Lecture Notes in Civil Engineering

Volume 27

Lecture Notes in Civil Engineering (LNCE) publishes the latest developments in Civil Engineering - quickly, informally and in top quality. Though original research reported in proceedings and post-proceedings represents the core of LNCE, edited volumes of exceptionally high quality and interest may also be considered for publication. Volumes published in LNCE embrace all aspects and subfields of, as well as new challenges in, Civil Engineering. Topics in the series include:

- Construction and Structural Mechanics
- Building Materials
- Concrete, Steel and Timber Structures
- Geotechnical Engineering
- Earthquake Engineering
- Coastal Engineering
- Hydraulics, Hydrology and Water Resources Engineering
- Environmental Engineering and Sustainability
- Structural Health and Monitoring
- Surveying and Geographical Information Systems
- Heating, Ventilation and Air Conditioning (HVAC)
- Transportation and Traffic
- Risk Analysis
- Safety and Security

To submit a proposal or request further information, please contact the appropriate Springer Editor:

- Mr. Pierpaolo Riva at pierpaolo.riva@springer.com (Europe and Americas);
- Ms. Swati Meherishi at swati.meherishi@springer.com (India);
- Ms. Li Shen at li.shen@springer.com (China);
- Dr. Loyola D'Silva at loyola.dsilva@springer.com (Southeast Asia and Australia/NZ).

More information about this series at http://www.springer.com/series/15087

Francesco Ricciardelli · Alberto Maria Avossa
Editors

Proceedings of the XV Conference of the Italian Association for Wind Engineering

IN-VENTO 2018

Springer

Editors
Francesco Ricciardelli
Department of Engineering
University of Campania "Luigi Vanvitelli"
Aversa, Italy

Alberto Maria Avossa
Department of Engineering
University of Campania "Luigi Vanvitelli"
Aversa, Italy

ISSN 2366-2557 ISSN 2366-2565 (electronic)
Lecture Notes in Civil Engineering
ISBN 978-3-030-12814-2 ISBN 978-3-030-12815-9 (eBook)
https://doi.org/10.1007/978-3-030-12815-9

Library of Congress Control Number: 2019932785

This Springer imprint is published by the registered company Springer Nature Switzerland AG
The registered company address is: Gewerbestrasse 11, 6330 Cham, Switzerland

Preface

This volume contains the full papers of most of the contributions presented at IN-VENTO 2018, the XV Conference of the Italian Association for Wind Engineering (ANIV), held at Castel dell'Ovo in Napoli, Italy, on September 9–12, 2018. The conference had 12 plenary sessions, aiming to provide a major forum for the exchange of ideas and discussion on the latest advances in the field of wind engineering, bluff-body aerodynamics, aeroelasticity, structural dynamics and reliability, wind modelling and forecast, wind tunnel and full-scale testing, computational fluid dynamics, wind energy. In particular, two of the session were devoted to the recent developments of wind energy exploitation in onshore and offshore environment. Invited lectures to the conference were Prof. Theodore Stathopoulos of the Concordia University, Dr. Andrew Allsop of ARUP UK, and Prof. Giovanni Solari of University of Genoa.

The previous fourteen editions of the IN-VENTO Conference series were held biennially since 1990 in Firenze, Capri, Roma, Trieste, Perugia, Genova, Milano, Reggio Calabria, Pescara, Cefalù, Spoleto, Venezia, Genova, and Terni; in its last four editions, the conference has become international, receiving attention from a large number of researchers worldwide.

IN-VENTO 2018 had an attendance of about 100 researchers, 40% of which coming from outside Italy. Out of the 76 papers presented in the conference, 55 were selected by the editors for publication in the proceedings book after a blind double peer-review process.

The editors thank all participants for their contributions to the conference programme and to this proceedings. We also acknowledge the support received by the Italian Association for Wind Engineering, by the Conference Scientific Committee and by the Local Organizing Committee.

Francesco Ricciardelli
Chairman of IN-VENTO-2018

Alberto Maria Avossa
Secretary of the Organizing Committee of IN-VENTO-2018

Organization

Scientific Committee

Alberto Maria Avossa	University of Campania "Luigi Vanvitelli", Italy
Gianni Bartoli	University of Florence, Italy
Marco Belloli	Politecnico di Milano, Italy
Claudio Borri	University of Florence, Italy
Luca Bruno	Politecnico di Torino, Italy
Massimiliano Burlando	University of Genoa, Italy
Luca Caracoglia	Northeastern University, USA
Yaojun Ge	Tongji University, China
Christos T. Georgakis	Aahrus University, Denmark
Massimiliano Gioffrè	University of Perugia, Italy
Vittorio Gusella	University of Perugia, Italy
Horia Hangan	Western University, Canada
Rudiger Hoffer	Ruhr-Universitaet Bochum, Germany
Ahsan Kareem	University of Notre Dame, USA
Gianpiero Manara	Permasteelisa Group, Italy
Francesco Petrini	"Sapienza" University of Rome, Italy
Maria Pia Repetto	University of Genoa, Italy
Francesco Ricciardelli	University of Campania "Luigi Vanvitelli", Italy
Daniele Rocchi	Politecnico di Milano, Italy
Giovanni Solari	University of Genoa, Italy
Ted Stathopoulos	Concordia University, Canada
Yukio Tamura	Tokyo Polytechnic University, Japan
Alberto Zasso	Politecnico di Milano, Italy

Organizing Committee

Alberto Maria Avossa
Stefania Di Ronza
Caterina Eramo
Giovanna Giugliano
Carmela Maienza
Francesco Ricciardelli (Chair)
Mustafa Vardaroglu

Conference Secretariat

Department of Engineering
University of Campania "Luigi Vanvitelli"
Via Roma, 9 – 81031 – Aversa, Caserta, Italy
e-mail: Invento2018@unicampania.it

Contents

Several studies provide recommendations for values of z_d and z_0. Grimmond and Oke (1999) have proposed a set of 14 categories to assess the urban roughness parameters (z_d and z_o) based on a comprehensive review of wind flow observations covering areas of different urban roughness densities in terms of non-dimensional roughness coefficients ($f_d = z_d/H$ and $f_o = z_o/H$). Values of $f_d = z_d/H$ range from 0.35 to 0.70 and values of $f_o = z_o/H$ range between 0.06 and 0.20.

Jian-Zhong et al. (2007) have introduced a new factor into Eq. (2), i.e. the wake factor α, to evaluate the mean wind speed ($\bar{u}_H$) at the mean roof height of the roughness elements (H), as follows:

$$\bar{u}_H = \alpha \cdot \frac{u_*}{k} \cdot \ln\left(\frac{H - z_d}{z_o}\right) \tag{3}$$

Also, the same study has provided an expression to estimate the roughness length, as given in Eq. 4:

$$z_o = H \cdot \left(1 - \frac{z_d}{H}\right) \cdot \exp\left[-\left(0.5 \cdot \frac{\alpha^2}{k^2} C_D \lambda_f\right)^{-0.5}\right] \tag{4}$$

in which C_D is the drag coefficient and λ_f is the frontal area density (ratio of total frontal area of roughness elements to the total area covered by the roughness elements).

On the other hand, it is hypothesized that the mean wind speed variation below the mean height of the roughness element (H) adopt an exponential profile (Heath et al. 2007), as provided by:

$$\bar{u} = \bar{u}_H \cdot \exp\left[\delta\left(\frac{z - H}{H}\right)\right] \tag{5}$$

in which δ is a constant which depends on building morphology such that for an array of cubes $\delta = 9.6 \cdot \lambda_f$.

The assessment of wind velocity of a bulk flow in urban environment, as discussed in this section, is not sufficient to rely upon for implementing micro-scale wind energy generation system; thus, estimating local urban wind resources must be viewed from a broader perspective.

3 Estimating Local Urban Wind Resources

The estimation of the wind speed in the urban environment remains a major obstacle in the successful development of micro-scale energy generation. This is due to the strong relation between the mean wind speed and the energy potential and the complex wind regime in built environment of highly stochastic roughness and the drag exerted by surface-mounted obstacles on the flow.

The most reliable method for the wind assessment in the urban environment is to directly measure the wind speed on-site (full-scale), ideally at the prospective location of the wind turbine. However, there are many difficulties that limit the possibilities of harnessing the wind for such applications. First of all, it is time consuming and expensive, it does not offer the possibility to consider different wind conditions, and moreover it is practically infeasible when the building is still at the design stage. On the other hand, several dependable methods can be conducted for preliminary assessment of wind resources for a specific site based on data measured by nearby meteorological stations, namely: probabilistic mathematical models (Weibull and Rayleigh; wind atlas data) and measurement methods such as atmospheric boundary layer wind tunnel testing and numerical simulation with Computational Fluid Dynamics (CFD).

Reviewing the previous studies, it is demonstrated that Weibull and Rayleigh probability functions are useful in case of offshore wind potential with high mean speeds (Jamil et al. 1995; Lun and Lam 2000; Seguro and Lambert 2000; Celik 2004; Carta and Ramirez 2007; Akdag et al. 2010; Rocha et al. 2012; Adaramola et al. 2014; Carneiro et al. 2016 and Wais 2017), which is indeed not the case of built environment, where the wind generally of complex nature.

Wind tunnel experimentation is still considered to be the most reliable, practical and effective approach, particularly for urban wind energy applications. However, there are impediments associated with the wind tunnel testing mainly owing to dynamic similarity and high blockage ratios. Recently, CFD has a great deal of light, and therefore, it has been widely utilized in several applications, mainly for the determination of mean flow conditions primarily for topics in environmental wind engineering such as pedestrian-level wind conditions, dispersion of pollutants, natural ventilation of buildings and wind-driven rain. Various simulation conditions and requirements such as the size of computational domain, mesh finesse sensitivity, flow profile and turbulence models, which have not all yet been fully understood limit the efficiency of the CFD applications (Yoshie et al. 2007). Examination and verification of results achieved through the CFD simulations is essential, as these are very sensitive to broad range of computational parameters.

Integrating wind turbines into buildings has been the subject of several case studies, where identifying the optimal location of the turbine was the pivotal issue. Unfortunately, most of the literature studies have been conducted on simple configurations of standalone cubes, prismatic buildings or groups of simple buildings. Cheng et al. (2007) carried out wind tunnel tests for uniform urban-type surfaces of two different area densities to investigate the effect of area density and array geometry on the aerodynamic characteristics of the considered surfaces. Heath et al. (2007) investigated computationally mean wind speeds at potential turbine mounting locations around a simple pitched-roof building surrounded by an array of similar houses for different wind directions using k-ε turbulence model. Lu and Ip (2009) studied the wind flows over three configurations of two tall buildings by considering parameters such as building height and roof shape and the spacing between the buildings using CFD. Ledo et al. (2011) examined the wind flow characteristics (wind velocity, turbulence intensity and flow patterns) above the roofs in three suburban areas of buildings with different roof shapes (pitched, pyramidal and flat) using CFD under several wind directions. Abohela et al. (2013) computationally investigated the effect of the roof shape, wind direction, building height and surrounding urban configurations on the above roof flow patterns to examine their effect on installing roof-mounted turbines. Wang et al. (2015) performed CFD tests to investigate the wind energy potential over roofs of two perpendicular buildings considering different lengths, widths, heights and corner spacing distances. Sari (2015) performed CFD simulations for isolated buildings to investigate the influence of the roof slope (20°, 30°, 40° and 50°) on the wind power density.

Also, analysing the literature shows that there are very limited research studies on the wind power evaluation and utilization over existing buildings in urban areas. Table 1 provides examples of recent actual building studies conducted by means of field measurements, wind tunnel experimentation and CFD simulations. As illustrated from comparison the prediction/errors of CFD and atmospheric boundary layer wind tunnel studies, appropriate utilization of atmospheric boundary layer wind tunnel approach can provide dependable approach to derive wind speed data, often from meteorological stations or airports, for a specific locality in urban area and to produce realistic estimates of associated wind energy potential within reasonable accuracy (Stathopoulos et al. 2018).

4 Wind Turbine Installations in Urban Areas

The assessment of wind flow in an urban locality is an indispensable measure for optimal installation of building integrated wind turbine. The assessment mainly involves the characteristics of wind speed, turbulence intensity and flow patterns. The boundary layer wind tunnels and CFD are feasible approaches to estimate realistic results of the wind energy potential.

Indeed, the prospective application of the turbine to be implemented points the way for choosing the size of the turbine and its generator type, the method of control, and how it is to be constructed and operated. For instance, wind turbines of open rural areas have generally a horizontal-axis rotor with three large blades. However, when it comes to wind turbines for urban environments, implementation is different for several reasons related primarily to the wind behaviour, aerodynamics and land limitations. Also, other constraints such as size and noise limitations, visual disturbances and low start-up wind speeds play a prominent role in the efficiency of the design.

Generally, wind turbines are divided into two main sets: horizontal axis and vertical axis wind turbines (Paraschivoiu 2002). The mechanism of wind energy conversion into mechanical power in the rotor axis is classified as drag-driven rotor, lift-driven rotor and hybrid rotor.

Vertical axis wind turbines (VAWTs) tend to be effective in their interactions with highly fluctuating winds, evoke lower noise levels and are aesthetically more pleasing. Moreover, the generator and gearbox of these wind turbines can be fixed at ground level making them easy to reach for operation services. These advantages, along with their smaller sizes, make VAWTs more suitable for application in urban environments (Toja-Silva et al. 2013). However, horizontal axis wind turbines (HAWTs) are still most prevalent in urban environment. This may be attributed, among other reasons, to the fact that HAWTs have a relatively low cut-in speed compared to VAWTs. Figure 3 shows some examples for HAWTs that can be integrated into buildings.

Table 1. Examples of recent studies conducted on actual site locations (Stathopoulos et al. 2018).

Study	Data collection approach	Site model (concerned building)	Simulated flow	Prediction/errors	Key findings
Lu and Sun (2014)	CFD	Lee Shau Kee building (97.4 × 38.7 × 69.9 m) (L, W, H). Kowloon, Hong Kong	RNG k-ε.	No validation.	Minimum height 1.4 m above the roof is recommended for turbine installations. Building proximity effects can increase the wind power significantly (1.5 to 5.5 times) at 4 m above building roof.
Tabrizi et al. (2014)	CFD Field measurements from one location.	Bunnings warehouse H = 8.4 m. Port-Kennedy, Western Australia. Radius of the surrounding: 200 m	Steady RANS with k-ε. Computational domain (H): 0.2 km	Wind speed: Error < 20%.	A small HAWT is recommended to be installed in the middle of the roof.
Al-Quraan et al. (2016)	Wind tunnel experiments. Field measurements from two locations.	EV Building (76 m high) Equiterre building (23 m high) Montreal, Quebec, Canada.	Homogeneous terrain Nonhomogeneous terrain	Wind energy: Error < 5%. Wind energy: 5% ≤ Error ≤ 20%	Accuracy of wind tunnel results depends on the upstream terrain.
Yang et al. (2016)	CFD Field measurements from 10 locations.	The building has dimensions of 68.9 × 62.5 × 33.5 m (L, W, H). Central metropolitan Taipei, Taiwan.	Standard k-ε. RNG k-ε. Realizable k-ε. Computational domain (L, W, H) 16 × 8 × 0.3 km	Turbulence intensity: Error < 56% (RNG k-ε model) Error < 20% (realizable k-ε model) Wind speed: Error < 20% (realizable k-ε model) Wind direction: Prediction of realizable k-ε is more reliable than RNG k-ε and standard k-ε.	Recommend increasing the hub height and install microturbines on the windward side of the building. Curved roof edges increase the power density.

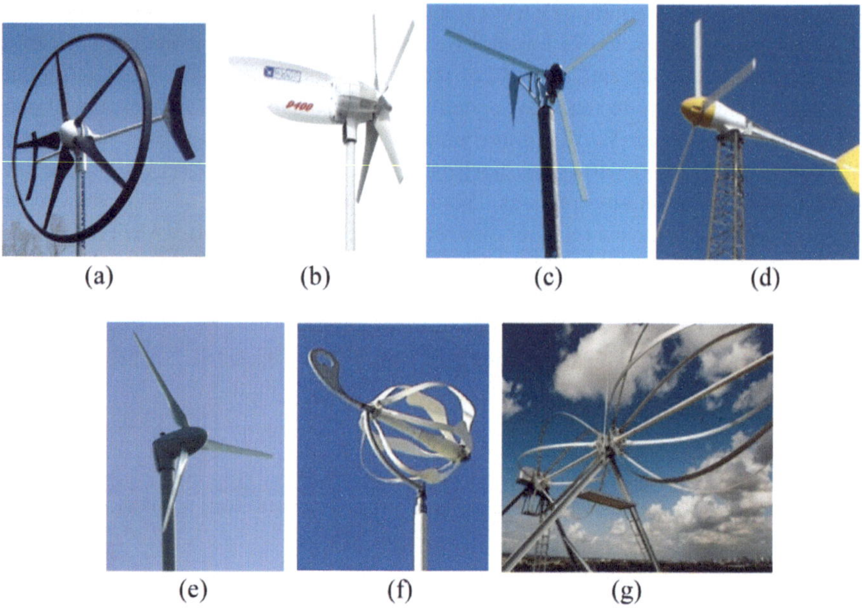

Fig. 3. Examples of HAWTs (a) Swift (https://www.ecopowershop.com) (b) Eclectic (https://boutique.marinagosselin.com/en/product/products) (c) Fortis Montana (https://www.fortiswindenergy.com) (d) Scirocco (http://archive.constantcontact.com/fs191/1102207134962/archive/1113940344041.html) (e) Tulipo (http://www.renugen.co.uk) (f) Energy Ball (http://www.wind-works.org;) and (g) WindWall (https://www.alibaba.com/product-detail/WindWall-3kW)

The wind aerodynamic industries have worked on those unfavorable characteristics of VAWTs and much improvements have been made. As a result, the use of VAWTs in urban applications has been increased, particularly for micro-scale urban operations due to their size. Currently, there are four main designs of vertical axis wind turbines utilized in urban environment, namely: Darrieus, Gorlov, Giromill and Savonius – as provided in Fig. 4.

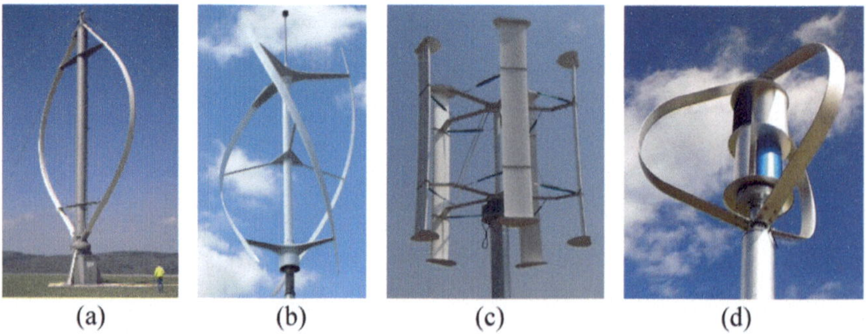

Fig. 4. Different types of VAWTs (a) Darrieus (https://en.wind-turbine-models.com) (b) Gorlov (https://www.quietrevolution.com) (c) Giromill (http://maketurbine.blogspot.com) and (d) Savonius (http://www.whirlopedia.com/)

Building-integrated wind turbines (BIWT) is a way of harvesting micro-scale energy, i.e., building-scale in urban environment. There is a growing number of implementations of BIWT systems on buildings in urban environment. Generally, the installations exist in two forms: wind turbines with large swept area attached to the rooftop, placed between two adjacent buildings or inside openings within the buildings; and small size wind turbine or a system of small wind turbines distributed on the roof or along the building height.

Retrofitting wind turbine into building requires consideration of urban surroundings on wind conditions around the turbines, wind conditions on optimal turbine placement and reliable prediction of wind power output. Along with these requirements, implementing wind turbines on the roof of the highest building as far as possible from the surrounding buildings may protect them from unexpected wind regimes. However, this measure would be conservative and in case of lacking detailed preliminary wind aerodynamic studies. Yang et al. (2016) confirmed that geometric configurations of the windward structures clearly affect the wind power on the roof of the building on which the turbine mounted.

For roof retrofitting wind turbines, it is recommended to be implemented outside the separation bubble at maximum attainable hub height and chamfering the roof edge of the prevailing wind direction in order to speed up the wind flow. Minimum hub height of 1.3 times the building height was recommended by Abohela et al. (2013) to avoid the region of high turbulence intensity inside the separation bubble. Lu and Sun (2014) found that installing the wind turbine at height 4 m above the roof enhanced wind power density by a factor of 1.3–5.4 times.

Increasing the swept area may have a positive effect on enhancing the energy yield. This can be achieved through many practical options, i.e., upscaling the rotor of the wind turbine by increasing the size of rotor blades, implementing multi-rotor turbine or multiple turbines at the same building. Upscaling always causes an unfavorable weight and loads on the turbines and on the supporting building (Sieros et al. 2012). Goltenbott et al. (2017) have investigated experimentally the aerodynamics of two configurations of diffuser-augmented wind turbines (DAWTs) of two and three rotors in laminar uniform flow. The wind turbines were placed in close proximity in the same vertical plane. It was reported that using two- and three-rotor configurations respectively enhanced the power efficiency by a factor up to 5% and 9% with respect to a single rotor turbine.

In addition to the mentioned requirements, retrofitting wind turbine into building may need to consider issues related to public safety and comfort, vibration, aesthetic and environmental impacts. In this regard, wind turbine developers have come up with innovative means. As a case in point, Chong et al. (Chong et al. 2012a) have introduced power-augmentation-guide-vane (PAGV) design encasing the *Sistan* rotor vertical axis wind turbine; and Chong et al. (2012a, b) have introduced an omni-directional-guide-vane (ODGV) to surround the vertical axis wind turbine (VAWT).

Keeping in mind that the theoretical maximum wind power is directly proportional to cube of the wind speed, a slight variation on the wind speed may cause substantial alteration on the wind power. In case of full integration of wind turbines together with the architectural form, a prior planning of design and construction on building aerodynamics shape is required, since it has a direct bearing in enhancing the power performance of wind turbines. There are three possible designs for integrating wind turbines (Mertens 2002): (1) Close to top or beside a building, (2) Between adjacent buildings, and (3) in an intubate path. Iconic examples of such installations are shown in Fig. 5.

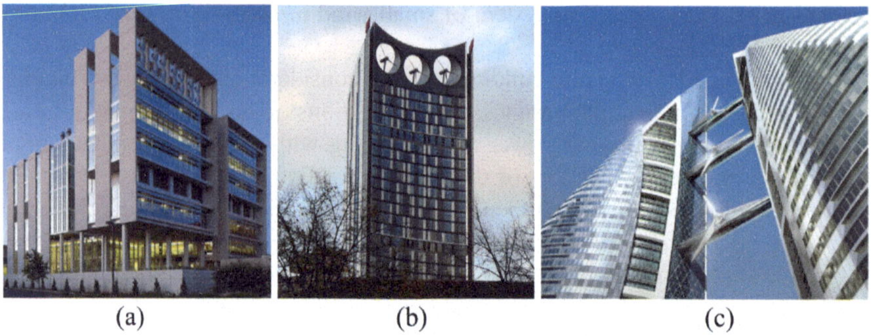

Fig. 5. Examples of BIWTs: (a) Oklahoma Medical Research Foundation (https://wordlesstech.com/omrf-roof-top-wind-farm) in Oklahoma City, USA (b) Strata tower in Southwark (http://www.urban75.org/blog/the-rarely-spinning-turbines-of-the-strata-tower-south-london), United Kingdom and (c) Bahrain World Trade Center (https://inhabitat.com/bahrain-world-trade-center-has-wind-turbines) in Manama, Bahrain

Wind turbine developers from both sectors, academia and industry, have come up with innovative aerodynamic solutions to enhance the upwind and integrating techniques. For instance, the objective building may be aerodynamically shaped to scoop and funnel wind into turbines to accelerate the upwind velocity. Chaudhry et al. (2014) have computationally studied the shape of Bahrain World Trade centre and found that the circular aerodynamic shape is optimal for wind turbines around high-rise buildings compared to square and triangular shapes, especially for regions with a prevailing wind direction.

Other thoughts on further enhancing the approaching wind flow have been the core subject of several studies, Diffuser Augmented Wind Turbines (DAWTs) is key cases in point, in which a brimmed diffuser structure encased the rotor blades can amplify the wind speed in a constricted flow path. Ohya and Karasudani (2010) investigated a wind turbine system consisting of a diffuser shroud with a broad-ring flange at the exit periphery and a wind turbine inside it. It was reported that the power production of the Diffuser Augmented Wind Turbine (DAWT) is factored by 4 to 5 times compared to conventional wind turbines. Recently, Dilimulati et al. (2018) have studied the aerodynamics of a flanged diffuser shroud mounted on a tall building in Montreal. It was found that although the diffuser does not impact the flow pattern, it enhances the flow at the windward wall of the building to be directed and accelerated into the diffuser. Figure 6 shows the differences between the velocity distribution over the building of the prospective diffuser with and without it.

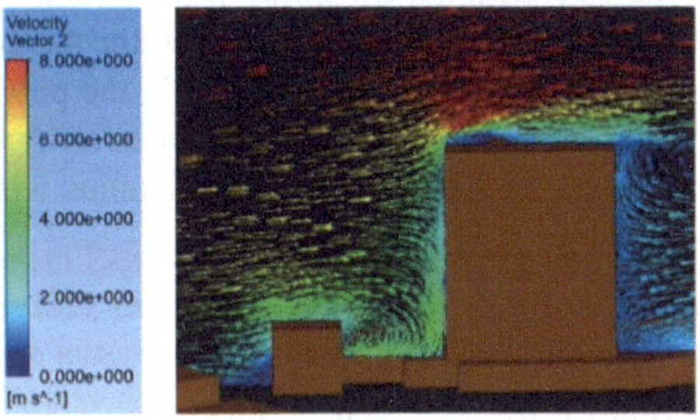

(a)

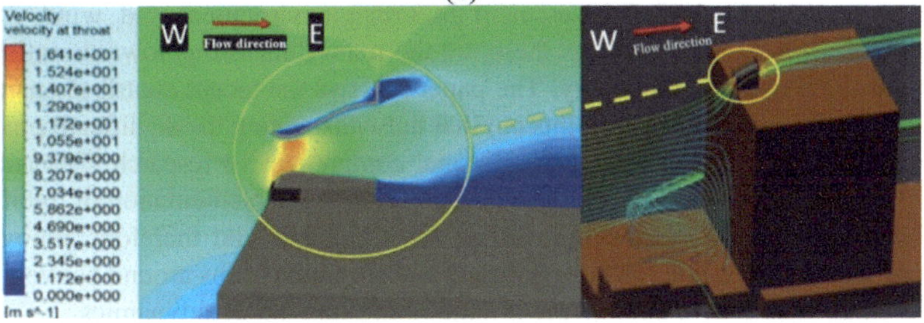

(b)

Fig. 6. (a) Velocity distribution around the objective building without the diffuser (b) Symmetry plane velocity distribution at the diffuser location (left) and velocity streamlines (right) with the diffuser (Dilimulati et al. 2018)

Stathopoulos et al. (2018) compared the power curve of a Building Integrated Wind Turbine (BIWT) (C_p = 0.381) derived from Park et al. (2015) with a conventional Savonius wind turbine (C_p = 0.09) based on results from Saha et al. (2008) – see Fig. 7.

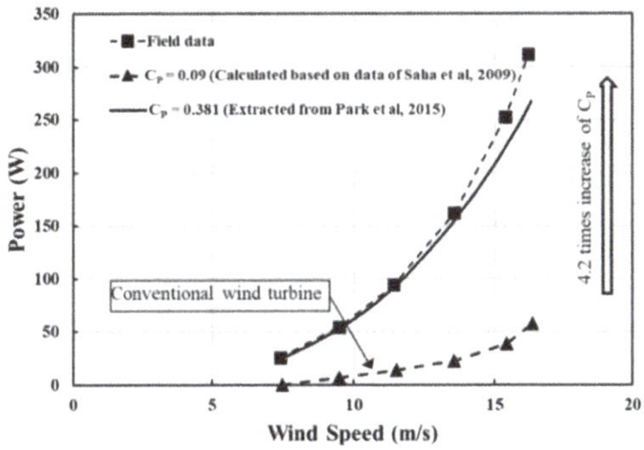

Fig. 7. Comparison of BIWT power curve with the respective conventional Savonius wind turbine Power curve comparison of BIWT based on data from Saha et al. (2008) and Park et al. (2015) (Stathopoulos et al. 2018)

De Gaetano P, Repetto MP, Repetto T, Solari G (2014) Separation and classification of extreme wind events from anemometric records. J Wind Eng Ind Aerodyn 126:132–143
Didden N, Ho CM (1985) Unsteady separation in a boundary layer produced by an impinging jet. J Fluid Mech 160:235–256
Duranona V, Sterling M, Baker CJ (2006) An analysis of extreme non-synoptic winds. J Wind Eng Ind Aerodyn 95:1007–1027
Elawadi A, Aboshosha H, El Damatty A, Bitsuamlak G, Hangan H, Elatar A (2017) Aero-elastic testing of multi-spanned transmission line subjected to downbursts. J Wind Eng Ind Aerodyn 169:194–216
Fujita TT (1985) Downburst: microburst and macroburst. University of Chicago Press, Chicago
Fujita TT (1990) Downburst: meteorological features and wind field characteristics. J Wind Eng Ind Aerodyn 36:75–86
Fujita TT, Wakimoto RM (1981) Five scales of airflow associated with a series of downbursts on 16 July 1980. Mon Weather Rev 109:1438–1456
Glauert MB (1956) The wall jet. J Fluid Mech 1:625–643
Goff RG (1976) Vertical structure of thunderstorm outflows. Mon Weather Rev 104:1429–1440
Gomes L, Vickery BJ (1976) On thunderstorm wind gusts in Australia. Civ Eng Tans Aust 18:33–39
Gomes L, Vickery BJ (1977/1978). Extreme wind speeds in mixed climates. J Ind Aerodyn 2:331–344
Gunter WS, Schroeder JL (2015) High-resolution full-scale measurements of thunderstorm outflow winds. J Wind Eng Ind Aerodyn 138:13–26
Haklander AJ, Van Delden A (2003) Thunderstorm predictors and their forecast skill for the Netherlands. Atmos Res 67–68:273–299
Hangan H, Refan M, Jubayer C, Romanic D, Parvu D, Lo Tufo J, Costache A (2017) Novel techniques in wind engineering. J Wind Eng Ind Aerodyn 171:12–33
Hjelmfelt MR (1988) Structure and life cycle of microburst outflows observed in Colorado. J Appl Meteorol 27:900–927
Hjelmfelt MR, Roberts RD, Orville HD, Chen JP, Kopp FJ (1989) Observational and numerical study of a microburst line-producing storm. J Atmos Sci 46:2713–2744
Holmes JD, Oliver SE (2000) An empirical model of a downburst. J Ind Aerodyn 22:1167–1172
Holmes JD, Hangan HM, Schroeder JL, Letchford CW, Orwig KD (2008) A forensic study of the Lubbock-Reese downdraft of 2002. Wind Struct 11:19–39
Huang G, Chen X (2009) Wavelets-based estimation of multivariate evolutionary spectra and its application to nonstationary downburst winds. Eng Struct 31:976–989
Huang G, Zheng H, Xu YL, Li Y (2015) Spectrum models for non-stationary extreme winds. J Struct Eng ASCE 141:04015010
Karman TV (1948) L'aerodynamique dans l'art de ingénieur. Mémoires de la Société des Ingenieurs Civils de France, pp 155–178
Kasperski M (2002) A new wind zone map of Germany. J Wind Eng Ind Aerodyn 90:1271–1287
Katsura J (1997) Some conceptions for the flow pattern formation around bluff bodies in natural winds. J Wind Eng Ind Aerodyn 66:1–15
Kim J, Hangan H (2007) Numerical simulations of impinging jets with application to downbursts. J Wind Eng Ind Aerodyn 95:279–298
Knupp KR (1989) Numerical simulation of low-level downdraft initiation within precipitating cumulonimbi: some preliminary results. Mon Weather Rev 117:1517–1529
Kwon DK, Kareem A (2009) Gust-front factor: new framework for wind load effects on structures. J Struct Eng ASCE 135:717–732
Le TH, Caracoglia L (2015) Reduced-order wavelet-Galerkin solution for the coupled, nonlinear stochastic response of slender buildings in transient winds. J Sound Vib 344:179–208

Letchford CW, Chay MT (2002) Pressure distributions on a cube in a simulated thunderstorm downburst. Part B: moving downburst observations. J Wind Eng Ind Aerodyn 90:733–753

Letchford CW, Mans C, Chay MT (2002) Thunderstorms – their importance in wind engineering (a case for the next generation wind tunnel). J Wind Eng Ind Aerodyn 90:1415–1433

Li CQ (2000) A stochastic model of severe thunderstorms for transmission line design. Probab Eng Mech 15:359–364

Lin WE, Savory E (2006) Large-scale quasi-steady modeling of a downburst outflow using a slot jet. Wind Struct 9:419–440

Lin EW, Orf LG, Savory E, Novacco C (2007) Proposed large-scale modelling of the transient features of a downburst outflow. Wind Struct 10:315–346

Lombardo FT (2014) Extreme wind speeds from multiple hazards excluding tropical cyclones. Wind Struct 19:467–480

Lombardo FT, Main JA, Simiu E (2009) Automated extraction and classification of thunderstorm and non-thunderstorm wind data for extreme-value analysis. J Wind Eng Ind Aerodyn 97:120–131

Lombardo FT, Smith DA, Schroeder JL, Mehta KC (2014) Thunderstorm characteristics of importance to wind engineering. J Wind Eng Ind Aerodyn 125:121–132

Lundgren TS, Yao J, Mansour NN (1992) Microburst modelling and scaling. J Fluid Mech 239:461–488

Marsh PT, Brooks HE, Karoly DJ (2009) Preliminary investigation into the severe thunderstorm environment of Europe simulated by the community climate system model 3. Atmos Res 93:607–618

Mason M, Letchford CW, James DL (2005) Pulsed jet simulation of a stationary thunderstorm downburst. Part A: physical structure and flow field characterization. J Wind Eng Ind Aerodyn 93:557–580

Mason MS, Wood GS, Fletcher DF (2009) Numerical simulation of downburst winds. J Wind Eng Ind Aerodyn 97:523–539

Matsumoto M, Shimamura M, Maeda, T, Shirato H, Yagi T, Hori K, Kawashima Y, Hashimoto M (2007) Drag forces on 2-D cylinders due to sudden increase of wind velocity. In: 12th international conference on wind engineering, Cairns, Australia, pp 1727–1734

McCann DW (1994) WINDEX – a new index for forecasting microburst potential. Weather Forecast 9:532–541

McConville AC, Sterling M, Baker CJ (2009) The physical simulation of thunderstorm downbursts using an impinging jet. Wind Struct 12:133–149

Mitchell KE, Hovermale JB (1977) A numerical investigation of the severe thunderstorm gust front. Mon Weather Rev 105:657–675

Mohr S, Kunz M, Richter A, Ruck B (2017) Statistical characteristics of convective wind gusts in Germany. Nat Hazards Earth Syst Sci 17:957–969

Nicholls M, Pielke R, Meroney R (1993) Large eddy simulation of microburst winds flowing around a building. J Wind Eng Ind Aerodyn 46–47:229–237

Nissen KM, Leckebusch GC, Pinto JG, Ulbrich U (2014) Mediterranean cyclones and windstorms in a changing climate. Reg Environ Change 14:1873–1890

Ogura Y (1963) The evolution of a moist convective element in a shallow, conditionally unstable atmosphere: a numerical calculation. J Atmos Sci 20:407–424

Oliver SE, Moriarty WW, Holmes JD (2000) A risk model for design of transmission line systems against thunderstorm down burst winds. Eng Struct 22:1173–1179

Orf LG, Anderson JR (1999) A numerical study of travelling microbursts. J Atmos Sci 127:1244–1258

Orf L, Kantor E, Savory E (2012) Simulation of a downburst-producing thunderstorm using a very high-resolution three-dimensional cloud model. J Wind Eng Ind Aerodyn 104–106:547–557

Oseguera RM, Bowles RL (1988) A simple analytic 3-dimensional downburst model based on boundary layer stagnation flow. NASA Technical Memorandum 100632
Pagon WW (1934, 1935) Aerodynamics and the civil engineers. Engineering News Records
Peng L, Huang G, Chen X, Kareem A (2017) Simulation of multivariate nonstationary random processes: hybrid stochastic wave and proper orthogonal decomposition approach. J Eng Mech ASCE 143:04017064
Piccardo G, Solari G (1998) Generalized equivalent spectrum technique. Wind Struct 1:161–174
Ponte J Jr, Riera JD (2010) Simulation of extreme wind series caused by thunderstorms in temperate latitudes. Struct Saf 32:131–137
Poreh M, Tsuel YG, Cermak JE (1967) Investigation of a turbulent radial wall jet. J Appl Mech Trans ASME 34:457–463
Proctor FH (1987a) The terminal area simulation system. I: theoretical formulation. NASA Contractor Report 4046
Proctor FH (1987b) The terminal area simulation system. II: verification cases, NASA Contractor Report 4047
Proctor FH (1988) Numerical simulations of an isolated micro burst. Part I: dynamics and structure. J Atmos Sci 45:3137–3159
Proctor FH (1989) Numerical simulations of an isolated microburst. Part II: sensitivity experiments. J Atmos Sci 46:2143–2165
Púcik T, Groenemeijer P, Rädler AT, Tijssen L, Nikulin G, Prein AF, Van Meijgaard E, Fealy R, Jacob D, Teichmanni C (2017) Future changes in european severe convection environments in a regional climate model ensemble. J Clim 30:6771–6794
Repetto MP, Burlando M, Solari G, De Gaetano P, Pizzo M (2017) Integrated tools for improving the resilience of seaports under extreme wind events. J Sustain Cities Soc 32:277–294
Repetto MP, Burlando M, Solari G, De Gaetano P, Pizzo M, Tizzi M (2018) A web-based GIS platform for the safe management and risk assessment of complex structural and infrastructural systems exposed to wind. Adv Eng Softw 117:29–45
Richter A, Ruck B, Mohr S, Kunz M (2018) Interaction of severe convective gusts with a street canyon. Urban Clim 23:71–90
Riera JD, Viollaz AJ, Reimundin JC (1977) Some recent results on probabilistic models of extreme wind speeds. J Ind Aerodyn 2:271–287
Riera JD, Nanni LF (1989) Pilot study of extreme wind velocities in a mixed climate considering wind orientation. J Wind Eng Ind Aerodyn 32:11–20
Sarpkaya T (1963) Lift, drag, and mass coefficients for a circular cylinder immersed in time dependent flow. J Appl Mech 30:13–15
Selvam RP, Holmes JD (1992) Numerical simulation of thunderstorm downdrafts. J Wind Eng Ind Aerodyn 41–44:2817–2825
Sengupta A, Sarkar PP (2008) Experimental measurement and numerical simulation of an impinging jet with application to thunderstorm microburst winds. J Wind Eng Ind Aerodyn 96:345–365
Shehata AY, El Damatty A, Savory E (2005) Finite element modelling of transmission line under downburst wind loading. Finite Elem Anal Des 42:71–89
Solari G (2007) The International Association for Wind Engineering (IAWE): progress and prospect. J Wind Eng Ind Aerodyn 95:813–842
Solari G (2014) Emerging issues and new frameworks for wind loading on structures in mixed climates. Wind Struct 19:295–320
Solari G, Repetto MP, Burlando M, De Gaetano P, Pizzo M, Tizzi M, Parodi M (2012) The wind forecast for safety and management of port areas. J Wind Eng Ind Aerodyn 104–106:266–277

Solari G, Burlando M, De Gaetano P, Repetto MP (2015a) Characteristics of thunderstorms relevant to the wind loading of structures. Wind Struct 20:763–791

Solari G, De Gaetano P, Repetto MP (2015b) Thunderstorm response spectrum: fundamentals and case study. J Wind Eng Ind Aerodyn 143:62–77

Solari G (2016) Thunderstorm response spectrum technique: theory and applications. Eng Struct 108:28–46

Solari G, Rainisio D, De Gaetano P (2017) Hybrid simulation of thunderstorm outflows and wind-excited response of structures. Meccanica 52:3197–3220

Steiner JT (1973) A three-dimensional model of cumulus cloud development. J Atmos Sci 30:414–435

Straka JM, Anderson JR (1993) Numerical simulations of microburst producing storms: some results from storms observed during COHMEX. J Atmos Sci 50:1329–1348

Twisdale LA, Vickery PJ (1992) Research on thunderstorm wind design parameters. J Wind Eng Ind Aerodyn 41:545–556

Vermeire BC, Orf LG, Savory E (2011) Improved modeling of downburst outflows for wind engineering applications using a cooling source approach. J Wind Eng Ind Aerodyn 99:801–814

Vicroy DD (1992) Assessment of micro burst models for downdraft estimation. J Aircr 29:1043–1048

Wang L, McCullough M, Kareem A (2013) A data-driven approach for simulation of full-scale downburst wind speeds. J Wind Eng Ind Aerodyn 123:171–190

Wang D, Chen X, Xu K (2017) Analysis of buffeting response of hinged overhead transmission conductor to nonstationary winds. Eng Struct 147:567–582

Wood GS, Kwok KCS, Motteram NA, Fletcher DF (2001) Physical and numerical modelling of thunderstorm downburst. J Wind Eng Ind Aerodyn 89:535–552

Xu Z, Hangan H (2008) Scale, boundary and inlet condition effects on impinging jets. J Wind Eng Ind Aerodyn 96:2383–2402

Zhang Y, Hu H, Sarkar PP (2013a) Modeling of microburst outflows using impinging jet and cooling source approaches and their comparison. Eng Struct 56:779–793

Zhang Y, Sarkar P, Hu H (2013b) An experimental study of flow fields and wind loads on gable-roof building models in microburst-like wind. Exp Fluids 54:1511

Zhang Y, Sarkar P, Hu H (2014) An experimental study on wind loads acting on a high-rise building model induced by microburst-like winds. J Fluids Struct 50:547–564

Zhang S, Solari G, De Gaetano P, Burlando M, Repetto MP (2018a) A refined analysis of thunderstorm outflow characteristics relevant to the wind loading of structures. Probab Eng Mech 54:9–24

Zhang S, Solari G, Yang Q, Repetto MP (2018b) Extreme wind speed distribution in a mixed wind climate. J Wind Eng Ind Aerodyn 176:239–253

Zhu S, Etkin B (1985) Model of the wind field in a downburst. J Aircr 22:595–601

Technical Papers

Comparison of High-Resolution Pressure Peaks in Closed and Open-Section Wind Tunnels

L. Amerio[1], G. Lamberti[2], G. Pomaranzi[3(✉)], A. Zasso[3], and C. Gorlé[2]

[1] Advanced Technology and Research Group, ARUP, 8 Fitzroy Street, London W1T 4BQ, UK
luca.amerio@arup.com
[2] Department of Civil and Environmental Engineering, Stanford University, 473 via Ortega, Stanford, CA 94305, USA
{giacomol,gorle}@stanford.edu
[3] Department of Mechanical Engineering, Politecnico di Milano, Via G. La Masa 1, 20156 Milan, Italy
giulia.pomaranzi@polimit.it, alberto.zasso@polimi.it

Abstract. Wind tunnel testing represents an established technique for the assessment of wind-induced pressure on cladding systems. Nonetheless, some physical events, such as the strong negative peaks of pressure that occur on a building's lateral facades, are not fully understood. These events can be catastrophic for cladding systems, which motivates their further investigation. The objective of the present work is to study the nature of extreme suction events, by comparing high-resolution pressure tap measurements on a high-rise building collected at two different facilities: the closed-circuit wind tunnel of Politecnico di Milano, and the open-circuit Wall of Wind facility at Florida International University. We first compare the statistics of the two atmospheric boundary layers, and subsequently present results for the mean and root mean square pressure distributions and negative peak pressure events. The experiments provide consistent results, and they both exhibit two types of negative peak pressure events: one is characterized by an extremely short duration and spatial extension, while the other impacts a larger portion of the facade.

Keywords: Wind tunnel · Cladding · Atmospheric boundary layer · Pressure peaks

1 Introduction

Glazed panels are often employed to cover the external facades of high-rise buildings. Correct estimation of wind loads on these components is critical, both from a safety and economic point of view. Building codes provide one approach to estimate the pressure acting on the panels; however, when the pressure-induced load is critical, common

F. Ricciardelli and A. M. Avossa (Eds.): INVENTO 2018, LNCE 27, pp. 35–48, 2019.
https://doi.org/10.1007/978-3-030-12815-9_3

practice is to rely on wind tunnel cladding tests. During such experiments, pressure time-series are recorded in several points on the building's surface by means of a synchronous multi-pressure sensing system (SMPSS) (Cluni et al. 2011). The area-averaged loads acting on the panels can then be computed by integration of these local pressure measurements or by filtering the pressure time-history using a moving-average filter (Lawson 1980). The pressure time-series measured on a building's lateral facades often exhibit extreme suction events that can be catastrophic for a glazed panel, and hence determine the dimensioning of the cladding system. In these regions, the pressure is characterized by strong non-Gaussian behavior; negative pressure peaks of up to 15 standard deviations can occur and result in large negative skewness (Rocchi et al. 2011); (Schito et al. 2013); (Amerio 2018). In this paper we present a comparison between experimental measurements that were performed in two different wind tunnels, with the objective to improve our understanding of the nature of the negative pressure peaks. The first experiment was performed at the wind tunnel of the Politecnico di Milano (PoliMi) and considered pressure measurements on a high-rise building. The focus was on measuring the pressure in the most critical regions on the building's surface, for an open-terrain exposure and different inflow directions. The high-rise building model was equipped with 448 closely-spaced pressure taps connected to high-frequency pressure scanners, in order to study the temporal and spatial extension of the peak events in detail (Amerio 2018). The experiment confirmed the presence of extreme suction events in several locations on the building's lateral facade; for these events the traditional assumption of correlation between duration and spatial extension (Lawson 1980) appeared to be incorrect (Amerio 2018). The same high-rise building model was then tested at the Wall of Wind (WoW) facility of Florida International University. The objective of these tests was to verify the conclusions from the PoliMi measurements. Hence, the tests were performed using a similar open-terrain exposure, considering a subset of critical inflow directions identified at PoliMi (Amerio 2018). In the following we first discuss the experimental set-up for the atmospheric boundary layer (ABL) velocity measurements and the pressure measurements in both wind tunnels. Subsequently, the comparison of the results is presented. To compare the incoming ABL statistics we consider the mean velocity, turbulence intensity, integral length scales, and spectra. The wind pressure measurements are compared in terms of the mean and root mean square values for the pressure coefficients, and time series are analyzed to investigate the intensity of the negative pressure peaks. Finally, the conclusions and areas for future work are presented.

2 Experimental Setup

This section describes the experimental set-up for the ABL velocity measurements and for the pressure measurements on the high-rise building in both wind tunnels. The PoliMi facility is a closed-circuit wind tunnel; the boundary layer test section has a cross section of 14 × 4 m^2 and is 35 m long. The WoW facility is an open-circuit wind tunnel with a 6.1 m wide and 4.3 m high test section.

2.1 Velocity Measurements

2.1.1 PoliMi

A detailed characterization of the ABL was carried out at the wind tunnel of PoliMi. 3D hot-wires with a sampling frequency of 2000 Hz were employed to measure the velocity components at 280 points distributed on a plane at the building location, as shown in Fig. 1. The outcome of the experiment consists of 20 s time-series of the three velocity components at 5 spanwise locations (0.6 m apart) and 56 vertical locations (43.7 mm apart below 0.75 m and 87.5 mm above).

Fig. 1. Setup of the experiment to characterize the ABL in the wind tunnel.

2.1.2 WoW

The ABL at WoW was characterized using TFI Cobra probes with a sampling frequency of 2500 Hz. 60 s time-series of the three components of velocity were recorded at 6 vertical locations in the center of the turntable. The spatial resolution of these measurements varied between 0.3 m and 0.6 m. Table 1 summarizes the parameters of the two experiments in terms of sampling frequency, total duration and reference velocity.

Table 1. Sampling frequency, total duration, and reference velocity at 1 m height for the velocity measurements performed in both facilities

	f_{samp} [Hz]	T [s]	$U_{ref@1m}$ [m/s]
PoliMI	2000	20	6.9
WoW	2500	60	30.8

2.2 Pressure Measurements

2.2.1 Building Model

The model of the high-rise building is a 1 m long, 0.3 m wide and 2 m high rectangular box, representative of a 100 m tall building in full-scale. The model was placed at the center of the turntable to allow testing at different inflow directions. In the present work we focus on three wind directions: 0°, 20° and 180°, following the convention defined in Fig. 2. The axis convention is also shown in Fig. 2, adopting x as along wind direction, y and z as vertical and transversal direction respectively.

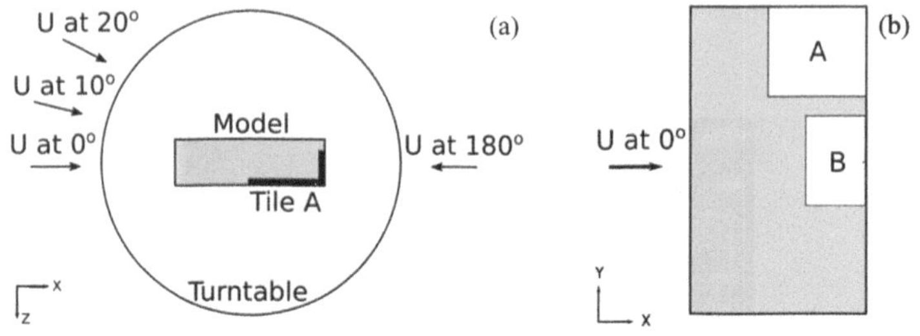

Fig. 2. Top view of the turntable, indicating the wind direction convention (a); side view of the building model (b).

The goal of the experiment was to perform a detailed study of the pressure distribution in the regions of the building where the highest pressure peaks are expected, i.e. near the corners and edges. Therefore, we designed two aluminum tiles containing 224 pressure taps each: tile A located on the top-corner of the model and tile B centered at 1 m height adjacent to the building edge (Fig. 2). The minimum tap resolution is 3 mm on both tiles; the spacing is progressively increased when moving away from the building edges. Figure 3 shows the distribution of the pressure taps on tile A, together with the set-up of the experiment in the two wind tunnel facilities.

2.2.2 PoliMi Pressure Measurement System

At PoliMi, the model was instrumented with 8 PSI ESP-32 HD high-speed pressure scanners, connected to a data acquisition system with a sampling frequency of 500 Hz. The outcome of each test consists of 300 s time-series of pressure measured at 446 taps. A Pitot tube, located 7 m upwind of the building, was employed during the experiments to measure the reference velocity (10.3 m/s) at 1 m height.

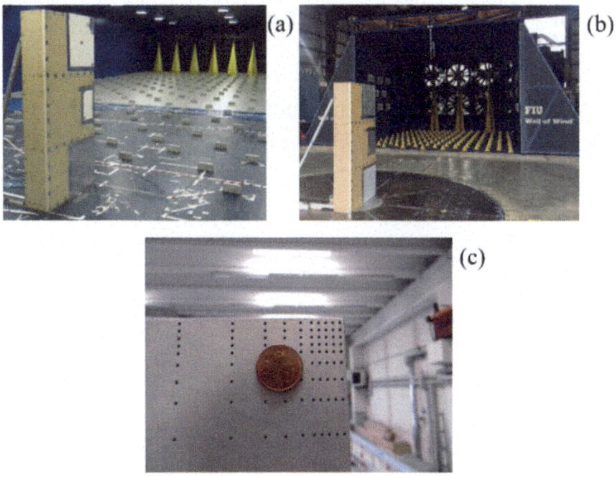

Fig. 3. View of the test section of the two wind tunnel facilities: PoliMi (a) and WoW (b); pressure taps distribution on tile A (c).

2.2.3 WoW Pressure Measurement System

The WoW measurements were performed using pressure scanners with a sampling rate of 520 Hz. As in the PoliMi wind tunnel, 300 s time-series of pressure were recorded at 446 taps. The tests were performed at the same Reynolds number, i.e. with a target reference velocity at 1 m height of 10.3 m/s. At the beginning of the experiment, TFI Cobra probes were employed to measure the velocity at two vertical locations at a lateral distance of 2 m from the building. The details of the two experiments on the high-rise building are summarized in Table 2.

Table 2. Sampling frequency, total duration and reference velocity at 1 m height for the pressure measurements performed in both facilities.

	f_{samp} [Hz]	T [s]	$U_{ref@1m}$ [m/s]
PoliMI	500	300	10.3
WoW	520	300	10.3

3 Results

In this section, we first compare the inflow statistics generated by the two wind tunnels, namely mean velocity profiles, turbulence intensities, integral time-scales and velocity spectra. Non-dimensional quantities are presented, since the measurements were performed for different reference velocities. Subsequently, we present the comparison of the pressure measurements both in terms of statistics, i.e. mean and root mean square values, and in terms of time-series and peak values.

3.1 Comparison of Velocity Measurements

In the following plots, the gray error-bars represent the variation of the velocity statistics measured at the different spanwise locations in the PoliMi wind tunnel and the gray circles indicate the corresponding spanwise-averaged values; the red dots represent the data measured at WoW.

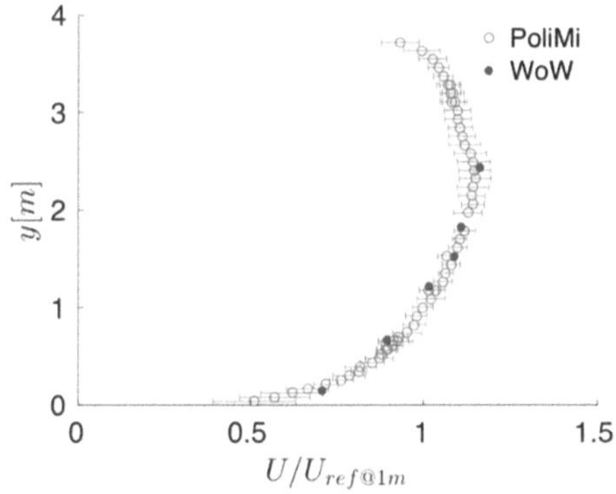

Fig. 4. Comparison of adimensional mean velocity between PoliMi and WoW data.

Figure 4 shows the profiles of mean streamwise velocity, adimensionalized by the reference velocity at 1 m height. The profiles agree well: the WoW measurements are within the spanwise variation of the PoliMi data at all available measurement heights. When fitting the logarithmic velocity profile typical of a neutral ABL to both sets of measurements, we obtain a roughness length of ~3 mm and ~2.5 mm for the PoliMi and WoW tests respectively.

The turbulence intensities, defined as the ratio of the root mean square of the turbulent velocity fluctuations and the mean velocity, are compared in Fig. 5. The streamwise turbulence intensity measured at the WoW is within the interval defined by the spanwise variation in the PoliMi experiments, except for the lowest and highest points. The highest discrepancy occurs close to the ground where the turbulence intensity in WoW is ~0.07 lower than the PoliMi spanwise-averaged value. The WoW profiles of vertical turbulence intensity are ~0.06 lower close to the ground and ~0.02 higher above 1.8 m. Similarly, the spanwise turbulence intensity computed from WoW data at 0.15 m height is ~0.045 lower than PoliMi spanwise-averaged value at the same height and above 1.8 m it is ~0.02–0.03 higher.

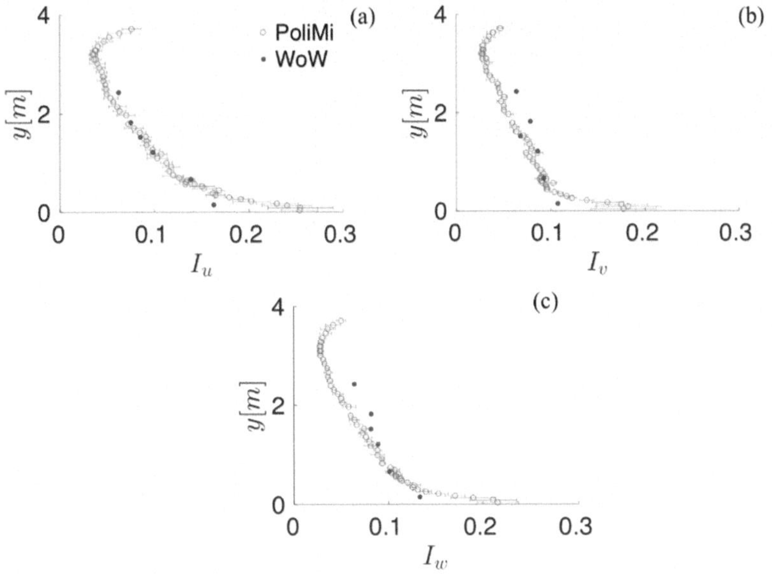

Fig. 5. Comparison of turbulence intensities between PoliMi and WoW data: streamwise component (a), vertical component (b) and spanwise component (c).

Figure 6 shows the comparison in terms of adimensional integral time-scales; the profiles are adimensionalized using the reference velocity at 1 m height and the height of the building. All three components show reasonable agreement, considering that a significant spanwise variation is observed in the PoliMi experiment. The main discrepancy between the two data sets again appears close to the ground in the vertical and spanwise profiles.

Finally, the power spectra of the streamwise velocity component are compared in Fig. 7. The longer time-series measured in the WoW test allows to capture larger scales compared to PoliMi experiment; this is evident from the lower frequencies of the spectrum computed from WoW data (Fig. 7). Conversely, even though the sampling frequency of the cobra probes is higher than the sampling frequency of the hot-wires, the spectrum measured at PoliMi contains higher adimensional frequencies compared to WoW data. This is a result of the fact that the Reynolds number of the WoW experiment is almost 5 times larger than the Reynolds number of the PoliMi test; therefore, the sampling frequency of 2500 Hz does not enable the measurement of scales as small as in the PoliMi wind tunnel.

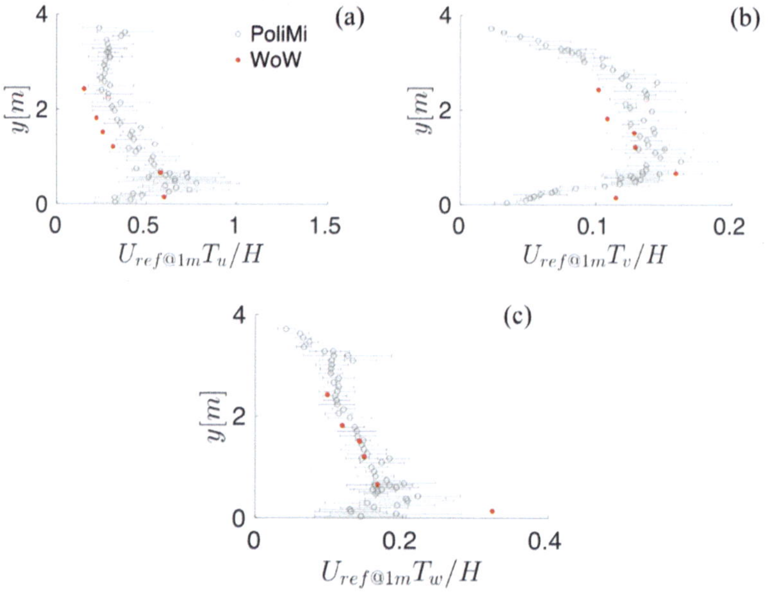

Fig. 6. Comparison of adimensional integral time-scales between PoliMi and WoW data: streamwise component (a), vertical component (b) and spanwise component (c).

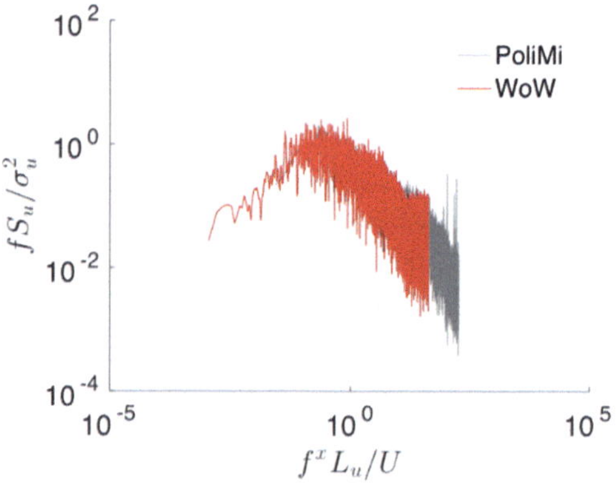

Fig. 7. Comparison of streamwise velocity spectra between PoliMi and WoW data at 1 m height.

In summary, it can be concluded that the PoliMi and WoW experimental set-ups have very similar incoming ABLs, both in terms of the time-averaged velocity and the turbulence statistics.

3.2 Comparison of Pressure Measurements

Figure 8 shows the distribution of the time-averaged pressure coefficient on the building's lateral facade at 0° (or equivalently 180°) exposure. The measurements recorded in both wind tunnels agree well qualitatively: the flow first separates at the windward edge generating an area of relatively strong negative pressure (red region in Fig. 8), then it reattaches in the rear part of the model (yellow region in Fig. 8). Nonetheless, from the contour plot it appears that the WoW data experience a less strong negative pressure, especially in the separation region.

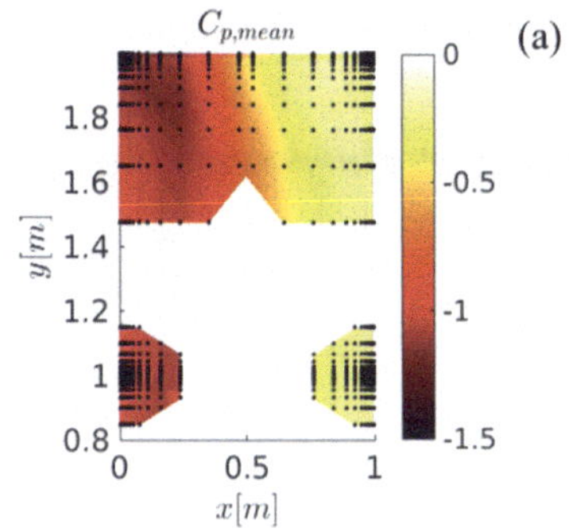

Fig. 8. Mean pressure coefficient at 0° exposure, measured at PoliMi (a) and WoW (b).

The distribution of the root mean square (rms) of the pressure coefficient is shown in Fig. 9. Both sets of measurements show high fluctuations in the region of flow separation and reattachment; the rms Cp computed at WoW is slightly lower than the PoliMi one, especially in the rear portion of the model, where the flow is attached.

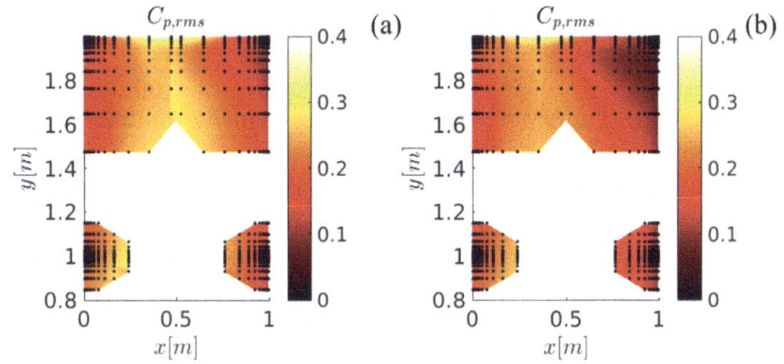

Fig. 9. Root mean-square pressure coefficient at 0° exposure, measured at PoliMi (a) and WoW (b).

Figure 10 compares the time-series of the pressure coefficients recorded on a pressure tap adjacent to the corner of tile A, at 20° exposure. We selected this particular flow direction and tap position because previous work observed the occurrence of extreme suction events (Amerio 2018). During a 50 s period, multiple negative peaks occur in both tests, namely 13 peaks stronger than Cp = −4 appear in both PoliMi and WoW time-series, indicated with a red arrow in Fig. 10. The largest negative value of pressure coefficient recorded at PoliMi is around −10, while at WoW it is around −8. Overall the two experiments seem to agree and, despite the fact that the negative peaks measured at PoliMi are slightly stronger than the WoW ones, a similar non-Gaussian behavior is found.

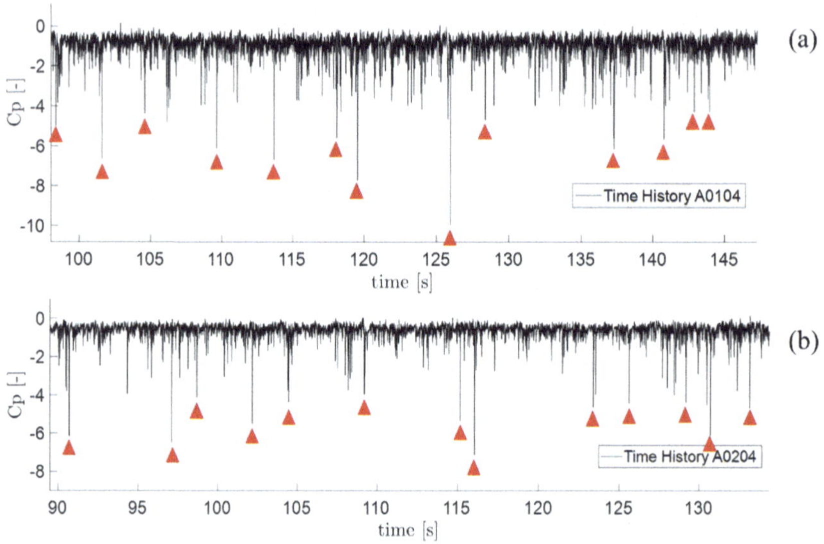

Fig. 10. Time history of a tap close to the building corner at 20° exposure, measured at PoliMi (a) and WoW (b). The red arrows mark the peaks lower than Cp = −4.

The high spatial resolution of the pressure taps enables further analysis on the spatial extension of the peaks. In Fig. 11 we selected two taps on tile A next to the corner, where both PoliMi and WoW tests experience extreme negative peaks of pressure. Figure 11 depicts a short time interval around the occurrence of the peaks and shows the corresponding spatial distribution of pressure at the time instants marked by the red dots. During both experiments, extremely concentrated peak events, both in time and space, are recorded in the same location, for the same exposure angle. These types of events that are limited in space and time are not critical for a typical glazed panel of size 3×2 m^2, visualized by the dashed red line in Fig. 11 (Amerio 2018).

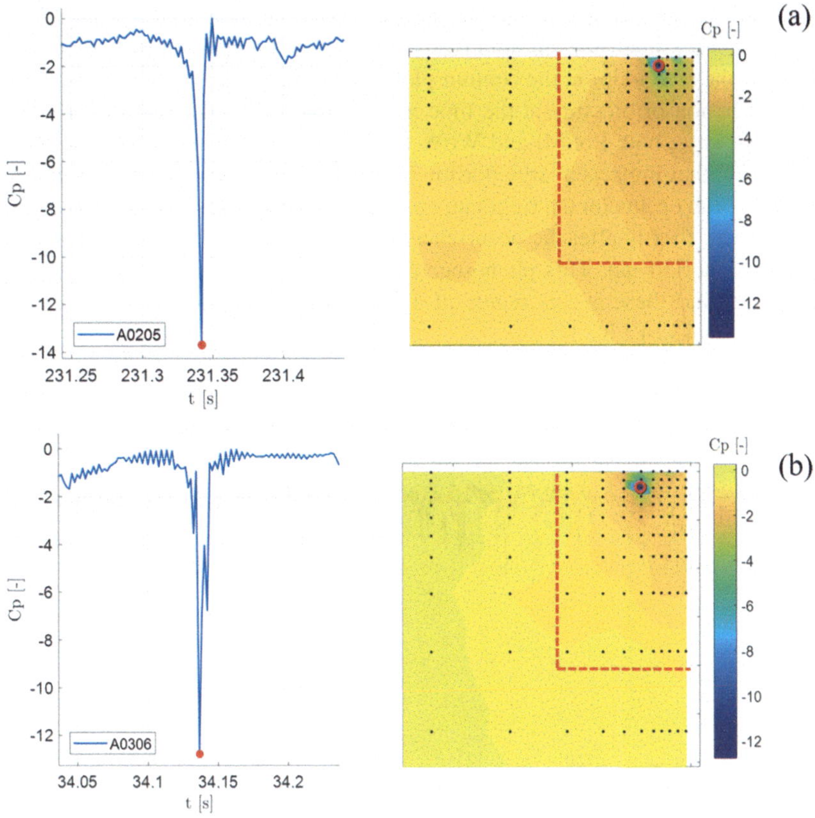

Fig. 11. Left: pressure coefficient time histories measured at a pressure tap on tile A at PoliMi (a) and WoW (a) at 20° exposure. Right: contour plot of the instantaneous pressure coefficient distribution at the top corner of the building on tile A, measured at PoliMi (a) and WoW (b), at the time indicated by the red dot in the time histories.

A very different phenomenon is observed when considering tile B for an inflow direction of 180°, when the strongest loads are recorded next to the model edge (Amerio 2018). Figure 12 compares pressure coefficient time histories recorded on tile A and tile B: the tile B signals (red lines in the figure) are no more characterized by sharp and concentrated peaks and such time-series appearance is smoother. This implies that the frequency content of the signals recorded on the tile B is limited in the high frequency range compared to the one measured on the tile A (Fig. 10). This different phenomenology of the peak events with respect to the ones previously analyzed is consistent between the Polimi and Wow experiments. The high spatial resolution allows for a spatial extension analysis on the tile B as well. Figure 13 on the left depicts a short time-series around the occurrence of the peaks at two adjacent taps, indicated by the red circles in the contour plots on the right. The contour plots show the spatial distribution of pressure at the time instants marked by the red dots in the time series. In this case, both PoliMi and WoW tests experience a strong negative pressure up to Cp = −4 that impacts a large portion of the tile. The time series reveal that when the negative peak occurs on the right tap, the left one experiences a peak in the opposite direction. Then, shortly after the occurrence of this first negative peak, a second one is measured by the left tap. This phenomenology appears to be consistent between the two data sets, and these events represent a more critical situation than the sharp and concentrated peaks of Fig. 11.

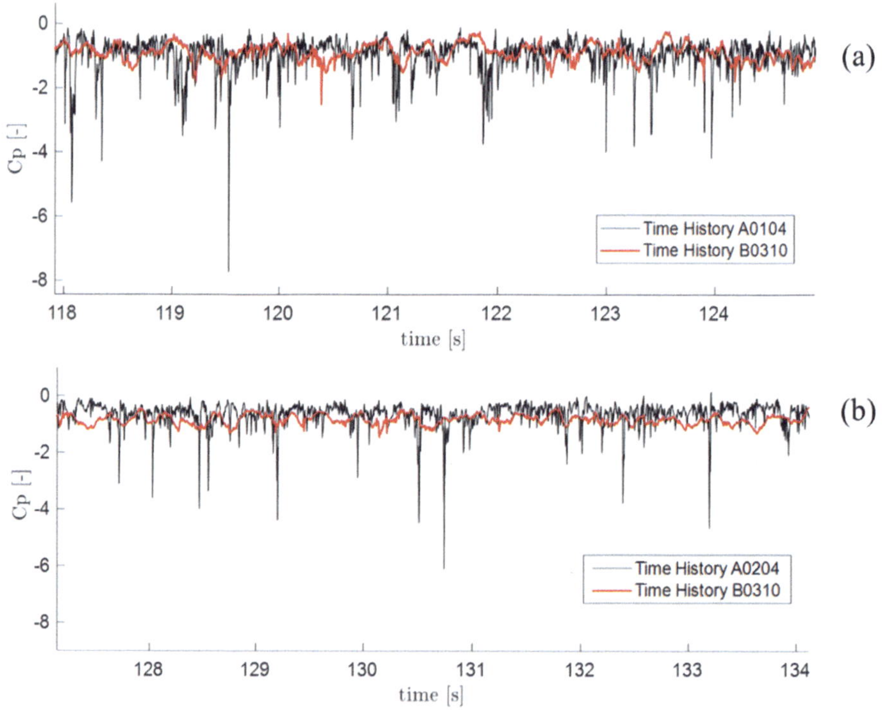

Fig. 12. Time history of a tap placed on the tile A (black line) compared to one recorded on tile B (red line) at 180° exposure, from Polimi data (a) and Wow experiment (b).

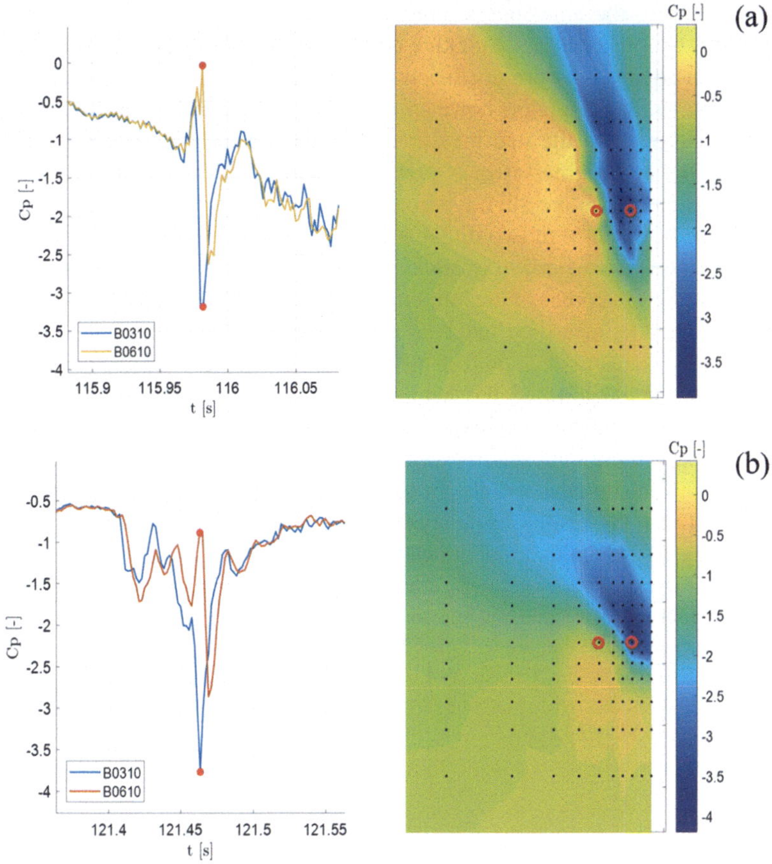

Fig. 13. Time history of a tap placed on the tile A (black line) compared to one recorded on tile B (red line) at 180° exposure, from Polimi data (a) and Wow experiment (b).

4 Conclusions and Future Work

In the present work we analyzed measurements on a high-rise building model, acquired in two different wind tunnels: the closed-circuit wind tunnel of Politecnico di Milano and the Wall of Wind open-circuit wind tunnel of Florida International University. The objective was to assess the consistency of both experiments and to investigate the extreme peak events that occur on the building's lateral facade (Amerio 2018).

First, we compared the statistics of the two atmospheric boundary layers. The mean velocity data recorded at WoW falls inside the interval defined by the spanwise variation of PoliMi measurements over the entire building height. The turbulence intensities and integral time-scales exhibit some differences close the ground and near the top of the building. The largest discrepancies appear near the ground, where the turbulence intensity in the WoW is up to 0.07 lower than the PoliMi value. Comparison of the power spectra of the streamwise velocity component further confirm good agreement between the atmospheric boundary layers generated in both experiments. Subsequently, we presented a qualitative comparison of the distributions of the mean and root mean square pressure coefficients on the building's lateral facade. The two data sets show similar behavior; however, the negative pressure measured at WoW is slightly less strong than the one measured at PoliMi. Similar considerations can be made by looking at the time-series of pressure: multiple extreme peaks occur, but the ones measured at PoliMi are slightly stronger.

Finally, the occurrence of two different kinds of negative peaks was shown in both experiments: one is characterized by an extremely short duration and spatial extension, measured near the top corner of the building on Tile A, while the other impacts a larger portion of the facade, measured mid height near the building edge on Tile B.

In summary, we can conclude that the two experiments provide consistent results, and the WoW data confirmed the nature of the negative pressure peaks that was observed at PoliMi, for the same exposure angles (Amerio 2018). In future work, we will further analyze the data sets by looking at the correlation between duration and spatial extension of the peak events. In addition, we will investigate the effect of the incoming ABL on the results by analyzing additional tests performed at the WoW considering higher Reynolds numbers and a different terrain exposure.

Acknowledgements. The second and last authors' contribution to this work was supported by the National Science Foundation under Grant Number 1635137. We gratefully acknowledge the contribution of the staff at the NHERI Wall of Wind facility who made the WoW experiments possible. In particular, we wish to thank Walter Conklin, Arindam Chowdhury, Peter Irwin, Ashkan Rasouli and Maryam Refan.

References

Cluni F, Gusella V, Spence SMJ, Bartoli G (2011) Wind action on regular and irregular tall buildings: Higher order moment statistical analysis by HFFB and SMPSS measurements. J Wind Eng Ind Aerodyn 99(6–7):682–690

Lawson TV (1980) Wind Effects on Buildings: Design Applications, vol 1. Spon Press, London

Rocchi D, Schito P, Zasso A (2011) Investigation on the relation between incoming wind characteristics and surface pressure distribution. J. Wind Eng. Ind. Aerodyn

Schito P, Giappino S, Zasso A, Gorlé C (2013) Uncertainty quantification of peak pressure loads on a rectangular building

Amerio L (2018) Experimental high resolution analysis of the pressure peaks on a building scale model facades. Ph.D. thesis, Italy

Aerodynamic Tailoring of a Bluff Deck Section Subjected to Inclined Winds Due to the Complex Orography of the Construction Site

T. Argentini[1(✉)], S. Omarini[1], A. Zasso[1], and M. Petrangeli[2]

[1] Department of Mechanical Engineering, Politecnico di Milano, Milan, Italy
tommaso.argentini@polimi.it
[2] Mario Petrangeli & Associati srl., Rome, Italy

Abstract. Bridges crossing steep valleys in mountainous regions may be subjected to large angles of attack of the incoming wind flow, because of the complex orography of the region where the structure is located. Therefore, the aerodynamic analysis of bridges in this kind of regions should account for this peculiarity. In this paper, the aerodynamic characterization of a bluff deck section of a railway bridge, with angles of attack ranging from −6 to +6° is discussed. It is shown that standard analyses that consider only angles of attack near 0° only are not sufficient to ensure optimal aerodynamic performances. Indeed, with such large angles of attack the flow field near the deck may change consistently, also leading to aeroelastic instabilities or strong vortex-induced vibrations. To mitigate the aeroelastic issues, without an abrupt change of the deck section, several aerodynamic appendices were tested, and a tailored section with acceptable behavior was selected.

Keywords: Bridge · Steep valley · Bluff deck section · Instability · Large angle of attack · Aerodynamic tailoring

1 Introduction

The bridge studied in this paper is a road railway cable-stayed bridge located in a mountainous region. The bridge crosses a steep valley (see Fig. 1a and the terrain contour map in Fig. 1b), therefore a wind tunnel study was performed in order to analyze the topographical effects on the wind flows which would affect the design of the proposed bridge (Rowan Williams Davies & Irwin Inc. (RWDI) 2006). The local mean flow approaches the deck with a mean angle of attack that ranges from almost −6° to +6°.

This fact is not common for standard bridges, that are generally subjected to nearly horizontal mean wind flows. Therefore, it may happen, as in this case-study, that a section with acceptable aerodynamic performances at 0-degree angle of attack may not behave properly, when large mean angles of attack are considered.

F. Ricciardelli and A. M. Avossa (Eds.): INVENTO 2018, LNCE 27, pp. 49–62, 2019.
https://doi.org/10.1007/978-3-030-12815-9_4

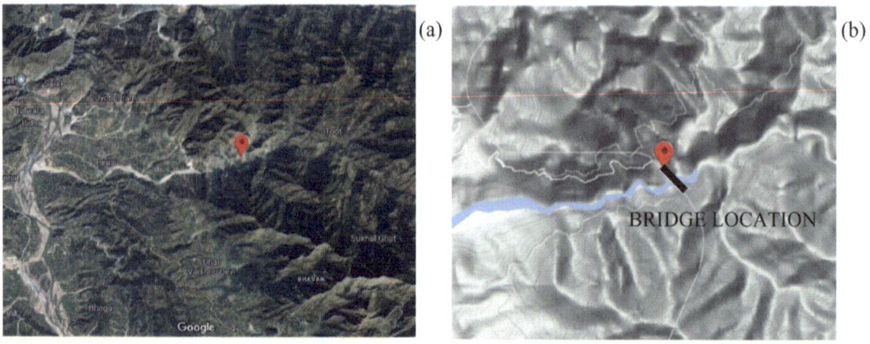

Fig. 1. Site plan (a) and zoom of the valley (b) where the bridge is located

Since the aerodynamic tests in wind tunnel were commissioned at a late stage of the design, only small modifications of the deck were possible, and to improve the aerodynamics small aerodynamics appendices were successfully tested, exploiting devices already tested on similar bridges (see Zasso et al. 2016, Amerio et al. 2017). A tailoring of the deck section geometry was carried out, finalized at a more streamlined shape, optimized to reach a positive slope of the aerodynamic static coefficients and positive aerodynamic damping on both the direct terms of the flutter derivative coefficients.

2 Bridge Location and Wind Characteristics

The structure in consideration is the Anji Khad bridge, a cable-stayed bridge proposed to connect Katra and Reasi Section of Jammu–Baramulla line in India. Since the bridge is located in a mountainous region, a study was performed by RWDI in order to analyse the topographical effects on the wind flows which would affect the design of the proposed bridge (RWDI 2006). A 1:3000 scale model including the bridge site as well as all major topographical features within a 7 km radius was constructed and tested in RWDI's boundary layer wind tunnel. The primary objectives of this study were to provide data on the variation of the mean wind speeds, flow direction and turbulence intensity at the proposed location of the bridge deck, and to determine design wind speeds at the bridge deck level for stability assessment and structural design.

The terrain surrounding the bridge deck is mountainous, except for a flat plateau to the west of the bridge site. For wind along the valley from the westward directions, the bridge is more exposed to the wind than it would be in other suburban terrain. Speedup effects were present in the topographical study for these wind directions.

The mean horizontal and vertical angles of attack are summarized in Fig. 2 at deck height only. These angles were obtained based on the mean speeds measured in the horizontal, lateral and vertical directions. The local wind directions are the middle set of values in Fig. 2a, and have been expressed in the same manner as the upper level wind directions which are the outer values. Figure 2a, which also shows the alignment of the bridge, suggests that the flow is channeled in a narrow range of local wind directions both up the valley ($\sim$230 to 260°) and down ($\sim$50 to 70°) and that these flow directions tend to be nearly normal to the bridge alignment.

The inner values in Fig. 2a show the vertical angle of attack. Large angles of attack can be seen for certain directions, especially over the mountains, these wind directions tend to have low mean wind speeds and are therefore unimportant for the bridge stability design (see Fig. 2b). For winds from the southwest, where the highest wind speeds were measured, the vertical angles of attack were nearly 5°, so standard analyses that consider only angles of attack near 0° only are not sufficient to ensure optimal aerodynamic performances for the bridge.

Note that a negative sign in Fig. 2 denotes downward flow (see conventions in Fig. 5) and that the angles of attack tend to follow the slope of the valley.

For flutter instability of the completed bridge, a very long return period needs to be considered because if instability occurs there is a very high probability of structural failure. A 10,000-year return period was chosen for the bridge studied in this paper, therefore the design wind speed for aerodynamic stability was calculated to be 54.9 m/s.

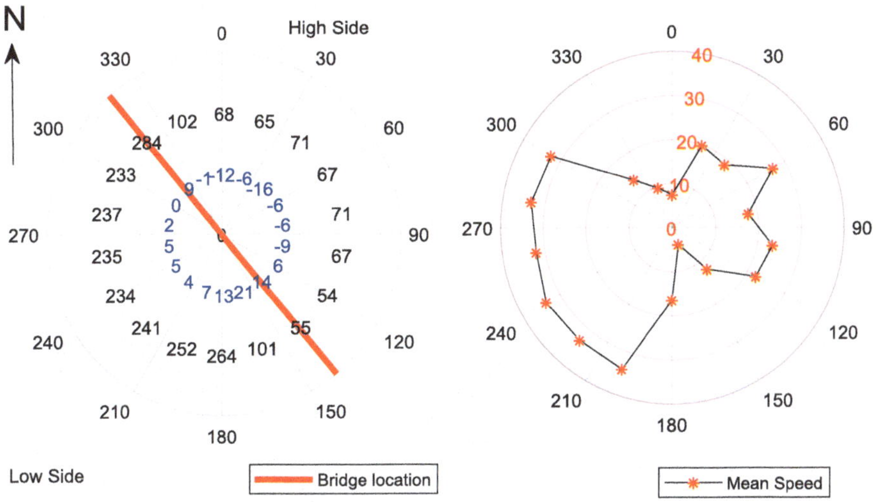

Fig. 2. (a) Mean vertical and horizontal angles of attack. Inner values (blue) are the vertical angles of attack, with negative angles denoting a downward flow. Middle values are the local wind horizontal directions, measured clockwise from North. (b) Mean wind speed at deck height, 50 year return period

3 Bridge Layout and Structural Dynamics

The deck cross section and the main dimension of the bridge studied in this paper are shown in Fig. 3. The structural properties are defined by a FEM model and the modal properties of the first vertical and torsional mode are reported in Table 1. Figure 4 shows the modal shapes in consideration, the most critical modes concerning instability issues.

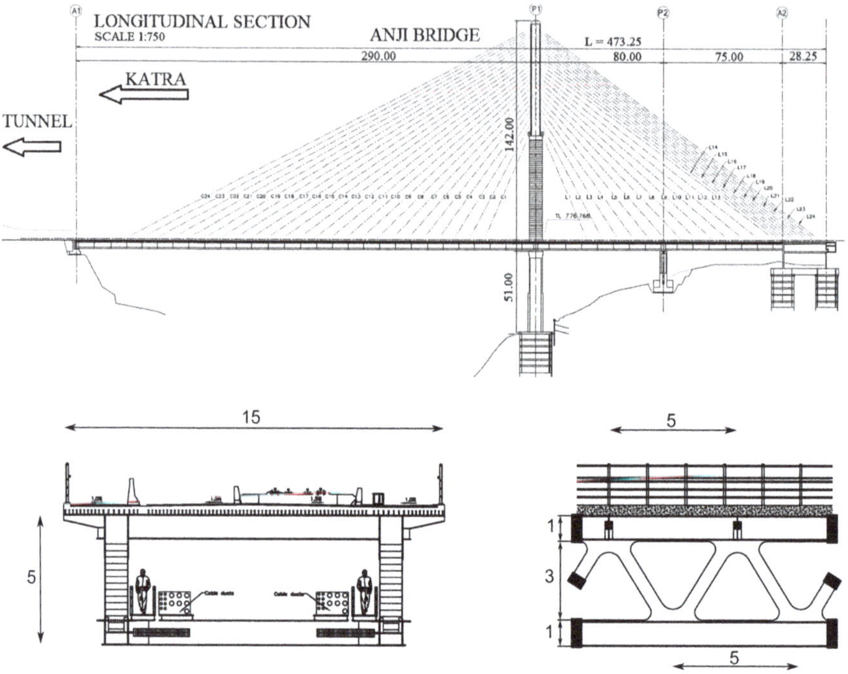

Fig. 3. Full bridge layout (top) and deck cross section (bottom-left), deck side view (bottom right)

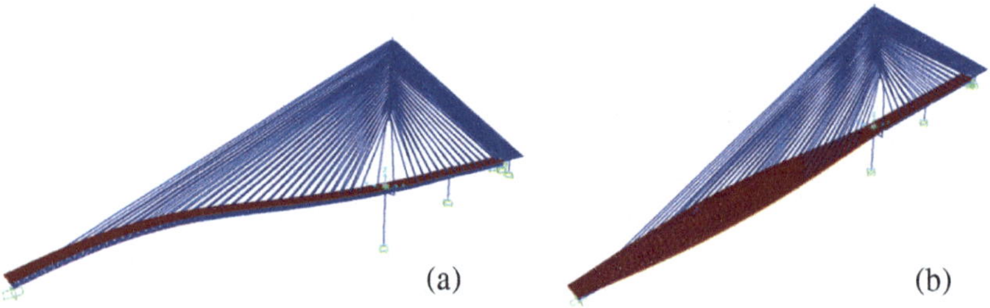

Fig. 4. (a) Modal shape of the 1st vertical mode and (b) modal shape of the 1st torsional mode

Table 1. Modal properties of the bridge

Mode	Effective modal mass per unit length	Frequency [Hz]	Damping ratio [-]
First vertical	36800 [kg/m]	0.4	0.003
First torsional	1297000 [kg m]	1.05	0.003

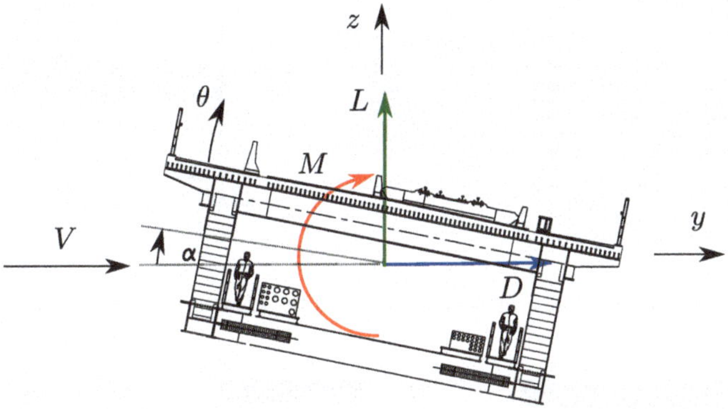

Fig. 5. Deck cross section and sign conventions

4 Wind Tunnel Tests on Sectional Model

Wind tunnel tests on the Anji Khad bridge deck section were carried out in the in the Boundary Layer test section of the Politecnico di Milano wind tunnel (GVPM), using a rigid sectional model. The primary objectives of this study were:

- Measurement (with two external 6-component balances) of the static aerodynamic coefficients on the sectional bridge model designed according to the nominal geometry
- evaluation of the Vortex Induced Vibrations (VIV) as a function of damping
- measurement of the aerodynamic derivatives (flutter derivatives h^*_{1-4} and a^*_{1-4} through free motion tests as a function of the reduced velocity V^*) and numerical estimation of the bridge stability.

The rigid sectional model of the bridge deck is in a geometrical scale of 1:20, with a chord-to-length ratio of 1/4. The deck chord is 0.75 m, depth 0.27 m and the model length is 3 m. The rigid sectional model of the bridge is made of a structural aluminium frame and honeycomb carbon and aluminium sandwich panel with plastic and wooden parts to reproduce the external aerodynamic shape. Figure 6a and b show a general overview of the schematic CAD model of the bridge deck that was tested.

For sectional model testing the test section width can be reduced by using an experimental test rig whose main characteristic is the presence of "End plates" used to confine the air flow allowing a 2-D fluid-structure interaction, see Fig. 7a. This test rig can be adopted both for static tests and dynamic tests (stability and VIV) using a suspension system with springs and tuneable dampers. A detailed description of the measurement setup is given in Amerio et al. (2017). Some pictures of the model in wind tunnel are shown in Fig. 7 (suspension system) and in Fig. 8.

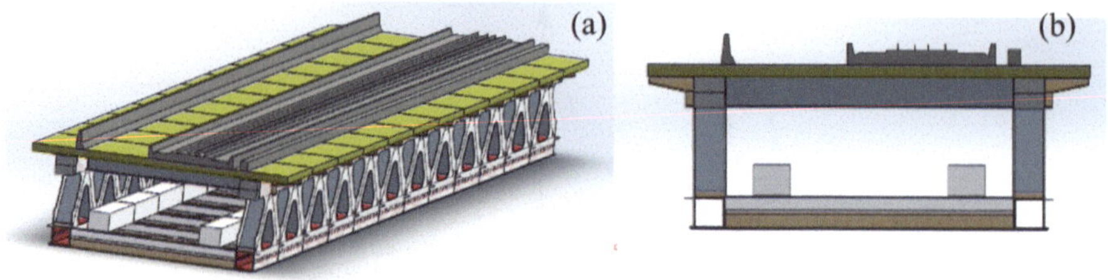

Fig. 6. Full bridge (a) sectional model composed by 12 modules, (b) deck cross section

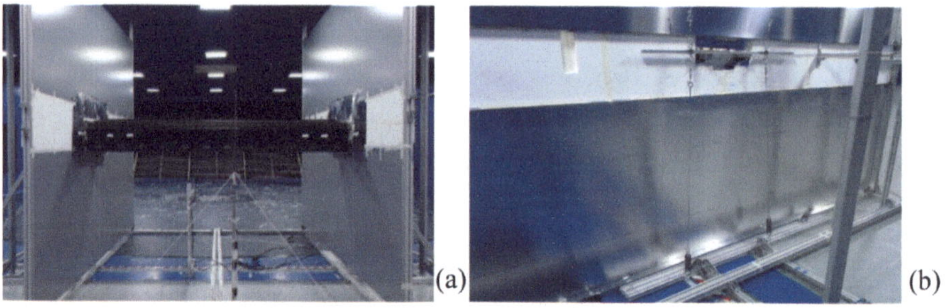

Fig. 7. (a) Overall view of the test rig, (b) suspension system with vertical springs

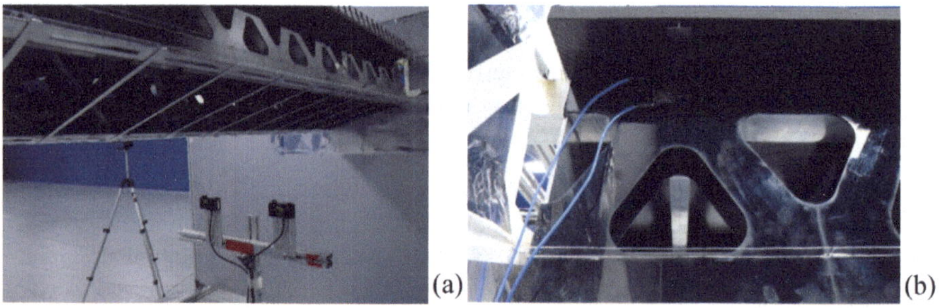

Fig. 8. (a) Laser pointing on the model to control the mean angle, (b) accelerometers installed on the model

5 Aerodynamic Behaviour of the Original Deck Section

Considering the topographical study of the construction site performed by RWDI, mean vertical angles of attack from −16° to +10° were investigated. In particular, for VIV tests four mean angles of attack were studied in the wind tunnel tests (−16, −6, +3, +6°), while for stability only three (−6, +3, +6°), since at −16° low wind speeds are expected (see Fig. 2). The angle +3° was chosen since the results showed negative slopes of the static lift and moment coefficients for positive mean angles of attack (as shown in the following) and slope of the lift coefficient, K_L, and of the moment, K_M, have a critical minimum at +3°.

5.1 Static and Dynamic Force Coefficients

The wind tunnel tests showed negative slopes of the static lift and moment coefficients for positive mean angles of attack (see Fig. 9). This negative slopes at positive angle of attack α are an index of likely 1-dof aeroelastic instability at relatively low reduced velocities, as it was verified during wind tunnel tests on suspended sectional model. The slope of the lift coefficient $K_L = dC_L/d\alpha$ and the moment coefficient $K_M = dC_M/d\alpha$ have a critical minimum at +3° (as reported in Table 2). This angle must be studied during the stability tests to verify the aerodynamic behavior of the bridge.

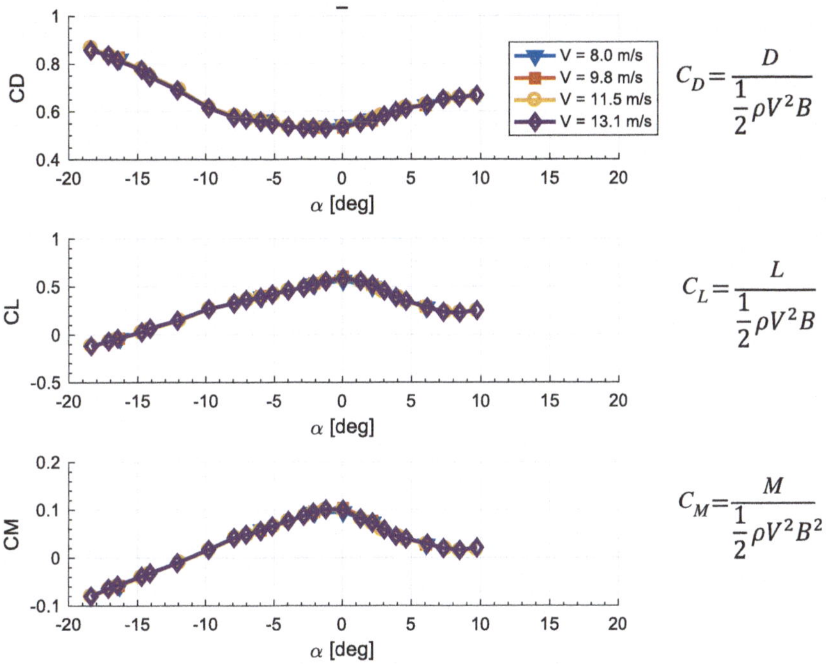

Fig. 9. Aerodynamic static force coefficients as a function of the mean angle of attack

Table 2. Slope of the lift and moment coefficients

Angle of attack α [deg]	K_L	K_M
−16	2.26	0.59
−6	1.78	0.51
−3	1.98	0.55
0	0.08	−0.43
+3	**−4.38**	**−0.96**
+6	−2.46	−0.42

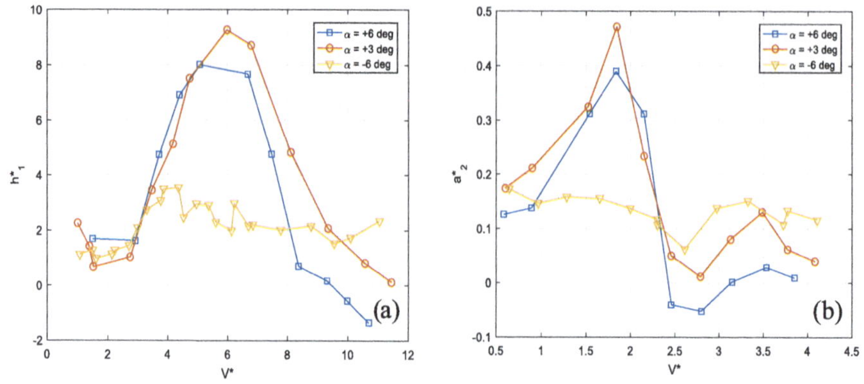

Fig. 10. Aerodynamic flutter derivatives (a) h_1^* and (b) a_2^* as a function of the mean angle of attack

Indeed, dynamic tests showed negative aerodynamic damping on both the direct terms of the flutter derivatives h_1^* and a_2^* at +6° of mean angle of attack, as shown in Fig. 10a and b. The aerodynamic self-excited forces are defined through the flutter derivatives as:

$$L_{se} = \frac{1}{2}\rho V^2 B\left(h_4^*\left(\frac{\pi}{2V_\omega^{*2}B}\right)z + h_3^*\theta\right) - \frac{1}{2}\rho VB(h_1^*\dot{z} + h_2^*B\dot{\theta}) \quad (1)$$

$$M_{se} = \frac{1}{2}\rho V^2 B\left(a_4^*\left(\frac{\pi}{2V_\omega^{*2}}\right)z + a_3^*\theta\right) - \frac{1}{2}\rho VB\left(a_1^*B\dot{z} + a_2^*B^2\dot{\theta}\right) \quad (2)$$

where L_{se} and M_{se} are the aerodynamic lift and the moment, ρ is the air density and $V_\omega^* = V^*/(2\pi)$. The reduced velocity is defined as $V^* = V/(fB)$, where V is the wind speed, f is the frequency considered and B the deck chord. The flutter derivatives h_4^* and a_3^* are the direct stiffness terms for the aerodynamic lift and the moment respectively, the flutter derivatives h_1^* and a_2^* are the direct damping terms, h_3^* and a_4^* are the non-diagonal elements for the stiffness terms, h_2^* and a_1^* for the damping terms.

A numerical analysis of the aeroelastic stability was performed using the effective modal parameters of the bridge. A 1-dof instability simulation was computed for the 1st vertical mode. If the quasi-steady theory is applied ($h_1^* = C_D + K_L$) the instability is expected at 32 m/s, considering a structural damping coefficient of the first vertical mode equal to 0.003. If the unsteady experimental coefficients h_1^* are used, the instability is expected at 61.8 m/s (see Fig. 11a), considering that the design 10-min mean wind speed is equal to 54.9 m/s.

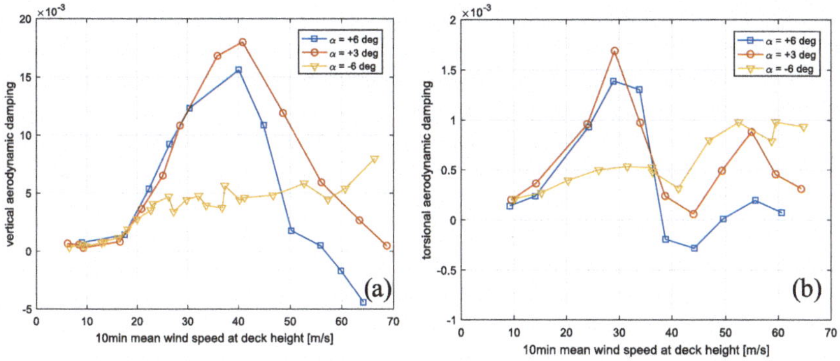

Fig. 11. (a) Vertical aerodynamic damping coefficient as a function of the mean wind velocity at deck height, (b) torsional aerodynamic damping coefficient as a function of the mean wind velocity at deck height.

A 1-dof instability simulation was computed also for the 1st torsional mode, but the critical instability onset does not occur even if the aerodynamic damping becomes negative at +6 deg mean angle of attack (see Fig. 11b).

Based on the experience and "best practices" rules accepted by the international wind engineering scientific and technical community, a deck section having negative aerodynamic damping on both the direct terms of the flutter derivatives h_1^* and a_2^* is not satisfactory. At this stage of the design, a revision of the deck section geometry finalized at a more streamlined shape, was performed and optimized to reach a positive slope of the aerodynamic static coefficients and positive aerodynamic damping on both the direct terms of the flutter derivatives h_1^* and a_2^*.

5.2 Vortex Induced Vibrations

Deck displacement due to vortex induced vibrations are shown in Fig. 12 as a function of the Scruton number and in terms of non-dimensional peak displacement. The Scruton number is calculated as $Sc_z = \frac{4\pi(m\zeta)}{\rho B^2}$ for the vertical oscillations, and $Sc_\theta = \frac{4\pi(J\zeta)}{\rho B^4}$ for the torsional ones. m and J are respectively the modal mass and modal mass moment of inertia, ρ is the air density, ζ is the non-dimensional damping and B the deck chord. In case of vertical oscillations, the non-dimensional displacement is z/B, while for torsional vibrations it is defined as z_{eq}/B, where z_{eq} is the equivalent displacement at the deck edge due to a deck rotation θ, calculated as $z_{eq} = B\theta/2$.

No VIV problems are expected for the original cross section of the bridge at the critical angles tested.

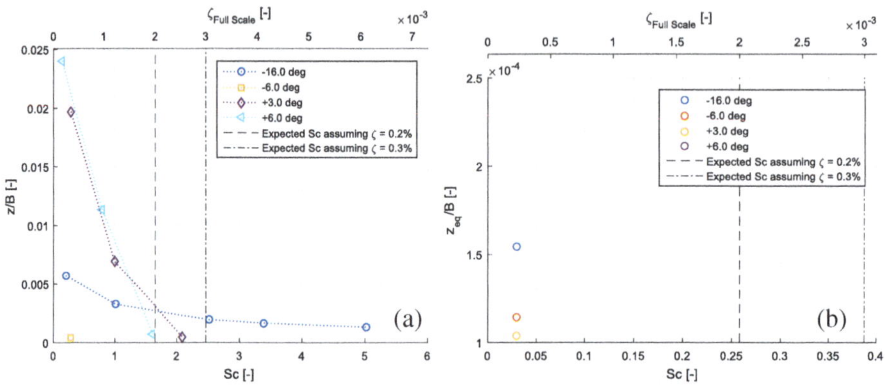

Fig. 12. Maximum vortex induced vibrations as a function of Sc (a) vertical motion, (b) torsional motion.

6 Tailoring of the Bluff Deck Section

An aerodynamic optimization of deck section was carried out to improve the aerodynamic stability of the deck section.

Six configurations were designed and tested in order to obtain a more streamlined "equivalent shape" of the deck section with leading and trailing edge noses, avoiding the stall of the moment and lift coefficients, see Fig. 13. The new deck sections are compared with the nominal one, called Configuration 00.

Configurations 01 and 02 evaluate the effect of the deck elongation on the static aerodynamic coefficients increasing the ratio of the section from 3:1 (Configuration 00) to 4:1, see Fig. 13a and b. The stall of the moment and lift coefficient is reached at larger angles, demonstrating that a more streamline shape is better, but they show again negative slopes of the lift and moment coefficients for positive mean angles of attack. Configurations 01 and 02 do not show any significant optimization of the lift and moment slopes (see Fig. 15).

Configuration 03 and 06 evaluate the effect of the length of the deck leading and trailing aerodynamic noses. Configurations 04 and 05 repeat configurations 03 and 06 respectively, adding a lower nose, so they evaluate the effect of the deck leading and trailing edge noses plus the effect of the lower noses, as shown in Fig. 13c–f.

Configurations 03, 04, 05 and 06 showed good results in terms of static coefficients.

The leading and trailing edge upper noses correct the aerodynamic behavior of the bluff deck section. These configurations show positive slopes of the lift and moment coefficients for positive mean angles of attack. Lower noses improve again the global behavior of the deck, but they add further changes and technical complexities to the bridge design. Table 3 reports the slope of the coefficients KL and KM as a function of the mean wind angle for the Configuration 06 compared with 01 and 00.

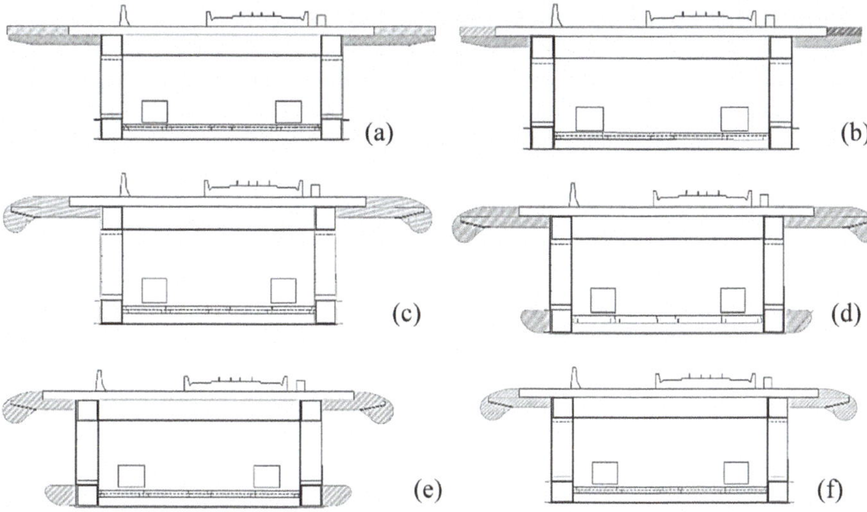

Fig. 13. Sketch of the cross section (a) Configuration 01, (b) Configuration 02, (c) Configuration 03, (d) Configuration 04, (e) Configuration 05, (f) Configuration 06

Fig. 14. (a) Sketch of the final cross section Configuration 06, (b) bridge model in the test section

Table 3. First derivative of lift and moment coefficients for Configuration 01 and Configuration 06

Angle of attack α [deg]	K_L Conf00	K_L Conf01	K_L Conf 06	K_M Conf00	K_M Conf 01	K_M Conf 06
−6	1.78	3.59	0.82	0.51	1.59	0.77
−3	1.98	3.60	0.55	0.55	1.39	0.79
0	0.08	3.68	−0.24	−0.43	0.89	0.54
+3	−4.38	1.10	−0.07	−0.96	−0.60	0.52
+6	−2.46	−4.83	0.81	−0.42	−1.54	0.60

7 Final Configuration Selection and Performance

Configuration 06 (see Fig. 14) was chosen in agreement with the bridge designer. With only short upper noses, Configuration 06 is the better solution in term of aerodynamic performance, technical complexities and costs. During dynamic tests the model of the bridge was suspended as done with the original Configuration 00. The flutter derivative h_1^* is always positive, the 1-dof instability never occurs considering the TQS and the dependence upon the reduced velocity V^* of h_1^*. Also for the 1st torsional mode, considering the dependence upon V^* of a_2^*, the 1-dof instability never occurs again and the aerodynamic damping is always positive.

A 2 DOF numerical estimation of the bridge stability was carried out. The aerodynamic lift and the moment are defined through the flutter derivatives, where the direct damping terms for the aerodynamic lift and the moment are defined through the flutter derivatives h_1^* and a_2^*. The direct stiffness terms for the aerodynamic lift and the moment are defined through the flutter derivatives h_4^* and a_3^*. h_3^* and a_4^* are the non-diagonal elements for the stiffness terms, h_2^* and a_1^* for the damping terms. The direct damping terms h_1^*, a_2^* and the direct stiffness terms h_4^* and a_3^* are identified using Fig. 16.

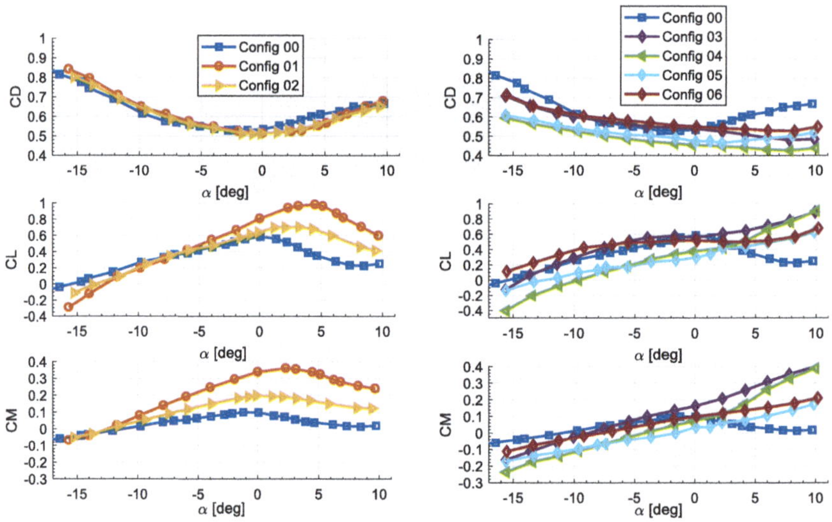

Fig. 15. (a) Aerodynamic static force coefficients for Configuration 01 and Configuration 02 with respect to Configuration 00. (b) Aerodynamic static force coefficients for Configuration 03, 04, 05 and 06 with respect to Configuration 00

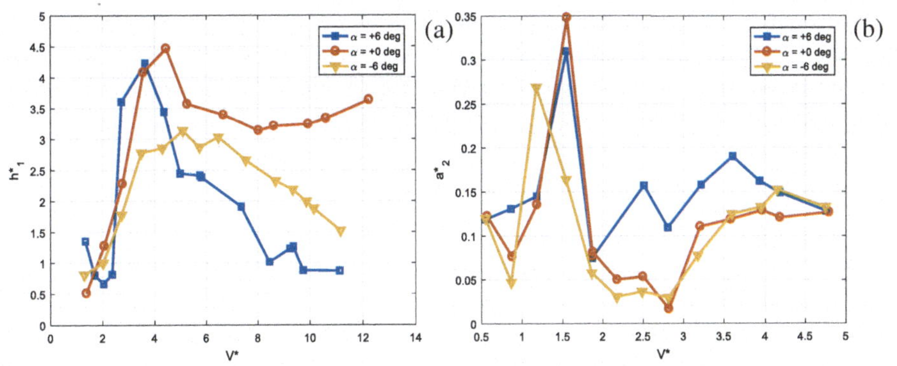

Fig. 16. Aerodynamic flutter derivatives (a) h_1^* and (b) a_2^* as a function of the mean angle of attack

DOF free motion decay tests. The non-diagonal coupling terms h_3^* and a_1^* are identified using 2 DOF free motion decay tests, through a best fitting of the eigenvalues trends (frequency and damping) and eigenvectors (mode shapes) function of the mean wind speed.

A 2 DOF numerical estimation (Argentini et al. 2014) of the bridge stability was carried out considering the identified flutter derivatives and the eigenvalues trend is shown in Fig. 17. The first vertical mode and the first torsional mode only are considered and a sensitivity analysis in carried out considering the mean angle of attack.

The coupling of mode shapes is weak, and no flutter speed is reached up to the design wind speed. The vertical and torsional damping are always increasing.

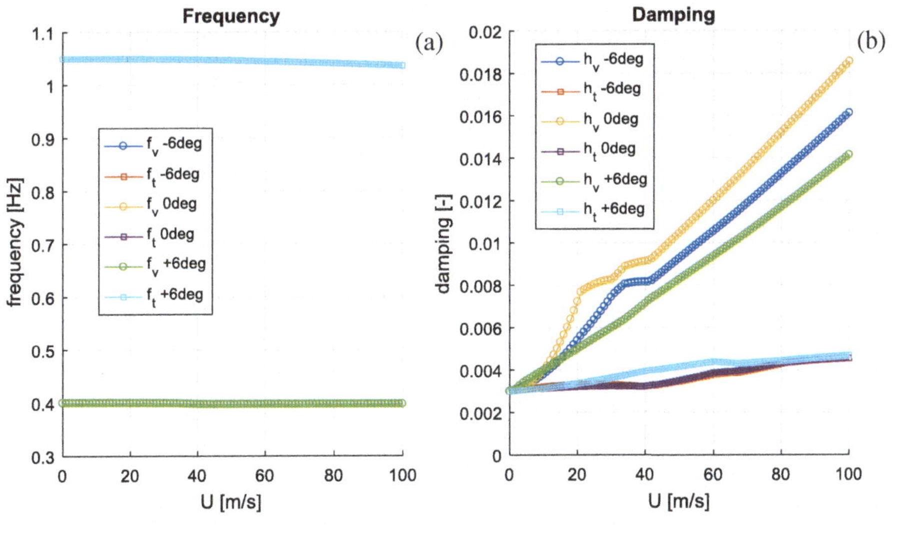

Fig. 17. Frequency (a) and damping ratio (b) as a function of the mean wind speed, considering the identified flutter derivatives

To conclude the aerodynamic study of the new cross section (Configuration 06), dynamic VIV tests were repeated for the more important angles −16, −6, +3, +6°. The short upper noses do not change the behavior of the bridge to VIV phenomena and no vortex induced vibrations are expected as for the original section of the bridge.

8 Conclusions

Bridges crossing steep valleys in mountainous regions may be subjected to large angles of attack of the incoming wind flow, because of the complex orography of the region where the structure is located. In this paper, the aerodynamic characterization of a bluff deck section of a railway bridge located in a steep valley, is performed. Special care on the analysis of the local winds and inclined flows must be kept in consideration, because large angles of attack may be seen for certain directions and where the highest wind speeds are measured, standard analyses that consider only angles of attack near 0° only are not sufficient to ensure optimal aerodynamic performances for the bridge. Without considering the large positive mean angle of attack, the original deck section would have performed satisfactorily for the bridge studied in this paper, but for positive angles of attack (+6°), critical conditions of instability are found. At this late stage of the design, only small modifications of the deck were possible, and to improve the aerodynamics small aerodynamics appendices were successfully tested, exploiting devices already tested on similar bridges. A tailoring of the deck section geometry is carried out, finalized at a more streamlined shape, trying to delay the stall as much as possible. The cross section is optimized to reach a positive slope of the aerodynamic static coefficients and positive aerodynamic damping on both the direct terms of the flutter derivative coefficients, also for low reduced velocities and a numerical verification of the full bridge response is computed with the complete set of flutter derivatives.

For this kind of bridges with a bluff section, upper noses seem to give a beneficial effect on the slope of the static coefficients and on the aeroelastic behavior. Moreover, for non-standard bridges and locations, the aerodynamic study of the deck section must be kept in consideration at an early stage of the design, because if instability occurs there is a very high probability of structural failure.

References

Zasso A, Rocchi D, Argentini T, Giappino S, Costantini T (2016) Experimental and numerical aerodynamic analysis of a concrete railway bridge in tandem arrangement with a truss road bridge. In: 8th international colloquium on bluff body aerodynamics and applications

Amerio L, Argentini T, Rocchi D, Zasso A (2017) Experimental and numerical aerodynamic optimization of a post-tensioned concrete railway bridge in tandem arrangement with a truss road bridge. In: 7th European African conference on wind engineering

RWDI (2006) Topographical model study Anji Khad arch bridge Jammu e Kashmir India, Final report

Argentini T, Pagani A, Rocchi D, Zasso A (2014) Monte Carlo analysis of total damping and flutter speed of a long span bridge: effects of structural and aerodynamic uncertainties. J Wind Eng Ind Aerodyn 128:90–104

A SCADA-Based Method for Estimating the Energy Improvement from Wind Turbine Retrofitting

D. Astolfi[1(✉)], F. Castellani[1], M. L. Fravolini[1], S. Cascianelli[1], and L. Terzi[2]

[1] Department of Engineering, University of Perugia, Via Duranti, 06125 Perugia, Italy
{davide.astolfi,francesco.castellani,mario.fravolini, silvia.cascianelli}@unipg.it
[2] Renvico srl, Via San Gregorio 34, 20124 Milan, Italy
ludovico.terzi@renvico.it

Abstract. Retrofitting of operating wind turbines has been spreading in the recent years in the wind energy industry, at the aim of optimizing the efficiency of wind kinetic energy conversion. This kind of interventions has material and labor costs and it is therefore fundamental to estimate realistically the production improvement. Further, the retrofitting of wind turbines sited in harsh environments, as for example complex terrain, might exacerbate the stressing conditions and therefore affect the residue lifetime. This work deals with a case of retrofitting: the testing ground is a multi-megawatt wind turbine from a wind farm sited in a very complex terrain. The blades have been optimized by installing vortex generators and passive flow control devices. A general method is proposed for estimating production upgrades from wind turbine retrofitting, basing on multivariate linear modeling of the power output of the upgraded wind turbine. Applying the model to the test case of interest, it arises that the upgrade increases the annual energy production of the wind turbine of an amount of the order of the 2%.

1 Introduction

The retrofitting of operating wind turbines has been spreading in the wind energy industry, in order to improve the efficiency of the conversion of wind kinetic energy. This kind of interventions has material and labor cost and producible energy is lost during the installation: the priority of wind farm owners dealing with wind turbine upgrades is therefore to estimate realistically the profitability. Furthermore, in perspective, the adoption of upgrades might exacerbate stressing mechanical conditions to which wind turbines are subjected: for example, a typical control system upgrade arising this kind of issues is the extension of the wind turbine operation above the cut-out wind speed [4,12]. For this reason, wind turbine upgrades is a relatively novel topic in wind energy scientific literature but it is attracting a certain attention.

F. Ricciardelli and A. M. Avossa (Eds.): INVENTO 2018, LNCE 27, pp. 63–72, 2019.
https://doi.org/10.1007/978-3-030-12815-9_5

Computing the energy improvement from a wind turbine upgrade is not straightforward because wind turbines operate under non-stationary conditions and therefore it makes little sense to compare the pre-upgrade production against the post-upgrade one. It makes much more sense to compare the post-upgrade production against a reliable simulation of the pre-upgrade production in the same conditions. Some recent relevant studies in the subject are [8] and [7]. In [8], a kernel plus method is adopted for computing wind turbine performance upgrades. The test cases are pitch angle control optimization and aerodynamic retrofitting. In [7], an academia-industry joint study is presented and the test cases are two onshore wind farms that have been upgraded through the installation of vortex generators. The approach is twofold: on one hand, SCADA data with 10 min of sampling time are employed; on the other hand, high-frequency power data are employed: the estimates of production improvement are shown to be similar.

This study also is a partnership between academia and industry: the involved company is Renvico[1], owning and managing 335 MW of wind turbines in Italy and France. One multi-megawatt wind turbine from a wind farm sited in a very complex terrain [2,5,6] has been upgraded through the installation of vortex generators [11] and passive flow control devices, increasing the lift and therefore the energy production. This has been a pilot test and the decision of adopting the retrofitting on the whole wind farm has been based on the quantification of the benefits by means of this study. One key point of this study is that the estimates of the undisturbed wind speed at the retrofitted wind turbine are unreliable: this implies that a straightforward power curve analysis is impossible and, moreover, the wind speed of the upgraded wind turbine can't be used as input to model its power output. The wind turbines nearby the upgraded one must therefore be used as reference for the wind conditions, but this is tricky because of the complexity of the terrain: the model must therefore be formulated with care. Some first developments about this problem have been achieved in [3] and [1] using an Artificial Neural Network approach. The objective of this work is overcoming the limitations of the results in [3] and [1] and providing a more methodological approach for studying in general the problem of quantifying the production improvement from wind turbine upgrades. On these grounds, an appropriate multivariate linear model has been searched for by means of stepwise regression. Stability and quality of the results have been the required benchmarks and therefore particular care has been devoted to the validation of the model.

Therefore, the outcome of this work is not only the computation of the production improvement (2% of the annual energy production is the estimated order of magnitude), but also the method for selecting appropriate input and validating the model. The model is generalizable to the study of whatever kind of wind turbine upgrades and having tested it for the study of an upgrade in complex terrain on a data set having a severe drawback represents a sort of crash course in its solidity and versatility.

[1] www.renvicoenergy.com.

The structure of the paper is therefore the following: in Sect. 2, the test case and the data sets are described. Section 3 is devoted to the description of the method; the results are collected in Sect. 4; conclusions are drawn and further directions are indicated in Sect. 5.

2 The Test Case and the Data Set

In Fig. 1, the layout of the wind farm is sketched. From the contour lines, it is possible to appreciate the complexity of the terrain. T7 (in red in Fig. 1) is the upgraded wind turbine. The wind turbines are multi-megawatt and the hub height is 80 m above ground level.

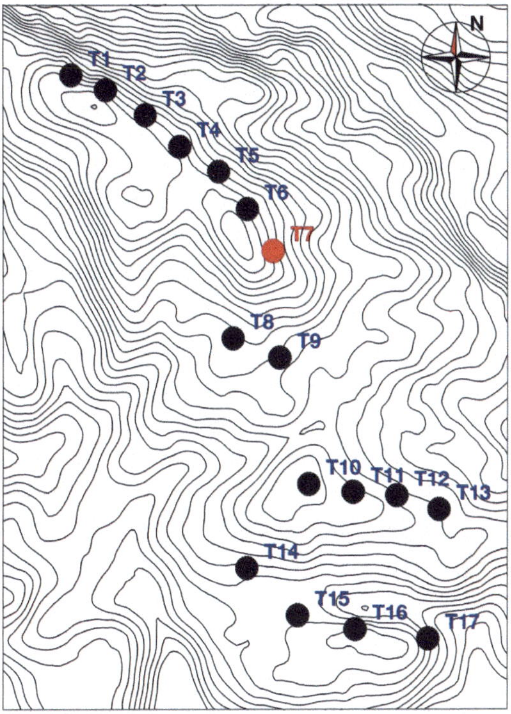

Fig. 1. The layout of the wind farm.

The data sets at disposal are the following:

- D_{bef} goes from 01/01/2016 to 01/07/2017. It is a period during which the standard blade configuration was adopted.
- D_{aft} goes from 01/09/2017 to 01/04/2018. It is a period during which T7 has been operating with the improved blade configuration.

The data sets have 10 min of sampling time, as typical in the SCADA technology for wind turbines.

3 The Method

Before discussing the method, it is interesting to motivate why a straightforward power curve analysis can't be performed for a qualitative assessment of production improvement. In Fig. 2, the power coefficient

$$C_p = \frac{P}{\frac{1}{2}\rho A v_\infty^3} \tag{1}$$

as computed from the SCADA data during the D_{aft} data set is reported as a function of the power production for the upgraded turbine T7 and for a sample wind turbine from the rest of the wind farm (T2). P is the measured power output, ρ is the air density on site, A is the blade swept area and v_∞ is the undisturbed wind speed as reconstructed from the nacelle wind speed through the nacelle transfer function. From Fig. 2, it arises that the C_p estimates for T7 are often implausible, because well above the Betz limit. This result is already sufficient to exclude the possibility of studying the power curve of T7 wind turbine after the upgrade.

Basing on the Eq. 1, the bias in the C_p estimates likely is in the reliability of v_∞. A further study has been performed in order to inquire if the unreliability of v_∞ estimates at T7 is connected to the installation of the flow control devices on the wind turbine blades. The fraction of C_p estimates above the Betz limit (0.593) is computed for T7 during data sets D_{bef} and D_{aft}: it arises that the 27% of C_p estimates are above Betz limit after the upgrade and the 3% are above Betz before the upgrade. For T2 (or whatever sample wind turbine from the wind farm), the percentage of C_p estimates above Betz doesn't significantly change in the data sets D_{bef} and D_{aft} and is of the order of few percents. On these grounds, it has been hypothesized that the unreliability of v_∞ estimates is connected to the blades retrofitting: the speculation is that the nacelle transfer function might haven't been updated and the wind turbine control systems fails in estimating reliably the undisturbed wind speed v_∞ from the wind speed measurements behind the rotor. This possible explanation has been deduced uniquely from the data at disposal and no information has been provided by the wind turbine manufacturer.

From the above results, it follows that it is necessary to adopt the wind turbines nearby T7 as references for constructing the model. The output y of the model is the power output of T7. The possible input data at disposal, from the remainder 16 wind turbines in the farm, are:

- nacelle wind speed,
- power output,
- individual blade pitch angles,
- rotor revolutions per minute,
- high speed rotor temperature.

The data are filtered on the conditions of power output production from all the wind turbines in the farm and of power production of T7 below the rated, because at rated power the upgrade has no visible effect.

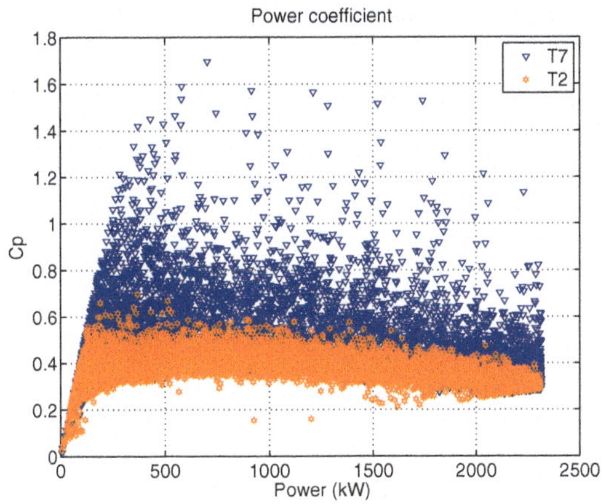

Fig. 2. The power coefficient C_p, as computed from the SCADA data, vs. nacelle wind speed: T7 and a sample wind turbine (T2), D_{aft} data set.

In the following, the procedure is reported for selecting the inputs non-ambiguously and optimally. In general, the measurements of the output y are assembled in a column vector $\boldsymbol{y}$ and the values of the inputs $\boldsymbol{x}$ are assembled in a matrix $\underline{x}$. A stepwise regression is performed: it is a systematic method for adding and removing terms from a multilinear model, based on their statistical significance in a regression. The method begins with an initial model and then compares the explanatory power of incrementally larger and smaller models. In this work, the models are distinguished by giving different values of the p_{remove} parameter: it is an exit tolerance for the probability of rejecting the hypothesis of a zero coefficient for a given input in the model. The tested values of p_{remove} are 10^{-p} with $p = 1, \ldots, 10$.

For each value of p_{remove}, the model is subjected to k-fold cross-validation [13]: the data set D_{bef} is divided J times randomly in (100-k)% of the data for training the model and k% for validating. k has been selected to be 10 for this study, because the objective is validating the model for very short folds, in order to test its robustness. J is selected to be 300.

For each value $j = 1, \ldots, J$ of the stepwise regression for a given p_{remove}, the most significant inputs are selected. The training input matrix $\underline{x}_{train_{j,p}}$, constructed with the selected inputs using the measurements selected for the training, is normalized and its pseudo-inverse is used to compute the weight matrix as

$$\underline{W}_{j,p} = \underline{x}^{-1}_{train_{j,p}} \cdot \boldsymbol{y}_{train_{j,p}}.$$

The simulated output on the validation data set is computed as

$$\hat{\boldsymbol{y}}_{valid_{j,p}} = \boldsymbol{x}_{valid_{j,p}} \cdot \underline{W}_{j,p}.$$

For each of the J runs, the mean absolute error

$$\bar{\delta}_{j,p} = \overline{|\hat{\boldsymbol{y}}_{valid_{j,p}} - \boldsymbol{y}_{valid_{j,p}}|}$$

is computed. Subsequently, it is possible to average the $\bar{\delta}_{j,p}$ over j and obtain the average mean absolute error for a given p:

$$\bar{\delta}_p = \sum_{j=1}^{J} \frac{\bar{\delta}_{j,p}}{J}$$

In Table 1, the results of the cross-validation are reported.

Table 1. Results of the model validation

p	$\bar{\delta}_p$ (kW)	Number of different regressor configurations
1	97	13
2	97	11
3	98	6
4	98	6
5	98	6
6	99	8
6	99	6
8	100	5
9	100	4
10	100	2

It arises that $\bar{\delta}_p$ is of the same order of magnitude for each value of p_{remove} ($\simeq$100 kW): the improvement in choosing $p_{remove} = 10^{-1}$ instead of $p_{remove} = 10^{-10}$ is averagely only 3 kW in δ_p, at the cost of employing many more regressors (more or less 20, against 3) and having a less stable model (the number of possible regressor configurations is much higher). Choosing $p_{remove} = 10^{-10}$, instead, one obtains for the highest number of times (291 out of 300) that the selected inputs are the same and they are:

- the power output of T6,
- the power output of T9,
- the rotor revolutions per minute of T8.

Therefore, the decision is modeling the power output of T7 as a function of the three inputs above. In the following Sect. 4, it shall be shown how to use this model to obtain an estimate of the energy improvement since the adoption of the T7 upgrade.

4 The Results

The data sets at disposal are employed as follows:

- D_{bef} is randomly divided in two subsets: D0 (1 year of data) and D1 (6 months of data). D0 is used for training the model and constructing the weight matrix $\underline{W}$, D1 is used for validating the model.
- D_{bef} (also named D2 for simplifying the notation in the following) is used to quantify the performance improvement.

The approach is based on the analysis, for the data sets D1 and D2, of the residuals between the measurement y and the estimation $\hat{y}$ as a function of the inputs $\boldsymbol{x}$ (power of T6, power of T9, rotor revolutions per minute of T8, as indicated in Sect. 3). In other words, the interest is in how the residuals vary after the upgrade with respect to before. Therefore, consider Eq. 2 with $i = 1, 2$.

$$R(\boldsymbol{x}_i) = y(\boldsymbol{x}_i) - \hat{y}(\boldsymbol{x}_i). \tag{2}$$

For $i = 1, 2$, one computes

$$\Delta_i = 100 * \frac{\sum_{\boldsymbol{x} \in Data_i} (y(\boldsymbol{x}) - \hat{y}(\boldsymbol{x}))}{\sum_{\boldsymbol{x} \in Data_i} y(\boldsymbol{x})} \tag{3}$$

Since Δ_i is constructed with the relative discrepancies of power data each having the same sampling time (10 min), the quantity $\Delta = \Delta_2 - \Delta_1$ provides a percentage estimate also of the energy improvement.

The above procedure has been repeated several times with several random choices of D0 and D1 for training and validating the model. In Table 2, statistics are reported for the discrepancy between model and measurement for the different choices of the D0 and D1 data sets: average residual between measurement and simulation (δ_{ave}), average mean absolute error ($\bar{\delta}_{ave}$), average standard deviation of the residuals (σ_{δ}).

Table 2. Statistic behavior of the residuals between measurement and simulation, for the different random choices of the D1 data set.

δ_{ave} (kW)	$\bar{\delta}_{ave}$ (kW)	σ_{δ} (kW)
−0.5	100	146

The above procedure allows to have not only an average estimate of the energy improvement, but also a standard deviation (and therefore reasonable lower and upper limits for the energy improvement).

In Figs. 3 and 4, some results are reported. The measured and estimated power of T7 is plotted, for a sample choice of D0 and D1 (and therefore for a sample run of the model). From these figures, a qualitative picture is obtained

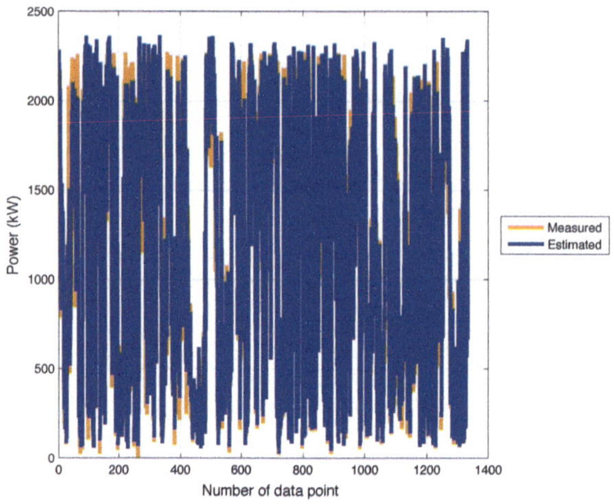

Fig. 3. The measured and estimated power of T7 before the upgrade, for a sample choice of D1 and a sample run of the model.

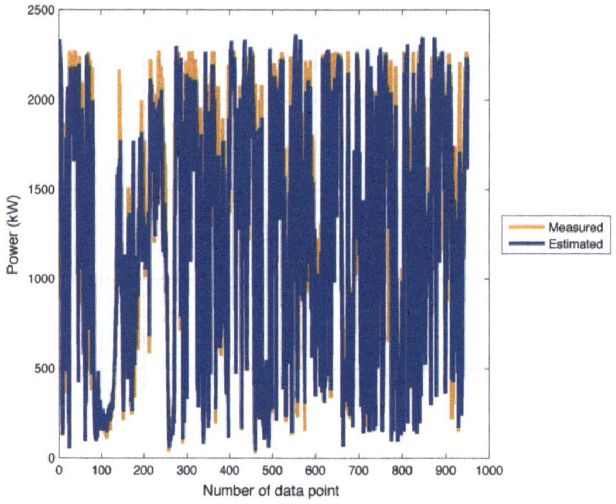

Fig. 4. The measured and estimated power of T7 after the upgrade, for a sample run of the model.

about the fact that, consistently, the difference between measurement and model increases in the post-upgrade period.

The average energy improvement is computed as $\Delta = 3.9\%$. In other words, the estimate is that, since the upgrade has been installed, T7 has produced below rated power the 3.9% more than it would have done without aerodynamic improvement. The standard deviation is computed as $\sigma_{\Delta} = 0.4\%$: therefore, reasonable upper and lower limits of the energy improvement are $\Delta_{+} = 4.3\%$

and $\Delta_{-} = 3.5\%$. This corresponds to an estimate of the improvement of annual energy production of the order of $\Delta_{AEP} = 2.0 \pm 0.2\%$.

5 Conclusions and Further Directions

In this work, a test case of wind turbine retrofitting through installation of vortex generators and passive flow control devices has been studied using SCADA data. The objective has been the computation of the production improvement since the adoption of the upgrade.

This kind of problems is non-trivial in general, because wind turbines are subjected to non-stationary operation conditions and therefore the most appropriate approach is comparing the post-upgrade production against a simulation of the pre-upgrade production under the same conditions.

In particular, this test case is challenging for two reasons: the first is that the data set at disposal has a severe limitation. Actually, the nacelle anemometer of the upgraded wind turbine is unreliable in the post-upgrade period. Therefore, the wind turbines nearby the upgraded one must be employed as reference of the external conditions and here comes the second peculiarity of the test case: the wind farm is sited in a very complex terrain and therefore it is difficult to select the appropriate inputs for modeling the power of the upgraded wind turbine. This has stimulated the formulation of a general method for studying wind turbine upgrades, basing on stepwise regression for selecting the most appropriate inputs for modeling a given output. In particular, it has been observed that in this complex-terrain test case the trickiness relies mainly in selecting a robust model.

One main result of this work is that the value of the upgrade has been estimated of the order of the 2% of the annual energy production. It is important to notice that it is of the same order of magnitude as the estimate provided by the wind turbine manufacturer, but it is non-negligibly lower. This can be explained by the fact that commonly wind turbine upgrades are estimated through simulation or field tests in conditions that are quite different from the ones to which wind turbines operating in real environments are subjected. This observation motivates a valuable further direction of this work: it is important to elaborate on the wake-terrain interaction in this environment and to study how the mechanical conditions of the upgraded wind turbine have changed and could possibly affect the residue lifetime.

Another main result of this work is the robust and generalizable method: actually, the study of wind turbine upgrades is conceptually and practically quite different from power curve modeling (a subject about which there is plenty of scientific literature: see for example [9,10,14]). Therefore, another valuable further direction of this work is extending to other test cases: different wind turbine upgrades, different environments, different peculiarities of the data sets. Furthermore, it is possible to explore non-linear approaches to the problem, assess the added value and investigate what kind of problems actually require non-linearity to be successfully tackled.

References

1. Astolfi D, Castellani F, Terzi L (2018) A SCADA data mining method for precision assessment of performance enhancement from aerodynamic optimization of wind turbine blades. J Phys Conf Ser 1037:032001
2. Astolfi D, Castellani F, Terzi L (2018) A study of wind turbine wakes in complex terrain through RANS simulation and SCADA data. J Sol Energy Eng 140(3):031,001
3. Astolfi D, Castellani F, Terzi L (2018) Wind turbine power curve upgrades. Energies 11(5):1300
4. Bossanyi E, King J (2012) Improving wind farm output predictability by means of a soft cut-out strategy. In: European Wind Energy Conference and Exhibition EWEA, vol 2012
5. Castellani F, Astolfi D, Burlando M, Terzi L (2015) Numerical modelling for wind farm operational assessment in complex terrain. J Wind Eng Ind Aerodyn 147:320–329
6. Castellani F, Astolfi D, Mana M, Piccioni E, Becchetti M, Terzi L (2017) Investigation of terrain and wake effects on the performance of wind farms in complex terrain using numerical and experimental data. Wind Energy 20(7):1277–1289
7. Hwangbo H, Ding Y, Eisele O, Weinzierl G, Lang U, Pechlivanoglou G (2017) Quantifying the effect of vortex generator installation on wind power production: an academia-industry case study. Renewable Energy 113:1589–1597
8. Lee G, Ding Y, Xie L, Genton MG (2015) A kernel plus method for quantifying wind turbine performance upgrades. Wind Energy 18(7):1207–1219
9. Lydia M, Kumar SS, Selvakumar AI, Kumar GEP (2014) A comprehensive review on wind turbine power curve modeling techniques. Renew Sustain Energy Rev 30:452–460
10. Ouyang T, Kusiak A, He Y (2017) Modeling wind-turbine power curve: a data partitioning and mining approach. Renewable Energy 102:1–8
11. Øye S (1995) The effect of vortex generators on the performance of the ELKRAFT 1000 kw turbine. In: 9th IEA symposium on aerodynamics of wind turbines. Stockholm. ISSN 0590–8809
12. Petrović V, Bottasso CL (2014) Wind turbine optimal control during storms. J Phys Conf Ser 524:012052
13. Refaeilzadeh P, Tang L, Liu H (2009) Cross-validation. In: Encyclopedia of database systems. Springer, pp 532–538
14. Shokrzadeh S, Jozani MJ, Bibeau E (2014) Wind turbine power curve modeling using advanced parametric and nonparametric methods. IEEE Trans Sustain Energy 5(4):1262–1269

Wind Effects on Containers Stability

A. Balbi(✉), M. P. Repetto, G. Solari, A. Freda, and G. Riotto

Department of Civil, Chemical and Environmental Engineering (DICCA), University of Genoa, Genoa, Italy
alberto.balbi@dicca.unige.it

Abstract. Due to their natural location, ports are very exposed to extreme weather conditions. Considering in particular terminal containers, the piled containers can fall and get damaged under strong wind. This paper describes a wide research activity aimed at the evaluation of wind effects on containers stability. The analytical formulation of stability conditions shows that they depend on three groups of parameters: the local wind field and turbulence parameters, the friction coefficient between the containers surface and the aerodynamic coefficients of piled containers. The friction coefficient has been analysed by means of in situ tests on the piled containers. The aerodynamic coefficients have been studied by means of wind tunnel tests on scaled models. The main results are shown and discussed.

Keywords: Aerodynamic coefficients · Containers · Seaports · Wind tunnel tests

1 Introduction

The seaports are critical infrastructures very exposed to extreme weather actions like wind; the consequences of these actions can have very destructive effects on many terminal activities (Lam et al. 2017). Terminal containers are crucial areas due to the strategic importance of the movement of goods. Containers, piled in the terminal yard, can fall and get damaged under strong winds, causing serious safety problems and economic losses. For this reason, companies usually adopt safety systems furnishing alert and regulating the terminal operability on the basis of the real time wind velocity measured in terminal areas.

The University of Genova (Italy) developed a wide research activity in this field, with reference to the European Projects "Wind and Ports" (Solari et al. 2012) and "Wind, Ports and Sea" (Repetto et al. 2017), handling the problem of the wind safe management and risk assessment of the North Tyrrhenian seaports. The projects developed a web-based GIS platform integrated into the operational management of the involved ports, making directly available in real time in situ monitoring wind velocities, wind and wave medium- and short-term forecasting, statistical analyses and wind climatological mapping (Repetto et al. 2018). However, in literature there is a lack of information about wind stability of the piled containers (Zhen et al. 2011); the critical wind velocity values at the base of the safety systems are still based on port operators experience and affected by many uncertainties.

F. Ricciardelli and A. M. Avossa (Eds.): INVENTO 2018, LNCE 27, pp. 73–86, 2019.
https://doi.org/10.1007/978-3-030-12815-9_6

This paper presents a step forward to a complete formulation of the safety management procedure against extreme wind in containers terminals. Firstly, the sliding, tilting and lifting stability problem is considered, identifying the main parameters governing the problem. Among them, the fundamental parameters are the local wind field and turbulence parameters, the friction coefficient between the containers surface and the aerodynamic coefficients of piled containers. The friction coefficient has been analyzed by means of in situ tests on the piled containers (Balbi et al. 2018). The results show a large variability, due to the surface conditions and correct positioning of the piled containers. The aerodynamic coefficients have been studied by means of wind tunnel test on scaled models. Several set-up conditions have been reproduced, considering seven different container storage configurations, two different incoming wind profiles and two different container's tiers, obtaining as many aerodynamic coefficient sets of values. The paper shows in detail the wind tunnel tests and the obtained aerodynamic coefficients in terms of pressure and force coefficients, discussing the variability of the values. The conclusions summarize the obtained results and describe the perspectives of the analysis.

2 Container's Stability Conditions

Considering the container as a rigid body placed at the top of a containers pile, let define a global reference system, *X, Y, Z* with origin *O* on the ground; *X*- and *Y*-axes are coplanar with ground; *Z* is directed upward; let define also a local reference system *x, y, z*, with origin *o* in the center of the container basis; *y* axis aligned with the container; *z* is directed upward (Fig. 1a, b).

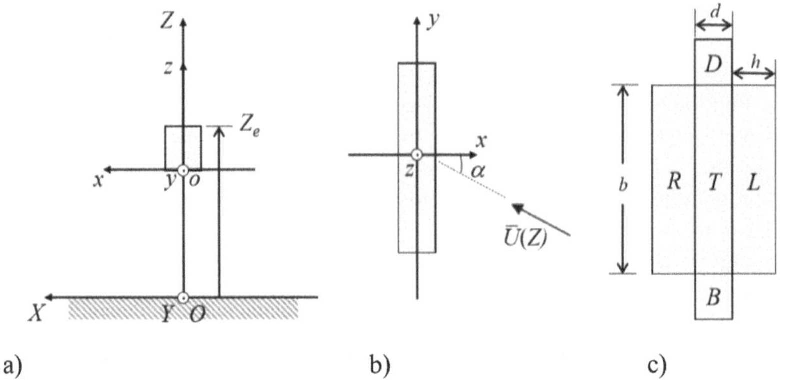

Fig. 1. (a) Global and local Cartesian reference systems: (a) Lateral view (b) Plan view; (c) Scheme and denomination of the sides of the container exposed to wind

The container is immersed into an Atmospheric Boundary Layer (ABL) wind field, described by a horizontal 10-min mean wind velocity with intensity $\bar{U}(Z)$ and direction (Fig. 1b) and a stationary, Gaussian, three-variate turbulence process, composed by the longitudinal (horizontal, aligned with $\bar{U}(Z)$), lateral (horizontal, orthogonal to $\bar{U}(Z)$) and vertical turbulence components. In the following, only the longitudinal turbulence effect is taken into account. The peak wind velocity at the site can be defined according to the relation:

$$\hat{U}(Z) = \bar{U}(Z)[1 + g_U(Z)I_U(Z)] \tag{1}$$

where I_U is the site turbulence intensity, g_U is the peak coefficient (CNR 2008).

Figure 1c represent the scheme of the container, with the denomination of the sides exposed to wind. It is worth noting that, when $\alpha = 0$, the wind is orthogonal to the container, hitting the lateral side L.

The wind induces aerodynamic peak pressures on the container surfaces are given by

$$p(x, y, z) = \frac{1}{2}\rho\hat{U}(Z_e)c_p(x, y, z) \tag{2}$$

where x, y, z are the local coordinates of the point on the surface, ρ is the air density, Z_e is the global reference height, assumed here as the maximum height of the container (Fig. 1a), c_p is the aerodynamic (external) pressure coefficient. Conventionally, pressure are positive when producing an overpressure, negative when producing a suction on the surface. Note that, assuming the container's doors locked, the internal pressures are disregarded.

The global aerodynamic loads on the container can be described by means of three forces, F_x, F_y, F_z, and three moments, M_x, M_y, M_z, with respect to the origin o of the local reference system. They can be obtained as resultant forces and moments of the pressure distributions in Eq. (2) (CNR 2008).

The critical conditions of the containers stability can be defined by imposing the wind-induced forces and moment to the body and applying the equilibrium equations. In particular, sliding critical condition occurs when the global horizontal aerodynamic force, $F_{xy,c}$ obtained as the vector sum of $F_{x,c}$ and $F_{y,c}$, equals the friction resistant force between the piled containers surfaces; the tilting critical condition occurs when the global vertical aerodynamic force, $F_{z,c}$, equals the container's weight force; the lifting critical condition occurs when the global aerodynamic overturning moment, $M_{y',c}$, calculated with respect to the y'-y' axis placed at the basis longitudinal corner of the container, equals the stabilizing moment due to the weight itself.

The sliding, tilting and lifting equilibrium conditions can be then expressed as:

$$F_{xy,c} = (P - F_z)C_a \tag{3}$$

$$F_{z,c} = P \tag{4}$$

$$M_{y',c} = P\frac{d}{2} \tag{5}$$

where P is container's weight force, C_a is the friction coefficient between the containers' surface and d is the transverse depth of the container (Fig. 1c). Expressing the global wind loads on the basis of Eq. (2) and substituting them into Eq. (3)–(5), it is possible to formulate explicitly the containers stability problem. As it can be shown, the critical wind velocity causing the container stability problems depends on wind direction α, turbulence intensity I_U, friction coefficients C_a and aerodynamic pressure coefficients c_p and of the piled containers. The turbulence intensity depends on the local roughness and the orography of the site. It can be obtained by analytical models, numerical models or on-site measurements (Burlando et al. 2013). The friction and aerodynamic coefficients are discussed in the following sections.

3 Friction Coefficients of Piled Container

The static friction coefficient between two stacked containers is a key parameter for assessing the stability of the container; however, the data provided in the literature present a considerable dispersion, inducing an important uncertainty in the analyses. Therefore, an on-site experimental campaign has been carried out to experimentally measure the friction coefficient between stacked containers (Balbi et al. 2018).

In the overall, 39 tests were considered, obtaining as many friction coefficient values; Fig. 2 shows the histogram of the obtained friction coefficient values. The mean value of the friction coefficient is 0.450 and the standard deviation is 0.0840. The variability of the obtained results is mainly related to the surface conditions of the containers and position of the test container respect to the below piled container, causing a different net surface support of the bodies. A deeper discussion is reported in (Balbi et al. 2018).

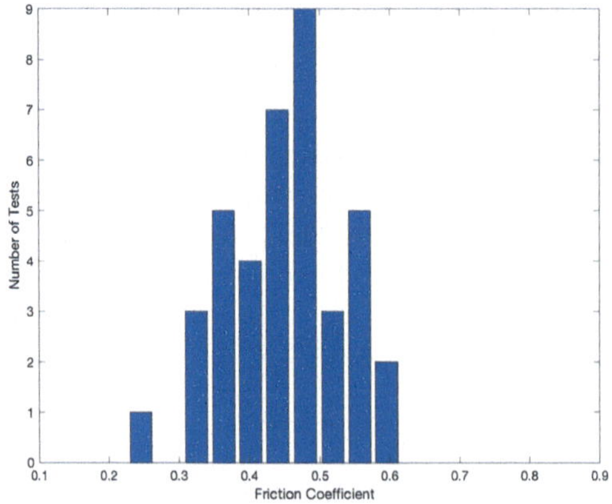

Fig. 2. Histogram of the values of the friction coefficient

4 Aerodynamic Coefficients of Piled Container

The aerodynamic coefficients for the container are not known in the literature and are difficult to assess using the current literature and standards, due to the peculiarities of the shape and the surface of the container, and the variability of the configurations that the stacked containers can take. In order to evaluate reliable values, the aerodynamic coefficients have been studied by means of wind tunnel tests performed in the wind tunnel (WT) of the Department of Civil, Chemical and Environmental Engineering (DICCA). The wind tunnel test concerned the 40' high cube container, the characteristics are shown in Table 1.

Table 1. Characteristics of a 40' high cube box

Width b (m)	12.19
Height h (m)	2.90
Depth d (m)	2.44

A single container model has been reproduced in scale 1:50 using the 3D printing technique, in order to model properly the particular geometric features of the container. The model is equipped with 225 external pressure taps (Fig. 3b): 15 are placed on each of the lateral smaller faces B and D, 65 are placed on each of lateral faces R and L, 65 are placed on the top face T (Fig. 1c).

a) b)

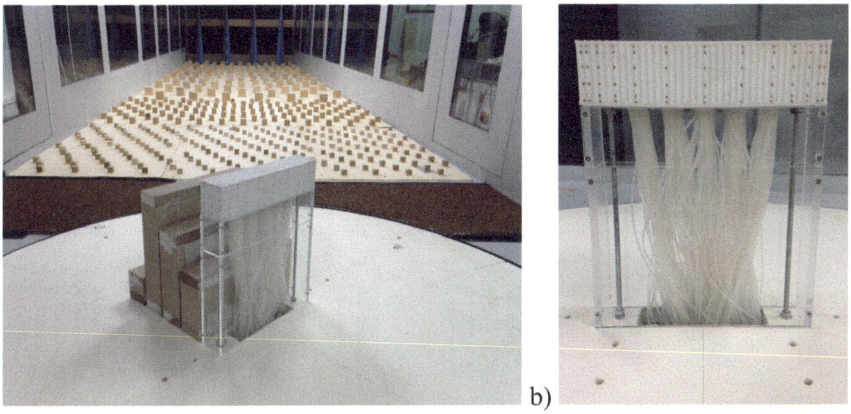

Fig. 3. (a) Test chamber set-up; (b) Instrumented container model

Seven set-up conditions have been reproduced, considering seven different container storage configurations (Fig. 4) of containers with the same properties. The isolated pile of containers is the standard reference case, which allows a qualitative comparison. The dummy containers models have been reproduced in medium density fibreboard, and brought close to the *L* side of the instrumented model. For each configuration, the instrumented container was placed in 4th and 5th tier, always considering it as the top container of the pile.

Two different wind profiles were considered for the wind tunnel test, corresponding to a less intense but more turbulent (*Urban profile*) and a more intense but less turbulent (*Sea profile*) wind field. The wind characteristics were reproduced in the wind tunnel through the insertion of passive devices at the entrance and on the floor of the test chamber (Fig. 3a). The container symmetry allows to repeat the pressure measurements just for 9 wind directions, starting from 0° up to 180° with steps of 22.5° (0°, 22.5°, 45°, 67.5°, 90°, 112.5°, 135°, 157.5°, 180°). It should be noted that, coherently with the reference systems already introduced, the 0° direction represents the incident wind perpendicular to the *L* face of the container; in this case the flow run over the dummy containers at the back of the instrument model; the 180° direction represents the incident wind perpendicular to the *R* face of the container; in this case the flow directly the instrumented model. Each test condition was analyzed in steady state, for a measurement period of 120 s (wind tunnel scale). The tests have been carried out for all the set-ups and for all the angles with the *Sea profile* and containers in the 5th tier. After a post-processing phase, additional tests were carried out with containers in the 4th tier both with *Sea profile* and *Urban profile*, considering only the most critical set-up and angle cases.

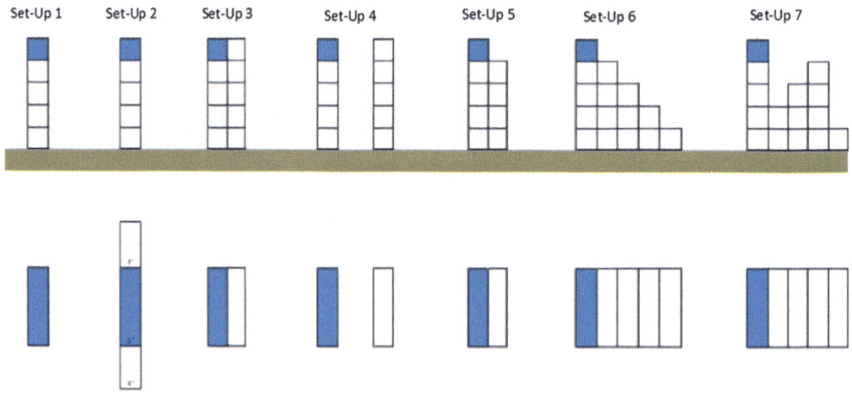

Fig. 4. Test set-up in 5th tier, the instrumented container is shown in blue

The pressure coefficient c_p at the point *x, y, z* (local reference system) of the test body surface at time *t*, is defined (CNR 2008) as the dimensionless value:

$$c_p(x, y, z, t) = \frac{p(x, y, z, t) - p_0}{\frac{1}{2}\rho U(Z_{wt})^2} \tag{6}$$

where $p(x,y,z,t)$ is the pressure measured at point (x,y,z) at time t, p_O is the static pressure in the test chamber, ρ the air density based on local test chamber conditions and $U(Z_{wt})$ the undisturbed average velocity at the Z_{wt} wind tunnel reference height 1.15 m found from the undisturbed artificial boundary layer used for the test. By averaging the pressure coefficients over the test time interval 120 s, the mean pressure coefficient is obtained for each test configuration.

A total number of 63 cases has been analyzed. In the following, the most relevant and critical cases are presented and commented, corresponding to set-up 1, set-up 3 and set-up 7. It is worth noting that, in all analyzed cases, the worst directions are in the range $90° \leq \alpha \leq 180°$, for which the flow hits directly the instrumented container model, while the presence of dummy containers at the back in different set-ups affects the values of the pressures.

Figure 5 shows the mean pressure coefficients map for $\alpha = 90°$ and set-up 1 (a) and (b) set-up 3 (c) set-up 7, with container in 5° tier, Sea Profile. At $\alpha = 90°$, the wind flow is orthogonal to the container's lateral side *B* The pressure distributions on the five sides of the container are very similar for the reference set-up 1 and for the set-up 3 and 7. This is due to the fact that, for this wind direction, the wind flow is almost unaltered by the containers at the back of the instrumented model in different set-up. On *B* face, all c_p values vary between 0.7 and 0.5 while on the *L* and *R* leeward faces c_p vary between −0.9 and −0.1.

At $\alpha = 135°$ (Fig. 6), the pressure distributions on *R* (upwind) face are very similar for the set-up 1, 3 and 7, with c_p varying between 0.7 and 0.1. On the contrary, the pressure distributions on *T* (roofing) and *L* (leeward) faces show suction values increasing significantly in set-up 3 and 7 with respect to the reference set-up 1 (c_p maximum values of −1.1, −2.2 and −1.8 and −0.7 for set-up 1, 3 and 7, respectively in *T* face and of −0.5, −0.8 and −1.1 for set-up 1, 3 and 7, respectively in *L* face), showing also a strong gradient on the surfaces, especially of the roofing. These effects are generated by the vortex wake phenomena induced by the presence of other containers piles behind the instrumented model in set-up 3 and 7.

At 180° (Fig. 7), the wind flow is orthogonal to the *R* face. The pressure distributions on *R* (upwind) face are very similar for the set-up 1, 3 and 7, with c_p varying between 0.7 and 0.3. On the contrary, the pressure distributions on *T* (roofing) and *L* (leeward) faces show suction values increasing significantly in set-up 3 and 7 with respect to the reference set-up 1 (c_p maximum values of −0.6, −0.8 and −0.8 for set-up 1, 3 and 7, respectively in *T* face and −0.5, −0.8 and −0.7 for set-up 1, 3 and 7, respectively in *T* face) showing a slow variation on the surfaces.

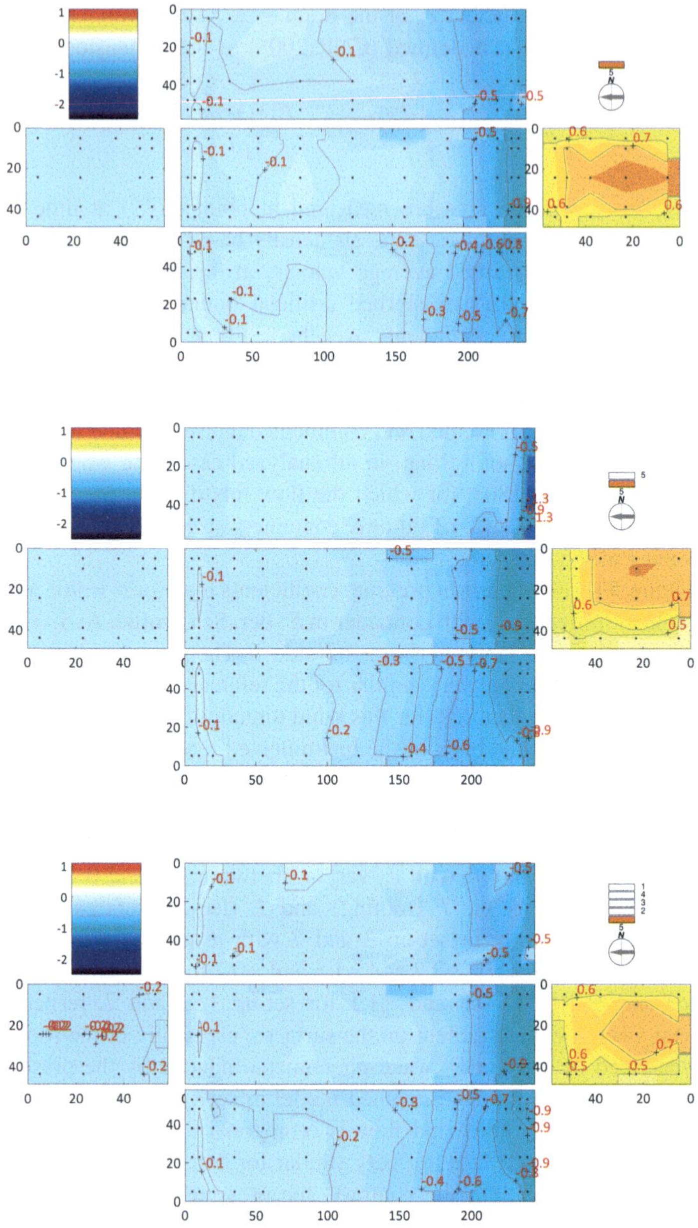

a)

b)

c)

Fig. 5. Mean pressure coefficients map for α = 90° and (a) set-up 1 and (b) set-up 3 (c) set-up 7, 5° tier, Sea Profile

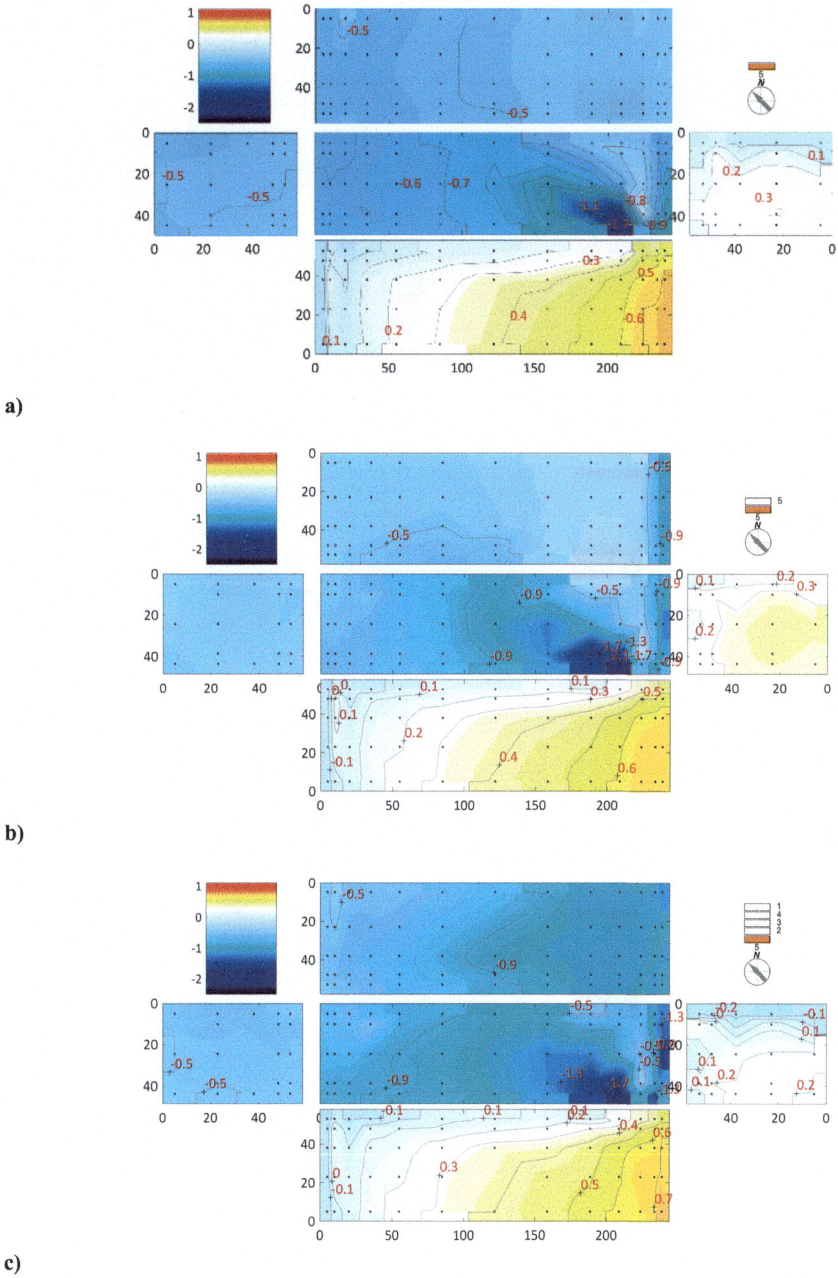

Fig. 6. Mean pressure coefficients map for α = 135° and (a) set-up 1 and (b) set-up 3 (c) set-up 7, 5° tier, Sea Profile

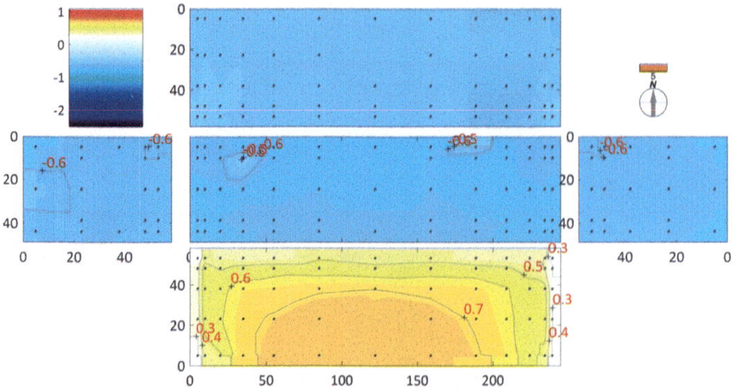

a)

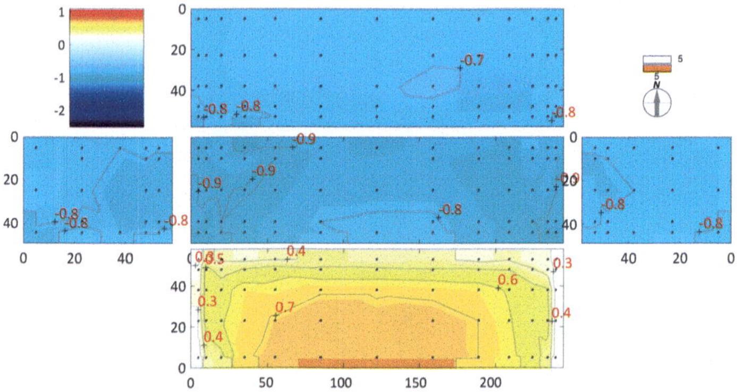

b)

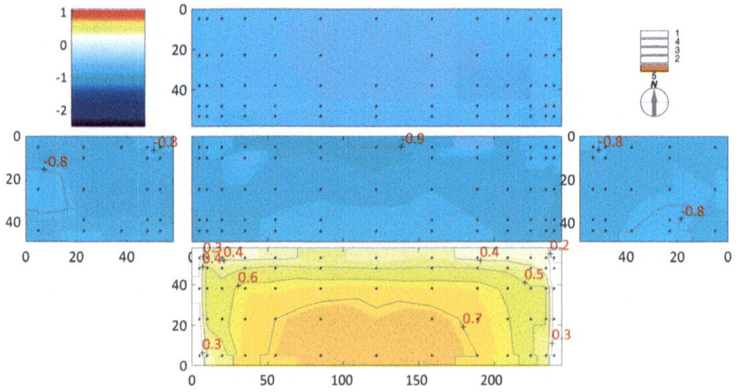

c)

Fig. 7. Mean pressure coefficients map for α = 180° and (a) set-up 1 and (b) set-up 3 (c) set-up 7, 5° tier, Sea Profile

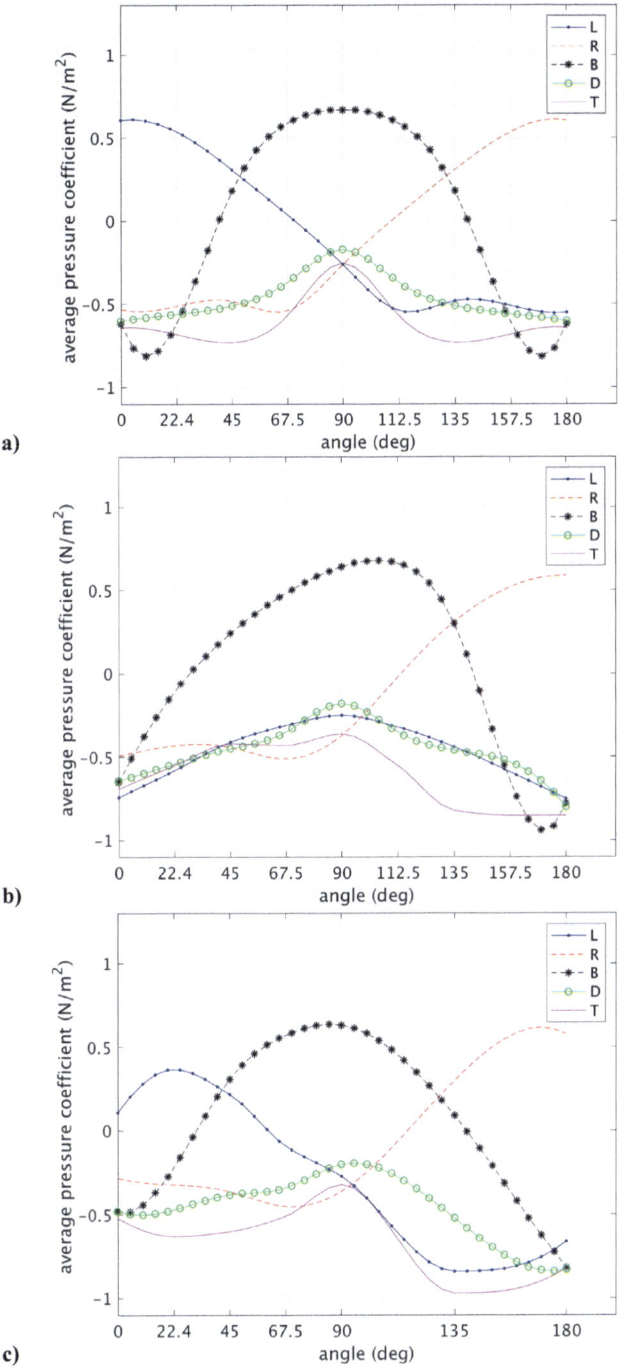

Fig. 8. Average pressure coefficient on the face $\delta = L, R, B, D, T$, 5° tier, *Sea Profile*, set-up 1 (a), set-up 3 (b) and set-up 7 (c)

By averaging the mean pressure coefficients over surface areas, the average pressure coefficient on the container faces are obtained. Figure 8 shows the average pressure coefficients on the face δ = *L, R, B, D, T*, considering the *Sea profile*, 5th tier, set-up 1 (a), set-up 3 (b) and set-up 7 (c). Overall, it results that the average pressures coefficient of the upwind face does not significantly change respectively to the other different set-ups if that face is directly invested by the wind flow, while the most significant changes are noted on the leeward and lateral faces.

Starting from the local pressures, the global forces F_{xy}, F_z, M_y, on the container have been obtained. In particular, the normalized values R_i between the F_{xy}, F_z, $M_{y'}$, and the peak dynamic pressure are expressed by:

$$R_i = \frac{F_i}{\frac{1}{2}\rho\hat{U}(Z_{wt})^2} \quad (8)$$

The corresponding values are presented respectively in Tables 2, 3 and 4. The values are obtained for each Set-up and for the angles between 0°–180°.

Based on these values, it results that the most critical values are those corresponding to set-up 3 and set-up 7. The most critical angles are 135° and 180°.

The most critical case for the F_{xy} force and M_y moment corresponds to the 180° angle of the set-up 3, while the most critical case for the F_z force corresponds to the 135° angle of the set-up 7.

Analyses have been repeated for instrumented container in 4th tier and *Urban Profile*, obtaining, in general lower values.

Table 2. Ratio *Rxy* between the horizontal force *Fxy* and the peak dynamic pressure, 5° tier, *Sea Profile*

α	Set-up 1	Set-up 2	Set-up 3	Set-up 4	Set-up 5	Set-up 6	Set-up 7
0°	40.32	38.29	8.61	3.29	27.85	29.66	13.96
22.5°	37.11	36.12	6.19	–	26.04	26.82	24.61
45°	28.45	29.77	4.98	22.13	22.58	21.18	21.22
67.5°	21.4	20.81	9.38	–	16.78	13.92	14.09
90°	5.98	5.87	7.23	–	5.66	6.29	6.76
112.5°	–	–	12.82	24.17	25.97	–	–
135°	–	–	27.1	36.01	32.82	34.51	40.67
157.5°	–	–	39.33	–	38.79	–	–
180°	40.32	–	47.42	43.91	42.86	42.06	43.71

It is not possible to identify a univocal result that represents the most penalizing case for container stability, since the equilibrium to the slide depends on both the horizontal resulting F_{xy} force and the F_z vertical force. However, the obtained results are general and can be used in any port location to characterize the critical wind conditions.

Table 3. Ratio R_z between the vertical force F_z and the peak dynamic pressure, 5° tier, *Sea Profile*

α	Set-up 1	Set-up 2	Set-up 3	Set-up 4	Set-up 5	Set-up 6	Set-up 7
0°	19.03	16.68	20.54	5.76	19.01	17.62	15.34
22.5°	20.19	−9.06	16.42	–	19.69	18.3	18.67
45°	21.68	23.08	12.42	20.09	20.32	18.75	17.98
67.5°	16.87	17.78	2.79	–	16.53	15.65	15.15
90°	7.54	6.11	10.58	–	9.25	10.16	9.45
112.5°	–	–	16.32	17	24.28	–	–
135°	–	–	24.46	25.52	24.12	24.81	28.92
157.5°	–	–	25.31	–	21.05	–	–
180°	19.03	–	25.4	22.58	21.24	20.76	24.48

Table 4. Ratio R_{My} between the overturning moment M_y and the peak dynamic pressure, 5° tier, *Sea Profile*

α	Set-up 1	Set-up 2	Set-up 3	Set-up 4	Set-up 5	Set-up 6	Set-up 7
0°	−34.26	−33.82	−11.23	−3.36	−25.57	−29.12	−4.65
22.5°	−27.69	−27.8	−10.52	–	−19.02	−21.53	−13.97
45°	−11.07	−12.33	14.29	−4.78	−7.62	−7.93	−6.83
67.5°	−6.53	−4.38	4.49	–	−1.61	1.19	0.78
90°	0.22	1.35	−5.58	–	−0.67	−4.17	−4.72
112.5°	–	–	−7.07	9.92	3.74	–	–
135°	–	–	4.53	15.97	14.86	16.01	19.64
157.5°	–	–	23.69	–	28.24	–	–
180°	34.26	–	36.01	33.87	33.99	32.88	31.83

5 Conclusions

This paper describes a research activity aimed at the formulation of wind effects on containers stability. In particular it describes the main parameters affecting containers stability and presents the results of experimental campaigns aimed at quantifying the friction coefficients and aerodynamic coefficients of piled containers. The former are obtained by means of in-situ measurements; the latter have been studied by means of extensive wind tunnel tests on scaled models, reproducing seven different container storage configurations, two different incoming wind profiles and two different containers tier, obtaining as many aerodynamic coefficient sets of values. The coefficients presented in this work are general and valid for any terminal location. They represent key parameters to obtain containers critical wind velocity values. Future work will develop the analytical formulation of the critical wind velocity, and will derive its probability distribution, considering the variability of the input parameters and taking into account local turbulence conditions and the actual orientation of the containers at the site. Relevant effect can also be related to local non-synoptic meteorological phenomena. A wide research project is addressed in modelling such phenomena and their effects on coastal structures and infrastructures (Zhang et al. 2018a, b; Solari et al. 2015).

Acknowledgements. The authors gratefully acknowledge the VTE management for technical and logistical assistance, with a special thanks to dr. Francesca Meloni and dr. Stefano Mordeglia. This research has been carried out in the framework of the Project "Wind monitoring, simulation and forecasting for the smart management and safety of port, urban and territorial systems", funded by Compagnia di San Paolo (grant number 2015.0333, ID ROL: 9820) in the period 2016–2018.

References

Balbi A, Repetto MP, Solari G, Freda A, Riotto G (2018) Critical wind velocity for harbor container stability. In: 2018 international conference progress in maritime technology and engineering (martech), pp 301–307

Burlando M, De Gaetano P, Pizzo M, Repetto MP, Solari G, Tizzi M. (2013) Wind climate analysis in complex terrains. J Wind Eng Ind Aerodyn 123:349–362

CNR (2008) Guide for the assessment of wind actions and effects on structures - CNR-DT 207/2008. National Research Council of Italy, Roma

Lam JSL, Liu C, Gou X (2017) Cyclone risk mapping for critical coastal infrastructure: Cases of East Asian seaports. Ocean Coast Manag 141:43–54

Repetto MP, Burlando M, Solari G, De Gaetano P, Pizzo M (2017) Integrated tools for improving the resilience of seaports under extreme wind events. Sustain Cities Soc 32:277–294

Repetto MP, Burlando M, Solari G, De Gaetano P, Pizzo M, Tizzi M (2018) A web-based GIS platform for the safe management and risk assessment of complex structural and infrastructural systems exposed to wind. Adv Eng Softw 117:29–45

Solari G, Burlando M, De Gaetano P, Repetto MP (2015) Characteristics of thunderstorms relevant to the wind loading of structures. Wind Struct 20:763–791

Solari G, Repetto MP, Burlando M, De Gaetano P, Pizzo M, Tizzi M, Parodi M (2012) The wind forecast for safety management of port areas. J Wind Eng Ind Aerodyn 104–106:266–277

Zhang S, Solari G, De Gaetano P, Burlando M, Repetto MP (2018a) A refined analysis of thunderstorm outflow characteristics relevant to the wind loading of structures. Probab Eng Mech 54:9–24

Zhang S, Solari G, Yang Q, Repetto MP (2018b) Extreme wind speed distribution in a mixed wind climate. J Wind Eng Ind Aerodyn 176:239–253

Zhen W, Xingqian P, Chunhui Z (2011) Overturning analysis of Harbor Containers based on wind tunnel test of rigid models. In: 2011 international conference on electric technology and civil engineering (ICETCE), pp 544–548. IEEE

Mooring System Optimization for a Spar-Buoy Wind Turbine in Rough Wind and Sea Conditions

G. Barbanti(✉), E. Marino, and C. Borri

Department of Civil and Environmental Engineering,
University of Florence, Via Santa Marta 3, 50139 Florence, Italy
gbarbanti@dicea.unifi.it

Abstract. In this paper the dynamic response of a spar-buoy floating offshore wind turbine (FOWT) is investigated in order to optimize the design of the deep water mooring system. Several mooring system design parameters are considered respecting the assigned constraints with the aim to minimize the platform motions, the normal stress in the mooring lines and the fatigue estimation for harsh environmental conditions. The influence of mooring line ballasting positioning and their amount along the cables is investigated and compared to the reference case.

Keywords: Wind turbines · Mooring lines optimization · Fatigue

1 Introduction

Floating offshore wind turbines have the potential to greatly expand the siting possibilities for offshore wind energy plants and thus become one of the most important source of sustainable renewable energy. In contrast to bottom-fixed offshore wind turbines, floating support platforms can be installed in water deeper than 40 m, opening new opportunities for installations in areas characterized by deep water (e.g. the Mediterranean Sea). In October 2017, Statoil built the first floating offshore wind park making the spar-buoy concept one of the most promising floating platform technology. Offshore wind turbines are subject to a variety of environmental conditions, which cause a significantly different response depending on the turbine state (operational or parked), making the hydro-aero-servo-elastic analysis not a trivial task.

Karimirad and Moan (2011) compared the response of the floating wind turbine in both survival and operational conditions and emphasized the importance of analyzing the structural response in survival conditions to obtain lifetime evaluation. Li et al. (2018) found that the wave-induced fatigue damage is greater than that induced by wind loads for a spar-buoy wind turbine. Haid et al. (2013) showed that fatigue lifetime estimation presents greater sensitivity to the method of counting unclosed cycles as compared to simulation length. The biggest increase found for fairlead fatigue estimation was about 3.5% for a simulation of 6 h compared to the reference 10 min' case. Fylling and Berthelsen (2011) presented an integrated design tool for the optimization of moored spar-buoy type support structures for wind turbines including the mooring

F. Ricciardelli and A. M. Avossa (Eds.): INVENTO 2018, LNCE 27, pp. 87–98, 2019.
https://doi.org/10.1007/978-3-030-12815-9_7

system. The optimization was performed in the frequency domain to determine the minimum material cost by the minimization of the objective functions, defined by the costs of spar buoy, mooring line, and cable using a gradient method. Brommundt et al. (2012) described a tool for the optimization of a catenary moored semi-submersible floating wind turbine using a frequency domain approach with the purpose to minimize the line lengths under site specific environmental loads. In the above studies a large number of different scenarios can be efficiently simulated due to the linear approximation made in the frequency domain analysis, but the inability to simulate the non-linear behaviour of the system is a limit for the optimization reliability. Kvittem and Moan (2012) showed substantial differences in the motion response of a floating wind turbine modelled with a linearized mooring line and a nonlinearmodel for different mooring line configurations. In Chen et al. (2017) the mooring line is optimized for a spar-buoy FOWT, limitating the analysis only for surge, heave and pitch motions and simplifying the aerodynamic and the overall structure's dynamic. The results showed that the spar dynamic response and the mooring line stress can be improved with respect to a non-optimized mooring system.

The purpose of the present study is to optimize the mooring system of a spar-buoy floating wind turbine in order to enhance the structural response to survival environmental conditions using a hydro-aero-servo-elastic time-domain modelling. Optimization is performed in order to find the mooring lines geometry that allows the minimization of pre-established target functions. In this study different structural responses are considered as objective functions: the average platform pitch motion and mooring lines stress and the fatigue loads estimation for the mooring lines and tower base. Firstly, a parametric comparison of the responses is performed to assess which parameters have the greatest influence on the optimization of the mooring lines configuration. Successively, the mooring ballast weights position and amount are considered in the optimization routine to evaluate their influence on the system response.

2 Methodology

In this study the OC3-Hywind spar-buoy featuring the NREL 5-MW (Jonkman 2010) baseline wind turbine is investigated using FAST, a fully coupled hydro-aero-servo-elastic time-domain simulator for horizontal axis wind turbines developed by the National Renewable Energy Laboratory (Jonkman and Buhl 2005). First in this study the strip-theory solution and the hybrid solution adopted by FAST are compared for the simulation of the OC3-Hywind FOWT dynamic response in rough environmental sea conditions. Successively an optimization procedure is developed to optimize the dynamic response of the wind turbine and the mooring lines system. Thus, the dynamic response of the optimized mooring system adopting mooring ballasts is studied and compared to the original reference case.

2.1 OC3-Hywind Spar-Buoy Floating Wind Turbine Description

In this study, the wind turbine properties, platform properties and mooring system properties are kept the same as described in the Offshore Code Comparison (OC3) for the Hywind concept (Jonkman 2010). This concept is derived from Statoil Hywind MW- scale prototype installed in Scotland in 2011, which is the world's first floating wind turbine installed, and is modified to create a benchmark case for codes comparison. The OC3 Hywind is a spar-buoy wind turbine, with three slack catenary mooring lines and equipped with the NREL 5 MW offshore reference wind turbine, a hypothetical wind turbine designed to enable a benchmark for the comparison of different modelling tools. The OC3-Hywind model includes the specification of additional hydrodynamic damping terms in the surge, sway, heave and yaw in order to match the results provided by experimental tests. The mooring lines properties are formulated with some simplification: a yaw stiffness value is adopted to reproduce delta lines connections, mooring lines damping is neglected and the multisegment lines are replaced with an equivalent homogeneous line, whose properties are obtained by averaging the mooring lines properties. To achieve a more accurate estimation of the mooring lines response a multi-segmented lumped-mass module MoorDyn is used (Hall et al. 2013). This module can model the mooring lines dynamics improving the mooring load predictions (Wendt et al. 2015). In particular, for fatigue analysis a dynamic approach is recommended since the quasi-static approach significantly under predicts the mooring loads (Davidson and Ringwood 2017).

2.2 Hydrodynamic Theory

The platform hydrodynamic loads are computed by Hydrodyn (Jonkman et al.), a time-domain hydrodynamics module coupled into FAST. Hydrodyn allows multiple approaches for calculating the hydrodynamic loads on a structure: a potential-flow theory solution, a strip-theory solution and a hybrid-combination of the two. The potential-flow solution involves the frequency-to-time-domain transformation, and requires the definition of the hydrostatic restoring coefficients, the added mass and damping contributions from linear wave radiation, and the incident-wave excitation (Froude-Kriloff forces). The hydrodynamic coefficients are frequency dependent and must be supplied by an external frequency-domain panel code.

The strip-theory solution is based on Morison's equation, one of the most reliable and adopted slender-body formulation for the computation of hydrodynamic forces on slender body. In offshore engineering a body can be considered slender (with respect of incident waves) if the ratio of the body length over the wavelength (D/λ) is less than 0.2 (Matha et al. 2011). According to Rainey (Rainey 1989), the errors using the slender body theory are of the order of $(D/\lambda)^3$ thus, the smaller the ratio D/λ, the better is the slender body theory approximation. Furthermore, this formulation is most suitable for the cases in which the inertia is dominant compared to drag force, i.e. when the Keulegan-Carpenter number is small, in general assumed less than 2 (Jonkman 2010). This implies that the surface profile is unaffected by the body itself, i.e. diffraction contribution is less important and the force on the body can be calculated from the drag and the inertia forces induced by incident waves. In the strip-theory the hydrodynamics loads are then evaluated by dividing the cylinder into a number of elements, and then evaluating the loads by integrating the contribution of each element according to Morison's equation (Faltinsen 1995).

The hybrid solution increases the hydrodynamic loads evaluated from the potential-flow theory with the contribution of the viscous-drag component of the strip-theory solution, combining the benefits of the two approaches together.

2.3 Test Cases

The wind turbine is considered in a parked state (shutdown) and exposed to harsh environmental conditions. The blades are feathered to 90° and the nacelle yaw is zero. According to the International Electrotechnical Commission (IEC 61400-3), the most unfavourable condition for offshore structures is when the incident angles of the wind and wave loads are the same. In this study, wind and wave loads are applied in the same direction on the structure, which are along the x-axis. For this reason, the analysed mooring line is that one aligned with the wind and wave direction. Given the axial-symmetry geometry of the spar-buoy platform concept, the assumption of unidirectional wind and wave loads leads to negligible out of plane motions (e.g. roll and yaw motions) and a maximization of the in-plane response (e.g. surge and pitch motions). Steady wind assumption is made for all the simulations since the turbulence doesn't affect the structural response reliability and a constant wind model can be used for the fatigue and ultimate limit state analysis with acceptable accuracy (Kamirad and Moan 2012). A hypothetical linear irregular sea state with a significant wave height of H = 4 m and a wave peak period T = 10 s is adopted with JONSWAP spectrum formulation. For this sea state the slender body formulation holds since the body ratio to wavelength D/λ is equal to 0.06 and the Keulegan-Carpenter number values are always less than 2. A steady wind speed of 30 m/s is adopted in order to simulate a realistic wind-wave case. An initial time interval of 400 s is discarded from the time histories of the response in order to remove the effects of the transient response.

2.4 Optimization Scheme

The purpose of this study is to find the geometry configuration of the mooring system which can improve the mooring lines and the wind tower responses in survival environmental condition considering different optimization targets: the average platform pitch motion and the mean mooring stress, and the fatigue estimation of the mooring lines and the tower base. Therefore, the optimization problem described aims to find a global minimum of a function with prescribed boundary conditions. For this reason, a hybrid constrained nonlinear optimizer method implemented in MATLAB is chosen, composed of a genetics algorithm (GA) followed by a local solver. Since the studied problem is not continuous and not smooth (i.e. not continuously differentiable) the pattern-search method was chosen (MathWorks 2003).

Five independent primary design parameters of the system are considered (see Fig. 1):

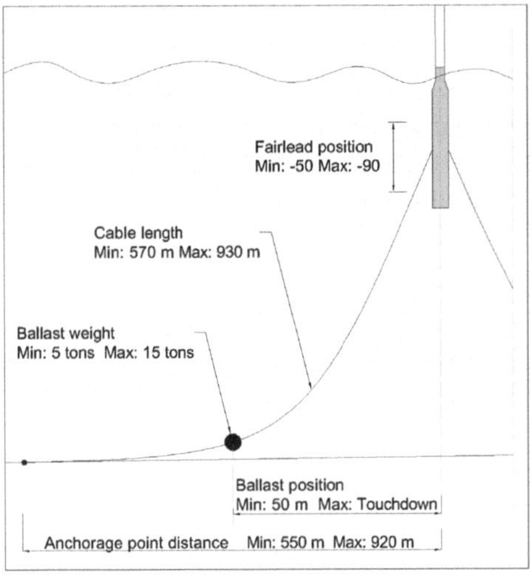

Fig. 1. Optimization's parameters

1. Overall length of the mooring system;
2. Radius distance to anchors from platform centreline;
3. Vertical fairlead position;
4. Clump weights amount;
5. Clump weights position along the catenary.

The mooring system is optimized varying the designed parameters within prescribed boundaries and given specific constraints. Several constraints are adopted in the simulations:

1. average stress in the hydrostatic equilibrium on the cables must not exceed the value of 100 tons;
2. drag embedded anchors reactions must be horizontal only;
3. mean surge motions of the floater must be within 6 m from the rest position;
4. avoiding clump weights touchdown.

Surge constraint threshold is adopted given that from the first simulations run a strong relationship was found between the maximum surge motion threshold and the pitch motion and mooring lines stress, i.e. all the optimal solutions found approached the maximum surge displacement allowed. For this reason, the adopted limit in surge displacement aims in making comparable results with the reference FOWT. Furthermore, since type and diameter of the chains are kept fixed during the simulations, a representative mooring line stress limit is chosen.

A routine is developed to restore the hydrostatic equilibrium position of the system varying the platform ballasting in order to balance the forces exerted by the cables (due to a modification of mooring lines geometry) and the mooring ballasts. Since the OC3-Hywind concept derives from the patented Statoil spar, only the global structural properties for the whole spar are available, i.e. there is no accessible information about the ballasting amount. For this reason, an iterative scheme was developed to obtain hypothetical platform steel and ballasting amount (density and total mass) matching the original mass, Center of Mass position (CoM) and platform pitch and roll inertias. This procedure increases the reliability of the results at the cost of small variations in the platform ballasting amount (generally less than 1% of the total spar mass). Then the routine calculates the updated platform properties, i.e. total mass, CoM position, pitch and roll inertias.

The overall optimization steps are the following ones:

1. GA generates random input parameters within prescribed constraints;
2. hydrostatic output control to check if the cable is lying on the seabed and if the mooring line stress is within the prescribed threshold;
3. mooring line stress is compared with Hywind's reference value and platform ballasting mass and properties are computed to restore the hydrostatic equilibrium;
4. simulation running;
5. iterations continue until GA's population size are not reached and function tolerances criteria are not respected.

Six optimization routines in total are performed in order to avoid local minima estimation and to make comparisons of the results.

2.5 Fatigue Load Assessment

A fatigue estimation analysis is computed using the Rainflow Cycle-Counting (RCC) method and S-N curves for damage accumulation. The S-N approach is based on the linear damage accumulation assumption (Palmgren-Miner rule). Fatigue curves are modelled with a power law, that stems from the linear regression of fatigue test results. According to DNV guidelines, a power exponential $m = 3$ is assumed for the mooring stud/studless chain (DNV-OS-E301 2010). A power exponential $m = 3$ is also assumed for the wind turbine tower welded steel. Rainflow counting method is commonly used to evaluate fatigue damage, as it realistically considers the fatigue damage caused by each individual load cycle. It identifies and counts the stress ranges corresponding to individual hysteresis loops of the component material. A disadvantage of the above method is the neglecting of sequence effects, i.e. the crack growth depends only on the current applied stress range and crack dimensions, and not on past loading. Here a rainflow counting method is applied to the time series of mooring lines axial stress for the number of stress cycles (corresponding to different mean stresses and stress ranges). N is the number of stress cycles that have been rainflow-counted and, in this study, N is taken to be the number of full cycles plus the number of half-cycles weighted by a factor of 0.5 (Sutherland 1999). For convenience an "equivalent fatigue load" (EFL) is adopted, defined as the constant-amplitude stress range which, over a fixed number of cycles, causes the same amount of damage as the original variable-amplitude stress time series (Marino et al. 2014; Ragan and Manuel 2007). We conducted fatigue lifetime estimation with simulations' length of 10 min.

3 Results

3.1 Cable Optimization with Clump Weights

The influence of the amount and position of the ballasts on the global platform response and on the stress of the mooring lines are now examined. Subsequently, an optimization routine is presented with the aim of minimizing the fatigue damage on the mooring lines and on the tower base due to the bending moment. For a mooring system with clump weights, the most influential design parameters are the size (mass) of the clump weights and their position along the cables. The emphasis is therefore placed on these two design parameters. In this optimization routine the influence of the quantity and position of the ballast is simulated in combination with the geometrical parameters of the mooring lines. A total of five independent parameters are considered: unstretched cable length, radius distance from fairlead to anchorage point, vertical fairlead position along the spar, weight and position of ballasts along the cables (Fig. 1). Further constraints are added for the clump weights: the ballasts must weigh between 5 and 15 tons and must be positioned on the catenary from a distance of 50 m to the seafloor touchdown (i.e. the weight must not lie on the seabed, see Table 1).

Table 1. Simulation specifications for the cable geometry with ballasting optimization routine

• Simulations performed	354
• Simulation time	1000 s
• Transition time discarded	400 s
• Unstretched cable length	Min. 570 m
	Max. 930 m
• Radius anchorage distance	Min. 550 m
	Max. 920 m
• Vertical fairlead position	Min. −90 m
	Max. −50 m
• Ballast weight	Min. 5000 kg
	Max. 15000 kg
• Ballast distance along the mooring lines	Min. 50 m
	Max. touchdown

Plotting the distribution frequency of the objectives for the simulations launched it is possible to see that the improvement for the mooring lines stress and platform pitch motion are small (for the first the reference value is close to the distribution tail, while for the latter the improvement is almost inconsistent), while it may be relevant for the fatigue in mooring lines (Fig. 2). Moreover, the optimized configurations of the mooring lines do not influence the fatigue of the tower base due to the bending moment. Thus, the implementation of the ballasting of the mooring lines has the potential to improve the system response more significantly for the EFL of the mooring lines and for the mean cables stress and in a negligible way for the EFL of the tower base (see Table 2).

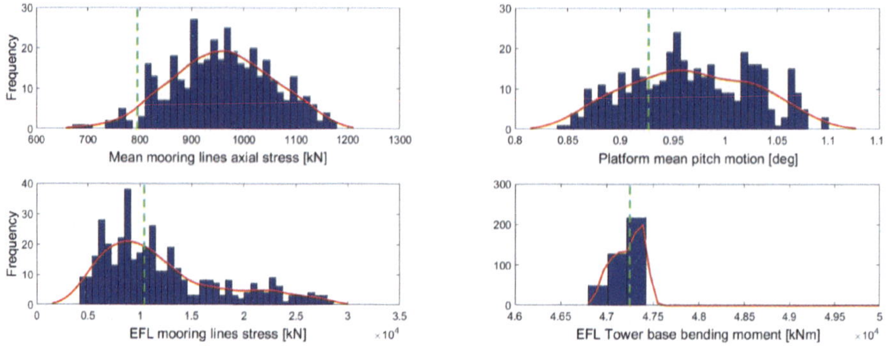

Fig. 2. Histogram charts of the simulation outputs' values (in green dashed line the reference values)

Table 2. Optimization results for the mooring lines with ballasts

Mean mooring lines stress optimization objective								
Case	CL (m)	FP (m)	CAD (m)	BW (kg)	BP (m)	MS (m)	MFMT (N)	%
M1	813	−79	757	10205	468	5.29	6.757e+5	−15
Reference	902	−70	853	/	/	5.52	7.948e+5	/
Mean pitch motion optimization objective								
Case	CL (m)	FP (m)	CAD (m)	BW (kg)	BP (m)	MFMT (N)	MP	%
P1	799	−60	745	160	73	9.053e+5	0.850	−7.6
Reference	902	−70	853	/	/	7.948e+5	0.923	/
Mooring lines equivalent fatigue load optimization objective								
Case	CL (m)	FP (m)	CAD (m)	BW (kg)	BP (m)	MFMT (N)	EFL_ML (N)	%
EFL1	833	−89	786	13175	73	8.634e+5	4.290e+3	−61
Reference	902	−70	853	/	/	7.948e+5	1.041e+04	/
Tower base equivalent fatigue load optimization objective								
Case	CL (m)	FP (m)	CAD (m)	BW (kg)	BP (m)	MFMT (N)	EFL_Twr(N)	%
EFL2	893	−60	838	301	13881	1.021e+6	4.688e+4	−1
Reference	902	−70	853	/	/	7.948e+5	4.725+04	/

CL = cable length, FP = fairlead position, CAD = cable anchorage distance, BW = ballasting weight, BP = ballasting position, MFMT = mooring fairlead mean stress, MS = mean surge motion, MP = mean pitch motion, EFL_ML = equivalent fatigue load mooring lines, EFL_Twr = equivalent fatigue load tower base

The optimization results found reveal shorter mooring catenaries for all the configurations found compared to those of the reference OC3-Hywind concept. A comparison study is then conducted for the optimizated configuration M1 in Table 2 with respect to the reference case (Table 3). From Fig. 3 it is possible to see that the optimized configuration adopting the ballasts can lead to a significant improvement in the reduction of the stress and fatigue of the mooring lines, but does not influence the response of the wind turbine keeping unchanged all the platform motions and reactions. With this optimized configuration a reduction of about 15% for the mean stress and of about 35% for the fatigue load of the mooring system can be achieved.

Table 3. Mooring system comparison for OC3-Hywind concept and the optimized configuration

Case	CL (m)	FP (m)	CAD (m)	BW (kg)	BP (m)	MFMT (N)	EFL_ML (N)
Cable and ballast opt.	813	−79	757	10205	468	6.757e+5	6.726e+03
Reference	902	−70	853	/	/	7.948e+5	1.041e+04

CL = cable length, FP = fairlead position, CAD = cable anchorage distance, BW = ballasting weight, BP = ballasting position, MFMT = mooring fairlead mean tension, EFL_ML = equivalent fatigue load mooring lines

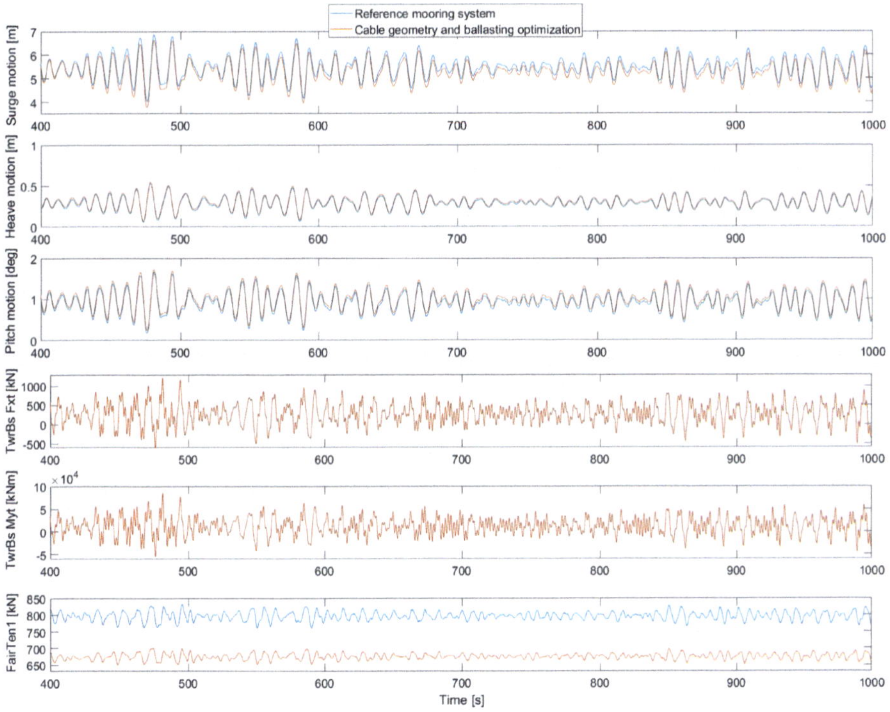

Fig. 3. OC3-Hywind concept and optimized configuration comparison

4 Conclusions

In this study the optimization of the mooring system of the OC3-Hywind concept was studied to withstand harsh environmental conditions. A dynamic hydro-aero-servo-elastic time-domain simulator with lumped-mass mooring line model is adopted. Firstly, strip-theory solution is compared with the hybrid solution for the computation of the hydrodynamic loads. For harsh sea conditions the two methods yielded comparable results for the spar-buoy concept, allowing faster calculation of optimization routines. The geometry of the mooring lines together with the ballasts along the catenaries are therefore optimized in order to minimize different targets: the average platform pitch motion, the cables stresses and the equivalent fatigue load (EFL) in the mooring lines and tower base. Optimization is performed using a genetics algorithm (GA) together with a pattern search algorithm. Several simulations have been performed and different statistic indicators are adopted, highlighting some interesting trends and confirmed the bibliography results.

We can conclude that it is possible to improve the concept of OC3-Hywind mooring system in harsh environmental conditions. In particular, the following considerations can be made:

- Mooring lines ballasting can improve the system response and reduce the fatigue damage in the mooring lines. A reduction in the mooring lines stress and in particular of their fatigue was found: with the optimized configuration a reduction of about 35% in the equivalent fatigue load was found. Besides, all optimized configurations need shorter cables length.
- Lowering the vertical fairlead position close to the center of mass reduces the mooring lines stress and consequently also the fatigue in the cables. Contrarily a reduction of the platform pitch motion can be made by moving the fairlead position upward close to the center of buoyancy. From these results we can conclude that the Hywind fairlead position along the spar seems a reasonable compromise in balancing the response between the mooring lines response and the wind tower response.
- Wind turbine platform responses proved to be almost insensitive to the mooring lines optimization. Only small improvements can be made by using ballasts. This point may be related to the fact that the wind turbine dynamics are more sensitive to wind loads than hydrodynamic loads.

It is important to note that these results refer to a parked wind turbine configuration. The constraint imposed to the platform surge motion turned out to be a very restrictive limitation that discarded most of the simulated configurations. In a future study, the relaxation of this parameter will be studied together with the consideration of a wind turbine in an operational state.

References

Brommundt G, Krause L, Merz K, Muskulus M (2012) Mooring system optimization for floating wind turbines using frequency domain analysis. Energy Proc 24:289–296

Chen D, Gao P, Huang S, Fan K, Zhuang N, Liao Y (2017) Dynamic response and mooring optimization of spar-type substructure under combined action of wind, wave, and current. J Renew Sustain Energy 9

Davidson J, Ringwood J (2017) Mathematical modelling of mooring systems for wave energy converters—a review. Energies 10:666

DNV-OS-E301 (2010) Offshore standard - Position Mooring. Standard. Det Norske Veritas [DNV]

Faltinsen OM (1995) Sea loads on ships and offshore structures. Cambridge University Press, Cambridge

Fylling I, Berthelsen P (2011) WINDOPT - an optimization tool for floating support structures for deep water wind turbines. In: Proceedings of the ASME 2011 30th international conference on ocean, offshore and arctic engineering, OMAE 2011. Rotterdam, Netherlands, p 1–10

Haid L, Matha D, Stewart G, Lackner M, Jonkman J, Robertson A (2013) Simulation-length requirements in the loads analysis of offshore floating wind turbines. In: 32nd international conference on ocean, offshore and arctic engineering, Nantes, France, 9–14 June 2013

Hall M, Buckham B, Crawford C (2013) Evaluating the importance of mooring line model fidelity in floating offshore wind turbine simulations. Wind Energy 17(12):1835–1853

IEC 61400-3 (2009) Ed. 1: Wind Turbines-Part 3: Design Requirement for Offshore Wind Turbines (International Electrotechnical Commission, 2009)

Jonkman J, Buhl M (2005) FAST user's guide. Technical report NREL/EL-500-38230

Jonkman J, Robertson AM, Hayman GJ HydroDyn user's guide and theory manual. Technical report (draft version)

Jonkman J (2010) Definition of the floating system for phase IV of OC3. Technical report NREL/TP-500-47535

Karimirad M, Moan T (2011) Extreme dynamic structural response analysis of catenary moored spar wind turbine in harsh environmental conditions. J Offshore Mech Arctic Eng 133

Kamirad M, Moan T (2012) Wave- and wind-induced dynamic response of a spar-type offshore wind turbine. J Waterw Port Coast Ocean Eng 138(1):9–20

Kvittem MI, Moan T (2012) Effect of mooring line modelling on motions and structural fatigue damage for a semisubmersible wind turbine. In: Proceedings of the twenty-second international offshore and polar engineering conference, Rhodes, Greece, pp 273–278

Li H, Hu Z, Wang J, Meng X (2018) Short-term fatigue analysis for tower base of a spar-type wind turbine under stochastic wind-wave load. Int J Naval Archit Ocean Eng 10(1):9–20

Marino E, Lugni C, Manuel L, Nguyen H, Borri C (2014) Simulation of nonlinear waves on offshore wind turbines and associated fatigue load assessment. In: Proceedings of the ASME 2014 33rd international conference on ocean, offshore and arctic engineering, San Francisco, California, USA

Matha D, Schilpf M, Cordle A, Pereira R, Jonkman J (2011) Challenges in simulation of aerodynamics, hydrodynamics, and mooring-line dynamics of floating offshore wind turbine. In: 21st international offshore and polar engineering conference, Maui, Hawaii, 19–24 June 2011

MathWorks (2003) Optimization toolbox user's guide for use with MATLAB

Ragan P, Manuel L (2007) Comparing estimates of wind turbine fatigue loads using time-domain and spectral methods. Wind Eng 31(2):83–99

Rainey RCT (1989) New equation for calculating wave loads on structures. J Fluid Mech 204:295–324

Sutherland HJ (1999) On the fatigue analysis of wind turbines. Sandia National Laboratories, SAND99-0089

Wendt F, Robertson A, Jonkman J, Hayman G (2015) Verification of new floating capabilities in FAST v8. In: Proceedings of AIAA SciTech 2015. American Institute of Aeronautics and Astronautics, Florida, USA, pp 1–11

Hybrid HIL Testing of Floating Wind Turbines Within LIFES50+ Project

I. Bayati[1(✉)], L. Bernini[1], A. Facchinetti[1], A. Fontanella[1], H. Giberti[2], A. Zasso[1], and M. Belloli[1]

[1] Department of Mechanical Engineering, Politecnico di Milano, Via la Masa 1, 20156 Milan, Italy
{ilmasandrea.bayati,luca.bernini, alan.facchinetti,alessandro.fontanella,alberto.zasso, marco.belloli}@polimi.it

[2] Dipartimento di Ingegneria Industriale e dell'Informazione, Università di Pavia, Via Ferrata 5, 27100 Pavia, Italy
hermes.giberti@unipv.it

Abstract. This paper describes the innovative experimental approach, introduced by the authors in the framework of the EU H2020 project LIFES50+, to perform scale model tests on floating offshore wind turbines (FOWTs). A 6-DOFs hardware-in-the-loop (HIL) system was designed and realized to reproduce the global FOWT response to wind and waves in the atmospheric boundary layer (ABL) of the Politecnico di Milano (PoliMi) wind tunnel. A 2-DOFs (surge and pitch) HIL system was used to perform preliminary tests, assessing the capabilities of the hybrid experimental methodology, and to gather data for the finalization of the 6-DOFs setup. Results from the first experimental campaign are discussed, showing the effect of aerodynamic loads on the coupled FOWT response.

Keywords: Floating offshore wind turbines · Wind tunnel testing · Hardware-in-the-loop · Unsteady aerodynamics · Control

1 Introduction

Floating offshore wind turbines were recently recognized as the key technology to exploit the favourable wind conditions of many coastal areas where the deployment of traditional fixed-bottom wind turbines is not cost-effective. FOWTs are nowadays designed relying on numerical codes, demanded to simulate the complex, interconnected phenomena involved in the wind turbine response to combined wind and wave loading. For this approach to be effective, it is necessary to calibrate numerical models and to assess their fidelity comparing simulation outputs with experiments. The high value of full-scale data is often balanced by the high costs, long testing time and the partial knowledge of the environmental conditions, that are not directly controlled. Scale model testing can support, or even replace, full-scale experiments, providing data affected by a lower degree of uncertainty at much lower costs.

F. Ricciardelli and A. M. Avossa (Eds.): INVENTO 2018, LNCE 27, pp. 99–109, 2019.
https://doi.org/10.1007/978-3-030-12815-9_8

In this contest, it was born the EU H2020 LIFES50+ project, with the aim of optimizing and qualifying two innovative substructure designs for the DTU 10 MW wind turbine (Bak et al. 2013) at a technology readiness level (TRL) of 5. One major project goal is to develop a novel experimental approach based on the HIL methodology, to overcome the scaling issues traditionally faced by ocean basin experiments and favour scale model tests on FOWTs.

In this work the hybrid/HIL experimental approach developed at PoliMi is introduced discussing its main peculiarities and advantages with respect to traditional ocean basin tests. The HIL methodology is explained, focusing on how the structure is divided into a physical and numerical subsystem. The experimental validation of the hybrid/HIL methodology is discussed, together with the main problematics linked to the reproduction of the FOWT coupled servo-aero-hydro-elastic response.

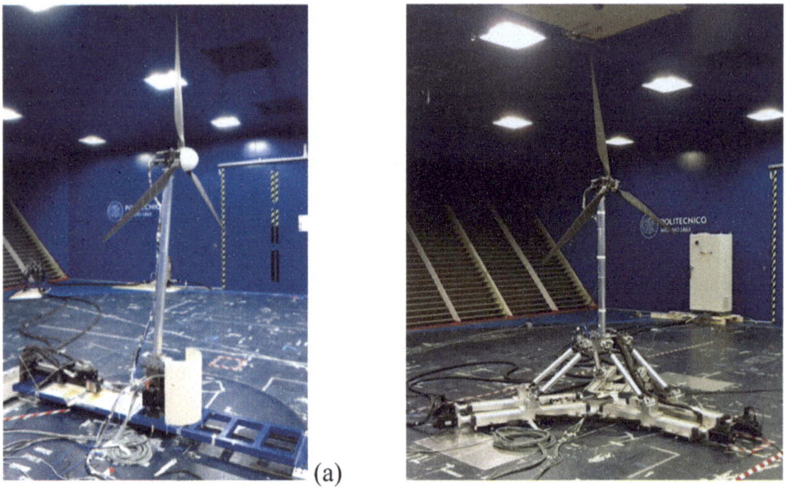
(a) (b)

Fig. 1. General view of the hybrid/HIL experimental setup: 2-DOFs (a) and 6-DOFs (b) configurations

2 Hybrid/HIL Testing

Performing scale model tests on FOWTs is a complex task since wind and wave loads, gravitational and buoyancy forces, mooring lines and flexible dynamics must be simultaneously reproduced. Ocean basin tests were recently performed under many research projects with the aim of investigating the coupled dynamics of different FOWT concepts (Goupee et al. 2014; Koo et al. 2014; Goupee et al. 2017; Stewart et al. 2012; Yu et al. 2017). These tests campaigns were based on an experimental methodology extensively used in the oil and gas industry, mainly developed to preserve hydrodynamic similarity (Martin et al. 2014). Ocean basin tests are traditionally designed according to Froude scaling, fixing the same Froude number ($Fr = U^2/gL$, where U is the wind speed, g the gravity constant and L a significative length) for the full-scale system and the scale model. This allows to achieve a unity scale factor for accelerations and correctly model gravity dependent loads (gravitational restoring, buoyancy and wave excitation forces). However, being the gravity acceleration fixed for the model and prototype, the velocity scale factor λ_v is function of the length scale factor only λ_L, as reported in Eq. (1).

$$\lambda_v = \sqrt{\lambda_L} \tag{1}$$

where the generic scale factor λ_x is defined as the ratio between a given quantity evaluated at prototype scale X_p and model scale X_m. Length scale factors between 50 and 100, generally required to fit a multi-megawatt wind turbine in the test facility, result in excessively low wind speeds and Reynolds numbers that are 350–1000 lower than those experienced by the full-scale machine. This makes the rotor scaling process exceptionally difficult and penalizes the correct reproduction of aerodynamic loads (Martin et al. 2014).

Hybrid testing is a potential solution for the scaling issues mentioned above. The floating wind turbine is divided into two separated and complementary subsystems that are demanded to model different phenomena (Sauder et al. 2016). In the HIL methodology developed by the authors, a wind turbine scale model operating in the ABL section of the PoliMi wind tunnel (the physical subsystem) is used to reproduce aerodynamic loads, whereas the floating structure response, hydrodynamic loads and mooring lines dynamics are computed in real-time from the integration of a numerical model (the numerical subsystem) (Bayati et al. 2018). The resulting platform rotations and translations are the set-point of a the HexaFloat robot that moves consistently the wind turbine scale model.

3 The Hybrid/HIL Experimental Methodology

The hybrid/HIL experimental methodology for wind tunnel tests on FOWTs developed at PoliMi within the LIFES50+ project is resumed by the scheme reported in Fig. 2. The leftmost block represents the numerical subsystem, responsible of modelling the response of the floating platform, set by the equation of motion of Eq. (2), directly used for real-time integration.

$$[M]_f \ddot{\underline{q}}_s + [K]_f \underline{q}_s = \left(\underline{F}_{hst} + \underline{F}_{moor} + \underline{F}_{hydro}\right) + \underline{F}_{aero} \tag{2}$$

where vector $\underline{q}_s = [x, y, z, \rho, \theta, \sigma]^T$ collects the absolute platform translations and rotations (surge, sway, heave, roll, pitch, yaw). These are the set-point of the actuation system (second block) that moves consistently the tower base of the wind turbine scale model. Aerodynamic loads developed by the wind turbine rotor (the physical subsystem) are continuously measured by a 6-components load cell mounted at tower base and are fed-back to the numerical subsystem.

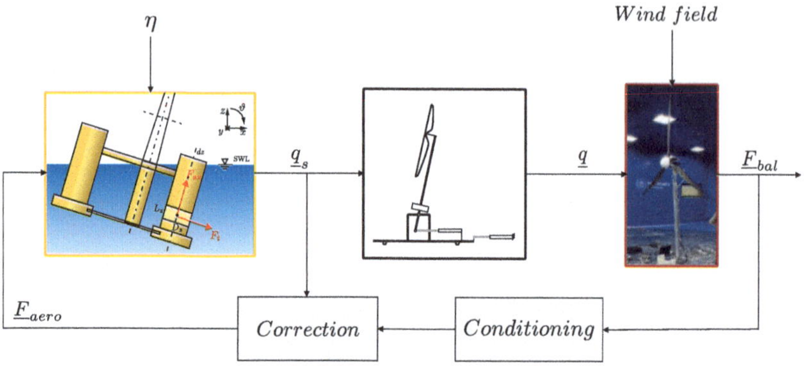

Fig. 2. Scheme of the hybrid/HIL experimental methodology

Hydrostatic restoring loads $\underline{F}_{hst}$, the hydrodynamic forcing due incident waves and free-body motion $\underline{F}_{hydro}$ are simulated by the numerical subsystem and are computed in real-time as function of the platform state $\underline{q}_s$ and the spectrum of the incident waves. Forces exerted by the mooring system on the floating platform $\underline{F}_{moor}$ are computed integrating in real-time a lumped-mass model of the mooring cables (Bayati et al. 2018).

3.1 Aerodynamic Loads

Aerodynamic loads are generated by the rotor of the wind turbine scale model when operating in the ABL test section of the PoliMi wind tunnel. Forces are measured, together with inertial and gravitational loads generated by the wind turbine scale model, by the 6-components load cell mounted at tower base. Total loads $\underline{F}_{bal}$ must be deprived of the inertial and gravitational components to extract the pure aerodynamic loads. Assuming the wind turbine scale model rigid, forces measured by the load cell are expected to be in the form of Eq. (3)

$$\underline{F}_{bal} = \underline{F}_{aero} + [M]_t\ddot{\underline{q}} + [K]_t\underline{q} \tag{3}$$

where $[M]_t$ and $[K]_t$ are the effective wind turbine scale model inertia and gravitational stiffness matrices and vector $\underline{q}$ collects the actual displacements and rotations of the wind turbine scale model. Small differences between $\underline{q}$ and $\underline{q}_s$ are introduced by the HexaFloat actuation system. The aerodynamic loads are then obtained in real-time subtracting from the load cell signal a correction force

$$\underline{F}_{corr} = [M]_t^*\ddot{\underline{q}}_s + [K]_t^*\underline{q}_s \tag{4}$$

where $[M]_t^*$ and $[K]_t^*$ are the inertia and gravitational stiffness matrices of the wind turbine scale model that are defined from the estimated WTM inertial and geometrical properties (Bayati et al. 2017c).

The aerodynamic loads calculation procedure was verified on a 2-DOFs HIL setup (Bayati et al. 2017c), designed to reproduce the FOWT surge and pitch motion being these DOFs the most affected by the aerodynamic forcefield action (Jonkman 2007). In Fig. 3 are analysed (at full-scale) the PSD of $\underline{F}_{bal}$, $\underline{F}_{corr}$ and the difference (reported as $\underline{F}_{res}$), that were measured during an irregular waves test (significative wave height $H_s = 7.1$ m, peak period $T_p = 12.1$ s) and no wind. When the system operates in still air, aerodynamic loads are almost negligible, and forces measured by the load cell are given only by inertial and gravitational effects.

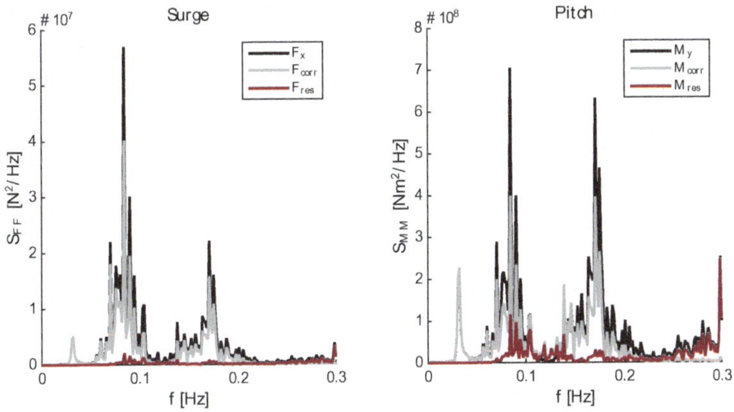

Fig. 3. PSD of surge force and pitch torque measured by the load cell, corresponding correction loads and residual forces.

As visible, a good matching between correction forces and measured loads is achieved across the frequency range of interest. An almost null residual force is obtained in the low-frequency range (below 0.1 Hz) where are located the platform natural frequencies and where aero-elastic effects are predominant. Some differences are present in the wave-frequency and high-frequency ranges, especially for the pitch DOF. This is caused by errors in the identification of the wind turbine structural parameters used for the computation of correction forces and by the scale model flexibility, neglected by Eqs. (3)–(4).

4 The Experimental Setup

The hybrid/HIL experimental methodology allows to separate the reproduction of Froude-dependent phenomena (hydrodynamic and gravitational loads) from Reynolds-dependent ones making possible to relax the scaling constraints for experimental tests and set independently the length and velocity scale factors. The length scale factor was set comparing the DTU 10 MW dimensions with the cross section of the PoliMi wind tunnel test section. A λ_L of 75 was chosen to avoid an excessive miniaturization of the model components, limiting at the same time the blockage effect. Moreover, this value ensures the correct reproduction of the wind profile typical of sea or coastal areas (e.g. Eurocode terrain category 0). The velocity scale factor λ_v was set to 3 to limit the Reynolds number reduction and have a reasonable frequency scale factor that sets the design requirements for actuators and aeroelastic components. Having fixed λ_L and λ_v, all the other scale factors for the wind turbine model and the hybrid/HIL experiments were derived from dimensional analysis and are listed in Table 1.

Table 1. Scale factors for the PoliMi wind turbine scale model

Scale	Expression	Value
Length	λ_L	75
Velocity	λ_v	3
Mass	$\lambda_M = \lambda_L^3$	75^3
Time	$\lambda_T = \lambda_L / \lambda_v$	25
Frequency	$\lambda_\omega = \lambda_v / \lambda_L$	1/25
Acceleration	$\lambda_a = \lambda_v^2 / \lambda_L$	$3^2/75$
Force	$\lambda_f = \lambda_L^2 \cdot \lambda_v^2$	$75^2 \cdot 3^2$

4.1 The Wind Turbine Scale Model

The PoliMi wind turbine model used for hybrid/HIL experiments is a scale model of the DTU 10 MW reference wind turbine (RWT) designed according to the scale factors of Table 1, as reported in (Bayati et al. 2016) and in (Bayati et al. 2017a). The scale factors of Table 1 lead to a Reynolds number 225 times lower than the one expected at full-scale ($\lambda_{Re} = \lambda_v \lambda_L$) that required to re-design the machine rotor. The WTM aeroelastic design is the result of an optimization process (Bayati et al. 2017b) aimed at matching the thrust coefficient, the scaled first flap-wise blade natural frequency and the scaled blade weight of the DTU 10 MW RWT.

A specific mechatronic system (visible in Fig. 4) was designed to control the PoliMi WTM with the same regulation principles adopted by real multi-megawatt wind turbines. An electronically commuted (EC) motor is connected to the high-speed side of the turbine drivetrain and used to control rotor torque (or alternatively rotor speed). A dedicated servomotor, housed on-board the hub, is connected to the root of each blade and used to set the blades full-span pitch angle. An embedded monitoring and control system based on *NI Veristand* makes up for the interface with the actuators and the measurement instruments on-board the WTM. Control algorithms developed in *MathWorks Simulink®* can be directly deployed on the real-time controller and effectively used to control the wind turbine scale model.

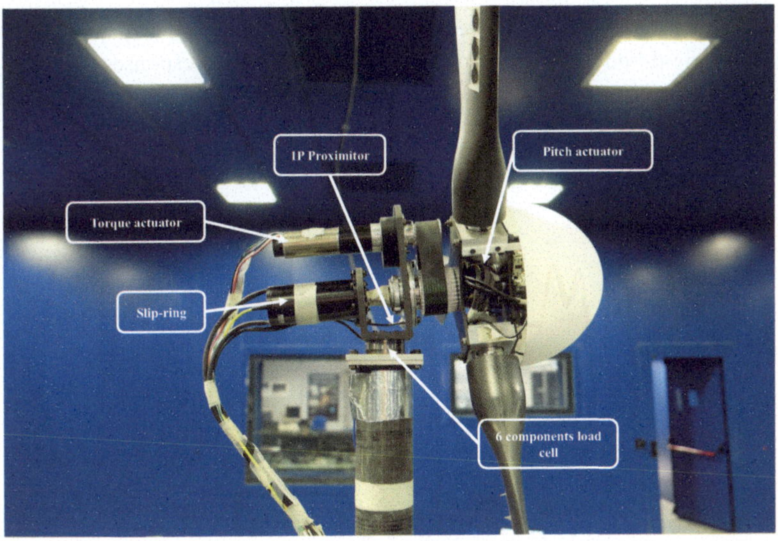

Fig. 4. Side view of the PoliMi WTM RNA and main mechatronic devices.

4.2 The Actuation System

The physical scale model is linked to the numerical subsystem by the actuation system that has to move the wind turbine tower base according to the outputs of real-time integration.

The 2-DOFs system (see Fig. 1a) used for verification tests was based on a long sled moved in the surge direction by a hydraulic actuator. A second actuator connected to a slider-crank mechanism on-board the sled was instead used to impose the pitch motion to the wind turbine base.

A 6-DOFs parallel kinematic machine (see Fig. 1b), the HexaFloat robot, was instead designed to move the wind turbine scale model during hybrid/HIL tests (Bayati et al. 2014). The robot structure was optimized to achieve a high stiffness and a low sensitivity of the end-effector position to error sources (Fiore et al. 2016).

5 Verification of the HIL Setup

A verification of the developed hybrid/HIL methodology was necessary in order to ensure a consistent reproduction of the FOWT motion in presence of real control signals, limited actuation and errors in the identification of the wind turbine model parameters. The stand-alone numerical model of the floating platform used for real-time integration was calibrated to match the still-air response of a FAST (Fatigue Aerodynamics Structures and Turbulence) (Jonkman et al. 2005) model of the floating wind turbine to be tested (Robertson et al. 2016). Free-decay and wave-only experiments were performed and compared to FAST simulations performed under equivalent environmental conditions to assess the effect of a non-ideal response of the actuation system, real measurements and flexible dynamics.

Free-decay tests in still air allowed to assess the natural frequency and damping of the platform rigid-body motion modes. The experimental responses for an initial imposed pitch rotation of ±8° are compared in Fig. 5 to the output of equivalent simulations performed in FAST.

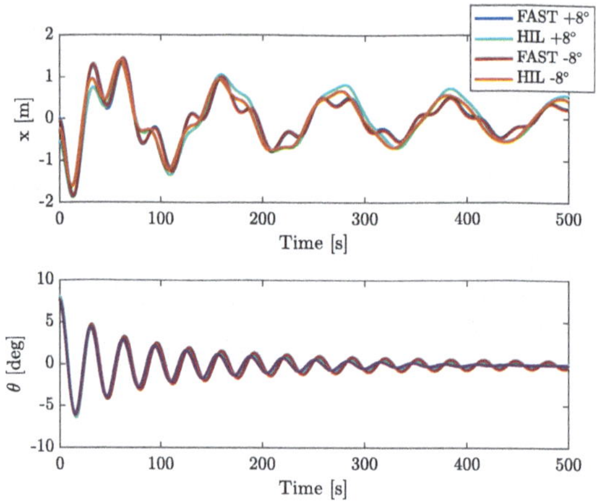

Fig. 5. Comparison between measured and simulated (FAST) free-decay response in still air for an initial pitch rotation of ±8°

As visible, the outcome of the HIL tests matches quiet well the target simulation, with the HIL response slightly more damped that the corresponding experiment.

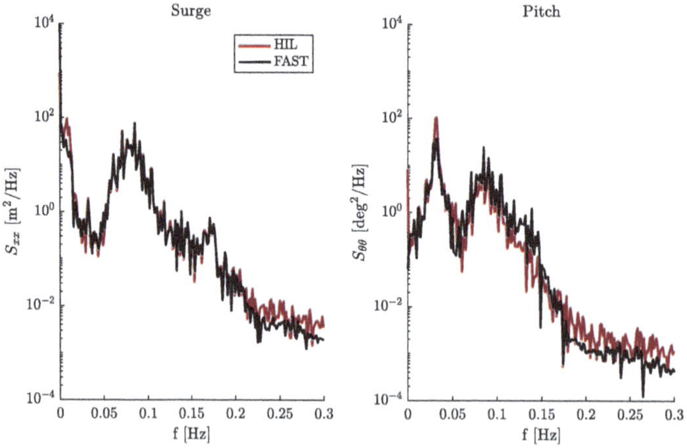

Fig. 6. PSD of surge and pitch response from HIL tests and FAST simulations for operational waves ($H_s = 7.1$ m, $T_p = 12.1$ s)

The FOWT response to irregular waves is assessed comparing the PSD of the platform response from HIL tests in still-air with the output of equivalent numerical simulations performed in FAST. In Fig. 6, the PSD of surge and pitch response to operational waves ($H_s = 7.1$ m, $T_p = 12.1$ s) is compared to the corresponding simulation output. The overall matching is good across the frequency range of interest, especially in the linear wave excitation range. Some minor differences can be seen in the low-frequency range and, for the pitch DOF only, for frequencies above 0.15 Hz.

6 Wind Tests

The aero-hydro-elastic response of the floating wind turbine can be assessed performing tests in laminar wind conditions. No active control logics were used, with rotor speed and pitch angle fixed to the values reported in Table 2, that were chosen to achieve the steady-state thrust force of the DTU 10 MW RWT.

Table 2. Wind turbine steady-state operating conditions for HIL tests

Operating point	U [m/s]	ω_R [rpm]	β [deg]
Below	9.3	7.2	−3.5
Rated	11.4	9.5	−1.0
Above	14.0	9.5	7.2

6.1 Decay Tests

The influence of the aerodynamic forcefield developed by the wind turbine rotor on the platform DOFs was assessed through free-decay tests.

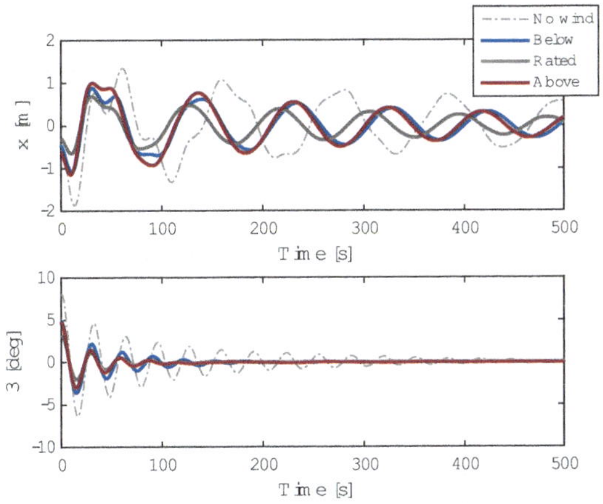

Fig. 7. Free-decay response of surge and pitch DOFs for an initial pitch rotation of 8° and different wind conditions

The coupled response of the surge and pitch DOFs measured during HIL experiments for an initial pitch rotation of 8° at the wind conditions of Table 2 is shown in Fig. 7 in terms of dynamic component only (the static response was removed from the signals shown in the figure). The natural frequency of the platform DOFs is affected by the positional component of the aerodynamic forcefield and slightly increases for increasing wind speeds. Damping of the surge and pitch DOFs is increased by the wind action.

6.2 Irregular Waves

The effects of combined wind and wave loading is investigated considering irregular waves ($H_s = 7.1$ m, $T_p = 12.1$ s) and the three different wind turbine operating points of Table 2. The PSD of surge and pitch DOFs is shown in Fig. 8. In the wave excitation range, the hydrodynamic loads prevail on aerodynamic loads and the platform response is almost unchanged for different wind conditions.

In the low frequency range, the effects of the aerodynamic forcefield are more evident, especially at the pitch natural frequency that shows a decreasing response amplitude for increasing wind speeds. The same effect is slightly visible also at the surge natural frequency, but it is less evident due to the low excitation provided by waves.

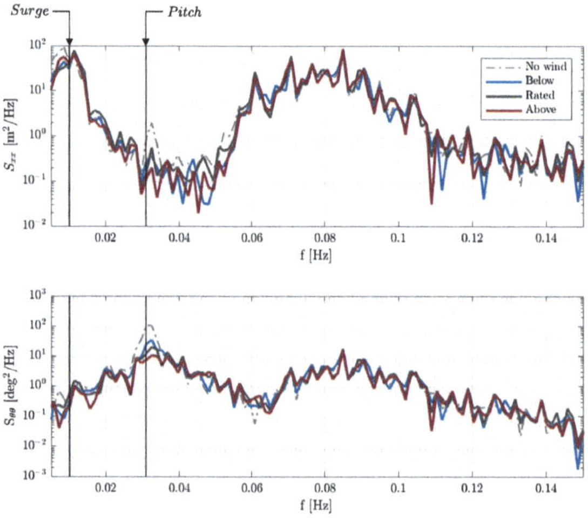

Fig. 8. PSD of surge and pitch response from HIL tests for operational waves ($H_s = 7.1$ m, $T_p = 12.1$ s) and different wind conditions

7 Conclusions

This paper presented the hybrid/HIL experimental methodology developed at PoliMi to perform wind tunnel tests on floating wind turbines. The physical and numerical subsystems are described, focusing on how the aerodynamic loads generated by the wind turbine rotor are measured and combined with hydrodynamic loads for real-time calculation of the floating structure motion. Results from a preliminary test campaign performed to assess the capabilities of the HIL system are discussed, showing how it is possible to investigate the aero-hydro-elastic response of a floating wind turbine.

Finally, the experimental setup is described, summarizing the scaling process and the design choices behind the wind turbine scale model and the HIL actuation system.

References

Bak C. et al (2013) Description of the DTU 10 MW reference wind turbine. DTU Wind Energy Report

Bayati I, Belloli M, Ferrari D et al (2014) Design of a 6-DoF robotic platform for wind tunnel tests of floating wind turbines. Energy Proc 53:313–323

Bayati I, Bernini L, Fiore E et al (2016) On the functional design of the DTU 10 MW wind turbine scale model of LIFES50+ project. J Phys: Conf Ser 735(5):052018

Bayati I, Belloli M, Bernini L et al (2017a) Scale model technology for floating offshore wind turbines. IET Renew Power Gener 11(9):1120–1126

Bayati I, Bernini L, Belloli M et al (2017b) Aerodynamic design methodology for wind tunnel tests of wind turbine rotors. J Wind Eng Ind Aerodyn

Bayati I, Belloli M, Facchinetti A (2017c) Wind Tunnel 2-DOF Hybrid/HIL tests on the OC5 floating offshore wind turbine. In: Proceedings of 36th international conference on ocean, offshore and arctic engineering, OMAE 2017

Bayati I, Facchinetti A, Fontanella A, Belloli M. (2018) 6-DOF hydrodynamic modelling for wind tunnel tests of FOWT: the real-time challenge. In: 37th international conference on ocean, offshore and arctic engineering, OMAE 2018. (forthcoming)

Fiore E, Giberti H (2016) Optimization and comparison between two 6-DoF parallel kinematic machines for HIL simulations in wind tunnel. In: MATEC Web of Conferences

Goupee AJ, Koo BJ, Kimball RW et al (2014) Experimental comparison of three floating wind turbine concepts. J Offshore Mech Arctic Eng 136:020906

Goupee AJ, Kimball RW, Dagher HJ (2017) Experimental observations of active blade pitch and generator control influence on floating wind turbine response. Renew Energy 104:9–19

Jonkman JM (2007) Dynamics modeling and loads analysis of an offshore floating wind turbine. Technical report NREL/TP-500–41958

Koo BJ, Goupee AJ, Kimball RW et al (2014) Model Tests for a Floaitng Wind Turbine on Three Different Floaters. J Offshore Mech Arctic Eng 136:020907

Martin HR, Kimball RW, Viselli AM et al (2014) Methodology for wind/wave basin testing of floating offshore wind turbines. J Offshore Mech Arctic Eng 136(2):020905

Jonkman J, Buhl M (2005) FAST user's guide. Technical report NREL

Robertson A, Jonkman J, Wendt F et al (2016) Definition of the OC5 DeepCwind semisubmersible floating system. Technical report NREL

Sauder T, Chabaud V, Thys M et al (2016) Real-time hybrid model testing of a braceless semi-submersible wind turbine. Part I: the hybrid approach. In: International conference on offshore mechanics and arctic engineering – OMAE

Stewart G, Lackner M, Robertson A et al (2012). Calibration and validation of a FAST wind turbine model of the DeepCwind scaled tension-leg platform. In: Conference paper, national renewable energy laboratory

Yu W, Lemmer F, Bredmose H et al (2017) The TripleSpar campaign: implementation and test of a blade pithc controller on a scaled floating wind turbine model. In: 14th deep sea offshore wind R&D conference EERA DeepWind

Wind-Tunnel Experiments on a Large-Scale Flettner Rotor

G. Bordogna[1(✉)], S. Muggiasca[2], S. Giappino[2], M. Belloli[2], J. A. Keuning[1], R. H. M. Huijsmans[1], and A. P. van't Veer[1]

[1] Section of Ship Hydromechanics, Delft University of Technology, Mekelweg 2, 2628 CD Delft, The Netherlands
{g.bordogna, riaan.vantveer}@tudelft.nl
[2] Department of Mechanical Engineering, Politecnico di Milano, Via la Masa 1, 20156 Milan, Italy
{sara.muggiasca, stefano.giappino, marco.belloli}@polimi.it

Abstract. Experiments on a large-scale Flettner rotor were carried out in the boundary-layer test section of Politecnico di Milano wind tunnel. The rotating cylinder used in the experimental campaign (referred to as Delft Rotor) had a diameter of 1.0 m and span of 3.73 m. The Delft Rotor was equipped with two purpose-built force balances and two different systems to measure the pressure on the rotor's outer skin. The goal of the experiments was to study the influence of different Reynolds numbers on the aerodynamic forces generated by the spinning cylinder. The highest Reynolds number achieved during the experiments was $Re = 1.0 \cdot 10^6$.

Keywords: Flettner rotor · Rotating cylinder · Magnus effect · Wind power · Wind assisted ship propulsion · Green shipping

1 Introduction

The Flettner rotor is a rotating cylinder that generates an aerodynamic force due to the Magnus effect. Invented by German engineer Anton Flettner, it was first used in 1925 on board the *Backau* ship as a form of propulsion. The *Backau* was equipped with two Flettner rotors and it successfully completed its first voyage across the Atlantic in 1926. The same year, following the success of the first rotor ship, the *Barbara* was launched. The vessel had three Flettner rotors and it served as a freighter in the Mediterranean between 1926 and 1929. Despite the proven concept, the rotor ship was fast abandoned since it could not compete with the increasing adoption of diesel engines and with the very low oil price of that time. In more recent years, however, the possibility to use wind energy as an auxiliary form of propulsion for commercial ships has again become of interest due also to the ever-stringent environmental regulations. Since its inception the Flettner rotor was seldom used for real-life applications in the maritime field, nonetheless the physical phenomena associated to rotating cylinders attracted the interest of many scientists over the past years.

F. Ricciardelli and A. M. Avossa (Eds.): INVENTO 2018, LNCE 27, pp. 110–123, 2019.
https://doi.org/10.1007/978-3-030-12815-9_9

Professor Lafay reported the earliest investigation on the Magnus effect in 1912 (Lafay 1912). The experiments comprised a set of two-dimensional tests during which the aerodynamic forces generated by a rotating cylinder were measured. The author also measured the pressure distribution on the cylinder by employing an external stationary probe placed near the cylinder's surface. Twelve years after, at the NASA Langley Memorial Aeronautical Laboratory, Reid (1924) carried out a systematic series of experiments on spinning cylinders at Reynolds numbers varying between $Re = 3.9 \cdot 10^4$ and $Re = 1.1 \cdot 10^5$. The results of Reid indicated that for the range considered, the Reynolds number had a marginal influence on the aerodynamic forces. The work of Thom represents a major contribution to the research on Flettner rotors. After the completion of his doctoral thesis (Thom 1926), the author worked on a series of reports (Thom 1931; Thom and Sengupta 1932; Thom 1934), that included a vast set of force and pressure measurements of rotating cylinders. The influence of surface roughness, aspect ratio and endplates was investigated. The Reynolds number effect was also studied, although no results for Reynolds numbers greater than $Re = 9.0 \cdot 10^4$ were published. Important to mention is also the pressure measurement system used by the author: an internal manometer tube connected to a spring valve, activated by a cam fixed to the spinning surface and by an external fixed pin. As the cylinder spun, the cam was lifted by the external pin that, by compressing the spring valve, was allowing the manometer tube to measure the pressure at a known position around the cylinder.

Several years after, Swanson (1961) provided a detailed summary of the experiments on rotating cylinders carried out until that time. Results of tests performed by the author for very low velocity ratios (ratio between the incoming flow velocity and the tangential velocity, defined as k), $0 < k < 1$, in the Reynolds number range $3.5 \cdot 10^4 < Re < 5.0 \cdot 10^5$ were also reported. As underlined by Swanson, a remarkable output of his work was the discrepancy in aerodynamic performance of rotating cylinders with different features, and of cylinders with similar features but tested at different Reynolds numbers.

During the oil crisis in the 1980's, wind assistance for ship propulsion was considered an appealing manner to cut operational costs and this led to several publications on the topic. Relevant to the aerodynamic performance of Flettner rotors, are the works of Clayton (1985) and Bergeson and Greenwald (1985). The former author carried out experiments for different numbers of endplates as well as for different Reynolds numbers. Tests were performed in the Reynolds number range $1.7 \cdot 10^4 < Re < 4.9 \cdot 10^4$, and the conclusion was that the Reynolds number had a marked influence on the aerodynamic performance of the Flettner rotor, being the drag coefficient the most affected. On the other hand, Bergeson and Greenwald mounted a 7.2 m tall rotor on aboard of a small motor yacht. Using a combination of mooring lines and spring dynamometers, the authors could measure the forces generated by the rotor for a variety of velocity ratios. The results showed the same trend of previous experiments, however, no details on the Reynolds numbers achieved were given.

Although a different topic, the research carried out on projectiles gives a useful insight on pressure measurement techniques applied on spinning cylinders. Particularly relevant are the studies of Miller (1976), Kayser et al. (1986), Rollstin (1990) and McLaughin et al. (1991). Miller used an internal stationary sliding O-ring seal to measure the pressure at a given position on the cylinder's surface. The tests were carried out on a cylinder of aspect ratio AR = 1.6 in the range $2.5 \cdot 10^5 < Re < 4.5 \cdot 10^5$ and for velocity ratios up to k = 2. On the other hand, Kayser et al., Rollstin and McLaughin et al. used a transducer attached to the internal surface of the spinning cylinder. The pressure signal was then passed to the readout instrumentation via a transmitter placed inside the cylinder (Kayser et al. and Rollstin) or via a slip ring (McLaughin et al.).

More recent is the effort of Badalamenti and Prince (2008), who performed a thorough investigation on rotating cylinders using different endplate sizes as well as using rotating or stationary endplates. A series of experiments for a cylinder with no endplate was performed for Reynolds numbers ranging between $Re = 1.9 \cdot 10^4$ and $Re = 9.5 \cdot 10^4$. The results showed that the Reynolds number influenced the lift coefficient for velocity ratios $k < 1$. At higher values of k, the difference became less pronounced. The use of endplates had a positive effect on the efficiency of the rotating cylinder, particularly in the velocity ratio range $2 < k < 4$.

Other than by experiments, rotating cylinders were also largely studied by means of CFD. In this context, the works of Craft et al. (2012), and Everts et al. (2014) appear to be the most relevant to the present research. Craft et al. performed simulations on a Flettner rotor with and without endplates at three different Reynolds numbers: $Re = 1.4 \cdot 10^5$, $Re = 8.0 \cdot 10^5$ and $Re = 1.0 \cdot 10^6$. The results indicated that the effect of the Reynolds number on the lift coefficient was marginal and no results for the drag coefficient were reported. On the other hand, Everts et al. completed a series of 2D CFD simulations on a rotating cylinder at $Re = 5.0 \cdot 10^5$, $Re = 1.0 \cdot 10^6$ and $Re = 5.0 \cdot 10^6$. The aim was to investigate the influence of the Reynolds number as well as the influence of different surface roughness coefficients on the aerodynamic forces. The results reported showed that both the Reynolds number and the surface roughness have a considerable influence on the lift and drag forces generated.

Other studies, both experimental and numerical, carried out on rotating cylinders and not included in this summary can be found in (Zdravkovich 2003).

The literature review presented shows that despite the numerous studies on rotating cylinders there is still disagreement on the influence of the Reynolds number on the aerodynamic forces generated. Moreover, the experimental efforts reviewed were typically undertaken at low Reynolds numbers, or at higher Reynolds numbers but at low velocity ratios or using cylinders with very low aspect ratios. In this context, the aim of the present research is to contribute to the understanding of the scale effects on the lift and drag forces generated by a Flettner rotor tested at high Reynolds numbers. To be noted that the current work focuses on the various experimental techniques employed during the testing campaign; only a small part of the experimental results is presented. A subsequent article will be published in which the complete set of results is provided with the aim to offer a useful benchmark for CFD validation.

2 Experimental Setup

Two series of tests were performed on the Delft Rotor: in December 2017 and in March 2018. The experiments were carried out in the boundary-layer test section of Politecnico di Milano wind tunnel. The test section is 13.84 m wide, 3.84 m high and 35 m long. The maximum reachable wind speed is 15 m/s and the turbulent index in smooth-flow condition is 2%. The Delft Rotor was mounted in the centre of the chamber, in correspondence with two large I-section beams used as a foundation (Fig. 1).

Fig. 1. The Delft Rotor in the boundary-layer chamber of Politecnico di Milano wind tunnel

The Delft Rotor is a rotating cylinder with diameter D = 1.0 m and span H = 3.73 m and it is comprised of three main parts: a static structure, a rotating frame and an outer skin. The static structure consisted of a lower and an upper steel assembly made of four M30 threaded bars screwed into four bases, that were bolted to the wind-tunnel ground and ceiling I-section beams. To improve the stiffness of the static assemblies, four oblique steel bars were connected to the threaded bars and the force balance (Figs. 3 and 5a).

Two aluminium purpose-built force balances were placed at the extremity of the threaded bars, on top of which the bearing housings were bolted (Figs. 2, 3 and 4). To compensate for a possible misalignment, it was chosen to use two self-aligning spherical roller bearings.

The reason to position the balances close to the bearings was to reduce to the minimum the bending moment. By doing so, it was possible to use a three-component force balance: only lift, drag and torque were measured.

Each force balance was carved out of 30 mm thick aluminium slab and it consisted of two concentric circles connected by four strips on which the lift, drag and torque extensometers were placed (Fig. 4). The force balances were purposely built for the Delft Rotor experiments in order to withstand the large aerodynamic forces expected to act on the cylinder ($Lift_{max} = 2500\ N$).

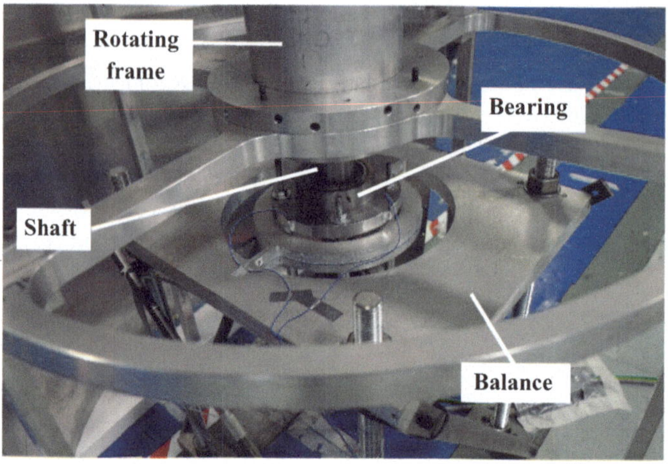

Fig. 2. Lower assembly of the Delft Rotor's static structure

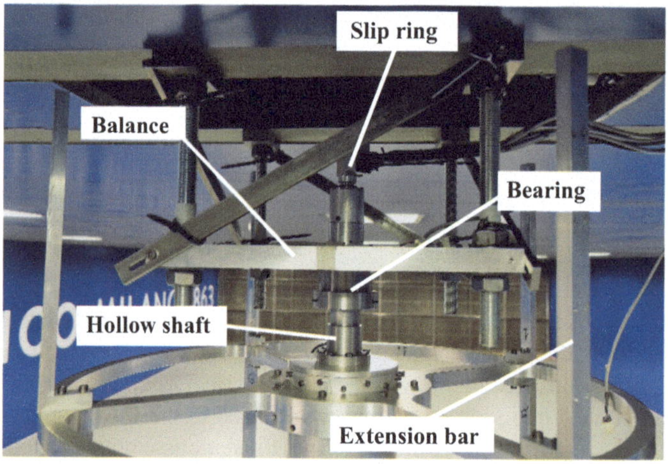

Fig. 3. Upper assembly of the Delft Rotor's static structure

The rotating frame of the Delft Rotor consisted of a 5 mm thick aluminium tube to which three equal aluminium wheels were bolted (Fig. 5a). The wheels were connected with eight vertical bars, and four lower bars (Fig. 5b) and four upper bars (Fig. 3) were used as an extension so that the internal frame matched the height of the outer skin.

The extremities of the internal tube were plugged with lids into which two steel shafts were screwed. The shafts were then fitted into the bearings (Figs. 2 and 3).

The Delft Rotor's outer skin was composed of four calendered aluminium sheets. The sheets were 2 mm thick and they were bolted to the internal frame with more than 250 countersunk head screws. The high number of screws used to secure the calendered sheets to the rotor's internal frame was necessary to achieve a safety factor of 2 while expecting a maximum centrifugal force of 15 kN (the total weight of the rotating components was 170 kg). The frame was rotated using an electric engine that was hung to the lower bearing housing by means of four threaded bars, and a flexible coupling was used to connect the engine shaft with the lower shaft of the internal frame (Fig. 5b).

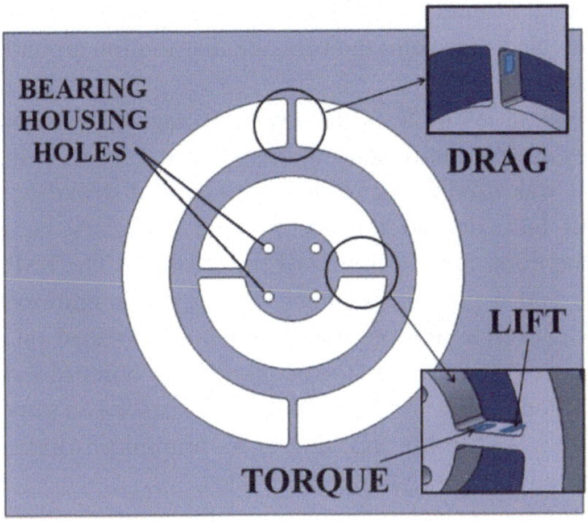

Fig. 4. Three-component force balance used during the experiments

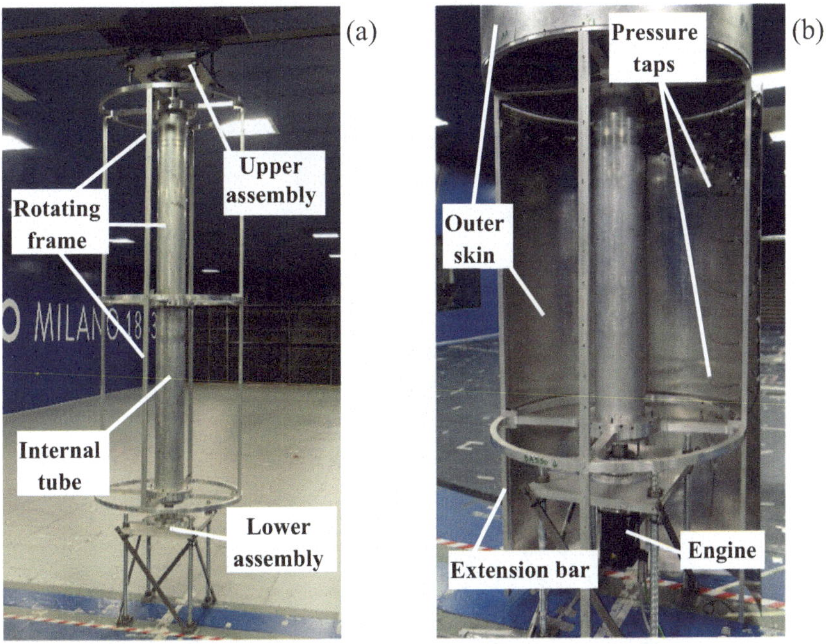

Fig. 5. Internal rotating frame without (a) and with (b) outer skin

The Delft Rotor was equipped with two different pressure measurement systems, namely one high sample-rate pressure scanner (PSI ESP-32HD) with 32 pressure sensors and one single AMS 4711 sensor. The AMS4711 sensor has a pressure range of ±2000 Pa and the ESP scanner of ±2500 Pa. The ESP scanner was mainly installed to be used during the static measurements, whereas during the dynamic experiments was used as redundancy system. During the first series of tests, all 32 pressure taps were equally distributed around the cylinder circumference at 1.85 m from the ground. Doing so, it was possible to measure the pressure distribution around a whole section at the cylinder mid-height.

During the second series of tests, however, 16 taps were removed from the circumference and they were equally spaced along a vertical line (Fig. 5b). This second type of arrangement was used to measure the pressure distribution on the cylinder at different heights. In fact, due to the wind-tunnel boundary layer, the flow speed decreases towards the floor and the ceiling of the chamber. The ESP system comprised a signal conditioner and a pressure scanner. The signal conditioner was fixed to the rotor's internal tube, whereas the pressure scanner was placed on the rotating frame middle wheel as close to the internal tube as possible in order to reduce the centripetal acceleration acting upon it (Fig. 6). The AMS4711 sensor is a very compact pressure instrument (32 × 25 × 13.5 mm) and, due to its minimized dimensions, it was possible to secure it directly to the cylinder inner face.

An advantage of doing so is that the length of the pressure tube was reduced to the minimum. The inner diameter and the length of the pressure tubes in fact have a considerable influence on the frequency response and this is particularly evident at high rotational speeds. Similarly to the ESP system, also the AMS4711 sensor was placed at 1.85 m from the ground (cylinder mid-height). The pressure systems installed on the Delft Rotor use a piezoresistive silicon chip as sensing element and both work as

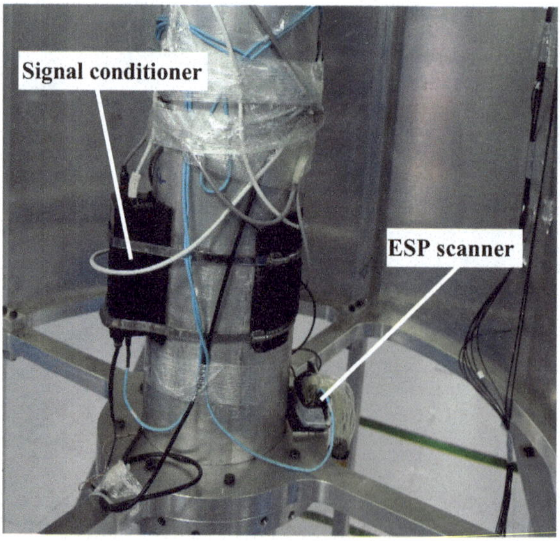

Fig. 6. Position of the ESP scanner and its signal conditioner on the rotating frame

differential transducers. This means that to know the actual pressure acting on the cylinder outer skin, it was necessary to measure also the static pressure in the interior of the cylinder. This was done with a single pressure tube located above the lower force balance. To pass the signal of both the ESP scanner and the AMS4711 sensor to the readout instrumentation, a slip ring connected to the upper hollow shaft was installed (Fig. 3).

The slip ring was also equipped with a transducer to measure the instantaneous velocity and angular position. The angular position of the cylinder was therefore in phase with the pressure measurements.

The experiments were conducted in smooth-flow conditions and the flow velocity was measured with a Pitot tube placed 5 m in front of the Delft Rotor at the height of 1.85 m from the wind-tunnel floor.

3 Pressure Measurement Correction

As mentioned in the previous section, the pressure tube length and inner diameter influence the pressure frequency response as they act as a filter on the signal. This effect is generally relevant in experiments in which the pressures on a moving object need to be measured, and it becomes more noticeable as the speed of the moving object increases. Since the tubes of the 32 pressure taps had to be connected to one single scanner, inevitably they had a considerable length. For this reason, the pressures measured by the ESP scanner had to be corrected to account for the frequency response of the tubing system. During the first series of tests, it was chosen to use tubes of 1.02 mm inner diameter whereas during the second experimental campaign larger tubes with an inner diameter of 1.50 mm were employed. From the analysis of the pressure data, it appeared that the tubes with larger diameter had a better frequency response when compared to the data of the AMS4711 sensor.

Another correction that needs to be applied is to consider the centripetal acceleration acting upon the air column enclosed inside the pressure tubes or acting directly upon the measurement instrument. In the first case, the effect of the acceleration on the air column is dependent on the length of the pressure tubes in the radial direction (not on the total length of the tubes). In the second case, the effect of the acceleration on the measurement instrument is dependent on the distance of the device from the axis of rotation. This means that, due to the way the pressure measurement systems were mounted on the Delft Rotor, the pressures measured by the ESP scanner need to be corrected to account for the effect of the acceleration on the air column enclosed in the pressure tube connecting the scanner and the pressure tap, while the pressures measured by the AMS4711 sensor need to be corrected for the effect of the acceleration acting on the sensor itself. As reported in (Pollack et al. 1972), the effect of the acceleration on the air column can be estimated according to Eq. (1):

$$\mathrm{p_A} - \mathrm{p_B} = \frac{\rho\omega^2}{2} \cdot \left(\mathrm{r_A^2} - \mathrm{r_B^2}\right) \tag{1}$$

where $p_A - p_B$ is the pressure difference between the measurement instrument and the pressure tap, ρ is the air density, ω is the angular velocity and r_A and r_B are the radial distances of the measurement instrument and the pressure tap from the axis of rotation. To provide an example of the effect of the centripetal acceleration on the pressure measurements, in Fig. 7 the pressures measured with the cylinder spinning at increasing angular velocities in still air are reported.

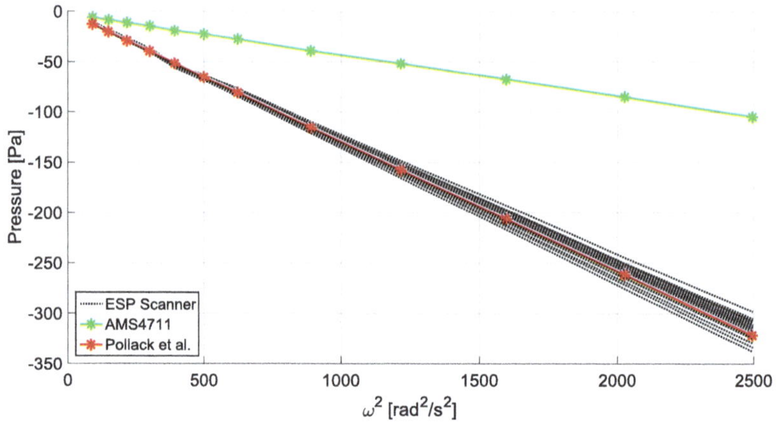

Fig. 7. Measured pressure in still air as a function of the angular velocity squared

The results of Fig. 7 show that the effect of the acceleration is similar for all 32 sensors of the ESP scanner, although the lengths of the pressure tubes are considerably different from one another (390 mm $< L_{tube} <$ 2500 mm). The results of the AMS4711 sensor on the other hand, do not follow the same trend as the measured pressure is due to the acceleration acting on the sensing membrane. This effect is described, for instance, in (Kurtz et al. 2003). The authors also provide a simple estimate of the acceleration effect on a generic piezoresistive pressure sensor depending on the magnitude of the acceleration and the full-scale pressure range of the device.

To overcome the issues concerning the correction of the pressure measurements here elaborated, during the experiments the cylinder was spun in still air at the corresponding angular velocity of each subsequent real measurement (rotation + wind switched on). This procedure was repeated for all performed tests. Thus, the pressures measured with both instrumentations were corrected according to:

$$\mathrm{p}_{\mathrm{corrected}} = [\mathrm{p}_{\mathrm{rot.\&wind}} - \Delta\mathrm{p}_{\mathrm{rot.\&wind}}] - [p_{rot.} - \Delta p_{rot.}] \tag{2}$$

where $\mathrm{p} - \Delta p$ is the pressure measured during experiments with wind switched on (*rot.&wind*) or with wind switched off and just the rotation (*rot.*), corrected for the difference in static pressure between inside and outside the cylinder for the corresponding test. By applying this correction, the effects of the acceleration were eliminated irrespective of whether they affected the tubing system or the measurement instrument itself. Nevertheless, using this correction method, also the pressure experienced by the cylinder due to the rotation itself in still air is discarded. The magnitude of this effect can be estimated by subtracting the averaged pressure measured in still air by the 32 ESP scanner sensors and the results of Eq. (1), (see Fig. 7). Using this procedure, it appears that the pressure due to the rotation itself is negligible as also reported in (Thom 1931).

4 Preliminary Results of the Experiments

The primary aim of the experiments was to investigate the influence of the Reynolds number on the aerodynamic forces generated by the Delft Rotor for different velocity ratios. Nonetheless, static and quasi-static experiments were also performed to check the accuracy of the measurement setup. While during the static tests the cylinder was not rotating, during the quasi-static experiments the cylinder was rotating at very low rpm. This type of experiments aimed at continuously measuring the pressures on the rotor's surface also using the AMS4711 sensor. During the quasi-static tests, the Delft Rotor was span at 2 rpm. This rotational speed was low enough to ensure that the quasi-static assumption with respect to the typical vortex shedding frequency of a cylinder at a given Reynolds number (in this case Strouhal number $St < 0.2$) held true for the entire range of wind speed tested.

As mentioned in the introductory section, the focus of the present work is to prove the validity of the experimental techniques adopted during the tests rather than on the experimental results themselves. For this reason, a comparison between the results obtained with the ESP scanner and with the AMS4711 sensor is provided, using both a static and quasi-static approach. In Figs. 8, 9 and 10 the results for $Re = 2.5 \cdot 10^5$, $Re = 5.2 \cdot 10^5$ and $Re = 9.8 \cdot 10^5$ are reported.

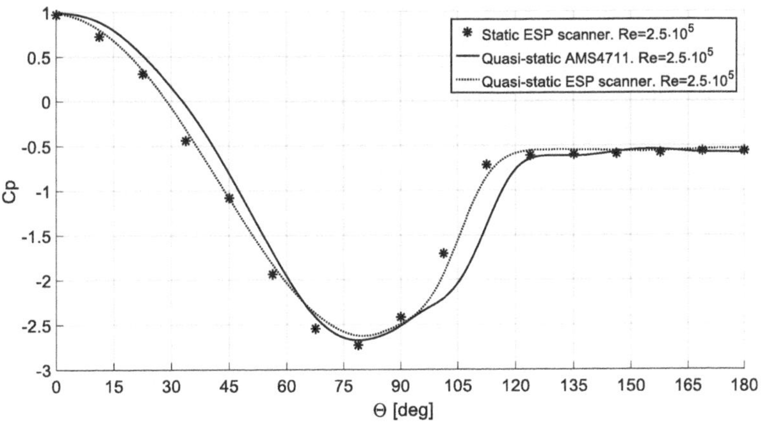

Fig. 8. Mean pressure distribution on the cylinder: static vs quasi-static approach. $Re = 2.5 \cdot 10^5$

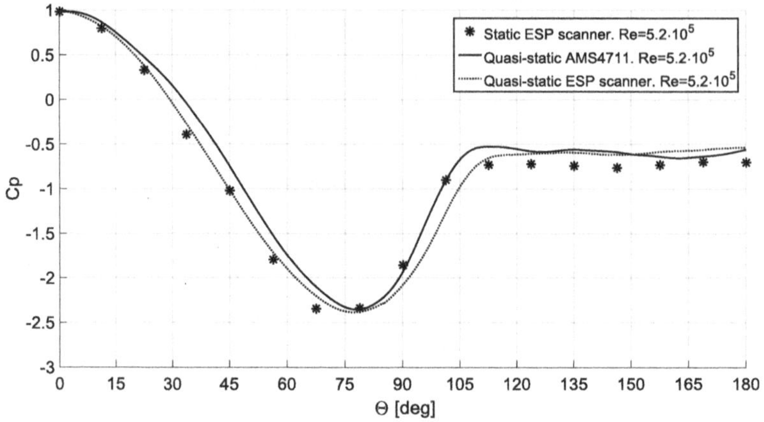

Fig. 9. Mean pressure distribution on the cylinder: static vs quasi-static approach. $Re = 5.2 \cdot 10^5$

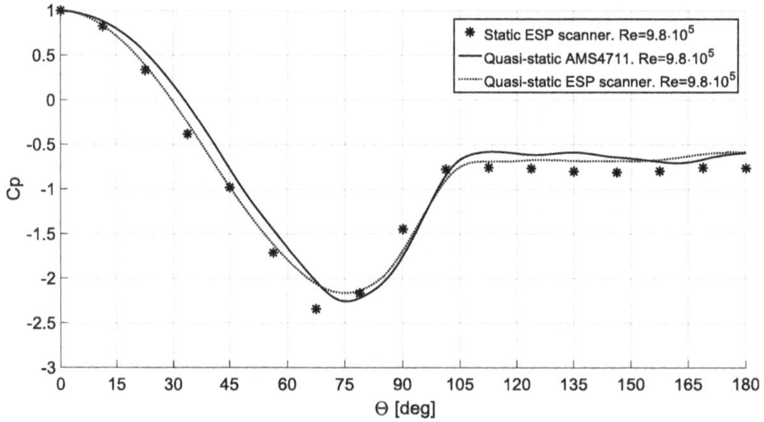

Fig. 10. Mean pressure distribution on the cylinder: static vs quasi-static approach. $Re = 9.8 \cdot 10^5$

The results of the comparison show that the two measurement instruments are in satisfactory agreement, both when a static or quasi-static approach is employed. The pressure distributions over the cylinder for the Reynolds numbers considered are also in line with data reported in (Zdravkovich 2003).

The lift and drag coefficient obtained at $Re = 3.6 \cdot 10^5$ for a set of various velocity ratios are given in Fig. 11. The forces were measured with the force balances as well as by integrating the pressures measured with the 16 sensors of the ESP scanner distributed over the span of the cylinder.

For the lift coefficient, it is shown that there is an excellent agreement between the forces measured by the balances and the results obtained by the pressure integration.

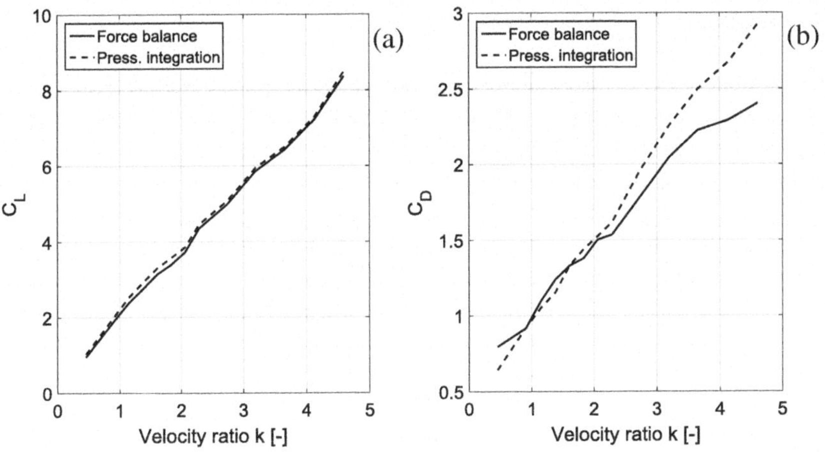

Fig. 11. Lift (a), drag (b) coefficients: measured forces vs pressure integration of ESP scanner data

For the drag coefficient, on the other hand, the agreement is less satisfactory, particularly at high velocity ratios. This is justifiable since the drag component is more sensitive to the orientation of the pressure distribution and, at high rotational speeds, achieving an accurate match between the pressure distribution and the angular position of the cylinder is a challenging task.

Although just a small set of results was published in this section, for the sake of completeness, a summary of all the dynamic experiments carried out on the Delft Rotor is given in Table 1.

Table 1. Summary of the dynamic experiments performed on the Delft Rotor

Date	Re [$\cdot 10^5$]	Velocity ratio k V = executed - = not executed												
		0.5	1.0	1.25	1.5	1.75	2.0	2.25	2.5	3.0	3.5	4.0	4.5	5.0
Dec.'17	1.8	-	V	V	V	V	V	V	V	-	-	-	-	-
Dec.'17	3.6	-	V	V	V	V	V	V	V	-	-	-	-	-
Dec.'17	5.5	-	V	V	V	V	V	V	V	-	-	-	-	-
Dec.'17	10.0	-	V	V	V	V	V	-	-	-	-	-	-	-
Mar.'18	1.8	V	V	-	V	-	V	-	V	-	-	-	-	-
Mar.'18	2.5	V	V	V	V	V	V	V	V	V	V	V	V	V
Mar.'18	3.6	V	V	V	V	V	V	V	V	V	V	V	V	V
Mar.'18	10.0	V	V	-	V	-	V	-	-	-	-	-	-	-

5 Conclusions

In the present work, a literature review on the experimental efforts conducted on Flettner rotors is provided. From the review, it appears that the effect of the Reynolds number on the aerodynamic forces generated by the rotor is still under discussion and that there is a lack of experimental data obtained at high Reynolds numbers. The experimental campaign on the Delft Rotor carried out at Politecnico di Milano wind tunnel is described in detail, with emphasis on the pressure measurement instruments used. The pressure measurements and the necessary correction to account for the effects of the centripetal acceleration represented one of the major difficulties of the described experiments. Nevertheless, the measurement instruments employed, and the correction methods applied, proved to provide meaningful results.

Acknowledgements. This research was supported by the Sea Axe Fund. The author would like to thank the research sponsor as well as all the staff at Politecnico di Milano wind tunnel for their kind help.

References

Badalamenti C, Prince SA (2008) The effects of endplates on a rotating cylinder in crossflow. In: 26th AIAA applied aerodynamics conference, Honolulu, HI, USA, 18–21 August

Bergeson L, Greenwald CK (1985) Sail assist developments 1979–1985. J Wind Eng Ind Aerodyn 19:45–114

Clayton BR (1985) BWEA initiative on wind assisted ship propulsion (WASP). J Wind Eng Ind Aerodyn 19:251–276

Craft TJ, Iacovides H, Johnson N, Launder BE (2012) Back to the future: Flettner-Thom rotors for maritime propulsion? In: 7th international symposium on turbulence heat and mass transfer, Palermo, Italy, 24–27 September

Everts M, Ebrahim R, Kruger JP, Miles E, Sharifpur M, Meyer JP (2014) Turbulent flow across a rotating cylinder with surface roughness. In: 10th international conference on heat transfer, fluid mechanics and thermodynamics, Orlando, FL, USA, 14–16 July

Kayser LD, Clay WH, Damico Jr WP (1986) Surface pressure measurements on a 155 mm projectile in free-flight at transonic speed. In: 14th aerodynamic testing conference, West Palm Beach, FL, USA, 5–7 March

Kurtz AD, Ainsworth RW, Thorpe SJ, Ned A (2003) Further work on acceleration insensitive semiconductor pressure sensor for high bandwidth measurements on rotating turbine blades. In: NASA propulsion measurement sensor development workshop, Huntsville, AL, USA, 13–15 May

Lafay A (1912) Contribution Experimentale a l'Aerodynamique du Cylindre. Reveus Mechanique 30:417–442

McLaughlin TE, Stephen EJ, Robinson MC (1991) Pressure measurements on a rotating circular cylinder. AIAA-91-3265-CP

Miller MC (1976) Surface pressure measurements on a spinning wind tunnel model. AIAA J 14:1669–1670

Pollack FG, Liebert CH, Peterson VS (1972) Rotating pressure measuring system for turbine cooling investigations. Technical report TM X-2621, NASA

Reid EG (1924) Tests of rotating cylinders. Technical report TN-209, NACA

Rollstin LR (1990) Measurement of in-flight base pressure on an artillery-fired projectile. J Spacecraft Rockets 27:5–6

Swanson WM (1961) The Magnus effect: a summary of investigations to date. J Basic Eng 83:461–470

Thom A (1926) The aerodynamics of a rotating cylinder. PhD thesis, University of Glasgow, UK

Thom A (1931) Experiments on the flow past a rotating cylinder. Technical report R&M No. 1410, Aeronautical Research Council

Thom A, Sengupta SR (1932) Air torque on a cylinder rotating in an air stream. Technical report R&M No. 1520, Aeronautical Research Council

Thom A (1934) Effects of discs on the air forces on a rotating cylinder. Technical report R&M No. 1623, Aeronautical Research Council

Zdravkovich MM (2003) Flow around circular cylinders, volume 2: applications. Oxford University Press Inc., New York

Self-excited Vibrations of a Bridge Deck with Single and Double Wind Barriers

A. Buljac[1(✉)], H. Kozmar[1], M. Macháček[2], and S. Pospíšil[2]

[1] Faculty of Mechanical Engineering and Naval Architecture, University of Zagreb, Ivana Lučića 5, 10000 Zagreb, Croatia
{abuljac, hkozmar}@fsb.hr
[2] Institute of Theoretical and Applied Mechanics Prague, Prosecká 809/76, 19000 Prague, Czechia
{machacek, pospisil}@itam.cas.cz

Abstract. Wind-tunnel experiments were carried out to study the influence of wind barriers on the flutter susceptibility of a bridge-deck section. This was performed on the model of a streamlined bridge-deck section equipped with various wind barriers. The wind barriers of various heights and constant porosity of 30% were placed at the (a) windward bridge-deck edge, (b) leeward bridge-deck edge, (c) both windward and leeward bridge-deck edges. The results indicate that the studied bridge deck remains dynamically stable with respect to the heave motion for all configurations of wind barriers. The configurations with wind barriers at both edges are more sensitive to torsional flutter when compared to configurations with single wind barriers placed either at the windward or at the leeward bridge-deck edge.

Keywords: Bridge deck · Self-excited vibrations · Flutter derivatives · Wind-tunnel experiments

1 Introduction

Strong winds that commonly blow on bridges may deteriorate dynamic stability of vehicles crossing the bridge. To prevent closure of these bridges in the events of strong cross-winds, wind barriers are commonly placed on bridges to protect vehicles from adverse cross-winds effects, e.g. Chu et al. (2013).

Wind barriers, often designed in various heights and porosities, proved to be successful in sheltering vehicles on bridges, e.g. Kozmar et al. (2012). However, long-span cable-supported bridges, characterized by low natural frequencies and low mechanical damping, are particularly sensitive to wind effects, and even a small change in the bridge-deck design may trigger dynamic instability, Jones et al. (1995).

Wind barriers significantly change the shape of the bridge-deck section, hence their influence on the dynamic behavior of the bridge should be carefully investigated. They may be placed in various arrangements at the bridge deck to correspond to wind and terrain characteristics on a certain geographic location. In the area where the wind blows predominantly from one direction, wind barriers are usually placed at the windward bridge-deck edge only. This is the most common arrangement of wind

F. Ricciardelli and A. M. Avossa (Eds.): INVENTO 2018, LNCE 27, pp. 124–132, 2019.
https://doi.org/10.1007/978-3-030-12815-9_10

barriers. However, the windward-placed wind barriers in various heights and porosities proved to significantly influence the aerodynamic and aeroelastic characteristics of bridges, Buljac et al. (2017).

In case the wind changes direction, the single windward-placed wind barrier may become leeward-placed wind barrier. This adverse configuration reduces the sheltering effects of the wind barrier, while at the same time it alters the dynamic behavior of the bridge deck; hence, it should be avoided.

In the areas with no predominant wind direction, wind barriers are commonly placed on both bridge-deck edges, e.g. He et al. (2016). For wind barriers placed both at the windward and leeward bridge-deck edges, the shear layer that separates from the windward wind-barrier top may remain captured at the bridge deck between the windward and leeward wind barriers, e.g. Avila-Sanchez et al. (2016). This issue is expected to influence a dynamic stability of long-span bridges.

The most common dynamic instability that affects long-span cable-supported bridges is flutter. It is a self-excited vibration, which may cause bridges to collapse due divergent and excessive amplitudes of oscillations, e.g. Xu (2013).

The flutter susceptibility is commonly analyzed using the dimensionless aeroelastic coefficients, i.e. flutter derivatives. The trends and values of these coefficients are considered as indicators of flutter instability, as they contribute to the generalized structural matrices of the dynamical system. Original method for extraction of flutter derivatives from the fluctuating response of a bridge deck is developed by Scanlan and Tomko (1971). The trends in flutter derivatives are considered to be dependent only on the shape of the bridge-deck section, Jones et al. (1995).

In the present study, an experimental study on the influence of various wind-barrier configurations on flutter susceptibility of a streamlined bridge-deck section was carried out using the custom-made experimental mechanism for measurements of complex aeroelastic phenomena.

2 Mathematical Model

In the present study, the heave and pitch degrees of freedom are considered, as these are commonly studied for long-span bridges, e.g. Xu et al. (2014).

The dynamic response of the bridge-deck moving in the two-degree-of-freedom (2DOF) system with the self-excited aerodynamic lift force and the pitch moment as external disturbances can be written as:

$$\begin{aligned} m\left[\ddot{h}(t)+2\xi_h\omega_h\dot{h}(t)+\omega_h^2h(t)\right] &= L_s(t),\\ I\left[\ddot{\alpha}(t)+2\xi_\alpha\omega_\alpha\dot{\alpha}(t)+\omega_\alpha^2\alpha(t)\right] &= M_s(t). \end{aligned} \tag{1}$$

The heave and pitch responses of the respective degrees of freedom are $h(t)$ and $\alpha\ (t)$; ω_h and ω_α are heave and pitch natural circular frequencies, respectively. System mass is m, I is mass of inertia, ξ_h and ξ_α are damping ratios for heave and pitch free decay. The equations of motion may be written in a symbolic way in the matrix form:

$$\mathbf{M}\ddot{\mathbf{x}}(t) + \mathbf{C}^m\dot{\mathbf{x}}(t) + \mathbf{K}^m\mathbf{x}(t) = \mathbf{F}(t). \tag{2}$$

M is mass matrix, **x**(t) is dynamic response vector, $\mathbf{C}^m$ and $\mathbf{K}^m$ are system damping and stiffness matrix, respectively, **F**(t) is vector of self-excited external disturbances.

Aerodynamic self-excited lift force $L_s(t)$ and the pitch moment $M_s(t)$ can be written using the dimensionless flutter derivatives and reduced frequency $K = B_\omega/v_\infty$, Scanlan and Tomko (1971):

$$\begin{aligned} L_s(t) &= \frac{1}{2}\rho v_\infty^2 BL\left[KH_1^*(K)\frac{\dot{h}(t)}{v_\infty} + KH_2^*(K)\frac{B\dot{\alpha}(t)}{v_\infty} + K^2H_3^*(K)\alpha(t) + K^2H_4^*(K)\frac{h(t)}{B}\right], \\ M_s(t) &= \frac{1}{2}\rho v_\infty^2 B^2L\left[KA_1^*(K)\frac{\dot{h}(t)}{v_\infty} + KA_2^*(K)\frac{B\dot{\alpha}(t)}{v_\infty} + K^2A_3^*(K)\alpha(t) + K^2A_4^*(K)\frac{h(t)}{B}\right]. \end{aligned} \tag{3}$$

Flutter derivatives H_i^* and A_i^* (i = 1, 2, 3, 4) contribute to the generalized stiffness and damping matrices of the system. Therefore, the self-excited external forces and moments modify the structural matrices of the system and alter its dynamic behavior.

The modified damping and stiffness matrices of a system are evaluated from the bridge-deck section dynamic response for various flow velocities using the modified unifying least-squares method. This numerical method proved to be quite accurate and robust for different flow velocities and responses, while it is only valid in the pre-critical state where assumption of linear behavior can be adopted, Král et al. (2014).

The flutter derivatives extracted from the diagonal terms of the system structural matrices are direct flutter derivatives (H_1^*, H_4^*, A_2^*, A_3^*). The direct flutter derivatives provide significant information on flutter susceptibility of bridge decks.

The indirect flutter derivatives (H_2^*, H_3^*, A_1^*, A_4^*) influencing the symmetry of the system matrices and thus the instability onset are extracted from the non-diagonal terms of the structural matrices. However, these derivatives are more difficult to be identified.

The flutter derivatives are calculated from the dynamic response of bridge deck section using the following equations:

$$\begin{aligned} H_1^*(K) &= -\frac{2m}{\rho B^2 L\omega_h}\left(C_{11}^e - C_{11}^m\right),\ H_2^*(K) = -\frac{2m}{\rho B^3 L\omega_\alpha}\left(C_{12}^e - C_{12}^m\right), \\ H_3^*(K) &= -\frac{2m}{\rho B^3 L\omega_\alpha^2}\left(K_{12}^e - K_{12}^m\right),\ H_4^*(K) = -\frac{2m}{\rho B^2 L\omega_h^2}\left(K_{11}^e - K_{11}^m\right), \\ A_1^*(K) &= -\frac{2I}{\rho B^3 L\omega_h}\left(C_{21}^e - C_{21}^m\right),\ A_2^*(K) = -\frac{2I}{\rho B^4 L\omega_\alpha}\left(C_{22}^e - C_{22}^m\right), \\ A_3^*(K) &= -\frac{2I}{\rho B^4 L\omega_\alpha^2}\left(K_{22}^e - K_{22}^m\right),\ A_4^*(K) = -\frac{2I}{\rho B^3 L\omega_h^2}\left(K_{21}^e - K_{21}^m\right). \end{aligned} \tag{4}$$

C_{ij} and K_{ij} (i, j = 1, 2, 3, 4) are components of the damping and stiffness matrices of the system, respectively. The subscript e denotes effective structural matrices obtained in the flow, while m denotes mechanical structural matrices which are obtained in the calm air without the flow. The damping and stifffness matrices are obtained from the dynamic response of the bridge-deck section using the modified unifying least-squares method, e.g. Král et al. (2014).

3 Experimental Setup

The wind tunnel experiments were carried out in the Climatic Wind Engineering Laboratory of the Institute of Theoretical and Applied Mechanics, Prague. The flutter susceptibility of the studied bridge deck was determined using the spring-supported portion of the bridge-deck section.

The studied bridge deck is a typical streamlined cross-section similar to the Great Belt Bridge cross-section. The bridge-deck section model was used together with 30% porous wind barriers in 3 m (WB 3_30), 5 m (WB 5_30) and 7 m (WB 7_30) height given at the full scale. The porosity of wind barriers is calculated as a ratio between the wind-barrier cavities and the frontal area of the entire wind barrier. This porosity of the wind barrier was used as it is anticipated to yield a good compromise between its sheltering efficiency and adverse modifications of the bridge-deck shape.

The bridge-deck section model and the wind-barrier models were manufactured in the 1:100 geometrical scale, Fig. 1.

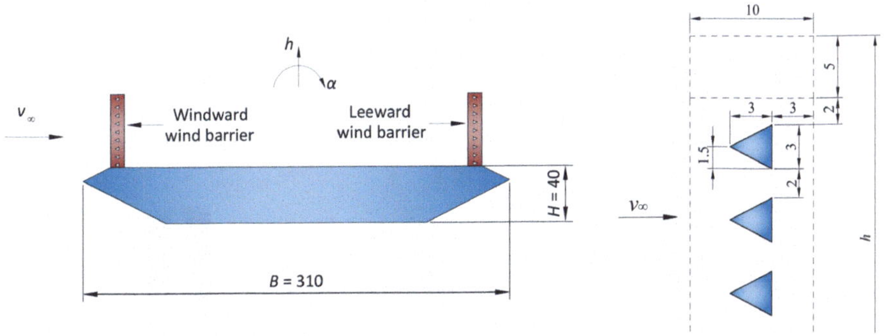

Fig. 1. Geometrical details of the studied bridge-deck section model and wind-barrier model. Dimensions are provided in mm in the wind-tunnel scale.

More details on the geometrical and mass properties of the studied bridge-deck section model and the wind-barrier models are provided in Buljac et al. (2017).

The wind-barrier models were placed on the studied bridge-deck section models in three different configurations: (a) single wind-barrier placed at the windward bridge-deck edge (WW configuration), (b) single wind barrier placed at the leeward bridge-deck edge (LW configuration), (c) double wind barriers placed at both windward and leeward bridge-deck edges (WW + LW configuration).

The studied bridge deck section was placed on the custom-made mechanism for measurements of complex aeroelastic phenomena. This mechanism allows for a dynamic motion of the bridge-deck section model in two-degrees of freedom (heave and pitch). It is designed to allow for large deflections with a linear behavior, while it allows for a precise natural frequency tuning in both the heave and pitch motions.

The undisturbed flow velocity was measured by the Prandtl-Pitot tube placed upstream of the bridge-deck section model in the freestream flow.

The dynamic response of the bridge-deck section model influenced by the airflow was obtained using the free-vibration method. This method allows for a natural interaction of the bridge-deck section model with the airflow, as it includes an initial displacement of the bridge-deck section model in both studied degrees of freedom (heave and pitch) using the trigger placed outside of the wind tunnel. Measurements of the free-decay oscillations were performed for various flow velocities.

The dynamic displacements of the bridge-deck section model in the heave and pitch motions were recorded using the magnetic transducers. More details on this mechanism are available in Král et al. (2016).

Experimental setup for measurements of a dynamic response of the bridge-deck section model is presented in Fig. 2.

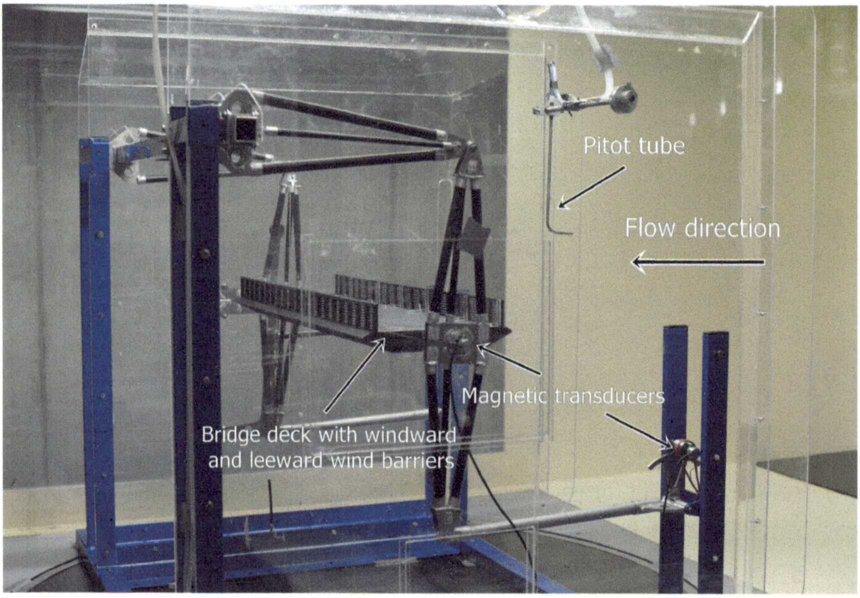

Fig. 2. Experimental setup for measurements of aeroelastic stability of the bridge-deck section model with wind barriers.

The experiments were carried out at the sampling rate of 1000 Hz and the time acquisition length of 30 s. The blockage of the studied bridge deck and the experimental mechanism was lower than the critical 5%, hence no corrections were applied to the results, Simiu and Scanlan (1996).

4 Experimental Results

Direct flutter derivatives H_1^*, A_2^* and A_3^* are reported in Figs. 3, 4 and 5 for various reduced flow velocities v_∞/fB. For the H_1^* derivative, the heave natural frequency of oscillations was used for normalization, while the pitch natural frequency was used for the A_2^* and A_3^* derivatives. Only these flutter derivatives are reported as they provide relevant insights in the dynamic stability with respect to bridge flutter. The legend presented in Fig. 1 is used for all subsequently reported results.

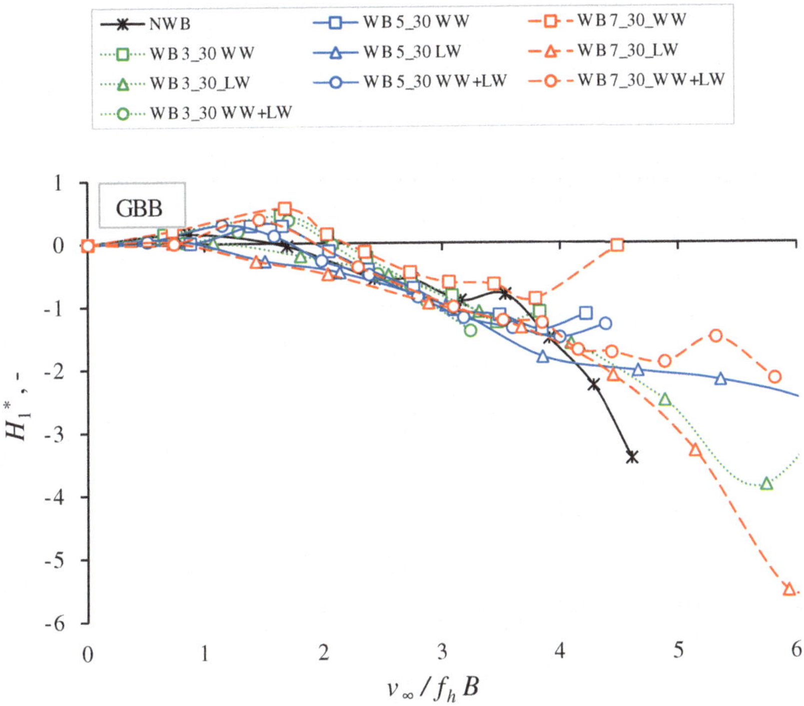

Fig. 3. The H_1^* derivative for various arrangements of wind barriers at various reduced flow velocities

The H_1^* derivative is important when studying the dynamic stability in the heave motion. It contributes to the generalized damping matrix of the system in the heave direction, as it describes the influence of the self-excited aerodynamic lift force on the heave motion. The H_1^* derivative has positive values for negative aerodynamic damping in the heave motion, which decreases overall damping and indicates a dynamic instability in the heave motion that may lead to the diverging oscillations of a bridge.

The long-span cable-supported bridges are not commonly sensitive to the instability in the heave motion. However, increasing the bluffness of the bridge equipped with the wind barriers may make those bridges sensitive to the heave instability.

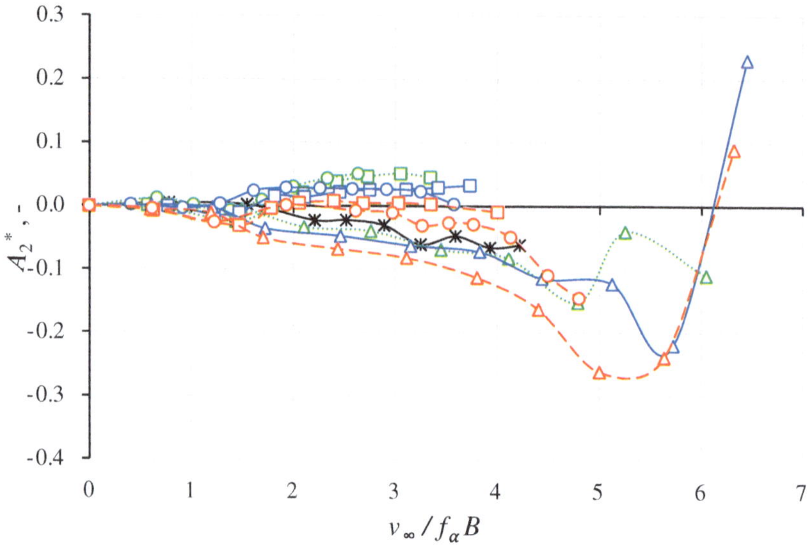

Fig. 4. The A_2^* derivative for various arrangements of wind barriers at various reduced flow velocities

The studied bridge deck is generally stable with respect to the heave motion, as the H_1^* derivative remained negative in a wide range of studied flow velocities. Positive values of the H_1^* derivative are observed only for the WW and WW + LW configurations of wind barriers in the range $v_\infty/f_h B = 1-2$. This is likely due to the fluctuating aerodynamic lift force with the frequency equal to the natural heave frequency of the bridge. As the flow velocity is increased, this phenomenon vanishes. This indicates that the WW and WW + LW configurations of wind barriers may trigger vortex-induced vibrations that adversely decrease the lifetime of bridge structural components, Ehsan and Scanlan (1990).

At relatively large flow velocities, only the configuration with the 7_30 WW wind barrier exhibits strong shift in the trend of the H_1^* derivative.

The A_2^* derivative is commonly considered as the most important in the analysis of a bridge-deck dynamic stability, e.g. Král et al. (2014). It contributes to the generalized damping matrix of the system in the torsional motion, as it describes the influence of the self-excited aerodynamic pitch moment on the pitch motion.

The A_2^* derivative has positive values for negative aerodynamic damping in the pitch motion. This deteriorates a dynamic stability in the pitch motion as it decreases the damping of the system. Excessive positive values of this derivative cause negative generalized damping and torsional flutter of the bridge.

Double wind barriers placed at both the windward and leeward bridge-deck edges prove to deteriorate the most the dynamic stability in the pitch motion, as the A_2^* derivative reaches positive values at the lowest reduced flow velocities for this arrangement of wind barriers.

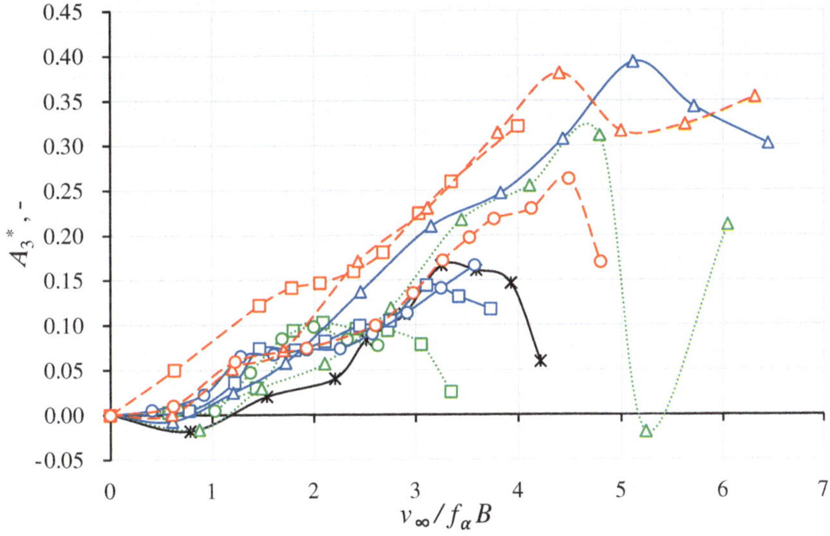

Fig. 5. The A_3^* derivative for various arrangements of wind barriers at various reduced flow velocities

The WW configuration of wind barriers yields positive values of the A_2^* derivative as well, which is observed at relatively low flow velocities. The LW configuration of wind barriers becomes dynamically unstable in the torsional motion at larger reduced flow velocities; hence it proved to be more dynamically stable.

The A_3^* derivative contributes to the generalized stiffness matrix of the system in the torsional motion, as it describes the influence of the pitch displacement on the self-excited aerodynamic pitch moment.

Positive values of this derivative are obtained for the negative aerodynamic stiffening, i.e. decreased system stiffness in the pitch motion.

The A_3^* derivative is generally positive suggesting a decrease in the damped frequency of the pitch motion.

5 Conclusions

The focus of the present study is on influence of various arrangements of wind barriers on the flutter susceptibility of a bridge-deck section. This was performed experimentally on the model of a streamlined bridge-deck section equipped with various wind barriers.

The wind barriers of various heights and constant porosity of 30% are placed at the (a) windward bridge-deck edge, (b) leeward bridge-deck edge, (c) both windward and leeward bridge-deck edges. The dynamic response of the bridge-deck section model was obtained using the free-vibration method for various reduced flow velocities.

The studied bridge generally remains stable with respect to the heave motion. However, the fluctuating aerodynamic lift force is observed for a single windward-placed wind barrier and double wind barriers in the range of investigated flow velocities.

Single windward-placed wind barrier and double wind barriers prove to deteriorate the most the dynamic stability in the pitch motion, while the configurations with a single leeward-placed wind barrier become unstable in the torsional motion at larger reduced flow velocities; hence they proved to be more dynamically stable. Increasing the bluffness of the bridge equipped with the wind barriers may make those bridges sensitive to the heave instability.

Acknowledgements. The activity presented in the paper is part of the research grant No. 17-26353 J by the Czech science foundation (GAČR).

References

Avila-Sanchez S, Lopez-Garcia O, Cuerva A, Meseguer J (2016) Characterisation of cross-flow above a railway bridge equipped with solid windbreaks. Eng Struct 126:133–146

Buljac A, Kozmar H, Pospíšil S, Macháček M (2017) Aerodynamic and aeroelastic characteristics of typical bridge decks equipped with wind barriers at the windward bridge-deck edge. Eng Struct 137:310–322

Chu CR, Chang CY, Huang CJ, Wu TR, Wang CY, Liu MY (2013) Windbreak protection for road vehicles against crosswind. J Wind Eng Ind Aerodyn 116:61–69

Ehsan F, Scanlan RH (1990) Vortex-induced vibrations of flexible bridges. J Eng Mech 116:1392–1411

He X, Shi K, Wu T, Zou Y, Wang H, Qin H (2016) Aerodynamic performance of a novel wind barrier for train-bridge system. Wind Struct 23:171–189

Jones NP, Scanlan RH, Sarkar PP, Singh L (1995) The effect of section model details on aeroelastic parameters. J Wind Eng Ind Aerodyn 54(55):45–53

Kozmar H, Procino L, Borsani A, Bartoli G (2012) Sheltering efficiency of wind barriers on bridges. J Wind Eng Ind Aerodyn 107(108):274–284

Král R, Pospíšil P, Náprstek J (2014) Wind tunnel experiments on unstable self-excited vibration of sectional girders. J Fluids Struct 44:235–250

Král R, Pospíšil S, Náprstek J (2016) Experimental set-up for advanced aeroelastic tests on sectional models. Exp Tech 40:3–13

Scanlan RH, Tomko JJ (1971) Airfoil and bridge deck flutter derivatives. J Eng Mech Div ASCE 97:1717–1733

Simiu E, Scanlan RH (1996) Wind effects on structures: fundamentals and applications to design. Wiley, New York

Xu YL (2013) Wind effects on cable-supported bridges. Wiley, Singapore

Xu F, Zhu L, Ge X, Zhang Z (2014) Some new insights into the identification of bridge deck flutter derivatives. Eng Struct 75:418–428

Wind Tunnel Experimentation on Stationary Downbursts at WindEEE Dome

M. Burlando[1(✉)], D. Romanić[2], H. Hangan[2], and G. Solari[1]

[1] Department of Civil, Chemical and Environmental Engineering (DICCA), University of Genoa, Via Montallegro 1, 16145 Genoa, Italy
{massimiliano.burlando,giovanni.solari}@unige.it
[2] Wind Engineering, Energy and Environment (WindEEE) Research Institute, Western University, 2535 Advanced Avenue, London, ON N6M 0E2, Canada
{dromanic,hmhangan}@uwo.ca

Abstract. In the context of the European Project THUNDERR a scientific collaboration between the Wind Engineering and Structural Dynamics (Windyn) Research Group of the University of Genoa (Italy) and the Wind Engineering, Energy and Environment (WindEEE) Research Institute of Western University (Canada) has been established to study experimentally at the WindEEE Dome facility how the main geometrical and mechanical properties of downbursts are affected by different cloud base outflows of stationary thunderstorms. At present, the analysis of the downbursts simulated experimentally is ongoing and some preliminary elaborations have been obtained concerning the qualitative and quantitative interpretation of the corresponding signals. Classical signal decomposition was applied to experimentally produced downbursts in the WindEEE Dome in order to study transient features of the time series. This study presents the results for two radial positions from downdraft centre and for twenty repetitions per radial position. Several prospects for further research are also discussed.

Keywords: Downbursts · THUNDERR project · WindEEE Dome · Signal analysis · Extreme winds · Impinging jets

1 Introduction

Downbursts are vigorous winds that descend from a thunderstorm and spread out rapidly once they encounter the ground. Strong downbursts can cause damage similar to that of an EF0 (105–38 m s^{-1}) or EF1 (38–49 m s^{-1}) tornados (McDonald et al. 2006). However, the dynamics of downbursts and tornadoes are completely different. In the initial stages of thunderstorm, a powerful updraft is the main feature of the cloud dynamics. The updraft is responsible for the influx of warm and moist air that "feeds" the thunderstorm. In this stage, the cumuliform cloud is typically a cumulus congestus, which will later evolve into a cumulonimbus. That is, as the growth of raindrops and hailstones continues in the mature cloud, eventually their weight is larger than the uplift caused by the updraft and they start falling to the ground. As the hydrometeors fall towards the ground two processes take place: (1) their evaporation and melting in the

F. Ricciardelli and A. M. Avossa (Eds.): INVENTO 2018, LNCE 27, pp. 133–142, 2019.
https://doi.org/10.1007/978-3-030-12815-9_11

subsaturated environment underneath the cloud, and (2) the drag on the ambient air caused by the falling hydrometeors, as described by the Stokes' law. In many cases, an inflow of dry air in the rear flank of the storm additionally amplifies evaporation and melting. The latent heat of evaporation/melting which is extracted from the surrounding air results in the cooling of the air in the precipitation region of thunderstorm, thus creating a downdraft of cold air that impinges on the surface. The downdraft and its radial spread upon hitting the ground are known as downburst. Meteorological aspects of downbursts and their evolution were studied in the pioneering studies by Simpson (1969), Charba (1974), Fujita (1981), Wakimoto (1982), Droegemeier and Wilhelmson (1987), and Sherman (1987), among many others.

In 2017, a first set of experiments has been carried out at WindEEE Dome (Hangan et al. 2017) to study the class of downbursts induced by stationary thunderstorm outflows (Solari et al. 2018). This class of phenomena comprehends all thunderstorms that are stationary in meteorological terms, which means that their vertical axis does not move with respect to the ground. However, even with the approximation of stationary thunderstorm, the resulting flow field is unsteady as it changes according to the time evolution of the cumulonimbus cloud as it passes from mature to dissipative stages.

In accordance with the above simplified assumption of real thunderstorms, the experiments presented in this paper were carried out in 2017 at WindEEE Dome and the focus was mainly on the simulation of downbursts generated by stationary but unsteady thunderstorm outflows. The main objective of this paper is to subject the measured time series to the classical signal decomposition techniques for transient winds (e.g., Holmes et al. 2008; Solari et al. 2015; Burlando et al. 2017a, b). Although these statistical techniques were applied to a number of full scale downburst events, it will be interesting to investigate their applicability to laboratory produced downbursts. Moreover, each downburst experiment in WindEEE Dome was repeated twenty times in order to build high statistical significance of the obtained results, as well as to account for the random nature of turbulent flow and its high dependency on initial conditions.

Downburst outflows produced in WindEEE Dome are effectively large impinging jets of various diameters. Over the last several decades, the impinging jet approach to simulate downburst has been established as a recognized laboratory technique to replicate full scale events, in particular for wind engineering purposes. Wood et al. (2001) used continues impinging jet to study downburst outflows over different topographies as well as flat surface. Two goals of their study were the investigation of the evolution of velocity profile at different radial distances from downburst centre and the comparison of speed-up factors between downbursts and atmospheric boundary layer (ABL) winds. Chay and Letchford (2002) and Letchford and Chay (2002) investigated the surface pressure distribution on a cube immersed in stationary and moving downbursts, respectively. Similar to Wood et al. (2001), their goal was also to compare the pressure distribution of stationary and translating downbursts to the one produced by ABL winds. For instance, they concluded that at lower translation speeds, translating downbursts can accurately be approximated as a stationary downburst. They also reported the largest similarity in pressure distributions between the uniform ABL winds and a stationary downburst at the radial distance $R/D = 1$, where R is the radial distance from downburst centre and $D = 0.51$ m is the jet diameter in their simulator. At a geometric scale of 1:3000, Mason et al. (2005) used pulsed impinging jet to simulate stationary downburst through flow visualization, velocity measurements, and surface pressures. Their jet was the same as the one used by Chay and Letchford (2002) and Letchford and Chay (2002). Xu and Hangan (2008) focused their analysis on dependencies of impinging jets on Reynolds number (i.e., scale), boundary conditions (geometry and surface roughness) and inlet conditions. They found that below the critical Reynolds number, the boundary layer height formed on the surface decreases while the maximum velocity increases by increasing the Reynolds number. Moreover, it was demonstrated that the flow is weakly dependent on the distance between the jet outlet and the surface for distances larger than the ring vortex formation length. However, their jet diameter was only $D = 0.0381$ m. McConville et al. (2009) investigated the transient nature of downbursts by directly controlling the simulator fans, as well as by using two different opening of the simulator. Their experiments were repeated ten times and the ensemble average was used for subsequent analyses. The jet diameter in their experiments was $D = 1$ m. It should be pointed out that they aligned the velocity signals from ten repetitions by assuming the time to be zero when the velocity in the ramp-up segment of time series is 40% of the peak velocity. At the end, it should be noted that the diameter of impinging jets analysed in the present paper is $D = 3.2$ m, which is by an order of magnitude or so larger than in all previous studies.

In the following, a short description of the experimental settings used to simulate downbursts at WindEEE Dome is reported in Sect. 2. An overview of decomposition methodology is provided in Sect. 3. The measured time series in WindEEE Dome and their analyses are discussed in details in Sect. 4. Lastly, the paper is closed with conclusions and prospects for further research in Sect. 5.

2 Experimental Settings

The Wind Engineering, Energy and Environment (WindEEE) Dome at Western University in Canada is a hexagonal wind tunnel with an inner diameter of about 25 m, designed to reproduce, at large scale and under controlled inflow and boundary conditions, time-dependent swirling winds such as tornadoes, downbursts, and gust fronts (Hangan et al. 2017). A schematics of the WindEEE Dome cross-section is depicted in Fig. 1. The total of one-hundred fans are distributed along the perimeter of the test chamber with one wall being equipped with sixty fans (so-called 60-fan wall). The 60-fan wall can also be used to produce ABL winds (Hangan et al. 2017). In addition six larger fans are installed in the upper plenum of the WindEEE Dome in order to produce either tornadoes or downbursts. The upper plenum is connected to the test chamber by a bell mouth with mechanical louvers that can be controlled remotely. The height of the chamber (H) is 3.75 m and the bell mouth opening used to generate downburst in this study is $D = 3.2$ m; thus $D \sim H$.

In terms of experimental inflow conditions, downbursts are generated producing a dynamic (sudden) vertical flow downward from the bell mouth mounted at the ceiling of the test chamber (Fig. 1). Since this study is interested in stationary downburst, the position of the bell mouth installed on a guillotine is kept fixed. However, using

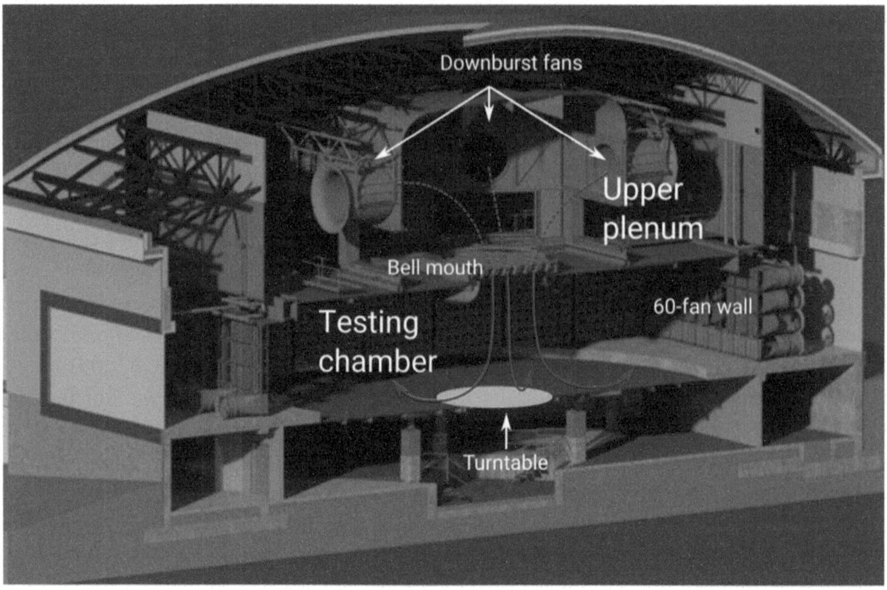

Fig. 1. Schematics of WindEEE Dome and downburst flow

various translation speeds of the guillotine, WindEEE Dome is capable of producing translating downbursts. In order to simulate as close as possible the transient nature of real downbursts, the time of louvres to open and close was approximately 10 s. Jet velocity at the exit of the bell mouth (i.e. outlet velocity) of $V_{jet} \approx 9\,\mathrm{m\,s^{-1}}$ is approximately 20% of the maximum rotation per minute (RPM) of the upper six fans in the WindEEE Dome.

The experiment setup is portrayed in Fig. 2. The data presented in this paper are obtained at two radial positions from downdraft centre, i.e., $R/D = 1$ and $R/D = 2$. In both cases, measurements were taken at 0.1 m above the surface. Velocity measurements are conducted deploying a Cobra Probe (Turbulent Flow Instrumentation 2015) with the sampling frequency of $f_s = 2500\,\mathrm{Hz}$. The simulated downbursts impinged on the bare floor which was not populated with any surface roughness elements or topography features. For each R/D position, experiments are repeated twenty times. This study investigates individual time series as well as their ensemble average, defined as:

$$\langle V(t_j) \rangle = \frac{1}{M} \sum_{i=1}^{M} V^{(i)}(t_j). \tag{1}$$

Here, $\langle V(t_j) \rangle$ is the ensemble-average of the jth wind speed (V) record in the time series ($j = 1, 2, \ldots, N$), N is the total number of records in the time series, and $i = 1, 2, \ldots, M$ is the ith member of the ensemble ($M = 20$), and t is the time.

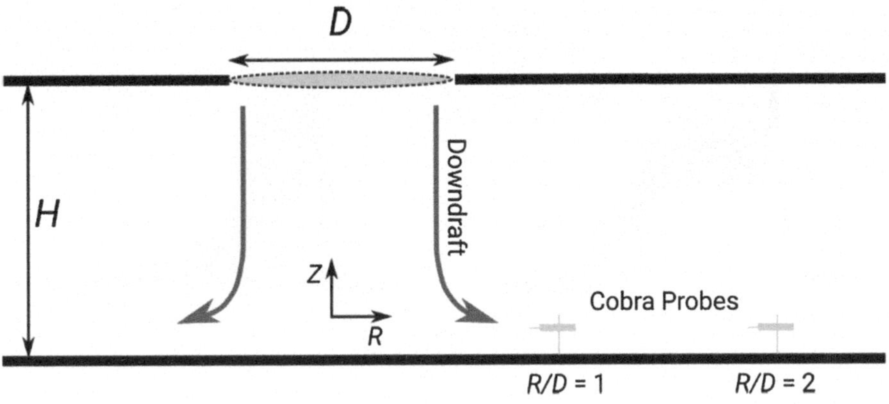

Fig. 2. Experiment setup

3 Brief Overview of Signal Analysis Methodology

The extension of Reynolds decomposition to transient wind velocity time series is applied to analyse experimentally produced downbursts in this study (e.g., Holmes et al. 2008; Kwon and Kareem 2009; Solari et al. 2015; Burlando et al. 2017a, b):

$$V(t) = \bar{V}(t) + V'(t), \tag{2}$$

where $t \in [0, \Delta T]$ is the time, ΔT is the period for which the velocity signals are examined, $\bar{V}(t)$ is the slowly-varying mean wind velocity (associated with low-frequency content) and $V'(t)$ is the residual fluctuation (associated with the high-frequency content of V).

By expressing the residual fluctuations as:

$$V'(t) = \sigma_V(t)\tilde{V}'(t), \tag{3}$$

where $\tilde{V}'(t)$ is the so-called reduced turbulent fluctuation, Eq. (2) can be cast in the following form:

$$V(t) = \bar{V}(t)\left(1 + I_V(t)\tilde{V}'(t)\right), \tag{4}$$

where $\sigma_V(t)$ is the slowly-varying standard deviation of V', and $I_V(t) = \sigma_V(t)/\bar{V}(t)$ is therefore the slowly-varying turbulence intensity. The analyzed features of the experimentally produced downbursts in WindEEE Dome are $V(t)$, $\bar{V}(t)$, $I_V(t)$, and $\tilde{V}'(t)$.

4 Results and Discussion

Figure 3 shows the instantaneous wind speed time series measured at $R/D = 1$ (left) and 2 (right). As pointed out earlier, the experiment has been repeated 20 times in order

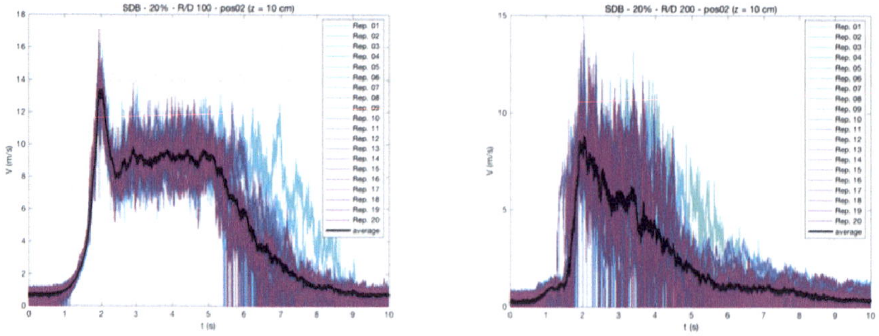

Fig. 3. Time series of the wind speed measured at $R/D = 1$ (left) and at $R/D = 2$ (right) and their ensemble mean.

to capture the variability of this transient phenomenon. The black line corresponds to the ensemble mean of all repetitions.

Both the mean time series (black lines), which correspond to the deterministic part of this phenomenon, as well as the statistical repeatability of the simulations show that the signal associated with downbursts consists of three main parts: (1) a sudden ramp-up followed by a rapid decrease of the wind speed, which determines the first maximum of the signal; (2) a plateau that can be more or less pronounced depending on the distance from the downburst centre; and (3) a tail during the downburst dissipation stage. Qualitatively, these two time series show the same transient footprint as the velocity records of full scale events (e.g., Sherman 1987; Holmes et al. 2008; De Gaetano et al. 2014; Burlando et al. 2017a, b).

In the following, the wind velocity decomposition described in Sect. 3 is reported. Figure 4 shows the slowly-varying mean wind velocity, $\bar{V}(t)$, obtained using a running mean with moving average period of 30 s (real time equivalent), chosen according to Solari et al. (2015).

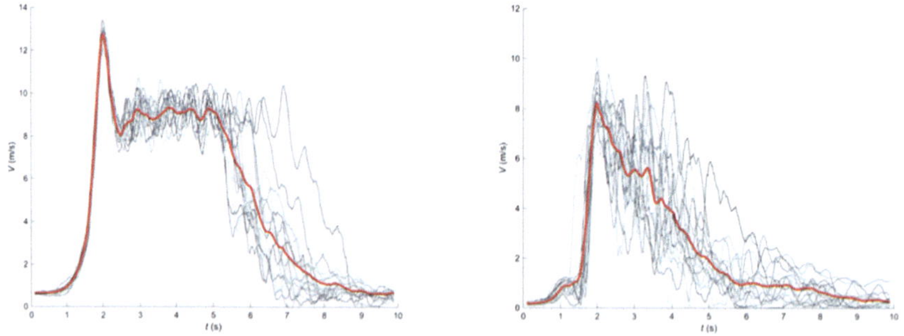

Fig. 4. Time series of the running mean of the wind speed measured at $R/D = 1$ (left) and at $R/D = 2$ (right), averaged over a period of 30 s. Red lines represent the ensemble mean over all the black lines.

Figure 5 shows the slowly-varying standard deviation, $\sigma_V(t)$, of the residual fluctuations. The turbulent fluctuations increase suddenly at the beginning of the downburst ramp-up, which is at about 1.5 s, then remain almost constant until the end of the aforementioned plateau, which lasts approximately up to 5 s and 3.5 s at $R/D = 1$ and $R/D = 2$, respectively. Fluctuations increase moving radially further from the downdraft, i.e. from around 0.7 m s^{-1} at $R/D = 1$ to around 1.2 m s^{-1} at $R/D = 2$, while at the same time the wind speed decreases. Consequently, this leads to an increase of turbulence intensity from about 0.1 to 0.2 at $R/D = 1$ and $R/D = 2$, respectively. Note that these values of I_V are comparable to the ones that are commonly found in the ABL (see for instance Pagnini et al. 2015).

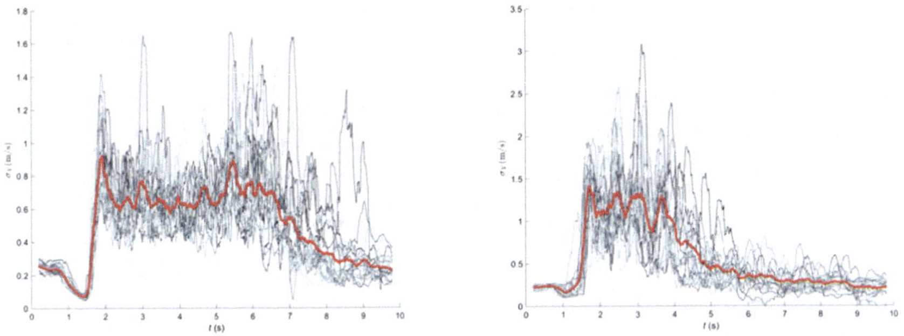

Fig. 5. Time series of the time-dependent standard deviations at $R/D = 1$ (left) and at $R/D = 2$ (right), considering a period of 30 s. Red lines represent the ensemble mean over all the black lines.

Figure 6 shows the reduced turbulent fluctuations, $\tilde{V}'(t)$. This parameter is usually dealt with as a rapidly-varying stationary Gaussian random process with zero mean and unit standard deviation (Chen and Letchford 2004). This approach is widely shared in literature, as reported for instance by Solari et al. (2015) and Zhang et al. (2017) and it holds also for the 20 replications analysed one by one through a global analysis. However, if the ensemble mean of $\tilde{V}'(t)$ is considered (i.e. red lines in Fig. 6), it seems that the assumption of statistical stationarity and in particular randomness of these signals does not hold, especially for the signal corresponding to $R/D = 1$ comparing the period during the ramp-up (1.5–2.5 s) and partly the plateau (up to 5 s) with respect to the rest. Closer inspection of Fig. 6 shows that this departure from Gaussian process is actually true for all the 20 replications, even though this behaviour is partially hidden by noise.

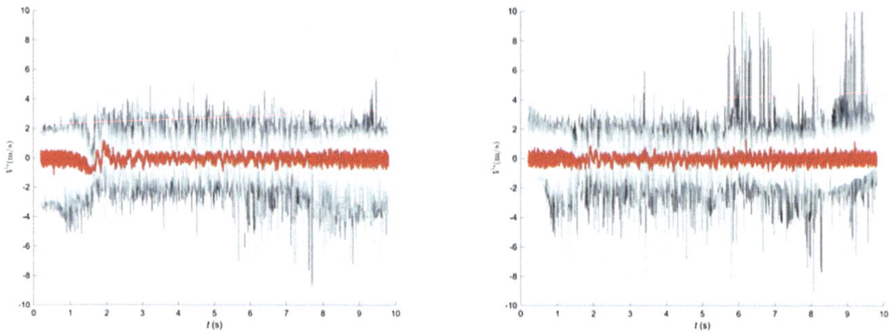

Fig. 6. Time series of reduced turbulent fluctuations at $R/D = 1$ (left) and at $R/D = 2$ (right). Red lines represent the ensemble mean over all the black lines.

5 Perspectives

This paper presented the first results of systematic analysis of the downburst time series measured in the WindEEE Dome. The results are shown for two radial distances from downdraft centre; namely $R/D = 1$ and $R/D = 2$ (R is the radial position and D is the diameter of the downdraft) and one height above surface ($Z = 0.1$ m). However, twenty repetitions of tests for each R/D are performed in order to analyse the transient nature of the phenomena. The analysis encompass the investigations of individual time series and their ensemble average. Using the signal analysis technique, all time series are decomposed into slowly-varying mean, slowly-varying turbulence intensity, and reduced turbulent fluctuations.

There are two great advantages of simulating experimentally-produced downbursts in a wind tunnel: firstly, the phenomenon can be generated under strongly controlled conditions; secondly, it can be replicated many times with exactly the same boundary conditions and forcing. This allows, on the one hand, to avoid any kind of environmental disturb, like background flows, which cannot be easily recognised and removed in real field measurements and, on the other hand, to distinguish deterministic and random parts of signals.

The downburst simulated at WindEEE was a transient stationary phenomenon without background flow. The lack of background flow allowed to analyse the deterministic part of the downburst, i.e. the ensemble mean $\langle V(t) \rangle$, without any external interference and to distinguish three phases of its evolution: ramp-up, plateau, and dissipation. Besides, the analysis pointed out that the randomness superimposed onto the deterministic part of the signal could be statistically non-stationary especially in the period of ramp-up and immediately afterwards. This is not confirmed by field measurements, however, in which the statistical characteristics of the reduced turbulence apparently do not change by time. During a real downburst event, the turbulence field obtained from direct measurements is almost never completely free from interactions with the boundary layer flow and its reduced turbulence is surely affected by the mixing with the background flow, which makes more difficult to compare experiments with field data.

By the end of 2018 a set of new experiments will be performed at WindEEE Dome in order to study more complex, but also more realistic, meteorological conditions associated with real downbursts, such as the case of background flows superimposed to thunderstorm outflows and translating downbursts.

Acknowledgements. This research is funded by European Research Council (ERC) under the European Union's Horizon 2020 research and innovation program (grant agreement No. 741273) for the project THUNDERR - Detection, simulation, modelling and loading of thunderstorm outflows to design wind-safer and cost-efficient structures – through an Advanced Grant (AdG) 2016 and by "Compagnia di San Paolo" for the "Wind monitoring, simulation and forecasting for the smart management and safety of port, urban and territorial systems" Project (grant number 2015.0333, ID ROL: 9820).

References

Burlando M, De Cio A, Pizzo M, Solari G (2017a) Analysis of wind vertical profiles of thunderstorm events in the mediterranean. In: 9th Asian-Pacific conference on wind engineering, Auckland, New Zealand, 3–7 December

Burlando M, Romanić D, Solari G, Hangan H, Zhang S (2017b) Field data analysis and weather scenario of a downburst event in Livorno, Italy, on 1 October 2012. Mon Weather Rev 145:3507–3527

Chay M, Letchford C (2002) Pressure distributions on a cube in a simulated thunderstorm downburst—part A: stationary downburst observations. J Wind Eng Ind Aerodyn 90:711–732

Charba J (1974) Application of gravity current model to analysis of squall-line gust front. Mon Weather Rev 102:140–156

Chen L, Letchford CW (2004) Parametric study on the alongwind response of the CAARC building to downbursts in the time domain. J Wind Eng Ind Aerodyn 92:703–724

De Gaetano P, Repetto MP, Repetto T, Solari G (2014) Separation and classification of extreme wind events from anemometric records. J Wind Eng Ind Aerodyn 126:132–143

Droegemeier KK, Wilhelmson RB (1987) Numerical simulation of thunderstorm outflow dynamics. Part I: outflow sensitivity experiments and turbulence dynamics. J Atmos Sci 44:1180–1210

Fujita TT (1981) Tornadoes and downbursts in the context of generalized planetary scales. J Atmos Sci 38:1511–1534

Hangan H, Refan M, Jubayer C, Romanic D, Parvu D, LoTufo J, Costache A (2017) Novel techniques in wind engineering. J Wind Eng Ind Aerodyn 171:12–33

Holmes JD, Hangan HM, Schroeder JL, Letchford CW, Orwig KD (2008) A forensic study of the Lubbock-Reese downdraft of 2002. Wind Struct 11:137–152

Kwon D, Kareem A (2009) Gust-Front factor: new framework for wind load effects on structures. J Struct Eng 135:717–732

Letchford CW, Chay MT (2002) Pressure distributions on a cube in a simulated thunderstorm downburst. part B: moving downburst observations. J Wind Eng Ind Aerodyn 90:733–753

Mason MS, Letchford CW, James DL (2005) Pulsed wall jet simulation of a stationary thunderstorm downburst, part A: physical structure and flow field characterization. J Wind Eng Ind Aerodyn 93:557–580

McConville AC, Sterling M, Baker CJ (2009) The physical simulation of thunderstorm downbursts using an impinging jet. Wind Struct 12:133–149

McDonald JR, Mehta KS, Mani S (2006) A recommendation for an enhanced Fujita scale (EF-scale). Wind Science and Engineering Center, Texas Tech University, Lubbock

Pagnini LC, Burlando M, Repetto MP (2015) Experimental power curve of small-size wind turbines in turbulent urban environment. Appl Energy 154:112–121

Sherman DJ (1987) The passage of a weak thunderstorm downburst over an instrumented tower. Mon Weather Rev 115:1193–1205

Simpson JE (1969) A comparison between laboratory and atmospheric density currents. Q J Roy Meteorol Soc 95:758–765

Solari G, De Gaetano P, Repetto MP (2015) Thunderstorm response spectrum: fundamentals and case study. J Wind Eng Ind Aerodyn 143:62–77

Solari G, Burlando M, Repetto MP (2018) THUNDERR: an ERC project for the "detection, simulation, modelling and loading of thunderstorm outflows to design wind-safer and cost-efficient structures". In: International workshop on wind-related disasters and mitigation, Sendai, Japan, 11–14 March

Turbulent Flow Instrumentation (2015) Getting started—series 100 cobra probe. Turbulent Flow Instrume Pty Ltd. https://www.turbulentflow.com.au/Downloads/Getting%20Started%20-%20Cobra%20Probe.pdf. Accessed 18 May 2018

Wakimoto RM (1982) The life cycle of thunderstorm gust fronts as viewed with Doppler radar and rawinsonde data. Mon Weather Rev 110:1060–1082

Wood GS, Kwok KC, Motteram NA, Fletcher DF (2001) Physical and numerical modelling of thunderstorm downbursts. J Wind Eng Ind Aerodyn 89:535–552

Xu Z, Hangan H (2008) Scale, boundary and inlet condition effects on impinging jets. J Wind Eng Ind Aerodyn 96:2383–2402

Zhang S, Solari G, De Gaetano P, Burlando M, Repetto MP (2017, in press) A refined analysis of thunderstorm outflow characteristics relevant to the wind loading of structures. Probab Eng Mech. https://doi.org/10.1016/j.probengmech.2017.06.003

Damage from Recent Thunderstorms in Romania

I. Calotescu(✉)

Department of Steel Structures, Management and Engineering Graphics (DCMMGI), Technical University of Civil Engineering Bucharest, Lacul Tei 122-124, 020396 Bucharest, Romania
ileana.calotescu@utcb.ro

Abstract. The paper presents an overview of wind-induced damage to the natural and the built environment caused by thunderstorms that have occurred in Romania in the past five years (2013–2017). To this aim, a vast online search of mass-media reports is performed in order to identify the date, location and damage type produced by each storm. Due to the lack of measured data, a thunderstorm is identified by means of descriptions of the event provided by interviewed witnesses, by images posted online by the newspaper or by the pattern of damage caused by the storm. The types of damage associated with the identified events include damage to buildings and transmission towers produced by strong winds, damage produced by large hail to buildings, cars as well as crops and flooding produced by heavy rainfall.

Keywords: Thunderstorms · Wind-related damage · Mass-media reports · Wind

1 Introduction

In this paper, an overview of damage produced by recent thunderstorms that have occurred in Romania in the past five years (2013–2017) is presented. Fort this purpose, a vast online search of mass-media reports was performed in order to identify the date, location and type of damage produced by each identified event. The objective of this study is, on one hand, to identify the part of the country where most thunderstorms occur and, on the other hand, to establish the damage type associated with these events.

Thunderstorms are localized events which last for a short period of time are identified by the presence of lightning and thunder, by a sudden drop of temperature, by strong winds as well as heavy rain and, in some cases, by large hail. Thunderstorm winds may be roughly divided into tornados and downbursts. A tornado is characterized by violently rotating winds whereas a downburst is defined as a strong downdraft which induces an outburst of damaging winds on or near the ground (Fujita 1981). Typically, thunderstorms occur during the summer months when atmospheric instability conditions are met. A study on thunderstorms that have occurred in Romania between 1961 and 1980 (Iliescu 1989) shows that thunderstorm season generally stars in March and ends in September, with the average annual number of thunderstorm days equal to approximately 20–40. Depending on their intensity, thunderstorms may

F. Ricciardelli and A. M. Avossa (Eds.): INVENTO 2018, LNCE 27, pp. 143–156, 2019.
https://doi.org/10.1007/978-3-030-12815-9_12

produce both structural as well as non-structural damage. High-velocity winds may induce the collapse of transmission towers and family houses (Mehta 1984; Tamura 2009; Demartino et al. 2016) as well as damage to large areas of forests (Dupont et al. 2015). Damage may also occur due to large hail which may affect claddings of buildings, cars as well as crops whereas heavy rain usually produces flooding. In some cases, thunderstorms may also generate tornadoes (Maeda 2009; Antonescu and Bell 2015), the largest tornado recorded in Romania being and F3 tornado that occurred in Făcăieni in 2002 (Lemon et al. 2003).

Unfortunately, measured data related to thunderstorms is not yet available in Romania. At most, the Romanian National Meteorology Administration may provide the maximum gust speed related to notable windstorms when journalists ask for this information. However, meteorological conditions related to particular events may be obtained from online databases such as Blitzortung, EUMETSAT (European Organization for the Exploitation of Meteorological Satellites) or ESWD (European Severe Weather Database) whereas wind data such as mean wind speed and gust speed may be obtained from the datasets collected by the National Oceanic and Atmospheric Administration (NOAA).

In this paper, available NOAA wind data corresponding to weather stations positioned in the proximity of the location of the studied wind storms was used in order to identify correlations between the damage type and the gust speed. However, wind data was not available for each of the events and, as such, only a rough relationship between gust speeds and wind-induced damage was established.

2 Methodology

The purpose of this paper is to investigate wind-induced damage from recent thunderstorms that have occurred in Romania in the past five years (2013–2017). To this aim, a vast online search of newspaper reports has been performed in order to identify those events that have produced considerable damage both to the natural as well as the man-made environment. Events were selected based on the amount of damage, only those causing at least damage to roof tiles of family houses or tree breakage being selected for investigation. Wind-related damage was of primary concern.

The methodology involved making use of the Google engine to perform a two-step search. First, dates of strong storms were identified by setting a custom search time frame such as the first and last day of each considered month (March–September) for each year (2013–2017). Then, for each identified event, a detailed search was performed and information about location, casualties and damage type from several newspapers was documented. Based on this methodology, approximately ten events were identified in 2017 and eleven events per year in 2013 to 2016 as seen in Table 1. Information about each storm together with images within this paper were taken from national newspapers and local newspapers.

Figure 1 shows the temporal evolution of the identified wind storms. It may be seen that, for each of the considered years, most storms occur during the summer months (June–August) where atmospheric instability conditions required for the occurrence of thunderstorms are met.

Table 1. Identified events (2013–2017)

2013	2014	2015	2016	2017
April 07	March 15–17	April 20	May 14	March 04
May 23	May 25–27	May 14	May 24	April 21
May 26	June 05	May 25	June 02	May 20
June 01	June 08	June 10	June 19	June 23
June 08	June 10	July 09	June 24	June 29
June 24	June 25	July 14	July 03	July 08
July 06	July 11	July 19	July 16	July 12
July 26	July 22	July 23	July 26	July 23
August 01	July 27	August 13	August 01	September 03
August 27	August 04	August 16	August 17	September 17
September 14	August 24	September 19	September 19	

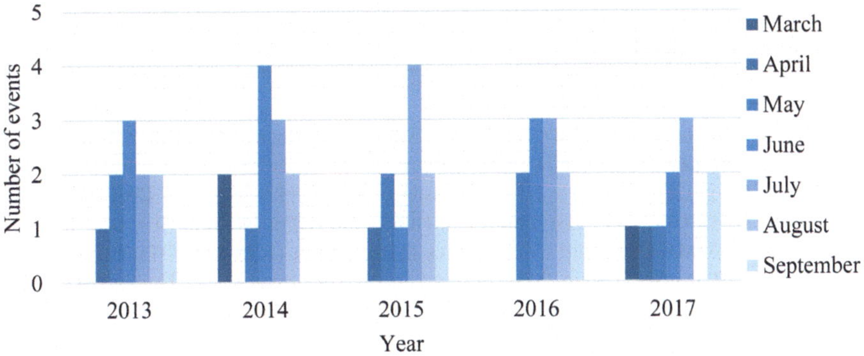

Fig. 1. Temporal evolution of identified events

3 Recent Thunderstorms

The data collected on each storm identified for this study is related to date, location, affected localities and counties as well as amount of damage. Figure 2 presents a map of Romania showing the division of the country: 41 counties plus the capital city as well as the precise location of the events presented in Table 1. The cities where damage has occurred within the past five years have been highlighted on the map, using different symbols for storms corresponding to each year. If a particular storm has affected several cities or villages within a certain county then all these points have been represented on the map. It may be seen from Fig. 2 that most storms have occurred in the western part of the country with 13 storms occurring in Arad (AR) and 10 storms in Timis (TM) within the past five years. This result agrees with (Iliescu 1989) which, based on daily observations at meteorological stations from 1951–1980, suggests that due to the predominant westerly direction of the wind in warm season, thunderstorm most frequently occur in the western part of the country.

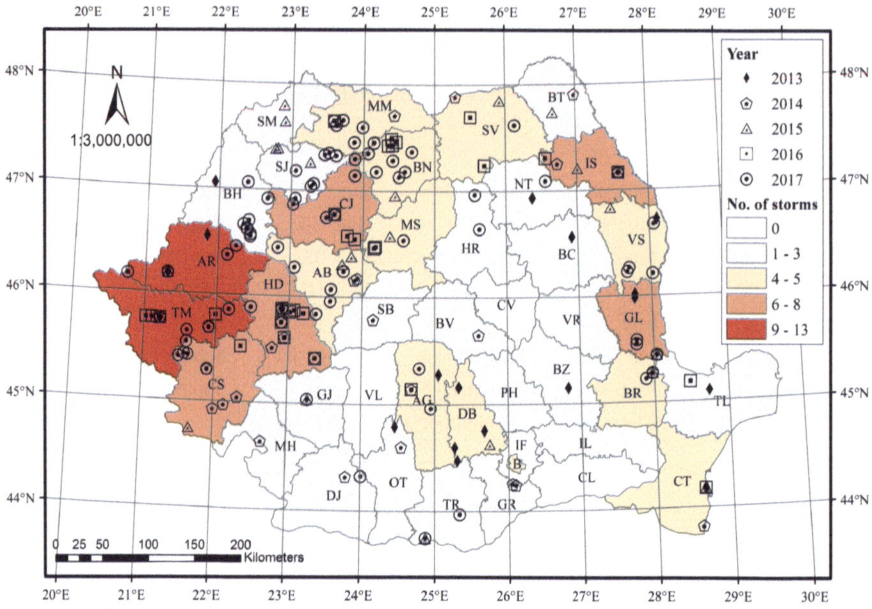

Fig. 2. Map of Romania: Alba (AB), Arad (AR), Argeş (AG), Bacău (BC), Bihor (BH), Bistriţa-Năsăud (BN), Botoşani (BT), Brăila (BR), Braşov (BV), Buzău (BZ), Călăraşi (CL), Caraş-Severin (CS), Cluj (CJ), Constanţa (CT), Covasna (CV), Dâmboviţa (DB), Dolj (DJ), Galaţi (GL), Giurgiu (GR), Gorj (GJ), Harghita (HG), Hunedoara (HD), Ialomiţa (IL), Iaşi (IS), Ilfov (IF), Maramureş (MM), Mehedinţi (MH), Mureş (MS), Neamţ (NT), Olt (OT), Prahova (PH), Sălaj (SJ), Satu Mare (SM), Sibiu (SB), Suceava (SV), Teleorman (TR), Timiş (TM), Tulcea (TL), Vaslui (VS), Vâlcea (VL) and Vrancea (VR) and Bucureşti (B), the capital city (Calotescu 2018)

3.1 Events Accompanied by Strong Wind

- On August 1st 2013 a thunderstorm that occurred in Dâmboviţa county produced wind gusts of 90 km/h as reported by the mass-media. Villages Răscăieţi and Şelaru were the most affected ones, the high winds damaging approximately 60 family houses.
- On July 22nd 2014 strong winds caused windthrow in the northern part of the country, Ulma village, Suceava county. In total 55 hectares of forest were affected resulting in 25000sqm of wood and a total estimated financial loss of 3,000,000 Romanian lei (approximately 650,000euro). Local relief conditions might have been an important factor as the village is located at the confluence of three deep valleys.
- On August 24th 2014 the roofs of over 250 family houses were damaged due to strong wind in Borlovenii Noi (100 houses) and Cornea (150 house) villages located in Caraş-Severin county.
- On June 19th 2016 a thunderstorm caused considerable damage to two neighboring villages in Mureş county: in Cuci the roofs of 153 houses were partially affected,

7 houses completely destroyed, the roofs of 2 churches collapsed and several high voltage and extra high voltage towers collapsed - 6 towers of 440 kV, 4 towers of 220 kV, 18 towers of 100 kV; in Dătăşeni 150 houses were partially affected and 5 houses were completely destroyed. The wind velocity was reported by the mass-media as reaching 122 km/h. A recent study describing the event in detail (Stăncescu and Reckerth, 2016) specify a maximum gust speed of 31 m/s (112 km/h) recorded at a nearby station (Târnăveni).

- On June 24th 2016 a thunderstorm caused windthrow in Broşteni, Suceava county, where 600 hectares of forest were destroyed, 500000 trees being either uprooted or broken.
- On July 3rd 2016 a strong wind event caused roof uplift of three blocks of flats, other ten being damaged in Haţeg, Hunedoara county. In total, 30 trees were either uprooted or broken damaging cars parked in their vicinity.
- On July 16th 2016, in Demăcuşa, Suceava county, strong winds destroyed 2335 hectares of forest in less than two hours. On 35 hectares of forest, all trees were windthrown.
- On August 17th 2016 in Susani, Timiş county, a strong wind storm produced damage to roofs of 50 houses in less than 20 min.
- On April 21st 2017 high winds and heavy snow produced the collapse of 8 high voltage towers, 400 medium voltage towers and 120 low voltage towers in Vaslui county (Murgeni-Bârlad). The interviewed witnesses described the event as a "summer thunderstorms with lightning and thunder. Wind velocities of 50–60 km/h up to 80 km/h were reported by the media.
- On June 23rd 2017 a strong storm that affected Arad, Arad county caused the damage of ten industrial buildings belonging to a bovine research facility located at the outskirts of the city, in Ceala village. In total, 10 workers were wounded. Over 25000 trees (39 hectares) from the Ceala forest were windthrown.
- On June 29th 2017 in Cenad and Sânnicolau Mare villages, Timiş county, 131 houses were damaged and 3 roofs were lifted off by wind which had an estimated velocity of 60–70 km/h, as reported by the mass-media.
- On September 17th 2017 a particularly strong event affected 212 localities from 15 counties. Due to the strong wind, 145 people were wounded and 8 people lost their lives, making this the most severe wind storm in recent history in Romania. The most affected city was Timişoara, Timiş county where 137 roofs were damaged and approximately 3500 trees were broken. Also, 3 steel towers and 5 concrete piles collapsed in Timiş county.

In Table 2 the location of damage, the county and the observed instantaneous gust speeds at nearby wind stations as obtained from NOAA (National Oceanic and Atmospheric Administration) corresponding to a series of the identified wind storms accompanied by strong winds are presented. Based on the information related to the wind-induced damage and the instantaneous gust speeds corresponding to the events presented in Table 2, the following conclusions may be drawn:

(i) Gust speeds over 12 m/s produced roof damage such as partial or total lift-off of steel sheeting and damage to roof tiles

(ii) Gust speeds over 20 m/s produced partial or total collapse of non-engineered family houses as well partial collapse of transmission lattice towers
(iii) Gust speeds over 30 m/s produced collapse of high-voltage transmission towers and damage to more than 130 family houses for each corresponding event

Table 2. Available NOAA wind data for events accompanied by strong winds

Date	Location of damage	County	Nearby station	Gust speed
19-Jun-2016	Cuci, Dătăşeni	Mureş	Sărmaşu	23 m/s
			Târnăveni	11 m/s
			Vidrasău	18 m/s
3-Jul-2016	Haţeg	Hunedoara	Deva	11 m/s
			Aurel Vlaicu	12 m/s
			Ţebea	13 m/s
			Caransebes	15 m/s
16-Jul-2016	Demăcuşa	Suceava	Poiana Stampei	14 m/s
17-Aug-2016	Susani	Timiş	Lugoj	12 m/s
21-Apr-2017	Murgeni	Vaslui	Bârlad	19 m/s
			Vaslui	10 m/s
			Tecuci	19 m/s
23-Jun-2017	Ceala	Arad	Arad	11 m/s
			Şiria	14 m/s
29-Jun-2017	Cenad; Sânnicolau Mare	Timis	Sânnicolau Mare	31 m/s
			Jimbolia	22 m/s
			Lugoj	15 m/s
			Banloc	17 m/s
17-Sep-2017	Timişoara	Timiş	Jimolia	34 m/s
			Lugoj	15 m/s
			Banloc	14 m/s
			Sânnicolau Mare	17 m/s

These conclusions are based on wind data recorded to stations located in the same county as the location of the event and, as such, the conclusions stated above may only be considered as rough correlations between damage pattern and gust velocities. However, they provide a general understanding on the events and the corresponding damage types.

3.2 Events Accompanied by Hail and Intense Rain

- On June 8th 2013 in the village Valea Grecului, Vaslui county, 6 houses were flooded and 30 hectares of agricultural land was damaged due to heavy rain and pea-size hail.
- On June 24th 2013, a thunderstorm produced 80 l/sqm of rain in 40 min flooding more than 100 family houses in 19 villages in the north-eastern part of Neamţ county. One person lost his life.

- On June 24th 2014 hail up to the size of ping-pong balls was responsible for the damage of 600 houses in the village Bozovici, Caraş-Severin county. The majority of houses suffered damage at roof level but also cladding damage was reported, including broken windows. Cars also got damaged due to the large hail.
- On May 25th 2015 hail the size of golf balls and larger damaged 600 houses and affected 400 hectares of strawberry fields in Ardud, Satu-Mare county. It also produced flooding. The storm lasted approximately 15 min.
- On May 24th 2016 a storm accompanied by marble size hail caused flooding in Cluj-Napoca, Cluj county. In the same day, flooding occurred also in a nearby village, Valea Drăganului where 120 houses were affected. In some places, the water reached half a meter, covering 15 hectares of agricultural land.

In Table 3 the location of damage, the county and the observed instantaneous gust speeds at nearby wind stations as obtained from NOAA (National Oceanic and Atmospheric Administration) corresponding to two of the identified wind storms accompanied by hail and flooding winds are presented. It may be noticed that large hail and flooding have produced damage to housing when nearby weather stations measured gust velocities between 11 m/s and 16 m/s.

Table 3. Available NOAA wind data for events accompanied by hail and flooding

Date	Location of damage	County	Nearby station	Gust speed
25-May-2015	Ardud	Satu Mare	Satu Mare	16 m/s
24-May-2016	Cluj-Napoca	Cluj	Cluj-Napoca	11 m/s
			Turda	13 m/s
			Câmpia Turzii	15 m/s

3.3 Events Accompanied by Tornados

- On May 23rd 2013 a strong storm completely destroyed one house and damaged another 29 in villages Drăguşeşti and Răscăieţi, Galaţi county. Witnesses described the event as a tornado that lasted for approximately three minutes.
- On September 14th 2013 winds that reached 100 km/h destroyed 5 houses and damaged other 40 in the village Cataloi, Tulcea county. Witnesses reported that the event was a tornado that lasted approximately 30–40 s.
- On June 24th 2014 a strong storm caused damage to 30 houses in Varlaam village, Giurgiu county. The storm duration was less than 15 min and interviewed people described the phenomenon as a tornado. The storm also produced damage to the nearby forest.
- On July 11th 2014 a tornado confirmed by meteorologists affected Mangalia in Constanţa county. Wind velocity reached 100 km/h partially destroying several houses. In Hăneşti, Botoşani county, another tornado formed partially destroying 37 houses and completely destroying 2 houses. It was described by the witnesses as a rotating motion of the wind coming from the east direction which took roofs of houses and displaced them on a distance of 200 m.

- On June 2nd 2016 in Niculițel, Tulcea county, an event described as a tornado by the interviewed people damaged the roofs of 12 houses and caused wind breakage. The event lasted for approximately 5 min and it had a west-east direction travelling approximately 60 m, as specified by the prefect of Tulcea.

4 Damage Associated with Recent Thunderstorms

In Sect. 3 the most damaging thunderstorms among those identified for the purpose of this study have been classified based on the damage caused by the accompanying hazardous weather event, i.e. strong winds, tornadoes, hail and flood. The type of storm could be identified only from descriptions provided by the interviewed witnesses, people who have seen the storm and, in some cases, have been affected by it. This methodology, of course, cannot be accurate. The mass-media tends to exaggerate facts and to present them in a manner which creates panic. For example, the majority of the events were described as "tornados". However, only in seldom cases description provided by locals could match the characteristics of tornados. For example, related the storm that occurred on June 24th 2014, the community priest said he heard "a strong howling sound" and saw a "whirlwind" from his window. His home was 70% damaged. An interviewed older lady said she heard a howling sound and saw "a circle rotating in the sky that took the roofs of people's houses" and, in another interview, one person described the tornado as "a big black balloon on the sky". All these descriptions could match the visual characteristics of a tornado.

4.1 Damage Caused by Wind

By far, the most common type of damage caused by strong winds was roof damage (Fig. 3). Roof cladding made of steel sheeting generally suffered partial or total lift-off, the latter one being reported for events with wind velocities larger than 90 km/h (where available). In numerous cases, the steel sheeting was lifted off along with roof purlins. Roofs covered with clay tiles or shingles generally suffered partial dislocation of tiles, without affecting the roof structure.

Another frequent type of damage that has occurred due to thunderstorm winds is damage to power lines (Fig. 4a–d). Reinforced concrete electricity poles suffered two types of damage, i.e. overturn and brakeage. The collapse of high voltage towers also reported in two instances. On June 16th 2016 when velocities reached 122 km/h, several high voltage and extra high voltage towers collapsed (6 towers of 440 kV, 4 towers of 220 kV, 18 towers of 100 kV) whereas on April 20th 2017 reported wind velocities of 50–60 km/h cause the collapse of 8 high voltage towers, 400 medium voltage towers and 120 low voltage towers. In the first case the collapse was caused by intense winds whereas in the latter case, the thunderstorm was accompanied by heavy snow which probably lead to the failure of the towers (Fig. 4d). Finally, in all of the identified events, thunderstorm winds caused tree breakage or uprooting which resulted in damaged cars or, sometimes, houses (Fig. 4e and f).

Fig. 3. Damage due to strong winds (a) damaged roof tiles August 17th 2016 (b) total roof uplift June 24th 2014 (c) uplift of steel sheeting June 23rd 2017 (d) uplift of roof sheeting and purlins August 24th 2014

Overturned trucks of various sizes were also reported (Fig. 5a–b). In forests, windthrow was caused by two of the identified wind storms (June 22nd 2014 and July 16th 2016), in both cases, the pattern of windthrow being similar to that caused by downbursts (Fig. 5c). High winds were also responsible for damage to crops (Fig. 5d).

The type of damage described herein and associated with strong winds produced by the identified events agrees to (Kaňák et al. 2007) who presented in detail a case study of a severe windstorm in Slovakia that occurred on May 9th 2003 as well as (Demartino et al. 2016) who presented a similar study on a thunderstorm that occurred on February 28th 2016 in Italy.

4.2 Damage Caused by Hail and Flooding

Thunderstorms support the formation of large hail, with sizes varying from "pea size" to "softball size", estimation based on real life objects according to NOAA. Quarter size hail is considered to be produced by severe thunderstorms.

(a) (b) (c) (d) (e) (f)

Fig. 4. Damage due to strong winds (a) collapsed electricity pole June 24th 2014 (b) collapsed electricity pole June 19th 2016 (c) collapsed transmission tower June 19th 2016 (d) bent transmission tower April 20th 2017 (e) damaged car July 3rd 2016 (f) roof damage due to tree breakage August 17th 2016

The most common type of damage caused by hail and flooding is damage to crops and agricultural land, having a strong impact on the overall costs of fruits and vegetables once they are on the market. Depending on its size, hail may also cause significant damage to buildings leaving dents in roofs as well as in wall claddings. This type of damage was caused by thunderstorms that occurred on June 24th 2014 as well as on May 25th 2015 (Fig. 6). Approximately 600 houses were damaged in both cases by hail as large as ping pong balls.

Fig. 5. Damage due to strong winds (a) overturned truck June 23rd 2017 (b) overturned pickup truck September 17th 2017 (c) windthrow June 22nd 2014 (d) damaged crops June 19th 2016

Fig. 6. Damage caused by the May 25th 2015 thunderstorm in Ardud (a) large-size hail (b) damaged house (c) damaged corn field

Related to flooding, the worst damage was caused by the storm that occurred on May 24th 2016 when approximately 120 houses were flooded in village Valea Drăganului in Cluj-Napoca county (Fig. 7), water reaching up to half a meter in some areas. Due to that storm, 15 hectares of agricultural land were flooded.

(a) (b)

Fig. 7. Flooding (a) May 24th 2016 (b) June 24th 2013

5 Conclusions

This paper presents the most frequent types of damage produced by thunderstorms as resulted from the investigation of mass-media reports related to events that occurred in Romania in the past five years (2013–2017). The search was performed online by making use of the Google search engine. Both national as well as local newspapers were consulted and information about the date, location, affected localities and counties as well as amount of damage were extracted.

Approximately eleven storms per year have been identified based on the methodology described in this paper, the selection of each storm being based on the amount of damage it has caused. The most severe wind storms were further detailed, categorizing them based on the accompanying hazardous weather event that caused the most damage, i.e. thunderstorms accompanied by strong winds, thunderstorms accompanied by hail and intense rain producing flooding and thunderstorms accompanied by tornados. Further on, damage types were categorized into damage caused by strong winds and damage caused by hail and flooding.

Results showed that strong intensity events produced the collapse of transmission towers as well family houses whereas the most common type of damage associated with medium intensity events was damage to the roofs of buildings. Overturned trucks have also been reported due to strong winds. Low intensity events generally produce breakage or uplift of trees damaging cars located nearby. Wind related phenomena such as large hail produced damage to roofs and claddings of buildings as well as damage to crops and cars. Most of the identified storms affected the country side, the damage being more significant than when the storm occurred in the cities. This might be due to the fact that, in Romania, family homes in the country side are generally not engineered structures and thus more prone to wind-induced damage.

Wind data obtained from the National Oceanic and Atmospheric Administration (NOAA) for a series of events was used in order to establish a relationship between instantaneous gust speeds recorded at weather stations positions nearby the wind storm location and the corresponding wind-induced damage. For events accompanied by strong winds it was found that gust speeds over 12 m/s produced roof damage such as partial or total lift-off and damage to roof tiles, gust speeds over 20 m/s produced partial or total collapse of non-engineered family houses as well partial collapse of lattice transmission towers and gust speeds over 30 m/s produced collapse of high-voltage transmission towers and damage to more than 130 family houses for each corresponding event.

To get a full understanding of the type of weather event and damage created by it, each storm should be deeply investigated, by carrying damage surveys as well as analyzing the synoptic situation corresponding to the date of the event (Kaňák et al. 2007; Stăncescu and Reckerth 2016; Burlando et al. 2017). This study is an empirical one, the objective being to identify some sort of damage pattern associated with thunderstorms as classified based on visual accounts of witnesses or on available images or videos posted online by newspapers. Its purpose is to give an overview on thunderstorms in Romania, in terms of most frequently affected areas as well as the damage associated to such events.

References

Antonescu B, Bell A (2015) Tornadoes in Romania. Monthly Weather Rev 143:689–701

Burlando M, Romanić D, Solari G, Hangan H, Zhang S (2017) Field data analysis and weather scenario of a downburst event in Livorno, Italy, on 1 October 2012. Monthly Weather Rev 145:3507–3527

Calotescu I (2018) Overview of recent wind-induced damage in Romania based on mass-media reports. In: International workshop on wind-related disasters and mitigation, Sendai, Japan, 11–14 March

Demartino C, Avossa A, Ricciardelli F (2016) Vulnerability of the urban and rural environment to extreme winds: the February 28 event in Italy. In: XIV conference of the Italian association for wind engineering, Terni, Italy, 25–28 September 2016

Dupont S, Pivato D, Brunet Y (2015) Wind damage propagation in forests. Agric For Meteorol 214–215:243–251

Fujita TT (1981) Tornadoes and downbursts in the context of generalized planetary scales. J Atmos Sci 38:1511–1534

Goliger AM, Retief JV (2007) Severe wind phenomena in Southern Africa and the related damage. J Wind Eng Indus Aerodyn 95:1065–1078

Iliescu MC (1989) Manifestări electrice atmosferice pe teritoriul României, Editura Academiei Republicii Socialiste România. (in Romanian)

Kaňák J, Benko M, Simon A, Sokol A (2007) Case study of the 9 May 2013 windstorm in southwestern Slovakia. Atmos Res 83:162–175

Lemon LR, Stan-Sion A, Soci C, Cordoneanu E (2003) A strong, long-track, Romanian tornado. Atmos Res 67–68:391–416

Metha KC (1984) Wind induced damage observations and their implications for design practice. Eng Struct 6:242–247

Maeda J (2009) Recent extreme wind events and damage assessment in Japan. Glob Environ Res 13:125–132

Stăncescu M, Reckerth U (2016) Fenomenele meteorologice severe din 19 iunie 2016, localitatea Cuci, Judetul Mureş, Revista Ştiinţifică a Administraţiei Naţionale de Meteoro-logie, 93–110. (in Romanian)

Tamura Y (2009) Wind-induced damage to buildings and disaster risk reduction. In: The seventh Asia-Pacific conference on wind engineering, Taipei, Taiwan, 8–12 November 2009
www.adevarul.ro
www.mediafax.ro
www.evz.ro
www.radiotimisoara.ro
www.aradon.ro
www.ziaruldambovita.ro
www.ziuaconstanta.ro
www.infocs.ro

Full-Scale Measurements of the Structural Response of a 63-Storey Mixed-Use High-Rise Under Wind Loading

S. Cammelli[1(✉)], H. Nguyen-Sinh[2], and J. García Navarro[3]

[1] WSP, London, UK
stefano.cammelli@wsp.com
[2] BMT Fluid Mechanics, New York City, USA
Hung.Nguyen-Sinh@bmtglobal.com
[3] Aecom, London, UK
jessica.garcianavarro@aecom.com

Abstract. Despite the hundreds of wind tunnel studies executed every year on tall- and super-tall buildings globally, the number of full-scale measurements conducted on these structures is still somewhat limited. Full-scale measurements, even for a short duration, have in fact tremendous value as they can be used to validate the structural frequencies predicted during the design stage through Finite Element Model (FEM) analysis and, at the same time, they can provide very valuable information on the level of damping these structures can inherently exhibit once built. In the assessment of the wind-induce response of tall- and super-tall buildings, damping does in fact constitute one of the most uncertain variables. In this technical paper, the findings of a 3-month full-scale monitoring campaign conducted on a 63-storey high-end residential tall building will be presented and discussed in detail.

Keywords: Full-scale monitoring · Modal identification · BFFTA · Frequency · Damping

1 Introduction

During the design of a slender 63-storey (230 m tall) high-rise in North America, wind tunnel studies highlighted the potential for wind-induced sway to cause discomfort to the residents of the uppermost occupied levels of the building. To mitigate such excessive vibrations, a Tuned Liquid Column Damper (TLCD) in the form of a 'U-tube' water tank with a total effective mass of ~410,000 kg was proposed and designed for this structure (Cammelli and Li 2016).

At the time of the topping-out of the structure, which took place in September 2017, the installation of the façade system was well progressed on site and the TLCD cast-in-situ but lying empty: this was the perfect timing for a short campaign of full-scale measurements to take place. This measurement campaign, the outcome of which will be presented within this technical paper, was primarily aimed at confirming some of the structural and modal parameters which were assumed during the design of the TLCD (primarily structural frequencies and damping).

F. Ricciardelli and A. M. Avossa (Eds.): INVENTO 2018, LNCE 27, pp. 157–164, 2019.
https://doi.org/10.1007/978-3-030-12815-9_13

2 Background

2.1 General

There is no doubt in the authors' mind that inherent damping, perhaps together with site-specific design wind speeds, represents one of the weakest "links" in the 'Alan G. Davenport Wind Loading Chain'. The uncertainty around the inherent level of damping of a tall building can in fact be rather large. Havilland (1976) estimated that the coefficient of variation of full-scale damping data is almost 70%: this means that, if the design value for the damping ratio is 2% (fraction of critical), the 'mean ± standard' deviation value can therefore range from 0.6% to 3.4% (Tamura 2012). This can translate in a potentially significant under-estimation of the wind-induced peak acceleration (by nearly a factor of 2), with profound impact on the exacted requirements of extra damping and size of damper. The accuracy of full-scale measurement techniques, as well as the robustness of the method used to identify the modal properties of a structure, are therefore incredibly important.

The number of modal identification methods, both in time and frequency domain, which have been developed during the last 60 years, is large. Amongst these, the Random Decrement (RD) technique (Cole 1973) has allowed modal properties of tall buildings, especially damping, to be robustly and reliably identified from full-scale data. Making use of a Multiple-mode Random Decrement (MRD) approach (Tamura et al. 2002, 2005) (Multi Degree Of Freedom (MDOF) method extracting Free Decay Damped Response (FDDR), to be precise), Tamura (2012) found that, as the magnitude of the wind-induced excitation increases, the natural frequency of tall buildings decreases, whilst the opposite trend was observed for the inherent damping of the structure. However, the damping of a tall building (in the non-damaged situation, of course), only increases up to a certain level of displacement of the structure, the so-called "critical tip drift ratio": beyond that point, the damping starts decreasing as the magnitude of the excitation increases. In his work, Tamura (2012) has also shown the "critical tip drift ratio" for a 200 m tall steel frame office building in Japan to be as low as $2 \cdot 10^{-5}$ (that is x_{HC}/H, where x_{HC} is the tip displacement of the structure and H is the height of the building).

2.2 The Bayesian Fast Fourier Transform Approach

In more recent years, Yuen and Katafygiotis (2003) developed a new modal identification technique: the Bayesian Fast Fourier Transform Approach (BFFTA). This technique makes use of the statistical properties of the Fast Fourier Transform (FFT) to derive not only the values of the modal properties of the structure but also their associated uncertainties. As described in Yuen and Katafygiotis (2003) and Au (2011, 2012), the measured acceleration on a structure is in fact modeled as the combination of the actual acceleration response which needs to be identified and a prediction error, primarily caused by measurement noise.

Modelling the prediction error as an independent and identically distributed Gaussian white noise and under the assumption of a linear-elastic and classically damped structure subjected to modal forces modelled as a stationary process with a constant spectral density matrix (i.e. independent from the number of samples per measuring channel), Yuen and Katafygiotis (2003) derived the Joint Probability Density Function (JPDF) for the augmented FFT vectors $\boldsymbol{Z}_k$ (collecting the real and imaginary parts of the FFT of the measured accelerations on the structure, where k = 1 to N and N is the number of samples per measurement channel) and applied the result to Bayesian modal identification. It can be shown that, choosing a non-informative prior distribution, the posterior Probability Density Function (PDF) of the set of modal parameters $\boldsymbol{\theta}$, given the FFT data, is proportional to the likelihood function $p(\boldsymbol{Z}_k|\boldsymbol{\theta})$. Using the FFTs in the selected frequency band for modal identification, the posterior PDF is given by (Au 2012):

$$p(\boldsymbol{\theta}|\boldsymbol{Z}_k) \propto p(\boldsymbol{Z}_k|\boldsymbol{\theta}) = (2\pi)^{-\frac{nN_f}{2}}\left[\prod_k \det \boldsymbol{C}_k(\boldsymbol{\theta})\right]^{-\frac{1}{2}} exp\left[-\frac{1}{2}\sum_k \boldsymbol{Z}_k^T \boldsymbol{C}_k(\boldsymbol{\theta})^{-1}\boldsymbol{Z}_k\right] \quad (1)$$

where N_f is the number of frequency values in the selected frequency band, n is the number of measured degrees of freedom (DOFs) and $\boldsymbol{C}_k$ is the covariance matrix of $\boldsymbol{Z}_k$. Equation (1) can also be conveniently written in term of the 'negative log-likelihood function' $L(\boldsymbol{\theta})$:

$$p(\boldsymbol{\theta}|\boldsymbol{Z}_k) \propto \exp[-L(\boldsymbol{\theta})] \quad (2)$$

where:

$$L(\boldsymbol{\theta}) = \frac{1}{2}\sum_k ln(det\,\boldsymbol{C}_k(\boldsymbol{\theta})) + \frac{1}{2}\sum_k \boldsymbol{Z}_k^T \boldsymbol{C}_k(\boldsymbol{\theta})^{-1}\boldsymbol{Z}_k \quad (3)$$

By minimizing $L(\boldsymbol{\theta})$, a set of modal parameters $\boldsymbol{\theta}$ can be found.

In general, as indicated in Au (2011), the BFFTA betters non-Bayesian methods (e.g. time domain modal identification methods used in conjunction with RD techniques) principally because the physical modelling assumptions are strictly obeyed, and its formulation allows virtually all information in the measured data to be utilised to make inferences about the modal parameters. However, this benefit comes at a price: the posterior statistics of the modal parameters is a non-trivial non-linear function, so the determination of the most probable value (MPV) of the modal parameters requires a multi-dimensional numerical optimisation problem to be solved. As indicated in Au (2011), this problem is computationally prohibitive and ill-conditioned because the number of parameters involved rapidly grows with the increasing number of measured DOFs and the optimisation function requires the inverse of an ill-conditioned matrix. To overcome this problem, Au (2011) developed an efficient algorithm to determine the posterior MPV and the covariance matrix which in turn enable a feasible application of the BFFTA in practical situations in which the number of measured DOFs can be moderate or large. This algorithm runs very efficiently if the signal-to-noise ratio is large and the modal frequencies of the structure are relatively well separated.

3 Monitoring Campaign

The full-scale monitoring campaign took place between October 2017 and December 2017. The instrumentation which was installed at the top of the building included (see Fig. 1):

Fig. 1. Instrumentations used during the 3-month full-scale monitoring campaign

- On the 62nd floor: two (2) three-axis accelerometers and tilt sensors mounted at approximately 1 m above the floor slab. One accelerometer was placed on the reinforced concrete (RC) core of the tower and another one was fixed onto one of the RC walls of the sloshing tank (see Fig. 2).
- On the 64th floor: a weather station equipped with an ultrasonic wind sensor to record and derive weather statistics (10-min averaging wind speed and direction;

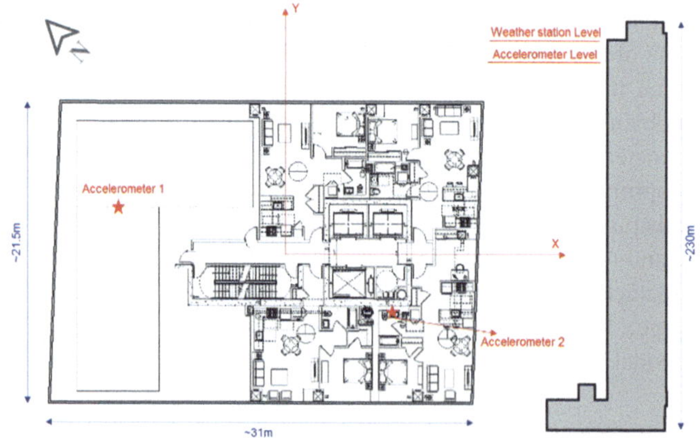

Fig. 2. Location of instrumentations within the building

3-s maximum/minimum wind speed and direction within each 10-min window; temperature; humidity; atmospheric pressure; rain and its duration) (see Fig. 2).

- On the 62nd floor: a 3G router installed near the sloshing tank to allow remote access to the data.

The accelerations (sampled at 20 Hz), angular accelerations (sampled at 10 Hz) and weather data were continuously and simultaneously recorded and stored on a cloud-based server.

During the monitoring campaign, a total of eleven (11) independent wind storms were identified and studied. The wind event which led to the largest structural response of the building peaked at 6.40am (EST) on the 24th October 2017. Anemometer (wind speed and directionality) and accelerometer data (*x* and *y* direction) recorded during this wind storm are presented in Fig. 3.

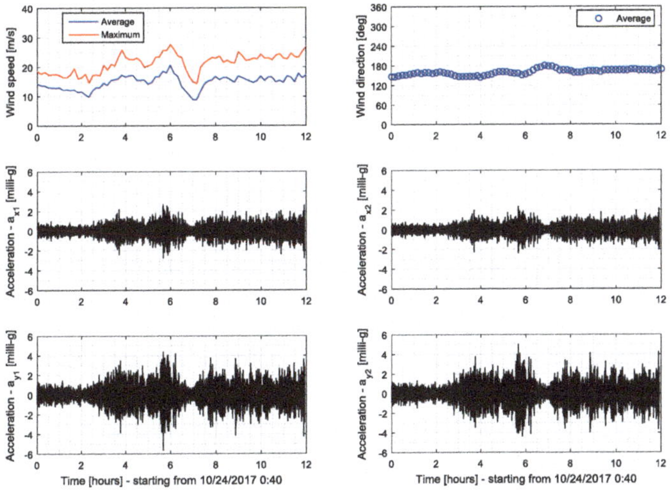

Fig. 3. Anemometer and acceleration recorded data during the 24th October 2017 wind event

4 Data Analysis and Results

The acceleration data for the six (6) hours prior and after the peak of each storm have been extracted for modal identification analysis. The various steps which were taken to obtain robust estimations of the structural frequencies and damping are listed below:

- Firstly, the PDF of the measured data for a given set of modal parameters were approximated by an appropriately selected multi-variate Gaussian distribution.
- Then, using the Bayes Theorem, the PDF of the modal parameters for a given set of actual measurements (usually referred as the 'likelihood function') were calculated based on the PDF of the actual measurements for a given set of modal parameters.
- By maximising the likelihood function, the most probable modal parameters were obtained.

For this study, the BFFTA for ambient data and well separated modes was used. Noting that, for ambient data, the potential for noise to introduce uncertainties in the measured data exists, the signal-to-noise ratios have been estimated and monitored to ensure the robustness of the results. The Power Spectral Densities (PSDs) obtained from accelerometer '1' (x and y direction) during the 24th October 2017 wind event are presented in Fig. 4.

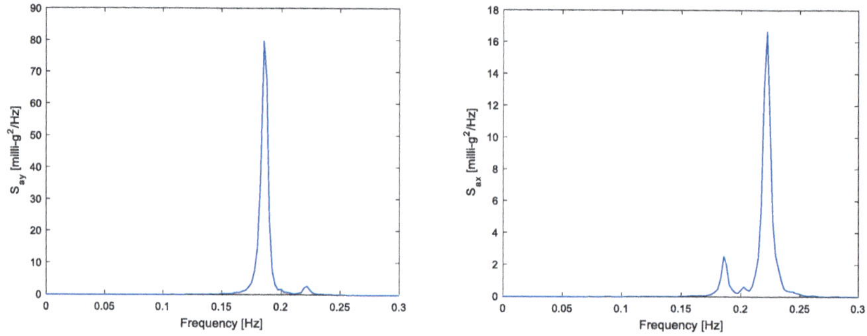

Fig. 4. PSDs of a_x (right) and a_y (left) during the 24th October 2017 wind event

For this specific storm, the structural frequencies for the first two modes of vibration were ~0.19 Hz and ~0.22 Hz, whilst the damping ~1.2% and ~1.4% (fraction of critical). Figure 5 also shows the variation of the damping for 'mode 1' as a function of the recorded wind-induced peak acceleration. The findings summarised in this chart – showing a near-linear relationship between damping and wind-induced acceleration – are well in line with what Tamura (2012) found on the 'ORC200 Symbol Tower' in Japan. It should be noted that the slight scatter in Fig. 5 is believed to be attributed to the variability in speed and direction of each single wind event.

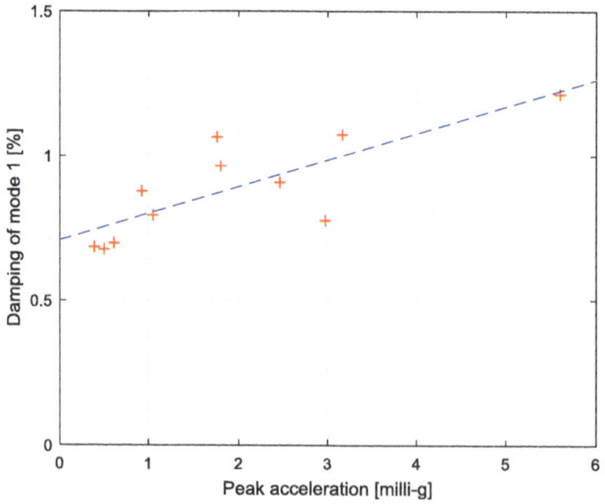

Fig. 5. Variation of damping ('mode 1') as a function of the wind-induced peak acceleration

It is important to note that, statistically speaking, the 24th October 2017 wind event was just shy of MRI = 1-yr and that the associated wind-induced response (~ 5.6 milli-g) was such that $x_{HC}/H \sim 1.7 \cdot 10^{-4}$, a figure that is likely to be very close to the "critical tip drift ratio" (Jeary 1998).

5 Discussion and Conclusions

Wind tunnel testing previously conducted on the tower, highlighted the need for the motion along the weak direction of the structure (*Y* in Fig. 2) to be mitigated by the introduction of a large TLCD near the top of the building. During the full-scale monitoring campaign conducted at the end of 2017, the structural frequencies of the building in the two principal directions have been detected to be sitting at ~ 0.19 Hz and ~ 0.22 Hz. The corresponding structural frequencies, numerically predicted by the structural engineers during design – and utilised in the sizing of the TLCD, were ~ 0.17 Hz and ~ 0.18 Hz (Cammelli and Li 2016). The authors of this technical paper believe that the two main reasons behind this discrepancy (full-scale vs. FEM) are:

- The Young's modulus of the concrete achieved at the site during construction is likely to be higher than the codified one assumed during the design of the structure.
- The inertia (and therefore the modal mass) of the building between the 'design scenario' and the condition of the tower at the end of 2017 is likely to be somewhat different (at the end of 2017, in fact, items such as finishes and partitions were not present, whilst machineries and crane were still in place).

Post-construction measurements would of course be needed to confirm these hypotheses. Unfortunately, at present (September 2018) there are still no plans to undertake further (and ideally more long-term) monitoring of the building to validate the actual performance of the structure (and TLCD) under wind loading.

Also, the damping assumed during design was $\sim 1.5\%$ (for all fundamental modes of the structure), whilst the values measured on-site were $\sim 1.2\%$ (first mode, weak axis) and $\sim 1.4\%$ (second mode, strong axis).

After performing a series of sensitivity analyses of the wind tunnel data in conjunction with the properties of the building measured on site, the authors of this technical paper could confirm the final arrangement of internal baffles of the TLCD.

Acknowledgements. The authors of this technical paper would like to thank Mr. Tom Krajcovic and Mr. David Hood for their invaluable support during the full-scale monitoring campaign.

References

Au S (2011) Fast Bayesian FFT method for ambient modal identification with separated modes. J Eng Mech 137(3):214–226

Au S (2012) Fast Bayesian ambient model identification in the frequency domain, part I: posterior most probable values. Mech Syst Signal Process 26:60–75

Cammelli S, Li YF (2016) Experimental and numerical investigation of a large TLCD. In: Proceedings of the institution of civil engineers: structures and buildings, no SB 00, pp 1–8
Cole HA (1973) On-line failure detection and damping measurement of aerospace structures by the random decrement signatures. NASA CR-2205
Havilland R (1976) A study of the uncertainties in the fundamental translational periods and damping values for real buildings. Massachusetts Institute of Technology, PB-253, p 188
Jeary AP (1998) The damping parameter as a descriptor of energy release in structures. In: Proceedings on CD-ROM, T193-4, structural world congress, San Francisco, vol 8
Tamura Y, Zhang, L-M, Yoshida A, Nakata S, Itoh T (2002) Ambient vibration tests and modal identification of structures by FDD and 2DOF-RD technique. In: Structural engineers world congress (SEWC), Yokohama, Japan, T1-1-a-1, p 8
Tamura Y, Yoshida A, Zhang L (2005) Damping in buildings and estimation techniques. In: Proceedings of the 6th Asia Pacific conference on wind engineering (APCWE-VI), Seoul, Korea, pp 193–214
Tamura Y (2012) Amplitude dependency of damping in buildings and critical tip drift ratio. Int J High-Rise Build 1(1):1–13
Yuen KV, Katafygiotis LS (2003) Bayesian fast Fourier transform approach for modal updating using ambient data. Adv Struct Eng 6(2):81–95

Mitigation of Structural Demand to Wind Turbines: Experimental Investigation of Three Control Strategies

N. Caterino[1,2(✉)], C. T. Georgakis[3], M. Spizzuoco[4], and J. Chen[5]

[1] Department of Engineering, University of Naples "Parthenope", Centro Direzionale, isola C4, 80143 Naples, Italy
nicola.caterino@uniparthenope.it

[2] Institute of Technologies for Construction, CNR, San G. Milanese, Milan, Italy

[3] Department of Engineering, Aarhus University, Inge Lehmanns Gade 10, 8000 Aarhus, Denmark

[4] Department of Structures for Engineering and Architecture, University of Napoli Federico II, Via Claudio 21, 80125 Naples, Italy

[5] Department of Structural Engineering, Tongji University, 1239 Siping Road, Shanghai 200092, China

Abstract. The adoption of wind turbines to produce electric energy nowadays represents one of the most promising alternatives to the use of the exhausting fossil fuel stocks. The actual tendency is toward the design of taller towers that can produce more power because excited by stronger winds. There is the need of designing these structures in a cost effective way, aiming to reduce the wind induced growing structural demand. Three different control systems are investigated and compared herein to this aim, on the basis of the experimental results gathered at the Structural Dynamics Laboratory of the Denmark Technical University. Two of these are passive (tuned rolling-ball damper, spherical tuned liquid damper), while the third one is semi-active and aims at realizing a time-variant base restraint. The experimental comparison of the three strategies, tested against two types of wind loads, allow to draw interesting conclusions and to provide useful hints to give rise to further developments of the technologies investigated.

Keywords: Wind turbines · Tuned mass damper · Tuned liquid damper · Semi-active MR devices · Shaking table test · Vibration control

1 Introduction

Wind turbines are becoming a highly capability and attractive mean of producing electricity, given that the energy crisis becomes more and more serious. Recently the wind turbine towers have grown from 40 m to 100 m, with consequent larger turbine power, because taller towers winds are stronger at higher altitudes. It is quite important to optimize the construction process of the towers, as it generally represents 20% of the final cost of a wind turbine. Taller wind turbines are characterized by complex vibration problems due to the interaction of the rotating blades with the tower. In case of offshore

F. Ricciardelli and A. M. Avossa (Eds.): INVENTO 2018, LNCE 27, pp. 165–178, 2019.
https://doi.org/10.1007/978-3-030-12815-9_14

wind turbines installed in coastal regions with high quality wind sources, the influence of wave loads makes their dynamic response even more complex, and the coupling of wind and wave loads on slender wind turbine towers may determine excessive vibrations that can inhibit the mechanical system of the nacelle and produce fatigue damage on the support system. Therefore, vibration control of wind turbine towers is a fundamental issue for guarantee the development of wind energy and reducing maintenance costs.

Nowadays passive vibration control is a mature technology and the tuned mass damper (TMD) is one of the simplest and the most reliable passive control devices: the application of TMD for tall buildings under wind or earthquakes has been extensively investigated.

Lackner and Rotea (2011) investigated the effectiveness of an optimal passive TMD and of a hybrid mass damper in reducing fatigue loads due to bending moment at the base of the tower, showing a percentage reduction of about 10% and 30% respectively due to each of the two proposed systems. Karimi et al. (2010) and Luo et al. (2011) proposed a SA control technique for floating wind turbines with TLCD: this device, generally used as a passive damper, turns into a SA device using a controllable valve; the orifice opening is real time adapted according to the structure response and loading conditions, by means of a control logic based on a H_∞ feedback methodology. Kirkegaard et al. (2002) have been the first to explore the use of magnetorheological (MR) dampers to control a wind turbine, assuming such type of smart device to be installed, in a vertical position, between the base and the top of the tower.

The authors Chen and Georgakis (2013) performed an experimental analysis of a prototype 1/20-scale wind turbine tower model, equipped with an innovative passive tuned rolling-ball damper (TMD), to reduce the structural vibrations. During the shaking table tests, two accelerograms were applied at the base of the tower, equivalent to two different wind load cases: "Extreme Operating Gust" (EOG) is a sharp increase and then decrease in wind speed which occurs over a short period of time while the turbine is operating; "Parking" refers to the load occurring when the turbine is parked and thus subjected to high velocity wind buffeting. The same authors (2015) assumed the same shaking table facility and dynamic inputs to test the above scaled wind turbine model equipped with a spherical tuned liquid damper (TLD). Moreover, on the base of the experience of the authors in the implementation of magnetorheological (MR) devices within semi-active (SA) control systems, they proposed the development of a time-variant base restraint for wind turbines, consisting of elastic springs and SA MR dampers in parallel (Caterino et al. 2014, 2015, 2016). The three different control systems proposed by the authors are compared herein on the basis of the experimental results achieved at the Structural Dynamics Laboratory of the Denmark Technical University, to the aim of derive useful considerations for designing these structures in a cost effective way to reduce the wind induced growing structural demand.

2 Case Study Wind Turbine

Figure 1 illustrates the reference full-scale wind turbine for the three experimental campaigns compared in the proposed work. The tower is 102.4 m tall, made of Q345 steel (Young's modulus of 206,000 MPa, Poisson ratio of 0.3, yield stress of 345 MPa), with a variable annular cross section with external diameter variable from 2.30 m (at the top) to 4.15 m (at the bottom). The tower body weights 3713 kN (including the flanges and internals), and the nacelle weights 1210 kN (including the rotor blades). For wind turbine towers, since the first vibration mode plays a dominant role in the dynamic analysis, Chen and Georgakis (2013, 2015) showed the dynamic equivalence of the case study structure with a generalized SDOF model. The tower was then simplified as a single degree of freedom system consisting in a tapered tubular cantilever beam with a concentrated mass at the top, representing the rotor and nacelle.

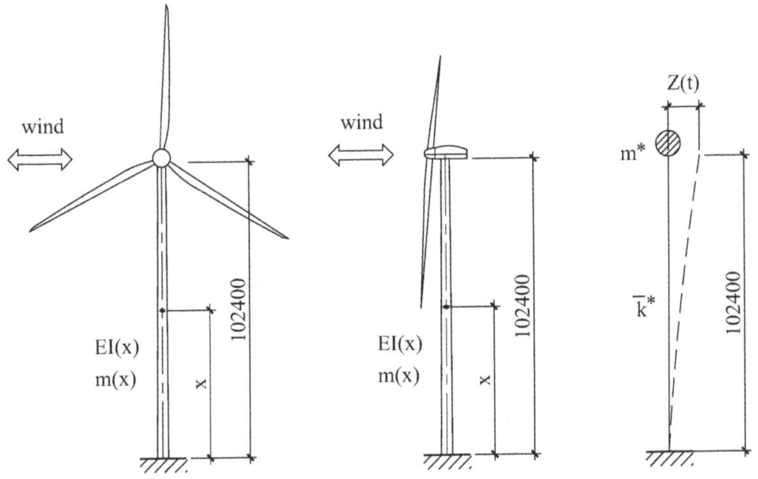

Fig. 1. Structural model of the reference wind turbine. (dimensions in mm)

2.1 The Scaled Tower Model

A 1/20-scale model of the prototype structure has been manufactured in the laboratory and tested on shaking table (Caterino et al. 2016): it consists in a 5.12 m high vertical tube with constant cross section Φ133/4, chosen according to the principle of the equivalent lateral stiffness (133 mm and 4 mm are the external diameter and the thickness of the tube), and with a lumped mass placed at the top to simulate the influence of the nacelle and blades on the generalized mass.

The test model has been installed on the shaking table (Fig. 2), available at the Laboratory of Structures of the DTU, which consists in an aluminium platform of 1.5 × 1.5 m in plan. It is able to move in a single horizontal direction through a 100 kN hydraulic actuator.

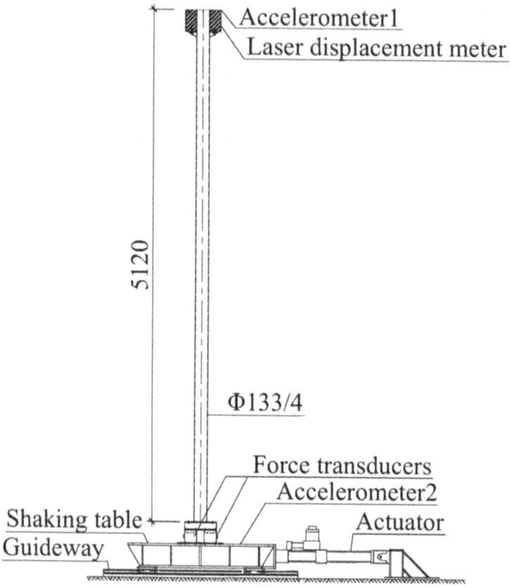

Fig. 2. Shaking table test setup. (dimensions in mm)

The base of the model practically is a rigid body that is connected to the center of the shaking table through a cylindrical steel hinge. At the beginning of the first experimental campaign, two force transducers have been set under the base of the test model to measure the base moment, while two accelerometers have been mounted to measure the absolute acceleration (one fixed on the top of the test model and the other one placed on the shaking table to monitor its motion). A laser displacement meter has been mounted on the shaking table to monitor its displacement. The data acquisition of the shaking table controller has been performed by using a real-time digital signal processor.

2.2 Equivalent Base Accelerations

During the three experimental campaigns, the tests have been carried out by applying to the shaking table different acceleration time histories, which have been defined by Chen and Georgakis (2013, 2015) using the wind turbine aeroelastic code HAWC2 (Horizontal Axis Wind turbine simulation Code, 2nd generation; Larsen and Hansen 2008) as the base inputs that would provide the same base moment and top mass response of the real fixed base structure subjected to the wind actions. Two considered load cases were: an extreme operating gust (EOG) occurring within a quick interval during turbine's operation; a high velocity wind buffeting ("parking" - PRK) occurring after a controlled shut-down. The equivalent base acceleration time histories imposed to the table through the actuator during the dynamic tests, are shown in Fig. 3.

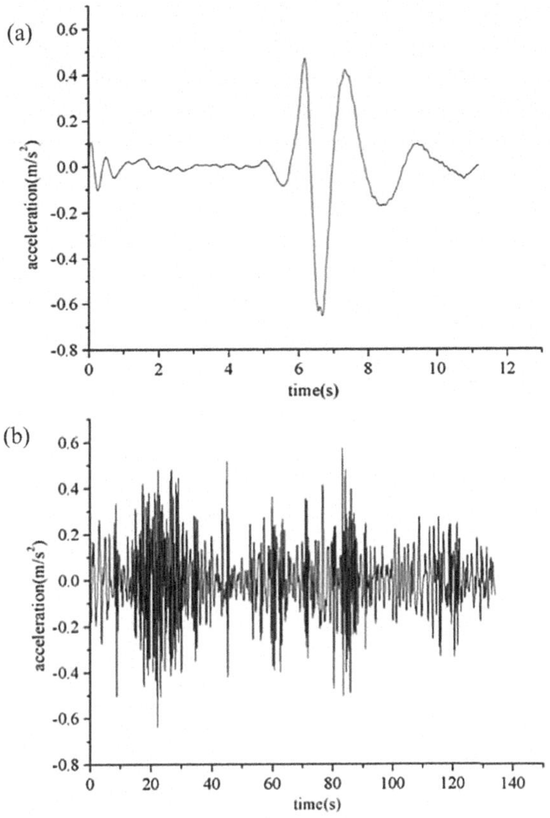

Fig. 3. Equivalent base accelerations corresponding to wind load cases: (a) EOG; (b) Parking

3 Control Systems: TMD, TLD, SA

The authors have performed three experimental tests campaigns relative to the implementation of different control techniques: two of them were passive systems (TMD, TLD) the third was SA. The following section shows the main results of the experimental campaign that involved all the three control techniques.

3.1 Tuned Rolling-Ball Dampers

Chen and Georgakis (2013) proposed to place a tuned rolling-ball damper on the top of the nacelle: it was characterized by single or multiple steel balls rolling on the inside surface of a spherical container (Fig. 4). The choice of ball radius and sphere radius was dependent on the optimal frequency of oscillation of the roller ball. (Warburton and Ayorinde 1980) proposed the optimal design formulas for the TMD system under different types of loads, such as harmonic forces, wind loads, and seismic loads. As the structural damping of the wind turbine tower is very low, the above proposed formulas allowed to evaluate the optimal frequency and damping ratio of the TMD attached to the undamped SDOF system subject to random excitation. The damping effect of the passive rolling-ball damper depends on the fact that the rolling of balls delays the main structural response by a phase angle of 90°, so that the elastic force transmitted by the rolling balls acts like a viscous force on the main structure. The vibration energy of the structural system is dissipated through the relative displacement between the top of the tower and the steel balls when their rolling frequency is tuned to the natural frequency of the turbine. The rolling-ball damper was rigidly attached to the top of the test model by means of a wood frame (see photo in Fig. 4c). Five damper's configurations were tested (Table 1): the first one consisted in a glass spherical container with one 4.1 kg steel ball inside (Fig. 4a); the others were characterized by two hemispherical containers with three to six 0.716 kg steel balls inside, respectively (Fig. 4b). The radiuses R of both the spherical and hemispherical containers were equal to 227 mm. A lubricant was used to reduce the rolling friction between the steel balls and the container.

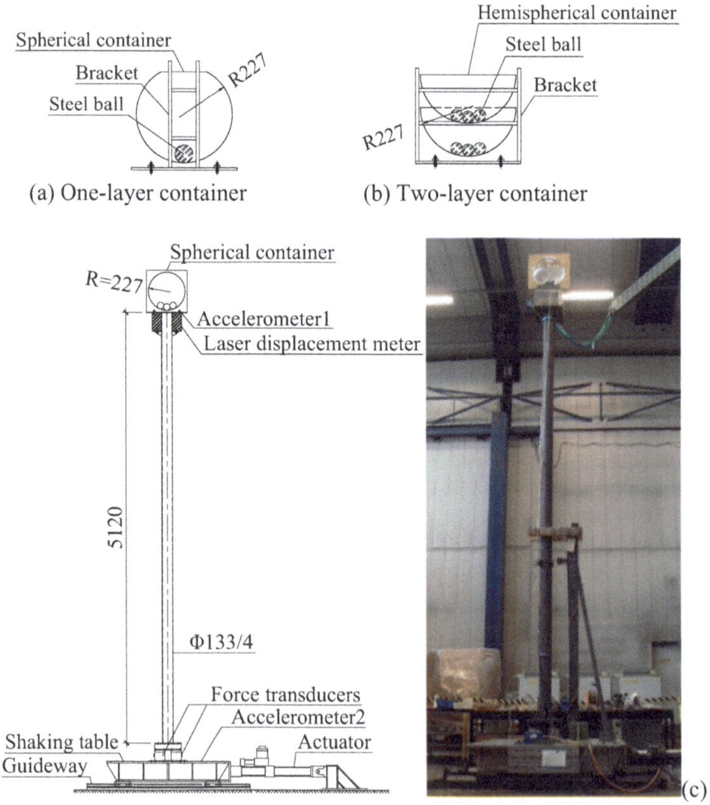

Fig. 4. Configurations including rolling-ball damper

Under EOG input, there was almost no vibration reduction in the first cycles of vibration, mainly because the mass of TMD needs sufficient time with the proper phase

Table 1. Parameters of TMD configurations

No.	Layers of container	Radius of balls (mm)	Number of balls	Total mass of balls (kg)	Frequency of damper (Hz)
TMD-A	1	49.6	1	4.1	1.000
TMD-B	2	27.8	6	4.3	0.943
TMD-C	2	27.8	8	5.7	0.943
TMD-D	2	27.8	10	7.2	0.943
TMD-E	2	27.8	12	8.6	0.943

before it can effectively reduce the structural motion (Fig. 5). But, all tested dampers could significantly reduce the standard deviation of the response. TMD-A had the best behavior in terms of peak reduction of base moment (−1.2%) and top displacement (−0.9%), while TMD-E was able to produce the largest decrease of the standard deviations (−44%) for both base moment and top displacement, with respect to the configuration without TMD. Under the PRK load case, the time histories of the base moment in the controlled cases were effectively reduced except for the period from about 80 to 120 s, probably due to the possible mistuning effects induced by the small deviation of the natural frequency of TMDs from the optimal design value. All the dampers could reduce the peak values of the dynamic response to about 80% of those without TMD: TMD-C determined the highest reduction of peak base moment (−24%) and peak top displacement (−25%). At the same time, the standard deviation of the dynamic response was suppressed significantly in spite of the effect of slight mistuning: TMD-B produced the highest reduction (−35%) for both the considered response quantities. The time histories of base moments under the long PRK input are not shown because the shape of input doesn't allow to clearly see the differences among the curves.

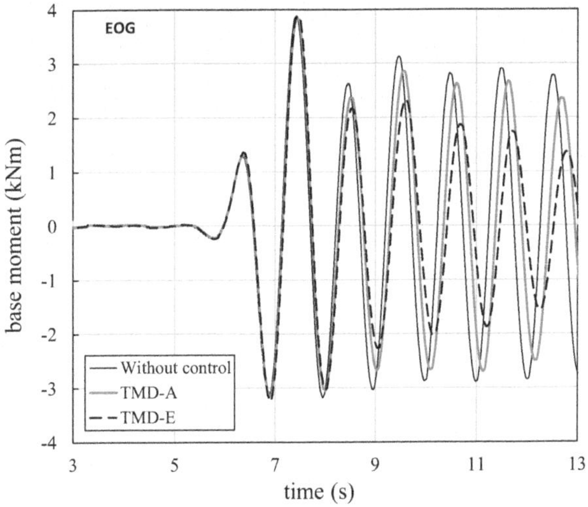

Fig. 5. Time histories of base moments without and with TMD under the EOG input

3.2 Spherical Tuned Liquid Dampers

Chen and Georgakis (2015) applied to the same wind turbine model the spherical tuned liquid damper in Fig. 6. It consists of two-layer hemispherical containers partially filled with water, which is able to slosh when the structure is subjected to dynamic accelerations. The effectiveness of this passive TLD is maximized if the liquid sloshing frequency is properly tuned to the natural frequency of the structure's vibration mode to be reduced. The TLD was rigidly mounted on the top of the model by means of the wood frame (Fig. 6). The plastic hemispherical containers were designed to have a radius of 227 mm and a wall thickness equal to 1 mm; the inside liquid was water with a density of 1000 kg/m^3. The depth of water in each container varied from 48 mm to 74 mm. The fundamental sloshing frequencies corresponding to different depths of water are given in Table 2. The floating particles and the bottom nylon net were used to reduce the beat phenomenon and improved the TLD damping.

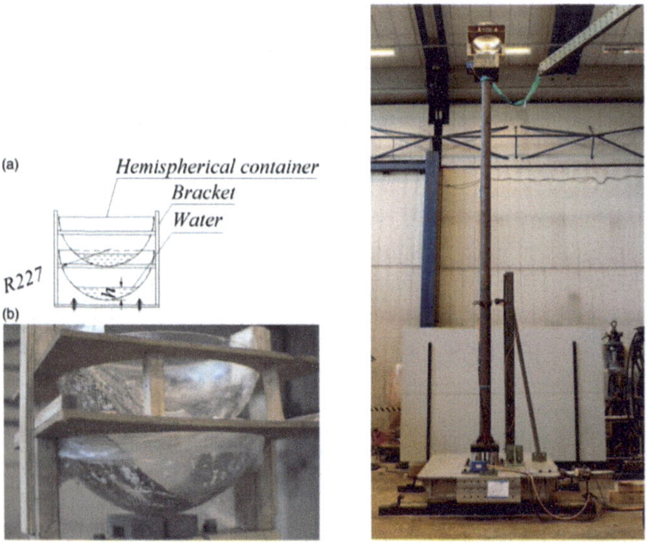

Fig. 6. Configurations including spherical tuned liquid damper

Under EOG load case, there was almost no vibration reduction produced by the TLDs in the first cycles of vibration, mainly because the liquid in the containers needs sufficient dynamic with the proper phase before it can effectively suppress the structure's motion (Fig. 7). Despite this, the dynamic responses with TLDs were suppressed rapidly. All TLDs could significantly reduce the standard deviation of the dynamic response although they could not reduce their peak values: TLD5 was characterized by the best peak reduction of base moment (−0.1%) and top displacement (−2.1%), and was able to determine the largest decrease of the standard deviations (−45%) with respect to the case without TMD. Under the PRK input, the time histories of the base moment with TLDs were effectively suppressed during the whole load process: all the

Table 2. Parameters of TLD configurations

No.	Layers of container	Water depth in each layer (mm)	h/R	Total weight of water (kg)	Frequency of damper (Hz)
TLD-1	2	48	0.21	2	1.084
TLD-2	2	56	0.25	4	1.091
TLD-3	2	62	0.28	5	1.097
TLD-4	2	68	0.30	6	1.103
TLD-5	2	74	0.33	7	1.108

dampers could reduce the peak values of the dynamic responses and, at the same time, their standard deviations. TLD5 produced the largest peak suppression (−35%) of those without TLD and the highest reduction of the standard deviations (−60%). It is worth to note that the experimental response in the case without control (Fig. 7) is a little bit different from that one shown in Fig. 5, due to small differences in the testing setup of the considered experimental campaigns.

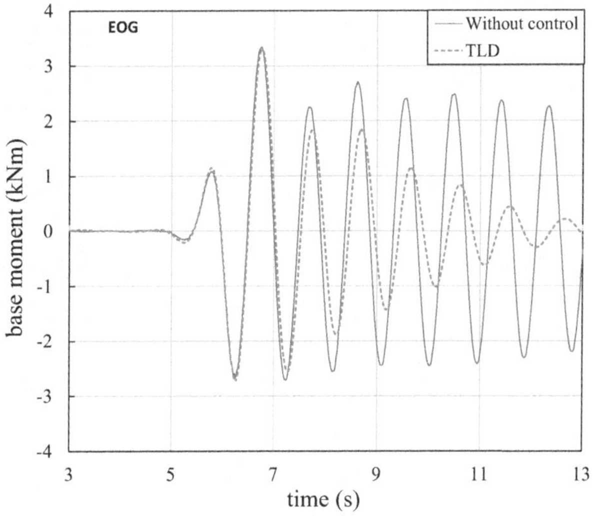

Fig. 7. Time histories of base moments without and with TLD under the EOG load case

3.3 Time-Variant Base Restraint

Caterino et al. (2014, 2015, 2016) focused on the realization of a time-variant restraint at the base of the tower model, consisting of elastic springs and SA MR dampers in parallel. The latters were driven in real time so as to change their mechanical behavior aiming at reducing the structural demand imposed to the tower. On both sides of the base, that is connected in the middle to the shaking table through a cylindrical steel hinge, one cylindrical spring (89 kN/m stiff) and one MR damper are installed, as shown in Fig. 8. The MR devices are two identical full-scale prototype dampers designed and manufactured by the German company Maurer Söhne. They can develop a maximum force of about 30 kN along the longitudinal axis, and have a stroke of ±25 mm. The current generated in the internal circuit is provided in the range 0÷3 A by a specific power supply. During a suitably designed promptness experimental tests, it was demonstrated that the response time of such type of SA MR damper can be bounded to 8–10 ms, when an adequate electric hardware is properly applied.

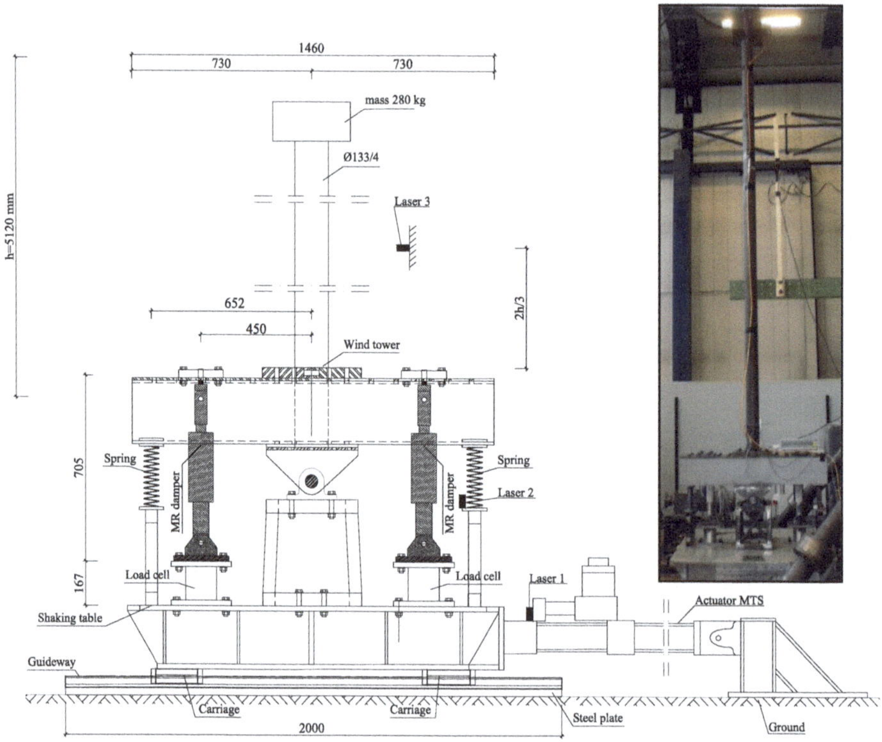

Fig. 8. Configurations including time-variant base restraint

The position of sensors was different with respect to the previous passive control systems: the horizontal displacements of the shaking table were measured through one laser transducer; the displacements of the tower were registered by a laser transducer located at 2/3 of total height of the tower and fixed to an external wood stick; the axial displacements of springs were measured by a third laser transducer attached on the steel plate where the East side spring was installed (rotation of the base and axial dampers' displacements were on-line derived directly starting from the above measures); two load cells mounted under each MR device allowed to measure their reacting force. A dedicated electronic equipment for acquisition and control has been used to drive the SA devices. It includes two operational power supplies able to provide the current needed to feed the circuitry inside the MR devices. A bang-bang control algorithm has been considered for semiactively controlling the wind turbine via MR dampers (Caterino et al. 2015, 2016). This controller switches back and forth from an "off" state (intensity of current $i = 0$) to an "on" state ($i = i_{max} = 1$A) according to a given logic designed in order to reduce the stress at the base, even aiming to bound the top displacement demand within given limits. To do that, the algorithm assumes a limit value for both base stress and top displacement (σ_{lim} and x_{lim} in the following, respectively).

Many tests have been performed with the above mentioned control algorithm under both types of base input, in order to evaluate and compare the effects, in terms of reduction of the structural response, of different couples of stress and displacement limits (σ_{lim} and x_{lim}, respectively). The different configurations, i.e. couples of values (σ_{lim}, x_{lim}), have been considered by varying the values of σ_{lim} and x_{lim} in the intervals [10, 40] MPa and [16, 46] mm, respectively.

Under EOG load case, the optimal calibration for the assumed controller resulted to be the configuration SA (σ_{lim} = 30 MPa, x_{lim} = 46 mm), leading to the maximum reduction of base stress (−29%) at the cost of a moderate increase of top displacement demand (+15%). Figure 9 shows the registration of base moments in case of the semiactive configuration, compared to the results of the fixed base test shown in Fig. 7 and the data recorded during the "passive ON" test carried out on the setup of Fig. 8 by feeding the MR dampers with a constant current equal to i_{max} = 3A. Under the PRK input, the configuration SA (σ_{lim} = 12 MPa, x_{lim} = 20 mm) resulted to be the best among the tests performed with the assumed control logic, both in terms of base stress suppression (−30%) and top displacement decrease (−11%).

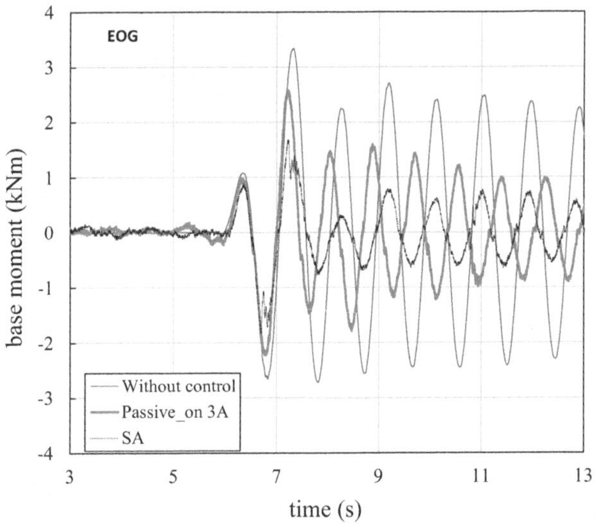

Fig. 9. Base moments without and with the SA control system under EOG input.

4 Comparison Among the Control Strategies

The following Table 3 compares the three control strategies in terms of capability of reducing the structural demand against the two load cases examined. Two configurations of the TMD have been selected for each load case. One is better for the higher reduction of the peak response, the other is better looking at the standard deviation of the response (that is in indirect measure of the fatigue phenomenon). For the TLD technique, the configuration TLD-5 resulted to be the best according to both criteria. For the SA strategy, where reduction of the peak response should be the first purpose, the configuration is that described in the previous section, and the standard deviation has not yet been assessed, it will be the subject of future work. However, it is worth noting that the reductions shown for the SA configuration in the following table are computed with respect to the "passive-ON", not to the uncontrolled case.

Table 3. Comparison among the best response reductions achieved by the three control systems, under EOG and PRK load cases ("M" and "x" stand for base moment and top displacement demand; "σ" stands for standard deviation; the sign minus indicates a reduction in respect to the uncontrolled configuration)

Control system	EOG load case		Control system	PRK load case	
	M/σ	x/σ		M/σ	x/σ
TMD-A	−1.2%/−23%	−0.9%/−23%	TMD-C	−24%/−29%	−25%/−30%
TMD-E	−0.2%/−44%	+0.9%/−44%	TMD-B	−20%/−35%	−21%/−35%
TLD-5	−0.1%/−45%	−2.1%/−45%	TLD-5	−.35%/−60%	−35%/−60%
SA	−29%	+15%	SA	−30%	−11%

Under a short input, pulse-like as EOG, both TMDs and TLD are only slightly able to reduce vibrations during the first strong cycles (around 1%). Actually, as known, such systems require a certain time to produce an effective suppression of the structural motion. Conversely, the SA control strategy is able to damp the response in the strong phase of such load case, causing a reduction of base moment up to 29%. However, the TMD and TLD techniques show the advantage of strongly reducing the fatigue demand and, above all, that of not causing significant change in the displacement demand in comparison with the uncontrolled configuration. The response of the semi-actively controlled system to the EOG load case instead is characterized by a significant increasing of the top displacement (15%) because, during a fast input like EOG, in correspondence of the stiffness reduction the displacement demand is significant and the system doesn't have sufficient time to reduce it through damping. Figure 10 shows the comparison among the base moment time-histories under this load case.

Different comments can be drawn looking at the performance of the three control systems against the longer, less impulsive PRK wind load. In this case, you can see that the three techniques lead about to the same reduction of bending moment demand. However the passive techniques result to be more able to reduce fatigue phenomenon and, above all, displacement demand, the latter even up to 35%. Finally, it is worth

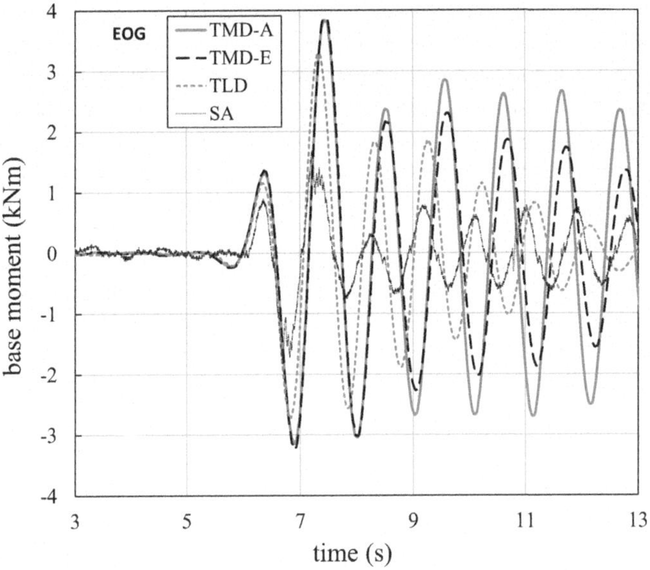

Fig. 10. Base moments measured during the best TMD, TLD and SA control systems

noting that the SA strategy also allows a reduction of top displacement demand. This reduction, while is smaller than the values registered for the passive cases, is still significant (11%), due to the amount of energy dissipated during the several cycles.

5 Conclusions

Three different control strategies have been proposed by the authors to reduce wind-induced vibrations of wind turbine towers. Two of them involve passive devices (TMD and TLD) mounted at the top of the structure. The third one mainly consists of semi-active dampers installed at the base of the tower and exploits the base rocking to dissipate energy. Three shaking table testing campaigns were carried out to assess the performance and potentialities of such technologies. Two strongly different load cases have been considered, the first shorter and impulsive (PRK), the second longer and almost stationary (PRK). Based on the results of all the tests, it can be conclude that:

- against the EOG case, the TMDs and TLDs were not able to suppress the peak value of the dynamic response, however they reduced the standard deviation significantly; under the PRK wind load, instead, both standard deviation and peak value of the dynamic response resulted to be strongly reduced;
- for the EOG load case, the MR SA control suppressed the peak value of the stress at the base of the tower, however with a moderate increase of the top displacement; for PRK load case, the SA system instead resulted to be able to significantly reduce both the dynamic responses.

In short, at least from the analyses conducted so far, it would seem that passive systems are able to reduce cyclic demand and are fail safe, in the sense that they rarely cause a worsening of the response compared to what would be of the uncontrolled system. In some cases, however, the reduction in response is very small.

On the other hand, the SA system results to always guarantee a significant reduction in the demand for the bending moment, even if, in case of impulsive actions, this can have as a side effect a significant increase in displacement at the top.

Acknowledgements. The Denmark Technical University (DTU) of Copenhagen is gratefully acknowledged for having financed the experimental activity and for the laboratory equipment, with special thanks to prof. C.T. Georgakis. The authors would like to thank the financial support of China Scholarship Council. The MR dampers have been designed, manufactured and provided for free by Maurer Söhne (Munich, Germany) that is also acknowledged for the support.

References

Caterino N, Georgakis CT, Trinchillo F, Occhiuzzi A (2014) A semi-active control system for wind turbines. In: Luo N, Vidal Y, Acho L (eds) Wind turbine control and monitoring (Advances in industrial control). Springer International Publishing Switzerland, ISBN 978-3-319-08412-1

Caterino N (2015) Semi-active control of a wind turbine via magnetorheological dampers. J Sound Vibr 345:1–17. https://doi.org/10.1016/j.jsv.2015.01.022

Caterino N, Georgakis CT, Spizzuoco M, Occhiuzzi A (2016) Design and calibration of a semi-active control logic to mitigate structural vibrations in wind turbines. Smart Struct Syst 18 (1):75–92

Chen J, Georgakis CT (2013) Tuned rolling-ball dampers for vibration control in wind turbines. J Sound Vibr 332(21):5271–5282

Chen J, Georgakis CT (2015) Spherical tuned liquid damper for vibration control in wind turbines. J Vibr Control 12(10):1875–1885

Karimi HR, Zapateiro M, Luo N (2010) Semiactive vibration control of offshore wind turbine towers with tuned liquid column dampers using H_∞ output feedback control. In: IEEE international conference on control applications, Yokohama, Japan

Kirkegaard H, Nielsen SRK, Poulsen BL, Andersen J, Pedersen LH, Pedersen BJ (2002) Semi-active vibration control of a wind turbine tower using an MR damper. In: Grundmann H, Schuëller GI (eds) Balkema Publishers, A.A./Taylor & Francis, Balkema/The Netherlands, pp 1575–1580

Lackner M, Rotea M (2011) Structural control of floating wind turbines. Mechatronics 21:704–719. https://doi.org/10.1016/j.mechatronics.2010.11.007

Larsen TJ, Hansen AM (2008) HAWC2 User Manual, Roskilde, Denmark: Risø National Laboratory, Technical University of Denmark

Luo N, Bottasso CL, Karimi HR, Zapateiro M (2011) Semiactive control for floating offshore wind turbines subject to aero-hydro dynamic loads. In: International conference on renewable energies and power quality, Las Palmas de Gran Canaria, Spain

Warburton GB, Ayorinde EO (1980) Optimum absorber parameters for simple systems. Earthq Eng Struct Dyn 8:197–217

Feasibility and Assessment of Real Time Monitoring Systems for Smart Structural Control of Wind Turbines

N. Caterino[1,2(✉)], G. Pugliano[1], M. Spizzuoco[3], and U. Robustelli[1]

[1] Department of Engineering, University of Naples "Parthenope", Naples, Italy
nicola.caterino@uniparthenope.it
[2] Institute of Technologies for Construction, CNR, San G. Milanese, Milan, Italy
[3] Department of Structures for Engineering and Architecture, University of Naples Federico II, Naples, Italy

Abstract. A semi-active (SA) control strategy has been proposed by the authors in the recent years to mitigate structural demand to high wind turbines against strong wind loads. Numerical as well as experimental analyses shown it is really promising and potentially useful to owners who want to optimize costs of realization and installation of such huge structures. The application to real cases of this technique, since based on the use of variable dissipative devices, requires the tower is instrumented for real time monitoring of structural response of the tower. This allows the control algorithm to make the decision about the optimal calibration, moment by moment, of the variable devices. Making reliable high frequency measurements of the horizontal displacement of points placed at a height of tens of meters can be not so trivial. With the aim of evaluating the efficiency and feasibility of Global Navigation Satellite System (GNSS)-based systems for structural control of wind turbines, this paper try to obtain insight into the characteristics (receiver type, type of observables, sampling data rate) and data processing techniques that can make the GNSS useful for such application. Finally, numerical investigations referred to a case study allow to discuss how the features of the measurement system may affect the performance of the proposed SA technique in reducing structural demand due to wind induced vibrations.

Keywords: Wind turbines · Semi-active control · GNSS · Real time monitoring

1 Introduction

The adoption of wind turbines to produce electric energy nowadays represents one of the most promising alternatives to the use of the exhausting fossil fuel stocks. The actual tendency is toward the design of taller towers that can produce more power because excited by stronger winds. There is the need of designing these structures in a cost effective way, aiming to reduce the wind induced growing structural demand. A semi-active (SA) control technique recently proposed by the authors (Caterino 2015; Caterino et al. 2016) is based on the use of a cylindrical hinge at the base of the wind tower and

F. Ricciardelli and A. M. Avossa (Eds.): INVENTO 2018, LNCE 27, pp. 179–191, 2019.
https://doi.org/10.1007/978-3-030-12815-9_15

magnetorheological (MR) dampers installed in parallel with linear springs. The aim is realizing a time-variant base restraint (Fig. 1). A purposely written control logic aims at reducing the base bending moment by relaxing - in selected intervals of time - the base restraint. This causes rocking of the base and dissipation of energy due to cyclic demand of deformation to MR devices. Such SA control strategy has been tested at the Structural Dynamics Laboratory of the Denmark Technical University by means of a shaking table facility (Fig. 1), leading to really encouraging results in terms of percentage reduction of structural demand in respect to the "fixed base" configuration.

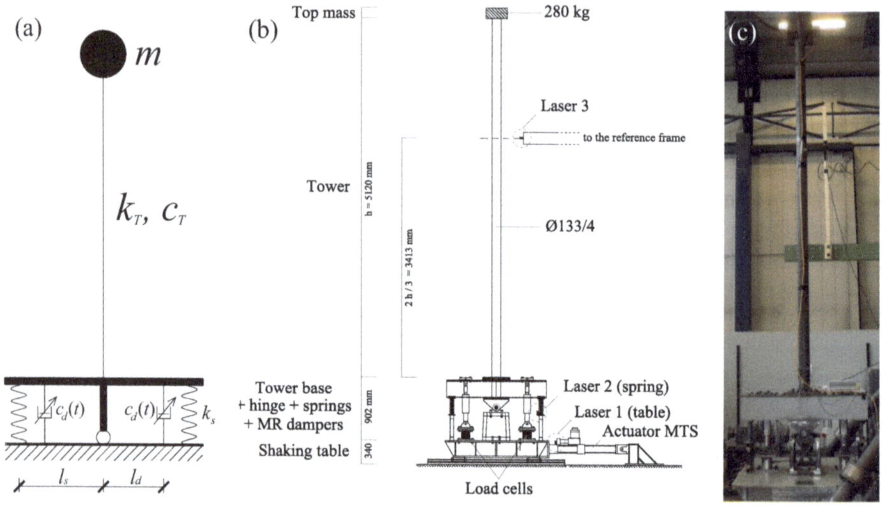

Fig. 1. Smart base restraint including semi-active MR dampers: (a) schematic idea, (b) set-up drawing and (c) picture

This control strategy requires that the tower is instrumented so as to monitor in real time the state of stress and deformation and to allow the algorithm to make the decision regarding the optimal calibration, moment by moment, of the variable devices. As regards, in particular, the displacement measurements, in the laboratory they can be carried out, on scaled models, easily with normal transducers and fixed external reference structures. However, for real applications, on- or off-shore wind turbines, making reliable high frequency measurements of the horizontal displacement of points placed at a height of tens of meters can be not so trivial.

Among the available geodetic technologies, very few are those can fully satisfy the required monitoring criteria for the above goal. Global Navigation Satellite Systems (GNSS) are among the most advanced for fully automated and continuous monitoring. Many GNSS-based applications for continuous monitoring of the dynamic behaviour of civil structures were developed in recent years, mainly for long-span bridges and tall buildings (Chmielewski and Górski 2015). GNSS actually has the advantage of directly measuring absolute 3-D position coordinates, thereby providing displacement measurements in real-time and full scale.

Usually, GNSS geodetic dual frequency receivers have been used as reliable tool for providing accurate and timely information on the actual status of the structure. The capability of measuring changes in position with high level of accuracy is extensively achieved by adopting the real-time kinematic (RTK) positioning.

With the aim of evaluating the efficiency and feasibility of GNSS-based systems for structural control of wind turbines, this paper try to obtain insight into the characteristics (receiver type, type of observables, sampling data rate) and data processing techniques that can make the GNSS useful for such application.

A case study structure – a real 3 MW, 102.4 m tall wind turbine - finally is considered to make numerical investigations and to evaluate how the specific features of the measurement system may affect the capability of the SA control system in reducing structural demand in respect to the conventional, fixed base configuration of the same wind turbine. The role of the sampling rate and that of the latency are mainly discussed, highlighting that they may significantly reduce the effectiveness of the SA system.

2 A Semi-active Rocking System for High Wind Turbines

A new idea of base restraint is proposed by Caterino et al. (2015) to reduce wind induced structural demand for high-rise wind turbine towers. It consists in SA rocking system, exploiting controllable fluid based devices, as schematically shown in Fig. 1: the uncontrolled wind turbine is fully restrained at the base and can be simulated by a single degree of freedom dynamic system having top mass m, stiffness k_T and inherent damping c_T. Figure 1(a) shows the proposed smart base restraint, whose stiffness can be instantaneously be modified during the motion, by mounting at the base of the tower a smooth hinge, two vertical linear springs (k_s) placed at a certain distance (l_s) from the hinge and two vertical SA magnetorheological (MR) dampers (c_d) at a distance l_d from the central hinge. The base control system is able to produce a real time modification of the system's stiffness by varying the MR dampers' mechanical properties according to a given control algorithm. A simple physical approach was formulated by the authors to reduce base stress, by restraining the increase of top displacements within certain limits to control second order effects: the objective was to achieve a trade-off between the contradictory purposes to limit the base stress σ_{lim} and the top displacement x_{lim}.

In other words, a bang-bang controller keeps 'stiffer' the base restraint (switches to an ON state by commanding a maximum current $i = i_{\max}$ to the dampers) until the stress doesn't exceed a limit value σ_{lim}. Conversely, the controller 'relaxes' the base restraint (switches to an OFF state by delivering a minimum current $i = i_{\min}$ to the dampers) when the limit stress is overpassed and the displacement falls within the limit of acceptability x_{lim}, so that the structure is able to convert its potential energy into kinetic energy. The following instructions are applied:

$$\begin{array}{lll} \text{(a)} & \text{if } |\sigma(t)| < \sigma_{\text{lim}} & \rightarrow i(t) = i_{\max} \\ \text{(b)} & \text{if } |\sigma(t)| \geq \sigma_{\text{lim}} \text{ and } |x(t)| < x_{\text{lim}} & \rightarrow i(t) = 0 \\ \text{(c)} & \text{if } |\sigma(t)| \geq \sigma_{\text{lim}} \text{ and } |x(t)| \geq x_{\text{lim}} \text{ and } \mathrm{x(t)} \cdot \dot{x}(t) > 0 & \rightarrow i(t) = i_{\max} \\ \text{(d)} & \text{if } |\sigma(t)| \geq \sigma_{\text{lim}} \text{ and } |x(t)| \geq x_{\text{lim}} \text{ and } \mathrm{x(t)} \cdot \dot{x}(t) \leq 0 & \rightarrow i(t) = 0 \end{array}. \quad (1)$$

where $\sigma(t)$, $x(t)$ and $\dot{x}(t)$ are respectively the value of stress at the base, top displacement and top velocity at the instant of time t. When both stress and displacement are beyond their threshold values, the controller switches the dampers to an ON state if the displacement is going towards a larger value (i.e., positive velocity), so trying to damp or invert the displacement's trend; otherwise it switches the MR devices to an OFF state to make them collaborating to both stress and displacement reduction. The springs have also the role to induce the recentering of the tower to the initial position at the end of a severe wind-induced excitation.

3 GNSS in Structural Monitoring

Although generally used in positioning applications such as in geodetic surveying and navigation, GNSS receivers have attracted much interest as sensors in the context of monitoring the deformation behaviour of structures under dynamic loads (Celebi 2000). Real-time measurement of dynamic response of engineering structures has been subject of intense research for structural health monitoring (SHM). Long span bridges, high-rise buildings and towers are examples of civil structures on which the SHM technology has been implemented. A typical SHM system involves assessment of the health state of in-service structures on a continuous real-time basis through several sensors. Relative displacements can be used to monitor the system performance and to carry out early damage detection.

Conventional monitoring methods, including accelerometers and robotic total stations, have been used for the purpose of measuring structural vibrations and dynamic displacements. Also, GNSS technology has been employed in the dynamic measurements of flexible structures for structural health monitoring (Fuggini 2009; Im et al. 2013; Kaloop et al. 2017). The first pioneering application of global positioning system (GPS) to measure the dynamic displacements of these structures, achieving a mm-level accuracy, was carried out in 1993 at the Calgary tower in Canada, collecting data at 10 Hz. The results showed a vibration frequency, under wind loading, of about 0.3 Hz in both north-south and east-west directions and the horizontal movement of ±15 mm was measured at 160 m above ground level in north-south direction (Lovse et al. 1995). The GNSS techniques have been broadly used for bridge monitoring. In the 1996, one of the earliest applications of real-time monitoring of a long span bridge by kinematic GPS was performed at the Humber Bridge in the United Kingdom and the results revealed that the resolution was ±1 mm horizontally and ±3 mm in height (Ashkenazi et al. 1996; Ashkenazi and Roberts 1997). More recently, Yu et al. (2014) employed GNSS technology in the dynamic monitoring of a medium span suspension bridge, by carrying out a series of experiments on the Wilford suspension bridge in Nottingham (UK). Three GNSS data processing modes, real-time kinematic (RTK), network RTK and post-processing kinematic (PPK), were considered. Using the multimode GNSS data and a multimode adaptive filtering algorithm the dynamic displacements of the bridge were identified with sub-millimeter level accuracy, showing the capability to monitor the vibration response of medium span bridges. Real-time monitoring on a full-scale of long-period structures, such as tall buildings and long-span bridges, has become increasingly popular due to advances in sampling rates of GNSS technology and developments in integrated systems. Both the quasi-static and dynamic responses are measured either as displacement or acceleration resulting from wind or seismic actions. The advantage over conventional monitoring using accelerometers is that relative displacements can be measured reliably in real-time and with sufficient accuracy.

Attention to the SHM technology has increased in recent years, also in the wind turbine sector. Although some research has applied GNSS to the wind turbines for the SHM application (Mostböck and Petryna 2014), the GNSS use for SA control of wind turbine structures is still rare and there is a lack of knowledge regarding measuring displacements for the SA control purpose using GNSS.

4 RTK-GNSS in Wind Turbine SA Control

Real-time kinematic (RTK) is the most attractive GNSS technique for high-accuracy instantaneous positioning, as the coordinate solutions are obtained in the field with a centimeter level accuracy and can be used immediately for both kinematic and time-critical static applications such as precise navigation, machine guidance and control. As emerging trend, RTK-based mechanic guidance systems are employed, in particular, in all phases of the agricultural operations to perform every activity on the crop, from soil preparation to harvesting, and they help farmers to choose consistent driving strategies enabling to increase accuracy and productivity and decrease time and operating costs.

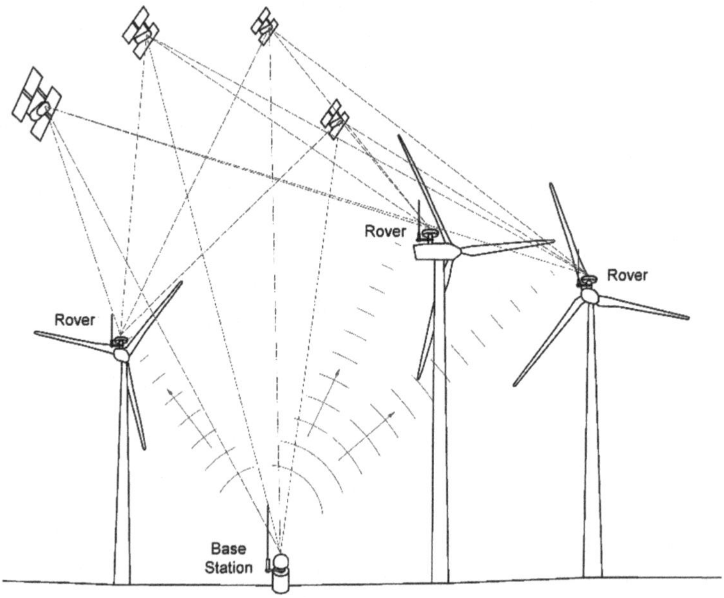

Fig. 2. Schematic of RTK-GNSS method

RTK positioning makes use of carrier phase measurements in the relative positioning mode. In addition to the rover receiver, used for the local positioning, one reference station receiver or a network, with a priori known coordinates, are employed, along with a data link to continuously transmit measurements from the base station to the rover, sometimes over the Internet using cellular or wireless service, or radio signal. The RTK solutions are derived from the current epoch instantaneously (Fig. 2).

The basis of RTK is the implementation of the on-the-fly (OTF) ambiguity resolution (AR) algorithms. As any kind of carrier phase solution, in order to obtain centimeter level positioning accuracy, an ambiguity-fixed solution is necessary, which indicates that the number of full phase cycles between the receiver and the satellite at the first instant of the receiver's lock-on are fixed at integer values. It is possible to perform the AR process with less than 10 s under the best circumstances, where the distance between the base and rover receiver is 10–15 km (which can be increased to 50–70 km if a network-based RTK service is available), the receivers are tracking a large constellation of satellites with low geometric dilution of precision (GDOP), the receivers are multi-frequency, the multipath and receiver noise are low.

However achieving reliable centimeter level accuracy with as few as possible data epochs in real time is not easy in the harsh environment encountered in wind turbines. Obstructions hindering satellite visibility (e.g. rotor blades) as also unexpected errors (e.g. multipath) related with GNSS difficult observation conditions could result in wrong ambiguity fixes degrading accuracy of the position even up to tens of centimeter. Therefore, to obtain an accurate and reliable RTK fix and to detect real displacements from the measurements a greater care has to be taken to characterize the RTK performance and to address each individual error present in GNSS data.

Multipath is one of the primary error sources in GNSS positioning that can degrade the position solutions, and its effect is difficult to be eliminated by current differencing techniques, since it is highly correlated with observation environments (Wang et al. 2017). The straightforward multipath mitigation method is to select a favourable environment to reduce the reflections surrounding the antenna, however this possibility is not feasible for wind turbine applications, since the reflection is inevitable due to the presence of blades placed near the tower. Therefore, the demand for high accuracy positioning, along with the improvement of the performance of ambiguity resolution, requires us to address multipath: wavelet analysis (Pugliano et al. 2016) and short-time Fourier transform (Robustelli and Pugliano 2018a) can be used to extract the time-varying frequency and magnitude contents of multipath.

GNSS receivers that are able to track multiple GNSS constellations can lead to improvements in the accuracy and reliability of RTK positioning, since the increased number of satellites enables the receiver to exclude satellites with poor signal. The GPS from the USA has led in the satellite positioning in the past decades, although lately a number of devices that use the Russian GNSS system named GLONASS have rapidly expanded, since the constellation expanded to 24 operational satellites. Moreover, many devices lately are compatible with the European GNSS system, Galileo, which is currently (April 2018) in its full operational capability (FOC) phase: 18 FOC satellites were launched up to the end of 2017 in addition to the 4 IOV launched between 2011 and 2012. The navigation signals of these satellites are transmitted on five frequencies E1, E5a, E5b, E5 and E6 (Robustelli and Pugliano 2018b).

In order to assist the SA control system in making better decisions by providing precise on-time information, the ability to detect and characterize displacements quickly and to update the position can make a crucial difference in the wind turbine SA control. The availability of low-latency data has a fundamental impact on how we perform the SA control, since the size of latency in practice results in wrong real time information. Latency is the time taken for the receiver to calculate and output a position. Lower latencies and higher position update rates mean that the SA control system is more responsive and can be operated without loss of performance. GNSS monitoring receivers are used typically gathering high-rate data at 10–20 Hz with an accuracy of the order of millimeters. Acquiring data at the maximum sampling rate, up to 50–100 Hz (ultra-high-rate), the highest performing receivers could identify the high structural vibration frequencies of up to 10 Hz (Psimoulis et al. 2008; Yi et al. 2013), with an equal level of accuracy and latencies of less than 20 ms. Table 1 exhibits up-to-date GNSS receivers, their accuracies for RTK positioning and the maximum sampling rate, based on the brochures provided by manufacturers.

Table 1. GNSS Receivers for RTK positioning based on brochures in 2018

Manufacturer	Receiver model	RTK position accuracy (RMS)		Max sampling rate (Hz)
		Horizontal	Vertical	
Leica	GM30	8 mm + 1 ppm	15 mm + 1 ppm	50
Javad	Delta-3	10 mm + 1 ppm	15 mm + 1 ppm	100
Navcom	SF-3050	10 mm + 0.5 ppm	20 mm + 1 ppm	100
Novatel	PwrPak7D	10 mm + 1 ppm	-	100
Septentrio	AsteRx SB	6 mm + 0.5 ppm	10 mm + 1 ppm	100
Sokkia	GNR5	5 mm + 0.5 ppm	10 mm + 0.8 ppm	100
Topcon	MR-2	5 mm + 0.5 ppm	10 mm + 0.8 ppm	100
Trimble	R9s	8 mm + 1 ppm	15 mm + 1 ppm	20

5 The Role of Frequency and Latency of Measures for SA Control of a Case-Study Wind Turbine

The case study structure is a prototype real wind turbine. It is a 3 MW wind turbine with horizontal power transmission axle, 102.4 m tall, with a variable hollow circular cross section whose external diameter is variable from 2.30 m (top) to 4.15 m (bottom), and whose thickness is variable from 14 mm (top) to 40 mm (bottom). A lumped mass of 111 t is placed at the top of the tower. The base of the prototype structure is highly stiff and is supported in the middle by a cylindrical steel hinge. On both sides of the base, one cylindrical spring (1417 kN/m stiff) and one SA MR damper are installed. The assembly "elastic springs + SA MR dampers", whose elements are placed in parallel at the base of the tower, just represents the smart base restraint proposed by the authors to control the dynamic behaviour of the structure. An extreme operating gust (EOG) loading has been considered in the following as reference wind action: a sharp increase, then decrease in wind speed within a short period of time. Chen and Georgakis (2013) defined an equivalent base acceleration time history (Fig. 3), that is the base input that would provide the same top mass response of the real fixed base structure subjected to the wind action. Values adopted to calibrate the SA control algorithm are x_{lim} = 900 mm and σ_{lim} = 10 MPa.

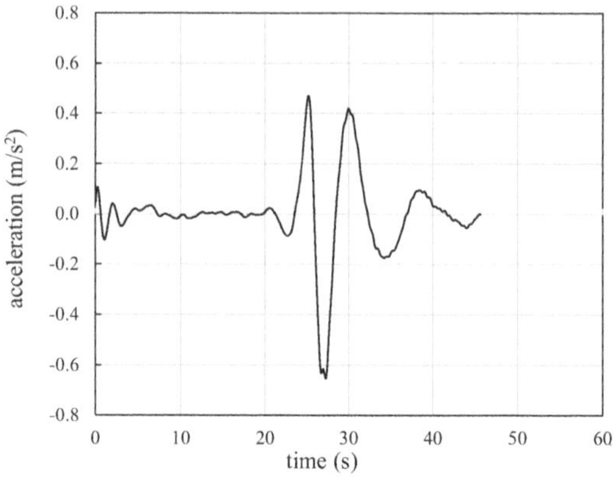

Fig. 3. Equivalent base acceleration time history for the EOG wind load case.

Table 2 and Fig. 4 allow the comparison of the structural response exhibited by the wind turbine under the same EOG wind load, in the case of fixed base and SA control. Five cases of SA control have been considered, each time assuming a different frequency of acquisition (10, 20, 50, 100, 1000 Hz) of the structural response data, those needed to the controller to make decision about the optimal calibration of MR Dampers. In other words, for each test has been assumed that measurements are made with different monitoring systems, with different sampling rate, fixed around the values valid for commercial devices (Table 1). The 1000 Hz case has been explored as an ideal case, hard to be achieved in case of real applications.

It is worth noting that the slower the acquisition, the lower the effectiveness of the SA control system in reducing the base stress demand (84% of reduction for command at 1000 Hz, 50% at 10 Hz), even if in the worst case (10 Hz) the peak value is still half that registered in fixed base configuration. This is due to the fact that the controller calibrates the MR dampers much less frequently when low sampling rate devices are adopted for measurements. This means that the configuration of MR dampers during the motion is far from optimal for most of the time (note number of operations commanded by the control logic: 62 at 10 Hz, 1648 at 1000 Hz). Finally it is worth observing from Fig. 4(b)-*zoom* that the demand of base stress over time is much more variable in case of lower sampling rate of measurement. This may causes higher risk of structural failure due to cyclic damaging and fatigue. Less evident is the effect of the sampling rate on the displacement demand. Considering the five cases, the reduction in respect to the fixed base case falls within 43–50%.

Table 3 and Fig. 5 allow to understand how the latency of the measurement can affect the performance of the SA control system. Two cases have been considered, i.e. latency of 20 ms and 50 ms, selected around common values for commercial GNSS Receivers. The ideal case of no latency (real time) has been explored for comparison.

It is worth noting that an high promptness of the measurement system is crucial to make really effective the SA control strategy under examination. Looking at Table 3, one can observe that a latency of 20 ms can worsen the ability of the SA system to reduce bending demand at the base of the tower by up to 15%. A latency of 50 ms even up to 60%. In the last case, the maximum top displacement (1019 mm) results to be definitely higher than in the other cases of SA control characterized by less latency (701–731 mm). It tends to be similar to that registered for the uncontrolled, fixed base configuration of the wind turbine (1320 mm).

Figure 5(b)-*zoom* finally shows that the demand of base stress over time is much more variable in case of higher latency of measurement. As said above, this causes fatigue and may lead to premature structural collapse of the tower.

As regards the accuracy of the data provided by the GNNS system, it has been found that the system's response is not strongly influenced by it when the accuracies of those devices in Table 1 are considered. However, it should be considered that this depends on the type of algorithm (there are several in the literature), the type of turbine analysed and, not least, the type of wind load.

Table 2. Peak response of the wind turbine considering different frequencies of measurements

	Maximum base stress [MPa]	Maximum top displacement [mm]	No. of state changes (on/off)
Fixed base	*90*	*1320*	–
SA @1000 Hz	14	752	1648
SA @100 Hz	22	731	366
SA @50 Hz	36	656	220
SA @20 Hz	40	684	104
SA @10 Hz	45	659	62

Table 3. Peak response of the wind turbine considering different latencies of measurements

	Maximum base stress [MPa]	Maximum top displacement [mm]	No. of state changes (on/off)
Fixed base	90	1320	–
SA – no latency	22	731	366
SA – lat. 20 ms	33	701	156
SA – lat. 50 ms	76	1019	114

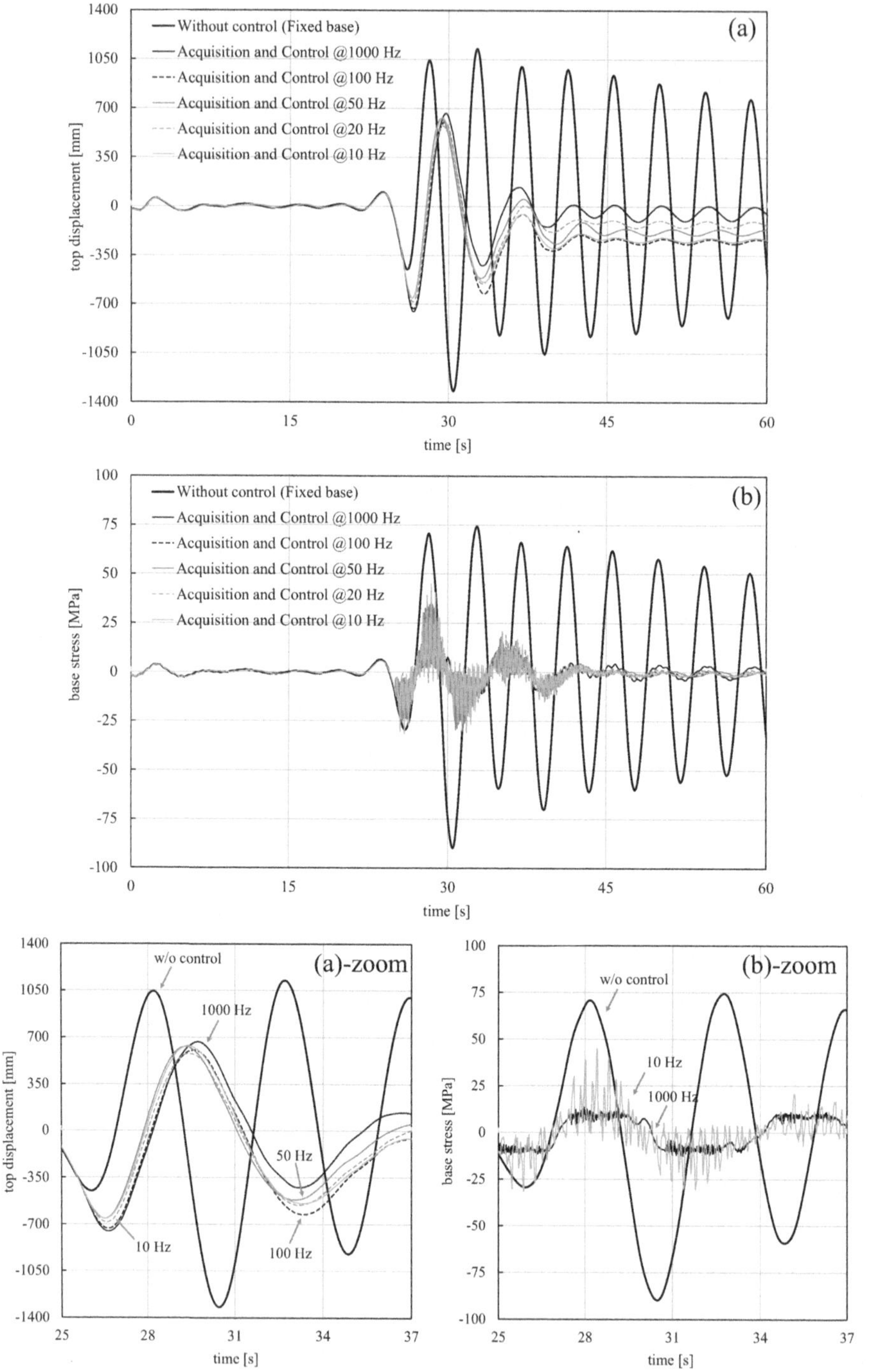

Fig. 4. Displacement (a) and stress (b) response for different frequencies of measures for control

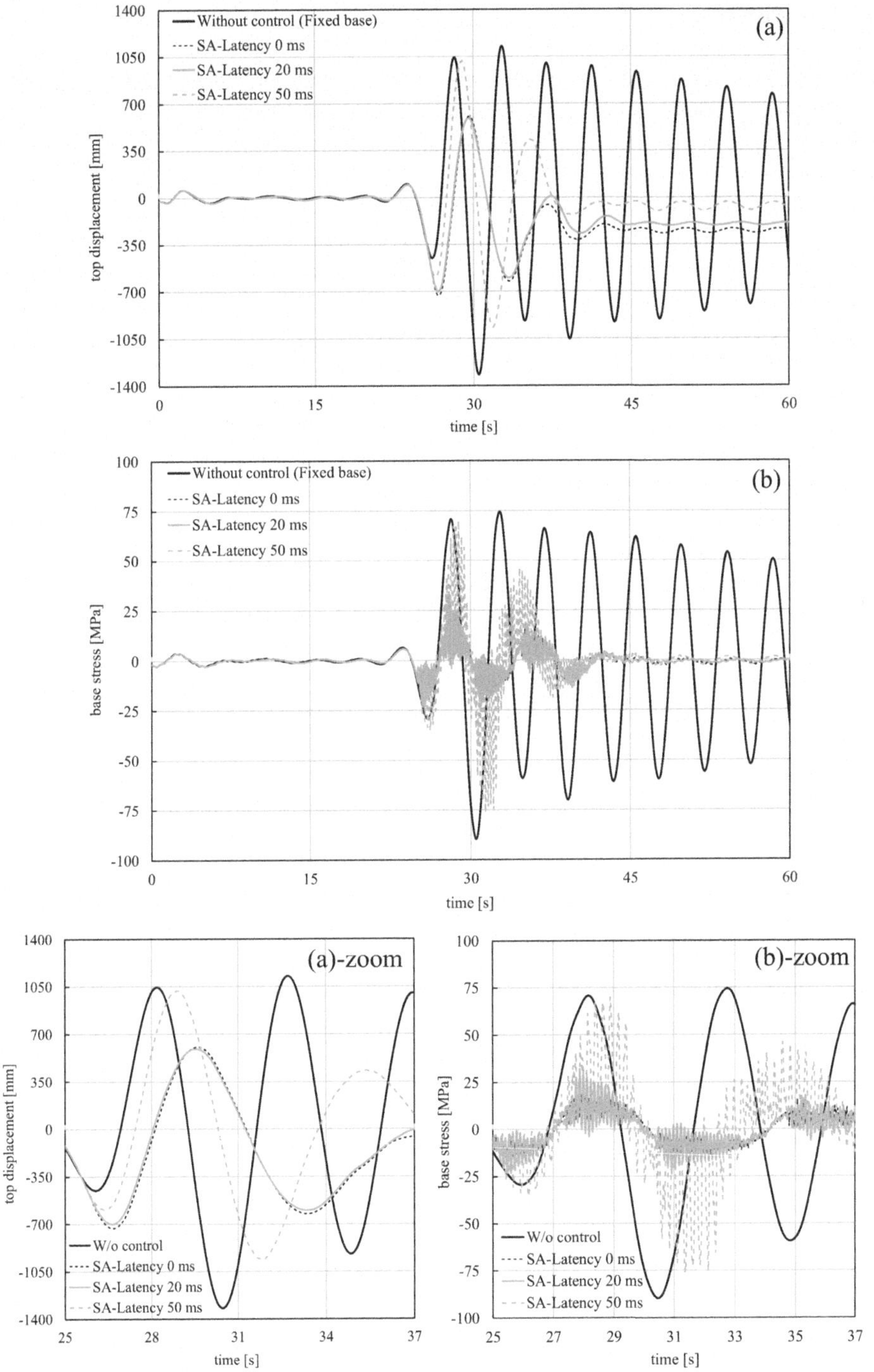

Fig. 5. Displacement (a) and stress (b) response for different latencies of measures for control

6 Conclusions

The main focus of this study was that of identifying the optimal real time monitoring system able to supply continuous data needed for the functioning of a smart wind turbine semi-active control system that the authors have proposed and experimented in recent years.

On the basis of the review of the current research and development of monitoring systems for displacements of structures under dynamic loads, RTK-GNSS measurements turned out to have the potential to be considered a suitable candidate to fulfill the aforementioned purpose. The efficiency and feasibility of such technique for the wind turbine SA control implementation have been investigated. First the main aspects have been discussed (receiver type, type of observables, sampling data rate), even comparing those of many products available on the market. In particular, to ensure that the capability of the proposed technique is matched to the required outcome of the SA control purpose, it is necessary to examine some real-world challenges, i.e. multipath error, tracking multiple GNSS constellations and data latency.

Then, numerical investigations with reference to a case-study wind turbine have been performed, highlighting that the technical characteristics of the monitoring system can significantly influence the operation of the vibration control system. In particular, measurements with a low sampling rate (<100 Hz) or with latency of the order of tens of ms can seriously reduce the effectiveness of said systems. Even the worsening can reach 84%, in the worst case. As regards the accuracy of the data provided by the GNNS system, it has been found that the system's response is not strongly influenced by it.

It is worth noting that the above results cannot be generalized, since they depends on the type of control algorithm considered, the model of turbine analysed and, not least, the type of wind load. They should rather be seen as preliminary analysis results for the feasibility study of the proposed monitoring system, starting point for the future developments that this activity can have, also - but not only - by carrying out field trials.

Acknowledgements. The research activity has been supported by the University of Naples "Parthenope" with a grant within the call "Support for Individual Research for the 2015–17 Period" issued by Rectoral Decree no. 793/2017. The above support is gratefully acknowledged.

References

Ashkenazi V, Dodson AH, Moore T, Roberts GW (1996) Real time OTF GPS monitoring of the Humber bridge. Surv World 4(4):26–28

Ashkenazi V, Roberts GW (1997) Experimental monitoring of the Humber bridge using GPS. In: Proceedings of the institution of civil engineers-civil engineering, vol 120, no Nov, pp 177–182

Caterino N (2015) Semi-active control of a wind turbine via magnetorheological dampers. J Sound Vib 345:1–17

Caterino N, Georgakis CT, Spizzuoco M, Occhiuzzi A (2016) Design and calibration of a semi-active control logic to mitigate structural vibrations in wind turbines. Smart Struct Syst 18(1):75–92

Celebi M (2000) GPS in dynamic monitoring of long-period structures. Soil Dyn Earthq Eng 20(5–8):477–483

Chen J, Georgakis CT (2013) Tuned rolling-ball dampers for vibration control in wind turbines. J Sound Vib 332:5271–5282

Chmielewski T, Górski P (2015) Full-scale investigations of civil engineering structures using GPS. In: Chang et al (eds) Advances in civil engineering and building materials IV. CRC Press, London, pp 171–175

Fuggini C (2009) Using satellites systems for structural monitoring: accuracy, uncertainty and reliability. Ph.D. thesis, Civil Engineering VIII Nuova serie (XXII Ciclo), Università di Pavia

Im SB, Hurlebaus S, Kang YJ (2013) Summary review of GPS technology for structural health monitoring. J Struct Eng 139(10):1653–1664

Kaloop M, Elbeltagi E, Hu J, Elrefai A (2017) Recent advances of structures monitoring and evaluation using GPS-time series monitoring systems: a review. ISPRS Int J Geo-Inf 6:382

Lovse JW, Teskey WF, Lachapelle G, Cannon ME (1995) Dynamic deformation monitoring of tall structure using GPS technology. J Surv Eng 121(1):35–40

Mostböck A, Petryna Y. (2014) Structural vibration monitoring of wind turbines. In: 9th international conference on structural dynamics, EURODYN 2014, Porto, Portugal

Psimoulis P, Pytharouli S, Karambalis D, Stiros S (2008) Potential of global positioning system (GPS) to measure frequencies of oscillations of engineering structures. J Sound Vib 318 (3):606–623

Pugliano G, Robustelli U, Rossi F, Santamaria R (2016) A new method for specular and diffuse pseudorange multipath error extraction using wavelet analysis. GPS Solutions 20(3):499–508

Robustelli U, Pugliano G. (2018a) GNSS code multipath short-time fourier transform analysis. Navig J Inst Navig https://doi.org/10.1002/navi.247

Robustelli U, Pugliano G (2018b) Code multipath analysis of galileo FOC satellites by time-frequency representation. Appl Geomat https://doi.org/10.1007/s12518-018-0241-3

Wang D, Meng X, Gao C, Pan S, Chen Q (2017) Multipath extraction and mitigation for bridge deformation monitoring using a single-difference model. Adv Space Res 60(2017):2882–2895

Yi T-H, Li H-N, Gu M (2013) Experimental assessment of high-rate GPS receivers for deformation monitoring of bridge. Measurement 46(2013):420–432

Yu J, Meng X, Shao X, Yan B, Yang L (2014) Identification of dynamic displacements and modal frequencies of a medium-span suspension bridge using multimode GNSS processing. Eng Struct 81(2014):432–443

Flow Distortion Recorded by Sonic Anemometers on a Long-Span Bridge

E. Cheynet[1(✉)], J. B. Jakobsen[1], and J. Snæbjörnsson[1,2]

[1] Department of Mechanical and Structural Engineering and Materials Science, University of Stavanger, Kjell Arholmsgate 41, 4036 Stavanger, Norway
etienne.cheynet@uis.no

[2] School of Science and Engineering, Reykjavík University, Menntavegur 1, 101 Reykjavík, Iceland

Abstract. Eight months of sonic anemometer records, collected above the deck of a long-span suspension bridge in complex terrain, have been used to assess possible distortion effect of the bridge girder on the monitored flow. For the two main wind directions observed on the bridge site, the velocity records from the downwind side of the deck, in particular, the vertical wind component, are systematically affected by the bridge deck. More precisely, the friction velocity and the variance of the vertical wind fluctuating component are underestimated on the downstream side of the deck. For a deck height H, both the anemometers located on the downwind side, on the same hanger, at a height of $3.6H$ and $2.2H$ show a substantial level of flow distortion. This indicates that flow distortion issues can be mitigated more efficiently using anemometers on both sides of the bridge instead of mounting them as high as possible above the deck. The influence of terrain-induced flow distortion is also found to be non-negligible, such that turbulence characteristics significantly different than observed in flat and homogeneous terrains are found.

Keywords: Full-scale · Flow distortion · Sonic anemometer · Turbulence · Suspension bridge · Complex terrain

1 Introduction

Measurements of wind turbulence above the deck of long-span cable-suspended bridges in full-scale can be traced back to the 1970s (Jensen and Hjort-Hansen 1977; Kristensen and Jensen 1979). The data gathered this way can be used to model the dynamic wind load acting on the deck (Hay 1984; Bietry et al. 1995; Toriumi et al. 2000; Miyata et al. 2002; Macdonald 2003; Xu and Zhu 2005), as well as to conveniently study the horizontal structure of wind turbulence without deploying multiple met-masts.

However, the possible distortion of the flow due to the presence of the bridge deck has rarely been studied in details. Jensen and Hjort-Hansen (1977) noted that for the case of the Sotra suspension bridge (Norway), which has a truss

F. Ricciardelli and A. M. Avossa (Eds.): INVENTO 2018, LNCE 27, pp. 192–206, 2019.
https://doi.org/10.1007/978-3-030-12815-9_16

girder, the wind measurements conducted 3 m above the upwind side of the deck were influenced by the bridge structure. This influence was, however, assumed small enough to be negligible for the purpose of their study. They found an average incidence angle of 7° and the turbulence statistics were thus studied in the streamline coordinate system, i.e. in a coordinate system where the vertical mean wind velocity is as close as possible to 0 ms^{-1}. Hay (1981, 1984) estimated the power spectral density (PSD) of the along-wind and vertical wind components in two positions above the Erskine Bridge (UK), which has a 31 m-wide and 3.2 m-high trapezoidal deck. The first position was on a 7 m-high boom on the upwind side of the deck, whereas the second one was on a 9 m-high lamp near the middle part of the deck. For the latter sensor, Hay (1984) found that the vertical wind spectrum was clearly disturbed, whereas the along-wind spectrum was much less affected by the presence of the girder. Frandsen (2001) used finite element flow simulations to access the expected flow disturbances by the Great Belt East suspension bridge deck. The analysis, considering the deck without the railings, indicated that the cup anemometer installed 2:3 m above the upwind side of the deck was expected to be located in the undisturbed mean free-stream velocity region. On the downwind side, a height of 2 m was considered sufficient.

The aforementioned observations underline the complex influence of the deck geometry on the flow measured by anemometers. They also reflect the need for improved understanding of flow distortion by a bridge deck in full-scale. In the present study, the bridge girder influence on the flow across the deck is investigated in the case of the Lysefjord suspension bridge (Norway), which has been used since 2013 as a full-scale laboratory to study wind effects on a long-span bridge in complex terrain (Snæbjörnsson et al. 2017). Since July 2017, the flow across the bridge has been recorded by 9 sonic anemometers installed on both the upwind and downwind side of the deck. By studying the differences in the recorded velocity components, it is possible to analyse the flow disturbance caused by the girder. The study is based on eight months of full-scale wind records from July 2017 through February 2018.

The present paper is organized as follows: In Sect. 2, the bridge instrumentation is presented as well as the data post-processing. Section 3 compares the one-point wind characteristics recorded by the sonic anemometers on both the upwind and downwind side of the girder. Section 4 provides some guidance regarding the installation of sonic anemometers on a cable-supported bridge with a streamlined hexagonal deck.

2 Instrumentation and Methods

2.1 Location of the Sonic Anemometers

The Lysefjord Bridge is a long-span suspension bridge with a main span of 446 m, oriented with an azimuth of −42°, located at the inlet of a narrow and long fjord. The bridge is surrounded by mountains and steep hills (Fig. 1), channelling the flow, such that two main wind directions are usually observed. The first one corresponds to a flow from north-northeast, i.e. from the inside of the fjord,

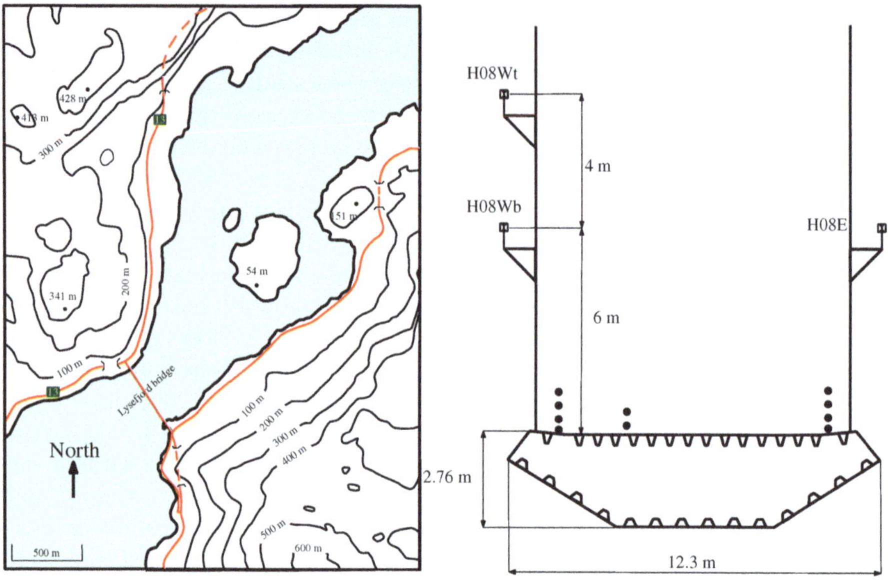

Fig. 1. Left: topographic map of the Lysefjord Bridge. Right: Sketch of the cross-section of the Lysefjord bridge deck on hanger 08.

at azimuth between 0° and 60°. The second one corresponds to a flow from south-southwest, i.e. a wind direction between 180° and 270°.

The Lysefjord Bridge has been instrumented with a wind and structural health monitoring system since November 2013 (Snæbjörnsson et al. 2017). In June 2017, the arrangement of the wind sensors was modified and three sonic anemometers were installed on the east side of the bridge. In Fig. 2, the position of the anemometers above the deck is defined using the hanger name HXY, where X is a digit between 08 and 24 indicating the hanger number, and Y denotes the west side (W) or east side (E) of the deck. Since two anemometers are mounted on the hanger west no. 08 (H08W), the notations H08Wb and H08Wt refer, in the following, to the sonic anemometer mounted 6 m (bottom) and 10 m (top) above the deck, respectively. Eight of the sonic anemometers are 3-D WindMaster Pro from Gill instruments (Lymington, UK), which can record the three wind velocity components and the sonic temperature with a sampling frequency up to 32 Hz. The last sonic anemometer, mounted on H10W, is a Weather Transmitter WXT520 from Vaisala (Helsinki, Finland), which monitors the horizontal wind components, relative humidity, pressure and absolute temperature with a sampling frequency up to 4 Hz. Note that the bridge instrumentation includes also accelerometers and Global navigation satellite displacement sensors, but these are not used in the present study.

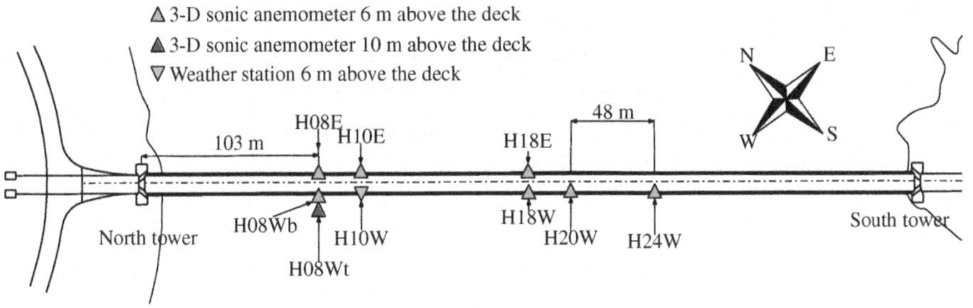

Fig. 2. Instrumentation of the Lysefjord Bridge since June 2017. The distance between each hanger is 12 m.

2.2 Data Processing

The along-wind, crosswind and vertical wind components are denoted u, v and w (positive z-axis), respectively. Each component is expressed as the sum of a mean component, denoted by an overline and a fluctuating component with zero mean denoted by a prime, defined as a Gaussian stationary random process, such that for $j = \{u, v, w\}$, one can write $j = \overline{j} + j'$.

The sonic anemometers deployed on the Lysefjord Bridge continuously monitor the flow and the amount of data gathered is, therefore, enormous. For this study, the data processing of the sonic anemometer records has been done as follows: (1) Wind records are de-spiked and decimated down to a frequency of 10 Hz to facilitate data handling. (2) Unphysical signals (3% of the data set), characterized by an absolute value of the skewness above 2 or a kurtosis above 8 are dismissed, as advised by Vickers and Mahrt (1997) and Stiperski and Rotach (2016). (3) The underestimation of the vertical wind component recorded by the sonic anemometers is corrected. Two correction algorithms have been evaluated, the first one is proposed by Nakai and Shimoyama (2012) and the second is provided by Gill Instrument. The selection criterion relies here solely on the ability of the correction to provide a ratio of the vertical wind spectrum over the along-wind spectrum that converges toward the theoretical value of 1.33 in the inertial subrange. Based on this criterion, the simple correction provided by Gill Instrument was chosen. Note that the anemometers located on H08E and H10E, which are more recent, do not need such a correction. (4) The coordinate system is rotated such that $\overline{v} \approx \overline{w} \approx 0\ \mathrm{ms}^{-1}$. The study in such a coordinate system can be done using the planar-fit (PF) algorithm (Paw et al. 2000; Wilczak et al. 2001), the double rotation method or triple rotation method (McMillen 1988). In the present case, the high complexity of the terrain justifies the application of the double rotation technique rather than a sectoral PF algorithm. (5) To select only stationary samples, two stationarity tests are combined and applied to the records of the along-wind component. The first one assesses the slope of the linear trend, such that the time series is considered as non-stationary if a difference between the extrema of the linear trend and its arithmetic mean is larger than

30%. If the sample passes the first step, the non-parametric test described in Bendat and Piersol (2011), also called reverse arrangement (RA) test, is applied with a 95% confidence level. Note that prior to the application of this test, each time series has been decimated down to a frequency of 0.1 Hz to ensure that the data points can be considered as statistically independent records. This assumption requires a time step larger than the integral time scale of turbulence T_u^x. The latter is assumed here to fulfil $T_u^x \leq 10$ s, which is supported by the integral time scale estimate from the EN 1991-1-4 (2005), which is lower than 6 s at a height of 60 m for a terrain category 2 to 4. (6) The measurement uncertainties in the variance and covariance of the wind velocity records are assessed as by Stiperski and Rotach (2016). For the wind velocity component $j = \{u, v, w\}$, the measurement uncertainty a_j for the variance is defined as:

$$a_{jj}^2 = \frac{4z}{\tau \overline{u}(z)} \left[\kappa_j - 1\right] \tag{1}$$

where τ is the averaging time and κ_j is the kurtosis of the velocity component j. Similarly, the measurement uncertainty for the covariance $\overline{u'w'}$ and $\overline{v'w'}$ are defined as:

$$a_{uw}^2 = \frac{z}{\tau \overline{u}(z)} \left[\frac{\overline{(u'w')^2}}{u_*^4} - 1 \right] \tag{2}$$

$$a_{vw}^2 = \frac{z}{\tau \overline{u}(z)} \left[\frac{\overline{(v'w')^2}}{u_*^4} - 1 \right] \tag{3}$$

where u_* is the friction velocity, calculated as advised by Weber (1999):

$$u_* = \left(\overline{u'w'}^2 + \overline{v'w'}^2 \right)^{1/4} \tag{4}$$

The averaging time τ is here equal to 20 min. Although the measurement uncertainties are reduced when $\tau = 60$ min is used, the number of non-stationary records increases, thereby reducing the amount of suitable data for the analysis. Therefore, a duration of 20 min was found better suited to collect a larger number of high-quality data. For that averaging time, the median value of a_{ii}, $i, j = \{u, v, w\}$, is ca. 25% whereas the median value for a_{uw} and a_{vw} is around 32%. Samples associated with a measurement uncertainty above 50% are disregarded (Stiperski and Rotach 2016), which accounts for ca. 14% of samples with a mean wind velocity above 6 ms^{-1}. On the Lysefjord Bridge, the traffic flow is low (less than one vehicle per minute on average), indicating that flow-distortion induced by the presence of moving vehicles upstream of the anemometers is negligible in the present case.

Turbulence characteristics are studied after removing the mean and any linear trend from the approved time series. The atmospheric stability is estimated using the non-dimensional Obukhov length ζ, local scaling (Nieuwstadt 1984; Sorbjan 1986) and assuming that the virtual potential temperature is well approximated by the sonic temperature:

$$\zeta = \frac{-g\kappa z \overline{w'\theta_v'}}{\overline{\theta_v} u_*^3} \tag{5}$$

where $g = 9.81\ \mathrm{ms}^{-2}$ is the acceleration of gravity; $\overline{\theta_v}$ is the average virtual potential temperature; $\overline{w'\theta_v'}$ is the flux of virtual potential temperature and $\kappa = 0.40$ is the von Kármán constant. Only samples characterized by $|\zeta| < 0.1$, which represents near neutral conditions, are considered in the following.

In addition to the one-point wind spectra and cross-spectra, the turbulence intensity I_j, where $j = \{u, v, w\}$, is studied at wind velocities large enough so that I_j depends less on the topography than at low wind velocities. We recall that I_j is defined as:

$$I_j = \frac{\sigma_j}{\overline{u}} \tag{6}$$

where σ_j is the standard deviation of the wind component j.

The PSDs of the velocity fluctuations are estimated using the periodogram method with a Hamming window and an averaging time of 20 min, allowing the study of the wind spectra down to a frequency of 0.8 mHz. The periodogram method is generally associated with a large random error, which is reduced by block-averaging each individual spectra into 60 bins equally spaced on a logarithmic scale. Then, each PSD estimate is normalized and ensemble averaged.

3 Results

Figure 3 displays the mean wind velocity and the along-wind turbulence intensity estimated on the Lysefjord Bridge from July 2017 to February 2018. The high turbulence intensity recorded, previously mentioned in Cheynet et al. (2016), is consistent with measurements from other Norwegian fjords (Harstveit 1996). It suggests that mechanically-generated turbulence dominates the buoyancy-generated turbulence. For the period considered, this is also indicated by the stability parameter ζ, for which 90% of the values are bounded between -0.06 and 0.05 for a wind from the north-northeast and between -0.02 and 0.12 for a wind from the south-southwest. The different sonic anemometers give values of ζ that are similar in average, but those corrected for the underestimation of the vertical wind component provides estimates with a considerable scatter, leading to a significant reduction of the data availability. For this reason, the value of ζ, used to select samples associated with near-neutral conditions, is estimated from the sonic anemometer on H10E only.

Whereas the flow from north-northeast is generally more turbulent than the one from south-southwest, the largest velocities are recorded for the latter sector.

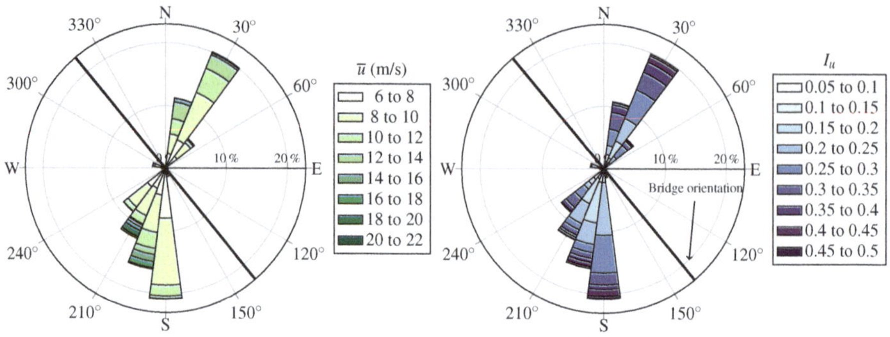

Fig. 3. Mean wind velocity (left) and along-wind turbulence intensity (right) recorded from every anemometer the Lysefjord Bridge between July 2017 and February 2018 (1.6×10^3 stationary records of 20 min duration with $\overline{u} \geq 6$ ms^{-1}).

Nevertheless, the flow from south-southwest is associated with a turbulence intensity above 0.20 when the wind direction is between 180° and 190°, for which the flow brushes past the west face of the mountain on the south side of the bridge.

3.1 Mean Flow Characteristics

A relative difference between the mean wind velocity estimated from a reference sonic anemometer and the others, denoted ϵ_u, is investigated in Table 1. The reference sensor is the one located on H18W or H18E for a wind from south-southwest or north-northeast, respectively. Note that the discrepancies due to the slightly different measurement heights are not corrected for as the wind shear is assumed small.

For a flow from south-southwest, Table 1 shows that, on average, the absolute relative difference is lower than 3%. For such a wind direction, it indicates that the bridge deck has a limited influence on the mean flow recorded on the downwind side of the bridge. The larger discrepancies observed between the mean velocity measurements on H18W and H24W may be due to terrain-induced flow inhomogeneity. This was previously suggested in Cheynet et al. (2017), where lidar measurements indicated the existence of a non-uniform mean flow on the south side of the bridge.

For a flow from north-northeast, the relative difference ϵ_u indicates a stronger flow inhomogeneity than for a flow from south-southwest. Values of ϵ_u larger than 7% on H08Wb and H08Wt cannot be simply due to effect of the terrain on the flow since, for example, the mean velocity estimated on H08E is almost identical to the one estimated on H18E. Different flow characteristic from the two wind sectors may be responsible for different levels of flow distortion by the bridge deck. Note that the large relative difference observed on H10W for a flow from north-northeast, where the weather station is located, may be due to a difference in sensor measurement accuracy under highly turbulent wind conditions.

The mean wind incidence angle $\overline{\alpha}$ recorded for the two wind directions considered are displayed in Table 1 for $\overline{u} \geq 6$ ms^{-1}. Whereas the south-southwest flow is characterized by $\overline{\alpha} \geq 3°$ in average on the upwind side, the sensors located on the downwind side of the deck monitor a lower incidence angle, reflecting a possible influence of the bridge girder on the vertical mean velocity. If a sonic anemometer is not perfectly levelled, a non-zero mean incidence angle is measured. For several anemometers monitoring a horizontal flow, this can lead to incidence angles that have various values as the levelling is different for each sensor. For a flow from north-northeast, which is more horizontal, such an imperfect levelling may explain the more heterogeneous values of the incidence angle recorded along the bridge deck.

Table 1. Average relative percentage wise difference ϵ_u in the recorded mean wind velocity and mean incidence angle α along the deck of the Lysefjord bridge for samples with $\overline{u} \geq 6$ ms^{-1} and a 20 min averaging time (599 samples for a flow NNE and 779 samples for a flow from SSW).

Sensors		Wind direction			
West side	East side	South-southwest		North-northeast	
		ϵ_u (%)	$\overline{\alpha}$ (°)	ϵ_u (%)	$\overline{\alpha}$ (°)
H08Wb	-	0.19	3.9	10	−0.39
H08Wt	-	0.08	3.3	7.7	−1.1
H10W	-	0.13	–	12	–
H18W	-	0	3.6	1.5	0.35
H20W	-	0	3.1	−0.31	1
H24W	-	−2.7	2.5	−2.9	0.59
-	H08E	1.7	1.6	3	0.56
-	H10E	1.2	0.82	1.4	1
-	H18E	3	1.7	0	3.9

3.2 One-Point Turbulence Characteristics

The flow from south-southwest is particularly complex to study as the turbulence intensity depends strongly on the wind direction. To study consistent turbulence characteristics on the bridge site, it is important that the turbulence intensity is more or less constant with the mean wind velocity and the wind sector selected. As the turbulence intensity was seen to stabilize as the mean wind approached 10 ms^{-1}, it was decided to consider only samples characterized by $\overline{u} \geq 10$ ms^{-1} in the present subsection, although it reduces considerably the number of samples available.

The one-point turbulence characteristics computed are displayed separately for a wind direction from south-southwest and north-northeast (Tables 2 and 3). Tables 2 and 3 shows that the discrepancies between the values estimated on the upwind and downwind side are particularly important for the vertical wind component whereas they are not clearly defined for the horizontal components.

Table 2. Turbulence characteristics estimated from every sonic anemometer on the Lysefjord Bridge, under near-neutral conditions, $\overline{u} > 10$ ms^{-1} for a wind from south-southwest (SSW) with 54 samples and north-northeast (NNE) with 120 samples.

Sector	Sensor location: West side	Sensor location: East side	$\frac{\sigma_w}{u_*}$	$\frac{\sigma_v}{u_*}$	$\frac{\sigma_u}{u_*}$	$\frac{\sigma_w}{\sigma_u}$	$\frac{\sigma_v}{\sigma_u}$
SSW	H08Wb	-	1.69 ± 0.54	2.82 ± 0.88	2.41 ± 0.91	0.72 ± 0.11	1.04 ± 0.17
	H08Wt	-	1.62 ± 0.30	2.45 ± 0.5	2.23 ± 0.58	0.70 ± 0.09	1.03 ± 0.16
	H10W	-	–	–	–	–	1.09 ± 0.18
	H18W	-	1.64 ± 0.44	2.26 ± 0.91	2.37 ± 0.78	0.74 ± 0.08	1.02 ± 0.06
	H20W	-	1.63 ± 0.27	1.90 ± 0.65	1.90 ± 0.41	0.71 ± 0.09	1.03 ± 0.09
	H24W	-	1.67 ± 0.48	2.26 ± 0.94	2.49 ± 0.64	0.73 ± 0.09	1.03 ± 0.16
	-	H10E	1.40 ± 0.18	3.03 ± 0.48	2.86 ± 0.47	0.48 ± 0.07	1.03 ± 0.16
	-	H08E	1.60 ± 0.18	3.11 ± 0.33	3.28 ± 0.53	0.47 ± 0.07	0.96 ± 0.13
	-	H18E	1.25 ± 0.14	2.46 ± 0.93	2.52 ± 0.13	0.50 ± 0.04	0.98 ± 0.32
NNE	H08Wb	-	1.36 ± 0.22	2.20 ± 0.42	2.77 ± 0.47	0.50 ± 0.05	0.81 ± 0.08
	H08Wt	-	1.44 ± 0.25	2.00 ± 0.41	2.47 ± 0.45	0.58 ± 0.05	0.80 ± 0.08
	H10W	-	–	–	–	–	0.79 ± 0.10
	H18W	-	1.55 ± 0.40	2.70 ± 0.79	3.22 ± 0.88	0.47 ± 0.06	0.85 ± 0.09
	H20W	-	1.60 ± 0.53	2.75 ± 1.10	3.20 ± 1.10	0.50 ± 0.05	0.86 ± 0.09
	H24W	-	1.55 ± 0.35	2.77 ± 0.67	3.30 ± 0.94	0.45 ± 0.05	0.80 ± 0.08
	-	H08E	1.38 ± 0.20	1.68 ± 0.34	1.98 ± 0.33	0.70 ± 0.07	0.86 ± 0.08
	-	H10E	1.38 ± 0.20	1.70 ± 0.33	1.94 ± 0.30	0.71 ± 0.07	0.87 ± 0.09
	-	H18E	1.50 ± 0.48	1.82 ± 0.56	2.18 ± 0.61	0.69 ± 0.07	0.84 ± 0.09

For a south-southwest flow, the ratio σ_w/u_* is slightly lower on the downwind side than on the upwind side, which must be interpreted with caution as the value generally found in the literature in flat terrain ranges from 1.2 to 1.3 (De Franceschi et al. 2009), whereas it corresponds here to a flow likely distorted by the presence of the deck. Jensen and Hjort-Hansen (1977) found a ratio σ_w/u_* between 1.4 and 1.6 on the upwind side of the Sotra Bridge (truss girder), which is consistent with the measurements conducted on the upwind side of the deck for the two wind sectors considered.

The ratio σ_w/σ_u estimated on the upwind side of the bridge is above 0.6, which is larger than the value $\sigma_w/\sigma_u \approx 0.5$, usually found for a flat terrain (Solari and Piccardo 2001). This means also that in mountainous terrain, a ratio σ_w/σ_u estimated from the deck of a suspension bridge that displays a value close to or below 0.5 may indicate flow distortion by the girder. On the other hand, only small differences are found for I_u and σ_v/σ_u between the downwind and upwind sides of the bridge deck, indicating that the horizontal components are not significantly affected by flow distortion from the girder. The unusually large ratio $\sigma_v/\sigma_u \approx 1$

Table 3. Same as Table 2, but for the turbulence intensity.

Sector	Sensor location		I_u	I_v	I_w
	West side	East side			
SSW	H08Wb	-	0.15 ± 0.05	0.16 ± 0.16	0.11 ± 0.04
	H08Wt	-	0.15 ± 0.05	0.16 ± 0.16	0.10 ± 0.04
	H10W	-	0.14 ± 0.05	0.16 ± 0.16	–
	H18W	-	0.20 ± 0.05	0.19 ± 0.19	0.14 ± 0.05
	H20W	-	0.27 ± 0.09	0.27 ± 0.27	0.22 ± 0.07
	H24W	-	0.21 ± 0.06	0.17 ± 0.17	0.13 ± 0.05
	-	H08E	0.18 ± 0.05	0.16 ± 0.16	0.08 ± 0.03
	-	H10E	0.15 ± 0.05	0.16 ± 0.16	0.07 ± 0.02
	-	H18E	0.16 ± 0.08	0.19 ± 0.19	0.08 ± 0.04
NNE	H08Wb	-	0.23 ± 0.05	0.19 ± 0.19	0.12 ± 0.02
	H08Wt	-	0.24 ± 0.06	0.20 ± 0.20	0.15 ± 0.03
	H10W	-	0.24 ± 0.06	0.19 ± 0.19	–
	H18W	-	0.24 ± 0.06	0.19 ± 0.19	0.11 ± 0.03
	H20W	-	0.24 ± 0.06	0.20 ± 0.20	0.12 ± 0.03
	H24W	-	0.23 ± 0.07	0.18 ± 0.18	0.10 ± 0.03
	-	H08E	0.25 ± 0.06	0.22 ± 0.22	0.18 ± 0.04
	-	H10E	0.26 ± 0.07	0.22 ± 0.22	0.18 ± 0.04
	-	H18E	0.25 ± 0.06	0.21 ± 0.21	0.17 ± 0.03

obtained for a wind from south-southwest may, therefore, be attributed to terrain-induced flow distortion. For a wind from north-northeast, a more common value $\sigma_v/\sigma_u \approx 0.8$ is found, although it is still larger than for the Sotra Bridge experiment (Kristensen and Jensen 1979), which was slightly lower than 0.7. The considerable scatter obtained for σ_v/u_* and σ_u/u_*, for both wind direction, may also be attributed to local terrain effects rather than the influence of the bridge structure. Since Table 3 shows that the presence of the deck does not clearly affect I_u, the larger values of σ_u/u_* on the downwind side of the deck are likely due to the underestimation of u_* due to flow distortion by the girder.

For a turbulent flow past a flat plate, Hunt and Graham (1978) predicted an increase of the variance of the horizontal component normal to the plate and a decrease of the variance of the vertical component downwind of the leading edge. In the present case, only the latter prediction is clearly observed, which may be due to the bridge deck shape, which is a bluff body and not a flat plate. More generally, wind records collected on the downwind side of a bridge deck may be characterized by an underestimation of the vertical wind component, and thus the underestimation of the vertical and torsional bridge buffeting response.

3.3 Spectral Ratios

The ensemble-averaged ratio of the PSD estimates of the vertical wind component over the along-wind component is displayed in Fig. 4 for a flow from south-southwest and in Fig. 5 for a flow from the north-northeast. These ratios are expressed as a function of the reduced frequency $f_r = fz/\overline{u}$. As predicted by the theory of local isotropy (Kolmogorov 1941), S_w/S_u should reach a value of 1.33 in the inertial subrange, whereas a lower value is expected if the flow is distorted by the bridge deck. On the Sotra bridge (truss girder), Jensen and Hjort-Hansen (1977) obtained a ratio S_w/S_u converging adequately toward 1.33 at $f_r > 2$ using sonic anemometers mounted 3 m on the upwind side of the deck, i.e. at a height of 60 m above the sea level, indicating undistorted flow conditions.

Figures 4 and 5 show that the ratio S_w/S_u obtained using the anemometers located on the downwind side of the bridge is significantly lower than upwind over the entire range of reduced frequencies. On the other hand, if the anemometers on the upwind side are used, the ratio S_w/S_u converges toward a value ranging

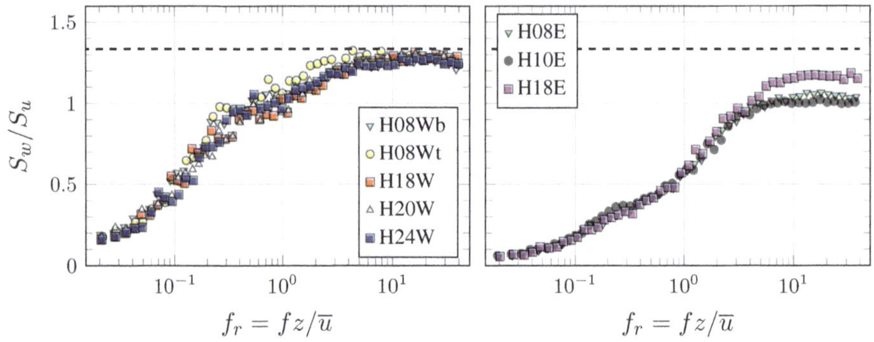

Fig. 4. Ratio of the PSD estimates of the vertical wind component over the along-wind component for a flow from south-southwest, near-neutral conditions and $\overline{u} > 6\ \mathrm{ms}^{-1}$ at the deck height (415 samples). The dashed line shows the theoretical ratio of 1.33.

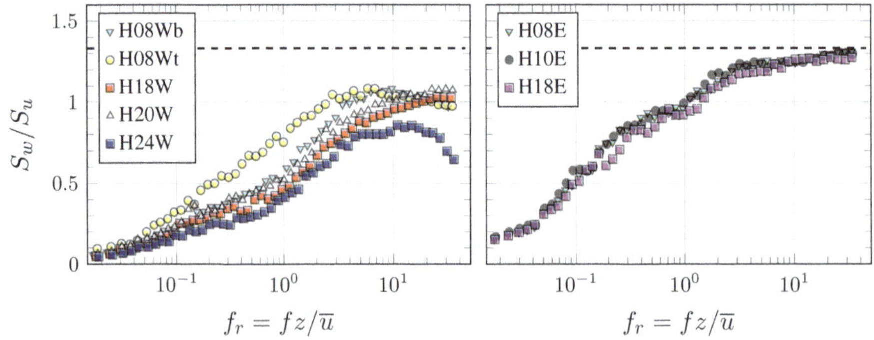

Fig. 5. Same as in Fig. 4 but for a wind from north-northeast (433 samples).

from 1.2 to 1.3 at $f_r > 5$, attesting that flow distortion is significant on the downwind side of the girder only. On the upwind side, the slight discrepancy between the estimated and theoretical values of the ratio S_w/S_u might indicate minor flow distortion from either the girder or the terrain upstream.

In addition, Fig. 5 shows clearly that flow distortion is not uniform along the girder. For a flow from north-northeast, the anemometer mounted on the southern part of the bridge records the strongest distortion with a ratio $S_w/S_u \approx 0.8$ at $f_r \approx 10$, which for $z = 60$ m and $\overline{u} = 10$ ms^{-1} corresponds to $f = 1.7$ Hz. For a flow from south-southwest, the three anemometers located on the East side of the bridge show a similar level of flow distortion, except at $f_r > 4$ where the ratio S_w/S_u is smaller on the northern side of the bridge. The non-uniformity of the flow distortion along the span may be due to the large yaw angles recorded, such that the anemometers installed furthest from the initial separation point are located in a more developed turbulent layer. Another contributing factor may also be the asymmetric layout of the bridge girder with respect to its midspan, associated with the 7.5 m altitude difference between the north and south end.

4 Influence of the Sensor Height on Turbulence Characterization

For a north-northeastern flow, the results presented in Sect. 3 indicate that on the downstream side of the girder, even the sonic anemometer mounted 10 m above the deck (H08Wt) is affected by bridge-induced flow distortion. Figure 6 compares the normalized PSD estimates of the vertical wind velocity component on hanger 8, on both sides of the deck. On the downwind side, a flat peak is observed, as by Hay (1984), and demonstrates the influence of the bridge girder on the vertical wind component. For a flow from north-northeast (left panel of Fig. 6), the two anemometers on the downwind side are those located

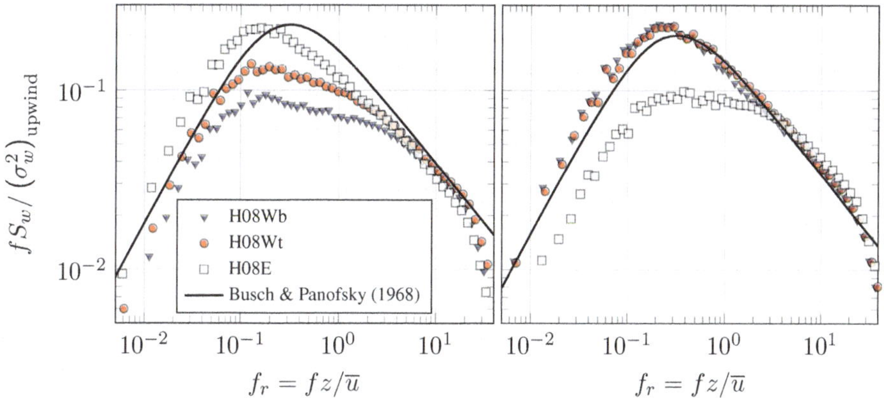

Fig. 6. Vertical wind spectra estimated on hangers 08 for a flow from north-northeast (left, 436 samples) and south-southwest (right, 415 samples), where σ_w is estimated on the upwind side only.

a height of 10 m (H08Wt) and 6 m (H08Wb) above the road. Their records show a reduced flow distortion at $f_r > 3$ only, which for $\overline{u} = 10\ \mathrm{ms}^{-1}$ and $z = 60$ m corresponds to $f = 0.5$ Hz, i.e. above the lowest eigen-frequencies of the Lysefjord Bridge. At lower frequencies, the PSD estimates on H08Wt and H08Wb show significant discrepancies with the one estimated from H08E, assumed representative of undisturbed flow conditions.

A viable strategy to avoid deck-induced flow distortion may, therefore, rely on the use of anemometers on both sides of the girder. It is still recommended to mount anemometers at heights corresponding to several deck height above the road. If they are installed on horizontal booms at the deck height, flow distortion is also expected. Such an instrumentation was applied for a streamlined closed-box girder in e.g. Xu and Zhu (2005), for which horizontal booms with a length of 9 m were mounted perpendicularly to a 41 m-wide deck. Using their fitted spectral model, the computed ratio S_w/S_u converges toward 0.7 in the inertial subrange, compared to 1.33 following local isotropy. The results displayed in Sect. 3 corroborate also those from Hay (1984), which showed that mounting a sonic anemometer at the centre of a deck (height $H = 3.2$ m and width $B = 31$ m), even at a height of several meters above the road, should be avoided as flow distortion from every direction can be expected.

The Lysefjord Bridge instrumentation aims primarily to study the turbulence characteristics along the bridge. It is unclear whether the coherence of the vertical wind component is significantly affected by flow distortion from the deck because the normalization of the cross-spectral density may attenuate the flow distortion effects. This question will be investigated in a further study.

5 Conclusions

Eight months of wind velocity data collected by sonic anemometers mounted above the deck of a suspension bridge crossing the inlet of a fjord in southwestern Norway have been investigated. The goal is to better understand to what degree the presence of the girder distorts the flow recorded by the anemometers. The data collected on the downstream side of the deck indicates that the vertical velocity component w is underestimated. Such a bias can lead to an underestimation of the vertical and torsional dynamic response of a long-span bridge to turbulent wind load.

On the other hand, the two horizontal wind components u and v are much more affected by the terrain than by the girder. Consequently, the turbulence characteristics established for both the flow from south-southwest and north-northeast can greatly deviate from the case of flat and homogeneous terrain. For the two main wind directions considered, the flow is not affected the same way by the topography, which reinforces the idea that turbulence characteristics observed on a bridge crossing a fjord or a canyon should be studied separately for different sectors.

For a hexagonal girder with a height H and width B such that $B/H \approx 4.5$, mounting a sonic anemometer at a height equal to $3.6H$ above the deck instead of $2.2H$ shows only minor improvements in flow distortion mitigation. To study turbulence characteristics from a long-span bridge, it is thus advised to mount sonic anemometers on both sides of the girder. Such an installation is particularly important in a mountainous area where the characterization of the wind conditions needs to be conducted after separation of the records in different wind sectors, where some of the anemometers will inevitably be located on the downstream side of the deck.

Acknowledgements. The support of the Norwegian Public Roads Administration is gratefully acknowledged, as well as their assistance during the installation and maintenance of the monitoring system.

References

Bendat J, Piersol A (2011) Random data: analysis and measurement procedures. Wiley series in probability and statistics. Wiley

Bietry J, Delaunay D, Conti E (1995) Comparison of full-scale measurement and computation of wind effects on a cable-stayed bridge. J Wind Eng Ind Aerodyn 57(2):225–235

Cheynet E, Jakobsen JB, Snæbjörnsson J (2016) Buffeting response of a suspension bridge in complex terrain. Eng Struct 128:474–487

Cheynet E, Jakobsen JB, Snæbjörnsson J, Reuder J, Kumer V, Svardal B (2017) Assessing the potential of a commercial pulsed lidar for wind characterisation at a bridge site. J Wind Eng Ind Aerodyn 161:17–26

De Franceschi M, Zardi D, Tagliazucca M, Tampieri F (2009) Analysis of second-order moments in surface layer turbulence in an Alpine valley. Q J R Meteorol Soc 135(644):1750–1765

EN 1991-1-4 (2005) Eurocode 1: Actions on structures–part1-4: General actions-wind actions

Frandsen J (2001) Simultaneous pressures and accelerations measured full-scale on the Great Belt East suspension bridge. J Wind Eng Ind Aerodyn 89(1):95–129

Harstveit K (1996) Full scale measurements of gust factors and turbulence intensity, and their relations in hilly terrain. J Wind Eng Ind Aerodyn 61(2–3):195–205

Hay JS (1981) The wind induced response of the Erskine Bridge. ICE Publishing, pp 81–90

Hay JS (1984) Analyses of wind and response data from the Wye and Erskine Bridges and comparison with theory. J Wind Eng Ind Aerodyn 17(1):31–49

Hunt J, Graham J (1978) Free-stream turbulence near plane boundaries. J Fluid Mech 84(2):209–235

Jensen N, Hjort-Hansen E (1977) Dynamic excitation of structures by wind-turbulence and response measurements at the Sotra Bridge. Technical report, Report No. STF71 A., SINTEF

Kolmogorov AN (1941) The local structure of turbulence in incompressible viscous fluid for very large Reynolds numbers, vol 30. Doklady Akademiia Nauk SSSR, pp 299–303

Kristensen L, Jensen NO (1979) Lateral coherence in isotropic turbulence and in the natural wind. Bound-Layer Meteorol 17(3):353–373

Macdonald JH (2003) Evaluation of buffeting predictions of a cable-stayed bridge from full-scale measurements. J Wind Eng Ind Aerodyn 91(12):1465–1483

McMillen RT (1988) An eddy correlation technique with extended applicability to non-simple terrain. Bound-Layer Meteorol 43(3):231–245

Miyata T, Yamada H, Katsuchi H, Kitagawa M (2002) Full-scale measurement of Akashi-Kaikyō Bridge during typhoon. J Wind Eng Ind Aerodyn 90(12):1517–1527

Nakai T, Shimoyama K (2012) Ultrasonic anemometer angle of attack errors under turbulent conditions. Agric For Meteorol 162:14–26

Nieuwstadt FT (1984) The turbulent structure of the stable, nocturnal boundary layer. J Atmos Sci 41(14):2202–2216

Paw UKT, Baldocchi DD, Meyers TP, Wilson KB (2000) Correction of eddy-covariance measurements incorporating both advective effects and density fluxes. Bound-Layer Meteorol 97(3):487–511

Snæbjörnsson JT, Jakobsen JB, Cheynet E, Wang J (2017) Full-scale monitoring of wind and suspension bridge response. IOP Conf Ser Mater Sci Eng 276(1):012,007

Solari G, Piccardo G (2001) Probabilistic 3-D turbulence modeling for gust buffeting of structures. Probab Eng Mech 16(1):73–86

Sorbjan Z (1986) On similarity in the atmospheric boundary layer. Bound-Layer Meteorol 34(4):377–397

Stiperski I, Rotach MW (2016) On the measurement of turbulence over complex mountainous terrain. Bound-Layer Meteorol 159(1):97–121

Toriumi R, Katsuchi H, Furuya N (2000) A study on spatial correlation of natural wind. J Wind Eng Ind Aerodyn 87(2):203–216

Vickers D, Mahrt L (1997) Quality control and flux sampling problems for tower and aircraft data. J Atmos Oceanic Technol 14(3):512–526

Weber R (1999) Remarks on the definition and estimation of friction velocity. Bound-Layer Meteorol 93(2):197–209

Wilczak JM, Oncley SP, Stage SA (2001) Sonic anemometer tilt correction algorithms. Bound-Layer Meteorol 99(1):127–150

Xu Y, Zhu L (2005) Buffeting response of long-span cable-supported bridges under skew winds. Part 2: case study. J Sound Vib 281(3):675–697

Influence of the Measurement Height on the Vertical Coherence of Natural Wind

E. Cheynet(✉)

Department of Mechanical and Structural Engineering and Materials Science,
University of Stavanger, Stavanger, Norway
etienne.cheynet@uis.no

Abstract. Two years of sonic anemometer records, collected on the offshore platform FINO1 in the North Sea are used to study the vertical coherence of the along-wind and vertical wind components under near-neutral conditions. The goal is to assess the influence of the measurement height on the coherence estimates. For the data set considered, a 3-parameter coherence model, which depends explicitly on the measurement height and accounts for the limited dimensions of the eddies, is found to be more appropriate than the Davenport model or the uniform shear model to describe the vertical coherence. This is partly because the latter two models do not take into account the blockage effect by the sea surface. The computation of the joint acceptance function of a line-like vertical structure with the Davenport model and the 3-parameter coherence model suggests that the use of the latter model may substantially improve the design of high-rise wind-sensitive structures such as wind turbines.

Keywords: Full-scale · Marine atmospheric boundary layer · Coherence · Turbulence · Surface layer

1 Introduction

The coherence, which according to Ropelewski et al. (1973), "can be thought as a correlation in frequency space", is widely used to describe the spatial structure of wind turbulence. Davenport (1961, 1962) is among the first wind engineers who used the coherence to model the dynamic wind load on large wind-sensitive structures. Nowadays, the coherence is a key element of the buffeting theory (Davenport 1964; Scanlan 1978). The wind coherence has been traditionally estimated using met-masts (Ropelewski et al. 1973; Panofsky et al. 1974; Panofsky and Mizuno 1975) or suspension bridges (Kristensen and Jensen 1979; Bietry et al. 1995; Toriumi et al. 2000; Miyata et al. 2002). The majority of the aforementioned studies have been conducted onshore or in coastal areas. Above the sea, at a distance of several kilometres from the coast, the wind coherence is not well known. Nevertheless, the ongoing development of large offshore wind turbines

F. Ricciardelli and A. M. Avossa (Eds.): INVENTO 2018, LNCE 27, pp. 207–221, 2019.
https://doi.org/10.1007/978-3-030-12815-9_17

(Thresher et al. 2007) emphasizes the need to investigate more in details the adequacy of the established coherence models, the majority of which are based on the so-called Davenport model (Davenport 1961).

In the standards IEC 61400-1 (2005) and IEC61400-3 (2009), used for the design of wind turbines, the coherence can be computed using a modified Davenport model or the uniform shear model (Mann 1994). Both models have fixed parameters, which may not be appropriate above the ocean (Eliassen and Obhrai 2016). In addition, the two coherence models in the IEC standards take into account the influence of the wind shear, but not the blocking effect by the surface. Consequently, the decay coefficient in the Davenport model may be height-dependant, as observed by e.g. Sacré and Delaunay (1992) at four different heights between 4 m and 40 m above the surface. A decay coefficient decreasing with the measurement height is consistent with the idea that eddies get larger further from the ground but is rather inconvenient for the design of tall wind-sensitive structures. Although Mann (1994) also proposed an improved uniform-shear model including the blockage effect by the surface, it is achieved at the cost of an increased complexity. For this reason, the simpler uniform shear model is generally used for engineering applications. A more direct way to take into account the blocking effect may be to introduce an explicit dependence of the coherence on the measurement height, as done in the studies of Bowen et al. (1983) and Iwatani and Shiotani (1984), conducted in open rural terrain and a coastal area, respectively. However, these models do not account for the limited dimensions of the eddies, such that for large crosswind separations, the coherence may be significantly lower than unity as the frequency approaches zero (Kristensen and Jensen 1979). In the present study, a coherence model that includes both the features of the models described by Bowen et al. (1983) and Kristensen and Jensen (1979) is introduced. The ability of such a model to characterize the vertical coherence is investigated using wind velocity data collected in 2007 and 2008 at heights above 40 m on the offshore platform FINO1 in the North Sea.

The present paper is organized as follows: Sect. 2 describes the instrumentation of the platform, the data post-processing and the coherence models investigated. Section 3 illustrates the influence of the measurement height on the coherence through a comparison between the estimated and fitted coherence functions. In Sect. 4, the Davenport model and the 3-parameter coherence model are compared through the joint acceptance function. The latter section discusses whether the use of a coherence model depending explicitly on the measurement height can improve the design of a tall wind-sensitive structure.

2 Instrumentation and Methods

2.1 Data Post-processing

The FINO1 platform is located in the North Sea (N 54°0′53.5″ E 6°35′15.5″), ca. 45 km North of Borkum, on the German coast (Fig. 1). A 81 m high steel lattice tower is mounted on the 20 m high jacket platform at 28 m water depth.

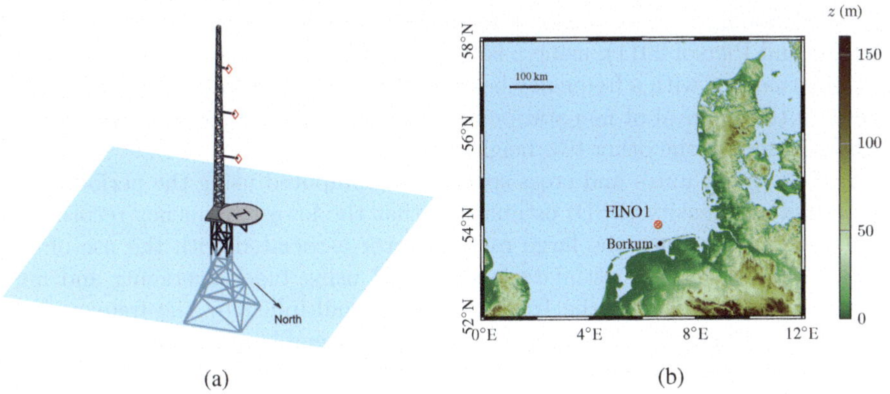

Fig. 1. Schematic of the FINO1 platform (a) and its location in the North Sea (b). For the sake of clarity, only the three booms supporting the sonic anemometers are displayed in the panel (a).

A detailed description of the dimensions of the towers and booms can be found in Neumann and Nolopp (2007) and Westerhellweg et al. (2012). Since 2003, the tower has been instrumented with numerous sensors, including 3 Gill R3-50 sonic anemometers (Gill Instruments Ltd., UK) at a height of 41.5 m, 61.5 m and 81.5 m above the sea level, which are the only sensors considered in the present study.

The sonic anemometers record the three wind velocity components and the sonic temperature with a sampling frequency of 10 Hz. They are mounted on booms oriented toward north-northwest, with an azimuth of 308° for the sensors at the two lowest levels and 311° for the sensor at 81.5 m. Only velocity records with a wind direction between 190° and 359° at $z = 81.5$ m are selected so that they are not significantly affected by flow distortion induced by the mast structure. This choice is supported by the study of Westerhellweg et al. (2012), and corresponds to a conservative approach with respect to the area effectively affected by flow distortion, which is between 90° and 160°.

The data set used in the present study corresponds to sonic anemometer records collected in 2007 and 2008, i.e. ca. 17×10^4 h of records for each sensor, which have been studied previously in Cheynet et al. (2018). To reduce the measurement uncertainties, the averaging time is set to 60 min. Samples with an hourly mean wind velocity between 5 ms^{-1} and 28 ms^{-1} at $z = 81.5$ m are selected as they correspond to the range of operational conditions of a large offshore wind turbine. Under strong wind conditions, a turbulence intensity around 10% is expected on the FINO1 platform at $z = 81.5$ m (Türk and Emeis 2010). Therefore, samples characterized by a turbulence intensity below 1% or above 20% were dismissed as they may correspond to non-physical signals.

The non-stationary wind fluctuations are disregarded by applying a two-step algorithm to each sample: (1) if the difference between the two extrema of the linear trend and its mean value is larger than 20%, the sample is considered as non-stationary; (2) each sample is subjected to the reverse arrangement test (Bendat and Piersol 2011), using a 95% confidence interval and considering only wind fluctuations with a frequency below 0.40 Hz. The resulting data availability, including the removal of non-stationary samples, is 35% (6204 h) at 61.5 m and slightly larger for the other two heights.

The one-point auto- and cross-spectra are computed using the periodogram power spectral density (PSD) estimate so that the lowest frequency recorded is ca. 0.3 mHz. To reduce the large random error associated with the use of the periodogram, each PSD estimate is smoothed using block averaging and normalized by the square of the friction velocity multiplied by the frequency. A single PSD estimate is afterwards obtained by ensemble averaging those corresponding to the same atmospheric stratification. The co-coherence is estimated using Welch's algorithm (Welch 1967) with 6 segments, 50% overlapping and a Hamming window, such that the lowest frequency analysed is ca. 17 mHz. The use of several overlapping segments aims to reduce the bias of the coherence estimate and the associated random error (Kristensen and Kirkegaard 1986). To reduce the measurement uncertainties as much as possible, the coherence models investigated here are compared to the ensemble-averaged co-coherence estimates only.

2.2 Assessment of the Atmospheric Stability

The three wind components are denoted u, v and w and refer to the along-wind (x-axis), the crosswind (y-axis) and the vertical (positive z-axis) components, respectively. For the sake of simplicity, the sonic temperature is assumed to be equal to the virtual potential temperature θ_v. One shall assume that u, v, w and θ_v can be decomposed into a mean component and a fluctuating component with zero mean, which is a stationary Gaussian random process. In Eqs. (1) to (4) the overline stands for the mean component whereas the prime refers to the fluctuating component:

$$u = \overline{u} + u' \tag{1}$$
$$v = \overline{v} + v' \tag{2}$$
$$w = \overline{w} + w' \tag{3}$$
$$\theta_v = \overline{\theta_v} + \theta_v' \tag{4}$$

In a flat and homogeneous area, it is often assumed that $\overline{v} = \overline{w} \approx 0\,\mathrm{ms}^{-1}$, which is established here using a sectoral planar fit (Wilczak et al. 2001) with an angular sector from 190° to 360°. The atmospheric stability is studied with the eddy covariance technique, for each sonic anemometer (Schotanus et al. 1983; Kaimal and Gaynor 1991). Using local scaling (Nieuwstadt 1984; Sorbjan 1986), the non-dimensional Obukhov length ζ is:

$$\zeta = \frac{-g\kappa z \overline{w'\theta_v'}}{\overline{\theta_v} u_*^3} \tag{5}$$

where $g = 9.81\,\mathrm{ms}^{-2}$ is the acceleration of gravity; z is the measurement height; u_* is the local friction velocity; $\overline{w'\theta'_v}$ is the local flux of virtual potential temperature and $\kappa = 0.40 \pm 0.01$ (Högström 1988) is the von Kármán constant. Here, u_* is calculated following the suggestion from Weber (1999). An additional reason to use Eq. (6) is given by e.g. Geernaert (1988) who observed that $\overline{v'w'}$ may not be negligible in an offshore environment due to the air-sea heat fluxes, a possible larger-scale forcing or the influence of the sea-state on the wind stress.

$$u_* = \left(\overline{u'w'}^2 + \overline{v'w'}^2\right)^{1/4} \tag{6}$$

In the following, a near-neutral atmospheric stratification corresponds to $|\zeta| < 0.05$, i.e. 1329 samples of 1 h duration.

2.3 Modelling of the Co-coherence

For vertical separations, the co-coherence γ_{ij}, where $i, j = \{u, v, w\}$, is defined as:

$$\gamma_{ij}(z_1, z_2, f) = \frac{\mathrm{Re}\left\{S_{ij}(z_1, z_2, f)\right\}}{\sqrt{S_{ii}(z_1, f)S_{jj}(z_2, f)}} \tag{7}$$

where $S_{ij}(z_1, z_2, f)$ is the cross-spectral density of the i and j components between heights z_1 and z_2; $S_{ii}(z_1, f)$ is the single-point spectrum of the i component measured at z_1 and $S_{jj}(z_2, f)$ is the single-point spectrum of the j component measured at z_2.

2.3.1 The Uniform Shear Model

The uniform shear model, introduced by Mann (1994), aims to describe the one-point auto spectra, the cross-spectrum S_{uw} as well as the associated coherence using only three adjustable parameters. The first parameter, denoted $\alpha\epsilon^{3/2}$, is a measure of the energy dissipation, where ϵ is the rate of viscous dissipation of specific turbulent kinetic energy and α is the three-dimensional Kolmogorov constant. The second parameter is a length scale of the spectral velocity tensor, which is here denoted L. The third parameter is called the "shear parameter" and is written Γ as it quantifies the anisotropy of the spectral tensor. In IEC 61400-1 (2005), the uniform shear model is considered, but its formulation differs slightly from the original model proposed by Mann (1994). The latter model is nevertheless adopted here so that more general conclusions can be drawn. Note that in Cheynet et al. (2017), which focused only on the one-point turbulence characteristics recorded on the FINO1 platform, the parameters $\alpha\epsilon^{3/2}$, L and Γ were found to be more or less constant with the mean wind velocity, such that in the present study, the one-point spectra are not separated into different velocity bins.

The co-coherence computed using the uniform shear model is expressed as a function of the wave number $k_1 = 2\pi f/\overline{u}$ and the vertical separation d_z:

$$\gamma_{ij}(k_1, d_z) = \frac{\mathrm{Re}\left\{\chi_{ij}(k_1, d_z)\right\}}{\sqrt{F_i(k_1)F_j(k_1)}} \tag{8}$$

$$\chi_{ij}(k_1, d_z) = \int_{-\infty}^{+\infty}\int_{-\infty}^{+\infty} \phi_{ij}(k_1, k_2, k_3)\exp\left(-ik_3 d_z\right) dk_2\, dk_3 \tag{9}$$

$$F_i(k_1) = \int_{-\infty}^{+\infty}\int_{-\infty}^{+\infty} \phi_{ii}(k_1, k_2, k_3)\, dk_2\, dk_3 \tag{10}$$

where ϕ_{ij} is the spectral velocity tensor and k_2 and k_3 are the wavenumbers in the lateral and vertical direction, respectively. More details on the coherence computed with the sheared spectral velocity tensor can be found in Mann (1994, 1998). The study of the coherence with $i \neq j$ is not considered here, such that the notation γ_{ij} is replaced in the following by γ_{ii} for the sake of simplicity.

2.3.2 Empirical Coherence Models

The Davenport coherence model (Davenport 1961) is one of the first empirical model used to describe the vertical co-coherence. For $i = \{u, v, w\}$, it is defined as:

$$\gamma_{ii}(z_1, z_2, f) = \exp\left(-\frac{c_z^i f|z_1 - z_2|}{\overline{U}(z_1, z_2)}\right) \tag{11}$$

$$\overline{U}(z_1, z_2) = \frac{1}{2}\left[\overline{u}\left(z_1\right) + \overline{u}\left(z_2\right)\right] \tag{12}$$

where c_z^i is a constant called "exponential decay" and z_1 and z_2 are two measurement heights.

At a height below 40 m above the surface, a possible dependency of the exponential decay c_z^i with the measurement height has been documented in e.g. Bowen et al. (1983) or Sacré and Delaunay (1992), which may reflect the increasing size of the turbulent eddies with the altitude and the blockage by the surface. Bowen et al. (1983) proposed an improved coherence model based on Eq. (11), where $c_z^i = c_1^i + 2c_2^i d_z/\left(z_1 + z_2\right)$:

$$\gamma_{ii}(z_1, z_2, f) = \exp\left(-\frac{c_1^i f|z_2 - z_1|}{\overline{U}(z_1, z_2)}\right)\exp\left(-\frac{2c_2^i f|z_2 - z_1|^2}{(z_1 + z_2)\,\overline{U}(z_1, z_2)}\right) \tag{13}$$

Equation (13) shows that the dependency of the coherence on z_1 and z_2 decreases as the height increases, such that sufficiently far from the surface, the coherence can be approximated by the Davenport model. In the present study, the coherence model proposed by Bowen et al. (1983) is modified to include an additional decay parameter c_3^i, which accounts for the limited size of the eddies in the vertical direction, such that the coherence cannot be equal to unity at zero frequency:

$$\gamma_{ii}(z_1, z_2, f) = \exp\left\{-\left[\frac{|z_2 - z_1|}{\overline{U}(z_1, z_2)}\sqrt{(c_1^i f)^2 + (c_3^i)^2}\right]\right\}\exp\left(-\frac{2c_2^i f|z_2 - z_1|^2}{(z_1 + z_2)\overline{U}(z_1, z_2)}\right) \tag{14}$$

3 Results

At heights between 40 m and 80 m, under neutral conditions, the wind shear is expected to be low enough so that the Davenport coherence model reduces to a single curve when expressed as a function of $k_1 d_z$. Otherwise, it indicates that additional environmental effects need to be accounted for. In the present section, the co-coherence is, therefore, expressed as a function of $k_1 d_z$ to evaluate the validity of the Davenport coherence model.

3.1 Coherence Computed with the Uniform Shear Model

For near-neutral conditions, the three parameters of the uniform shear model (Mann 1994) are estimated simultaneously using a non-linear least-square fit to the one-point spectra and the real part of the cross-spectrum S_{uw}. Whereas the uniform shear model is limited to turbulent fluctuations, the estimated spectra may cover a larger frequency range. The fit is thus performed for a reduced frequency $f_r = fz/\overline{u}$ bounded between 0.006 and 5.1. The co-coherence computed with the uniform shear model between the heights z_1 and z_2 is obtained using the values of $\alpha\epsilon^{3/2}$, Γ and L averaged between z_1 and z_2. This allows accounting for the evolution of these three parameters with the height. The corresponding fitted and estimated one-point PSD estimates are shown in Fig. 2 at the height of 61.5 m above sea level, whereas the values of the fitted parameters are summarized in Table 1.

Table 1. Parameters of the uniform shear model fitted to the estimated one-point spectra S_u, S_v, S_w and co-spectrum Re (S_{uw}) on the FINO1 platform, for a neutral atmospheric stratification.

Height above sea level (m)	$\alpha\epsilon^{3/2}$ (s^{-1})	Γ	L (m)
81.5	0.025	3.2	52
61.5	0.028	3.5	44
41.5	0.036	3.8	32

In the absence of the blockage effect by the surface, the uniform shear model predicts that the imaginary part of the cross-spectrum S_{uw} is equal to zero. Figure 2 shows that even at 60 m above the surface, this is not the case, especially

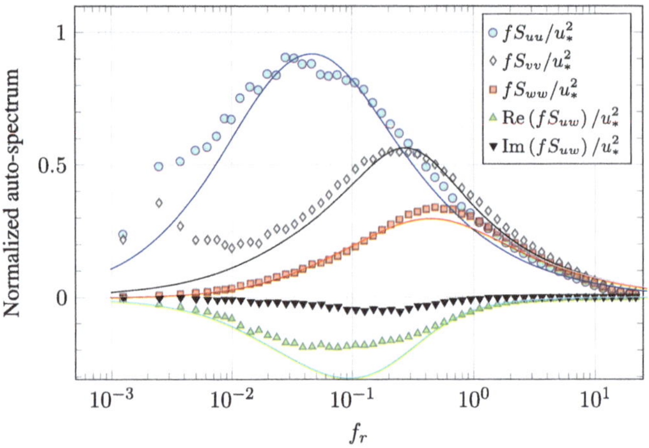

Fig. 2. PSD estimates (scatter plot) of the turbulent wind velocity components at the height $z = 61.5$ m and a stability parameter $|\zeta| \leq 0.05$ superposed to the fitted spectra (solid lines) computed with the uniform shear model.

at $f_r \approx 0.2$, where the ratio of the imaginary part over the real part is equal to ca. 0.3, which is also found at $z = 80$ m. This observation indicates that, in the present case, the spatial structure of the turbulence may still be affected by the surface at a height up to 80 m. It is also possible that the flow distortion by the structure slightly amplifies the value of the imaginary part of the cross-spectrum. The in-depth examination of the latter possibility is, nevertheless, out of the scope of the present study.

There is a fairly good agreement between the estimated value of $\alpha\epsilon^{3/2}$ with those estimated by De Maré and Mann (2014, Fig. 3–5) on a met-mast at the Rødsand II offshore wind farm. In their study, the estimated values of Γ and L are, however, slightly lower than those displayed in Table 1. The latter values differ also slightly from those found in Cheynet et al. (2017) because the atmospheric stratification is here assessed much more accurately. Figure 3 shows that the computed co-coherence for the u and w components provides conservative estimates, except at $k_1 d_z > 1$. If a lower value of L is used, the agreement between the computed and estimated co-coherence is improved whereas the discrepancies between the computed and estimated one-point spectra increase.

Although the fitted and estimated one-point spectra agree reasonably well in Fig. 2, they exhibit discrepancies that may partly explain those observed between the estimated and computed co-coherence:

- In the first 30 m above the surface, the distortion of the eddies by the surface is such that the normalized spectrum $fS_u(f)/u_*^2$ is expected to exhibit a plateau instead of a spectral peak (Tchen 1953; Hunt and Morrison 2000). Such a behaviour has been observed in full-scale (Högström et al. 2002; Mikkelsen et al. 2017), but is not predicted in the uniform shear model. The plateau is expected to occur in the wavenumber range corresponding to $\Lambda^{-1} \ll k_1 \ll$

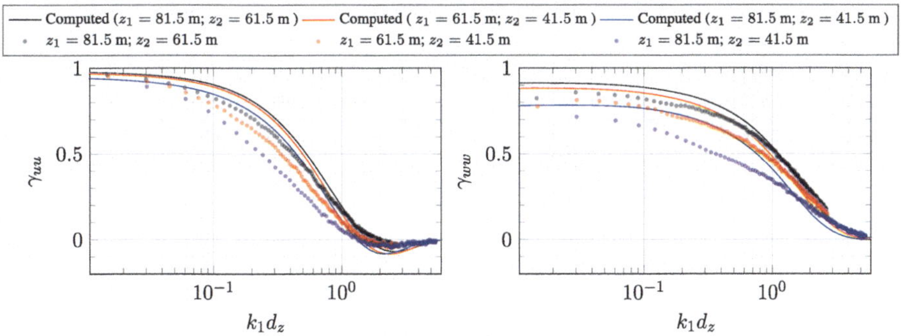

Fig. 3. Estimated (scatter plot) co-coherence compared to the one computed using the uniform shear model (solid line) for near-neutral conditions in 2007 and 2008 (1329 samples of 1 h duration).

z^{-1}, where Λ is the largest horizontal eddy (Högström et al. 2002). Although the anemometers are located at heights above 40 m, Fig. 2 shows that a rather flat spectral peak is observed in the along-wind PSD estimate for $0.02 < f_r < 0.1$. The value $f_r = 0.1$ corresponds here to a wavenumber of $0.01\,\mathrm{m}^{-1}$, i.e. slightly below the upper limit where one would expect to observe the k^{-1} spectral scaling. Note that a flat peak is also observed for $\mathrm{Re}\,(fS_{uw})$ but is not modelled by the uniform-shear model.

- The uniform shear spectral model assumes that the ratio S_w/S_u approaches the theoretical value of 1.33 in the inertial subrange. In the present case, the estimated ratio S_w/S_u reaches a value around 1.2, which may be due to flow distortion by the mast structure and/or possible local anisotropy (Smedman et al. 2003).

3.2 Coherence Computed with the Empirical Coherence Models

In Fig. 4, the co-coherence estimates of the u and w components are superposed to the Davenport model, computed with the fitted coefficients $c_z^u = 12.9$ and $c_z^w = 5.3$. For a wind turbine with a hub height at 80 m above the surface, the exponential coherence model from the IEC standard (IEC 61400-1 2005, Eq. B.16) is almost identical to the Davenport model with a decay coefficient equal to 12, which is remarkably close to the value $c_z^u = 12.9$ found from the full-scale data. In the data set considered, the low wind shear measured above 40 m leads to a co-coherence computed with the Davenport model that almost collapses into a single curve when expressed as a function of $k_1 d_z$, as expected. This is not the case for the full-scale data, which is attributed to the blocking effect by the surface.

In Fig. 5, the application of the 3-parameter coherence model (Eq. (14)) shows an excellent agreement with the estimated values of γ_{uu} at every frequency. For the w component, the computed co-coherence agrees also well with the estimated

Table 2. Parameters of the 3-parameter coherence model corresponding to the computed co-coherence in Fig. 5.

Decay parameter	c_1^u	c_2^u	c_3^u (s^{-1})	c_1^w	c_2^w	c_3^w (s^{-1})
Value	6.0	17.8	0.02	2.7	4.0	0.16

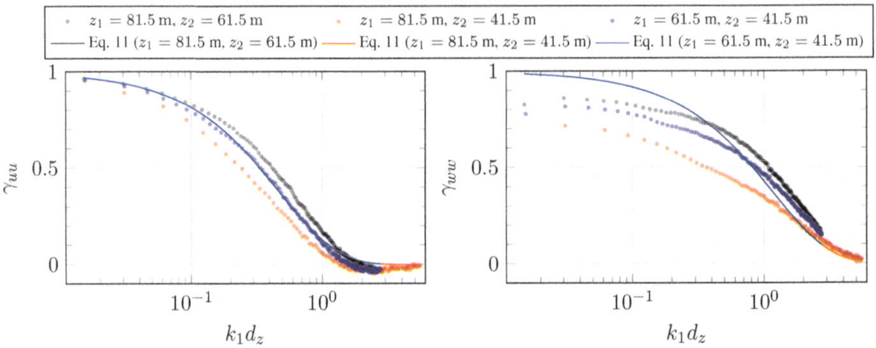

Fig. 4. Estimated co-coherence (scatter plot) and fitted Davenport model (solid lines) for near-neutral conditions in 2007 and 2008 (1329 samples of 1 h duration).

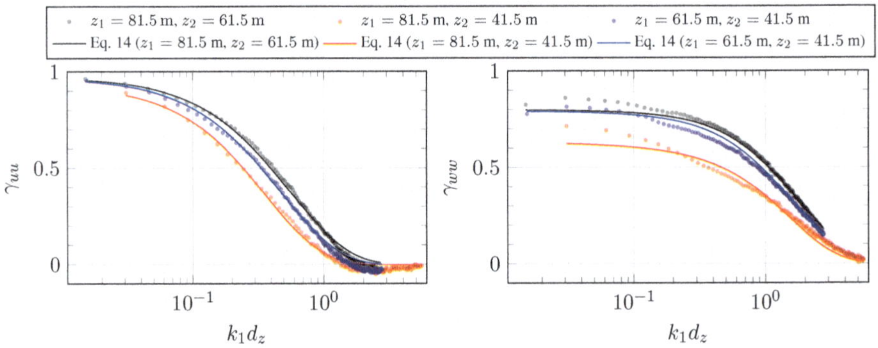

Fig. 5. Estimated (scatter plot) and fitted (solid lines) co-coherence for near-neutral conditions in 2007 and 2008 (1329 samples of 1 h duration). The fitted parameters are displayed in Table 2.

one, except at $k_1 d_z < 0.1$, where it is slightly lower than estimated from the full-scale data. In Table 2, which displays the fitted parameters of the 3-parameter coherence model, the low value of the coefficient c_3^u shows that the co-coherence of the along-wind component approaches unity when the frequency becomes small, such that Eq. (13) may directly be used instead of Eq. (14). However, for the w component, the parameter c_3^w cannot be neglected, as γ_{ww} reaches values significantly lower than 1 at low frequencies.

4 Discussions

Since the co-coherence estimates do not collapse into a single curve when expressed as a function of $k_1 d_z$ (Figs. 3, 4 and 5), the approximation:

$$\gamma_{uu}(d_z, f) \approx \gamma_{uu}(z_1, z_2, f) \tag{15}$$

needs be re-assessed by evaluating whether a coherence model accounting for an explicit dependency on the height z can substantially improve the estimation of the wind load on a large wind-sensitive structure. This is in particular important for a high-rise structure spanning from the surface, for which the blocking effect of the ground may increase at lower heights, although the most important load effects are applied further away from the surface.

A detailed investigation in terms of sensitivity of the wind-induced response of a structure with the coherence model used is out of the scope of the present study. Nevertheless, a preliminary analysis can be conducted using the joint acceptance function. For the along-wind component and the kth vibration mode, it is defined as:

$$J_k^2 = \frac{1}{\mathrm{A}} \int_0^H \int_0^H \phi_k(z_1)\gamma_{uu}(z_1, z_2, f)\phi_k(z_2)\, \mathrm{d}z_1\, \mathrm{d}z_2 \tag{16}$$

$$\mathrm{A} = \int_0^H \int_0^H |\phi_k(z_1)\phi_k(z_2)|\, \mathrm{d}z_1\, \mathrm{d}z_2 \tag{17}$$

where ϕ_k is the kth mode-shape of the structure and H its height.

The mode shapes are computed here for the simple case of a vertical cantilever beam with a height H of 80 m and a constant circular cross-section with a diameter equal to $H/20$. The mode shapes and eigenfrequencies are computed using the classical beam theory where the characteristic equation is numerically solved. Note that Eq. (16) and the computed mode shapes are independent of the eigenfrequencies. For the case considered here, if the first eigenfrequency is set to 0.3 Hz, this leads to a second and third eigenfrequency equal to 1.9 Hz and 5.3 Hz, respectively. In this case, the response of the structure to wind turbulence will be dominated by the first mode. Although this numerical model is simplistic, the computed mode shapes are consistent with those identified with full-scale vibrations data from a wind turbine (Oliveira et al. 2018) or more complex numerical models of towers (Murtagh et al. 2004). For the sake of simplicity, only the Davenport and the 3-parameter coherence model are here compared using the decay coefficients estimated in Sect. 3.

Figure 6 shows the computed joint-acceptance function for the three mode shapes selected. In each panel, the inset on the top-right shows the corresponding mode shape ϕ_k, where $k = \{1, 2, 3\}$ is the mode number. For the simple case considered, the joint acceptance function of the first vibration mode is

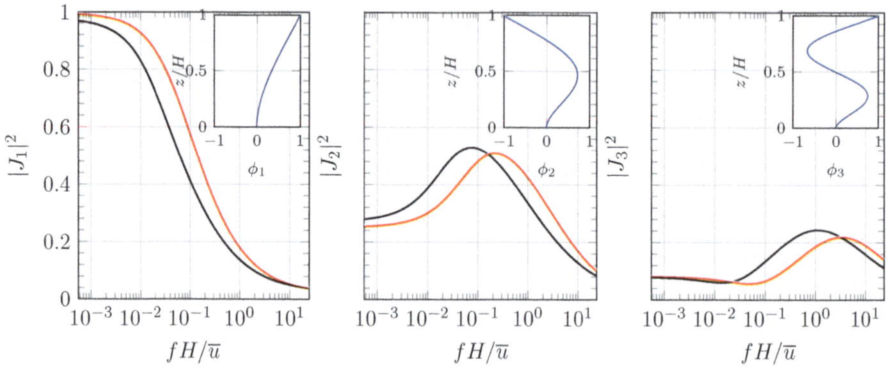

Fig. 6. Joint acceptance function computed for the first three modes of vibration, displayed in the insets, for the Davenport model (——) and the 3-parameter coherence model (——).

systematically larger if the Davenport model is used, except for high frequencies where the coherence becomes small in both cases. The integral of $J_1(f)$ computed with the Davenport model is 18% larger than the one computed with the 3-parameter coherence function. For the example considered, it indicates that the dynamic wind load is overestimated for the first mode of vibration. For the other two modes, which may play a minor role in the wind-induced response, the discrepancies depend on the frequency considered and their interpretation is more complex. Such a comparison illustrates, nevertheless, the limits of the Davenport model to describe the vertical coherence of the along-wind component. In a different context, the influence of the measurement height on the estimated co-coherence implies that for the design of a long-span suspension bridge, field measurements should be conducted at the deck height to properly estimate the turbulent wind load on the girder.

5 Conclusions

The vertical co-coherence of the along-wind and vertical wind components has been studied for near-neutral conditions using ca. 1.3×10^3 h of records collected on the FINO1 platform in the North Sea. The focus is on the proper modelling of the height dependence of the coherence, reflecting the influence of the blockage by the surface. The latter is responsible for the wind shear and the deformation of the eddies as they approach the ground or the sea. The velocity records show that the influence of the surface on the coherence and the one-point spectra is still detectable at a height above 60 m.

At the heights considered, i.e. between 40 m and 80 m above the sea level, where the wind shear is low, the Davenport model fails to include the dependency of the decay coefficient with the measurement height, even though it provides a fairly good approximation of the vertical coherence of the along-wind component.

On the other hand, the 3-parameter coherence model manages to capture well this height-dependency, whereas the co-coherence computed with the uniform shear model provides more conservative estimates of the coherence than the other two models investigated.

The computation of the joint acceptance function of a simple line-like vertical cantilever beam indicates that a more in-depth investigation of the coherence model depending explicitly on the measurement height is pertinent to improve the design of a tall wind-sensitive structure.

Acknowledgements. The author would like to acknowledge the Federal Ministry for Economic Affairs and Industry and the Projektträger Jülich for funding the FINO project, and UL DEWI for providing the sonic data. Prof. Jasna Bogunović Jakobsen is gratefully acknowledged for her review of the manuscript. Thanks are also due to Prof. Joachim Reuder for his useful advice regarding the wind data analysis.

References

Bendat J, Piersol A (2011) Random data: analysis and measurement procedures. Wiley series in probability and statistics. Wiley

Bietry J, Delaunay D, Conti E (1995) Comparison of full-scale measurement and computation of wind effects on a cable-stayed bridge. J Wind Eng Ind Aerodyn 57(2–3):225–235

Bowen AJ, Flay RGJ, Panofsky HA (1983) Vertical coherence and phase delay between wind components in strong winds below 20 m. Bound-Layer Meteorol 26(4):313–324

Cheynet E, Jakobsen JB, Obhrai C (2017) Spectral characteristics of surface-layer turbulence in the North Sea. Energy Procedia 137:414–427

Cheynet E, Jakobsen JB, Reuder J (2018) Velocity spectra and coherence estimates in the marine atmospheric boundary layer. Bound-Layer Meteorol:1–32

Davenport A (1964) The buffeting of large superficial structures by atmospheric turbulence. Ann N Y Acad Sci 116(1):135–160

Davenport AG (1961) The spectrum of horizontal gustiness near the ground in high winds. Q J R Meteorol Soc 87(372):194–211

Davenport AG (1962) The response of slender, line-like structures to a gusty wind. Proc Inst Civ Eng 23(3):389–408

De Maré M, Mann J (2014) Validation of the Mann spectral tensor for offshore wind conditions at different atmospheric stabilities. J Phys Conf Ser 524:012106

Eliassen L, Obhrai C (2016) Coherence of turbulent wind under neutral wind conditions at FINO1. Energy Procedia 94:388–398

Geernaert G (1988) Measurements of the angle between the wind vector and wind stress vector in the surface layer over the North Sea. J Geophys Res Oceans 93(C7):8215–8220

Högström U (1988) Non-dimensional wind and temperature profiles in the atmospheric surface layer: a re-evaluation. Bound-Layer Meteorol 42(1):55–78

Högström U, Hunt JCR, Smedman AS (2002) Theory and measurements for turbulence spectra and variances in the atmospheric neutral surface layer. Bound-Layer Meteorol 103(1):101–124

Hunt JC, Morrison JF (2000) Eddy structure in turbulent boundary layers. Eur J Mech-B/Fluids 19(5):673–694

IEC 61400-1 (2005) IEC 61400–1 Wind turbines–Part 1: Design requirements
IEC61400-3 (2009) Wind Turbines–Part 3: Design Requirements for Offshore Wind Turbines
Iwatani Y, Shiotani M (1984) Turbulence of vertical velocities at the coast of reclaimed land. J Wind Eng Ind Aerodyn 17(1):147–157
Kaimal J, Gaynor J (1991) Another look at sonic thermometry. Bound-layer Meteorol 56(4):401–410
Kristensen L, Jensen N (1979) Lateral coherence in isotropic turbulence and in the natural wind. Bound-Layer Meteorol 17(3):353–373
Kristensen L, Kirkegaard P (1986) Sampling problems with spectral coherence. Risø National Laboratory. Risø-R-526
Mann J (1994) The spatial structure of neutral atmospheric surface-layer turbulence. J Fluid Mech 273
Mann J (1998) Wind field simulation. Probab Eng Mech 13(4):269–282
Mikkelsen T, Larsen SE, Jørgensen HE, Astrup P, Larsén XG (2017) Scaling of turbulence spectra measured in strong shear flow near the Earth's surface. Phys Scr 92(12):124,002
Miyata T, Yamada H, Katsuchi H, Kitagawa M (2002) Full-scale measurement of Akashi-Kaikyo Bridge during typhoon. J Wind Eng Ind Aerodyn 90(12):1517–1527
Murtagh P, Basu B, Broderick B (2004) Simple models for natural frequencies and mode shapes of towers supporting utilities. Comput Struct 82(20–21):1745–1750
Neumann T, Nolopp K (2007) Three years operation of far offshore measurements at FINO1. DEWI Mag 30:42–46
Nieuwstadt FT (1984) The turbulent structure of the stable, nocturnal boundary layer. J Atmos Sci 41(14):2202–2216
Oliveira G, Magalhães F, Cunha Á, Caetano E (2018) Continuous dynamic monitoring of an onshore wind turbine. Eng Struct 164:22–39
Panofsky HA, Mizuno T (1975) Horizontal coherence and pasquill's beta. Bound-Layer Meteorol 9(3):247–256
Panofsky HA, Thomson D, Sullivan D, Moravek D (1974) Two-point velocity statistics over Lake Ontario. Bound-Layer Meteorol 7(3):309–321
Ropelewski CF, Tennekes H, Panofsky H (1973) Horizontal coherence of wind fluctuations. Bound-Layer Meteorol 5(3):353–363
Sacré C, Delaunay D (1992) Structure spatiale de la turbulence au cours de vents forts sur differents sites. J Wind Eng Ind Aerodyn 41(1–3):295–303
Scanlan R (1978) The action of flexible bridges under wind, II: Buffeting theory. J Sound Vib 60(2):201–211
Schotanus P, Nieuwstadt F, De Bruin H (1983) Temperature measurement with a sonic anemometer and its application to heat and moisture fluxes. Bound-Layer Meteorol 26(1):81–93
Smedman AS, Högström U, Sjöblom A (2003) A note on velocity spectra in the marine boundary layer. Bound-Layer Meteorol 109(1):27–48
Sorbjan Z (1986) On similarity in the atmospheric boundary layer. Bound-Layer Meteorol 34(4):377–397
Tchen C (1953) On the spectrum of energy in turbulent shear flow. J Res Nat Bur Stan
Thresher R, Robinson M, Veers P (2007) To capture the wind. IEEE Power Energy Mag 5(6):34–46
Toriumi R, Katsuchi H, Furuya N (2000) A study on spatial correlation of natural wind. J Wind Eng Ind Aerodyn 87(2):203–216

Türk M, Emeis S (2010) The dependence of offshore turbulence intensity on wind speed. J Wind Eng Ind Aerodyn 98(8–9):466–471

Weber R (1999) Remarks on the definition and estimation of friction velocity. Bound-Layer Meteorol 93(2):197–209

Welch P (1967) The use of fast Fourier transform for the estimation of power spectra: a method based on time averaging over short, modified periodograms. IEEE Trans Audio Electroacoust 15(2):70–73

Westerhellweg A, Neumann T, Riedel V (2012) FINO1 mast correction. Dewi Mag 40:60–66

Wilczak JM, Oncley SP, Stage SA (2001) Sonic anemometer tilt correction algorithms. Bound-Layer Meteorol 99(1):127–150

Modeling and Simulation of Non-stationary Thunderstorm Outflows

M. Ciano[1(✉)], M. Gioffrè[1], V. Gusella[1], and M. Grigoriu[2]

[1] Department of Civil and Environmental Engineering, University of Perugia, Via G. Duranti 93, 06125 Perugia, Italy
{matteo.ciano,massimiliano.gioffre,vittorio.gusella}@unipg.it
[2] Department of Civil and Environmental Engineering, Cornell University, Ithaca, NY 14853, USA
mdg12@cornell.edu

Abstract. Wind action on structures is commonly estimated under the stationary assumption that is related to the phenomena that occur in neutral atmospheric conditions at the synoptic scale with velocity profiles in equilibrium with the atmospheric boundary layer. Nevertheless, when thunderstorm outflows are concerned the wind induced load description should take into account its inherent non-stationary features. In this paper a non-stationary model is proposed to describe wind velocity fluctuations of a typical thunderstorm in the northern Italian coast. This model can be obtained by modulating both amplitude and frequency of a stationary process. First, experimental time histories are used to calibrate the non-stationary model, then a parametric dynamic response study to stationary and non-stationary wind loads is finally investigated.

Keywords: Thunderstorm outflows · Wind-excited structures · Non-stationary excitation · Simulation methods · Monte Carlo simulation

1 Introduction

Thunderstorms are non-stationary phenomena at the mesoscale that occur in convective conditions with totally different velocity profiles with respect to synoptic winds. This phenomenon might significantly affects the estimation of the structural random response peak values [1–3]. In the last few years the wind engineering scientific community has focused attention on this class of non-stationary phenomena given the increase of severe damages to civil structures observed during this kind of thunderstorms. Nevertheless, this topic is still fragmented and not completely assessed.

A time domain parametric approach to predict the dynamic response of a structure affected by thunderstorm loads was proposed in [5,8]. Chen et al. proposed an empirical numerical model for downburst wind fields in [6], while Chen introduced a frequency domain analytical approach to quantify the along-wind

F. Ricciardelli and A. M. Avossa (Eds.): INVENTO 2018, LNCE 27, pp. 222–233, 2019.
https://doi.org/10.1007/978-3-030-12815-9_18

response of tall buildings to non-stationary winds [7]. Kwon et al. presented a gust-front factor-based framework to evaluate the non-stationarity in terms of mean and standard deviation of the wind velocity fluctuations [13]. The effect of thunderstorms on tall buildings structural response was investigated in [12] with a time domain approach using multiple wind speed time history samples and in [14] with an alternative and efficient method based on wavelet-Galerkin methods. Aboshosha and Damatty developed a closed form solution to estimate the forces exerted by transmission line conductors to the supporting towers under down-burst loads [4]. Very recently, the study of such phenomena has been addressed in [20] where the maximum response of vertical slender structures affected by thunderstorm outflows was estimated using the wind velocity data obtained from a significant experimental work in the northern Italian coast [18].

In this work a probabilistic model to describe the non-stationary wind velocity fluctuations of a typical thunderstorm outflow recorded in the northern Italian area is proposed. This model is based on the one presented in [10] to describe seismic loads in the time-frequency domain using suitable time varying functions for both amplitude and frequency coefficients. First, experimental data in [18] are used to calibrate the non-stationary probabilistic model that is then used to generate virtual samples of the wind velocity thunderstorm time histories. Second, the effect of the wind velocity non-stationary features on the structural response is estimated using single degree of freedom systems.

2 Non-stationary Model and Simulation

A typical one-hour record of wind velocity time history before, during and after a thunderstorm outflow, $U(t)$, is reported in Fig. 1 (top panel). It is clear that the time series shows inherent non-stationary features at least in the mean and variance components, superimposed to a non negligible synoptic flow. Indeed, it is possible to process the sample time history with a moving average to obtain the series, $U_m(t)$, shown in Fig. 1 (middle panel) and to estimate the zero-mean wind velocity fluctuations

$$u(t) = U(t) - U_m(t) \tag{1}$$

reported in Fig. 1 (lower panel) where it is clear that the variance is also time dependent. Figure 2 shows the estimated spectrogram of the time series $u(t)$ where it is also rather evident a change in the frequency content with time. Nevertheless, it is possible to consider few time intervals, m, where the frequency content of $u(t)$ can be considered constant with time. Sampling frequency of the recorded data is $f_s = 10\,\text{Hz}$.

Such a behavior can be described using the model presented in [10] for earthquake non-stationary time series. Let $x(t) = u(t)$ be a sample time history of the non-stationary process

$$X(t) = c(t)Y(\phi(t)), \tag{2}$$

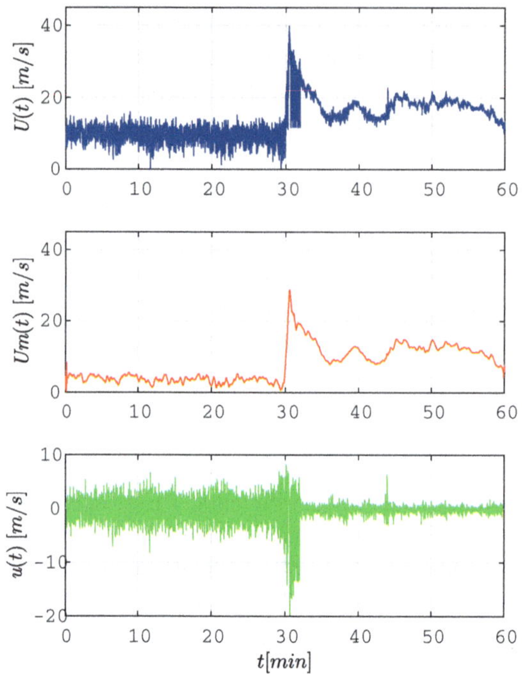

Fig. 1. Thunderstorm outflow experimental sample: wind velocity time series $U(t)$ (top panel), mean component $U_m(t)$ (middle panel), meanless component $u(t)$ (bottom panel).

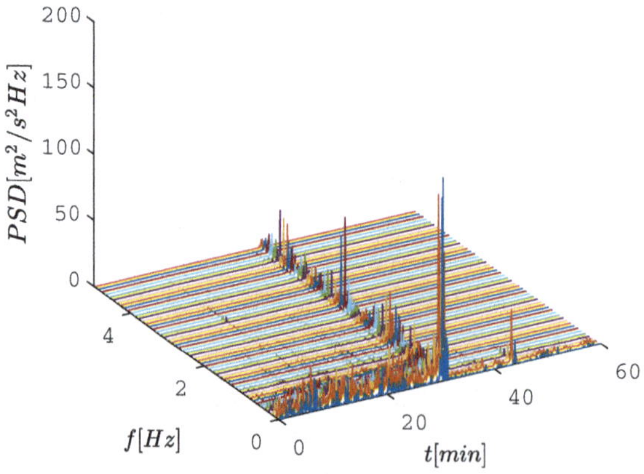

Fig. 2. Spectrogram of $u(t)$.

where $Y(t)$ is a real-valued zero-mean wide sense stationary process with variance σ_Y^2 and one sided spectral density $G_Y(f)$, $c(t)$ and $\phi(t)$ are amplitude and frequency modulating functions, respectively. In particular, $c(t)$ is a slowly varying real-valued deterministic function modulating the variance of $X(t)$, while $\phi(t)$ is a positive, real valued function that satisfy the conditions $\phi(0) = 0$ and $\dot{\phi}(t) > 0$ for $t \geq 0$. In this way the process in (2) is derived from $Y(t)$ by modulating its amplitude and frequency.

It can be demonstrated that the variance of $X(t)$ at time t has the expression [10]

$$\sigma^2(t) = c^2(t)\sigma_Y^2. \tag{3}$$

while the time-dependent one-sided spectral density of $X(t)$ can be defined as

$$G_X(t,f) = c^2(t)G_Y(f\frac{t}{\phi(t)}) \tag{4}$$

The modulating function $\phi(t)$ determines the rate at which the spectral density of $X(t)$ changes with time. If $\phi(t) = t$ one obtains

$$G_X(t,f) = c^2(t)G_Y(f) \tag{5}$$

and the process $X(t)$ in (2) is a uniformly modulated process.

Given the main features of the estimated spectrogram in Fig. 2 it is possible to use only a few uniformly modulated processes where the shape of the spectral density function can be assumed to be constant with time. The one sided spectral density $G_Y^{(j)}(f)$ in each of these m intervals can be modeled using the k-modal function

$$G_Y^{(j)}(f) = \sum_{i=1}^{k^{(j)}} d_i^{(j)} \exp[-A_i^{(j)}(2\pi f - \Omega_i^{(j)})^2] \tag{6}$$

with $j = 1, 2, ..., m$, where $k^{(j)}$ can be chosen in each interval in order to describe the numbers of dominant modes in the spectral density function. In the case study of this paper $k^{(j)} = k = 12$ is assumed. The parameters $d_i^{(j)}$, $A_i^{(j)}$, $\Omega_i^{(j)}$, $j = 1, 2, ..., m$, $i = 1, 2, ..., k^{(j)}$ can be calibrated using the average spectral density estimated in each of the m intervals.

The variance modulating function $c(t)$ in (3)–(5) can be estimated from data using Eq. (3).

Figure 3 shows the amplitude modulating function $c(t)$ computed from the time series $u(t)$ in Fig. 1, where $\sigma_Y^2 =$ was estimated using (6).

Virtual time histories of the non-stationary process in (2) can be generated using the approximation

$$X_Q(t) = c(t) \sum_{q=1}^{Q} \sigma_q [V_q \cos\omega_q(t)t + W_q \sin\omega_q(t)t] \tag{7}$$

where $\omega_q(t) = \omega_q\phi(t)/t$, while V_q and W_q are uncorrelated random variables with zero mean and unit variance. It can be demonstrated that the process

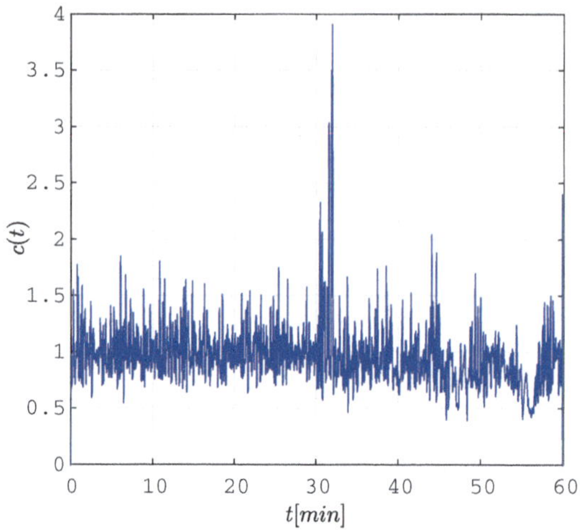

Fig. 3. Amplitude modulation function $c(t)$

$X_Q(t)$ in (7) has energy of magnitude $c^2(t)\sigma_q^2$ at the time dependent frequencies $\omega_q(t) = 2\pi f_q(t)$, $q = 1, 2, ..., Q$. It is worth noting that the time variation in this case study is discrete and is limited to the m time intervals which define Eq. (6) where it was assumed $\phi(t) = t$. Furthermore, V_q and W_q are assumed to be Gaussian.

Figure 4 shows the experimental sample $u(t)$ used to calibrate the non-stationary model (top panel) and a sample $x_Q(t)$ of $X_Q(t)$ generated using (7) (bottom panel). Figures 5 and 6 show estimated local variances and spectrograms, respectively. Figure 5 reports the local variance $\sigma_u^2(t)$ of the experimental time history $u(t)$ used to calibrate the non-stationary model (top panel) compared to $\sigma_{X_Q}^2(t)$ estimated from the simulated time series $x_Q(t)$ of Fig. 4 (bottom panel).

Figure 6 compares the spectrogram model resulting from the calibration of Eq. (6), using the experimental $u(t)$, (top) and the mean evolutionary spectral density estimated from $n_s = 1000$ samples of X_Q generated using (7) (bottom). Spectral density estimates at the three time instants 30,50 min, 31 min and 31,50 min are compared in Fig. 7. In particular, the calibrated model of Eq. (6) is compared with the estimation from the experimental $u(t)$ and from two simulated time series, $x_Q^{(1)}(t)$ and $x_Q^{(2)}(t)$.

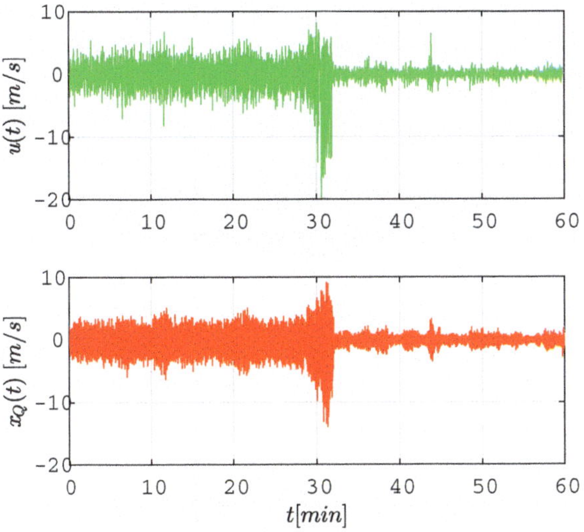

Fig. 4. Experimental time series $u(t)$ in Fig. 1 (top) and a sample $x_Q(t)$ of $X_Q(t)$ (bottom).

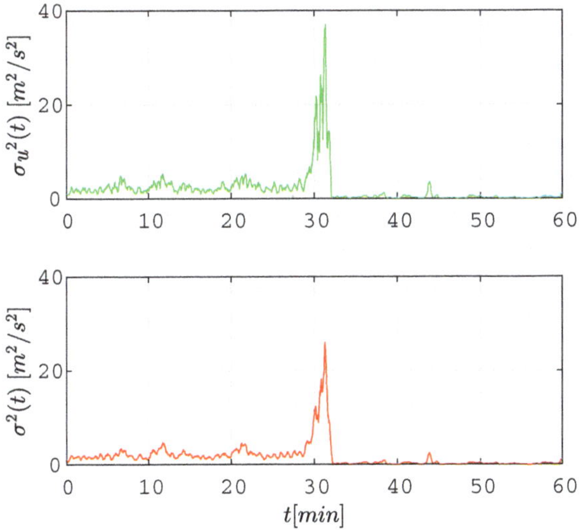

Fig. 5. Local variance of $u(t)$ (top) and of a sample of $X_Q(t)$ (bottom).

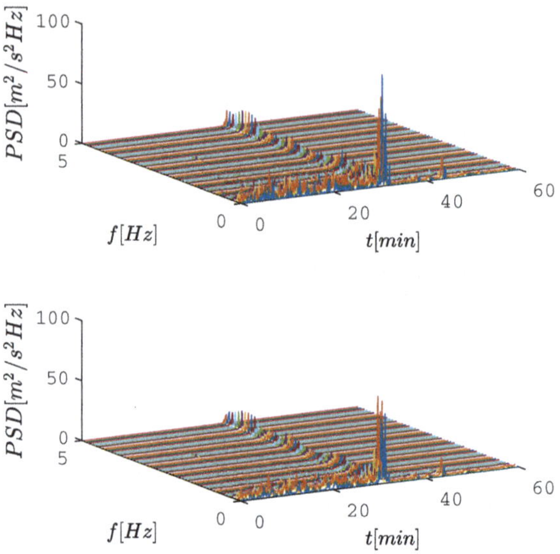

Fig. 6. Evolutionary spectral density model (top) and mean spectrogram from $n_s = 1000$ generated samples (bottom).

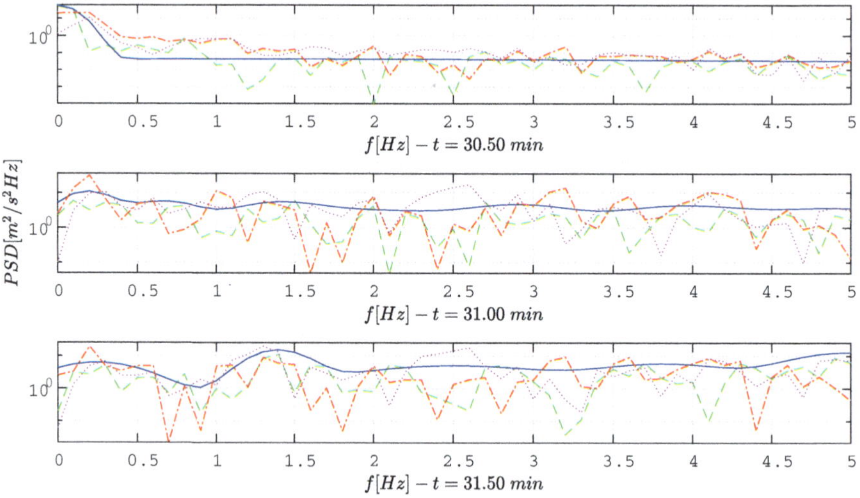

Fig. 7. Spectral density estimates at the three time instants: experimental (dashed line), model (continuous line), sample $x_Q^{(1)}(t)$ (dash-dotted line), sample $x_Q^{(2)}(t)$ (dotted line).

3 Influence of Non-stationarity on Structural Response

In the previous Section the main non-stationary features of the wind velocity fluctuations during a thunderstorm outflow were modeled and an effective algorithm to generate virtual samples was proposed. In this Section, the influence of these non-stationary properties on the structural response is investigated using Monte Carlo simulation.

For sake of comparison a stationary Gaussian process $Z(t)$ with zero mean and spectral density

$$S_Z(f) = \frac{\alpha\, \sigma_Z^2}{(1+50\, f)^{\frac{5}{3}}}. \tag{8}$$

representing the typical synoptic wind velocity fluctuation is considered [16]. The variance σ_Z^2 is calibrated on the interval $0 < t < 29$ minutes of the time series $u(t)$ in Fig. 1 assuming the average value of the variance fluctuations in Fig. 5 (top) where the process has limited variance fluctuations, i.e. $\sigma_Z^2 = E[\sigma_u^2(t)] = 2.47\,\text{m}^2/\text{s}^2$.

Equation (7) and the classical method in [9] are used to generate $n_s = 1000$ samples of $X_Q(t)$ and $Z(t)$, respectively. These samples are then used to obtain wind forces of the kind

$$f_{x_i}(t) = 0.5\rho C_d A x_i^2(t), \quad i = 1, ..., n_s \tag{9}$$

$$f_{z_i}(t) = 0.5\rho C_d A z_i^2(t), \quad i = 1, ..., n_s \tag{10}$$

where ρ is the air mass density, $C_d = 2.06$ is the drag coefficient and $A = 1$ is the tributary area.

Classical numerical integration [15] is used to obtain the response time series $u_{x_i}(t)$ and $u_{z_i}(t)$, $i = 1, 2, ..., n_s$, of linear sdof systems with natural frequency ω and damping ratio ξ affected by the wind loads in (9) and (10). These responses are then compared trying to investigate the effect of the sudden increase of the wind velocity fluctuations induced by the thunderstorm outflow, with respect to a stationary wind with properties calibrated on the first 29 min of $u(t)$ in Fig. 1. In particular, the main focus is on the effects of changes with time of both variance (Fig. 5) and spectral density function on the extremes of the system response. Figure 8 compares the assumed models for the spectral density of $X(t)$ (Eq. (6), continuous lines) and $Z(t)$ (Eq. (8), dashed line).

In order to compare the system responses the following ratio is considered

$$r_i = \frac{\max_{a \le t \le b} |u_{x_i}(t)|}{\max_{a \le t \le b} |u_{z_i}(t)|}, \quad i = 1, ..., n_s \tag{11}$$

where $a = 25.5$ min and $b = 35.5$ min define the 10-min interval including the thunderstorm outflow. Figure 9 reports the $n_s = 1000$ samples of the ratio in Eq. (11) for four linear sdof systems defined by damping ratio $\xi = 0.02$ and natural frequencies $f_0 = 1.02, 1.59, 2.82, 4.06$ Hz. The horizontal lines show the average ratio value estimated on the n_s samples. The response to the non-stationary forces is increasing with the sdof natural frequency. This result was expected and

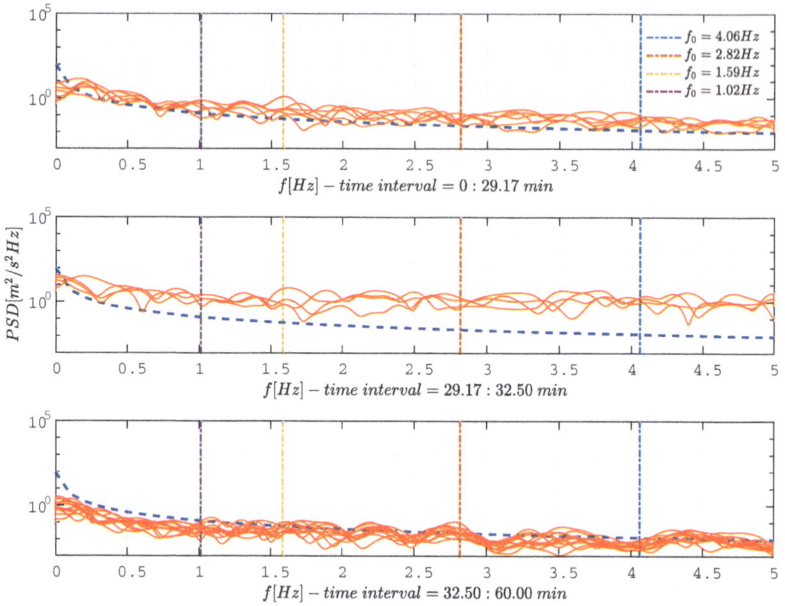

Fig. 8. Spectral density models in the tree main time intervals: stationary model in (8) (dashed line) and non-stationary model in (4) (continuous lines). The vertical lines indicate the frequencies of the sdof systems investigated in the comparison.

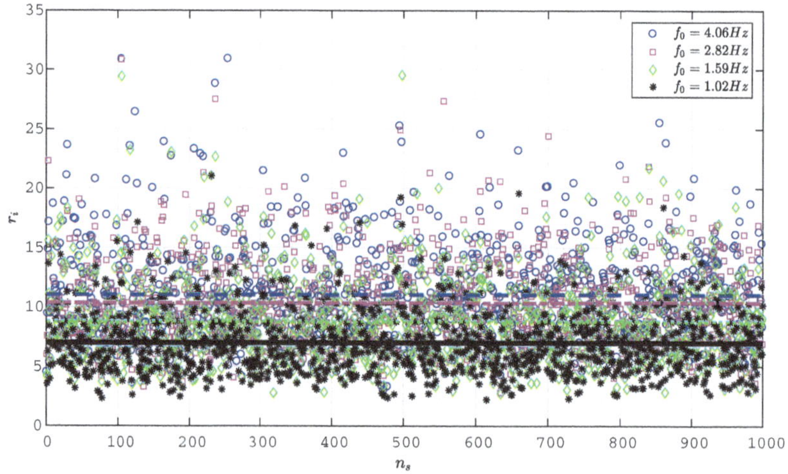

Fig. 9. Ratios r_i, $i = 1, 2, ..., n_s = 1000$, for sdof systems with damping ratio $\xi = 0.02$ and different values of natural frequencies f_0. The horizontal lines correspond to the four ratio average values: $f_0 = 4.06$ Hz (dashed), $f_0 = 2.82$ Hz (dashdotted), $f_0 = 1.59$ Hz (dotted), $f_0 = 1.02$ Hz (continuous).

it is consistent with the frequency content of the wind load fluctuations reported in Fig. 8 (both stationary and non-stationary). The vertical lines indicate the four natural frequencies of the considered sdof systems.

In order to extend the investigation described above the whole response spectra [11,17] is estimated by Monte Carlo simulation starting from the generation of the stationary and non-stationary wind velocity fluctuations $Z(t)$ and $X(t)$.

First, the samples of the stationary $f_{z_i}(t)$ and non-stationary $f_{x_i}(t)$ wind loads, $i = 1, 2, ..., n_s = 1000$, described above are used to compute the sdof responses for each of the frequencies in the range $0 < f \leq 5$ Hz, with damping ratio $\xi = 0.02$. Second, the average response spectra

$$\bar{S}_{d_Z}(f) = \frac{\sum_{i=1}^{n_s} \max_{a \leq t \leq b} |u_{z_i}(t)|}{n_s} \tag{12}$$

$$\bar{S}_{d_X}(f) = \frac{\sum_{i=1}^{n_s} \max_{a \leq t \leq b} |u_{x_i}(t)|}{n_s} \tag{13}$$

in the interval $[a, b]$ are estimated. Finally the mean pseudo-velocity response spectrum and mean pseudo-acceleration response spectrum are estimated using

$$\begin{aligned} \bar{S}_{v_Z}(f) &= 2\pi f \bar{S}_{d_Z}(f) \\ \bar{S}_{v_X}(f) &= 2\pi f \bar{S}_{d_X}(f) \end{aligned} \tag{14}$$

and

$$\begin{aligned} \bar{S}_{a_Z}(f) &= (2\pi f)^2 \bar{S}_{d_Z}(f) \\ \bar{S}_{a_X}(f) &= (2\pi f)^2 \bar{S}_{d_X}(f) \end{aligned} \tag{15}$$

This tree step algorithm is repeated for sdof systems with damping ratio $\xi = 0.25$. Figure 10 shows the results of the estimations obtained with Eqs. (12), (13), (14) and (15) in the top, middle, lower panels, respectively. This simple investigation gives clear information: the change with time of variance and frequency content of the spectral density function caused by thunderstorm outflow seems to have greater influence as the system natural frequency increases. This would indicate a greater sensitivity to thunderstorms for higher modes of structures such as tall cranes, posts and power line supports that experienced severe damages and/or failures as reported by the research group in [19].

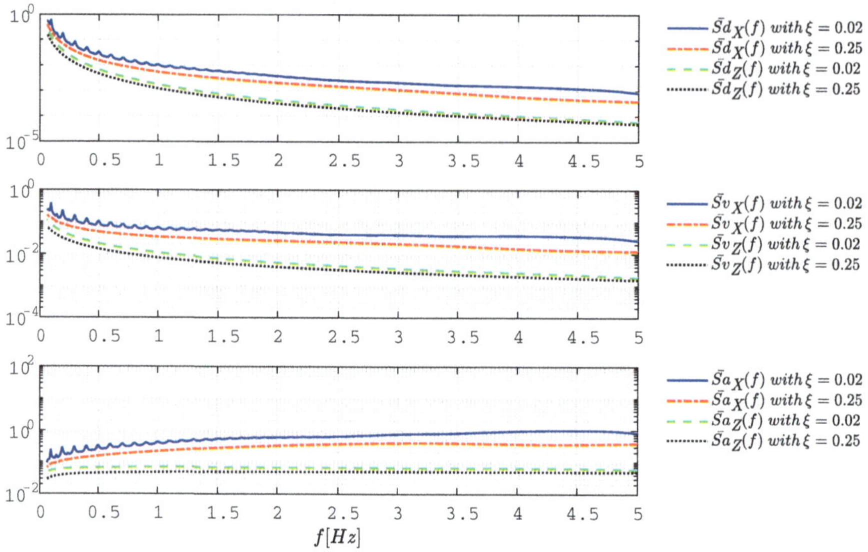

Fig. 10. Mean response spectra (top), mean pseudo-velocity response spectra (middle) and pseudo-acceleration response spectra (bottom) estimated for linear sdof systems with $\xi = 0.02$ and $\xi = 0.25$.

4 Conclusions

In this paper a simple and effective non-stationary model was proposed to describe wind velocity fluctuations of a typical thunderstorm in the northern Italian coast. The model was introduced in the literature to describe earthquake time series modulating both amplitude and frequency of a stationary process. Experimental data were used to calibrate the probabilistic model and an algorithm to generate non-stationary wind velocity fluctuations was developed. The obtained virtual time histories were used to investigate the effect of the non-stationary wind load fluctuations on the structural response using Monte Carlo simulation. Response spectra were estimated for both stationary and non-stationary wind loads and then compared. The obtained results seem to indicate a greater sensitivity of higher modes in tall slender structures to thunderstorm outflows. There is a significant influence given by the time variations of the wind velocity variance and the spectral density function.

Work is in progress to investigate into the influence of linear damping and to identify the single contribution of the change in frequency content of the time variant wind spectral density function during a thunderstorm.

Acknowledgments. These Authors are indebted to Prof. Giovanni Solari for providing the wind velocity time series used in this work. The support of Ing. Stefano Fioravanti is also gratefully acknowledged.

References

1. Gioffrè M, Gusella V (2007) Peak response of a nonlinear beam. J Eng Mech 133(9):963–969
2. Cluni F, Gioffrè M, Gusella V (2013) Dynamic response of tall buildings to wind loads by reduced order equivalent shear-beam models. J Wind Eng Ind Aerody 123:339–348
3. Ouyang Z, Spence SMJ (2019) A performance-based damage estimation framework for the building envelope of wind-excited engineered structures. J Wind Eng Ind Aerodyn 186:139–154
4. Aboshosha H, El Damatty A (2015) Engineering method for estimating the reactions of transmission line conductors under downburst winds. Eng Struct 99:272–284
5. Chen L, Letchford C (2004) Parametric study on the along-wind response of the caarc building to downbursts in the time domain. J Wind Eng Ind Aerodyn 92(9):703–724
6. Chen L, Letchford C (2007) Numerical simulation of extreme winds from thunderstorm downbursts. J Wind Eng Ind Aerodyn 95(9–11):977–990
7. Chen X (2008) Analysis of alongwind tall building response to transient nonstationary winds. J Struct Eng 134(5):782–791
8. Choi EC, Hidayat FA (2002) Dynamic response of structures to thunderstorm winds. Prog Struct Eng Mat 4(4):408–416
9. Deodatis G (1996) Simulation of ergodic multivariate stochastic processes. J Eng Mech 122(8):778–787. https://doi.org/10.1061/(ASCE)0733-9399(1996)122:8(778)
10. Grigoriu M, Ruiz SE, Rosenblueth E (1988) The mexico earthquake of september 19, 1985 – nonstationary models of seismic ground acceleration. Earthq Spectra 4(3):551–568
11. Hadjian A (1981) Seismic response of structures by the response spectrum method. Nucl Eng Des 66(2):179–201
12. Huang G, Chen X, Liao H, Li M (2013) Predicting tall building response to nonstationary winds using multiple wind speed samples. Wind Struct 17(2):227–244
13. Kwon DK, Kareem A (2009) Gust-front factor: new framework for wind load effects on structures. J Struct Eng 135(6):717–732
14. Le TH, Caracoglia L (2015) Reduced-order wavelet-galerkin solution for the coupled, nonlinear stochastic response of slender buildings in transient winds. J Sound Vib 344:179–208
15. Newmark NM (1959) A method of computation for structural dynamics. J Eng Mech Div 85(3):67–94
16. Simiu E, Scanlan R (1996) Winds effects on structures: fundamentals and applications to design. A Wiley-Interscience Publication, Wiley. https://books.google.it/books?id=DcSZTIMZObEC
17. Solari G (1989) Wind response spectrum. J Eng Mech 115(9):2057–2073
18. Solari G, Burlando M, De Gaetano P, Repetto MP (2015) Characteristics of thunderstorms relevant to the wind loading of structures. Wind Struct 20(6):763–791
19. Solari G, De Gaetano P, Repetto MP (2015) Thunderstorm response spectrum: fundamentals and case study. J Wind Eng Ind Aerodyn 143:62–77
20. Solari G, Rainisio D, De Gaetano P (2017) Hybrid simulation of thunderstorm outflows and wind-excited response of structures. Meccanica 52(13):3197–3220

On the Aerostructural Design of Long-Span Cable-Stayed Bridges: The Contribution of Parameter Variation Studies with Focus on the Deck Design

M. Cid Montoya(✉), F. Nieto, S. Hernández, and J. Á. Jurado

Structural Mechanics Group (GME), School of Civil Engineering, University of La Coruña, Campus de Elviña s/n, 15071 La Coruña, Spain
{miguel.cid.montoya, fnieto, hernandez, jjurado}@udc.es

Abstract. The design of long-span bridges consists of dealing with a large number of design variables and it is conditioned by responses of structural and aeroelastic nature. A deep knowledge about the influence of these variables governing the bridge responses is crucial to achieve efficient and safe designs. Apart from heuristic rules, numerical approaches, such as parameter variation studies, sensitivity analysis and optimization algorithms, can provide reliable information to improve designs. This work studies the effects on the flutter and structural responses of a cable-stayed bridge when the mechanical, mass, aerodynamic and aeroelastic properties of a streamlined mono-box deck are modified. These results are used to understand qualitatively and quantitatively the effects caused by the variation of the deck plate thickness and cross-section shape on the bridge responses.

Keywords: Long-span bridges · Bridge aerodynamics · Aeroelasticity · Aerostructural design · Parameter variation studies · Sensitivity analyses

1 Introduction

Parameter variation studies and sensitivity analyses are instrumental techniques for designers in all disciplines. A deep knowledge of the effect in the structure's responses of possible variations in the design variables considered is pivotal. Particularly, in the case of long-span bridges, the high number of possible design variables and the many responses to consider, make preliminary parameter variation studies a compulsory step in the design process. This is also very useful when selecting the most adequate design variables for the formulation of an optimization problem (see the authors' previous works by Nieto et al. 2009, and Cid Montoya et al. 2018a, b).

These techniques have been previously implemented to analyse the responses of cable-stayed bridges. For instance, in Zhang and Sun (2003), and Zhang (2006), parameter variations studies were conducted to study the influence of geometrical variations of the layout of the bridge in the flutter response. In Jurado et al. (2008), the aeroelastic performance of several cable arrangement was studied.

F. Ricciardelli and A. M. Avossa (Eds.): INVENTO 2018, LNCE 27, pp. 234–246, 2019.
https://doi.org/10.1007/978-3-030-12815-9_19

In the case of suspension bridges, parametric studies were carried out by Nieto et al. (2011), or Wang et al. (2014), where the influence in flutter of some relevant mechanical properties of the deck and towers were studied. In Øiseth et al. (2010), the influence of each flutter derivative is studied. In Andersen et al. (2016), the role of the frequency ratio is analysed. In Argentini et al. (2014), the effect of variations in both mechanical and aerodynamic properties are systematically analysed. In Tao et al. (2017), a parametric study of the buffeting responses of a suspension bridge is reported.

The results reported in this piece of research are part of the output of an ongoing research in which variations in a broader set of design variables are studied. This work reports some of the results of the parameter variations studies obtained when modifying design variables related with the bridge deck. The goal of the research is to carry out sensitivity analysis using finite differences in order to identify the most relevant design variables and their influence in the bridge structural and aeroelastic responses of super-long-span cable-stayed bridges.

2 Description of the Super-Long-Span Cable-Stayed Bridge Model

The application case used in this work to carry out the parameter variation studies is a super-long-span cable-stayed bridge with a main span of 1316 m, described in the authors' previous works Cid Montoya et al. (2018a, b). The bridge layout is similar to the Russky or the Sutong Bridges, and an scheme is shown in Fig. 1. The baseline geometry of the deck cross-section is the well-known Scanlan's G1 section, which is similar to the box deck cross-section of the Great Belt Bridge, with the following initial dimensions: $B_0 = 40$ m, $H_0 = 5.6$ m and $t_0 = 2.5$ cm. Hence, the mechanical properties of the deck cross-section are $A_x = 2.06$ m^2, $J_x = 41.36$ m^4, $I_y = 12.52$ m^4 and $I_z = 284.70$ m^4. The deck non-structural mass and mass moment of inertia are $M_x = 8$ T/m and $I_M = 2300$ Tm/m. The stays supporting system consist of two parallel planes of 40 stays. The cross-section area of the backstays is $A^B = 0.5$ m^2 and the area of the other stays is $A^s = 0.05$ m^2. The structural damping is set to $\xi = 0.01$.

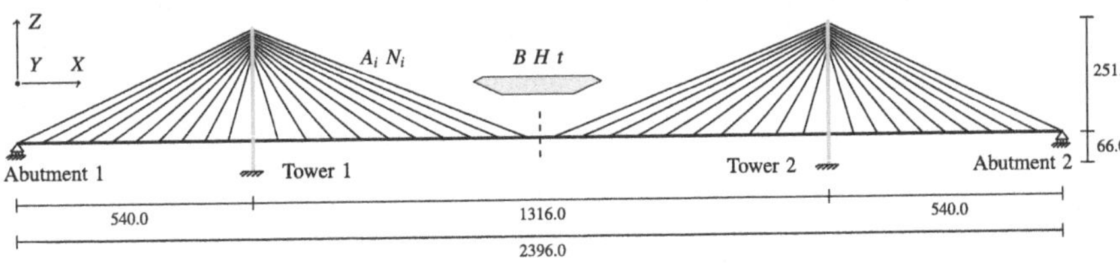

Fig. 1. Layout of the cable-stayed bridge model.

The most relevant natural frequencies of the bridge for flutter analysis are reported in Table 1 and they are in accordance with those of the Sutong Bridge. The flutter velocity of the bridge was obtained using the numerical approach developed in the author's previous work Cid Montoya et al. (2018a). It consists of estimating of the force coefficients and their slopes of the deck cross-section using a surrogate model that was built from the result of CFD analyses, obtaining the flutter derivatives taking advantage of the formulation derived from the quasi-steady theory and assessing the flutter velocity by the so-called multi-modal analysis, as described, for instance, in Jurado et al. (2011).

Table 1. Natural frequencies, modal shapes and participations of the main modes in flutter.

Mode #	f [Hz]	Shape	Contribution [%]
5	0.203	V1s	33.545
8	0.351	V3s	8.060
10	0.419	L3s & T1s	12.762
11	0.438	V5s	11.620
12	0.461	T1s	21.880
15	0.518	V5s	6.238
16	0.533	L3s & T1s	1.530

The flutter velocity obtained by means of this numerical procedure is 126.6 m/s, and the participation of each mode is shown in Table 1.

3 Influence of Deck Mechanical Properties on Flutter Response

In this section the influence of the deck cross-section mechanical properties in the flutter response is analysed. Assessing the changes in the bridge responses is of utmost interest in order to understand how it will perform when shape and size design variables are modified. The mechanical properties are modified by multiplying their original value by a set of factors ranging from 0.1 to 3, focusing on values close to the original baseline design. This wide range of variation was chosen aiming to explore all possible designs alternatives even though they could be considered outside the range of common values used in bridge engineering practice. This approach will be used all along this work in order to analyse the performance of both conventional and non-conventional designs. Sensitivity studies of the influence of the deck mechanical properties can be found in the works by Nieto et al. (2011), and Jurado and Hernández (2004).

3.1 Deck Cross-section Area A_x

Modifying the deck cross-section area affects to the stiffness of the deck, and more importantly, to its structural translational mass. Consequently, the vertical and lateral mode shapes are drastically affected, as it can be seen in Fig. 2. It can be appreciated how the vertical modes (namely, 5, 8, 11 and 15) reduce the value of their natural frequencies, while the pure torsional mode (12) keeps its value constant. However, mode 10, which presents a combined torsional-lateral shape, decreases its value due to the existence of a lateral component. The flutter velocity slightly increases its value in the neighborhoods of the original design as A_x grows, given that the torsional to flexural frequency ratio (modes 12 and 5) $\gamma_{\omega 12-5}$ grows with A_x.

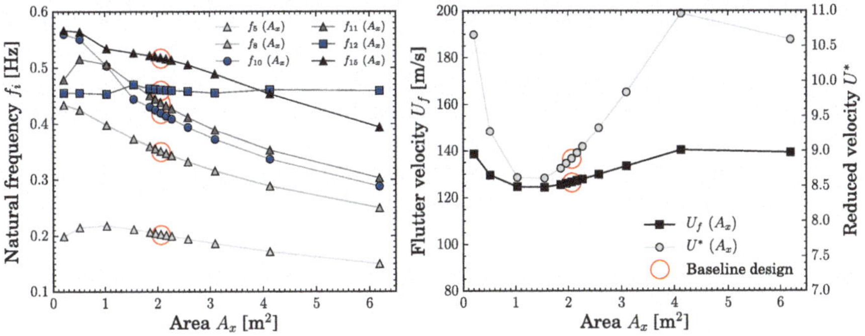

Fig. 2. Influence of A_x in the dynamic and flutter responses of the bridge.

3.2 Deck Cross-section Vertical Bending Moment I_y

Since the structural mass moment of inertia is given by $I_M = \rho(I_y + I_z)$, the vertical bending moment affects to I_M, although its influence is much lower than the one of I_z because of its lower absolute value. Therefore, the torsional modes slightly decrease the value of the natural frequency when increasing I_y. The influence of this property on the bridge vertical stiffness is reduced since the major contribution is provided by the cable supporting system. This is clearly shown in the variation of the natural frequency of the first vertical mode when I_y is modified, as shown in Fig. 3. However, the influence of I_y is larger in higher vertical modes, since those mode shapes are more dependent on the deck stiffness as the number of nodes of the shapes grow. On the other hand, the frequency ratio is almost constant in the range of values of I_y considered, which leads to an almost non-sensible response of the flutter velocity of the bridge. Similar conclusions were reported in Wang et al. (2014).

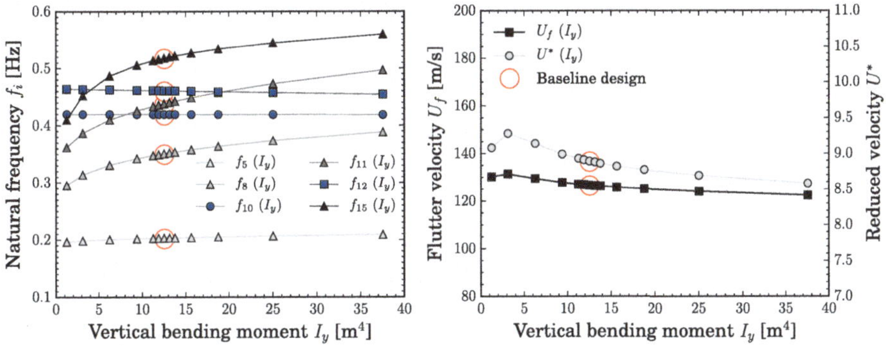

Fig. 3. Influence of I_y in the dynamic and flutter responses of the bridge.

3.3 Deck Cross-section Lateral Bending Moment I_z

Given that the absolute value of I_z is much higher than I_y in deck cross-sections, the variation induced in I_M is larger when I_z is modified. Hence, the variations introduced in the natural frequencies of the torsional modes are higher than those induced by variations on I_y. It can be appreciated in Fig. 4 that the frequency of the pure torsional mode 10 decreases its value as the value of I_z, and subsequently I_M, grows. However, the combined lateral-torsional modes, namely 12 and 16, increase their values of frequency with I_z, given that the influence of the stiffness is dominant over the influence of I_M. Alternatively, it can be appreciated that the vertical modes are insensitive to changes in I_z. The flutter velocity is considerably affected since the variations on the torsional modes noticeably affects the frequency ratio $\gamma_{\omega_{12-5}}$.

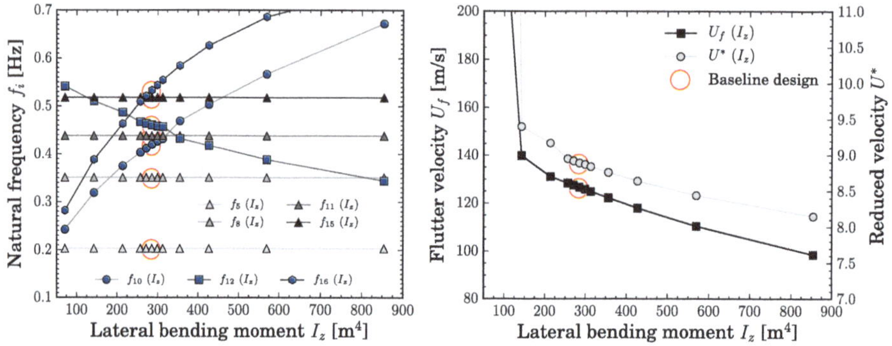

Fig. 4. Influence of I_z in the dynamic and flutter responses of the bridge.

3.4 Deck Cross-section Torsional Stiffness J_x

Variations in J_x are very influent in the value of the natural frequency of the pure torsional mode, while other modes are insensitive to these variations. This affects drastically to the frequency ratio and subsequently to the flutter velocity (see Fig. 5), which presents the most dramatic variation in the critical flutter speed when any of the properties of the deck are modified. Similar results are reported in Wang et al. (2014).

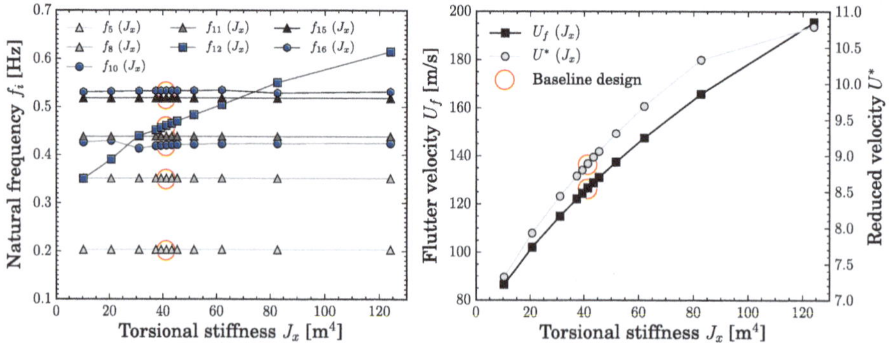

Fig. 5. Influence of J_x in the dynamic and flutter responses of the bridge.

3.5 Deck Translational Non-structural Mass Per Unit of Length M_x

The total mass of a bridge deck is composed by the structural mass and the non-structural mass. The influence of the non-structural mass in the dynamic and flutter response of the bridge is obviously the same as the structural mass. The interest in the non-structural mass is that it can be modified independently to the deck shape and size, while the structural mass is inherently dependant on the mechanical properties.

Focusing on the translational non-structural mass, it can be seen in Fig. 6 that the influence on the dynamic response of the bridge is similar to the influence of modifications in A_x. Lateral and vertical modes reduce their value of the natural frequency, while pure torsional mode keeps the value of the natural frequency almost constant. This generates slight variations in the flutter response of the bridge.

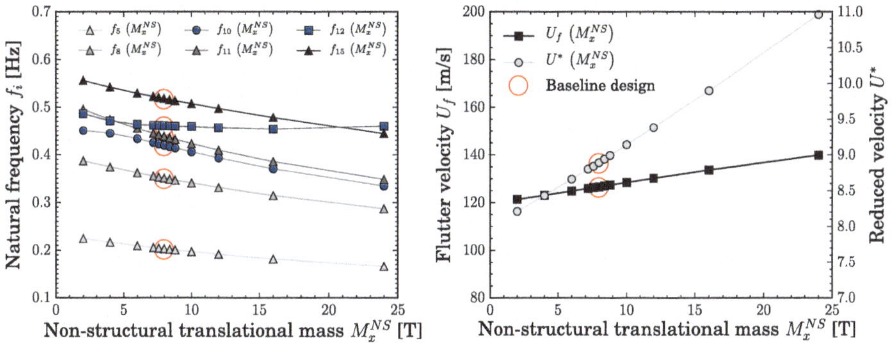

Fig. 6. Influence of M_x in the dynamic and flutter responses of the bridge.

3.6 Deck Rotational Non-structural Mass Per Unit of Length I_M

The effect of the rotational mass in the dynamic responses of the bridge is the opposite of the effect caused by the translational mass. It can be seen in Fig. 7 that the translational vertical and lateral modes are insensitive to variations in the value of I_M, while the pure torsional mode is deeply affected by changes in I_M. This motivates a drastic change in the flutter response of the bridge. It is also worthy to mention that when I_M takes values higher than 3500 Tm, the shapes of modes 10 and 12 are switched, changing mode 12 from pure torsional to torsional-lateral shape and mode 10 from torsional-lateral shape to pure torsional. This explains why for values of I_M higher than 3500 Tm mode 12 became insensitive to variations in I_M, while mode 10 became very sensitive to these variations.

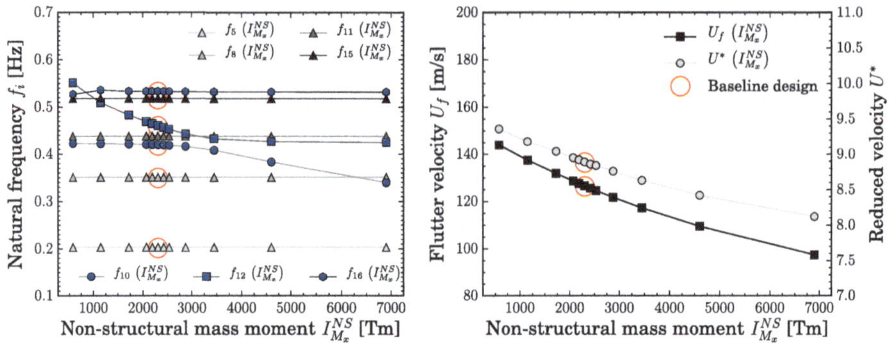

Fig. 7. Influence of I_M in the dynamic and flutter responses of the bridge.

4 Influence of Natural Frequencies on Flutter Response

In this section the influence of variations in the value of the most relevant natural frequencies (see Table 1) on the flutter response of the bridge is studied. Figure 8 reports the results obtained in terms of flutter velocity and reduced velocity when each natural frequency is modified independently. These analyses help to stablish qualitatively and quantitatively which are the most influential modes on the flutter response, and they are very useful to understand the conclusions raised in the previous section. It can be appreciated that both torsional modes, especially the pure torsional model (12), and the first vertical symmetrical mode are the most influential ones, as expected. This highlights the role of the frequency ratio $\gamma_{\omega_{12-5}}$ to ascertain the flutter response of the bridge (see Matsumoto et al. 2010). Mode 12 causes the most dramatic variations on the flutter responses, which explains that the deck properties responsible for the most important variations on the flutter responses reported in Sect. 3 were J_x and I_M.

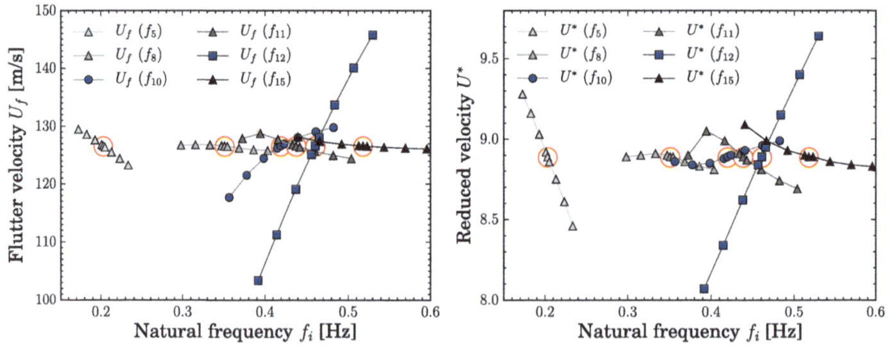

Fig. 8. Influence of each natural frequency in the flutter response of the bridge.

5 Influence of Deck Plate Thickness *t* on Flutter and Structural Responses

In this section the role of the uniform deck plate thickness t in the flutter response of the bridge is studied. The effect of variations in the width B and depth H for this bridge model were studied in the authors' previous work Cid Montoya et al. (2018b).

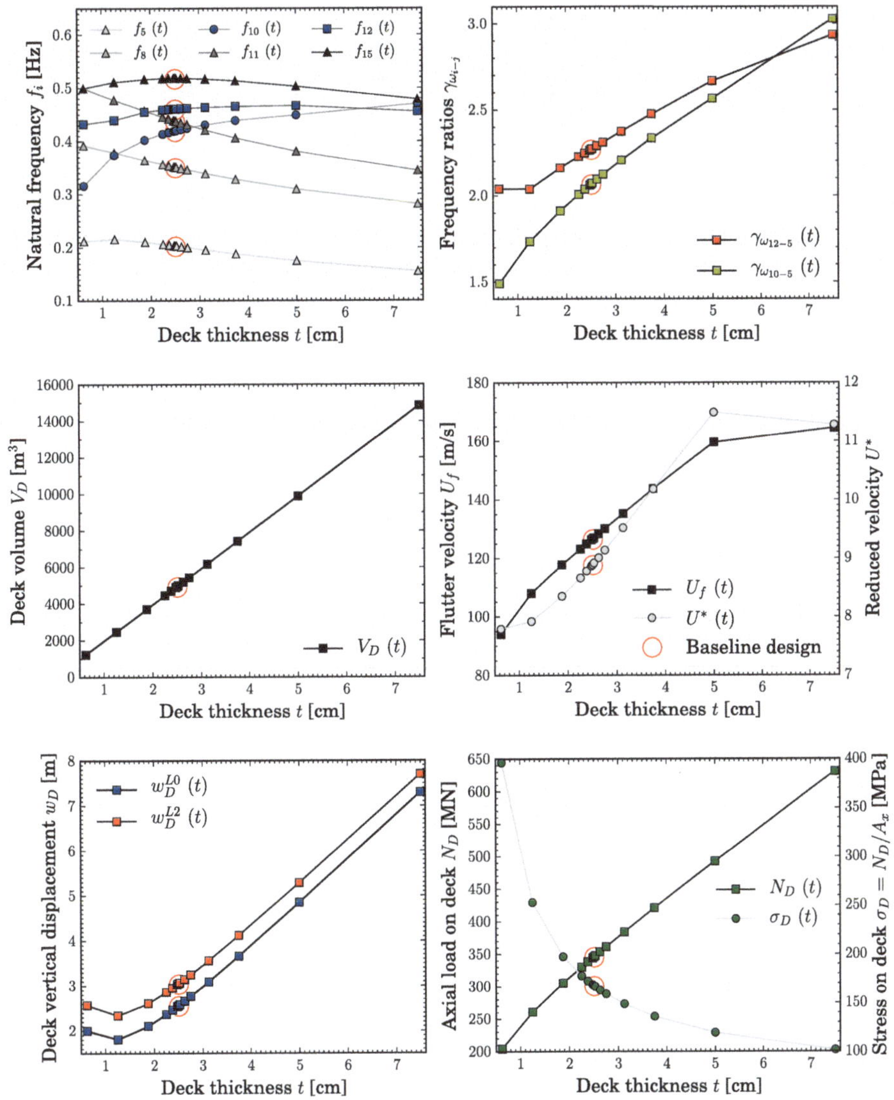

Fig. 9. Influence of t in flutter and structural responses of the bridge: natural frequencies f_i, torsional to vertical frequencies ratios $\gamma_{\omega i-j}$, flutter velocity U_f and reduced velocity U^*, deck vertical displacement w_D under load cases L_0 and L_2 and axial load N_D and stress σ_D on deck under load case L_0.

Variations in the value of t affects to all mechanical properties, generating linear changes in their values in all cases. This influences the dynamic responses as shown in Fig. 9. The value of the natural frequencies of all the vertical translational modes is reduced as the thickness is increased because of the growth in the translational mass, while the natural frequencies of the torsional modes slightly increase their values, affected by the simultaneous growth of J_x and I_M, which as studied in Sect. 3, generates contradictory effects on the value of the natural frequencies of those modes. As a consequence, the resulting frequency ratios $\gamma_{\omega_{12-5}}$ and $\gamma_{\omega_{10-5}}$ present a growing trend with the increase of t. Subsequently, the flutter velocity of the bridge grows with t.

Some of the most relevant structural responses of the bridge are also reported in Fig. 9. The structural analyses were carried out considering two load cases: L_0 and L_2. L_0 consists on a uniform overload along the whole deck and self-weight, while L_2 only considers the uniform load in the main span and self-weight (see Cid Montoya et al. 2018b for more details). It can be appreciated in Fig. 9 that the deck vertical displacement at the center of the main span of the bridge grows with t, because the growth in the translational mass is more influential than the increase in the vertical bending stiffness, as it is reflected in the values of the natural frequencies of the vertical modes. Alternatively, the stress level in the deck cross-section at the tower decreases with the growth of t. This represents a trade-off in the structural design of cable-stayed bridges.

6 Influence of Stays Transversal Separation Ω on Flutter Response

Another key aspect in the design of long-span bridges is the separation of the stays supporting the deck, as shown in Fig. 10. This parameter drastically affects the torsional stiffness of the deck, and subsequently to the natural frequency of the torsional mode, which is the most influential one in the flutter response, as studied in Sect. 4.

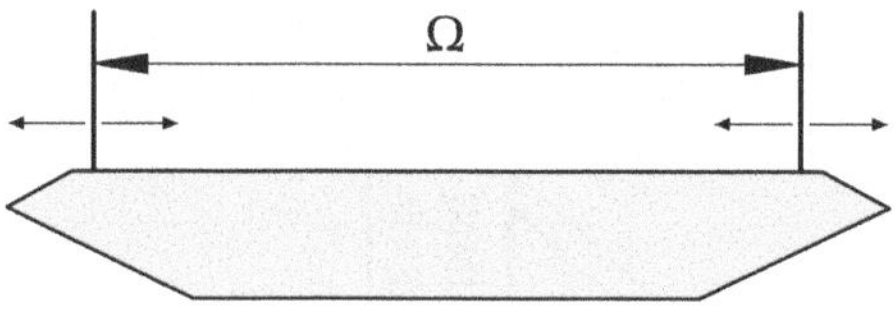

Fig. 10. Representation of the stays transversal separation Ω.

It can be seen in Fig. 11 that mode 12 is drastically affected by the variation of Ω, and subsequently the flutter velocity dramatically modifies its value, as well as the reduced velocity. Similar studies were reported by Zhang (2006). It must be noticed

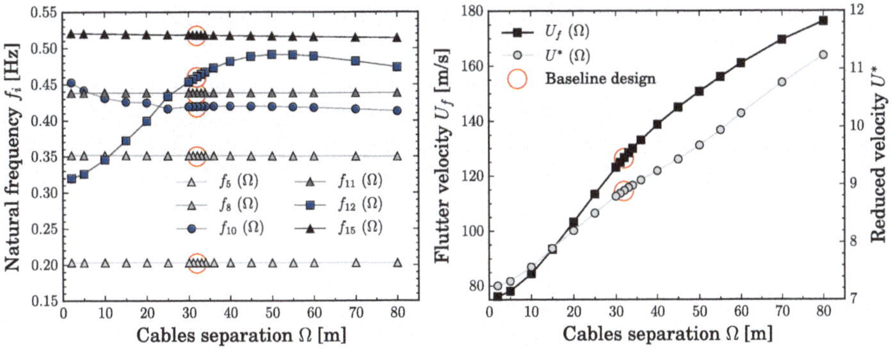

Fig. 11. Influence of Ω in the dynamic and flutter responses of the bridge.

that when Ω tends to 0, all the torsional stiffness is provided by the deck itself. In the case of mono-box deck cross-sections, the deck can provide enough torsional stiffness. Alternatively, for other deck cross-sections a more drastic reduction of the natural frequency of the torsion mode may take place, leading to frequency ratios even lower than 1. This is studied, for instance, in Andersen et al. (2016).

7 Influence of Structural Damping ξ on Flutter Response

The structural damping is a parameter with a relevant effect in the value of the flutter velocity. Figure 12 shows the values of flutter velocity and reduced velocity obtained for a set of structural damping values in order to ascertain its influence on the bridge under study. The flutter velocity is almost linearly increased as the value of the structural damping grows. It is interesting to mention that when the damping value tends to 0, the flutter velocity converges to a constant value, in this case to 120 m/s.

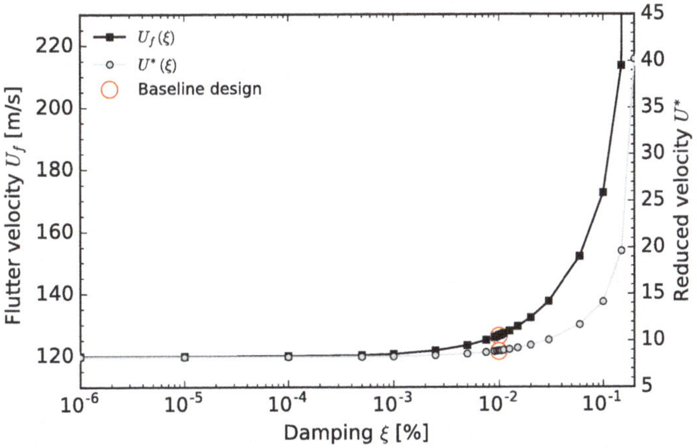

Fig. 12. Influence of ξ in the flutter response of the bridge.

8 Influence of Flutter Derivatives on Flutter Response

The aerodynamic and aeroelastic properties of the deck cross-section can also have a dramatic influence on the aeroelastic responses of the bridge (see Matsumoto et al. 1996). In this section, the influence of the most relevant flutter derivatives in the flutter response is studied. Table 2 reports the percentage changes of the flutter velocity when each flutter derivative is independently multiplied by a given factor λ. It can be seen that A_1^*, A_2^*, A_3^*, and H_3^* are more influential in the flutter response, being the A_3^* the most influential in this case. Similar results were obtained by Øiseth et al. (2010) for the Handanger Bridge, and in Argentini et al. (2014), for the Messina Strait Bridge project.

Table 2. Percentage change of the flutter velocity when each flutter derivative has been scaled separately by the factor λ.

Factor λ	H_1^*	H_2^*	H_3^*	A_1^*	A_2^*	A_3^*
0	−7.3%	−3.3%	-	-	−29.6%	77.4%
0.25	−5.0%	−2.5%	-	-	−20.7%	43.1%
0.50	−3.3%	−1.7%	19.2%	15.5%	−12.5%	21.7%
0.75	−1.8%	−0.9%	7.3%	6.2%	−5.7%	9.0%
0.95	−0.4%	−0.2%	1.3%	1.1%	−1.1%	1.6%
1	0%	0%	0%	0%	0%	0%
1.01	0.4%	0.2%	−1.2%	−1.0%	1.0%	−1.5%
1.25	2.1%	0.9%	−5.4%	−4.7%	5.0%	−6.8%
1.50	4.8%	1.8%	−9.7%	−8.4%	9.6%	−12.2%
2	12.9%	3.6%	−16.4%	−14.1%	18.0%	-
2.5	-	5.6%	−21.2%	−18.2%	26.0%	-

9 Influence of the Slopes of the Force Coefficients on Flutter Response

Since in this work the flutter derivatives are estimated using the formulation derived from the quasi-steady theory, it is interesting to study how the flutter response is affected by the simultaneous variation of the flutter derivatives that are dependent in this formulation, or in other words, how the flutter velocity responds to variation in the slopes of the lift and moment coefficients. Figure 13 shows the values of flutter velocity and reduced velocity obtained for a set of values of C_L' and C_M' in the usual range that these quantities can adopt for mono-box deck cross-sections. The results reflect the already known concept of the inverse relationship of the slopes of the force coefficients with the aeroelastic performance of the sections when the values of the slopes are positive. It can be also appreciated that in this particular case the flutter velocity is more sensible to variations of the slope of the lift coefficient than to variations of the slope of the moment coefficient.

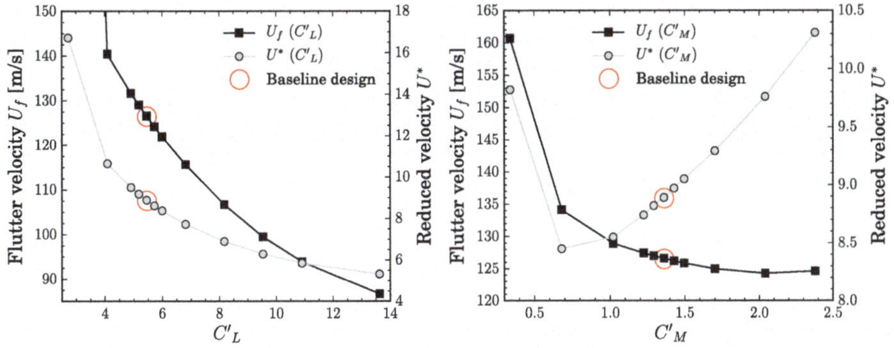

Fig. 13. Influence of C'_L and C'_M in the flutter response of the bridge.

10 Conclusions

From the results obtained for the cable-stayed bridge with mono-box deck cross-section under study some conceptual design guidelines can be stablished.

From the structural point of view, it has been shown that the most effective way to alter the flutter response of the bridge was to modify the torsional stiffness of the deck or the mass moment of inertia. An interesting way to achieve this is to modify the separation of the stays, since it allows to drastically increase the torsional stiffness without increasing the mass moment of inertia of the deck cross-section. It must be noted that the relocation of the anchorages can cause a relevant effect on the mass moment of inertia.

Regarding to the interpretation of the results, it has been seen that the flutter response can be qualitatively anticipated by analysing the frequency ratio.

From the aerodynamic point of view, the flutter derivatives more influential in the flutter phenomenon of this particular bridge have been identified, as well as the effect of the slopes of the lift and mass coefficients.

However, in order to obtain a deeper understanding for bridge design, sensitivity analyses must be conducted. This will be carried out in the following step of this ongoing research.

Acknowledgements. The research leading to these results has received funding from the Spanish Minister of Economy and Competitiveness (MINECO) with reference BIA2016-76656-R. The first author has been also founded by the Fundación Pedro Barrié de la Maza and the University of La Coruña. The authors fully acknowledge the support received.

References

Andersen MS, Johansson J, Brandt A, Hansen SO (2016) Aerodynamic stability of long span suspension bridges with low torsional natural frequencies. Eng Struct 120:82–91

Argentini T, Pagani A, Rocchi D, Zasso A (2014) Monte Carlo analysis of total damping and flutter speed of a long span bridge: effects of structural and aerodynamic uncertainties. J Wind Eng Ind Aerodyn 128:90–104

Cid Montoya M, Nieto F, Hernández S, Kusano I, Álvarez AJ, Jurado JÁ (2018a) CFD-based aeroelastic characterization of streamlined bridge deck cross-sections subject to shape modifications using surrogate models. J Wind Eng Ind Aerodyn 177:405–428
Cid Montoya M, Hernández S, Nieto F (2018b) Shape optimization of streamlined decks of cable-stayed bridges considering aeroelastic and structural constraints. J Wind Eng Ind Aerodyn 177:429–455
Jurado JÁ, Hernández S (2004) Sensitivity analysis of bridge flutter with respect to mechanical parameters of the deck. Struct Multidiscip Optim 27(4):272–283
Jurado JÁ, Nieto F, Hernández S, Mosquera A (2008) Efficient cable arrangement in cable stayed bridges based on sensitivity analysis of aeroelastic behaviour. Adv Eng Softw 39:757–763
Jurado JÁ, Hernández S, Nieto F, Mosquera A (2011) Bridge aeroelasticity: sensitivity analysis and optimal design. WIT Press, Southampton
Matsumoto M, Kobayashi Y, Shirato H (1996) The influence of aerodynamic derivatives on flutter. J Wind Eng Ind Aerodyn 60:227–239
Matsumoto M, Matsumiya H, Fujiwara S, Ito Y (2010) New consideration on flutter properties based on step-by-step analysis. J Wind Eng Ind Aerodyn 98:429–437
Nieto F, Hernández S, Jurado JÁ (2009) Optimum design of long-span suspension bridges considering aeroelastic and kinematic constraints. Struct Multidiscip Optim 39:133–151
Nieto F, Hernández S, Jurado JÁ, Mosquera A (2011) Analytical approach to sensitivity analysis of flutter speed in bridges considering variable deck mass. Adv Eng Softw 42(4):117–129
Øiseth O, Rönnquist A, Sigbjörnsson R (2010) Simplified prediction of wind-induced response and stability limit of slender long-span suspension bridges, based on modified quasi-steady theory: a case study. J Wind Eng Ind Aerodyn 98:730–741
Tao T, Wang H, Wu T (2017) Parametric study on buffeting performance of a long-span triple-tower suspension bridge. Struct Infrastruct Eng 14:381–399
Wang H, Tao T, Zhou R, Hua X, Kareem A (2014) Parameter sensitivity study on flutter stability of a long-span triple-tower suspension bridge. J Wind Eng Ind Aerodyn 128:12–21
Zhang X-J, Sun B-N (2003) Study of design parameters on flutter stability of cable-stayed bridges. Wind Struct 6(4):279–290
Zhang X-J (2006) Study of design parameters on flutter stability of cable-stayed-suspension hybrid bridges. Wind Struct 9(4):331–344

Structural Optimization of a Reduced-Scale Model of a Wind Mill Tower for Wind Tunnel Testing

M. Cid Montoya(✉), S. Hernández, C. López, A. Álvarez, L. Romera, and J. Á. Jurado

Structural Mechanics Group (GME), School of Civil Engineering, University of La Coruña, Campus de Elviña s/n, 15071 La Coruña, Spain
{miguel.cid.montoya, hernandez, carlos.lopez.rodriguez, antonio.jose.alvarez, lromera, jjurado}@udc.es

Abstract. This paper describes a research carried out for the design of a wind mill tower reduced-scale model to be built and tested in the boundary layer wind tunnel at the University of La Coruña. It is well known that one of the many requirement of a reduced-scale model is its capability of reproduce the natural frequencies and vibration modes of the real structure. In this research, this is achieved by using a technique developed by the authors in a previous work and adapted to this particular case. This technique is based on the use of optimization algorithms, allowing the minimization of the error between these dynamic properties of the reduced-scale model and the target values. The numerical results are presented.

Keywords: Wind mill towers · Aeroelastic scaling · Structural dynamics · Reduced-scale models · Wind tunnel · Optimization

1 Introduction

Although in the early years of wind mill towers steel was the material used to build these constructions, more recently several companies have created designs using prefabricated segments of reinforced or prestressed concrete for towers that can be up to 120 m tall in order to reduce costs and improve their position in such a competitive market. During the construction phases and after completion, but before installing blades, they can be subjected to aeroelastic phenomena, for instance, vortex shedding excitation (Simiu and Scanlan 1996; Zasso et al. 2008; Muggiasca et al. 2018). Hence, wind tunnel tests using reduced-scale models can be very convenient to anticipate the performance of the real constructions (Davenport and Isyumov 1967; Irwin 2008, 2009).

During the construction phases, depending on the approach adopted, these structures can present different structural behavior. The case studied in this work consists of a wind mill tower made of four precast reinforced concrete segments. The construction procedure consist in locating each one of the precast segments, prestressing them altogether after placed in their final position and finally sealing the joints between the concrete segments. This procedure leads to a non-linear structural response of the tower

F. Ricciardelli and A. M. Avossa (Eds.): INVENTO 2018, LNCE 27, pp. 247–259, 2019.
https://doi.org/10.1007/978-3-030-12815-9_20

before it is finished, caused by the non-linear behavior of the joints between each precast segment. This complicates the design of the reduced-scale model for wind tunnel testing.

This work describes the procedure adopted for the aeroelastic design of a reduced model of the wind mill tower made of precast reinforced concrete segments to be built and tested in the boundary layer wind tunnel at the University of La Coruña, which is currently in progress. The goal of the process is to maintain dynamic similarity with the non-linear real construction, whose non-linear dynamic responses are anticipated by using finite element models (FEM). The procedure was developed in the authors' previous work Cid Montoya et al. (2017) and adapted for this particular case.

2 Procedure for the Design of Reduced-Scale Aeroelastic Models Maintaining Dynamic Similarity

The procedure used in this work consists of designing a reduced-scale model by means of the traditional design process (Phase 1), following with the model optimization in order to accurately reproduce the dynamic responses of the real structure, specifically, the natural frequencies (Phase 2). In this manner, the resulting design would guarantee the structural similarity which is of utmost relevance for the accuracy of the wind tunnel tests. The procedure is sketched in Fig. 1, and described in the following sections.

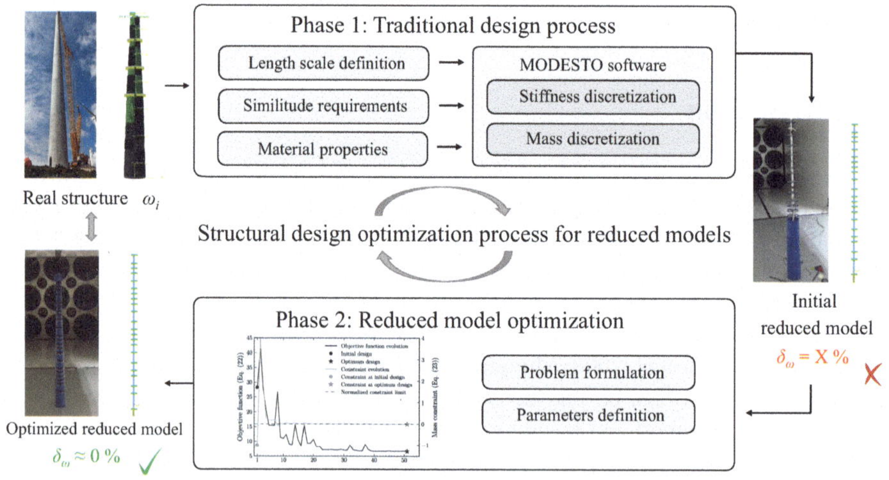

Fig. 1. Conceptual scheme of the design process adopted in this work.

2.1 Similitude Requirements

Similitude requirements are a set of design patterns aiming to achieve a design of the reduced-scale aeroelastic model that reproduces the exact structural dynamic response of the real structure when tested in a wind tunnel. Geometric similarity is accomplish by scaling the outside geometry of the real structure by using a length scale factor λ_L. Dynamic similarity is achieved with a number of non-dimensional quantities. The purpose of these quantities is to keep in the same ratio the inertia, gravitational and viscous forces acting on the reduced-scale model as it happens in full scale. This is widely described in literature (Cermak (1975); Dyrbye and Hansen (1999); Barlow et al. (1999)), and a summary of the most relevant is presented below:

- Cauchy number $\left(Ca = EI/\rho V^2\right)$: defines the ratio of the elastic forces of the bridge to the inertia forces of the flow, where E is the Young's modulus, I is the inertia of the cross-section of the model, ρ is the air density and V is the wind flow velocity. This ratio is used to define the scale factors related to the material and section properties.
- Reynolds number $(Re = VL/v)$: establishes the relationship between the flow velocity V, a representative dimension of the model L and the viscosity of the flow v. Keeping this similarity is not practical and test are usually carried out without enforcing it.
- Froude number $\left(Fr = gL/V^2\right)$: defines the ratio of the gravitational to the inertial forces, where g is the acceleration due to gravity. Time λ_T and velocity λ_V scale factors are obtained from this relationship.
- Damping factor $(\xi_{\text{modelo}}/\xi_{\text{real}})$: relates the damping of the reduced-scale model with the damping of the real structure. It is important to be preserved because of its influence on resonant motions.
- Density ratio $(\rho_{\text{modelo}}/\rho_{\text{real}})$: defines the ratio between the mass of the model with respect to the air in the wind tunnel.

From these relations all scale factors can be obtained, giving place to:

$$\begin{array}{llll} \lambda_L = L_m/L_p & \lambda_\rho = \rho_m/\rho_p & \lambda_V = V_m/V_p & \lambda_m = \lambda_\rho/\lambda_L^2 \\ \lambda_M = \lambda_\rho/\lambda_L^3 & \lambda_i = \lambda_m/\lambda_L^2 & \lambda_I = \lambda_M/\lambda_L^2 & \lambda_T = \lambda_L/\lambda_V \\ \lambda_a = \lambda_V/\lambda_T & \lambda_\xi = \xi_m/\xi_p & \lambda_{EI} = \lambda_V^2/\lambda_L^4 & \lambda_{EA} = \lambda_V^2/\lambda_L^2 \\ \lambda_f = \lambda_V^2/\lambda_L & \lambda_F = \lambda_V^2/\lambda_L^2 & \lambda_{BM} = \lambda_V^2/\lambda_L^3 & \lambda_{CWE} = \lambda_V^2/\lambda_L^6 \end{array} \quad (1)$$

where subindexes m and p indicate reduced-scale model and prototype, or real structure, respectively, λ_ρ is the density scale factor, M is the mass and m represents the mass per unit of length, I and i represent the mass moment of inertia and the mass moment of inertia per unit of length, respectively, T is time, a is acceleration, ξ indicates damping, EI is the bending stiffness while EA is the tensile stiffness, F and f indicate force and force per unit of length, BM is the bending moment and CWE the buckling stiffness.

2.2 First Phase: Traditional Design Process

2.2.1 Model Construction and Limitations

Reduced-scale model design consist of designing an equivalent structure at the desired scale that represents the dynamic response and geometry of the real structure, as well as all the similarity requirements enumerated above (Davenport et al. 1969; Isyumov 1999). In the case of linear structures or those composed by elements that can be modelled as 1D elements, such as wind mill towers or cable-stayed bridges, the reduced-scale model can be designed by a set of 1D bar elements. The conceptual design of this kind of structures is shown in Fig. 2, where it can be seen that the prototype or real structure is modelled by a 1D structural element, the spine, which reproduces the stiffness of the element. The outside geometry is reproduced by a set of cladding modules which also contribute to emulate the mass of the prototype.

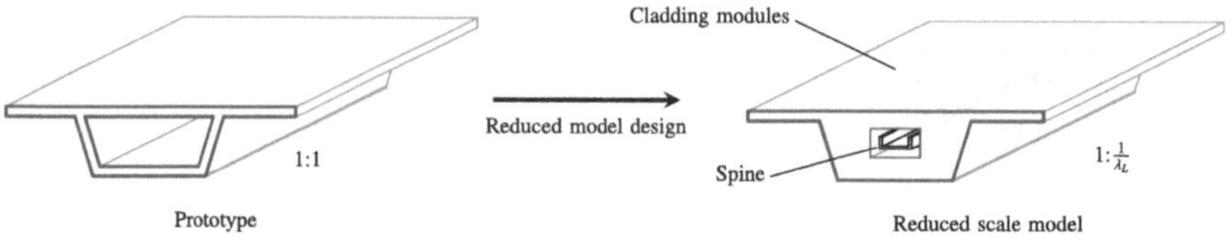

Fig. 2. Conceptual scheme of the design of a 1D element.

However, the scale reduction must be considered in the design in order to avoid the constructive limitations that can affect the capability of the model to reproduce the stiffness and mass of the real structure. Hence, the design must be adequately discretized in order to obtain a feasible design from the constructive point of view.

2.2.2 Model Discretization

The model design entails two levels of discretization: the stiffness discretization and the mass discretization. The stiffness discretization is related to the spine design and consists of defining the segments with uniform cross-sections in which the spine will be divided to reproduce the stiffness properties of the real bridge in a way that the spine construction is feasible. The mass discretization consist of redistributing the mass of the real structure into the cladding modules. This is carried out by defining the location and length of each cladding modules, taking into account that the mass properties of each module acts as a lumped mass in the connection of the cladding module with the spine.

The procedure for defining the dimensions of the cross-section of each spine segment obtained from the stiffness discretization is based on the beam bending theory, which states that the bending rotation of a point of the 1D element of the prototype or real structure is given by

$$\theta_{y,p} = \int_0^{L_p} \frac{M_{y,p}(x)}{E_p I_{y,p}(x)} dx = \frac{M_{y,p}}{E_p} \int_0^{L_p} \frac{1}{I_{y,p}(x)} dx, \tag{2}$$

where M_y stands for bending moment, I_y is the inertia of the cross-section, L_p is the length of the spine segment, and it is assumed that the bending moment is uniform along the length of the segment, as well as the Young's modulus. It must be noticed that the inertia along the segment of the real structure is not uniform along the length of the segment of the reduced-scale model, and consequently is the property to discretize. After some manipulations (see Cid Montoya et al. (2017) for full theory explanation), the stiffness properties of the reduced-scale model can be obtained from the properties of the prototype assuming that the bending rotation is equal, giving place to:

$$\theta_{y,m} = \theta^{m}_{y,p} \rightarrow \frac{L_m}{E_m I_{y,m}} = \frac{1}{E_p}\sum_{l=1}^{n}\frac{L_l\lambda_L}{I_{y,p,l}\lambda_{EI}} \rightarrow I_{y,m} = \frac{E_p}{E_m}\frac{L_m}{\sum_{l=1}^{n}\frac{L_l\lambda_L}{I_{y,p,l}\lambda_{EI}}} \tag{3}$$

In the same manner, the torsional stiffness can be estimated as:

$$J_{x,m} = \frac{E_p}{E_m}\frac{L_m}{\sum_{l=1}^{n}\frac{L_l\lambda_L}{J_{x,p,l}\lambda_{EI}}} \tag{4}$$

where E_p/E_m is the contribution of the possible change of material from the prototype to the reduced-scale model.

2.2.3 Automatization of the Design

Once the stiffness properties of each segment in which the spine is discretized are obtained, the dimensions can be estimated taking into account the geometry of the spine cross-section considered for the reduced-scale model. This is automatically conducted by the in-house software MODESTO developed in the authors' previous work (Cid Montoya et al. 2016). The software obtains the set of dimensions for each spine segment that replicates the set of stiffness properties obtained from Eqs. (3) and (4) by using optimization algorithms (Arora 2011; Henández and Fontán 2002). This procedure is schematized in Fig. 3 for the cross-section studied in this work. Since the goal is to study vortex shedding excitation of a wind mill tower, the rotational mass and stiffness properties are not required to be reproduced. Therefore, the reduced-scale model can simulate the aeroelastic behavior of the tower with just one dimension for each segment of the spine (its radios R) and one mass property (the translational mass M).

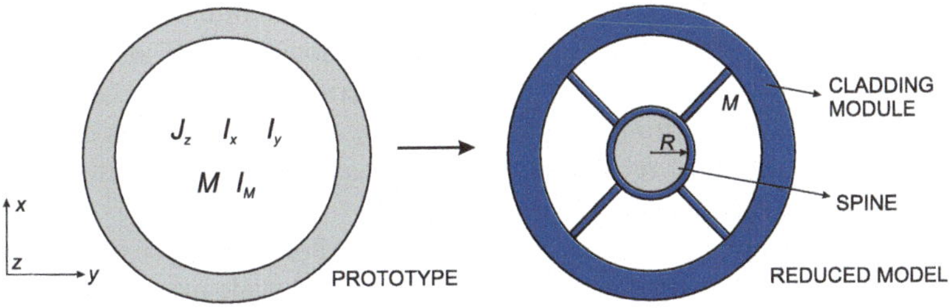

Fig. 3. Simplification of the process for obtaining the dimensions of the spine and mass properties of the cladding modules of the reduced-scale model.

Hence, the formulation adopted by MODESTO software to estimate the dimensions of each segment that compose the reduce model is given by

$$f = \lambda_{Iy}\delta_{Iy} + \lambda_{Iz}\delta_{Iz}, \quad (5)$$

where f is the objective function to minimize, which consists of the sum of relative differences in the mechanical properties, in order to ensure that when the optimization process ends, the mechanical properties of the prototype and the reduced-scale model are the same. The relative errors δ_I of each mechanical property (Iy and Iz are the bending inertias with respect to x and y axis as shown in Fig. 3) are obtained following the expression:

$$\delta_I = \frac{|I_m - I_p|}{I_m}, \quad (6)$$

while λ are weighting factors to assign the desired relative importance to each property when it is not possible to reproduce all values.

2.2.4 Sources of Inaccuracies

The procedure described before is based on the traditional approach for designing reduced-scale models, including some modifications to automatize the process reducing the time dedicated to the design. This procedure is based on some assumptions that leads to some possible sources of inaccuracy, as described in the authors' previous work Cid Montoya et al. 2017, which could lead to errors in the main natural frequencies about 10% (see, for instance, Reinhold et al. 1992).

Furthermore, the procedure described before presents another intrinsic assumption, which is the linear behavior of all elements of the structure. The representation of non-linear element in a reduced-scale model would require to create a non-linear element in reduced scale that reproduces the exact response of the real structure. This is extremely difficult as the scale adopted is higher from a constructive point of view. This is of utmost interest for the case under study in this work, which shows a strong non-linear response in the joints between the tower models, as it will be studied in Sect. 3. Therefore, a different approach must be considered in order to design a reduced-scale model capable of reproduce the dynamic response of a non-linear real structure.

2.3 Second Phase: Reduced-Scale Model Optimization Approach

An alternative procedure to reproduce the dynamic response of the real structure is the optimization approach developed in the authors' previous work (Cid Montoya et al. 2017). This procedure was developed to avoid the errors in the reduced-scale model dynamic properties derived from the traditional design process. This process achieves remarkable reductions in the relative errors of the main dynamic properties, including natural frequencies and generalized masses.

In this work, this approach is applied to a problem where the source of discrepancies in the dynamic properties of the prototype and the reduced-scale model comes from the non-linear behavior of the real structure. Instead of trying to reproduce the non-linear properties in the reduced-scale model, which is considered very difficult from the constructive point of view, the problem is tackled by defining a linear reduced-scale model that reproduces the dynamic responses of the non-linear prototype for the wind velocity under interest of the study. In this manner, the reduced-scale model obtained will reproduce the exact dynamic behavior only for the wind velocity of interest. The formulation of the problem to obtain this linear optimized reduced-scale model is

$$\min F\left(\mathbf{R}, \frac{1}{\mathbf{M}}\right) = \min \sum_{i=1}^{N_\omega} \left| \delta_i^\omega \left(\omega_{m,i}^2, \omega_{p,i}^2 \left(\mathbf{R}, \frac{1}{\mathbf{M}} \right) \right) \right| w_i, \tag{7}$$

where F is the objective function to minimize, $\mathbf{R}$ and $\mathbf{M}$ are the design variables, namely the set of radios of each segment in which the spine is divided and the set of values of the mass of each cladding module, respectively. In this manner, both stiffness and masses are considered to achieve the desired dynamic behavior of the reduced-scale model. The objective function is formulated as the sum of relative errors between the natural frequencies δ_i^ω of the considered set of modes N_ω of the reduced-scale model $\omega_{m,i}$ and the prototype in the reduced scale $\omega_{p,i}$. Relevance of each mode can be weighted by the term w_i. It must be noted that even though only the natural frequencies are considered in the objective function, other dynamic properties such as generalized masses and mode shapes of the optimized design are checked later to guarantee that the dynamic behavior obtained is the one desired. This properties are not included in the optimization process since their inclusion worse the numerical performance of the algorithm and, on the other hand, it is not expected that these dynamic responses increase their difference with the prototype properties after the optimization process, as studied in the authors' previous work Cid Montoya et al. 2017.

The initial design considered for the optimization process is the reduced-scale model design obtained from the traditional process described in Sect. 2.2. The optimization problem is constrained to keep constant the total mass of the model by the following expression:

$$g\left(\frac{1}{\mathbf{M}}\right) = \frac{\delta^M\left(\frac{1}{\mathbf{M}}\right)}{\delta_{tol}^M} - 1 \le 0, \tag{8}$$

where δ^M is the relative variation of the total translational mass of the model, and δ_{tol}^M is the maximum variation allowed, which is commonly set to a value around 1%. Furthermore, the maximum variation allowed to the lumped masses of the cladding modules and the dimensions of the spine are also limited to a maximum. This is given by

$$R_{j,\min} \le R_j \le R_{j,\max}, \tag{9}$$

where R_j is the radius of the segment j of the spine, and $R_{j,\min}$ and $R_{j,\max}$ are the minimum and maximum values allowed for each segment of the spine, and the constraint related to the masses can be written as

$$\frac{1}{M_{k,\max}} \le \frac{1}{M_{k,}} \le \frac{1}{M_{k,\min}}, \tag{10}$$

where M_k is the translational mass of the cladding module k, and $M_{k,\min}$ and $M_{k,\max}$ are the minimum and maximum values allowed for the translational mass of each cladding module. All these constraints allow to obtain a reduced-scale model that reproduces the dynamic behavior of the real structure introducing slight modifications on the reduced-scale model obtained from the traditional design process. In this manner, the mass distribution of the model and consequently the Scruton number are not significantly modified in the optimization process.

3 Wind Mill Tower Application

3.1 Description of the Structure and Construction Process

The wind mill tower under study in this piece of research is composed by 4 precast modules of reinforced or prestressed concrete (see Fig. 4). During the construction process, these modules are simply placed one on top of the other, waiting to do the connections between segments until all of them are in their final position. However, before the connections between segments are done, the tower shows a non-linear static and dynamic behavior (see Fig. 5). The resulting dynamic behavior subject to the wind conditions under study must be reproduced by the reduced-scale model used for the aeroelastic wind tunnel test. The dynamic responses of the prototype under the wind conditions that will be tested in the wind tunnel were analyzed using finite element models using ABAQUS software in order to know beforehand the dynamic properties that the reduced-scale model must represent. This allows to design a linear reduced-scale model capable to reproduce the dynamic properties of interest for the given wind conditions, even though the structure intended to be simulated (prototype) presents a non-linear dynamic behavior.

Fig. 4. Real tower and finite element model.

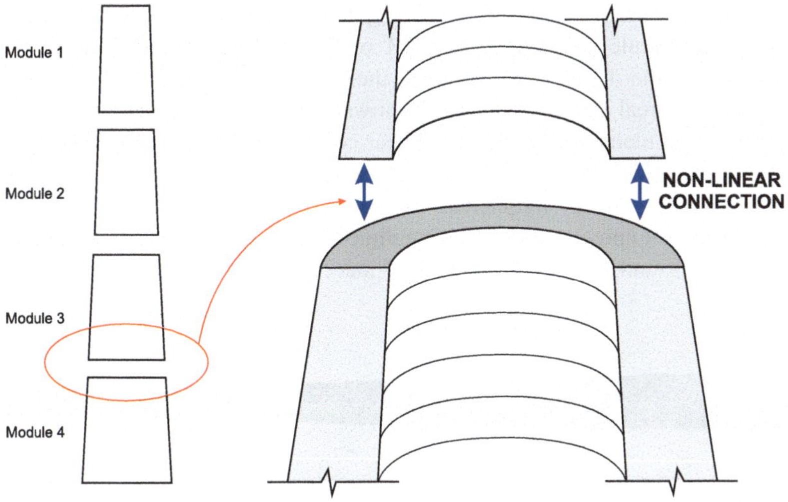

Fig. 5. Conceptual representation of the non-linear connections between the four modules (precast segments of reinforced or prestressed concrete).

3.2 Reduced-Scale Model Design. Traditional Approach

The procedure described in Sect. 2.2. was carried out to obtain an initial design of the reduced-scale model, taking advantage of the in-house MODESTO software (Cid Montoya et al. 2016). Figure 6 shows a screenshot of the program module for carrying out the stiffness discretization and define the dimensions of the spine. A similar procedure is carried out for the mass discretization required to define the mass properties (translational and rotational) assigned to each cladding module of the reduced-scale model.

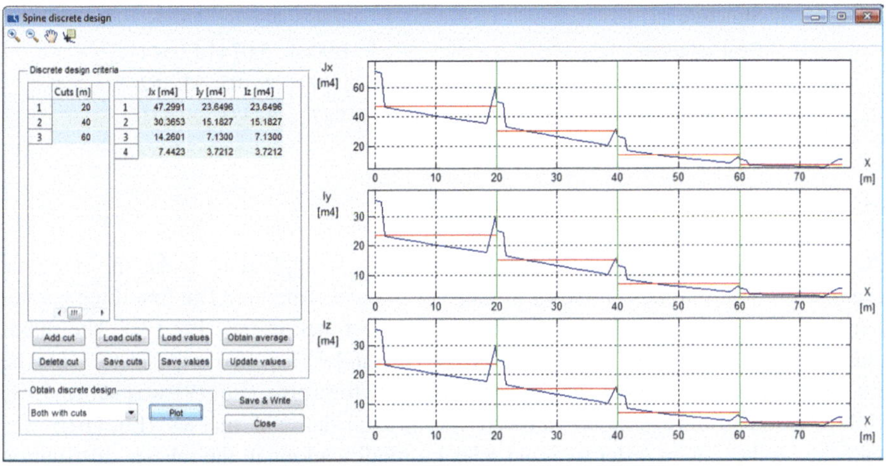

Fig. 6. Mechanical properties discretization using MODESTO software.

Once the dimensions of all the spine segments cross-sections and the masses of each cladding module are defined, a FEM of the resulting reduced-scale model is created to obtain the dynamic responses of the model. These responses are compared with those of the real structure. Figure 7 shows the FEM of the real tower, which is modelled by shell elements and the non-linear connections are modelled in detail, and the FEM of the initial design of the reduced-scale model made of bar elements. The comparison of the natural frequencies obtained for each structure is presented in Table 1. It can be appreciated that non-negligible differences in the first three natural frequencies are found, due to the sources of inaccuracy commented in Sect. 2.2.4.

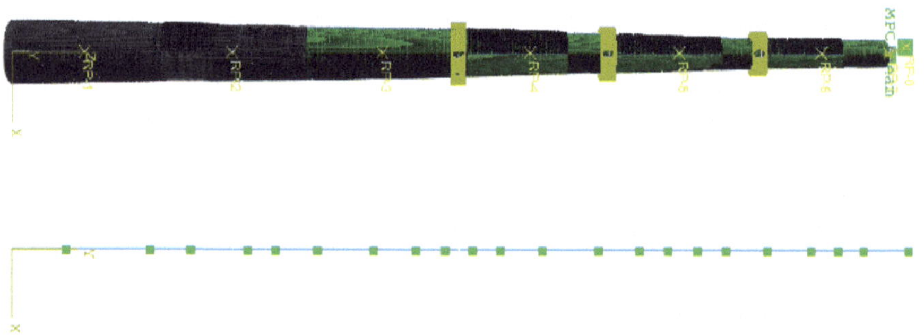

Fig. 7. FEM of the real tower modelled by shell elements and the reduced-scale model composed by bar elements.

Table 1. Comparison of the natural frequencies of the real structure and the initial design of the reduced-scale model. The values are normalized to the value of the first natural frequency of the real structure. The frequencies shown for the real structure have been scaled to the reduced model scale.

Structure	ω_1/ω_1	ω_2/ω_1	ω_3/ω_1
Real structure	1.000	5.074	13.544
Initial design of the reduced-scale model	1.149	6.105	17.562
Relative error of the initial design	14.872	20.331	29.671

3.3 Reduced-Scale Model Optimization

The optimization approach described in Sect. 2.3. is applied to make the dynamic response of the initial design closer to those of the real structure. The first three natural frequencies are considering to define the objective function (see Eq. (7)). The maximum variation allowed of the total mass of the model (Eq. (8)) was set to 1%. The maximum variation allowed for the dimensions of the spine (Eq. (9) was set to 15%, while the maximum variation allowed for the translational masses (Eq. (10)) was stablished to be 5%, in order to avoid relevant modifications in the masses distribution of the structure and Scruton number.

The optimum design obtained from the optimization process (see Fig. 8) allowed to reduce the differences in the natural frequencies between the real structure and the reduced-scale model to relative errors in the first and second modes to less than 0.2%, and a maximum relative error in the third frequency of 1.695%, which guarantees that the optimum design of the reduce model reproduces the dynamic response of the real structure, and the only possible sources of inaccuracy can only be found in the construction process (Table 2).

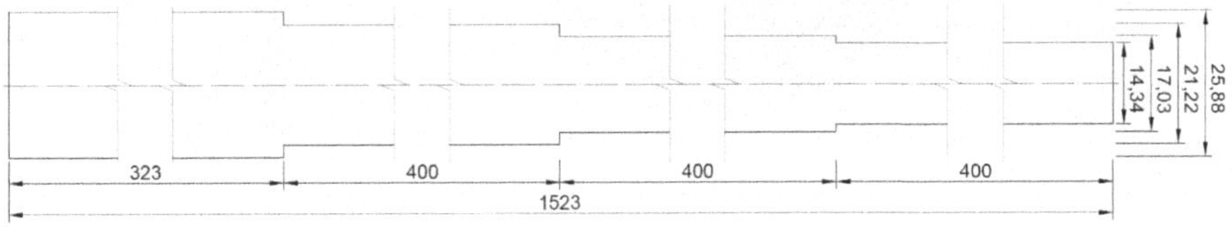

Fig. 8. Layout of the optimum design of the spine

Table 2. Numerical results of the optimization process. The values are normalized to the value of the first natural frequency of the real structure. The frequencies shown for the real structure have been scaled to the reduced model scale.

Structure	ω_1/ω_1	ω_2/ω_1	ω_3/ω_1
Real structure (optimization target)	1.000	5.074	13.544
Initial design of the reduced-scale model	1.149	6.105	17.562
Relative error of the initial design	14.872	20.331	29.671
Final design (optimized model)	0.998	5.064	13.773
Relative error of the final design	0.150	0.197	1.695

3.4 Reduced-Scale Model Construction

The spine of the reduced-scale model was design and build with aluminum of circular cross-section, as shown in Fig. 9(a). The cladding modules where built with a 3D printer in order to guarantee their exact outside geometry (see Fig. 9(b)). The mass of each cladding module was achieved by adding lead pieces to each plastic cladding module. The modules were set up from the bottom to the top, as shown in Fig. 9(c). The construction and wind tunnel tests of the resulting reduced-scale model are currently in progress.

(a) (b) (c)

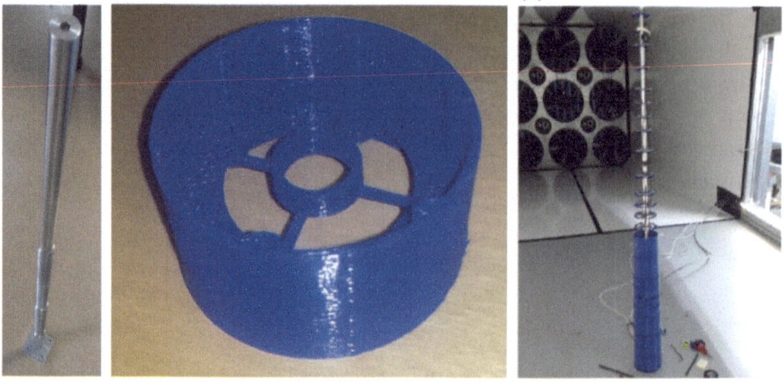

Fig. 9. Reduced-scale model of wind will tower. (a) Spine (b) Cladding module (c) Construction of the model

4 Conclusions

This paper briefly reports the process of designing a full reduced-scale model of a wind mill tower for wind tunnel testing in order to study possible vortex-shedding excitations. First, the traditional process for designing reduced-scale models is described and their limitations are discussed. Second, the optimization process developed in an author's previous work to achieve a dynamic response of the reduced-scale model closer to the real structure is outlined. Then, these two approaches are applied to design a reduced-scale model of a wind mill tower that presents a non-linear behavior during the construction phase, allowing to obtain a design of the reduce model that faithfully represents the set of natural frequencies considered in this study.

Hence, the optimization approach developed in the author's previous work can be also used to define reduced-scale models with linear behavior that reproduce the dynamic response of real structures with non-linear dynamic behavior in the conditions of the wind tunnel tests.

Acknowledgements. The research leading to these results has received funding from ACCIONA Windpower, the Spanish Minister of Economy and Competitiveness (MINECO) with reference BIA2016-76656-R. The first author has been also founded by the Fundación Pedro Barrié de la Maza and the University of La Coruña. The fourth author have been founded by grant BES-2014-068418. The authors fully acknowledge the support received.

References

Arora JS (2011). Introduction to optimum design, 3rd edn. Elsevier/Academic Press

Barlow B, Rae WH, Pope A (1999) Low-speed wind tunnel testing. Wiley-Interscience, New York

Cermak JE (1975) Applications of fluid mechanics to wind engineering - a freeman scholar lecture. J Fluids Eng 97(1):393–409

Cid Montoya M, Hernández S, Nieto N, King JPC, Kong, L (2016) A user friendly software for the design of full bridge reduced models in wind tunnel testing. In: AFM2016, 11th international conference on advances in fluid mechanics, September 5–7, Ancona, Italy

Cid Montoya M, King JPC, Kong L, Nieto F, Hernández S (2017) A method for improving the dynamic response of full bridge reduced-scale models in aeroelastic wind tunnel tests by using optimization algorithms. J Wind Eng Ind Aerodyn 167:198–216

Davenport AG, Isyumov N, Fader DJ, Bowen CFP (1969) A study of wind action on a suspension bridge during erection and on completion. BLWTL-3-69, University of Western Ontario

Davenport AG, Isyumov N (1967) The application of the boundary layer wind tunnel to the prediction of wind loading. In: International Research Seminar: Wind Effects on Buildings and Structures

Dyrbye C, Hansen SO (1999) Wind loads on structures. Wiley, Chichester

Hernández S, Fontán AN (2002) Practical applications of design optimization. WIT Press, Southampton

Irwin PA (2008) Bluff body aerodynamics in wind engineering. J Wind Eng Ind Aerodyn 96:701–712

Irwin PA (2009) Wind engineering challenges of the new generation of super-tall buildings. J Wind Eng Ind Aerodyn 97:328–334

Isyumov N (ed) (1999) Wind tunnel studies of buildings and structures. ASCE Manuals and Reports on Engineering Practice No. 67

Muggiasca S, Belloli M, Diana G (2018) Specific power input: comparison among rigid and flexible models. J Wind Eng Ind Aerodyn 173:180–186

Reinhold TA, Brinch M, Damsgaard A (1992) Wind tunnel test for the great belt link. In: Proceedings of the first international symposium on aerodynamics of large bridges

Simiu E, Scanlan RH (1996) Wind effects on structures. Wiley, New York

Zasso A, Belloli M, Giappino S, Muggiasca S (2008) Pressure field analysis on oscillating circular cylinder. J Fluids Struct 24:628–650

Probabilistic Assessment of the Galloping Stability of Ice-Accreted Bridge Hangers

C. Demartino[1(✉)] and F. Ricciardelli[2]

[1] College of Civil Engineering, Nanjing Tech University, Puzhu South Road, 30, Nanjing 211816, People's Republic of China
cristoforo.demartino@me.com
[2] Department of Engineering (DI), Università della Campania "Luigi Vanvitelli", Via Roma 9, 81031 Aversa, Italy
fricciaunicampania.it

Abstract. Galloping vibrations have recently been identified as a potential problem for ice-accreted bridge hangers. In this study, starting from wind tunnel measurements of the aerodynamic coefficients of an ice accreted HDPE cable hanger, the nature of the ice-accretion aerodynamics is shown. Then, a framework based on Montecarlo simulations is applied for the probabilistic assessment of the minimum structural damping required to prevent galloping of bridge hangers based on the output of a 2-DoFs sectional quasi-steady aeroelastic model. All the variables required to define the hanger dynamics, the sheath aerodynamics and the local wind climate are considered. The results highlight the advantages of the probabilistic procedure in terms of reliability quantification, compared to the deterministic approach.

Keywords: Bridge hangers · Quasi-steady aerodynamics · Galloping · Ice accretion cables · Reliability

1 Introduction

Wind-induced cable vibrations have been the topic of many studies carried out in the last decades. These mainly focus on the unstable behaviour deriving from a variety of causes and phenomena such as vortex shedding, galloping, shape modifications due to rain and ice. In particular, undesirable wind-induced vibrations of bridge cables can occur when atmospheric conditions are such to generate ice accretion on the surface of the cable (Demartino et al. 2013a, b; Demartino 2014; Demartino et al. 2015). These vibrations are generated by the complex aerodynamics of circular cylinders that is particularly sensitive to small variations of the superficial characteristics (Demartino and Ricciardelli 2017b; Demartino and Ricciardelli 2018b).

Following the approach proposed by Den Hartog (1932), in EN1991 Part 1-4 (CEN 2005) the onset wind velocity, U_{CG}, is expressed as a function of the mean drag and lift coefficients (C_D and C_L) function of the angle of attack (α), the nominal acrosswind

F. Ricciardelli and A. M. Avossa (Eds.): INVENTO 2018, LNCE 27, pp. 260–271, 2019.
https://doi.org/10.1007/978-3-030-12815-9_21

dimension of the structure (D), of the Scruton number (Sc), of the acrosswind fundamental frequency of the structure ($f_{1,y}$), and of the factor of galloping instability (a_G), which depends on cross-section geometry:

$$U_{CG} = \frac{2 \cdot Sc}{a_G} f_{1,y} \cdot D \text{ with } a_G = -\left(C_D(\alpha) - \frac{\partial C_D(\alpha)}{\partial \alpha} \right) \tag{1}$$

EN1991 Part 1-4 (CEN 2005) defines U_{CG} as the velocity at which galloping starts, reporting that normally the amplitude of oscillation rapidly increases with increasing wind velocity. A structure is considered stable against galloping, if U_{CG} is at least 25% larger than the mean wind velocity evaluated at the height where galloping excitation is expected to occur, likely to coincide with the point of maximum amplitude of oscillation (usually located in the middle of the cable). This serviceability problem is usually controlled by varying the structural damping. In Bridge Engineering, it is well known that different values of the damping can strongly affect the dynamic response under dynamic actions (e.g., Ricciardelli and Demartino 2016; Avossa et al. 2017a, b; Avossa and Pianese 2017c; Demartino et al. 2018).

In some circumstances, code approaches are not sufficiently accurate and more sophisticated analyses must be undertaken. In this context, a number of models have been derived for the prediction of galloping instability of bridge cables based on the QS theory; these were reviewed and compared in Demartino and Ricciardelli (2017b) and in Piccardo et al. (2015).

However, although many authors demonstrated the role of uncertainties in the galloping assessment of ice accreted bridge cables, the development of a probabilistic framework has never been explored.

In this study, a procedure for the probabilistic assessment of the minimum structural damping required to prevent galloping of ice-accreted bridge hangers is presented. The same framework is proposed in Demartino and Ricciardelli (2019). The procedure is based on the calculation of the minimum structural damping required to prevent galloping of bridge hangers using a 2-DoFs sectional QS aeroelastic model. The aerodynamic data of ice-accreted stay cable used in the calculations derived from wind tunnel tests performed at the DTU/Force Technology Collaborative Climatic Wind Tunnel (CWT) in Lyngby, Denmark (Demartino 2014; Demartino et al. 2015; Demartino and Ricciardelli 2015). The probability distribution of the minimum required structural damping is evaluated through Monte Carlo simulations. This framework allows for realistic worst case conditions among different scenarios, and their probabilistic definition.

By way of example to show the application of the framework, it is chosen a prototype fictitious suspended bridge located in Copenhagen since the climatic conditions are characterized by a large probability of ice formation on the cables. Using a normal, a log-normal and a deterministic model of capacity, the probability of failure is calculated. In the definition of the framework, all the variables required to define (*i*) the hanger dynamics, (*ii*) the ice accreted sheath aerodynamics and (*iii*) the local wind climate are considered. Finally, using the deterministic model of capacity, a simplified closed form equation for the evaluation of the minimum structural damping required to prevent galloping in a probability-based performance approach is proposed.

2 Aerodynamic Coefficients of the Ice Accreted Bridge Hanger

Static wind tunnel tests were performed in the DTU/Force Technology Climatic Wind Tunnel (CWT) in Lyngby, Denmark. The wind tunnel is a closed circuit, with test section dimensions of (length × width × height) 5 m × 2 m × 2 m with a maximum wind speed in the test section of 30 m/s and the minimum temperature is –5 °C. This section summarizes results of tests carried out using this facility and reported in Demartino et al. (2015). The following climatic conditions were employed for the ice accretion: mean wind speed $U = 11$ m/s; temperature $T = -5$ °C; accretion time: $t = 1$ h. In this condition, the windward accretion was characterized by a hard rime with a narrow core and by the absence of lateral rivulets (Fig. 1). This is referred as climatic condition V1 in Demartino et al. (2015).

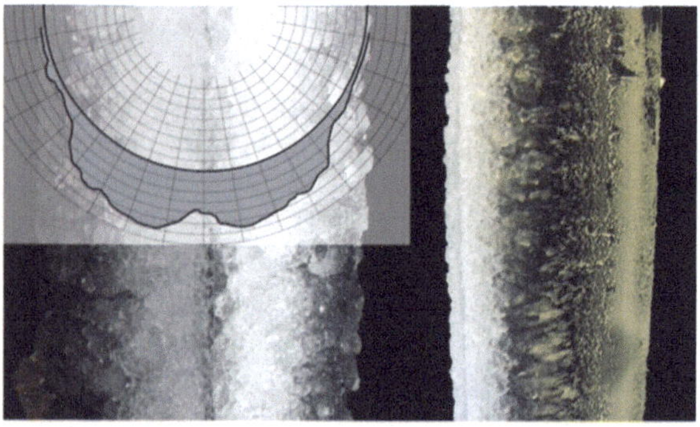

Fig. 1. Front and lateral view and contour tracing of the cable after ice accretion (mean wind speed U = 11 m/s; temperature T = –5 °C; accretion time: t = 1 h). After Demartino et al. (2015).

The mean aerodynamic force coefficients are evaluated with respect to the mean wind tunnel speed and the reference area, i.e. the product of the nominal diameter and the length of the model. The variation of the reference area due to the accretion was not considered. The mean drag C_D and lift C_L coefficients are presented in Fig. 2 for different values of the angle of attack ($0° \leq \alpha \leq 360°$) and Reynolds number ($1 \leq Re \cdot 10^{-5} \leq 3.5$ - corresponding to the wind speed range $6 \leq U \leq 30$ m/s). Drag coefficients were corrected for blockage following the Maskell III Method. Since the aerodynamic coefficients were measured in the range $0° \leq \alpha \leq 180°$ by Demartino et al. (2015), the adopted set of aerodynamic coefficients (i.e., $0° \leq \alpha \leq 360°$) was obtained by assuming symmetry, i.e., $C_D(\alpha) = C_D(\alpha + 180°)$ and $C_L(\alpha) = C_D(\alpha + 180°)$.

From Fig. 2, it can be seen that there are certain ranges in which the aerodynamic coefficients do not depend on wind speed and others in which they do. The values of the aerodynamic coefficients depend on the position of the separation point, which is determined by the ice accretion producing large asymmetry of the original circular cross-section. The position of the separation point, otherwise dependent on wind speed

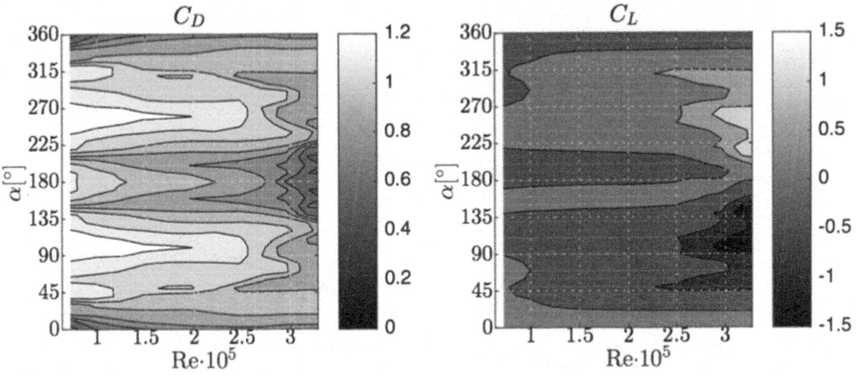

Fig. 2. Drag (left) and lift (right) coefficients maps in the Re - α plane. After Demartino et al. (2015).

for a plain cable, can become independent of it when geometric perturbations are created by the accretion, hence determining separation points. The aerodynamic coefficients are only little dependent on wind speed angles of attach near to 0° and 180°, whereas this dependency increases for different angles (Fig. 2). The value of C_D decreases with increasing wind speed as the effect of Reynolds critical behaviour while the opposite is usually observed for the absolute value of C_L. These variations (with respect to the angle of attack and to the Reynolds number) are the main cause of the galloping instability.

3 Aeroelastic Model

The 2-DoFs modal Equations of Motion (EoMs) of the hanger of length L in the (x, y) reference system (Fig. 3) at height z is:

$$\begin{aligned}\int_0^L \boldsymbol{\Phi}(z)^T \boldsymbol{M}\boldsymbol{\Phi}(z)dz \cdot \ddot{\boldsymbol{\eta}}(t) + \int_0^L \boldsymbol{\Phi}(z)^T \boldsymbol{C}_s\boldsymbol{\Phi}(z)dz \cdot \dot{\boldsymbol{\eta}}(t) + \int_0^L \boldsymbol{\Phi}(z)^T \boldsymbol{K}_s\boldsymbol{\Phi}(z)dz \cdot \boldsymbol{\eta}(t) \\ = \int_0^L \boldsymbol{\Phi}(z)^T \boldsymbol{F}_a(U(z), \boldsymbol{\Phi}(z)\dot{\boldsymbol{\eta}}(t), \alpha_s(t), \alpha(t), z)d\end{aligned} \tag{2}$$

where $\boldsymbol{M} = \mathrm{m}\boldsymbol{I}$, $\boldsymbol{C}_s$, $\boldsymbol{K}_s$ are the diagonal 2 × 2 structural mass, damping and stiffness sectional matrices, respectively, m being the mass per unit length and $\boldsymbol{I}$ being the identity matrix. The parameters in the matrices $\boldsymbol{M}$, $\boldsymbol{C}_s$, $\boldsymbol{K}_s$ are referred to a unit length of the cable and are considered constant with z. $\boldsymbol{q}(z,t) = \left[u(z,t) \quad v(z,t)\right]^T = \boldsymbol{\Phi}(z)\boldsymbol{\eta}(t)$ is the vector containing the sectional displacements of hanger being $\boldsymbol{\Phi}(z)$ the 2 × 2 mode shape matrix and $\boldsymbol{\eta}(t) = [\xi(t) \quad \eta(t)]^T$ the modal coordinate vector.

$\boldsymbol{F}_a$ indicates the non-linear sectional aerodynamic force vector (i.e., per unit length) in the x and y directions. $U(z)$ is the component of $\boldsymbol{U}(\beta, z)$ in the x' direction, i.e., its magnitude. The problem is governed by two angles, defining the aerodynamics (angle

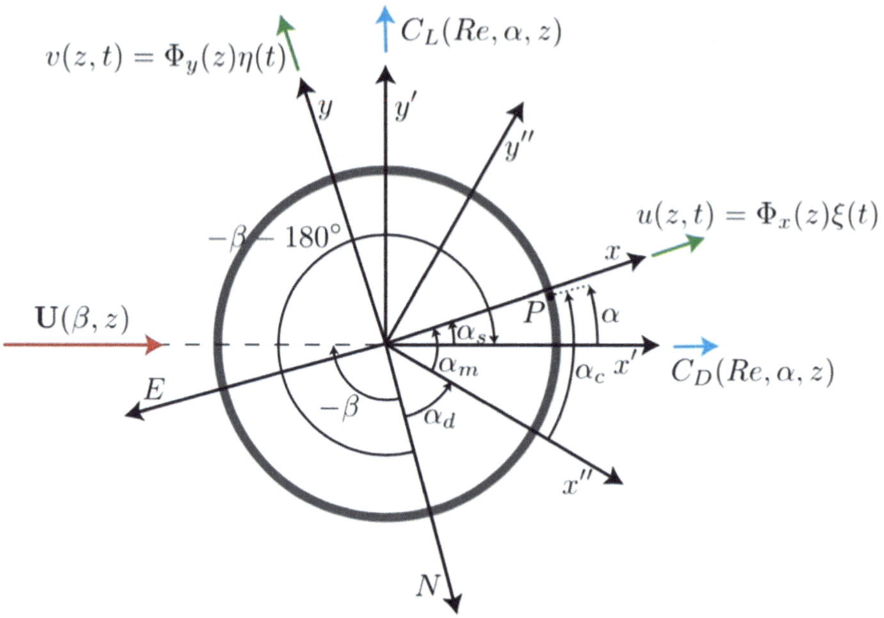

Fig. 3. Schematic of the cable cross-section with flow direction, reference systems definition, modal DoFs and aerodynamic forces. (*N,E*): Geographic reference system; (*x″,y″*): Bridge reference system; (*x′,y′*):Wind reference system; (*x,y*): Hanger reference system; *P*: characteristic point of the aerodynamic coefficients defying the angle of attack.

of attack, α) and the dynamics (structural angle, α_s) of the hanger. These two angles can be expressed as a function of the other angles involved (α_d, α_c, α_m, β) as:

$$\alpha = \alpha_d + \alpha_c + \beta - 180° \tag{3a}$$

$$\alpha_s = \alpha_d + \alpha_m + \beta - 180° \tag{3b}$$

The angles α_d (deck direction), α_c (cable cover location), α_m (directions of vibration) and β (wind direction with respect to *N*) are defined in Fig. 3 together with the definition of the different reference system adopted.

For a perfect cable, the first two frequencies and mode shapes coincide, and depend only on the cable length and tension. Indeed, perfect cable conditions are rarely found in bridge hangers, since the actual connections to the cable and to the deck make the boundary conditions deviate from those of the perfect cable, making the two mode shapes and natural frequencies slightly different from one another (e.g., Demartino and Ricciardelli 2015).

For the purpose of investigating stability, only the first term of the Taylor expansion of the non-linear aerodynamic forces of Eq. 2 around the equilibrium point is retained:

$$\boldsymbol{F}_a \cong \boldsymbol{F}_m(U(z), \alpha_s, \alpha, z) - \boldsymbol{C}_a(U(z), \alpha_s, \alpha, z)\boldsymbol{\Phi}(z)\dot{\boldsymbol{\eta}}(t) \tag{4}$$

where the first term is the mean action and the second term is the aeroelastic, self-excited action depending on cross-section velocity. C_a is the aerodynamic damping matrix, which in the cross flow case is given as (Macdonald and Larose 2008):

$$\begin{aligned} \boldsymbol{C}_a &\cong \boldsymbol{R}_z(\alpha_s)\bar{\boldsymbol{C}}_a(U(z),\alpha,z)\boldsymbol{R}_z^T(\alpha_s) \\ &= \frac{\rho D U(z)}{2}\boldsymbol{R}_z(\alpha_s)\begin{bmatrix} 2C_D + \frac{\partial C_D}{\partial Re}Re & -C_L + \frac{\partial C_D}{\partial \alpha} \\ 2C_L + \frac{\partial C_D}{\partial Re}Re & C_D + \frac{\partial C_D}{\partial \alpha} \end{bmatrix}\boldsymbol{R}_z^T(\alpha_s) \end{aligned} \tag{5}$$

where ρ is the air density, $\boldsymbol{R}_z$ is the rotation matrix, and $\bar{\boldsymbol{C}}_a$ is the aerodynamic damping matrix corresponding to $\alpha_s = 0°$, i.e., written in the wind reference system (Fig. 3). The diagonal terms of $\bar{\boldsymbol{C}}_a$ correspond to the aerodynamic damping appearing in the models of Martin et al. (1981) and of Den Hartog (1932).

As the aerodynamic damping is not constant over the hanger length due to the variation of the mean wind speed and of the mean aerodynamic coefficients, then the integrals in Eq. 2 must be calculated. However, some assumptions are usually doing in the common applications (Nikitas and Macdonald 2013): (*i*) uniform wind velocity profile and (*ii*) aerodynamic coefficients constant with the height. Under these assumptions, the EN1991 Part 1-4 (CEN 2005) approach can be used to evaluate the mean wind velocity at the reference height, chosen as that where the maximum amplitude of oscillation is expected, i.e., $z_e = z_d + z_{ref} = z_d + L/2$, z_d being the base of the hanger. Following this approach, the modal matrix becomes a unit matrix and the modal coordinate vector and the sectional displacements vector evaluated at z_{ref} becomes the same. The sectional EoM can be simplified in:

$$\boldsymbol{M}\cdot\ddot{\boldsymbol{\eta}}(t) + \left(\boldsymbol{C}_s + \boldsymbol{C}_a\left(\boldsymbol{U}\left(z_{ref}\right),\alpha_s,\alpha,z\right)\right)\dot{\boldsymbol{\eta}}(t) + \boldsymbol{K}_s\cdot\boldsymbol{\eta}(t) = 0 \tag{6}$$

Susceptibility to galloping is assessed by solving an eigenvalue problem. The solution of Eq. 6 is written in the form $\boldsymbol{\eta}(t) = \boldsymbol{w}exp(\lambda t)$, being $\boldsymbol{w}$ the eigenvectors and λ the eigenvalues, giving rise to the characteristic polynomial equation. By setting the determinant of the characteristic polynomial equation to zero, the stability is studied:

$$f(\lambda) = \det\left[\lambda^2\boldsymbol{M} + \lambda\left(\boldsymbol{C}_s + \boldsymbol{C}_a\left(U\left(z_{ref}\right),\alpha_s,\alpha,z\right)\right) + \boldsymbol{K}_s\right] = 0 \tag{7}$$

The system is stable if the real parts of all of the eigenvalues are negative. The stability boundaries can be found by imposing that the eigenvalues have zero real parts (Macdonald and Larose 2008).

4 Probabilistic Assessment

Under the assumption that the capacity and the demand are independent variables, the probability of failure is the product of the two marginal Probability Density Functions (PDFs) $f_C(\xi_s)$ and $f_D(\xi_c)$:

$$p_f = \iint\limits_{C \le D} f_C(\xi_s)\cdot f_D(\xi_c)d\xi_s d\xi_c \tag{8}$$

where ξ_c and ξ_s are defined using the mean values of the mass, $\bar{m}$, and of the natural frequency, $\bar{f}_x$.

Starting from Eq. 7, the PDF of the minimum structural damping required to prevent galloping is obtained through the convolution of the conditional PDF:

$$f_D(\xi_c) = \int_0^{360^\circ}\int_0^{360^\circ}\int_0^{\infty}\int_{-\infty}^{\infty}\int_{-\infty}^{\infty}\int_0^{\infty} f_D(\xi_c|f_x,\kappa,\boldsymbol{C},U,\alpha,\alpha_s)df_x d\kappa d\boldsymbol{C} dU(z_{ref})d\alpha d\alpha_s \quad (9)$$

Only m, D, ρ, and v have been considered deterministic assuming their mean value and omitted in the integral. The integration with respect to $\boldsymbol{C}$ is done with respect to C_D and C_L separately. The PDFs of α and α_s in Eq. 9 are the result of the convolution of other angles according to Eqs. 3a and 3b.

The capacity is assumed with three different distributions, namely normal, log-normal and deterministic:

$$f_C(\xi_s) = \begin{cases} \mathcal{N}(\bar{\xi}_s, CV_C\bar{\xi}_s) \text{ Normal distribution} \\ \ln\mathcal{N}(\mu,\sigma) \text{ Log}-\text{normal distribution} \\ \delta(\xi_s-\bar{\xi}_s) \text{ Deterministic} \end{cases} \quad (10)$$

where $\bar{\xi}_s$ is the mean structural damping available and CV_C its coefficient of variation of the normal distribution; μ and σ are the parameters defying the log-normal distribution and δ is the Dirac delta function in the deterministic model. In order to obtain similar capacity models using the normal and log-normal distributions, in the following, an equivalence is established between the first two moments (mean and standard deviation) of the distributions.

Because of the large uncertainties and of the non-linearity of Eq. 7, the distribution of the minimum structural damping required to prevent galloping is obtained by Monte Carlo simulations.

5 An Application to a Bridge Hanger Located in Copenhagen

By way of example, a hanger of a prototype fictitious suspended bridge in the area of Copenhagen is considered. Copenhagen is chosen because of the low temperature and large probability of ice accretion (e.g., Demartino and Ricciardelli 2014). The parameters involved in the assessment of the structural damping required to prevent galloping are considered using a probabilistic description. A summary of these parameters is given in Table 1. In the probabilistic approach, the assumed COVs are representative of typical values for a bridge.

The hangers have a diameter of 0.16 m. In this example, z_e is 100 m and the mass per unit length of the hanger is m = 100 kg/m. f_x and κ (detuning parameter: ratio of the natural frequency in the y to the x direction) are normal random variables with mean values coincident with that used for the deterministic calculations, and coefficients of variation $CV_{f_c} = CV_\kappa = 0.03$. The direction of vibration is uniformly distributed in the range of $\alpha_m = 0^\circ$ to 360°, corresponding to a detuning occurring in any arbitrary direction.

Table 1. Parameters governing the assessment of the structural damping required to prevent galloping. $\mathcal{N}(\mu, \sigma)$ indicates a normal distribution with mean μ and standard deviation σ. *U(a,b)* a circular uniform distribution in the range from *a* to *b*.

Hanger dynamics and orientation				
m	f_x	κ	α_m	
[kg]	[Hz]	[-]	[°]	
100	$\mathcal{N}(1,0.03)$	$\mathcal{N}(1,0.03)$	*U(0°,360°)*	
Cable cover aerodynamics				
$\mathbf{C}(\mathrm{Re}, \alpha)$	α_c	D	ρ	υ
[-]	[°]	[m]	[kg/m³]	[m²/s]
$\mathcal{N}(\boldsymbol{C}, 0.03\boldsymbol{C})$	*U(0°,360°)*	0.16	1.25	1.59x10⁻⁵
Local wind climate				
$U(z_e)$	z_e	v_b	β	z_0
[m²/s]	[m]	[m/s]	[°]	[m]
Eq. 11	100	Eq. 12	*U(0°,360°)*	0.003

The aerodynamic coefficients are taken as normally distributed with mean value equal to the value measured in the wind tunnel and coefficient of variation $CV_C = 0.03$. This accounts for errors in the wind tunnel measurements and for the variability of the aerodynamic coefficients for different cables. For wind speeds outside of the measurement range, the aerodynamic coefficients are taken equal to the value at the closest measured.

As the ice-accretion is considered to occur with an arbitrary angle (i.e., wind direction), it is assumed α_c as uniformly distributed in the range of 0° to 360°. The air density and the kinematic viscosity are taken as $\rho = 1.25$ kg/m^3 and $\upsilon = 1.59 \times 10^{-5}$ m^2/s, respectively, independently of atmospheric temperature, pressure and humidity.

The local wind climate is described considering the variation of the wind speed with height (e.g., Demartino et al. 2017a). The wind speed at the reference height is evaluated using EN1991 Part 1-4 (CEN 2005) that defines the wind profile using a reference velocity, v_b, defined as the 10-min omnidirectional mean velocity at 10 m above ground in a category II exposure. When no orography effects are considered, the mean wind profile is defined as (CEN 2005):

$$U(z_e) = v_b \left[0.19 \left(\frac{z_0}{0.05} \right)^{0.07} \right] \ln\left(\frac{z_e}{z_0} \right) \tag{11}$$

where z_0 is the roughness length considered independent of the wind direction and equal to 0.003 m, corresponding to a terrain category 0 (sea, costal area exposed to open sea). β is uniformly distributed in the range [0;360] and v_b is described by the marginal PDF of the wind speed. This assumption is related to the probability distributions assumed for α_c and α_m. In fact, it can be demonstrated that if the PDFs of α_c and α_m are uniform then the PDFs of α and α_s (Eqs. 3a and 3b) are also uniform distribution in (0°,360°), independently of the PDF of β (Mardia 1972). In the majority of synoptic wind climates, the marginal PDF of the mean wind speed is reasonably well represented by a Weibull distribution whose PDF is (e.g., Chiodi and Ricciardelli 2014):

$$f_W(v_b|c,k) = \frac{k}{c}\left(\frac{v_b}{c}\right)^{k-1}\exp\left[-\left(\frac{v_b}{c}\right)^k\right] \quad (12)$$

where k and c are the shape and the scale parameters, respectively. The model parameters for the wind speed and direction were estimated using the anemometric wind data coming from the Copenhagen Kastrup Airport, (ICAO: EKCH) for the period 1929–2017 (88 years). The estimated parameters are $k = 5.919$ and $c = 2.148$ m/s.

The solution of Eq. 8 requires the calculation of the PDF of the demand, $f_D(\xi_c)$. Montecarlo simulations are carried out for the two probabilistic cases (Table 1), by generating samples of 10^7 combinations of the input parameters. Figure 4 shows the empirical CDF, $F_D(\xi_c)$, for the probabilistic case derived from the generated samples. It can be seen that the empirical distribution is characterized by large values of the PDF (i.e., the derivative of the CDF) in the negative range of ξ_c. Although negative values of ξ_c have no physical meaning, they indicate a stable behavior even in the absence of structural damping.

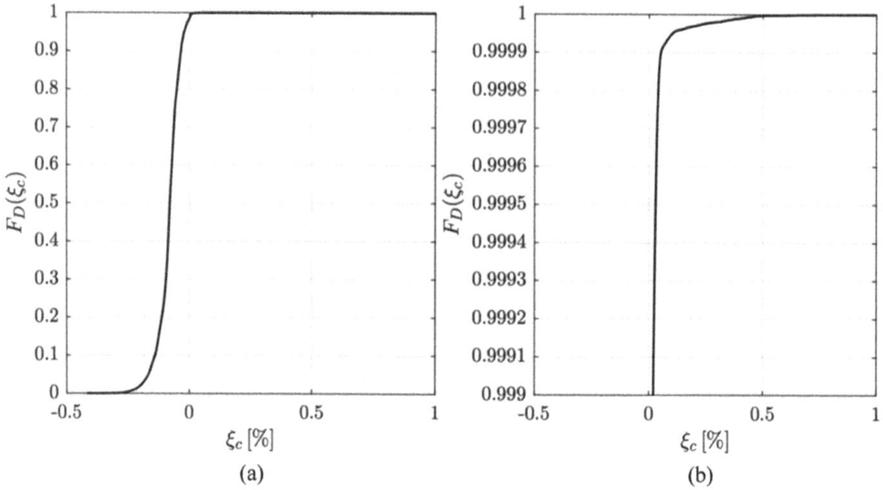

Fig. 4. Empirical CDF of the minimum structural damping required to avoid galloping in the range of 0 to 1 (a) and of 0.99 to 1 (b).

p_f (Eq. 8) was calculated with varying $\bar{\xi}_s$ in the range of 0% to 1% and reported the semi-log plan in Fig. 5a. Calculations were carried out using a deterministic, a normal and a log-normal capacity model with $CV_C = 30\%$ (Pagnini and Repetto 2012; Pagnini et al. 2016). As expected, increasing $\bar{\xi}_s$ the probability of failure rapidly decreases, and

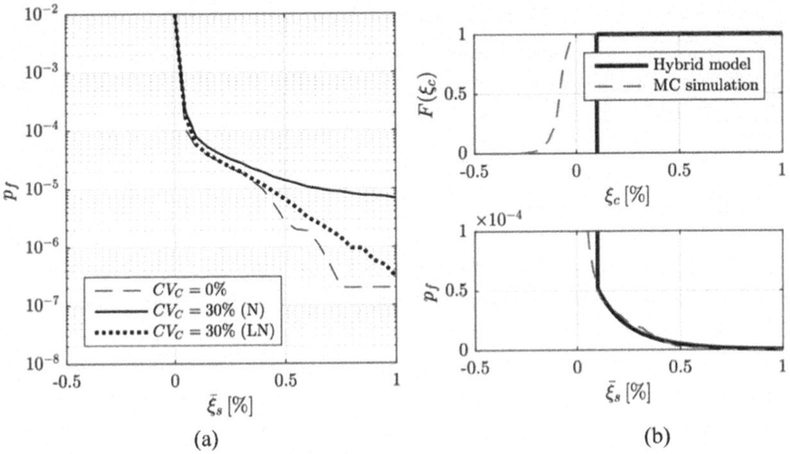

Fig. 5. Probability of failure for the two probabilistic cases for CV_C equal to 0% and 30% in the semi-logarithm plane (a). $CV_C = 0\%$ denotes a deterministic capacity model. N refers to a normal distribution. LN refers to a log-normal distribution. Comparison of the CDF of the structural damping required to avoid galloping and p_f as a function of $\bar{\xi}_s$ estimated using Montecarlo simulations and the hybrid model for $tr = 0.1\%$ (b).

it slightly increases with increasing CV_C. The normal capacity model always leads to larger values of p_f than the log-normal and deterministic ones.

Finally, a simplified closed-form equation (hybrid model) can be written with the failure probability expressed as:

$$p_f = 1 - \left\{ F_0(tr) + (1 - F_0(tr)) \left[1 - \exp\left(- \left(\frac{\bar{\xi}_s - tr}{c'} \right)^{k'} \right) \right] \right\} \text{ with } \bar{\xi}_s \geq tr \quad (13)$$

where $F_0(tr)$ is the fraction of damping values below the threshold tr, and k' and c' are the shape and scale parameters of a Weibull distribution, respectively, regressed to the data above tr. In Eq. 13, it is neglected all the data below tr since considered too low for to be representative of real damping values. The following parameters were regressed from the Montecarlo simulations using $tr = 0.1\%$ (this value is considered as a compromise between the two goals to have a lower value of tr and to have a good approximation of the failure probability beyond tr): $F_0(tr) = 0.999$; $c' = 0.1632$; $k' = 0.9482$. Considering the large simplification introduced, it can be seen a good agreement between the regressed model and the empirical data for $\bar{\xi}_s > tr = 0.1\%$ obtained through Montecarlo simulations (see Fig. 5b).

Finally, Eq. 13 can be used to calculate the minimum required damping to avoid structural vibration, $\bar{\xi}_s$, given a target probability failure $p_{f,t}$. ISO 2394 (ISO 1998) reports reliability requirements based on the expected social and economic consequences. For instance, assuming moderate costs of safety measures and great consequences of failure, it suggests $p_{f,t} = 10^{-5}$; using this value, it is found $\bar{\xi}_s = 0.35\%$. If it is low costs of safety measures and great consequences of failure ($p_{f,t} = 10^6$), then it is found $\bar{\xi}_s = 0.79\%$. This example showed that the required structural damping to prevent galloping can be designed according to a reliability approach as a function of the desired allowed risk.

6 Conclusions

Starting from a QS 2-DoFs model, this paper proposes a procedure for the probabilistic assessment of the minimum structural damping required to prevent galloping of ice-accreted bridge hangers. The framework is applied to a hanger of a prototype fictitious suspended bridge in the area of Copenhagen. The results highlight the advantages of the probabilistic procedure in terms of reliability quantification, compared to the deterministic approach. Moreover, the proposed procedure allows for a probabilistic safety check, based on reliability requirements given by International Standards, and it can be incorporated into performance-based design frameworks.

References

Avossa AM, Demartino C, Ricciardelli F (2017a) Probability distribution of footbridge peak acceleration to single and multiple crossing walkers. Procedia Eng 199:2766–2771

Avossa AM, Demartino C, Ricciardelli F (2017b) Design procedures for footbridges subjected to walking loads: comparison and remarks. Baltic J Road Bridge Eng 12(2):94–105

Avossa AM, Pianese G (2017c) Damping effects on the seismic response of base-isolated structures with LRB devices. Ing Sismica 34(2):3–29

CEN (2005) Eurocode 1: Actions on structures. Comité Européen de Normalisation

Chiodi R, Ricciardelli F (2014) Three issues concerning the statistics of mean and extreme wind speeds. J Wind Eng Ind Aerod 125:156–167

Demartino C (2014) Aerodynamics and aeroelastic behaviour of ice-accreted bridge cables. Ph.D. thesis. University of Naples Federico II University of Naples Federico II - Department of Structures for Engineering and Architecture

Demartino C, Avossa AM, Ricciardelli F (2018) Deterministic and probabilistic serviceability assessment of footbridge vibrations due to a single walker crossing. Shock Vib 26

Demartino C, Avossa AM, Ricciardelli F, Calidonna CR (2017a) Wind profiles identification using wind lidars: an application to the area of Lametia Terme. In: EACWE 2017-European-African conference on wind engineering, pp 1–10

Demartino C, Georgakis C, Ricciardelli F, (2013a) Experimental study of the effect of icing on the aerodynamics of circular cylinders - part II: Inclined flow. In: 6th European and African conference on wind engineering, Robinson College, Cambridge, UK

Demartino C, Koss H, Georgakis C, Ricciardelli F (2015) Effects of ice accretion on the aerodynamics of bridge cable. J Wind Eng Ind Aerod 138:98–119

Demartino C, Koss H, Ricciardelli F (2013b) Experimental study of the effect of icing on the aero dynamics of circular cylinders - part I: cross flow. In: 6th European and African conference on wind engineering, Robinson College, Cambridge, UK

Demartino C, Ricciardelli F (2014) Prediction of the buffeting response of ice-accreted stay cables. In: XIII Italian conference on wind engineering, Genova University Press

Demartino C, Ricciardelli F (2015) Aerodynamic stability of ice-accreted bridge cables. J Fluids Struct 52:81–100

Demartino C, Ricciardelli F (2017b) Aerodynamics of nominally circular cylinders: a review of experimental results for civil engineering applications. Eng Struct 137:76–114

Demartino C, Ricciardelli F (2019) Probabilistic vs. deterministic assessment of the minimum structural damping. ASCE J Struct Eng. (Accepted)

Demartino C, Ricciardelli F (2018b) Assessment of the structural damping required to prevent galloping of dry HDPE stay cables using the quasi-steady approach. ASCE J Bridge Eng 23 (4):1–17

Den Hartog J (1932) Transmission line vibration due to sleet. Electrical Eng 51:413

ISO (1998) ISO 2394: General principles on reliability for structures. International Organization for Standardization

Macdonald J, Larose G (2008) Two-degree-of-freedom inclined cable galloping part 1: general formulation and solution for perfectly tuned system. J Wind Eng Ind Aerod 96:291–307

Mardia K (1972) Statistics of directional data. Academic Press, London

Martin W, Naudascher E, Currie I (1981) Streamwise oscillations of cylinders. J Eng Mech Div 107:589–607

Nikitas N, Macdonald J (2013) Misconceptions and generalizations of the Den Hartog galloping criterion. J Eng Mech 140

Pagnini L, Freda A, Piccardo G, (2016) Uncertainties in the evaluation of one degree-of-freedom galloping onset. Eur J Environ Civ Eng 1–21

Pagnini L, Repetto M (2012) The role of parameter uncertainties in the damage prediction of the along wind-induced fatigue. J Wind Eng Ind Aerod 104–106:227–238

Piccardo G, Pagnini L, Tubino F (2015) Some research perspectives in galloping phenomena: critical conditions and post-critical behavior. Continuum Mech Therm 27:261–285

Ricciardelli F, Demartino C (2016) Design of footbridges against pedestrian-induced vibrations. J Bridge Eng C4015003:1–13

Experimental and Numerical Study of the Wind Tunnel Blockage Effects on the Behaviour of a Horizontal Axis Wind Turbine

A. Eltayesh[1(✉)], M. Burlando[2], F. Castellani[3], and M. Becchetti[3]

[1] Mechanical Engineering Department, Benha Faculty of Engineering, Benha University, Benha 13512, Egypt
abdelgalil.youssef@bhit.bu.edu.eg

[2] Department of Civil, Chemical, and Environmental Engineering, University of Genoa, Via Montallegro 1, 16145 Genoa, Italy
massimiliano.burlando@unige.it

[3] Department of Engineering, University of Perugia, Via G. Duranti 93, 06125 Perugia, Italy
{francesco.castellani,matteo.becchetti}@unipg.it

Abstract. This study presents an experimental and numerical investigation of the effects of tunnel blockage, defined as the ratio of the free portion of a wind tunnel cross-section with and without the rotor of a small-size (diameter 2 m) horizontal-axis turbine (HAWT). Experimental measurements were performed on three rotors with different number of blades, namely three, five and six, and different tip speed ratios, in the closed-loop closed-test chamber wind tunnel of the University of Perugia. Numerical wind tunnel simulations were performed at the University of Genoa through a steady-RANS method with SST k-ω turbulence model by means of the CFD solver OpenFOAM. The wind turbine was represented numerically by the actuator disc method. Simulations have been compared to experimental results for validation.

Keywords: Small-size wind turbines · Wind tunnel tests · CFD simulations · Wind tunnel blockage

1 Introduction

In wind tunnel (WT) experiments of small-size wind turbines, the most important factor affecting measurements is the blockage effect. The flow field in a wind tunnel differs from the one that would occur in unbounded free air under the same operating conditions due to the presence of wind tunnel walls. Therefore, it is usually required to correct measurements for blockage (Glauret 1935; Maskell 1963).

Some investigations on wind turbine performance and aerodynamics were carried out with no blockage correction, as for example by Schreck et al. (2007) and Hirai et al. (2008), who performed measurements on wind turbine performance in wind tunnel with blockage ratio around 10%.

F. Ricciardelli and A. M. Avossa (Eds.): INVENTO 2018, LNCE 27, pp. 272–285, 2019.
https://doi.org/10.1007/978-3-030-12815-9_22

Different approaches exist in the literature, however, to take into account blockage effects. Sørensen et al. (2006), for instance, derived a correction model for the effect of blockage of the German–Dutch wind tunnel's (DNW), which is an open test section WT, using the one dimensional momentum approach. A series of computational fluid dynamics (CFD) simulations using the Navier-Stokes/Euler actuator disc (AD/NS) model were performed to validate the model. The results showed that the corrections from the model were in very good agreement with the computations.

Bahaj et al. (2007), who performed experimental measurements of power and thrust of an 800 mm diameter model of a marine current turbine in a cavitation tunnel and a towing tank, corrected their measurements using the blockage model developed by Barnsley and Wellicome (1990), which is based on the actuator disc model as well.

Chen and Liou (2011) studied the effect of tunnel blockage on the power coefficient of horizontal-axis wind turbines (HAWT). The experimental work was carried out in an open-circuit wind tunnel with test section of 1.3 m times 1.3 m. They used rotor blade lengths of 30, 24 and 14.5 cm which corresponds to tunnel blockage ratio of 28.3%, 20.2% and 10.2%, respectively. The results indicated that the blockage effect was strongly dependent on tip speed ratio TSR, blockage ratio ε, and rotor pitch angle β. Blockage effect is smaller at larger TSR and ε values. Also, blockage correction is less than 5% for a ε of 10% and for TSR less than 1.5. Again, they corrected their results using the same method adopted by Bahaj et al. (2007).

Ryi et al. (2015) carried out an experimental work to investigate the blockage effect on small scale wind turbine using both closed and open circuit wind tunnel. The small-scale rotor was tested under the same operating conditions (rotating speed, rotor torque, speed of wind tunnel, advance ratio) for both wind tunnel. The experiments were performed at blockage ratios of 48.08%, 17.97%, and 8.08%. Blockage ratio was defined as the ratio of the project area of the model and the test section area. The correction model proposed by Glauret (1935) was used for comparison with both wind tunnel experiments. The results showed that Glauert's methods works well for all condition with low thrust.

Sarlak et al. (2016) studied the effect of wind tunnel blockage on the characteristic of wake and power of a HAWT using the large eddy simulation (LES) technique and an actuator line model. Numerical simulations were performed using different blockage ratios of 0.2, 0.09, 0.05 and 0.02. Numerical investigations were carried out using tip-speed ratios (TSR) varying from 1 to 12. Results showed that, when TSR is much lower than the optimal value, blockage effects are insignificant on both power and thrust coefficient as well as wake characteristics. Conversely, when TSR is close to or above the optimal value, whatsoever blockage ratio larger than 0.05 affects power and thrust.

Zaghi et al. (2016) analyzed the blockage effects for wind tunnel experiments on large-scale wind turbines. In this study, a number of simulations were carried out to investigate the blockage effects by approximate tunnel geometry. Two circular wind tunnel were tested: firstly, the radius was chosen to keep the same cross section area as actual wind tunnel; secondly, the radius was set to be equal to half of the smallest cross-section dimension. The results showed that no unique equivalent section exists that allow both global coefficients and local flow features to be matched.

Kinsey and Dumas (2017) presented a numerical study to investigate the effect of channel blockage on the performance of axial and cross flow hydrokinetic turbines. They also study different distances of vertical and lateral boundaries which lead to the same blockage ratio. The results show that the blockage correction based on simple momentum actuator disc approach is good for axial and low solidity cross flow. In addition, the performance of axial and cross flow turbines depend only on blockage ratio unless symmetry differences are greater than 3.

The aim of the present study is to investigate the effect of a closed-loop open-chamber wind tunnel blockage on the performance of a small scale wind turbine using experimental measurements as well as numerical calculations. In particular, the equations for blockage correction by Kinsey and Dumas (2017) are applied to calculate the real power coefficient of the wind turbine.

2 Experimental Setup

The experiments were carried out in the wind tunnel “R. Balli” of the University of Perugia. This is a closed-loop configuration as shown in Fig. 1, driven by a 375 kW electric motor so that the wind can be adjusted up to a maximum speed of 47 m/s. The inlet section of the open test chamber has an area of 5 m^2 while the recovery section is 7 m^2. The inflow conditions are measured through two pitot tubes and a cup anemometer in the inlet section while all the atmospheric parameters (static air pressure, temperature and relative humidity) are collected on a station point in the testing room. An honeycomb upstream the test section is able to keep turbulence level below 0.4%. Figure 2 shows the frontal and lateral view of the five blade HAWT that has been used in this study. The rotor diameter is 2 m, and the turbine blade were attached to a hub whose diameter is 26 cm. the blade has constant pitch angle with varying rotational speed from 100 to 700 RPM. The description of wind turbine rotor is summarized in Table 1. More details are presented in Scappatici et al. (2016). During each test the velocity were measured in the wind tunnel with and without wind turbine rotor.

3 Numerical Setup

Numerical simulations were carried out using the CFD solver OpenFOAM to study the flow field and the wind tunnel blockage effect on the wind turbine performance. Fluid flow characteristics were obtained by solving the governing equations for momentum, continuity and turbulence. The flow was assumed steady and incompressible. According to Bardina et al. (1997), who analysed the performance of different turbulence models, the SST k-ω model can predict properly the flows with strong adverse pressure gradients and separation, so that this model was employed to represent the turbulent flow.

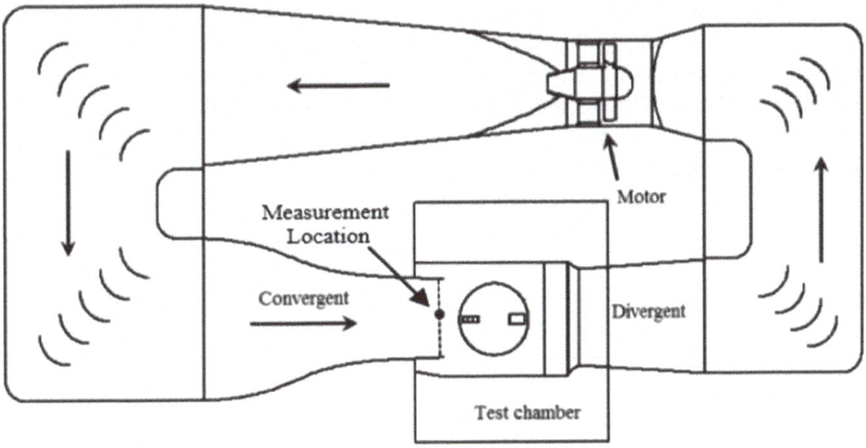

Fig. 1. A schematic of the wind tunnel.

Fig. 2. Frontal (left) and lateral view (right) of the wind turbine in wind tunnel.

Table 1. Wind turbine characteristics.

Wind turbine characteristics	
Maximum power	2800 W
Cut-in and cut-off wind speed	3, 15 m/s
Blades number	3 and 5

The three-dimensional structured grid was generated using the preprocessor software ICEM 14. Figure 3 illustrates the 3D computational domain and the corresponding boundary conditions. The computational domain was designed to match the exact shape of the wind tunnel in order to represent properly the experimental setup and to account for blockage effects. The mesh was generated in a multi-block topology. The domain was divided into fifty-one blocks and the grid was generated in each block individually. The appropriate distribution of grid coordinates is very important for the quality of the solution in terms of numerical accuracy and stability, so only heaxahedral cells were used (Hirsch et al. 2002). At each side of the domain a boundary layer region was employed. The size of the first cell near wall was chosen in order to obtain dimensionless wall distance y^+ in the logarithmic layer region (30–300). The first cell centre was around 0.0003 m from the wall and the growth factor of the grid was kept below 1.2 everywhere in the computational domain. The boundary layer had 16 rows along every side wall, which is far beyond the minimum number of ten mentioned in the best practice guidelines (Franke et al. 2007; Tominaga et al. 2008).

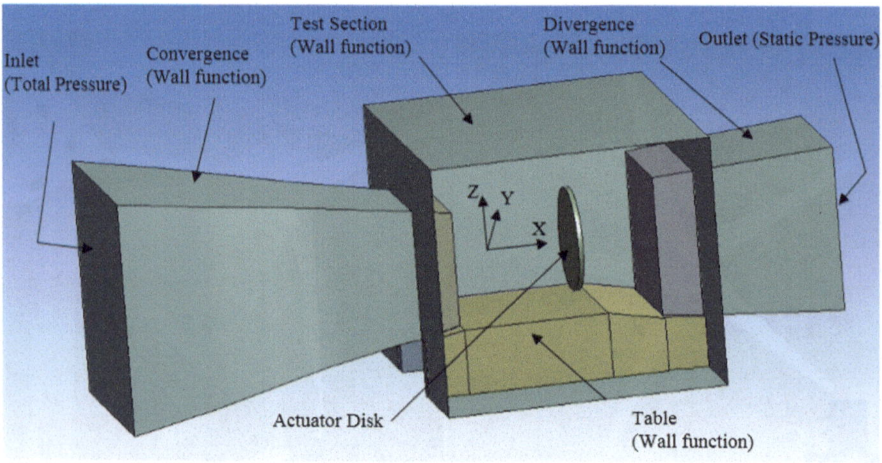

Fig. 3. Three-dimensional computational domain and boundary conditions used for the numerical simulations.

The total pressure of the incoming flow was obtained from the current experimental measurements and was applied at the inlet boundary for different velocities considered in this study. The non-slip boundary condition was considered at the walls. The flow was assumed fully developed at the outlet boundary and zero static pressure was imposed.

The actuator disk (AD) region was modelled as a very thin disk and the properties of the wind turbine were defined in terms of power and thrust coefficients according to the experimental measures. The velocity at convergence exit was calculated during each simulation for wind tunnel with and without actuator disc, which used to calculate

Fig. 4. Three-dimensional computational mesh.

predicted blockage factor. The final three dimensional mesh is shown in Fig. 4, which consists of about 15.8 million cells.

The grid dependency of the solution was examined using an inlet total pressure of 25.93 Nm^{-2} Three grids were used to check the grid dependency with a total number of cells of about 8, 10 and 15 millions, respectively. The effect of the grid size on velocity was examined: Fig. 5, which shows the comparison between the longitudinal (y-direction) wind speed values behind the actuator disc at 0.5 disc diameter along the transversal direction, indicates that there is no significant change in velocity when the mesh size increases above 10-million cells. Therefore, this grid was considered sufficiently refined for the wind tunnel and it was used in the current study. Furthermore, the vertical velocity profile inside the wind tunnel was measured in case of without turbine rotor and are compared to the numerical calculation as shown in Fig. 6. The velocity was measured at about 2 m from the exit of inlet convergence. The figure shows that good agreement was obtained between numerical predictions and experimental measurements.

4 Results

In this paper the measured power coefficient is corrected using the equations for blockage correction by Kinsey and Dumas (2017), which are mainly based on the definition of a so-called blockage factor (BF) as follows:

$$BF = \left(\frac{U}{U'}\right) \tag{1}$$

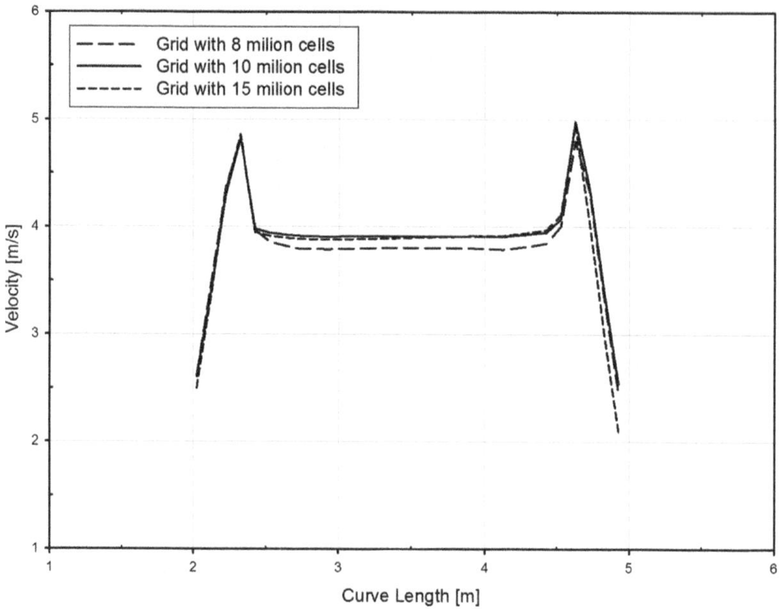

Fig. 5. Transversal velocity distribution of the longitudinal wind speed component for different grid sizes.

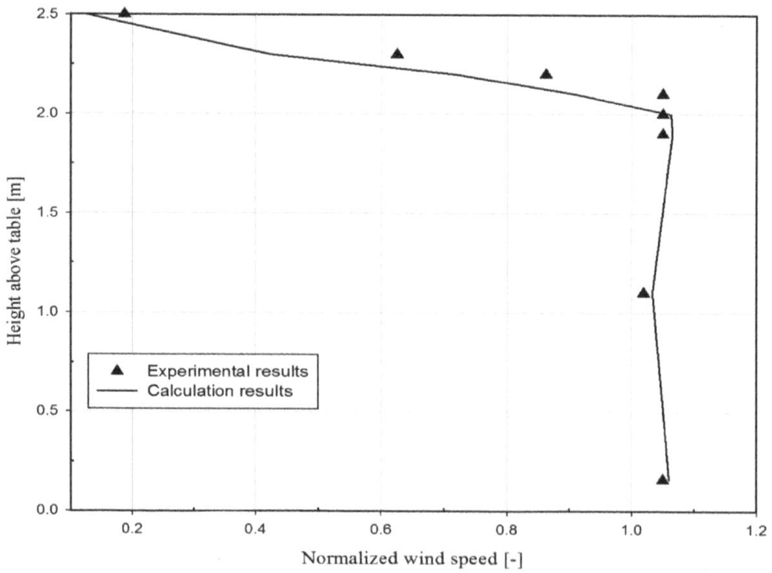

Fig. 6. Transversal velocity distribution of the vertical wind speed.

where BF is the ratio between the freestream wind speed in the wind tunnel with rotor, U, and the freestream wind speed in the wind tunnel without rotor, U'.

Then, power coefficient Cp as well as tip-speed ratio (TSR) are defined as a function of to scale parameter BF.

$$Cp' = Cp\left(\frac{U}{U'}\right)^3 = Cp \cdot BF^3 \tag{2}$$

$$TSR' = TSR\left(\frac{U}{U'}\right) = TSR \cdot BF \tag{3}$$

where Cp' and TSR' are the correction of non-dimension power coefficient and tip-speed ratio.

Figure 7 shows the experimental measurements and numerical calculations for the distribution of the tunnel blockage factor for the inlet free stream wind speed U' = 7.3 m/s, as a function of the TSR for the case of five-blade wind turbine. The horizontal and vertical axis represents TSR and BF, respectively. As show from the figure the measured blockage factor decreases as TSR increases until TSR of 5. The reason for this is that, while increasing the TSR, the blades rotation reduces the air flow through the turbine's rotor. After this value of TSR the turbine blades rotate so fast to act like a partially-solid disc, the blockage factor become independent on TSR and it assumes a constant value which is around 0.895 from experiments.

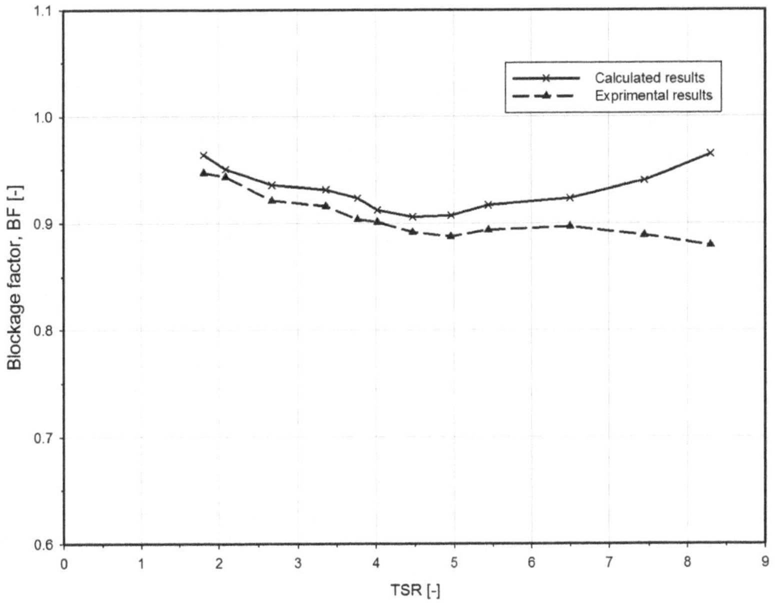

Fig. 7. Relation between BF and TSR, at free wind speed equal to 7.3 m/s, five blade.

The results from numerical calculation using AD shows a reasonable agreement with measured blockage factor until TSR of 5. After this value the calculated blockage factor start to increases again, and the difference between measure and calculation blockage factor grows. The reason for this is that rotational speed of the wind turbine is not considered in the numerical calculation, AD depends on power and thrust coefficient of wind turbine: as the TSR exceeds 5 the power and thrust coefficient start to decrease, as shown in Fig. 8. Therefore, the AD is not valid for TSR greater than 5.

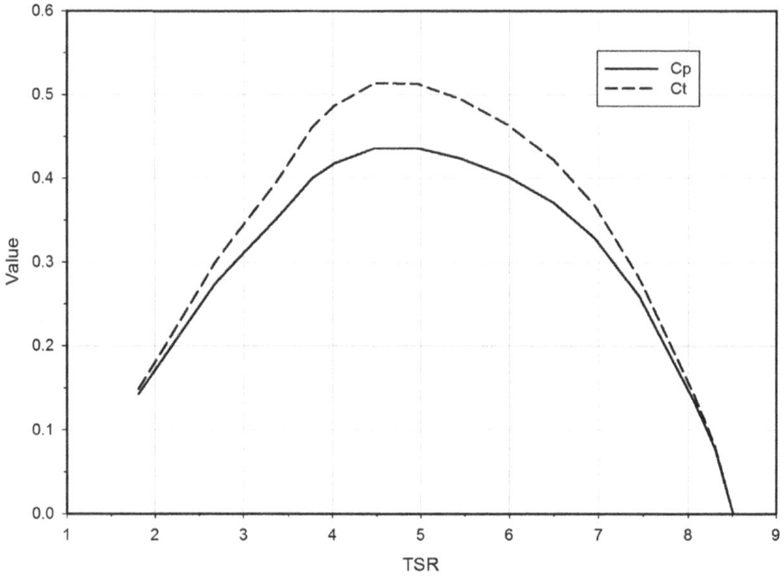

Fig. 8. Relation between uncorrected measurements of Cp, Ct and TSR, at free wind speed equal to 7.3 m/s.

Figure 9 illustrates the relationship between the measured and calculated blockage factor and TSR for the inlet free stream wind speed $U' = 7.3$ m/s, for the case of three-blade wind turbine. The figure indicates that the maximum blockage factor was predicted at TSR of 0.7. Then, the blockage factor gradually decreases as TSR increases. At low the air flow impede less with the turbine rotor as they pass through the rotor. As a result, the rotor effect on air flow is small and the tunnel blockage effect is small as well. When the TSR gets larger, the air flow impede more with the turbine rotor and lose more of kinetic energy. In this case, the rotor effect on air flow increases, and the tunnel blockage effect is also large. Generally, lager blockage factor the smaller TSR. The blockage factor became small when TSR about 6, and the blockage factor become independent on TSR, and approach a constant value of 0.92. Decline rate of blockage factor for three blade rotor are smaller than five blade rotor at the same TSR. The blockage factor at TSR of 3 is 0.95 for three blade rotor. The blockage factor at TSR of 3 is 0.92 for 5 blade rotor.

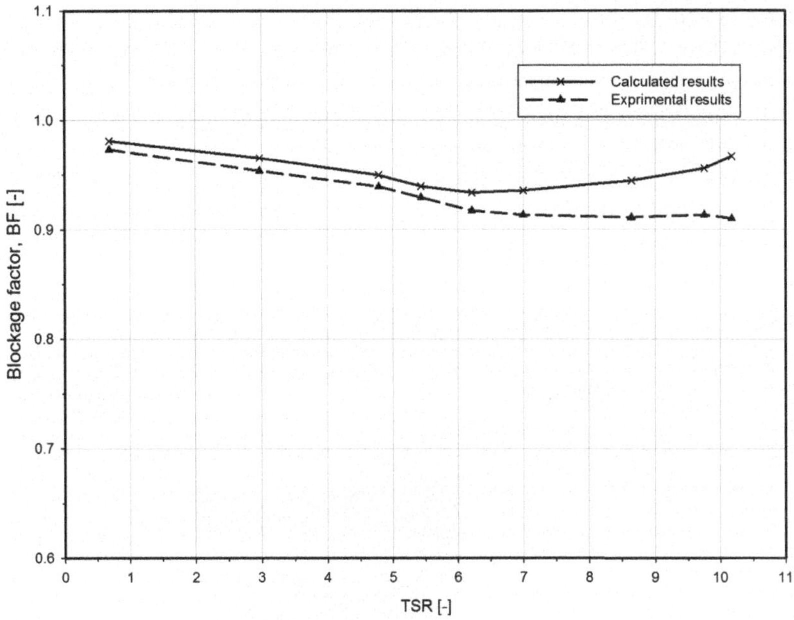

Fig. 9. Relation between BF and TSR, at free wind speed equal to 7.3 m/s, three blade.

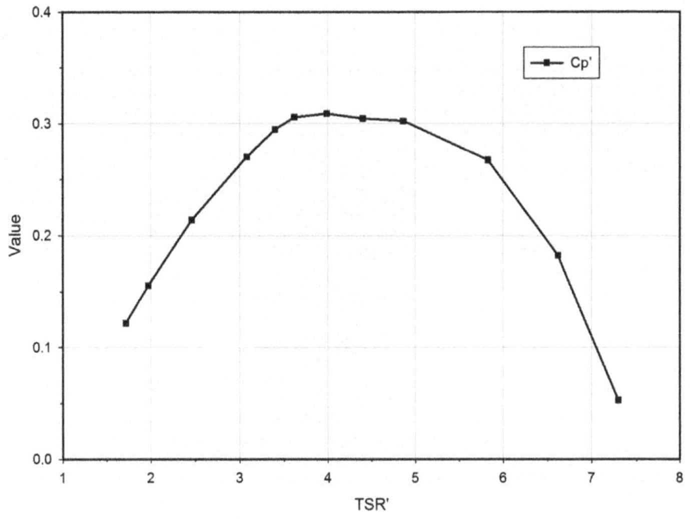

Fig. 10. Corrected Cp′, TSR′, at free wind speed of 7.3 ms^{-1} using $\left(\frac{U}{U'}\right)^3$.

Figures 8 and 10 show the uncorrected and corrected measured power coefficient and tip speed ratio for five blade rotor. The higher values of Cp with respect to its corrected value Cp′ demonstrate that the peak of power coefficient moves to higher TSR as the blockage ratio increases. The uncorrected maximum power coefficient is located at TSR around 4.5, while the corrected one is at 3.5.

Figure 11a and b show the velocity magnitude for two values of TSR corresponding to 1.8 and 5, respectively, simulated in the horizontal section at the centre of the wind tunnel, which is also in the centre of the AD. Generally, the wind speed

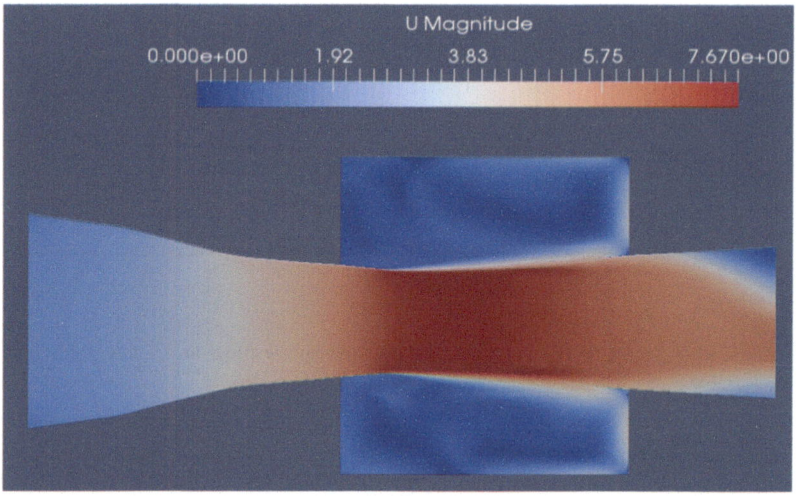

(a)

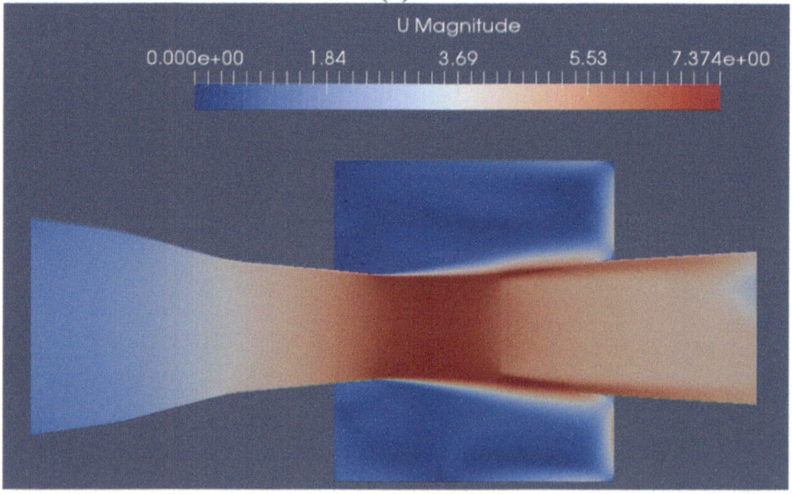

(b)

Fig. 11. Horizontal section velocity magnitude (a) TSR of 1.8, CP of 0.143 and CT of 0.149, and (b) TSR of 5, CP of 0.436 and CT of 0.513.

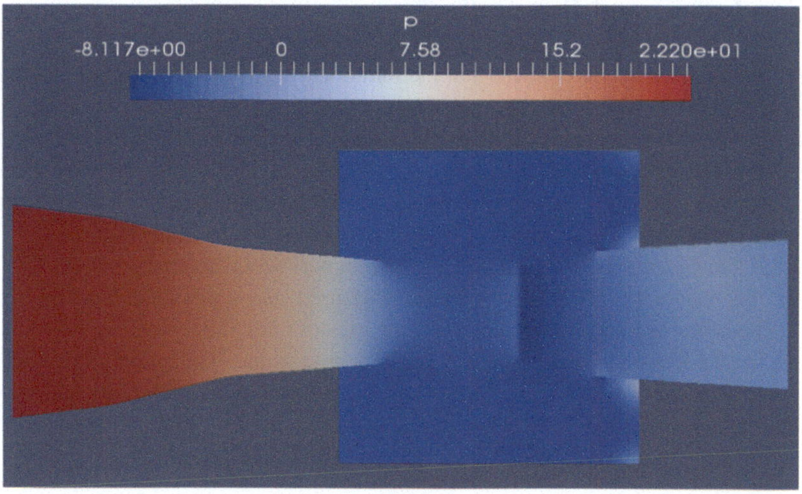

(a)

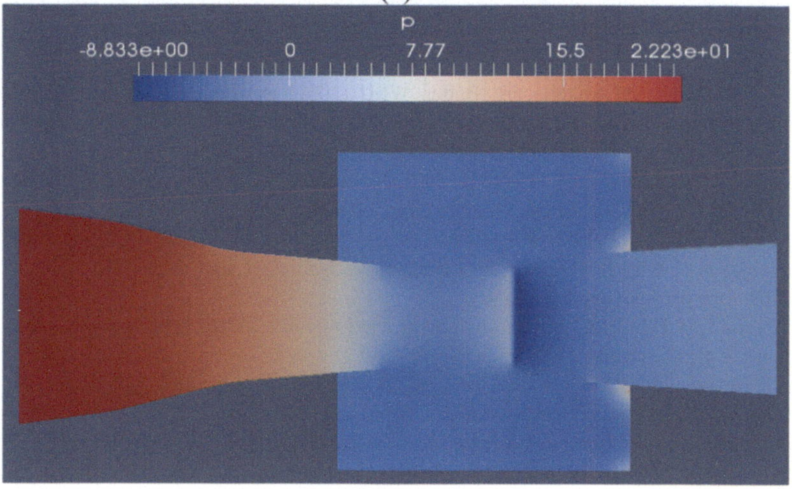

(b)

Fig. 12. Horizontal section pressure contour (a) TSR of 1.8, CP of 0.143 and CT of 0.149, and (b) TSR of 5, CP of 0.436 and CT of 0.513.

decreases before the AD due to the appeared force, while behind it the flow deviates from the AD along a stream tube which determines a higher velocity region. Increasing TSR to 5 the thrust and power coefficient increases. This causes the resistance force to increase. Further downstream, low velocity zone can be observed behind AD as shown in Fig. 11b.

Figure 12a and b show the behaviour of static pressure in the same sections represented in Fig. 10.

5 Conclusions and Perspectives

The results obtained so far confirm that the numerical wind tunnel simulation with AD of the facility in Perugia can be considered suitable for evaluating numerically the blockage effects of small wind turbines. In general, Experimental results, in particular, showed that the measured blockage factor decreases as TSR increases until a certain value, after this value the blockage factor becomes independent on TSR and it assumes a constant value which is around 0.895 for five blade rotor. Numerical calculations using AD showed reasonable agreements with experimental measurements until TSR 5, after this value the difference between measured and calculated blockage factor grows. Therefore, the AD is not valid for TSR greater than 5, because the AD model is not able to take into account the rotation of the wind turbine. The model, however, will be validated more extensively and further measurements and simulations are ongoing considering different kind of wind turbines, using three, five and six rotor blades in order to create a benchmark for future analyses.

References

Bahaj AS, Molland AF, Chaplin JR, Batten WMJ (2007) Power and trust measurements of marine current turbines under various hydrodynamic flow conditions in a cavitation tunnel and a towing tank. Renewable Energy 32:407–426

Bardina JE, Huang PG, Coakley TJ (1997) Turbulence modeling validation, testing and development. NASA Technical Memorandum 110446

Barnsley M, Wellicome J (1990) Final report on the 2nd phase of development and testing of a horizontal axis wind turbine test rig for the investigation of stall regulation aerodynamics. Technical report 5A/CON5103/1746

Chen T, Liou L (2011) Blockage corrections in wind tunnel tests of small horizontal axis wind turbines. Exp Therm Fluid Sci 35:565–569

Franke J, Hellsten A, Schlünzen H, Carissimo B (2007) Best practice guideline for the CFD simulation of flows in the urban environment. COST Action 732, quality assurance and improvement of microscale meteorological models

Glauert H (1935) Airplane propellers. In: Aerodynamic theory. Springer, Heidelberg, pp 169–360

Hirai S, Honda A, Kariromi K (2008) Wind loads investigations of HAWT with wind tunnel tests and site measurements. In: Wind Power Asia

Hirsch C, Bouffioux V, Wilquem F (2002) CFD simulation of the impact of new buildings on wind comfort in an urban area. In: Augsti G, Borri C, Sacrè C (eds) Impact of wind and storm on city life and built environment. CSTB, Nantes, pp 164–171

Kinsey T, Dumas G (2017) Impact of channel blockage on the performance of axial and cross-flow hydrokinetic turbines. Renewable Energy 103:239–254

Maskell EC (1963) A theory of blockage effects on bluff bodies and stalled wings in closed wind tunnel. ARC R&M 3400

Ryi J, Rhee W, Hwang UC, Choi J-S (2015) Blockage effect correction for a scaled wind turbine rotor by using wind tunnel test data. Renewable Energy 79:227–235

Scappatici L, Bartolini N, Castellani F, Astolfi D, Garinei A, Pennicchi M (2016) Optimizing the design of horizontal-axis small wind turbines: from the laboratory to market. J Wind Eng Ind Aerodyn 154:58–68

Sarlak H, Nishino T, Martinez-Tossas LA, Meneveau C, Sørensen JN (2016) Assessment of blockage effects on the wake characteristics and power of wind turbines. Renewable Energy 93:340–352

Schreck SJ, Sorensen NN, Robinson MC (2007) Aerodynamic structures and processes in rotationally augmented flow fields. Wind Energy 10:159–178

Sørensen JN, Shen WZ, Mikkelsen R (2006) Wall correction model for wind tunnels with open test section. AIAA J 44:1890–1894

Tominaga Y, Mochida A, Yoshie R, Kataoka H, Nozu T, Yoshikawa M (2008) AIJ guidelines for practical applications of CFD to pedestrian wind environment around buildings. J Wind Eng Ind Aerodyn 96:1749–1761

Zaghi S, Muscari R, Di Mascio A (2016) Assessment of blockage effects in wind tunnel testing of wind turbines. J Wind Eng Ind Aerodyn 154:1–9

Wind Field Variability in Complex Terrain: Lessons from the Hardanger Bridge

A. Fenerci(✉) and O. Øiseth

Department of Structural Engineering,
Norwegian University of Science and Technology,
Richard Birkelands vei 1A, 7034 Trondheim, Norway
{aksel.fenerci,ole.oiseth}@ntnu.no

Abstract. Along the Coastal Highway E39 in the western coast of Norway, Norwegian Government is planning to build several extreme bridges spanning from 1.5 to 5 km. The region is typically mountainous with deep fjords seeping inland. Here, experience gained from a 5-year monitoring campaign on the Hardanger Bridge in Norway is summarized relating to this ambitious project. The analysis of data provided valuable knowledge on the wind characteristics, which can be generalized for the whole region. Insight has also been gained on the dynamic behaviour of the bridge and how it is influenced by the wind conditions. The results are presented and discussed here with the future bridges in mind.

Keywords: Suspension bridge · Turbulence characteristics · Buffeting response · Complex topography · Probabilistic modelling

1 Introduction

Structural health monitoring of large or lifeline structures has been extensively carried out in the last two decades. Long-span cable-supported bridges in particular received considerable attention such that almost every major bridge in the world has a monitoring system installed in it. Although the main purpose of such systems are to monitor the structural health and identify any damage or unexpected phenomena, valuable data on environmental conditions and bridge behaviour is collected as a by-product. Such data are commonly used by researchers to improve the state-of-the-art techniques for structural design and assessment such as the wind-resistant design of long-span cable supported bridges.

In 2013, one such project was initiated on the Hardanger Bridge (HB), to this day the longest suspension bridge in Norway, for the sole purpose of research. The data acquired during the monitoring campaign is used to study the wind characteristics and the terrain effects, validity of common assumptions in wind engineering of such structures, performance of buffeting response predictions and also the state-of-the-art design methodology. In this paper, conclusions drawn from the monitoring campaign and their implications on future bridge designs in Norwegian fjords are discussed.

F. Ricciardelli and A. M. Avossa (Eds.): INVENTO 2018, LNCE 27, pp. 286–295, 2019.
https://doi.org/10.1007/978-3-030-12815-9_23

2 Hardanger Bridge Monitoring Project

Hardanger Bridge Monitoring Project (MBMP) has been initiated in 2013 by the Norwegian University of Science and Technology (NTNU) with financial support from the Norwegian Public Roads Administration (NPRA, Statens Vegvesen). It was aimed that the project and the full-scale observations would provide information and insight for the future bridge projects along the E39 Coastal Highway. In particular, the wind characteristics in the fjords, how are they influenced by the particular terrain and in return how they influence the bridge behaviour, were questions of interest.

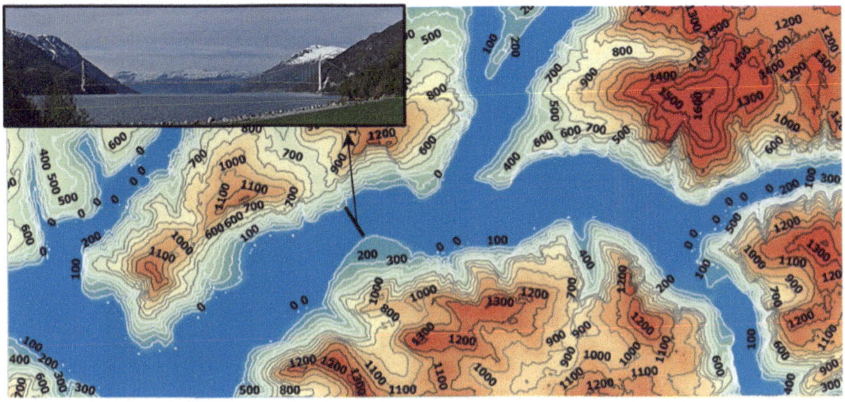

Fig. 1. The Hardanger Bridge and its surroundings

The Hardanger Bridge (Fig. 1) is the longest suspension bridge in Norway as of today with its main span of 1308 m. The bridge is located in western Norway and it is surrounded by a complex terrain, consisting of steep mountains and fjords in between. Such features makes the HB a very interesting case study, especially when the E39 project is concerned. The bridge was opened for traffic in 2013 and a structural monitoring system was installed on it shortly after. The monitoring system consists of 20 accelerometers, 9 anemometers and several other equipment for data communication and storage. An overview of the monitoring system and sensor locations are presented in Fig. 2 and the pictures of wind and acceleration sensors are shown in Fig. 3. For more information on the monitoring system, the reader is kindly referred to the published papers (Fenerci et al. 2017; Fenerci and Øiseth 2017).

The monitoring system measures wind velocities and accelerations at the HB site continuously. However, the data is recorded only when a threshold wind speed of 15 m/s (averaged over 1 min) is exceeded in any of the sensors. In that case, a 30-min recording is taken. The monitoring system has recorded data following such a triggering rule during the 5-year measurement period between 2013 and 2018. Occasionally, the system has also been triggered manually to include the low wind speed data in the database.

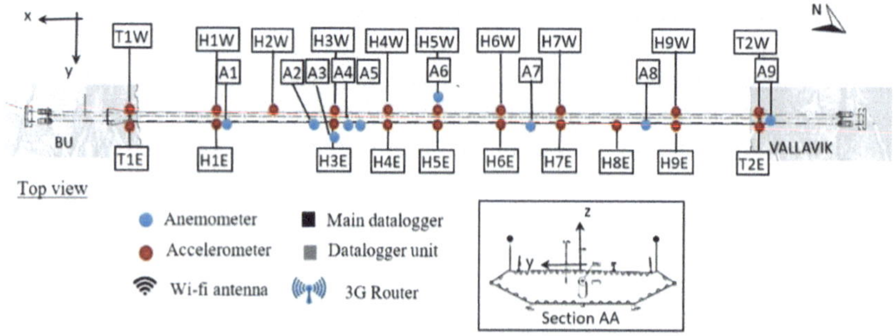

Fig. 2. Sensor layout

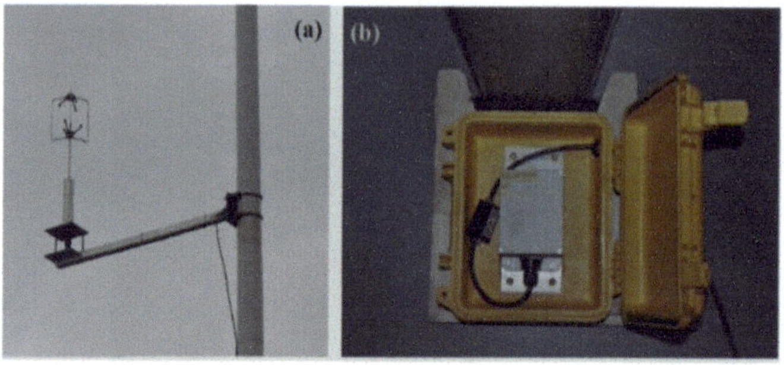

Fig. 3. Sensors: (a) anemometer attached to hanger, (b) accelerometer inside deck

Immediately after a recording is taken at the site, the data is first logged in the main logger unit located at the bridge and then transferred to designated servers at NTNU. A customized computer program then detects the new recording and starts processing data. Initially, the data is filtered, downsampled and transformed to appropriate coordinate systems and adjusted for errors. Both the raw and adjusted data are then stored in a database in the Matlab environment. The organization structure of a single recording in the database is shown in Fig. 4. The data are then processed to extract the interesting statistics concerning wind and response measurements, separately. An averaging interval of 10 min, which is rather common in wind engineering, is used to evaluate the statistics of both wind and response. The extracted statistics are then collected in a separate database than the raw and adjusted data. Loading such a database to common scientific computation software such as Matlab or Python, the wind and acceleration statistics for all recordings can be accessed easily. An overview of all 10 min recordings that has been taken during the monitoring period is given in Fig. 5.

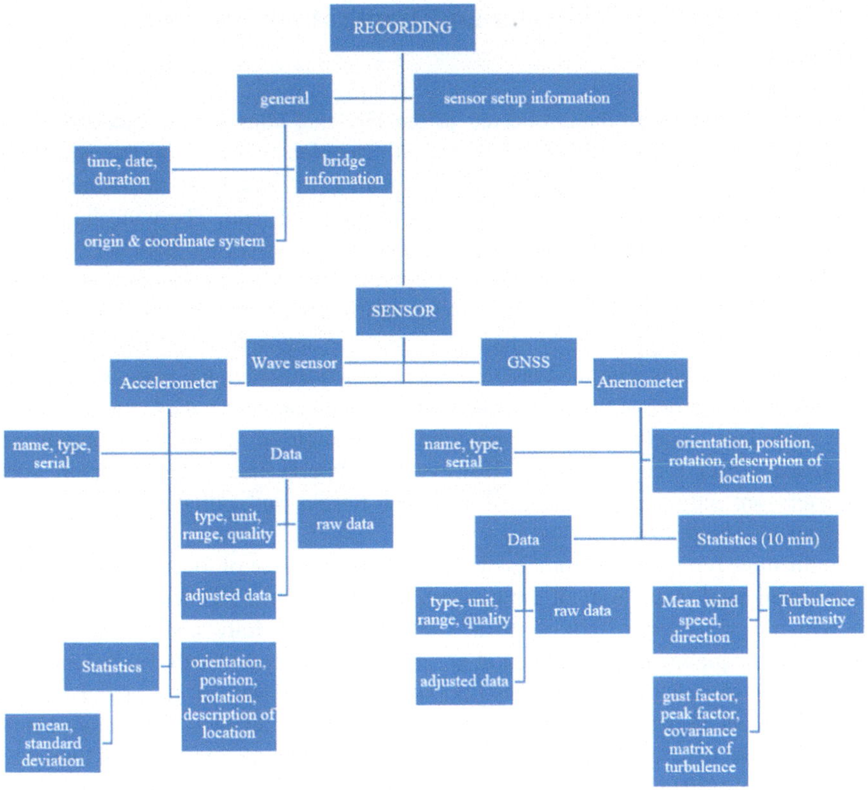

Fig. 4. Organization structure of the database for single recording

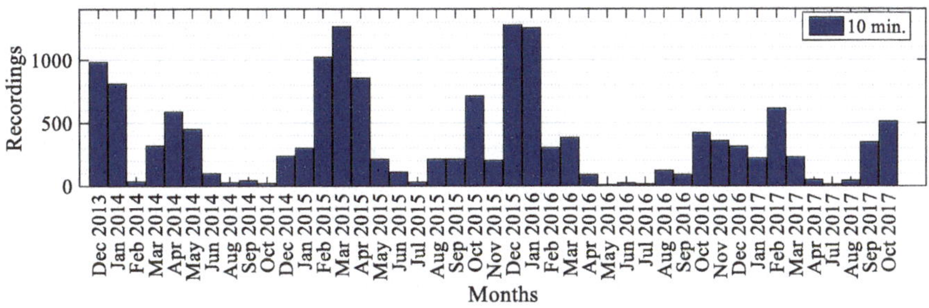

Fig. 5. Overview of 10 min recordings

3 Randomness in Field Data and Implications on Bridge Design

Examining the long-term data collected at the HB site, considerable randomness in wind and response characteristics is detected. Detailed analyses of the wind and acceleration data was conducted in Fenerci et al. (2017) and the randomness was attributed to a large extent to the randomness in the wind field. The wind field characteristics were calculated using traditional approaches, assuming stationarity of the time signals during a 10-min averaging interval. The inherent randomness in the resulting wind characteristics was found to be significant even for the high wind speed events, where the wind is fairly stationary and the atmosphere is stable. Such an observation contradicts with the common design practice, where the variable nature of the statistical properties of the turbulence such as turbulence intensities, integral length scales and decay coefficients for spanwise correlation of turbulence components (Davenport 1961) are overlooked. However, it appears that randomness in such parameters are significant in case of complex terrain and influence the dynamic behaviour to a large extent.

To illustrate how such randomness would affect the design of long-span suspension bridges, analytical predictions of the dynamic response of the HB deck is carried out. The dynamic response of the deck is calculated using fully coupled multimode approach in frequency domain (Chen et al. 2001; Fenerci and Øiseth 2017; Jain et al. 1996). A time-domain approach is also possible, but nonlinearities in case of the HB are not profound, allowing a much faster frequency domain analysis. The aerodynamic properties of the deck was obtained through forced vibration tests in the wind tunnel by Siedziako et al. (2017) and directly incorporated into the analysis, where the modal properties of the bridge were obtained through finite element (FE) analysis. The static force coefficients of the section are given in Table 1, where the natural frequencies of the bridge are given in Table 2. More detailed information concerning the bridge and the details of the analytical approach can be found in Fenerci and Øiseth (2017).

Table 1. Vibration modes of the Hardanger Bridge

Lateral		Vertical		Torsional	
Mode no	Freq. (Hz)	Mode no	Freq. (Hz)	Mode no	Freq. (Hz)
1	0.05	3	0.11	15	0.36
2	0.098	4	0.14	26	0.52
5	0.169	6	0.197		
10	0.233	7	0.21		

Table 2. Steady-state force coefficients of the Hardanger Bridge deck

$\bar{C}_D$	C'_D	$\bar{C}_L$	C'_L	$\bar{C}_M$	C'_M
1.05	0	−0.363	1.789	−0.017	0.654

* D = drag, L = lift, M = moment (bar denotes mean value and apostrophe denotes derivative)

Buffeting loads on a line-line structure such as the HB deck is commonly represented in frequency domain with the help of a cross-power-spectral-density (CPSD) matrix. In design of long-span suspension bridges; however, turbulence characteristics that defines the CPSD matrix of turbulence are treated as deterministic parameters. The parameters that define the spectra can be estimated using mast measurements at the site, wind tunnel terrain model tests, computational fluid dynamics (CFD) tools combined with mesoscale atmospheric models or on-site measurements or directly using code recommendations. In any case, the parameters would be average values and will result into a deterministic spectra, neglecting the inherent randomness of the turbulence field described earlier. Following such an approach, the turbulence field is modelled with the following formulae:

$$\frac{S_{u,w}f}{\sigma^2_{u,w}} = \frac{A_{u,w}f_z}{(1+1.5A_{u,w}f_z)^{5/3}}, \quad f_z = \frac{fz}{U}, \quad C_{uu,ww} = \exp(-K_{u,w}\frac{f\Delta x}{U}) \tag{1}$$

where $\sigma_{u,w}$ are the standard deviations of the along-wind (u) and vertical (w) turbulence components. $A_{u,w}$ are the parameters to be fitted. In the expression, $S_{u,w}$ denote the auto-spectral densities, f denotes frequency in Hz and z denotes the height above ground (68 m for the midspan) and U denotes the mean wind velocity. $C_{uu,ww}$ denote the normalized cross spectra of turbulence at two points separated by Δx and $K_{u,w}$ are commonly referred to as decay coefficients, which are also needed to be fitted to data. Therefore, for an estimation of the CSPD matrix of turbulence in modeling of the buffeting actions on the HB, the parameters $\sigma_{u,w}$, $A_{u,w}$ and $K_{u,w}$ need to be estimated from the available data. Here, most probable values (modes) of each parameters were used to model the turbulence field. The modes of each parameter were obtained by fitting lognormal distributions to the field data. Using, such input for the buffeting analysis, the resulting comparison of the measured and analytical dynamic response if given in Fig. 6. The figure shows that the analytical predictions lie somewhere around in the middle of the scatter of the measured response for all of the response components (lateral, vertical and torsional acceleration responses). Clearly, the design curves obtained in such a way will be exceeded by many measurement points since the randomness in the wind field is neglected. Thus, such an observation points out the need for a probabilistic description of the above mentioned turbulence spectra, and a coherent approach for evaluating the corresponding dynamic response.

4 Probabilistic Modelling of the Wind and Turbulence Field

Previous investigations showed that the turbulence field at the HB site can be modelled with reasonable accuracy using relatively simple expressions for the CPSD matrix with inclusion of only six turbulence-related parameters. These are namely the turbulence standard deviations ($\sigma_{u,w}$) which give the power of the turbulence components, spectral parameters ($A_{u,w}$) that relate to the spectral content of turbulence and decay coefficients ($K_{u,w}$) which represent the spanwise correlation information of turbulence. The information regarding these parameters gives the full turbulence field along the bridge deck.

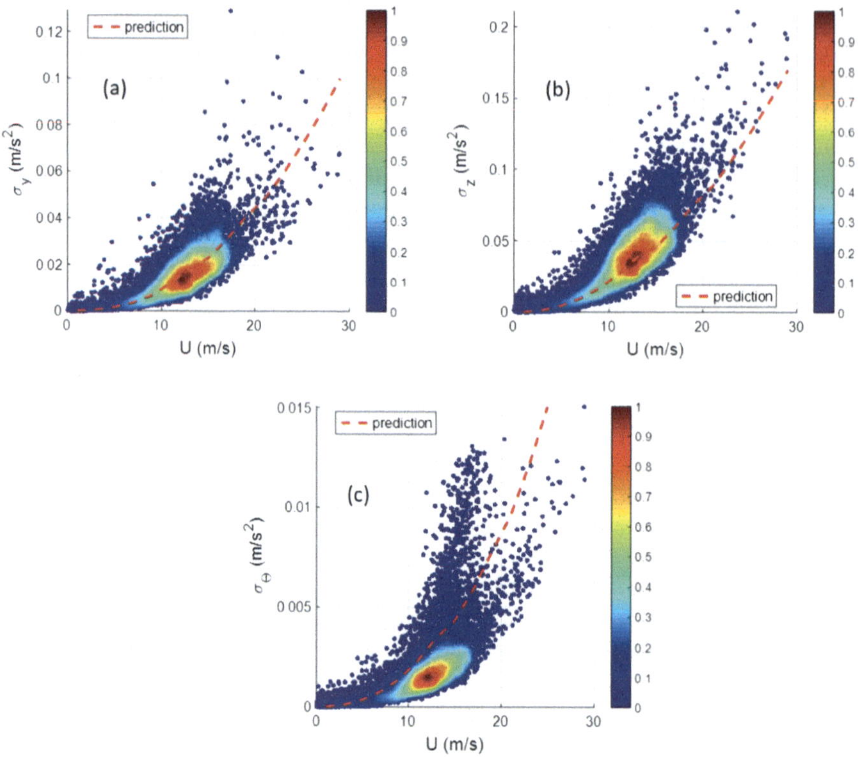

Fig. 6. RMS response at the midspan: (a) lateral (b) vertical and (c) torsional acceleration response

It should be noted that if the buffeting loading on other components of the structure, such as bridge cables or pylons are going to be considered, more parameters are needed.

It was observed that all of these parameters accommodate inherent randomness, which can be modeled reasonably well using lognormal probability distributions. Considering the nature of the problem and observing the long-term data, it is also expected that these parameters also depend on the mean wind speed and direction during the averaging period considered. Moreover, the correlation of different parameters can also be significant, for example in the case of parameters for different turbulence components. Therefore, a probabilistic description of the turbulence field along the whole structure requires definition of six lognormally distributed correlated random variables that are conditional on wind speed and direction. This is achieved by dividing the data into two wind directions (easterly and westerly winds) and several wind velocity intervals. The data is then fitted with lognormal distributions and the correlation coefficient matrix of all parameters are calculated. The resulting statistical properties of the six parameters are given in Table 3.

Table 3. Statistical properties of the turbulence parameters

	East			West		
	$\tilde{\mu}$	$\tilde{\sigma}$	ρ	$\tilde{\mu}$	$\tilde{\sigma}$	ρ
σ_u	0.122 + 0.039U	0.28	0.754	0.122 + 0.039U	0.28	0.772
σ_w	−0.657 + 0.032U	0.278		−0.657 + 0.032U	0.278	
A_u	2.67 + 0.0248U	0.456	0.15	2.407 + 0.048U	0.556	0.327
A_w	0.725	0.456		1.247	0.556	
K_u	1.938	0.275	0.267	2.11	0.275	0.459
K_w	1.833	0.415		2.213	0.415	

After establishing the statistical properties of the turbulence parameters, one can simulate many wind fields using Monte Carlo simulations to represent the randomness in the wind field, which was observed in measurements. As an illustrative example, 10 simulations of the one-point spectra and the normalized cross spectra were made for the design wind speed of HB (U = 39 m/s, 50 year return period) in such way using random number generators. The resulting spectra are shown in Fig. 7 for the easterly winds and Fig. 8 for the westerly winds. The difference in the turbulence fields between the two wind directions is also noteworthy and implies the importance of the incoming wind direction on the turbulence structure in case of such complex terrain. Such a probabilistic model is useful especially when a frequency domain approach is adopted. That way, many simulations of the wind field and the analytical response is possible to carry out avoiding exhaustive computation.

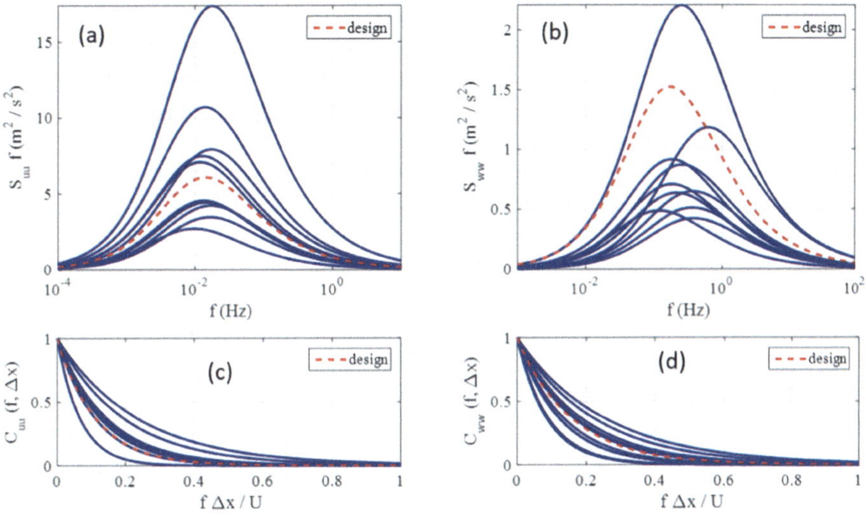

Fig. 7. Simulations of turbulence field for easterly winds under design wind speed (a) One-point spectra of along-wind and (b) vertical turbulence, (c) Normalized cross-spectra for the along-wind and (d) vertical turbulence

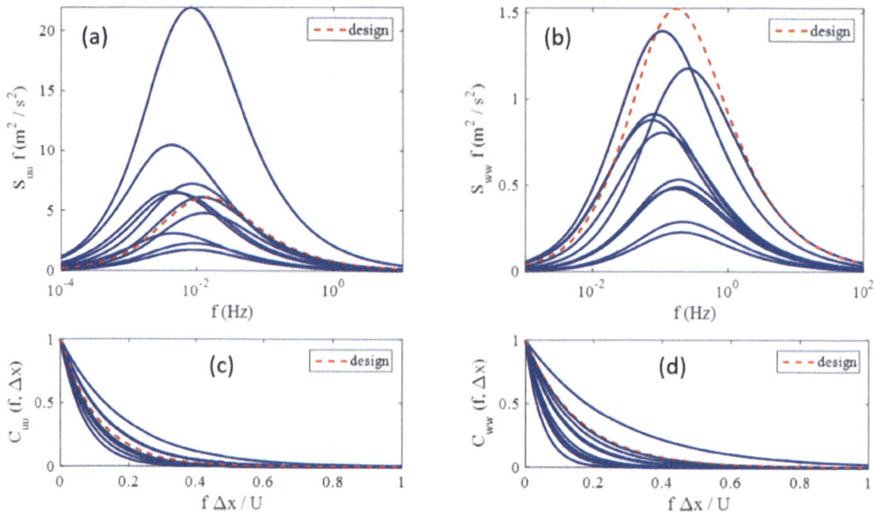

Fig. 8. Simulations of turbulence field for westerly winds under design wind speed (a) One-point spectra of along-wind and (b) vertical turbulence, (c) Normalized cross-spectra for the along-wind and (d) vertical turbulence

5 Insights, Challenges and Prospects

Traditional structural design against buffeting actions requires prediction of the buffeting response for the design wind speed, where other turbulence parameters such as the turbulence intensity, length scale and coherence of turbulence are modelled deterministically. On the contrary, a look of the data from the HB showed significant variability of the turbulence parameters, which was also reflected into the dynamic response of the bridge (Fenerci and Øiseth 2017). Therefore, it is suggested that the turbulence field along the bridge is modelled probabilistically, taking into account the variations in turbulence parameters.

The data from the HB were also used to evaluate the analytical predictions using state-of-the-art methods. For such applications, the wind field was modelled with maximum possible accuracy using the on-bridge measurements. Despite this, significant discrepancy was found between measured and predicted responses. Identifying the sources of such discrepancies and improving the predictions poses a great challenge and more research is needed on this topic, exploiting the vast amount of data available.

The wind measurements also showed nonstationary and span-wise non-uniform features (Fenerci and Øiseth 2018). Although they seem to be of secondary importance, possibility of inclusion of such effects into the design should be investigated. Moreover, when the bridge is not in place such as in the design process, the transfer of full probability distributions of turbulence characteristics from measurement locations to the bridge location should be handled with care (Lystad et al. 2017).

References

Chen X, Kareem A, Matsumoto M (2001) Multimode coupled flutter and buffeting analysis of long span bridges. J Wind Eng Ind Aerodyn 89:649–664. https://doi.org/10.1016/S0167-6105(01)00064-2

Davenport AG (1961) The spectrum of horizontal gustiness near the ground in high winds. Q J R Meteorol Soc 87:194–211. https://doi.org/10.1002/qj.49708737208

Fenerci A, Øiseth O (2018) Strong wind characteristics and dynamic response of a long-span suspension bridge during a storm. J Wind Eng Ind Aerodyn:116–138. https://doi.org/10.1016/j.jweia.2017.10.030

Fenerci A, Øiseth O (2017) Measured buffeting response of a long-span suspension bridge compared with numerical predictions based on design wind spectra. J Struct Eng (United States) 143. https://doi.org/10.1061/(asce)st.1943-541x.0001873

Fenerci A, Øiseth O, Rønnquist A (2017) Long-term monitoring of wind field characteristics and dynamic response of a long-span suspension bridge in complex terrain. Eng Struct 147:269–284. https://doi.org/10.1016/j.engstruct.2017.05.070

Jain A, Jones NP, Scanlan RH (1996) Coupled flutter and buffeting analysis of long-span bridges. J Struct Eng 122:716–725. https://doi.org/10.1061/(ASCE)0733-9445(1996)122:7(716)

Lystad TM, Fenerci A, Saetran L, Øiseth O (2017) Wind field characteristics at the Hardanger Bridge site: comparison of wind tunnel terrain mode tests with full-scale measurements. In: 7th european and African conference on wind engineering (EACWE 2017), Liege, Belgium

Siedziako B, Øiseth O, Rønnquist A (2017) An enhanced forced vibration rig for wind tunnel testing of bridge deck section models in arbitrary motion. J Wind Eng Ind Aerodyn 164:152–163. https://doi.org/10.1016/j.jweia.2017.02.011

Validated Numerical Simulation of Aerodynamic and Aeroelastic Characteristics of Rhein-Crossing Bridge in Leverkusen

G. Ferri[1(✉)], A. Chiaracane[1], C. Borri[1], R. Höffer[2], F. Lupi[2], and U. Winkelmann[2]

[1] Department of Civil and Environmental Engineering (DICEA), Università di Firenze, Via Santa Marta 3, 50139 Florence, Italy
{giulio.ferri,alessandro.chiaracane}@stud.unifi.it, cborri@dicea.unifi.it

[2] Faculty of Civil and Environmental Engineering, Ruhr-Universität Bochum, Universitätstraße 150, 44780 Bochum, Germany
{ruediger.hoeffer,Ulf.Winkelmann}@rub.de, Francesca.Lupi@ruhr-uni-bochum.de

Abstract. This paper presents the results of numerical investigation of the aeroelastic characteristics of an aged bridge deck. The study is developed on the hand of the Rhein-crossing bridge in Leverkusen (Germany), which is undergoing major investigations and will be likely demolished after 53 years of service. Static coefficients and instationary flutter derivatives are determined by means of both forced vibration tests at WIST wind tunnel (Ruhr-Universität Bochum, Germany) and CFD analysis by performing this last as two-dimensional simulations. Pressure measurements in a cross-section in the middle of the bridge model are performed in the wind tunnel experiments, to analyse the wind flow pattern. The numerical model uses the Finite Volume discretization and the turbulence is simulated by the k-ω-SST model. Force and pressure measurements from wind tunnel tests are used for validating the numerical model. The main and original contribution of the study consists in assessing the aerodynamic, as well as aeroelastic, behaviour of an aged bridge deck (with rather unusual "old-fashion" profile) and to verify the performance of CFD-URANS method to determine the bridge flutter derivatives. The aeroelastic forces are obtained by calculating the vectoral difference between the two sets of measurements (Šarkić et al. 2017; Lupi et al. 2018).

Keywords: Wind tunnel experiments · CFD model · Flutter derivatives

1 Introduction

The complex interaction between air flow and solid structures has always been a very important issue. The basic idea of fluid-structure interaction is that a body immersed in a fluid flow is affected by steady and unsteady loads, which tend to elastically deform the structure inducing vibration. If these deformations are larger enough to influence

F. Ricciardelli and A. M. Avossa (Eds.): INVENTO 2018, LNCE 27, pp. 296–309, 2019.
https://doi.org/10.1007/978-3-030-12815-9_24

the flow field around the body, the forces change. This fluid-structure interaction (FSI) is regarded as aeroelasticity and can lead to different aeroelastic phenomena. Theodorsen (1934) provided the first description of motion-induced forces at a flat plate. However, the physical flow around the real oscillating bridge decks contains viscosity and turbulence effects. A step forward in the study of FSI was taken by Scanlan and Tomko (1971). Introducing aerodynamic parameters, called flutter derivatives, they studied the phenomenon defining a linear aeroelastic subsystem with instationary force coefficients, in which the aeroelastic lift and moment can be expressed as follows (Eqs. 1 and 2).

$$L_{ae} = \frac{1}{2}\rho U^2 B\left[KH_1^*\frac{\dot{z}}{U} + KH_2^*\frac{B\dot{\alpha}}{U} + K^2H_3^*\alpha + K^2H_4^*\frac{z}{B}\right] \tag{1}$$

$$M_{ae} = \frac{1}{2}\rho U^2 B^2\left[KA_1^*\frac{\dot{z}}{U} + KA_2^*\frac{B\dot{\alpha}}{U} + K^2A_3^*\alpha + K^2A_4^*\frac{z}{B}\right] \tag{2}$$

where K is the reduced frequency, H_i^* and A_i^* are the flutter derivatives related to Lift and Moment. Flutter derivatives are determined experimentally in the wind tunnel, by means of forced vibration tests. They are evaluated either by means of force measurements and by means of pressure integration.

2 Identification of Flutter Derivatives from Wind Tunnel Tests

The procedure of calculation by means of force signals was developed by Neuhaus, more details regarding that and further information about the excitation mechanism can be found in Neuhaus et al. (2009). To separate the small signals of the aeroelastic force acting on the bridge deck from the larger one due to inertial forces, two measurements are performed: a reference measurement with forced vibration in still air, then a measurement with the same oscillation frequency under the action of the wind tunnel flow. Since the motion is applied harmonically, the measured forces are assumed to be harmonic too and they can be expressed as:

$$F_0 = \hat{F}_0 \cdot e^{i(\omega t + \varphi_0)} \quad F_w = \hat{F}_w \cdot e^{i(\omega t + \varphi w)} \tag{3}$$

where F_0 and F_w are the force amplitudes of the measurement signals, while φ_0 and φ_w are the phase shifts of the force signals with respect to the motion. The aeroelastic force is calculated as the vectorial difference between the two measurements:

$$\Delta F = F_w - F_0 \tag{4}$$

This difference can be represented in the complex plane as it is shown in Fig. 1.

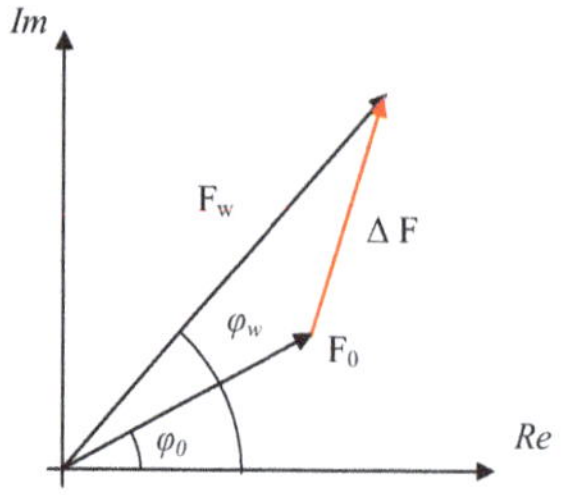

Fig. 1. Identification of the aeroelastic forces in the complex plane

Considering Eq. 4, it can be shown that flutter derivatives related to torsional motion can be obtained from Eq. 5:

$$\begin{aligned}\Delta L^{\alpha}(K) &= \frac{1}{2}\rho K^2 U^2 B\hat{\alpha}\left[H_3^*(K) + iH_2^*(K)\right]\\ \Delta M^{\alpha}(K) &= \frac{1}{2}\rho K^2 U^2 B^2\hat{\alpha}\left[A_3^*(K) + iA_2^*(K)\right]\end{aligned} \quad (5)$$

where $\Delta M^{i}(K), i = z, \alpha$, are the differences for the aeroelastic moment. A set of 8 flutter derivatives is presented in Eq. 6:

$$\begin{aligned}H_1^*(K) &= \frac{2\mathrm{Im}(\Delta L^{z}(K))}{\rho K^2 U^2 \hat{z}} & A_1^*(K) &= \frac{2\mathrm{Im}(\Delta M^{z}(K))}{\rho K^2 U^2 B\hat{z}}\\ H_2^*(K) &= \frac{2\mathrm{Im}(\Delta L^{\alpha}(K))}{\rho K^2 U^2 B\hat{\alpha}} & A_2^*(K) &= \frac{2\mathrm{Im}(\Delta M^{\alpha}(K))}{\rho K^2 U^2 B^2\hat{\alpha}}\\ H_3^*(K) &= \frac{2\mathrm{Re}(\Delta L^{\alpha}(K))}{\rho K^2 U^2 B\hat{\alpha}} & A_3^*(K) &= \frac{2\mathrm{Re}(\Delta M^{\alpha}(K))}{\rho K^2 U^2 B^2\hat{\alpha}}\\ H_4^*(K) &= \frac{2\mathrm{Re}(\Delta L^{z}(K))}{\rho K^2 U^2 \hat{z}} & A_4^*(K) &= \frac{2\mathrm{Re}(\Delta M^{z}(K))}{\rho K^2 U^2 B\hat{z}}\end{aligned} \quad (6)$$

Instead, pressure measurements on a cross-section allow to evaluate the aeroelastic force as a sum of contributions of the forces associated with each pressure tap:

$$L^{\alpha} = \sum_{j=1}^{Ntaps} L_j^{\alpha} = \sum_{j=1}^{Ntaps} p_j^{\alpha}\Delta x_j = \frac{1}{2}\rho K^2 U^2\hat{\alpha}\sum_{j=1}^{Ntaps}\left[H_{3,j}^*(K) + iH_{2,j}^*(K)\right]. \quad (7)$$

In the same way as motion-induced forces, flutter derivatives can be treated as a sum of contributions:

$$H_{2|3}^*(K) = \sum_{j=1}^{Ntaps}\left[h_{2|3,j}^*(K)\right] \quad (8)$$

Since the imposed motion is harmonic, flutter derivatives must be harmonic too. Therefore, they can be related to the *unsteady pressure amplitude* (or *self-excited pressure amplitude*), Eq. 9:

$$\hat{C}_{p,j} = \frac{\hat{p}_j^{\alpha}}{\frac{1}{2}\rho U^2 \hat{\alpha}} \tag{9}$$

The contribution of each pressure taps to the Flutter derivatives, can be expressed as:

$$h_{2,j}^*(K) = \frac{1}{K^2 B}\hat{C}_{p,j}\sin(\varphi_j)\Delta x_j \quad h_{3,j}^*(K) = \frac{1}{K^2 B}\hat{C}_{p,j}\cos(\varphi_j)\Delta x_j \tag{10}$$

This method gives a different source of validation for the numerical simulation (theoretically closer to the CFD), which provides a deeper comprehension the fluid-structure interaction. On the other hand, pressure signals become more and more disturbed increasing the reduced velocity (namely increasing the mean flow velocity and reducing the motion frequency) and the source of error rises.

3 Wind Tunnel Experiments

The experimental campaign was conducted at the WIST Boundary Layer Wind Tunnel at Ruhr-Universität, Bochum. It is an open circuit wind tunnel with a total length of 9.4 m, 1.8 m width and 1.6 m height. All the tests are performed with a turbulence intensity of 3.5%. A tubing system connects the pressure transducers to the bores at the model surface and it makes possible to measure pressures. A digital filter is applied to remove the effect of resonance of the tubing system. Force balances, placed at each side of the wind tunnel, allow to measure flow-inducted forces. Displacement are calculated by means of two lasers sensors placed at the force balances. The bridge model has a total width B of 0.251 m, a height H of 0.033 m and a length L of 1.8 m. Inclined struts are placed at a regular interval of 0.04 m. The total mass of the model is 1.959 kg. Only the bare deck configuration is investigated. The model is instrumented with 26 pressure taps, arranged in the middle of the span (Fig. 2).

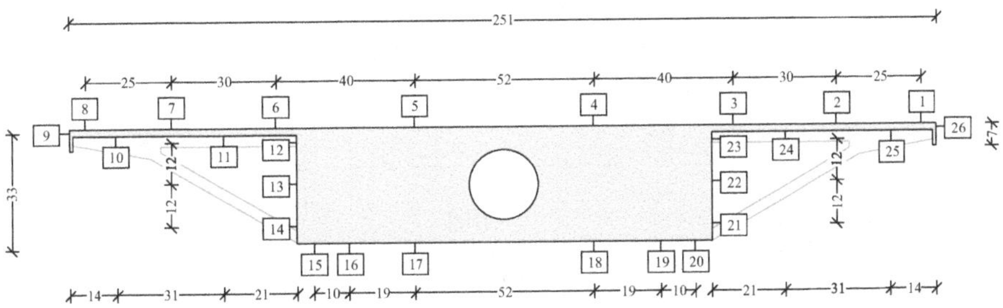

Fig. 2. Rhein-crossing bridge model, dimensions are in millimetres - pressure taps position

Two different type of wind tunnel tests are performed: fixed model tests (at 0° and at different angles of attack) and forced vibration tests.

3.1 Fixed Model Experiments

The main purpose of fixed model tests at 0° angle of attack is to evaluate the Strouhal number and the critical velocity (lock in velocity). These two parameters are essentials to avoid any resonance (synchronization between the model eigenfrequency and the vortex shedding), in forced vibration tests. Strouhal number is experimentally identified from U = 2.15 m/s to U = 20 m/s, following the equation:

$$S_t = \frac{f_s H}{U}. \tag{11}$$

where H is the characteristic length of the phenomenon (bridge model height). According to Eq. 11, a change in the mean flow velocity, U, results (out of the lock-in region) in a change of the vortex shedding frequency, f_s, so that S_t remains constant. The vortex shedding frequency is estimated trough the spectral analysis of the measured lift force, as the frequency corresponding to peaks different from the eigenfrequency peak.

As it is possible to observe in Fig. 3a, b, no Strouhal peak can be identified at 0° angle of attack. Instead, for positive angles of attack (Fig. 3c), St becomes more visible. Vortex shedding is very weak for this bridge section and not even visible until +8° (when the attack stall condition is reached). It does not disturb the identification of flutter derivatives. Stationary force coefficients are evaluated from different angles of attack tests, from −10° to +10° (positive nose up), two mean flow velocities are chosen: 4.80 m/s and 8.30 m/s. Neglecting fluctuation due to turbulence, Stationary force coefficients c_L, c_D and c_M, are estimated following Eq. 12:

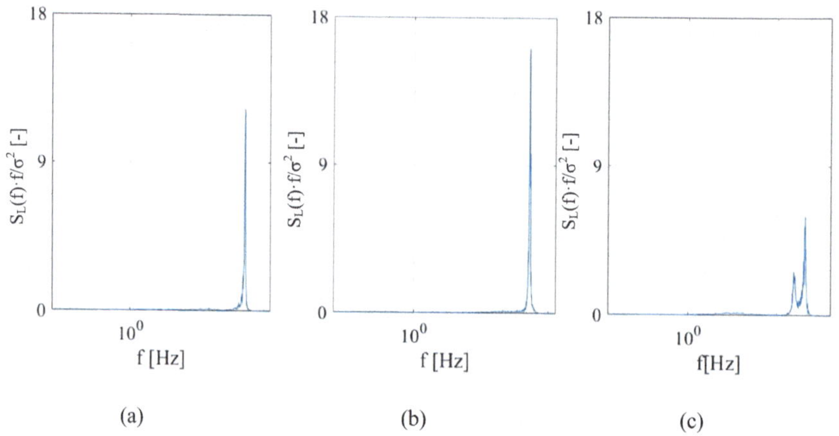

Fig. 3. Spectral analysis of lift force - (a) U = 4.80 m/s α = 0°; (b) U = 6.60 m/s α = 0°; U = 6.60 m/s α = 8°

$$c_D(\alpha) = \frac{D}{q_0 BL} \quad c_L(\alpha) = \frac{L}{q_0 BL} \quad c_M(\alpha) = \frac{M}{q_0 B^2 L} \tag{12}$$

where $q_0 = 0.5\rho U^2$ is the dynamic pressure of the wind flow. Stationary force coefficients are estimated by means of force and pressure. They are compared and used for the numerical validation. Results are presented in Fig. 4.

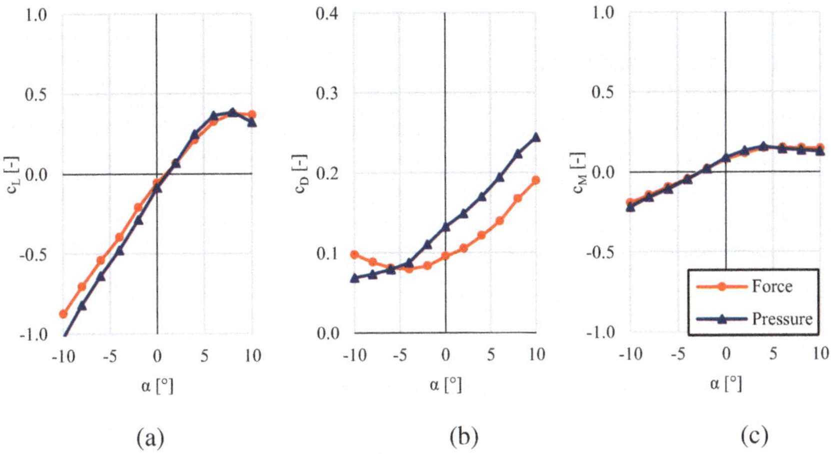

Fig. 4. Stationary force coefficients c_L (a), c_D (b), c_M (c), angles of attack are positive nose up

c_D patterns obtained with the two methods (Fig. 4b) are very distant. This could be due to the lack of pressure integration sections, a single cross section instrumented with pressure taps could not be very representative of the model global response, also because of non-uniformities of the approaching flow along the span. The effect is less important for the lift coefficient, because the flow is separated. Considering c_L, the attack stall condition occurs for angles of attack higher than +6°.

Another aeroelastic instabilities investigated is galloping. It is related to the gradient of the lift coefficient, $dc_L/d\alpha$. It may occur when lift coefficient has such a strong negative slope that it vanishes the drag coefficient, it can be evaluated following the *Den Hartog criterion*:

$$\frac{dc_L(\alpha)}{d\alpha} + c_D(\alpha) > 0 \tag{13}$$

Since c_L has a light negative slope only for positive angle up +8°, while c_D assumes its higher values, it is possible to state that Rhein-crossing bridge cross section is not prone to galloping.

3.2 Forced Vibration Experiments

Forced vibration tests are organized in heave and pitch vibration tests. The range of mean flow velocities goes from 2.15 m/s to 6.57 m/s, while the oscillation frequency goes from 4.5 Hz to 0.7 Hz. The range of reduced frequency investigated goes from 2 to 39. The set of eight flutter derivatives will be presented and commented in Sect. 6.1 together with the ones obtained from the numerical simulation.

4 Numerical Procedure

The flow regime in the simulation domain is modelled in according with the Averaged Navier-Stokes equations. In URANS approach the flow is separated into mean and fluctuating components, the time averaging velocity component is defined as (Ferziger and Peric 1996):

$$\overline{U}(x_i) = \lim_{T\to\infty} \frac{1}{N} \sum_{n=1}^{N} U(x_i, t) \tag{14}$$

where N is the number of ensemble members and must be large enough to eliminate the effects of fluctuations. Substituting it to the incompressible Navier-Stokes equations results in the ensemble averaged Navier-Stokes equations:

$$\overline{U}(x_i) = \lim_{T\to\infty} \frac{1}{N} \sum_{n=1}^{N} U(x_i, t) \tag{15}$$

$$\begin{cases} \frac{\partial \overline{U}_i}{\partial x_i} \\ \frac{\partial \overline{U}_i}{\partial t} + \frac{\partial \overline{U}_i \overline{U}_j}{\partial x_i} = \frac{\partial \overline{P}}{\partial x_i} + \frac{\partial}{\partial x_j}\left(\nu(\frac{\partial \overline{U}_i}{\partial x_j} + \frac{\partial \overline{U}_j}{\partial x_i}) - (\overline{u_i' u_j'})\right) \end{cases} \tag{16}$$

These formulas contain the averaged velocities $\overline{U}_i$, the fluctuating flow velocities $\overline{u}_i$, the averaged pressure $\overline{P}$ and the kinematic viscosity ν of the fluid. t stands for the time and x_i denotes the spatial coordinates. Due to replacement of the averaged product of the velocity components, appears a new unsteady term, the Reynolds stress tensor $\overline{u_i' u_j'}$, must be modelled separately.

For this aeroelasticity study, a forced motion of the bridge deck is performed by adapting the mesh of the domain by moving the boundaries of the bridge deck section. The motion is characterized by the spacing between nodes, which changes by stretching and squeezing. The equations of cell motion are solved based on the Laplacian of the diffusivity and the cell displacement.

The fluid domain is discretized using the Finite-Volume Method separating the computational domain into a finite number of elements known as control volumes.

The modelling of the Reynolds stress tensor is performed by the k-ω-SST model introduced by Menter (1993), more details can be found in Menter et al. (2003).

The cited RANS equations and the turbulence model are discretized with second order accurate central differencing schemes with a limiter in space for the convection term. For time integration a steady state scheme is used.

As inflow boundary condition a uniform velocity U_∞ is chosen to match with the mean oncoming velocity used in the wind tunnel. For the turbulence intensity and the turbulence length scale 3.5% and 3 cm are adopted. The cinematic viscosity chosen is up to $15*10^{-6}$. Top and bottom boundaries are modelled as symmetry plane.

To keep the mesh size realizable, no special refinement around the wall boundaries is performed. The used software is the CFD code OpenFOAM which operates with the Finite-Volume Method. The mesh is set up as a structured two-dimensional grid with up to $7*10^6$ cells created with ANSYS ICEM. The cells stretching ratio is settled to 1.3 in the far field and maintained constant round the body. The normalized wall distance y^+ obtained with the finest mesh is between 0.5 and 4. This y^+ means that the first cell off the wall lies within the laminar sub-layer, and therefore no special wall treatment is necessary. Definition of the non-dimensional time-step for fixed model simulations, performed with a RANS approach, is not required. Instead for the forced motion simulations is $\Delta s = \Delta t \ x \ U_\infty/B = 5 \ x \ 10^{-4}$, equivalent of resolving each period of vortex shedding with 600 time steps. The mentioned mesh is used for all the numerical simulations. The motion done by using the motion solver dynamicFvMesh. In the case of flow simulations over the fixed bridge deck with different angles of flow attack, the bridge deck is numerically rotated up to the desired angle while the grid is distorted.

5 Validation of the Numerical Simulation

Great importance is given to the choosing of the most fitting model in order to lead the upcoming simulations. Stationary coefficients are used for this proposal.

Secondly the force coefficients are compared in relation with different angle of attacks from −10° to 10° using a RANS approach.

Finally using a URANS simulations forced motion configuration is compared for the evaluation of flutter derivatives trends and also for the believed Strohual number.

5.1 Mesh Sensitivities

To accept the reliability of the model, a mesh sensitivities campaign is leaded with 0° angle of attack. Force coefficients and y^+ values are evaluated each refining mesh step. The mesh, set up as a structured two-dimensional grid with about $2.5*10^5$ cells for the coarse mesh up to $7*10^6$ for the finest one. During each simulation has been necessary to use the function "mapFields" to initialize the new finer mesh with the previous final values of pressure, velocity and turbulence parameters.

Figure 5 shows the trends for each force coefficient, calculated following Eq. 12, in relation with the refinement of the mesh.

At the beginning all values change repentantly after a while they remain constant or certainly smoother. This, added to the y^+ values, under the acceptance limits, means that the third mesh is enough fine to be compared with the wind tunnel tests.

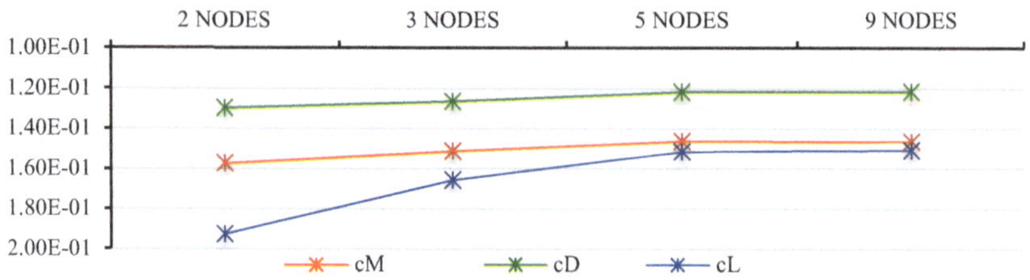

Fig. 5. Force coefficients trends in relation with different mesh qualities

Table 1 shows the comparison between the numerical and the experimental results. The number of nodes is referred to the tooth at the edge of the body.

Table 1. Comparison between the force coefficient values evaluated for model 1 (obtained with the finest mesh) and for experimental tests

c_L			c_D			c_M		
WT [-]	CFD [-]	Rel. Err [-]	WT [-]	CFD [-]	Rel. Err [-]	WT [-]	CFD [-]	Rel. Err [-]
0.086	0.146	0.412	0.11	0.122	0.095	−0.071	0.151	3.13

The relative error distance is quite big for the momentum value and mostly the lift one, that is positive instead in the experimental tests is negative.

Observing the wind flow field around the body it seems that the numerical approach is not able to catch the reattachment of the flow field in the bottom area, refer to Fig. 6, few differences lead to two completely different behaviors.

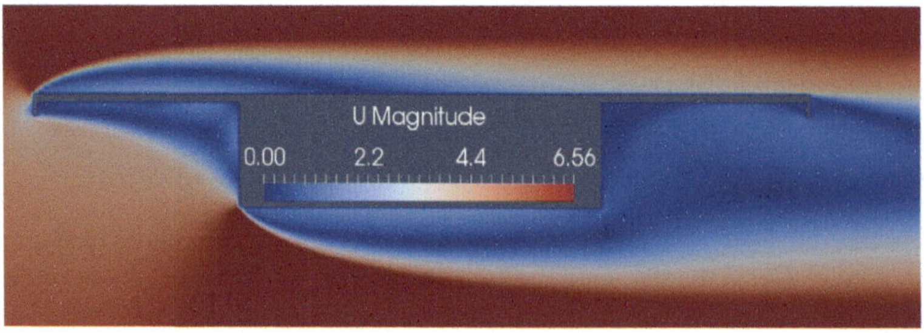

Fig. 6. Computed instantaneous stream wise velocity fields (in m/s) around the bridge cross-section

For the drag coefficient value some discrepancies could be connected to the presence of two diagonal constant elements that modify the aero-dynamicity of the body.

5.2 Simulations on the Fixed Model

Selected the fittest mesh, the analysis moved to the comparison of the stationary force coefficients as a function of the angle of a flow attack, refer to Fig. 7. Due to the reducing of the velocity for these tests from 6.56 m/s to 4.84 m/s, the y^+ value is under tolerability limit also using the 5 nodes mesh.

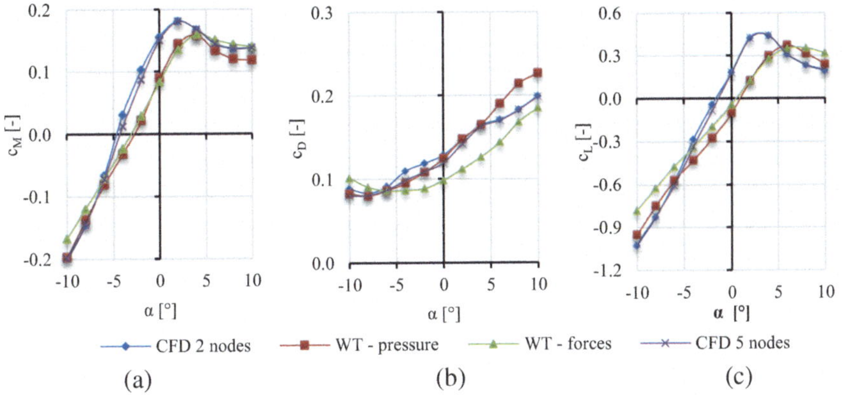

Fig. 7. Stationary force coefficients comparison - c_L (a), c_D (b), c_M (c), angles are positive nose up

All curves follows the same trend for highest and lowest angle of attack, instead in the mid part, closer to the zero, the distance is larger. This fact underlines a possible incorrect distribution of the pressure taps. However, it is not so elevated, and it proves the applicability of this technique for the evaluation of aeroelastic forces.

Considering the relative error in relation with the angle of attack, it is possible to see an increasing of it going close to the 0° angle of attack, this fact underlines the influence of the two beam elements, basically unconsidered for high angles of attack.

5.3 Simulations in Forced Vibration

All forced vibration simulations are performed forcing the bridge cross-section to move periodically, using the same vertical and torsional motion of the wind tunnel tests. The periodic motion acts with an initial angle of a flow attack of 0° with U = 4.87 m/s. The motion-induced force components are extracted from the measured signal following the mentioned identification procedure of flutter derivatives. This is done considering that the simulation reaches a converged periodic state after a few cycles of forced motion, refer to Fig. 8. The initial non-physical response of about 2 s is omitted. To ensure that the selected period is long enough, the whole time of the simulation is divided in two parts, the flutter derivatives are estimated for both of them and also for the whole period of the simulation. These differences are negligible. To reduce the disturbing effects in the filtered force coefficients particulars care are taken to the influence of the Courant number and the boundary conditions.

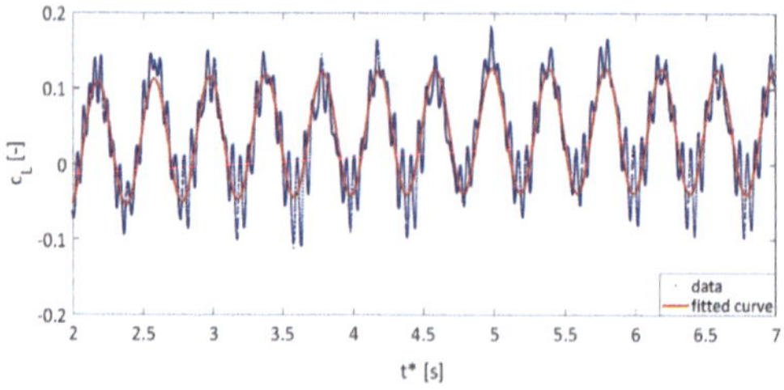

Fig. 8. Example of one of the comparison of simulated and filtered lift in the time history.

The amplitude spectrum of lift force highlights the presence of a peak other then motion frequency. This is close to the vortex shedding frequency, around f_{st} = 25 Hz, that leads to a Strouhal number normalized with the depth of St = 0.1705. The third peak is obtained considering a higher velocity of 6.57 m/s, refer to Fig. 11.

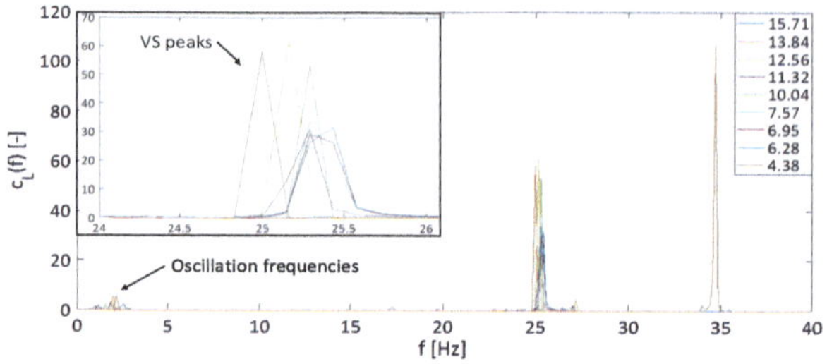

Fig. 9. Comparison between the simulated lift amplitude spectrum for U = 4.87 m/s.

6 Results and Discussion

The flutter derivatives evaluated from URANS simulations are compared with the experimental results. Results obtained from the integration of pressure measurement are also compared considering domain is up to U_{red} = 39.

6.1 Flutter Derivatives – Experiments and Simulations

Vertical and pitch related flutter derivatives are identified from both approaches, refer to Fig. 9. For what regards the determination of each point of the numerical simulation, represented by each blue star in the Fig. 10, it has been led by Eq. 8, considering that lift and momentum coefficients are already evaluated by the simulations, and so it has not been necessary to integrate the pressures along a specified area.

Focusing the attention on $A_2^*(K)$, since it has a negative slope, it is possible to state that Rhein-crossing bridge is prone to classical flutter (2DOF flutter).

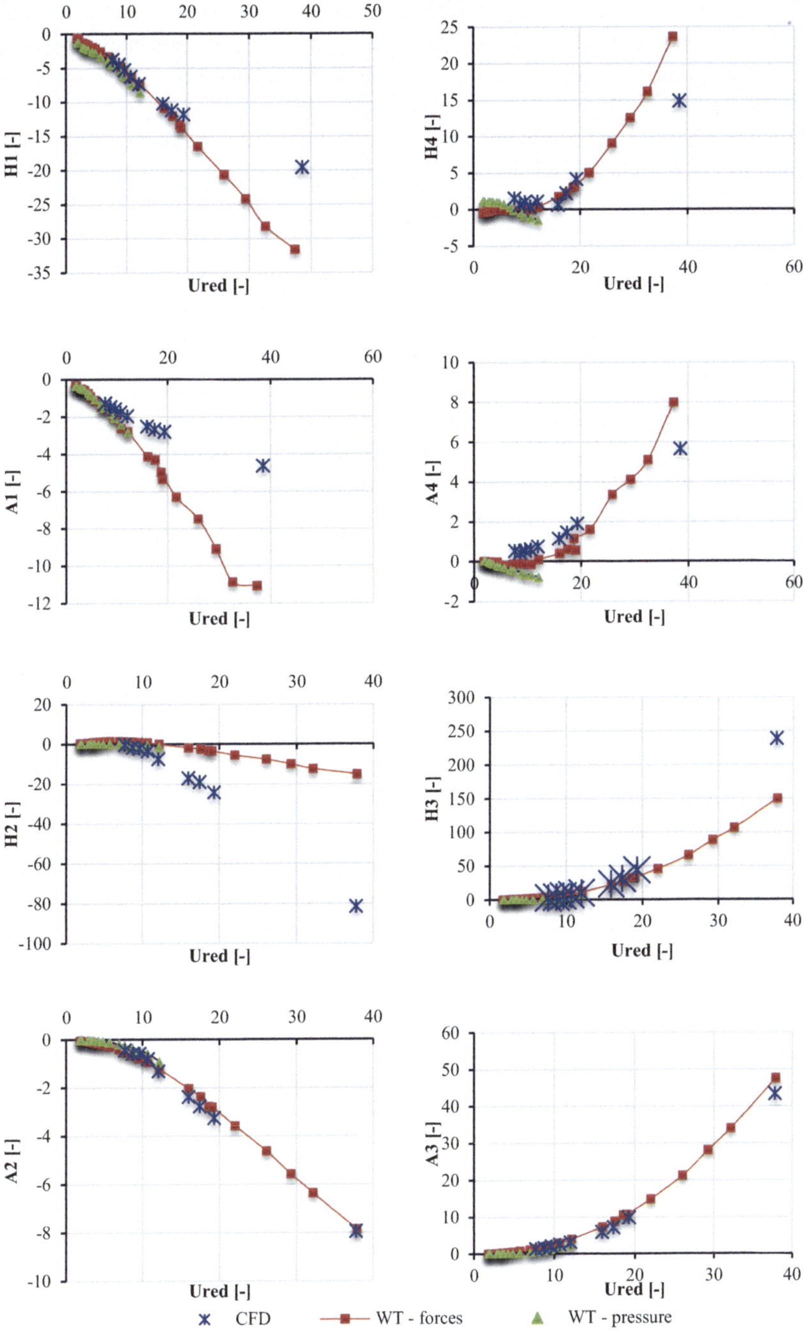

Fig. 10. Rhein-crossing bridge flutter derivatives - comparison between CFD and WT tests

CFD and wind tunnel flutter derivatives show the same trends, as it is possible to see in the case of H_1^*, H_3^*, H_4^* and A_4^*. These results are interesting because usually, due to the low amplitude of motion, errors rise with the increasing of $U_{red.}$ Instead, in the case of H_2^* and A_1^*, differences are more pronounced.

Considering force measurements and pressure integration, the higher differences are identified in $H_4^*(K)$ and $A_4^*(K)$ derivatives. Similar discrepancy is identified by Šarkić (2012) and Sarkar et al. (2009).

7 Conclusion

The main purpose of this paper is to prove the reliability of the aeroelastic characteristics of a bridge deck model obtained by means of computational simulations. A source of validation, namely an experimental campaign is required. A block structural mesh is built with the 2-equation k-ω-SST turbulence model to simulate the flow condition. For the fixed model, pressure distributions give good results in terms of stationary force coefficients for angles of attack far from 0°. Considering forced vibration simulations, the comparison of flutter derivatives proves that the 2D URANS method is able to handle these motion-induced pressure variations accurately over the dominant part of the section surface. Numerical approach allows an easier identification of flutter derivatives, because it is not affected by any inertial effect. Furthermore, it gives detailed information about the flow distribution and the pressure distribution at the bridge surfaces, not affected by any instrumentation error. Moreover, it can exceed the physical bounds like the restricted geometry dimension or wind speeds. In addition, there are no resonance effects that have to be avoided, leading to limitations in forced motion test cases. On the other hand, the 2D URANS approach shows limitations for the strong influence of the mesh quality and the accuracy such as the representation of separation bubbles near to edged corners. This specific limitation is shown in fixed model tests results for 0° angle of attack. Finally, this work shows that for a chosen bridge deck section good prediction of flutter derivatives for the range of moderate reduced velocities can be obtained using the computationally affordable numerical approach.

Acknowledgements. The activity presented in the paper is part of the Erasmus traineeship at Ruhr Universität Bochum (RUB) provided by the University of Florence during the years 2017-2018. The authors like to thank wind engineering and fluid mechanics department of RUB and Dr.-Ing. Claudio Mannini for the useful advices.

References

Neuhaus Ch, Roesler S, Höffer R, Hortmanns M, Zahlten W (2009) Identification of 18 flutter derivatives by forced vibration tests – a new experimental rig. In: Proceedings of the European and African conference on wind engineering, Florence

Šarkić A, Fisch R, Höffer R, Bletzinger K-U (2012) Bridge flutter derivatives based on computed, validated pressure fields. J Wind Eng Industr Aerodyn 104–106:141–151

Šarkić A, Höffer R, Brĉić S (2017) Identification of flutter derivatives by forced vibration tests. Gradevinar 69:267–280

Lupi F, Niemann H-J, Höffer R (2018) Aerodynamic damping model in vortex-induced vibrations for wind engineering applications. J Wind Eng Industr Aerodyn 174:281–295

Winkelmann U, Glumac A, Lupi F, Höffer R (2017) Vortex induced vibrations on circular structures: numerical modeling of aerodynamic loads. In: Proceedings of the European African conference for wind engineering, Liege

Sarkar PP, Caracoglia L, Haan FL, Sato H, Murakoshi J (2009) Comparative and sensitivity study of flutter derivatives of selected bridge deck sections: Part 1: analysis of inter-laboratory experimental data. Eng Struct 31(1):159–169

Menter FR (1993) Zonal two-equation k–ω turbulence models for aerodynamic flows. AIAA Paper 93-2906

Menter FR, Kuntz M, Langtry R (2003) Ten years of industrial experience with the SST turbulence model. Turbul Heat Mass Transf 4:625–632

Full-Aeroelastic Model of CAARC Building: Iterative Design Procedure and Wind Tunnel Tests

G. Frison[1(✉)], A. M. Marra[2], G. Bartoli[2], and R. Scotta[1]

[1] CRIACIV, Department of Civil and Environmental Engineering (DICEA), University of Padova, Via Francesco Marzolo, 9, 35131 Padova, Italy
{giovanni.frison,roberto.scotta}@dicea.unipd.it

[2] CRIACIV, Department of Civil and Environmental Engineering (DICEA), University of Florence, Via Santa Marta 3, 50139 Firenze, Italy
{antoninomaria.marra,gianni.bartoli}@unifi.it

Abstract. A first step of a broader framework, concerning the investigation of tall buildings' dynamic response to wind load, is presented. In order to accomplish this challenging task, a well-known benchmark has been selected: the CAARC (Commonwealth Advisory Aeronautical Research Council) standard tall building. Decades of literatures on the subject make it the perfect candidate to tackle the complexity of wind-structure interaction phenomena. The present work concerns the design of a full-aeroelastic multi degrees of freedom model. The configuration proposed by J. Holmes at International Conference on Wind Engineering (2007), accounting for both flexural and torsional decoupled modes of vibrations, is adopted. A design theory present in literature is re-proposed, from a different prospective, in a semi-automated fashion through a Python script. At last, preliminary results are shown, demonstrating the method's efficiency and applicability, not only to tall buildings, but more generally to structures with continuous distribution of mass.

Keywords: Tall buildings · CAARC · Fluid-structure interaction · Full-aeroelastic model · Wind tunnel · Python

1 Introduction

In the last decades, high-rise buildings acquired a strong symbolic, social and political meaning, beyond the architectural and technical merit. Structures onces devoted exclusively to offices, nowadays represents the main source of residential units in metropolis around the world. From 2000 to 2020, the average height of the 20 tallest building in the world jumped from 375 m to 598 m, with the leading of Kingdom tower surpassing the 1000 m (CTBUH 2018). One of the most engineering challenging task in the construction of such buildings, relies on the

F. Ricciardelli and A. M. Avossa (Eds.): INVENTO 2018, LNCE 27, pp. 310–323, 2019.
https://doi.org/10.1007/978-3-030-12815-9_25

correct analysis of wind action. This represents the design load and an hidden danger to habitability of structure. Indeed, high-rise buildings are characterized by low eigenfrequencies and therefore prone to wind-structure interaction dynamic effects, potentially devastating.

The present study is part of a broader framework that aims to give a deeper insight on wind-induced dynamic effects on tall buildings and their consequences from a structural point of view. It's a belief of the Authors that, in order to get useful informations on such a complex theme, it is mandatory to reduce the degree of uncertainty to a minimum. Therefore, a well know benchmark has been selected as research topic: the CAARC (Commonwealth Advisory Aeronautical Research Council) standard tall building. Decades of literature on the subject provides a solid database and make it the perfect candidate to a deeper investigation on wind effects on high-rise structures. Since Wardlaw and Moss proposal in 1970 of a "simple building model" (Wardlaw and Moss 1970), several experimental campaign have been carried out, analyzing many aspects with different approaches. A thorough review on CAARC's literature can be found in Wardlaw and Moss (2009). However, common wind tunnels practice is often limited to "pivot" models. Therefore, wind-induced response is restricted to linear and uncoupled modes. There are few examples in literature, concerning full-aeroelastic models of tall buildings, able to adequately simulates the complete dynamic of a real structure (e.g. Blackmore 1985).

As a first step, this work present the design of a 1:360 multi-degrees-of-freedom (MDOF) full-aeroelastic model. It will be the subject of extensive experimental campaign. The design theory reported in Templin and Cooper (1981) is re-proposed from a different prospective, introducing ad hoc modifications, functional to CAARC idealization. The scaling procedure is semi-automated through a modular Python code that proved to be potentially useful not only for tall buildings application but, more generally, for elongated structures with a continuous distribution of mass. The last section shows preliminary laboratory testings and wind tunnel results, obtained at the Inter-University Research Center on Building Aerodynamics and Wind Engineering (CRIACIV).

2 Design Theory

Templin and Cooper (1981) presented a model design procedure that allows reproducing complex dynamics of multi-storey structures. As a matter of fact, wind tunnel simulations, in order to be realistic, need to allow for modes that are likely to be coupled at some extent. The approach is based on a lumped-mass approximation, having 3 degrees of freedom for each rigid floor. Applying Lagrange's equation, the model is designed to have the correctly scaled kinetic and potential energies. Using the concept of generalized properties, two equations of motion are formulate: one for translational and one for rotational mode shapes. Indeed, for two systems to be dynamically similar, the equation of motion's coefficients must be identical. Therefore, if scaled model's number of subdivision is large enough, it is possible to find a good approximation of the relevant modes,

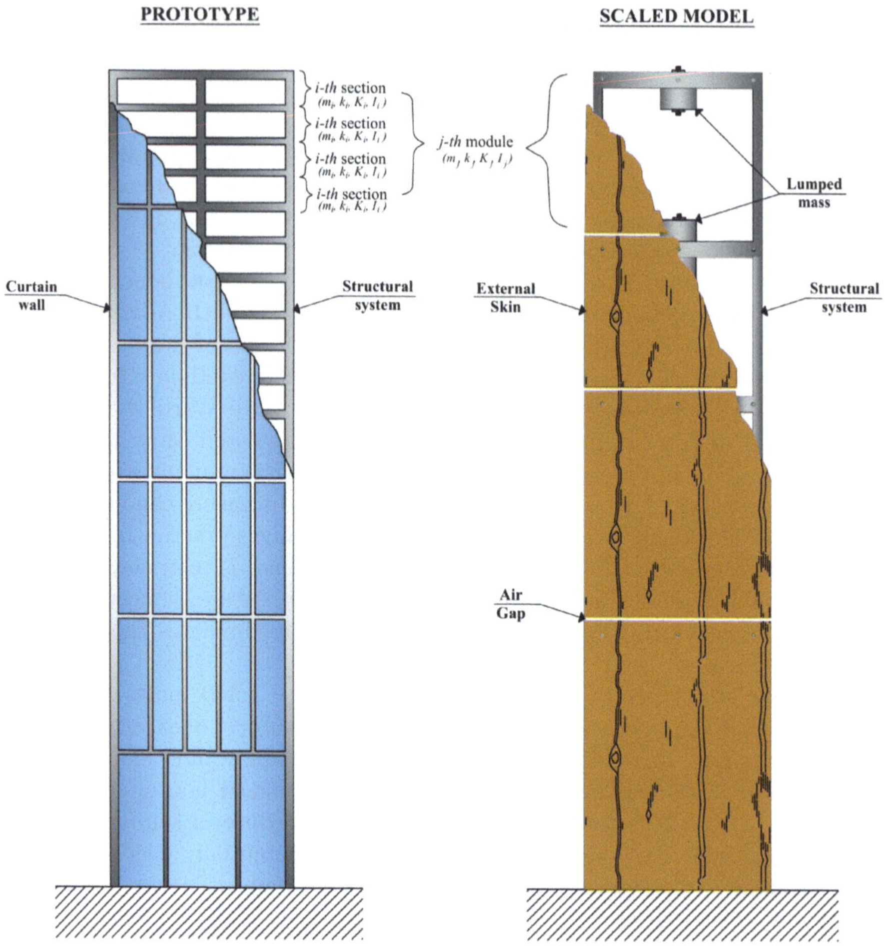

Fig. 1. Prototype's properties are condensed in a scaled model

linking i-sections of the real structures, hereafter denoted as the *prototype*, to j-modules of the model (Fig. 1).

For each mode shape that is to be reproduced, the following equations have to be formulated:

$$m_j \, {\phi_{kj}}^2 = \sum_{i=i_1}^{i_2} m_i \, {\phi_{ki}}^2 \tag{1}$$

$$k_j \, \Delta\phi_{kj} = \sum_{i=i_1}^{i_2} k_i \, \Delta\phi_{ki} \tag{2}$$

$$I_j\,{\phi_{k\theta j}}^2 = \sum_{i=i_1}^{i_2} I_i\,{\phi_{k\theta i}}^2 \tag{3}$$

$$k_j\,d_j\,\Delta\phi_{kj}\,\Delta\phi_{k\theta j} = \sum_{i=i_1}^{i_2} k_i\,d_i\,\Delta\phi_{ki}\,\Delta\phi_{k\theta i} \tag{4}$$

$$(k_j\,d_j^2 + K_j)\,\Delta{\phi_{k\theta j}}^2 = \sum_{i=i_1}^{i_2} (k_i\,{d_i}^2 + K_i)\,\Delta{\phi_{k\theta i}}^2 \tag{5}$$

where m_j and m_i are respectively the mass of model's *j-th* module and corresponding prototype's mass; ϕ_{kj} is the *k-th* translational modal shape coordinate of *j-th* module, ϕ_{ki} is the prototype's corresponding coordinate of the *i-th* section. k_j and k_i are lateral stiffnesses, the former refer to the structural system of *j-th* module, the latter to *i-th* prototype's section; $\Delta\phi_{kj} = \phi_{kj} - \phi_{kj-1}$ and $\Delta\phi_{ki} = \phi_{ki} - \phi_{ki-1}$ represent the linear variation, at consecutive coordinates, of translational mode shapes. I_j is the rotational mass linked to *j-th* module, I_i is the corresponding associated to *i-th* prototype's section; $\phi_{k\theta j}$ is the *k-th* torsional modal shape coordinate of *j-th* module, $\phi_{k\theta i}$ is the prototype's corresponding coordinate of the *i-th* section. d_j is the distance between center of mass and rigidity of the *j-th* module and d_i the corresponding of *i-th* prototype's section; $\Delta\phi_{k\theta j} = \phi_{k\theta j} - \phi_{k\theta j-1}$ and $\Delta\phi_{k\theta i} = \phi_{k\theta i} - \phi_{k\theta i-1}$ represent the linear variation, at consecutive coordinates, of torsional mode shapes. K_j and K_i are torsional stiffnesses, the former refer to *j-th* module, the latter to *i-th* prototype's section. Summarizing, left hand side of Eqs. (1) to (5) represent the model properties of the *j-th* module. Right hand side says that it has to duplicate the behavior of an arbitrary section of the prototype, ranging from i_1 to i_2, the lower and upper section considered.

2.1 Readjustment to CAARC Model

Templin and Cooper developed their design procedure aiming to simulate the dynamic response of an actual tall building (the Commerce Court West office tower in Toronto). Having in their hand a full description of the tower and a series of experimental data regarding masses and stiffnesses distributions, computation of Eqs. (1) to (5) was straightforward. Conversely, the CAARC is an idealized structure, created as a "standard tall building model for the comparison of simulated natural winds in wind tunnels" (Wardlaw and Moss 1970). Therefore, there have never been need to specify any structural property except eigenfrequencies, mode shapes and distributed mass. This has direct consequences, in particular, on the fulfillment of Eqs. (2), (4) and (5), where the prototype's lateral and torsional stiffness are directly computed into the model's properties. Since the CAARC configuration chosen in the present paper, proposed by Holmes at the International Conference on Wind Engineering in 2007 (Holmes 2014), considers only linear and uncoupled mode shapes, Eq. (4) is pointless and Eq. (5) redundant. However, a workaround is needed for Eq. (2).

2.1.1 Fooling Modal Mass "on average"

Scaling procedures, applied to a full-aeroelastic model of a tall building, are not particularly complex per se. Formulated the right equations, the computation involves simple math. However, they become a real challenge when tied together with the mechanical manufacturing. Indeed, aside from some rare exceptions, researchers deal with model scale ranging from 1:200 to 1:500. This implies working with a precision of the tenth of a millimeter and therefore, employing good amount of effort designing a structural system actually feasible. In the next section, a 1:360 scale model is presented. It is a four level, lumped-mass, shear-type approximation of the CAARC building structural system, having 12 degrees of freedom. Undoubtedly, the most time-consuming step in the design process has been the manufacturing. This led to an ad hoc readjustment of the design theory.

In the application of Templin and Cooper procedure, the first step concern Eq. (1). In this case, transfer the mass of continuum structure to a discrete one. Having linear mode shapes with a constant distribution of mass, require a dropping stiffness along the model. This can be accomplished reducing storeys' height or columns cross section. Due to fabrication issues, particularly related to top floors, the second solution has been chosen, keeping constant the inter-storey height. The aim is to find the lumped mass needed in order to obtain the correct modal mass. Under this assumption, Eq. (1) reads:

$$m_j = \frac{m \int_{h_i}^{h_{i+1}} \phi_k(z)^2 \, dz}{\phi_{kj}{}^2} \qquad j = 1,2,3,4 \quad i = 0,1,2,3,4 \tag{6}$$

where the summation has become an integral, along the height (z), due to the lack of storeys definition, m is the mass per unit length that, being constant, can be take out from the integral, while j represents the number of model's floor and h_i the coordinate of corresponding prototype's sections.

Equation (6) leads to determine the exact lumped mass of the *j-th* module, with proper modal mass. However, lumping mass at floors location introduce an approximation error that become increasingly relevant for a coarse discretized structures (four storeys for a whole high-rise). In particular, the mass associated to the *j-th* module performs a modal displacement proportional to the effective location of the floor. For example, let's say that floors of the model are attached in the middle of modules; m_j associated to the module is lumped at floor location, thus having its same displacement. In this way, the lower part of the module is developing more modal mass than it should, while the opposite is happening to the upper part. However, since the relation between modal mass and modal displacement is quadratic, the error produced by the lumped approximation is not compensated. Thus, since m_j, structural mass, is obtained imposing a prescribed modal shape, the correct modal mass is achieved inevitably employing lower structural mass then prescribed. Two main drawbacks have to be highlighted, one technical and the other practical:

- the mass-damping parameter (Scruton number) will not be respected, since the model's mass now differ from prototype's;

- having less structural mass, with fixed eigenfrequencies, traduces in lower stiffnesses, i.e. thinner columns.

Indeed, for j large enough, a very good approximation of the prototype is reached, the correct structural mass is obtained, but the model's feasibility is jeopardized.

The solution to this problem is found with an additional observation in Eq. (6) based on the following reasoning. Wind load does not takes into account the structural system, rather it act on the exterior walls/façades. If the height of modules is considered as a design variable, the storeys' area of influence can be varied without introducing any modification to structure. At the same time, structural mass, associated to *j-th* module, will keep moving with the same modal coordinate relative to the floor. Thus, the required modal mass is achieved on average, while structural mass is iteratively improved until convergence with prototype's requirements.

Eventually, Eqs. (1) and (2) are adjusted as follows:

$$m_j = \frac{m \int_{f_i}^{f_{i+1}} \phi_k(z)^2 \, dz}{\phi_{kj}{}^2} \qquad j = 1,2,3,4 \quad i = 0,1,2,3,4 \tag{7}$$

$$I_j = \frac{\int_{f_i}^{f_{i+1}} I_{f_i} \, \phi_{k\theta}(z)^2 \, dz}{\phi_{k\theta j}{}^2} \qquad j = 1,2,3,4 \quad i = 0,1,2,3,4 \tag{8}$$

where f_i and f_{i+1} represents the lower and upper coordinates of model's exterior skin.

2.1.2 Establish Stiffness Distribution

So far, linear mode shapes have been assumed to be property of the model and, thanks to Eqs. (7) and (8), its translational and rotational modal masses are defined; its total weight close enough to prototype's prescriptions. However, the elements involving Eq. (2) are still unknown quantities. To overcome this fact, an iterative procedure is required. In particular, both mode shapes and eigenfrequencies depend on the geometric features of columns. Therefore, first attempt values have to be computed. With all the elements needed to solve an eigenvalue problem, a computational formulation constitute of two steps is implemented.

First, has to be assured that ϕ_{kj}^2 and $\phi_{k\theta j}^2$ in Eqs. (7) and (8) are the actual modal shapes that produce the desired distribution of modal mass. This can be achieved by manual tuning of the columns cross sections along the height or, more elegantly, by choosing one of the infinite equations that fit the first prototype's mode shape $f(\phi_{j_i})$ (mode shapes have no units and are defined up to a constant), imposing its transition through modal displacement of one storey, and iteratively modifying the columns cross section until all the other storeys' modal displacements fit the curve properly. The procedure is applied for both translation directions and the torsional mode shape is satisfied accordingly.

Second, tuning of eigenfrequencies is needed. An iterative procedure over the firsts translational eigenmodes is required. Having the correct modal behavior of the model traduces in a proportionality vector coefficient $\vec{\gamma_s}$ of columns' stiffnesses along the height. Therefore, the subject of iteration is the lateral stiffness of only one storey (i.e. its columns cross section), which will be varied until convergence.

2.2 Python Code

The model design procedure presented in Sect. 2.1 is perfectly suited to be automatized in a computer code. A script is written in Python 3.X (Rossum 1995). Figure 2 shows the flowchart involving various step delineated in the previous sections. Project Parameters contain all initial data provided by the prototype, like the geometrical features of the building and target modal properties, together with some user inputs regarding the model, like number of storeys, properties of materials employed for fabrication, etc. Thereafter, first main block "Matching modal mass" follows the instruction given in Sect. 2.1.1. Here, the user have to essentially tune the effective height of the modules' façades in order to match the correct structural and modal mass of the prototype. Finally, the second main block follows the instruction given in Sect. 2.1.2. Thanks to two iterative procedure, the best fit of mode shapes and eigenfrequencies is found.

Although not tested, the Authors believe that the presented procedure, developed for tall buildings, could be easily adapt to any kind of structure presenting a continuous distribution of mass and stiffness.

3 Scaled Model Details and Fabrication

Even if a clear understanding of theoretical background is achieved and there is a working script able to automatize the scaling process, the manufacturing of a full-aeroelastic model remains a challenging, time-consuming task. Evaluate the structural system, the materials to employ, together with the physical limitation of the wind tunnel facility, establishes a first crucial step.

3.1 Prototype Description

The CAARC building, in the configuration provided by Holmes (2014), is a tall building with a rectangular cross section measuring 35×40 m and a total height of 180 m. The average building density is $160\,\mathrm{kg/m^3}$. The sway frequencies are 0.20 Hz and 0.23 Hz respectively along the short and long side. Twist frequency is 0.4 Hz. Mode shapes are linear. Structural damping is set to 1%. No structural system is specified. Required longitudinal turbulence intensity (I_z) at rooftop of building is 14.3% (±10%). Required integral turbulence length scale (L_z) at rooftop height of building is 175 m (±10%).

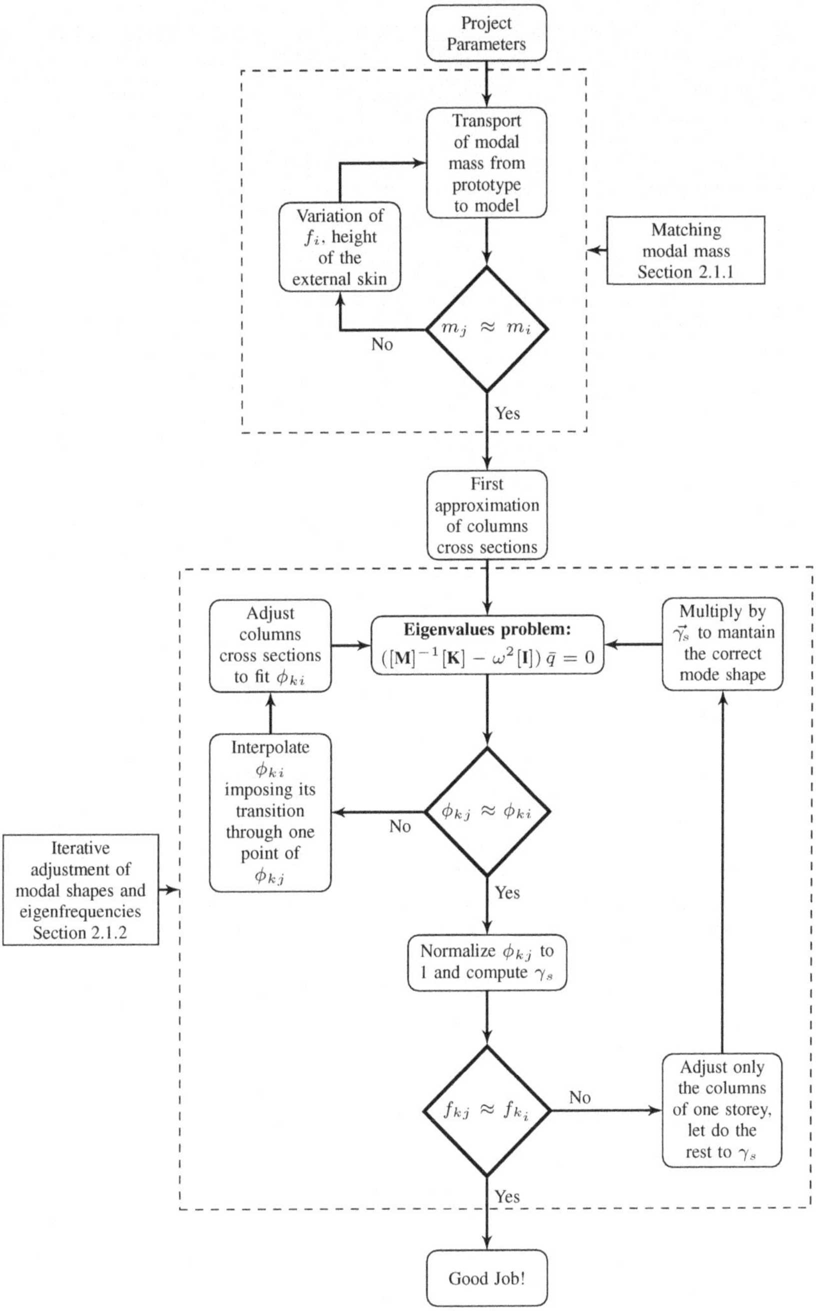

Fig. 2. Flowchart of Python's code

3.2 Model's Scaling Boundary Conditions

In the present work, the most tightening parameters results in I_u and L_u. As a matter of fact, wind tunnel's length and cross section determine the maximum height at which prescribed turbulence properties are still satisfied. Therefore, length scale is set to 1:360, resulting in a model 50 cm high. Secondly, the velocity scale is chosen in order to preserve integrity of model, in other words, lo limit its eigenfrequencies. Therefore, velocity scale is set to 1:8. A list of aeroelastic similarities and modeling laws adopted is shown in Table 1.

Table 1. Modelling parameters

Property	Scaling parameters	Value used
Length	l_m/l_p	1/360
Velocity	V_m/V_p	1/8
Time	$t_m/t_p = V_p l_m / V_m l_p$	1/45
Density	ρ_m/ρ_p	1
Mass	$m_m/m_p = \rho_m l_m^3 / \rho_p l_p^3$	$(1/360)^3$
Damping	ψ_m/ψ_p	1

3.3 Model Fabrication

Once the model's scaling boundary conditions are established and the python script has performed the condensation of modal parameters in a reduced numerical model (Sect. 2.2), a final edges shaping is required in order to proceed with fabrication. In order to keep tolerances in the tenth of a millimeter, wire erosion processing technique is employed. Due to mass scaling, the total weight has to be below 1 kg; therefore, aluminum is chosen as main material. In order to reduce at minimum uncertainties deriving from numerical assumptions, shear-type building deflection is assumed, disregarding relative stiffness between columns and floors. Therefore, the structural system is composed by two 4-storeys bi-dimensional steel frame, with two identical rectangular leg per module, ranging from 2.0 × 1.9 mm to 1.2 × 1.1 mm. These are connected by means of aluminum plates, appositely design to be several order of magnitude stiffer than the legs. These are connected by means of M2 × 6 bolted connections. The exterior shape of CAARC building is provided by 2.5-mm thick balsa sheet, connected to rigid floors by M2 × 16 screws. Each module is separated from the adjacent by 1.5 mm gap, in order to let free relative movement of storeys. The whole structure is attached to a rigid 30-mm steel plate (Fig. 4). Two instrumentation columns have been designed to let any interferences between the instrumentation cabling and the structural system be removed. To obtain the correct torsional mode shape, most of the mass is allocated at the center of rigid floors, in the form of rectangular steel blocks (Fig. 3).

(a) Structural system (b) Exterior façade

Fig. 3. Photographs of the model: (a) structural system, (b) exterior façade

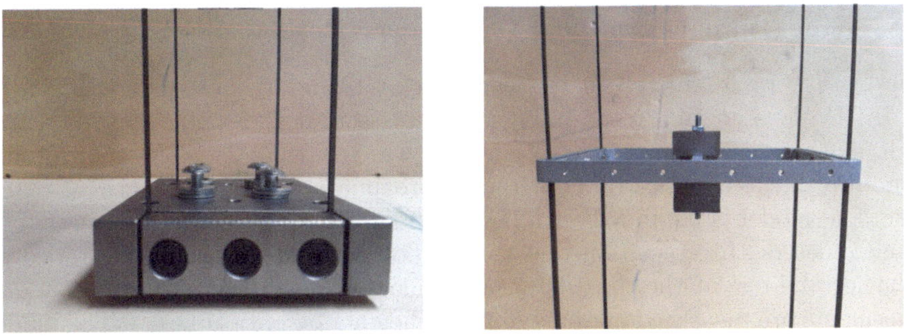

(a) Base steel plate (b) Typical structural floor

Fig. 4. Model's detailing

4 Preliminary Results

4.1 Modal Identification

Prior to wind tunnel testing, a series of laboratory measurements are performed in order to verify that the prescribed model's dynamic properties are achieved. The model is assembled inside the wind tunnel and bolted to the force balance, in order to consider the testing configuration. Measured frequencies are below prediction in respect to designed velocity scale (1:10), most probably due to shear-type idealization. However, for tall structures the Froude Number is not

Table 2. Measured modal properties

Modes	Model scale		Equivalent at full scale		Relative error [%]
	Measured	Benchmark	Measured	Benchmark	
1	8.33 Hz	9.00 Hz	0.185 Hz	0.20 Hz	–7.44
2	9.59 Hz	10.35 Hz	0.213 Hz	0.23 Hz	–7.39
3	15.59 Hz	18.00 Hz	0.346 Hz	0.40 Hz	–15.6
1/2	0.868	0.869	0.868	0.869	–0.001
1/3	0.534	0.500	0.534	0.500	+6.36

binding since gravitational forces do not affect stiffness, therefore the velocity scaling will be derived from length and frequency scales. Table 2 reports a list of the measured properties, showing, in fourth and fifth row, that ratio between modes has a good match. Figure 5 shows a comparison of measured and benchmark modules, derived from the modal displacement of rigid floors; very good agreement is found for translational modes while torsional behavior of the fourth floor requires some corrections. Damping is below 1% for all modes but dependent on amplitude of oscillation and the position of instrumentation's cables.

4.2 Wind Tunnel Setup and Preliminary Results

The tests were conducted in the open-circuit boundary layer wind tunnel of CRIACIV located in Prato, Italy. The test section is 2.4 m wide and 1.6 m high, so that the maximum blockage ratio was less than 2%. The wind profile reproduce a suburban terrain with an approximate roughness length $z_0 = 0.2$ mm. The model is equipped with a set of four mono-axial piezoelectric ICP® accelerometers, three positioned at the top floor in order to fully characterize its behavior and one at the third floor in the along-wind direction. Cross-flow displacements are measured at three different heights by optical lasers. The model is rigidly attached to a 5-component strain gauged balance (HFFB - High Frequency Force Balance) that allows the measurement of 2 shear forces, F_x, F_y, 2 bending moments, M_x, M_y, along two orthogonal directions, and of the torque component, M_z. The full wind tunnel experimental campaign is still to be performed; however, first results are presented for the main directions. Figure 6 shows the root-mean-square of cross-flow displacement against wind velocity, at full scale, up to 60 m/s. For two different configuration, namely 0° and 90°.

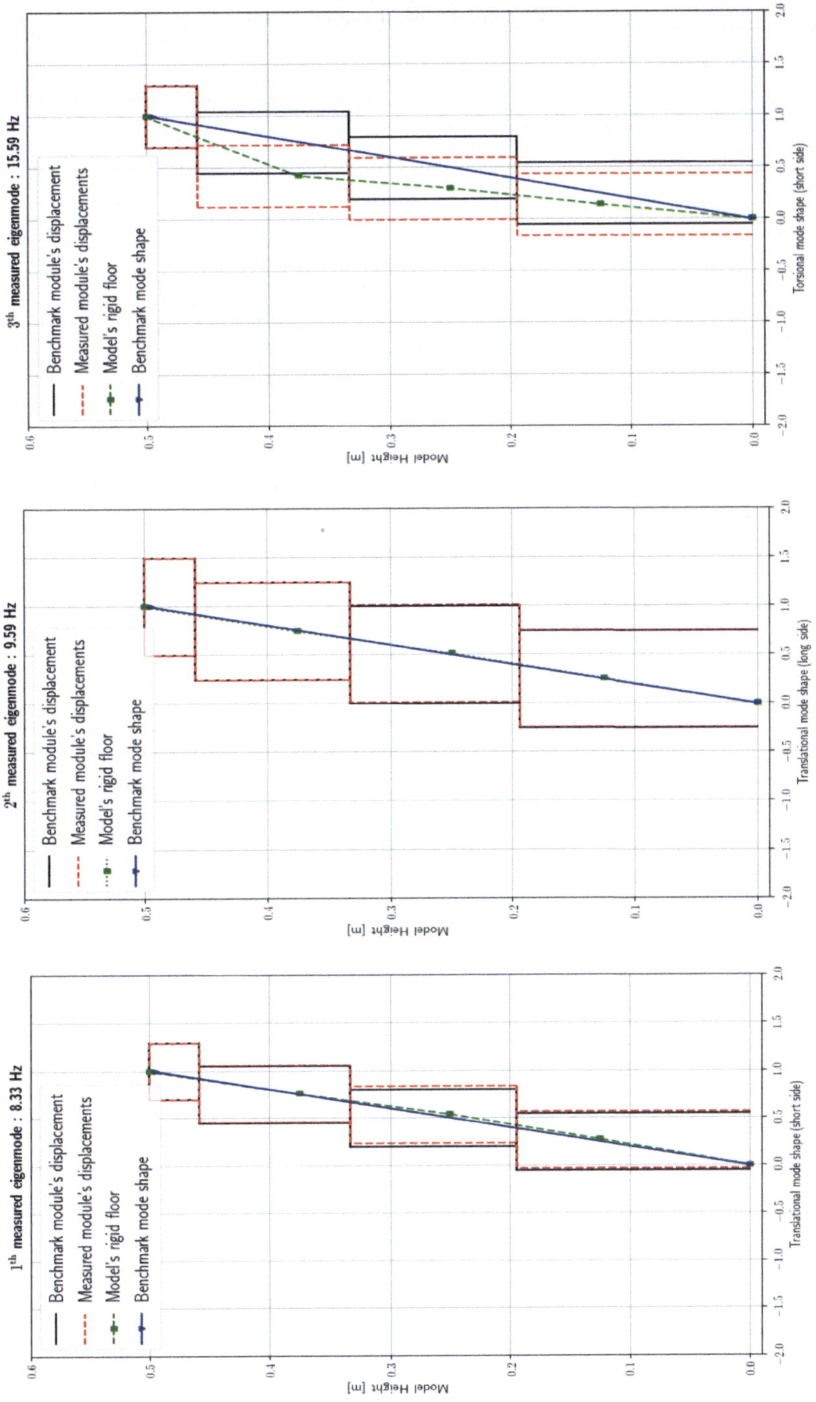

Fig. 5. Modal identification: measured and benchmark module's displacement

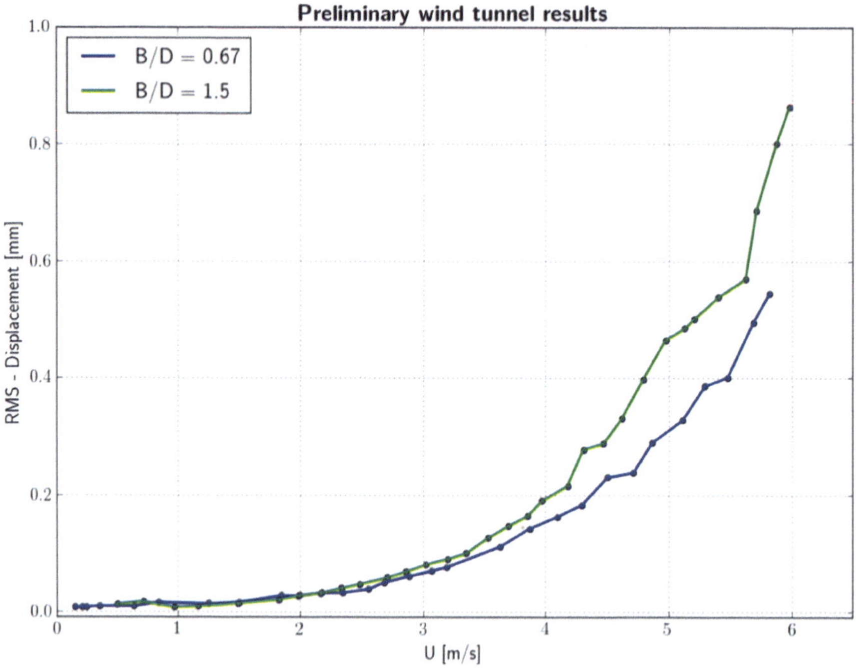

Fig. 6. Preliminary wind tunnel results

5 Conclusions

An iterative design method for multi degrees of freedom building model is presented. Its ability in extrapolating discrete properties out of continuum structures is assessed. A Python script is programmed in order to automatically transfer the generalized properties of the prototype in a reduced number of modules, each having correctly modal property. A full-aeroelastic model of the CAARC building is designed, consisting of 4 modules and 12 degree of freedom, considering both translational and torsional modes. Modal identification tests are presented, showing a good agreement with benchmark values, verifying the proposed design procedure. A glimpse of the preliminary wind tunnel results is shown, although a complete test schedule is to perform in the foreseeable future.

References

Blackmore P (1985) A comparison of experimental methods for estimating dynamic response of buildings. J Wind Eng Ind Aerodyn 18(2):197–212

Braun AL, Awruch AM (2009) Aerodynamic and aeroelastic analyses on the CAARC standard tall building model using numerical simulation. Comput Struct 87(9–10):564–581

CTBUH (2018) Years in review: tall trends of 2017

Holmes JD (2014) International high-frequency base balance benchmark study. Wind Struct 18(4):457–471
Rossum G (1995) Python reference manual. Technical report, Amsterdam, The Netherlands
Templin J, Cooper KR (1981) Design and performance of a multi-degree-of-freedom aeroelastic building model. J Wind Eng Ind Aerodyn 8(1–2):157–175
Wardlaw R, Moss G (1970) A standard tall building model for the comparison of simulated natural winds in wind tunnels. CAARC, CC 662m Tech, 25 January

Aerodynamic Instability of Cables with Circular Appendages

S. Giappino[2], A. Manenti[1], and S. Muggiasca[1](✉)

[1] Department of Mechanical Engineering,
Politecnico di Milano, Via G. La Masa 1, 20156 Milan, Italy
{alessandra.manenti, sara.muggiasca}@polimi.it
[2] GVPM Wind Tunnel, Politecnico di Milano,
Via G. La Masa 34, 20156 Milan, Italy
stefano.giappino@polimi.it

Abstract. The present paper describes the aerodynamic study performed on a cable fitted with a circular tube (pipe) of comparable diameter. Experimental tests were carried out in the high velocity low turbulence test section of the Politecnico di Milano wind tunnel. Typical galloping instability was observed, due to asymmetry introduced by the fitting, but also a strong torsional instability occurred. This phenomenon is not generally significant for cables, but it can appear when fittings change substantially the shape of the cable: for this reason, the study included experimental tests on this instability.

Keywords: Cable dynamics · Galloping · Wind tunnel · Torsional instability

1 Introduction

Cables are structures very sensitive to the wind actions and they may suffer from different kinds of flow-structure interaction: vortex shedding and galloping (Belloli et al. 2015; Belloli et al. 2012; Williamson and Govardhan 2004; Matsumoto et al. 2010). These problems are often observed on cables characterized by low structural frequencies, small structural damping and inclination with respect to the wind as stays of a cable-stayed bridge, hangers of suspended bridges or spoke cables of an observation wheel. Vortex shedding on circular cylinders is a well-known phenomenon that induce vibration perpendicular to the flow and it is generally experienced at low wind velocity. At higher wind velocities other aerodynamic phenomena may occur as ice-galloping, rain-wind induced vibration, high-speed vortex excitation and dry inclined cable galloping. When fittings are applied on this kind of flexible cables, giving rise to a geometrical asymmetry, instability can occur: asymmetries generate lift force on the cable and if the slope of the lift force as a function of the angular position is negative this can be indicative of a possible instability (Den Hartog 1932). Torsional instability due to negative slope of the moment coefficient with respect of the angle of attack is generally not relevant for cables but when fittings with dimensions comparable to the dimensions of the cable, are coupled to the cable itself, also the moment coefficient must be considered. The presented condition differs from the typical wake induced vibrations that occurs when two cylinders are arranged in tandem (Sumner 2010; Zhou and Mahbub

F. Ricciardelli and A. M. Avossa (Eds.): INVENTO 2018, LNCE 27, pp. 324–335, 2019.
https://doi.org/10.1007/978-3-030-12815-9_26

Alam 2016) because fittings and cable are closer than in the literature cases and especially because they are mechanically connected representing a single body i.e. no separated vibrations of the two circular cylinders can be observed. Fittings with relevant dimension are, for example, dehumidification air pipes applied to hangers of suspended bridges or pipes that contain power cables for illumination system for spokes of observation wheel. In the present paper particular attention was devoted to galloping and torsional instability for a cable coupled to with a cylindrical fitting, by means an experimental study performed in the Politecnico di Milano wind tunnel.

2 Model and Experimental Set-Up

The experimental campaign was performed on a prototype (scale 1:1) of the spoke cable of a large observation wheel equipped with a LED system. The cable model was an aluminum tube covered by an HDPE sheath with a helical fillet as the real spoke cable (external diameter D = 0.095 m). The LED system included the lamps and the power cables collected in a cylindrical pipe connected to the spoke cable. Four different configurations were considered changing the pipe diameter, the gap between the two cables and the power cable surface (see Sect. 2.3). Both static and dynamic tests were performed in the low turbulence high velocity tests section of the Politecnico di Milano wind tunnel. This is a 4 m × 4 m section characterized by turbulence intensity lower than 0.15%, smooth wind velocity profile and maximum wind speed of V_{Max} = 55 m/s.

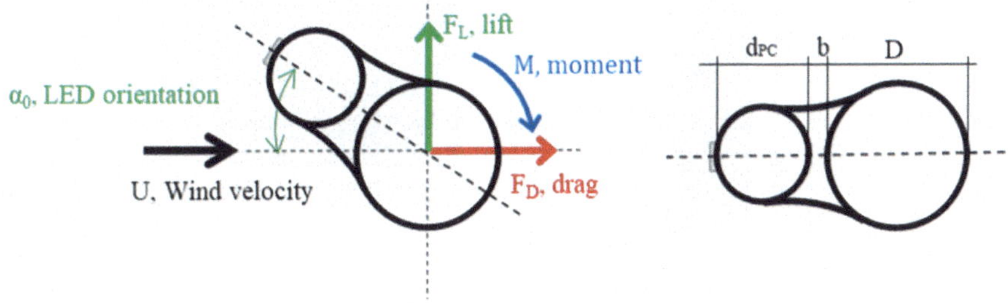

Fig. 1. Model section characteristics and aerodynamic forces definition

2.1 Static Tests Layout

Static tests were performed to measure the aerodynamic force coefficients as a function of the angle of attack α. The model was placed in vertical position in the test section. The layout is that of a 2-dimensional model that is placed between the test section walls (Fig. 2a). The turntable on the floor allows for an easy variation of the angle of attack.

The aerodynamic forces were measured using two six-component force balances (RUAG SG–Balance 192–6), placed at the model ends. Each balance measures the strain produced by loads using strain-gauges arranged in a Wheatstone bridge and calibrated using dead weights and a high precision data acquisition system by the manufacturer. The model was also equipped by two accelerometers to check possible vibrations in X and Y directions.

2.2 Dynamic Tests Layout

Dynamic tests were performed to define the dynamic behavior of the spoke cable + LED system. The main data of the cable model, compared with the full scale layout, are summarized in Table 1. The model was placed in horizontal position in the test section (Fig. 2b) and the constrain system was designed to allow for the oscillation in cross-flow i.e. vertical direction (rigid translation of the cable model) and the rotation around the estimated center of gravity of the system cable + LED (Fig. 3a). The model was suspended using calibrated springs: the suspension equipment provided vertical and torsional stiffness to obtain vertical and torsional frequencies close to the estimated first natural frequencies of the lateral and torsional motion of the real spoke cable + LED.

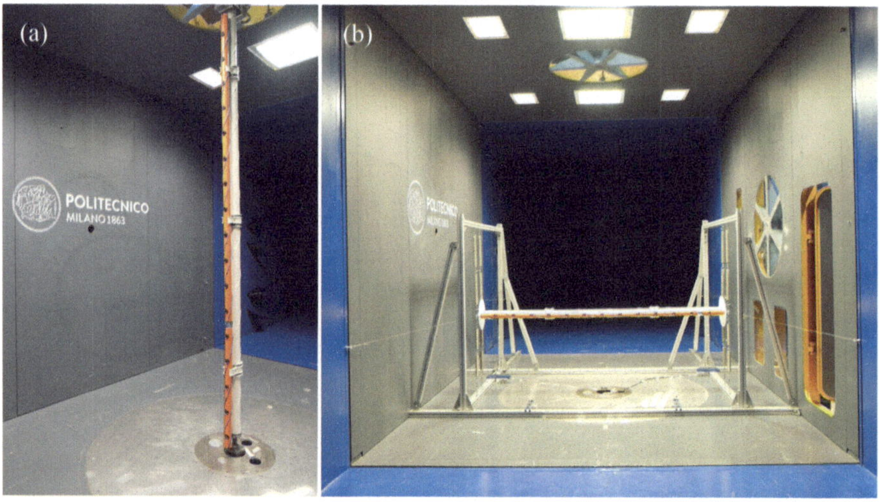

Fig. 2. Spoke cable model layout (a) Static layout (b) Dynamic layout

The vibration level was measured trough accelerometers placed at the model end (Fig. 3b). Three accelerometers are used to measure vertical vibrations and to check for undesired horizontal vibrations.

Table 1. Main model parameters

Parameter		Unit	Model	Real cable
Length	L	m	2.8	105
Cable diameter	D	mm	95	95
Linear mass	m_L	kg/m	7.47	40
Linear mass moment of inertia	J_L	kg m^2/m	0.033	0.08
Vertical frequency	f_v	Hz	1.2	1.2
Torsional frequency	f_t	Hz	2.5	2.5
Vertical damping ratio	ζ_v	‰	0.1	
Torsional damping ratio	ζ_t	%	2	
Scruton Number (vertical)	Sc_v	–	4.33	
Scruton Number (torsional)	Sc_t	–	42.4	

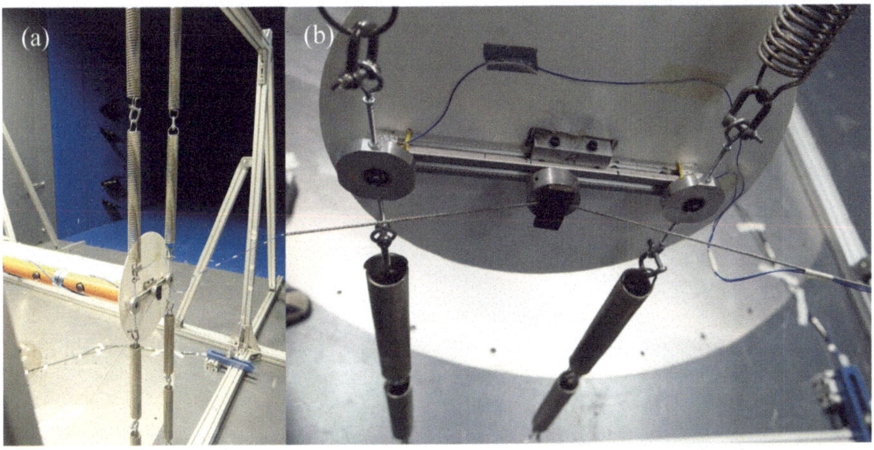

Fig. 3. Model set-up (a) Suspension system and end plates (b) Position of vertical accelerometers

2.3 Tested Configurations

The cable and the LED system were arranged in different ways during the tests, in particular were modified the diameter of the power cable, the distance between the cables and the surface of the power cable (see Fig. 1 for the characteristics parameter). Table 2 summarized all the considered configurations: all these were tested statically while only CONF 4 was also tested in dynamic condition.

Table 2. Tested configurations

Name		Power cable diameter d_{PC} [mm]	Distance between cables D [mm]	Power cable surface
CONF 1	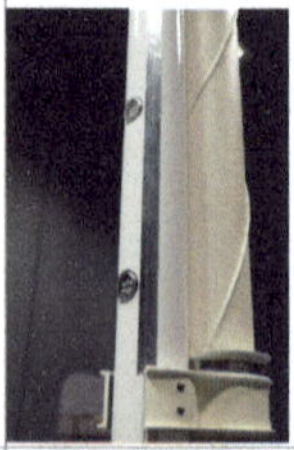	75	8	Smooth
CONF 2		63	8	Smooth
CONF 3	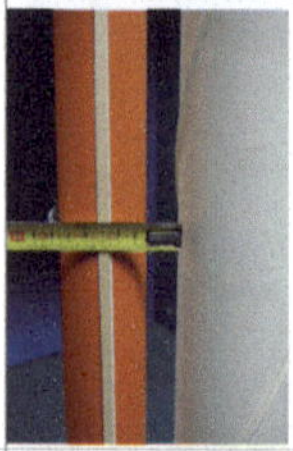	63	24	With lateral rectangular profiles
CONF 4	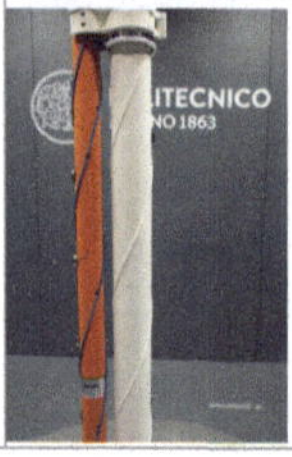	63	24	With helical fillet

3 Results

3.1 Static Tests

Aerodynamic coefficients are calculated from static tests according to the reference frame system defined in Fig. 1. The LED orientation angle α_0 is equal to 0 having the wind perpendicular to the LED.

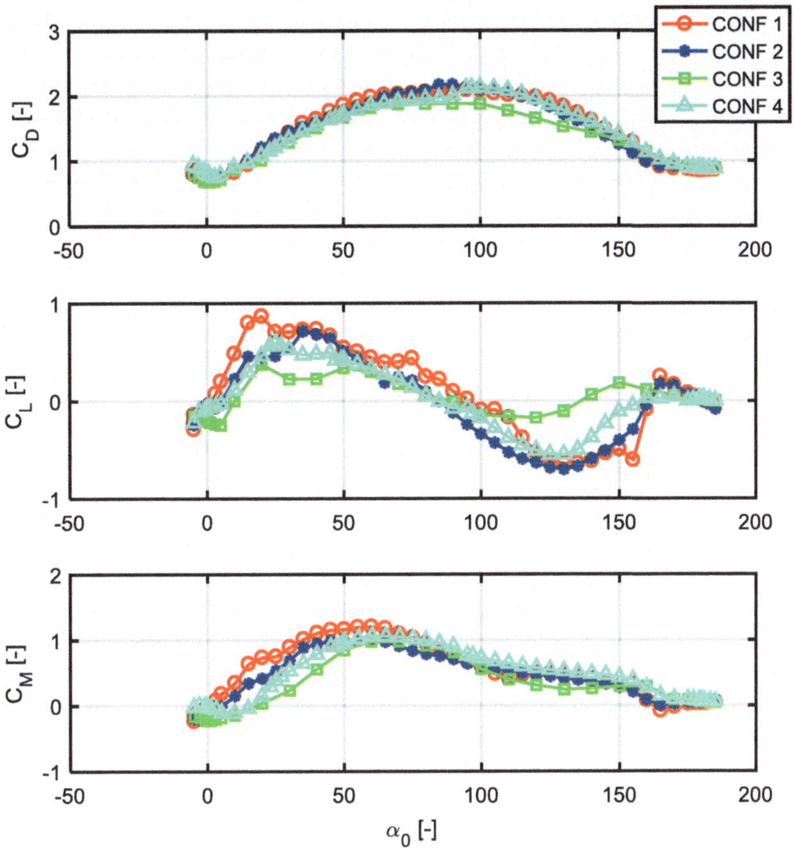

Fig. 4. Aerodynamic coefficient for the tested configurations as a function of the angle of attack

Aerodynamic coefficients were defined as:

$$C_D = \frac{F_D}{1/2\rho V^2 DL}. \quad (1)$$

$$C_L = \frac{F_L}{1/2\rho V^2 DL}. \quad (2)$$

$$C_M = \frac{M}{1/2\rho V^2 D^2 L}. \quad (3)$$

where F_D, F_L and M are drag force, lift force and moment respectively, ρ the air density, V the mean wind velocity, D the external cable diameter and L the cable length.

The coefficients concerning the tested configurations are reported in Fig. 4. It is possible to note that the variations in the LED system characteristics mainly affect lift and moment coefficients while the drag curves do not change significantly. Focusing on the lift curves, the reduction on the power cable diameter (from CONF 1 to CONF 2) produces a smoother lift curve as a function of the angle of attack from −5° to 160°. The increase in the cables gap (from CONF 2 to CONF 3) reduces the lift coefficient while the application of a helical fillet on the power cable surface permits to obtain an intermediate lift curve between CONF2 and CONF 3.

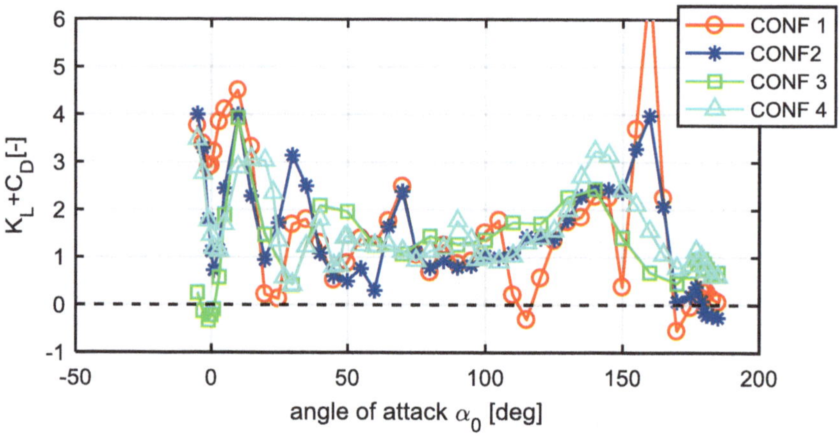

Fig. 5. $K_L + C_D$ as a function of the angle of attack

The analysis of the static aerodynamic coefficients it is very important to predict possible galloping instability: galloping refers to structural vibrations in a direction almost perpendicular to the wind direction (cross wind) mainly due to negative aerodynamic damping. Using a simple quasi-steady approach, the condition for the occurrence of galloping can be expressed by the so called Den Hartog criterion (Hartog 1932):

$$\left(\frac{\partial C_L}{\partial \alpha} C_D\right)\bigg|_{\alpha=\alpha_0} < 0. \tag{4}$$

where $(\partial C_L/\partial \alpha) = K_L$ is the lift derivative with respect to the LED orientation angle.

The application of the Den Hartog criterion to the curves obtained for the configurations 1, 2 and 3 highlighted galloping instability at the tests wind velocity, V = 30 m/s. As it possible to see in Fig. 5, CONF 1 does not satisfy Den Hartog criterion for different angular positions (115°, 170° and for 25° $K_L + C_D$ is very close to 0), CONF 2 can suffer from galloping at about $\alpha_0 = 180°$ and CONF 3 at $\alpha_0 = 0°$.

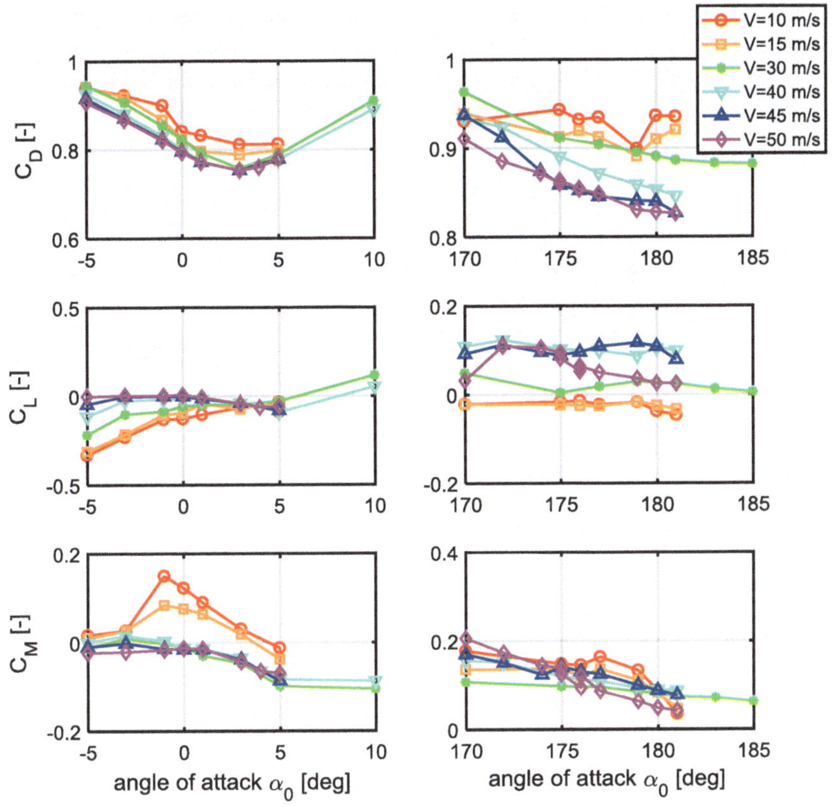

Fig. 6. Aerodynamic coefficients for CONF 4 at different wind velocities

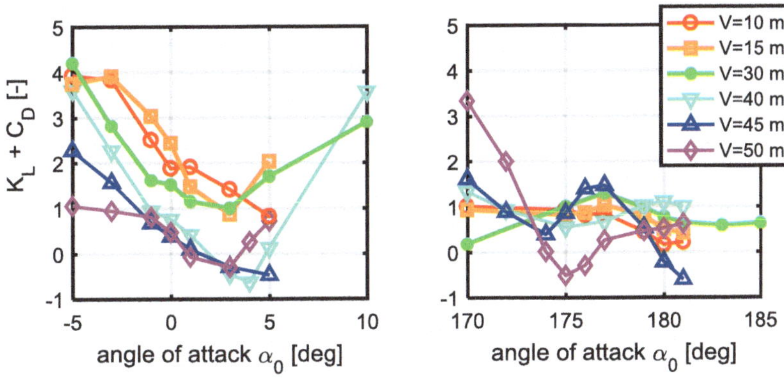

Fig. 7. $K_L + C_D$ for CONF 4 at different wind velocities

CONF 4 is the only one that does not experience instability for wind velocities up to 30 m/s. However, as shown in Fig. 6, its aerodynamic coefficients in the region around $\alpha_0 = 0°$ and $\alpha_0 = 180°$ are strongly dependent on Reynolds Number and, for wind velocities higher than 30 m/s, $K_L + C_D$ becomes negative, Fig. 7.

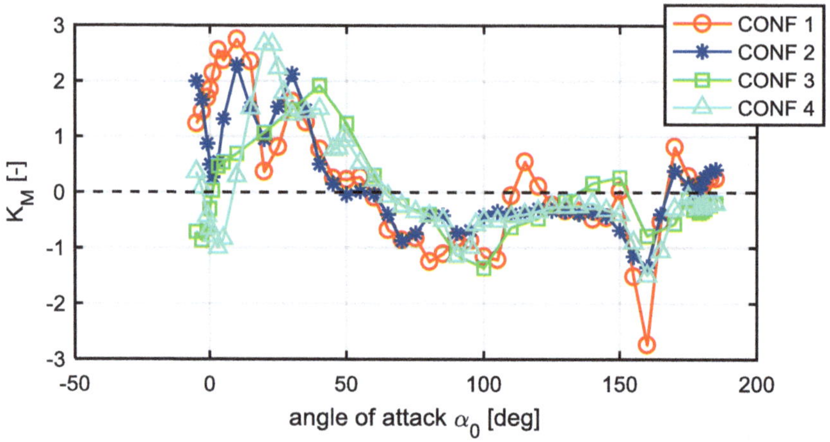

Fig. 8. K_M as a function of the angle of attack

Moreover, looking at Fig. 8 where the derivative of the moment coefficient with respect to the LED orientation angle α_0 is reported ($\partial C_M/\partial\alpha) = K_M$), it is possible to note that this value becomes negative for a large range of angle of attack: this means that torsional instability may occur. Applying Quasi Steady theory, the aerodynamic damping can be written as:

$$r_a = \frac{1}{2}\rho D^2 B_\alpha LVK_M. \tag{5}$$

If we assume B_α equal to D, the non-dimensional aerodynamic damping would be:

$$\zeta_a = \frac{\rho D^3 VK_M}{4\omega_t J_L}. \tag{6}$$

where ω_t is the circular frequency of the torsional mode and J_L is the linear mass moment of inertia. It is possible to note that when the K_M value is negative, a negative damping is introduced in the mechanical system and if this negative aerodynamic damping is in absolute value grater than the structural damping, instability can occur. In order to assure structure stability, the required structural damping should be:

$$\zeta_{str} \geq \frac{\rho D^3 VK_M}{4\omega_t J_L}. \tag{7}$$

Figure 8 shows that for all the tested configurations, negative values of the derivative K_M were measured: CONF 1 is characterized by the most negative value ($K_M = -2.73$) while the other configurations all reached $K_M = -1.5$. This means that small variations in the power cable layout do not permit to obtain significant advantages in terms of moment derivative. For CONF 4 the most negative value was $K_M = -1.47$ resulting in an aerodynamic damping for the real cable (see data included in Table 1) equal to 1% at V = 30 m/s. This means that the structural torsional damping must be higher than 1% to avoid torsional instability.

Due to the critical behavior came out of static tests, the model was tested also dynamically, in particular, dynamic tests were performed on CONF 4, the one that showed better static coefficients concerning the classical galloping instability.

3.2 Dynamic Tests

The spoke model equipped with the LED system in CONF 4 was elastically suspended in the test section: mass per unit length, mass moment of inertia and stiffness were adjusted to obtain vertical and torsional frequencies of the model close to the expected first natural frequencies of the lateral and torsional motion of the real spoke cable (see Table 1). Natural frequencies and structural damping ratios have been identified by means of free motion decay tests in still air. The damping coefficient was evaluated performing a Hilbert transform to damped free responses of the aeroelastic models. The analysis is performed on the acceleration measurements.

Measured damping ratios resulted in two Scruton numbers calculated, from values reported in Table 1, as:

$$Sc_v = \frac{2\pi m_L \zeta_v}{\rho D^2}. \tag{8}$$

$$Sc_t = \frac{2\pi J_L \zeta_t}{\rho D^4}. \tag{9}$$

Dynamic tests were performed for a wind angle of attack equal to 0°. Wind velocity has been increased and accelerations were acquired: standard deviation of vertical accelerations measured during the tests are reported in Fig. 9 both for vertical and torsional modes. It possible to note that vertical mode was excited by galloping instability for wind velocities lower than the ones expected from static tests results: this can be explained considering the interaction between vertical and torsional modes and considering that the vertical damping is much lower than the torsional one as reported in Table 1.

In order to check the torsional behavior of the model, vertical displacement was controlled during torsional tests: for a wind velocity equal to V = 16.7 m/s the cable + LED became torsionally unstable, as can be seen in Fig. 9 that shows an abrupt increase in acceleration signal. The amplitude reached in this condition points out that torsional instability is predominant with respect to classical galloping for this kind of structures. Figure 10 shows the rotation angle obtained from accelerometer signals during the build-up test at 16.7 m/s: the oscillation amplitudes show a limit cycle of

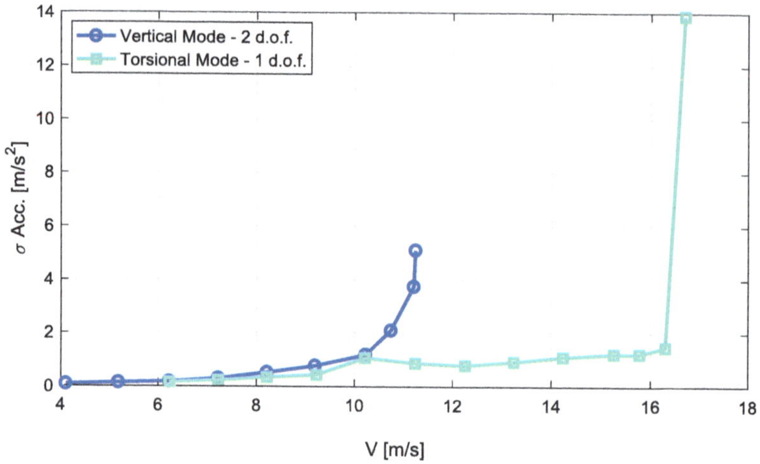

Fig. 9. Vertical acceleration standard deviation as a function of the wind velocity for vertical and torsional modes

about 40° at this wind speed probably due to the increase of the structural damping related to the amplitudes increase.

These tests confirmed the considerations already made on the base of the static tests, i.e. the solution with two parallel cables, one close to the other, is unstable.

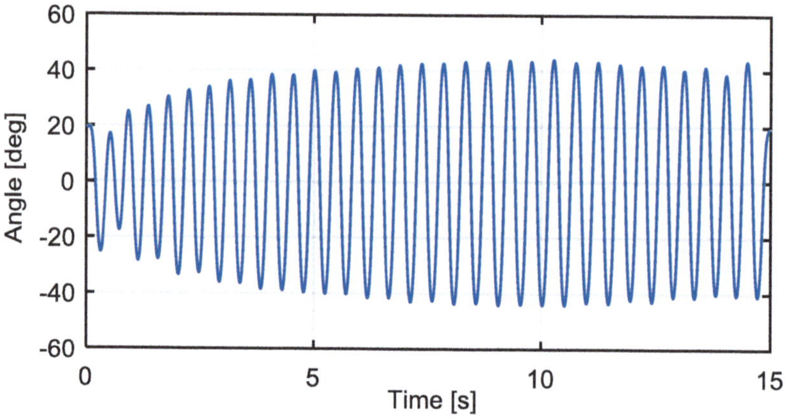

Fig. 10. Filtered angular displacement at V = 16.7 m/s

4 Conclusions

The present paper describes an aerodynamic study performed on a cable fitted with a circular tube (pipe) of comparable diameter: a spoke cable of a large observation wheel equipped with LED system was tested in the Politecnico di Milano wind tunnel. Four different configurations were considered changing the LED equipment i.e. the characteristics of the pipe used to collect LED power cables. The aerodynamic behaviour was firstly investigated through static tests to define the forces coefficients trends as a function of the wind angle of attack: these highlighted a strong sensitivity to Reynolds Number and a possible instability due to classical galloping but also a torsional instability of the two cables. Dynamic tests confirmed the critical behaviour of a cable fitted with a circular tube with particular reference to torsional instability.

References

Belloli M, Giappino S, Morganti S, Muggiasca S, Zasso A (2015) Vortex induced vibrations at high Reynolds numbers on circular cylinders. Ocean Eng 94:140–154. https://doi.org/10.1016/j.oceaneng.2014.11.017

Belloli M, Giappino S, Muggiasca S, Zasso A (2012) Force and wake analysis on a single circular cylinder subjected to vortex induced vibrations at high mass ratio and high Reynolds number. J Wind Eng Ind Aerodyn 103:96–106

Den Hartog JP (1932) Transmission line vibration due to sleet. Trans AIEE 51:1074–1076

Matsumoto M, Yagi T, Hatsuda H, Shima T, Tanaka M, Naito H (2010) Dry galloping characteristics and its mechanism of inclined/yawed cables. J Wind Eng Ind Aerodyn 98 (2010):317–327

Sumner D (2010) Two circular cylinders in cross-flow: a review. J Fluids Struct 26(6):849–899. https://doi.org/10.1016/j.jfluidstructs.2010.07.001

Williamson CHK, Govardhan R (2004) Vortex-induced vibrations. Annu Rev Fluid Mech 36:413–455

Zhou Y, Mahbub Alam M (2016) Wake of two interacting circular cylinders: a review. Int J Heat Fluid Flow 62:510–537. https://doi.org/10.1016/j.ijheatfluidflow.2016.08.008

Cyclist Aerodynamics: A Comparison Between Wind Tunnel Tests and CFD Simulations for Helmet Design

S. Giappino[1(✉)], S. Omarini[1], P. Schito[1], C. Somaschini[1], M. Belloli[1], and M. Tenni[2]

[1] Department of Mechanical Engineering, Politecnico di Milano, Via G. La Masa 1, 20156 Milan, Italy
{stefano.giappino, simone.omarini}@polimi.it
[2] MET Helmets, Talamona, Italy

Abstract. During the last decades, a particular attention has been paid to the aerodynamic optimization of professional bike riders and their equipment, with a large use of experimental investigation as well as numerical simulations. Extensive studies have been carried out in wind tunnels all over the world by professional cycling teams to study the best set-up for time trial competitions. The standard approach is to compare different parts of the equipment or different positions and find out the optimized configuration by a trial and error procedure. This paper presents an attempt of helmet design based on CFD calculations and experimental validation of numerical results with a new experimental set-up able to measure the surface pressure on a real helmet.

Keywords: Cyclist aerodynamics · CFD calculations · Wind tunnel tests · Helmet design · Time trial competitions

1 Introduction

It is well established that wind resistance is responsible for most of the metabolic cost of cycling in level ground. Aerodynamic drag is about 80% of the total resistive force in road racing at 30 km/h and up to 94% in time trial competitions at 50 km/h, so that it becomes important to reduce it to improve cycling performance (see Padilla et al. (2000); García-López et al. (2008); Gibertini et al. (2011); Belloli et al. (2016) and Belloli et al. (2014)).

Different techniques are used nowadays to evaluate aerodynamic drag in cycling: wind tunnel tests, track and road testing and CFD calculations. On one side track testing allows a natural athlete's behaviour while, on the other side, wind tunnel testing is the most accurate and reliable technique because of the high repeatability of the tests. In case of wind tunnel testing, particular attention should be given to the simulation of the pedalling with an adequate resistance, since significant differences are found between static and in effort tests, see Gibertini et al. (2011). Wind tunnel results showed a good agreement with track values considering also that the rolling resistance was not included in the wind tunnel measurements.

The original version of this chapter was revised: The correction in author's name has been incorporated. The correction to this chapter is available at https://doi.org/10.1007/978-3-030-12815-9_56

F. Ricciardelli and A. M. Avossa (Eds.): INVENTO 2018, LNCE 27, pp. 336–349, 2019.
https://doi.org/10.1007/978-3-030-12815-9_27

Blocken et al. (2013) performed CFD simulations of the riders to study rider positions and drafting effects as a function of the distance between bikes. CFD calculations showed good agreement with wind tunnel experiments but using as reference tests on mannequins, because spinning wheels and pedalling are very challenging to be simulated using a computational approach.

All these comparisons are made in terms of integral force, drag and side forces when side wind effects are included. CFD presents a great advantage: it can be used as a design tool, because it defines the flow field and the local pressure on the whole surface of the considered body, while in wind tunnel tests it is very complicated to measure the pressure on the rider, bike and equipment.

The work presented in this paper tries to overcome this issue by proposing a new experimental set-up able to measure the pressure distribution on a real helmet, worn by an athlete during wind tunnel testing. This local measurement is useful to validate the CFD results in terms of local pressure distributions.

2 Helmet Design and Instrumentation

Nowadays, the equipment design has been developed having aerodynamic issues in mind. The position of the rider and its size compared to the rest of the system mean that the helmet choice can have a significant effect on the net drag force that the rider must overcome, see Crouch et al. (2017). Some studies have shown that the difference between well-performing helmets and poorly performing helmets can be greater than the difference between fast and slow wheels, see Sidelko (2007). Modern time-trial helmets are designed for speed over comfort and, more recently, the design has achieved the development of hybrid helmets that attempt to reduce drag without compromising ventilation and mobility.

The helmet studied in this paper is an aerodynamic long-tailed one suitable for time trial and triathlon competitions, that cover the rider's head in an elongated fairing. The numerical analysis is conducted using the geometry of a professional cyclist and the CAD model of the helmet and of the visor. The rider geometry is obtained by means of a three-dimensional scan. As can be seen in Fig. 1, the geometry reproduces the upper-body of a real athlete, while some minor details, as the hair (which will be shaded by the helmet), the hands, the legs and the bicycle are neglected.

The helmet and the visor are placed correctly on the rider head and this geometry is used as a reference for the CFD analysis as can be seen in Fig. 1.

Several improvement steps are done with CFD calculations and the final version of the helmet design is tested in the Low Turbulence Test Section of the Politecnico di Milano wind tunnel (GVPM). The comparison between wind tunnel tests and CFD simulations is realized on an instrumented version of the real helmet provided by MET. The helmet is instrumented with 32 pressure taps, simultaneously acquired in order to measure the pressures distribution, as shown in Fig. 2. Pressure measurements are performed using a high-speed pressure scanning equipment with ESP miniature pressure scanners. These scanners consist of an array of 32 silicon piezoresistive pressure sensors, one for each pressure point, whose analog outputs are multiplexed and amplified within the scanner.

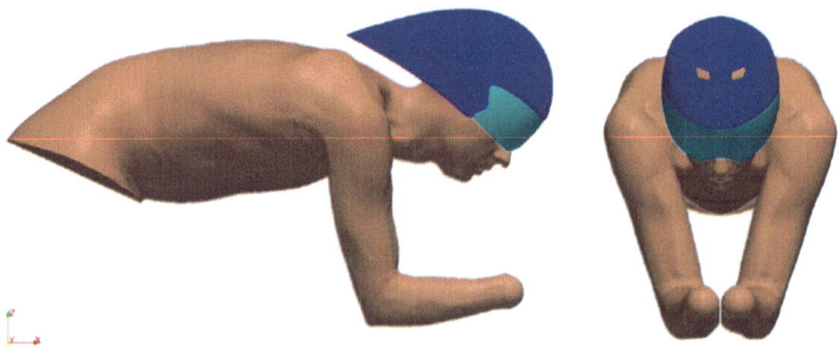

Fig. 1. Geometry of the rider and the helmet: side view (left), front view (right).

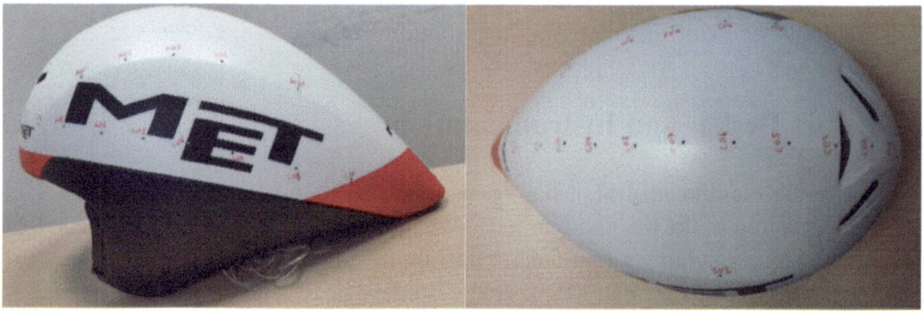

Fig. 2. Instrumented helmet for pressure measurements

A PVC-tube connects each measuring point on the surface of the helmet (afterwards called tap) to the scanner. The helmet is divided into sectors in which the pressure taps are positioned. The pressure taps are named with a code that has a first letter that recalls the positioning line and a progressive number that grows from the visor to the tail of the helmet. A central line has 14 taps from c01 above the visor to c14 on the tail, a medium line has 5 taps from m01 to m05, a low line 9 taps from l01 to l09 and finally 2 taps e01 and e02 are made on the black part that cover the ears. Two more taps are positioned on the other side of the helmet and they are called as "symmetric": s02 replicates the medium tap m02 and s06 replicates the low tap l06. Table 1 summarizes the pressure taps distribution and designation, which are shown in Fig. 3.

Table 1. Pressure taps distribution

Taps	From	To
Central line	c01	c014
Medium line	m01	m05
Low line	l01	l09
Ear line	e01	e02
Simmetric medium	s02	s02
Simmetric low	s06	s06

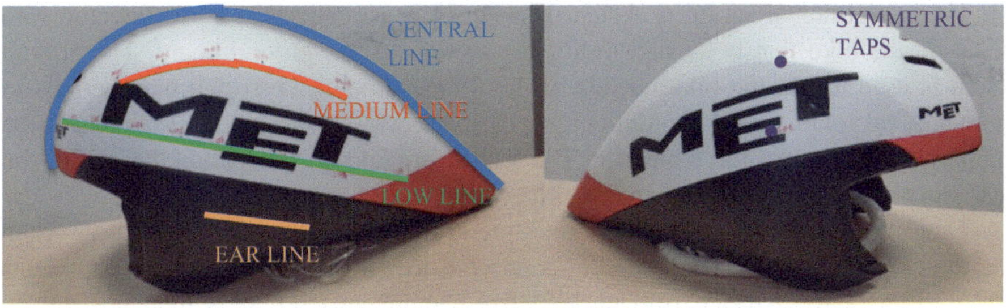

Fig. 3. Helmet with pressure taps distribution highlighting the positioning lines.

The small dimensions of the scanner allow to place it in the back of the cyclist under the clothes as shown in Fig. 4b, close to the measuring points but without troubling the flow. The PVC tubes are only 0.8 m long, because reducing the length of the tubes the phase-frequency distortion on the signals is smaller.

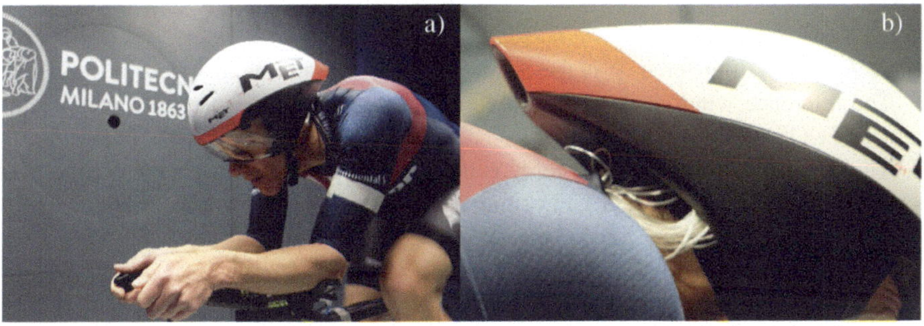

Fig. 4. Athlete in the test section of the wind tunnel (a) details of the helmet and PVC-tubes (b)

The particular geometry of the rider's head and upper back could limit the comparison between wind tunnel tests and CFD simulations. Different geometries and postures may influence the pressure distribution on the helmet, therefore the final helmet was tested in the wind tunnel with two different athletes, trying to replicate as much as possible the reference geometry used for the CFD analysis. Figure 4a shows the athlete in the test section with the instrumented helmet, while pedalling. The position and the size of the athlete are chosen in order to replicate the reference CFD position.

3 Wind Tunnel Set-up

The wind tunnel tests have been carried out at the Politecnico di Milano wind tunnel (GVPM) in the Low Turbulence Test Section. The dimensions of the test section (4×4 m) make it possible to perform tests with the athlete without the need of model scale testing. The blockage effects are negligible and the flow turbulence is very low ($I_t = 0.15\%$).

The experimental set-up is sketched in Fig. 5. It is realized to allow the athlete to pedal on the bike that is located in the test section over a ground board. The ground board is used to put the bike outside of the floor boundary layer and then permits to simulate real relative-wind conditions on the bike. Moreover, it permits to shield the support frame and the force balance. The wheels are laid on rollers and the axis of the rear wheel hub is fixed to a framework.

Thanks to a belt both wheels are spinning at the same speed during the tests when the athlete is pedaling. The frame is connected to a force balance, a 6-components RUAG Aerospace balance (model 192) fixed to the ground. It is then possible to measure the overall loads on the global system: bike and cyclist.

The drag force (along-wind direction force) is measured by the force balance. In cycling aerodynamic drag is often expressed in the form of a drag area (C_dA) as the ratio between the drag force and the wind kinetic pressure. The drag area, defined in Eq. (1), has the dimensions of an area [m^2] and can be interpreted as the product of a drag coefficient and the frontal area of the cyclist. The drag area is calculated as:

$$C_dA = \frac{F_D}{\frac{1}{2}\rho V^2}\left[m^2\right]. \tag{1}$$

where F_D is the drag force, ρ is the air density and V is the wind speed.

The reference wind speed is measured by a differential pressure micromanometer Furness FCO510.

The measurement of the drag force is not sufficient in order to have a comparison between wind tunnel tests and CFD simulations, therefore a new experimental set-up able to measure simultaneously the pressure distribution on the real helmet is developed.

The wind tunnel tests were performed at 14 m/s (50 km/h) with the main idea of investigating more than one athlete in order to check different geometries and postures, consequently studying more than one rider's head position based on the inclination of the leading edge, simulating the pedalling with an adequate resistance and also verifying the effects of the lateral wind yawing the test section turntable (clockwise and counter clockwise).

A group of two subsequent records (runs) sampled without stopping the wind is done for each yaw angle tested. A single run is 20 s long and a mean value is calculated. The pressure value reported in the results is the average of the mean values at each run. Fourteen yaw angles are tested: 0, 2.5, 5, 7.5, 10, 15, 20° and the equivalent negative ones. The same position/configuration at 0° of yaw angle was repeated at least two times.

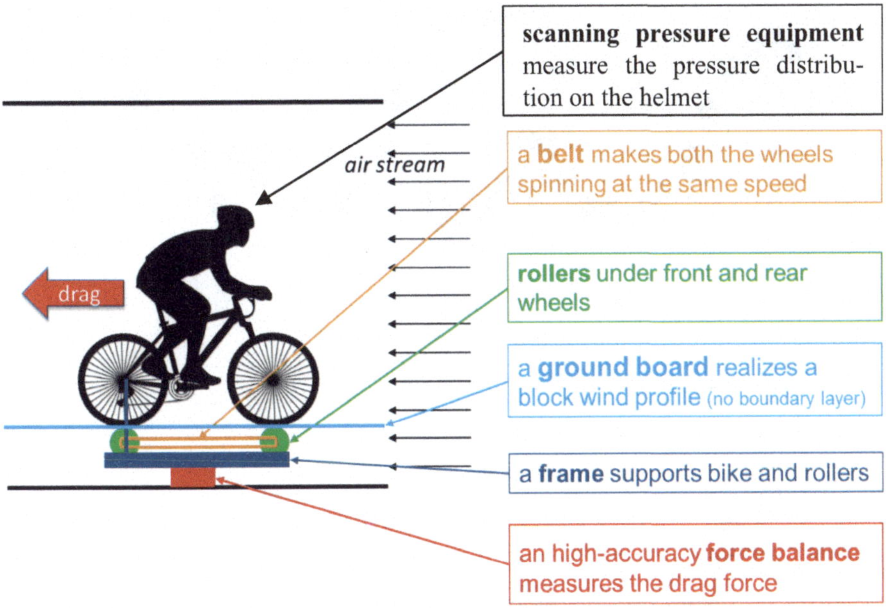

Fig. 5. Set-up sketch highlighting the main components

4 Wind Tunnel Experimental Results

In this section the wind tunnel results are presented. The pressure data are given as non-dimensional pressure coefficients, defined as the ratio of the difference between the local pressure and the reference one (the free-stream static pressure) and the kinetic pressure. Figure 6 shows the pressure distribution recorded for Athlete 1 (reference position). As mentioned above, the experimental tests were done with two athletes in order to consider different geometries and postures that may influence the pressure distribution on the helmet. Figure 6a shows the pressure distribution as a function of the pressure taps for positive angles: from 0° to +20°, and Fig. 6b the pressure distribution as a function of the pressure taps for negative angles: from 0° to −20°. The pressure values reported in the results are the average of the mean values at each run. Figure 7 shows the pressure distribution recorded for Athlete 2 for positive angles and negative ones respectively (reference position).

The behaviour is quite clear: varying the yaw angle the pressure distribution changes appreciably for the taps above the visor and along the tail of the helmet for the central line of pressure taps. This trend is clear for both the athletes but remarkable for the first one. In both cases the pressure distribution changes dramatically for the lateral pressure taps positioned in medium line and in the low line.

The pressure taps are only on one side, consequently positive and negative yaw angles are compulsory to describe the aerodynamic behavior of the helmet. The "symmetric" taps s02 and s06 reproduce correctly the values for the m02 and l06 taps for the opposite yaw angle. Symmetric taps were also used for a proper positioning of the cyclist head in the wind tunnel: a particular attention has been paid before the tests

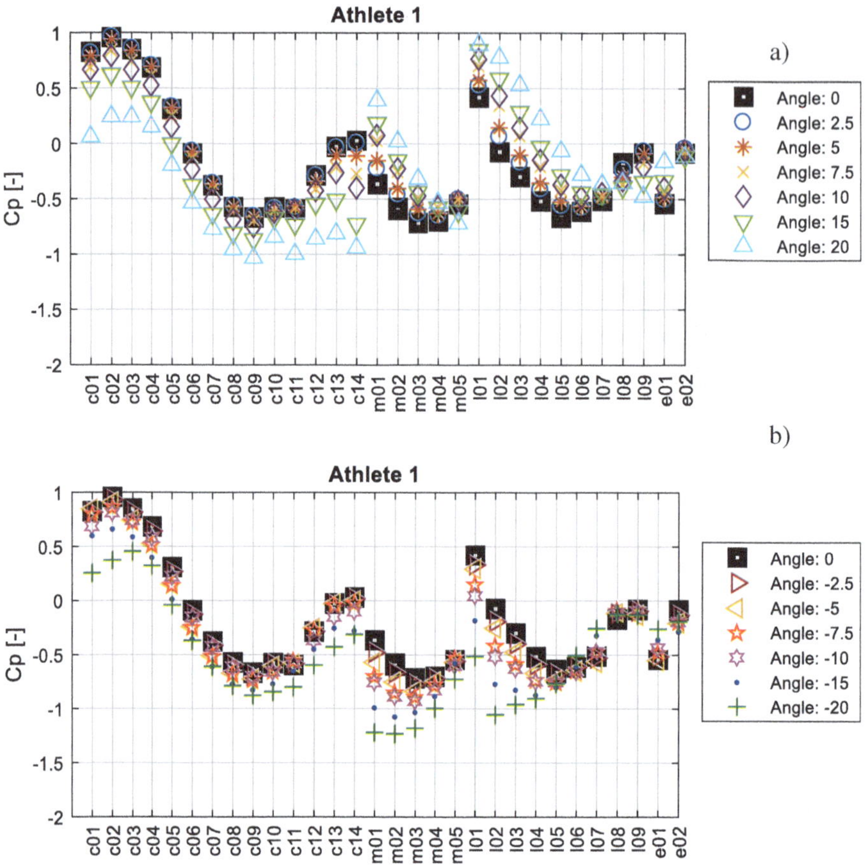

Fig. 6. Non-dimensional pressure distribution for Athlete 1 (a) positive yaw angles (b) negative yaw angles

in order to have the helmet with a perfect alignment at 0°. Authors also believe that the athletes may change the head's rotation during the tests at high yaw angle due to the lateral wind speed. This is why the values for the tap m02 and the "symmetric" s02 and the taps l06 and the "symmetric" s06 were checked at each run and the average values are reported in Figs. 6 and 7.

Figure 8 shows the non-dimensional pressure coefficients for Athlete 1 at 0°, considering different head positions based on the inclination of the leading edge. Varying head's angle or the global position of the athlete, the results change appreciably. It seems that a change of the head position is more important than small variations of the yaw angle, this is why the discussion of the results at 0° is carried out in a more exhaustive way. In particular, the black points in Fig. 8 show the reference position (also reported in Fig. 6), the red ones display a lower performance position with the head up, the blue triangles show an aggressive position with rider's head down (so with the tail of the helmet up) and finally, green triangles display a more aggressive position with a normal position of the tail. Analysing the results, it is visible a clear behaviour as a function of the head's angle.

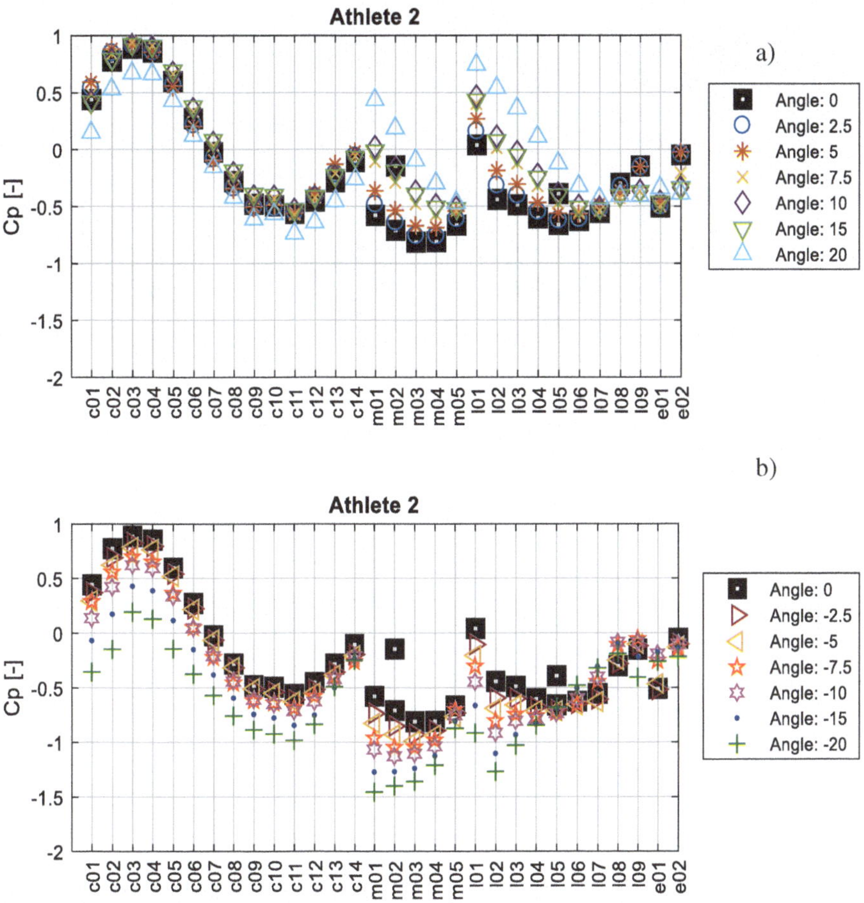

Fig. 7. Non-dimensional pressure distribution for Athlete 2 (a) positive yaw angles (b) negative yaw angles

Turning the head down and therefore creating a different angle of attack between the helmet and the wind, a shift of the stagnation point is visible. The maximum of the pressure shifts from tap c01 to tap c05 for more aggressive configurations indeed. This behaviour is equivalent to perform a shift of the pressure taps along the x-axis.

In the legend of Fig. 8, the C_dA values of the global system (bike, cyclist and helmet) are also reported. The position giving the best performance is the "aggressive" one with a normal condition for the helmet's tail. On the contrary, an aggressive position looking down is not recommended because of the angle reached by the helmet. In this condition, the tail is up and it creates more drag. The flow does not change too much for the pressure taps near the visor but deviates dramatically for the pressure taps positioned in the medium, low and so-called ear line, close to the helmet's tail. Looking at the results it is visible that this aggressive position with the head down is worse than

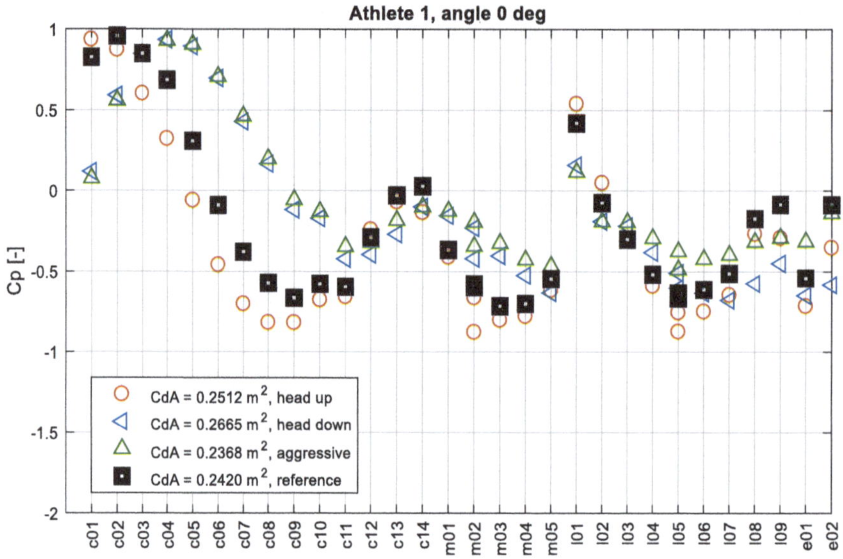

Fig. 8. Pressure distribution for Athlete 1 testing different head positions

a more comfortable configuration with the head up. This implies a greater effort for the athlete with a lower performance.

5 CFD Simulations

A CFD model of a cyclist has been performed using a three-dimensional scan of a real athlete with the helmet placed correctly on the rider head, as can be seen in Fig. 1.

The geometry of the numerical domain reproduces a box that has sizes of 15 m in the streamwise direction, is 6 m wide and 4 m high. The cyclist is positioned at 5 m from the inlet of the domain, with the head at 1.3 m from the ground (see Fig. 9). The numerical framework where the work has been developed is the open-source fluid-dynamic software OpenFOAM. The incompressible Navier-Stokes equations are solved using the Reynolds average approach. The flow is considered steady state. This is possible since we solve the mean flow explicitly and we model the unsteady fluctuations using a turbulence closure model. This approach is consistent with the problem in analysis, since the measured quantities are averaged over time. The turbulence closure used is the k-ω SST, since its performance is good on this kind of problems.

The mesh is realized using the OpenFOAM pre-processor snappyHexMesh. The mesh size goes from the minimum size of 0.5 mm on the surface of the helmet to a size of 5 mm on the cyclist body to a maximum size of 0.16 m far away. The total number of cells realized is about 11 million cells. A visualization of the mesh realized can be seen in Fig. 10.

In order to solve the problem some boundary conditions and numerical methods must be defined. The boundary conditions are indicated on the surfaces in Fig. 9. The boundary conditions that are applied to the domain are: a velocity of the flow

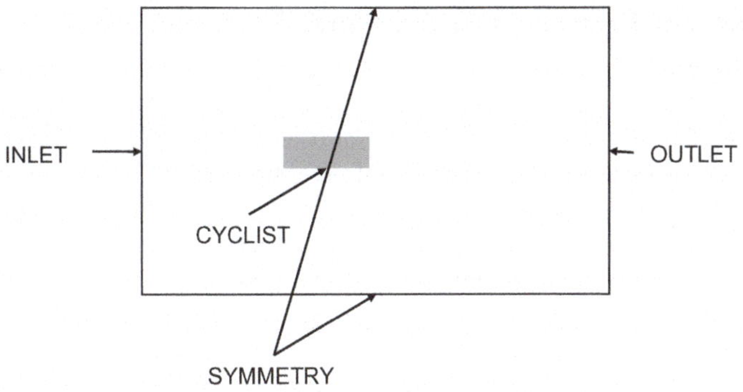

Fig. 9. Domain size for the CFD analysis

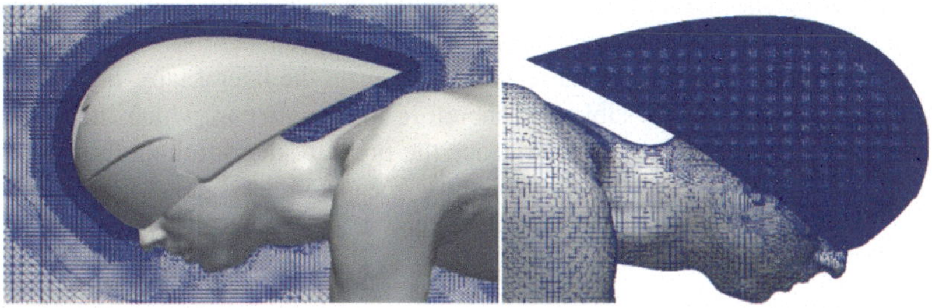

Fig. 10. Mesh used for the CFD analysis: mesh refinement around the helmet (left) and on the athlete and helmet surface (right)

(V = 14 m/s) with suitable direction with respect to support symmetry axis is defined on the inlet; the static pressure is imposed on the outlet of the domain; no slip condition with standard wall functions (V = 0 m/s) for the support surface; slip condition is imposed on the symmetry boundary. Results can be seen as pressure distribution on the athlete and the helmet (see Fig. 11) or as velocity field around the cyclist geometry (see Fig. 13a).

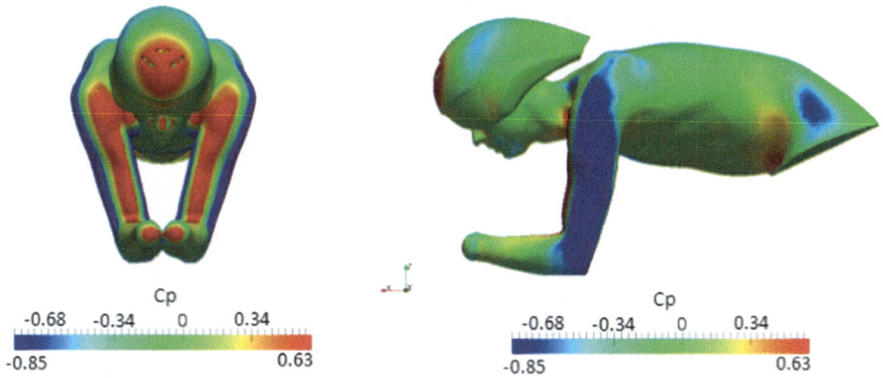

Fig. 11. Pressure distribution around the athlete and helmet geometry.

6 Comparison Between Wind Tunnel Tests and CFD Simulations

Numerical simulations were carried out only at 0° considering the reference geometry. CFD results are analysed as non-dimensional pressure coefficients in order to compare the experimental and numerical results. The values corresponding to the pressure taps are selected accordingly. The results at 0° of yaw angle are compared against the wind tunnel tests considering both the athletes and the several positions tested. Figure 12 shows the comparison in terms of non-dimensional pressure coefficients. As can be seen, the numerical pressure distribution (red asterisk) is close to the experimental values, therefore there is a good validation of numerical data. The numerical results are between the reference position and the aggressive one for Athlete 1. The reference posture for both athletes is quite similar and not so far from CFD simulations. This means that during the wind tunnel tests the helmet was correctly worn by the cyclist and the position taken is comparable to the reference in the CFD model. The deviation between numerical and experimental results is bigger in the lateral taps. In this region, large pressure gradients are visible and therefore the manner in which the helmet is worn, and the posture of the athlete influence the pressure distribution; also, the exact position of the pressure taps may influence the measurement since where large gradients are present, a small location difference may affect the result comparison.

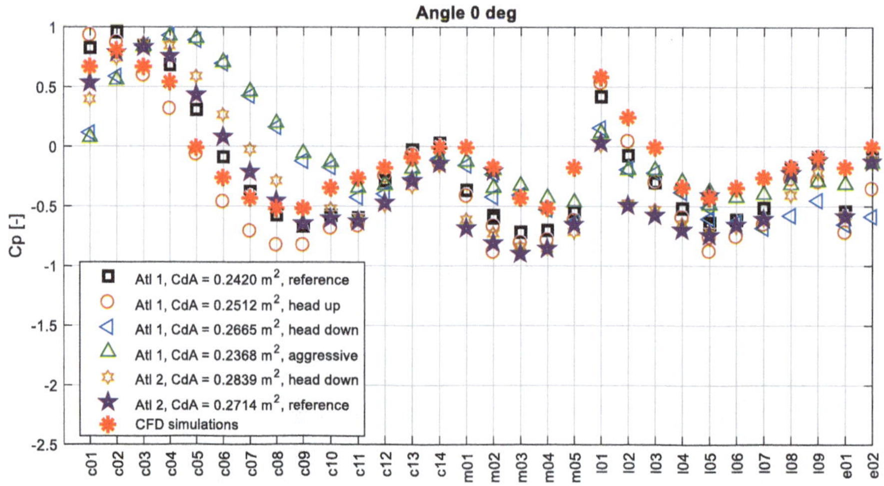

Fig. 12. Pressure distribution for both the Athletes in the wind tunnel and CFD simulations.

Figure 13 shows an example of pressure distribution around the helmet in CFD simulations, compared to experimental data. The general trend of pressure distribution is well captured by numerical calculation. When looking with more detail the results, it is possible to analyse the differences between experimental and numerical data. The region where the largest differences are seen is the one of the lateral taps (medium and low lines) where high pressure gradients are present (see Fig. 13b). Here the pressure values change dramatically in few millimetres and therefore depending on the actual pressure tap position, the numerical results move from the reference position and are closer to the aggressive position for Athlete 1. Flow separation occurs on the rear part of the helmet, as can be deduced from Fig. 13. This is also visible on the measured pressures where the behaviour of the central line of pressure taps is as a function of the head's angle. Changing the inclination of the leading edge of the helmet, the flow separation occurs in different positions troubling the complete pressure distribution downwind. In the last part of the tail instead, the flow is already separated, and the pressure values are quite similar for all the athletes and positions. The numerical simulations confirm the experimental data: a different head position based on a different inclination of the leading edge of the helmet modifies significantly the whole pressure distribution making evidence of strong pressure gradients. Therefore the experimental-numerical validation must be performed as a careful process where the posture of the cyclist and the position of the helmet on the head in the wind tunnel experimental tests have to replicate as much as possible the numerical position (or viceversa).

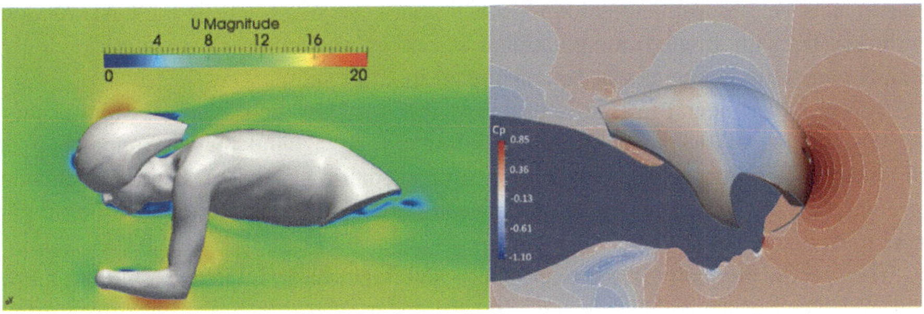

Fig. 13. Velocity field around the cyclist symmetry plane (left) and pressure distribution around the helmet in CFD simulations (right).

7 Conclusions

Wind tunnel testing and, more recently, CFD simulations have taken over as the primary analysis techniques for studying cycling aerodynamics due largely to their accuracy, reproducibility, see Timothy et al. (2017). Nowadays the cycling equipment is being developed having aerodynamic issues in mind. The helmet design becomes crucial because it is the first impact of the flow and affects the flow on the rest of the cyclist; moreover, its size relative to the rest of the system is relevant. This means that the choice of the helmet can have a significant effect on the net drag force that the rider must overcome.

The helmet studied in this paper is an aerodynamic long-tailed one provided by MET, that covers the rider's head in an elongated fairing. The real helmet is instrumented with 32 pressure taps, simultaneously acquired to measure the pressure distribution on the helmet external surface and to validate CFD results by means of local static pressure comparison. The geometry of the rider's head and upper back is used for conducting CFD analysis, while the motion of legs and the bike has been neglected and not reproduced in the numerical analysis. Different postures as well as the way the helmet is worn influence the pressure distribution, therefore the final helmet was tested in the wind tunnel with two different athletes, varying different head positions based on the inclination of the leading edge.

The results show strong interaction between the posture of the rider and the pressure distribution on the helmet. Comparing a more upright posture with a low aggressive position, the pressure distribution changes dramatically on the helmet and the global drag area is smaller. The angle of the rider's head modifies the pressure distribution on the helmetmore than the change of the yaw angle. Extremely high helmet inclination angles show higher drag area, showing that more comfortable positions are more efficient.

CFD simulations are computed considering a reference rider's geometry and posture. The flow field and the local pressure are computed on the whole surface of the considered body, therefore it is possible to optimize the performance by contouring the helmet in order to fit the profile of a rider's back. On the other hand, one of the main limitations of the numerical approach is that geometries of professional cyclists are not easy to achieve, because a three-dimensional scan is needed to create the CAD model. Furthermore, with limited computational resources and optimization purposes only limited and static positions can be considered in CFD simulations.

During experimental tests it is observed that the flow field changes significantly over the course of the pedal stroke, therefore during a wind tunnel campaign it is important to test more than one athlete while pedalling and to test more than one configuration to check the correlation between the pressure distribution and the rider's head position.

The specific geometry of the rider's head and upper back, limits the ability to make generalisations about helmet design. Streamlined shapes are a clear advantage, but small geometric details may positively or negatively influence the performance on the helmet, depending on the rider's posture and shape. In this paper CFD is used as a design tool and a first attempt of validation against wind tunnel tests is done. The comparison between the results shows a good agreement, moreover the experimental-numerical validation must be performed as a careful process where the posture in the wind tunnel has to replicate as much as possible the numerical reference position.

Acknowledgements. The activity presented in the paper is part of the research activity developed with MET helmets. We thank the athletes Davide Fumagalli and Emilio Previtali.

References

Padilla S, Mujika I, Angulo F, Goiriena JJ (2000) Scientific approach to the 1-h cycling world record: a case study. J Appl Physiol 89:1522–1527

García-López J, Rodrígue Marroyo JA, Juneau C, Peleteiro J, Martnez AC, Villa JG (2008) Reference values and improvement of aerodynamic drag in professional cyclists. J Sports Sci 26:277–286

Gibertini G, Campanardi G, Grassi D (2011) Cycling aerodynamics: a complex problem. In: 13th international conference on wind engineering (ICWE 2013)

Blocken B, Defraeye T, Koninckx E, Carmeliet J, Hespel P (2013) CFD simulations of the aerodynamic drag of two drafting cyclists. Comput Fluids 71:435–445

Crouch TN, Burton D, LaBry ZA, Blair KB (2017) Riding against the wind: a review of competition cycling aerodynamics. Sports Eng 20:81–110

Sidelko S (2007) Benchmark of aerodynamic cycling helmets using a refined wind tunnel test protocol for helmet drag research. Ph.D. thesis, Massachusetts Institute of Technology

Belloli M, Giappino S, Robustelli F, Somaschini C (2016) Drafting effect in cycling: investigation by wind tunnel tests. In: 11th conference of the international sports engineering association, ISEA 2016, pp 38–43

Belloli M, Cheli F, Bayati I, Giappino S, Robustelli F (2014) Handbike aerodynamics: wind tunnel versus track tests. In: 10th conference of the international sports engineering association, ISEA 2014, pp 750–755

Belloli M, Giappino S, Schito P, Tenni M (2014) Numerical and experimental optimization of a time-trial cycling helmet. In: INVENTO 2014, June 2014, Genova, Italy

Towards the Second Generation Eurocodes: Evolution of EN 1991-1-4 Wind Actions

S. O. Hansen[1(✉)], R. Höffer[2], J. Rees[3], F. Ricciardelli[4], and P. Spehl[5]

[1] Svend Ole Hansen ApS, Sct. Jørgens Alle 5, 1615 Copenhagen, Denmark
soh@sohansen.dk

[2] Faculty of Civil and Environmental Engineering, Ruhr-Universität Bochum, Universitätstrasse 150, Building IC 02-165, 44780 Bochum, Germany
ruediger.hoeffer@ruhr-uni-bochum.de

[3] COWI, Building Two, Riverside Court, Chipping Sodbury, Bristol BS37 6JX, UK
ejr@cowi.com

[4] Department of Engineering, Università della Campania "Luigi Vanvitelli", via Roma 9, 81031 Aversa, Italy
friccia@unicampania.it

[5] SECO Belgium, Rue d'Arlon, 53, 1040 Brussels, Belgium
http://www.groupseco.com

Abstract. In December 2012, through Mandate M/515, the European Commission asked CEN to develop new standards, or new parts of existing standards. This is to include the incorporation of new performance requirements and design methods, the introduction of a more user-friendly approach in several existing standards, and a technical report on how to adapt the existing Eurocodes and the new Eurocode for structural glass to take into account the relevant impacts of future climate change. Within Phase 2 of the Mandate, in 2017 Project Team SC1.T3 EN1991-1-4 was appointed to draft the new EN1991-1-4 on wind actions on structures. This paper summarises the first revised draft prepared in the first year of activity of the Project Team, providing preliminary information on the possible modifications that might appear in the second generation of the document, expected to be completed in 2020.

Keywords: Codes of practice · Eurocodes · Wind actions · Wind models · Aerodynamic coefficients · Dynamic response · Aeroelastic instabilities

1 Introduction

The European standard EN1991-1-4 Wind actions is one of the 58 parts of the Structural Eurocodes prepared since 1989 by CEN (European Committee for Standardization) under successive Mandates of the European Commission.

A pre-standard ENV1991-2-4 was first published in 1995, based on rules in use in several European countries, on (ECCS 1987), which contains a first European map of extreme wind velocities, and a set of pressure coefficients for buildings given in (Cook 1990).

F. Ricciardelli and A. M. Avossa (Eds.): INVENTO 2018, LNCE 27, pp. 350–364, 2019.
https://doi.org/10.1007/978-3-030-12815-9_28

Each Member State was then invited by the European Commission to prepare a National Application Document, in order to identify the changes and additions needed for this pre-standard to be used in each European country.

On basis of the information collected, of new publications such as (Dyrbye and Hansen 1997), and of further National comments to the successive drafts, the first edition of the European standard EN1991-1-4 was prepared and published in 2005, to replace the existing National standards on the same subject.

In the standard, notes indicate values or procedures which are left open for national choice, known as NDPs (Nationally Determined Parameters), and the National Standards implementing EN1991-1-4 have a National Annex containing the NDPs to be used for the design of buildings and civil engineering works constructed in the relevant countries.

A second generation of Eurocodes is now in preparation under mandate Mandate M/515 of the European Commission. A "*Climatic Actions*" working group has first been established within CEN/TC250/SC1 "*Actions on structures*" to prepare in a more coordinated way the parts on snow, wind and thermal actions, including European maps of basic data, and to oversee the revision of these parts to be prepared by new Project Teams.

Project Team SC1.T3 EN 1991-1-4 started in June 2017 to prepare the second edition of EN1991-1-4 and delivered its first revised draft at the end of April 2018. The aim of this revision is to improve the first edition based on a systematic review of the National comments, National Annexes to the first edition, and comments received from other SCs and WGs of CEN TC250.

2 Proposed Structure of New EN1991-1-4

The general task of the Project Team on wind actions was that of: (*a*) improving the consistency of the basis for calculations and new basis for calculations, (*b*) harmonizing the many different approaches used in the present Eurocodes with National Annexes, (*c*) potentially harmonizing the different Eurocode Parts and CENELEC standards for overhead electric lines, as well as other CEN or ISO standards, and (*d*) possibly reducing global wind loads leading to more economical structures. Such general task was then organised into the following subtasks:

1. reduction of National Choices (NDPs);
2. enhancement of the ease of use;
3. drafting of a European extreme wind map, emphasizing differences across borders;
4. improvement of the wind models for calculation of wind velocities and peak velocity pressures including roughness categories and transition between roughness categories;
5. review of the aerodynamic coefficients based on the current state of the art;
6. review of the methods for evaluating dynamic response, vortex-induced vibrations, aeroelastic effects, bridge response;
7. make editorial improvements.

A first effort was devoted to the rearrangement of the structure of the document, so to reduce the size of the main text by limiting this to the general information used for all or almost all design situations. The main text is structured in accordance with the wind-load chain (Fig. 1).

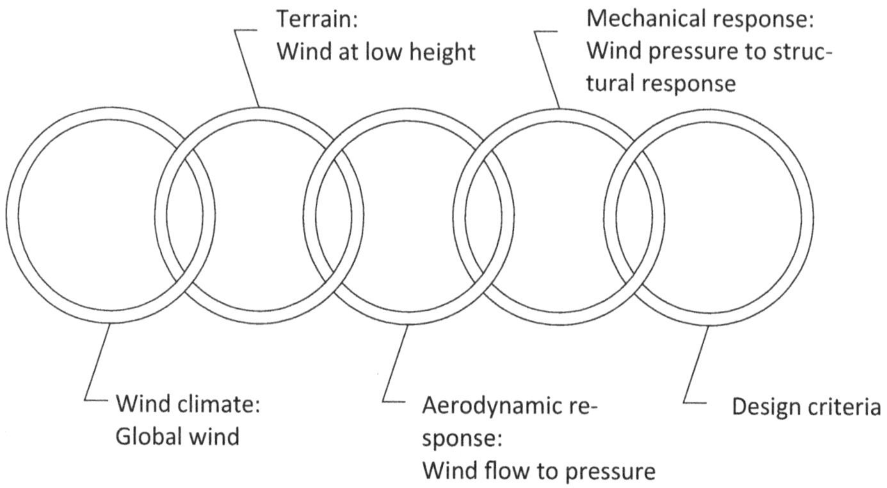

Fig. 1. Alan G. Davenport wind-load chain (Dyrbye and Hansen 1997).

All specific and detailed information will be contained in annexes. These will be in a greater number and larger than those in the current version of EN1991-1-4. Nevertheless, they will be used only when required, and for some of them (e.g. those containing the values of the aerodynamic coefficients) only a very limited number of pages will be necessary for a given structure. To guide the user through the document, a number of flow charts will be inserted at the beginning of the main text. These are aimed at setting a path through the document for different design situations, starting from the most common ones needing a very limited use of the annexes, to the more complex ones, needing considerable use of annexes and possibly requiring judgement by the user.

In the main text, simple sketches show the structures for which dynamic response and vortex-induced vibrations may become significant. This means that the majority of users do not need to consult the annexes that cover these more complicated calculations.

To date, 13 Annexes have been considered for inclusion, as follows:

- Annex A, containing a European wind map based on the data currently available in the National Annexes.
- Annex B, giving illustrations of the terrain categories and a wind model applicable for heights up to 300 m above terrain. It provides rules for the effects of orography including displacement height, roughness change, influence of landscape and influence of neighbouring structures.

- Annex C, giving pressure coefficients to calculate the pressures on a surface.
- Annex D, giving net pressure coefficients to calculate the pressure difference across a skin.
- Annex E, giving force coefficients to calculate the forces or forces per unit length acting on a structure or structural component.
- Annex F, giving a procedure for calculating the structural factor $C_s C_d$
- Annex G, giving procedures for calculation of cross-wind and torsional equivalent static actions and acceleration response of rectangular plan buildings.
- Annex H, giving rules for vortex-induced response and some guidance on other aeroelastic effects.
- Annex I, giving information concerning the dynamic characteristics of structures with linear behaviour.
- Annex J, giving a probabilistic description of wind actions.
- Annex K, giving basic information concerning wind tunnel tests and CFD simulations.
- Annex L, giving basic information on how to build wind maps based on measurements from metrological stations.
- Annex M, giving procedures for the calculation of response of steel lattice towers and guyed masts.

Annex J is new, and is meant to provide the information required, in conjunction with EN1990, for a probability based assessment of structures exposed to wind actions. The annex will be used as basis for the calculation of partial factors for wind actions included in EN1990. Furthermore, similar annexes in other climatic Eurocode parts, e.g. the Eurocode on snow load, enable characteristic load values and partial factors to be produced on a consistent basis and to provide the same structural reliability independently of the dominant climatic action.

Annex K is also new and it is meant to give the designer information concerning the fields of application of wind tunnel testing and of CFD simulations for the assessment of the wind loading of structures. It is also meant to provide the designer with the information required to commission wind tunnel tests or CFD simulations, to set the minimum requirements of the tests and of the simulation, and of the results to be delivered.

3 Wind Map and Wind Models

The wind map contains the basic wind velocity describing the wind climate of the site, and the wind models specify the influence of the terrain.

3.1 Wind Map

At present, wind maps are contained in the National Annexes. A combined European wind map has been prepared by the Project Team and included in the first draft of the revised EN1991-1-4 (Annex A). The map identifies a number of national borders where there are discontinuities in wind speeds. There are many possible reasons for these steps, such as: (*i*) Different methods may have been used to predict extreme winds from anemometer records. (*ii*) Data sets may cover different periods and so different confidence intervals may apply. (*iii*) Wind speeds at a border may be an extrapolation from anemometer sites within a country, rather than interpolation between anemometer sites across the border. (*iv*) Wind speeds may have been adjusted for non-technical reasons.

It is intended that Member States on a bilateral basis will initiate activities to harmonize basic wind velocities across borders and that a more harmonized wind map will be included in the second edition of EN1991-1-4. However, wind maps in National Annexes will still be permitted and take precedence.

It is an aspiration to have a single harmonised European wind map, but this requires a concerted effort to ensure all European nations have produced their national maps on a consistent basis. The revised version of EN1991-1-4 will provide guidance on how to generate wind maps from raw anemometer data, to support this process of convergence, see Annex L described above.

3.2 Wind Models

Mean Wind Speeds

EN1991-1-4 currently uses a simplified reduction of the Deaves & Harris wind model (Cook 1986) for the mean wind speed profile. Equation (1) gives the variation of mean wind speed with height and is accurate from the displacement height to the gradient height:

$$v_m(z) = \frac{u_*}{\kappa}\left[ln\left(\frac{z-d}{z_0}\right) + 5.75a - 1.875a^2 - \frac{4a^3}{3} + \frac{a^4}{4}\right] \tag{1}$$

Where $a = (z-d)/z_g$, $v_m(z)$ is the mean wind speed at height z, u_* is the friction velocity, κ is the von Karman constant (= 0.4), d is the displacement height, z_0 is the aerodynamic ground roughness and z_g is the gradient height.

Within approximately 300 m of the ground surface, the last three terms of Eq. (1) are insignificant, so it can be reduced to:

$$v_m(z) = \frac{u_*}{\kappa}\left[ln\left(\frac{z-d}{z_0}\right) + 5.75a\right] \text{for } z \leq 300\,\text{m} \tag{2}$$

Within approximately 30 m of the ground surface, the linear correction is not required and Eq. (2) can be reduced to:

$$v_m(z) = \frac{u_*}{\kappa}\left[ln\left(\frac{z-d}{z_0}\right)\right] \text{for } z \leq 30\,\text{m} \tag{3}$$

EN1991-1-4: Eq. 4.3 defines wind speed as (ignoring orography, so c_o = 1):

$$v_m(z) = c_r(z).v_b \tag{4}$$

And

$$c_r(z) = k_r.ln\left(\frac{z}{z_0}\right) \tag{5}$$

EN1991-1-4 does not include displacement height in the formulation for mean wind speed, so it can be seen that Eqs. (3) and (4) are identical when $k_r = 2.5u_*$. However, EN1991-1-4 currently allows Eq. (5) to be applied up to heights of 200 m and this is considered too high.

Turbulence

EN1991-1-4 currently uses the simplification that wind turbulence (σ_v) is constant with height. EN1991-1-4, Eq. 4.6 states: $\sigma_v = k_r.v_b.k_I$; all three terms in this equation are constants. The Deaves & Harris model contains a more accurate representation of the variation of turbulence with height and this is illustrated in Fig. 2.

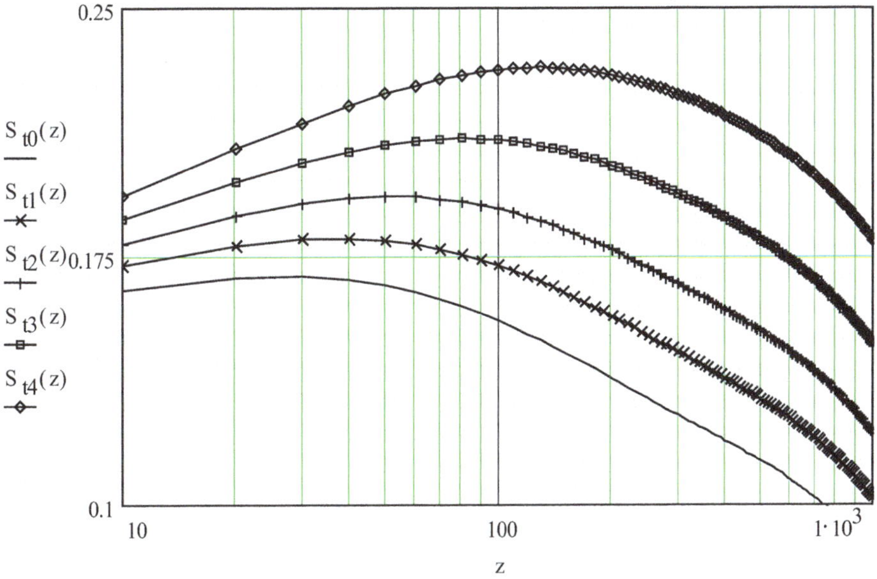

Fig. 2. Variation of turbulence factor for 5 typical aerodynamic ground roughnesses with height, z [m].

Peak Wind Pressure

EN1991-1-4 currently uses a linearization in the calculation of peak velocity pressures. This results in an underestimation of peak wind pressures in rough terrain.

Proposed Changes
For the most part, the wind model currently in the main body of EN1991-1-4 will remain unchanged, except it will only apply to structures up to 100 m high (it should give conservative values between 30 m and 100 m). Also, the calculation of peak velocity pressure will use the peak wind velocity, defined as:

$$v_p(z) = v_m(z).(1 + k_v.I_v) \quad (6)$$

This will remove the linearization.

EN1991-1-4 is used for a wide range of structures – from simple buildings through to complex, tall, wind sensitive structures (e.g. chimneys and guyed masts). To cater better for the more complex structures, a more refined wind model will be included in Annex B. This will be based on the Deaves & Harris model and be applicable for structures up to 300 m in height. It may also be used for structures up to 100 m if the user requires a more detailed wind model.

4 Aerodynamic Coefficients

Section 7 of the first revised draft lists the different loading models, and guides the user through the choice of the relevant aerodynamic coefficients. The coefficients are listed in annexes C through E. In particular, pressure coefficients to be used for buildings and other structures are given in annex C. Net pressure coefficients to be used for canopies, for porches and balconies, for walls and for signboards are given in annex D. Force coefficients per unit length for elongated structures or structural members, force coefficients for point-like structures and for lattice structures and force coefficients for bridges are given in annex E.

These three annexes are based on the coefficients available in the current version of EN1991-1-4, which have been complemented with additional data available from National Annexes or other Codes of Practice. The aim is that of providing the Engineering community with the largest possible aerodynamic dataset, with the idea that it should allow the use of EN1991-1-4 for the design of a large variety of structural shapes.

In assembling the dataset, it became clear that aerodynamic coefficients have been derived on different bases. It is beyond the scope of the EN to explain the basis of each coefficient. However, the PT will prepare an accompanying supporting document that gives as much background data on the source and basis of each group of coefficients as reasonably possible. This will assist users of the EN and inform those responsible for its ongoing maintenance.

Some of the coefficients added have been directly provided by Member States through their representatives in TC250 SC1/WG1. Conversely, in other cases specific requests have been made to the Project Team concerning aerodynamic coefficients; in the latter cases, the PT has tried to fulfil the request or, when unable to do so, has asked WG1 and in particular the requesting Member State, to provide data.

In annex C, external pressure coefficients $c_{pe,1}$ and $c_{pe,10}$ as already available in the current version of EN1991-1-4 for rectangular plan buildings are given; however, a different definition of the reference height is provided. In particular, a pattern for the choice of the reference height is given that should facilitate the calculation of the floor forces to be used in the Finite Element modelling of buildings. This is done by allowing an arbitrary choice of the height of each strip of constant reference height in which the building is divided; this will allow the choice of strips that match the inter-storey height or a multiple of that. By taking the reference height equal to the maximum height of the strip, the loading patterns turn out to be always on the safe side, and the larger the number of strips, the higher the complexity of the load pattern, but lowering the total load.

In addition to that, a third set of external pressure coefficients has been added, termed global external pressure coefficients. The latter are meant for use in the evaluation of the actions on principal structural members and on foundations; they do not necessarily give the exact pressure distribution, but they give rise to simplified loading patterns, whose resultant is generally conservative. Global external pressure coefficients apply to vertical walls and to flat, monopitch, duopitch and gable roofs; in addition, criteria are given for the use of global coefficients for multi-span roofs.

A new approach to the evaluation of pressure coefficients for walls or roofs with more than one skin is currently under preparation, based on data available to the industry.

Completely new to EN1991-1-4, a section has been added in annex C giving guidance on how to use global external pressure coefficients available for rectangular plan buildings in the case of irregular buildings. In particular buildings of uniform height having a floor plan which is a combination of rectangular elements, and buildings having a rectangular floor plan but made of elements of different height or recessed bodies, are considered. In addition, global external pressure coefficients for Y-shaped buildings are given.

A specific section has been devoted to global external pressure coefficients for structure with circular plan. Pressure coefficients are given for the walls of cylindrical buildings, towers and chimneys and for those of hyperbolic cooling towers; the latter are new to EN1991-1-4. The same section hosts also the global external pressure coefficients for domes, as they apply to the same structures for which the vertical walls are considered. The difference is highlighted between the meaning of the external pressure coefficients given in this section and the coefficients of force per unit length for circular members, given in annex E. In fact, the confusion is not uncommon between the latter and the integral of the global external pressure coefficients.

Finally, a new approach is proposed for the evaluation of internal pressure coefficients. Three cases are considered. A first case is that in which, for at least two sides of the building (facades or roof), the total area of openings on each side is more than 30% of the area of that side. In this case, the actions on the structure should not be calculated from the rules given for buildings, but rather by using the rules applying to canopies or to free-standing walls, depending on whether the large openings are on the vertical walls or on the roof, respectively. The second case is that already considered in the current version of EN1991-1-4, applying to buildings with a dominant face, to which no changes have been made. Finally, the third case is that of buildings without a dominant face, for which the internal pressure coefficient should be determined as a linear combination of the positive and negative averaged external pressure coefficient, weighted through the squares of the areas to which positive and negative external pressures apply. This is a linearized version of the solution using discharge and continuity equations given by (Cook 1986), proposed by (Vickery 1994). Linearization brings a violation of the continuity equation, but allows a closed form solution of relatively simple implementation.

In annex D, the net pressure coefficients and forces for monopitch, duopitch and multibay canopy roofs already available in the current version of EN1991-1-4 are given. To these, net pressure coefficients for porches and for balconies have been added. Finally, the same net pressure coefficients for free-standing walls, parapets, fences and signboards already available are proposed again.

Annex E contains the force coefficients for elongated structures and structural elements already available in EN1991-1-4, which have been complemented with values applying to structural metallic elements with sharp edged sections. In addition to the data for smooth or mildly rough circular sections, those applying to stranded cables have been added.

Completely revised is the section containing the force coefficients for square and triangular lattice structures. These have been transferred from EN1993-3-1 and are derived from wind tunnel tests undertaken in the UK in the 1980s on a range of lattice towers and masts, with and without ancillary items (e.g. dishes, ladders, cables, …); the background to these tests is given in BS8100: Part 2: 1986). Although developed for a specific form of structure, they are generally applicable to lattice structure with solidity ratios of up to 0.6.

Finally annex E will contain also the aerodynamic coefficients for bridges, which in the current EN1991-1-4 are given in a separate section with respect to the remaining aerodynamic coefficients. Drafting of this part is still in progress, and it is anticipated that a different approach will be proposed with respect to the available one, based on existing European codes and available data from wind tunnel test and numerical simulations.

5 Dynamic Response

The current EN1991-1-4 gives two alternative procedures to calculate the dynamic along-wind response of structures. The two procedures provide predictions of the same order of magnitude, and both procedures assume that:

- the wind load is determined from the undisturbed wind field;
- the structure is linear-elastic with viscous damping;
- the along-wind mode considered is uncoupled from other modes.

According to assumption 1 above, the pressure correlation is assumed to be identical to the correlation of the longitudinal turbulence in the undisturbed wind field. This assumption may lead to an overestimation of the load, due to lack of correlation between wind load on the structure front and on the rear face. On the other hand the load will be underestimated due to the fact that pressures on the structure are better correlated than the longitudinal turbulence in the undisturbed wind field.

The two above-mentioned effects, which to some extent counterbalance each other when predicting loads, are not taken into account in the current EN1991-1-4 due to rather limited data available. However, one possibility in the present revision will be to prepare a new format, which serves two purposes: the procedure should predict results in the same order of magnitude as the current two procedures, and the format should enable predictions based on more advanced methods in form of wind tunnel testing. Establishing a consistent format will facilitate that the more advanced methods over the coming years will provide the data needed for a codified full application of the more advanced procedure.

The new procedure may define a cross-sectional admittance function, which gives the conversion between the spectrum of the incoming air flow and the spectrum of the wind action on a cross section. The cross-sectional admittance function is subsequently combined with the correlation of the air flow for longitudinal turbulence separated along the axis of the structure. For a line-like structure the correlation between cross sectional forces will be described well by the correlation of the incoming air flow with the same separation.

The procedure described above has been applied successfully in buffeting response analyses for long span cable-supported bridges. It is believed that the same approach may turn out to give a consistent and operational description of buffeting wind actions on buildings. For along-wind response the cross-sectional admittance function describes the conversion of longitudinal turbulence to longitudinal forces, for across-wind response the cross-sectional admittance function describes the conversion of across-wind turbulence to across-wind forces.

One of the advantages of the new procedure above is that the actual correlation of pressures and forces is taken into account via the cross-sectional admittance function. The current assumption of equivalence between pressure correlation and velocity correlation do not provide consistent results, and this non-consistency will be removed by the new approach proposed.

The new procedure will relatively easily accommodate codified extensions in form of mode shapes changing sign, across-wind buffeting response and torsional buffeting response.

The detailed calculation of dynamic along-wind response is specified in Annex F in form of a determination of the structural factor $c_s c_d$. The majority of users do not need to apply Annex F. Figure 3 in the main text provides information on when Annex F becomes relevant (Table 1).

6 Aeroelastic Phenomena

Section 10, Aeroelastic phenomena, and annex H, Vortex shedding and aeroelastic instabilities, deal with wind effects due to the reciprocal interaction of structural displacements and the nearfield surround-flow during strong winds, i.e. high Reynolds numbers *Re*. Annex H is based on Section 10. Both parts contain calculation procedures for (with the exception of one case) isolated, flexible structures with either compact cross sections, such as slender rectangular or circular cross sections, or with cross-sections of an elongated shape, having a mass which is more or less

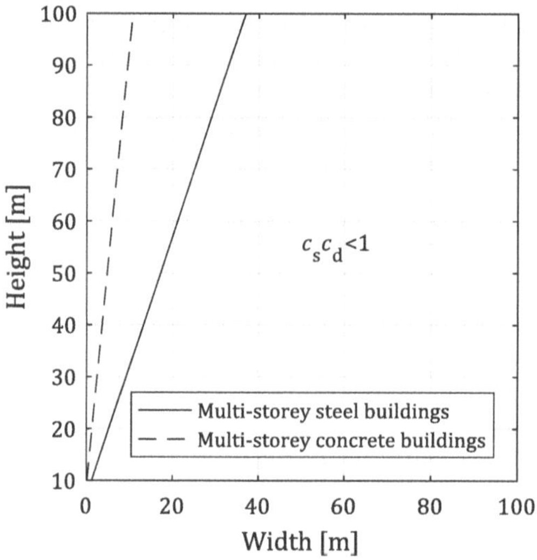

Fig. 3. $c_s c_d$ = 1,0 for multi-storey steel and concrete buildings and rectangular floor plan and vertical external walls, with regular distribution of stiffness and mass

Table 1. Reference data for Fig. 3

	Symbol	Value
Roughness length	z_0	0.05 m
Basic wind velocity	v_b	28 m/s
Logarithmic decrement of aerodynamic damping	δa	0.0
Logarithmic decrement of structural damping for steel	δs	0.05
Logarithmic decrement of structural damping for concrete	δs	0.10

homogeneously distributed over the length of the structure. The following aeroelastic phenomena is dealt with:

- galloping, being self-induced vibrations of line-like structures with compact, non-circular cross sections in cross wind bending modes including:
  (a) cross-sections affected by icing,
  (b) interference effects at closely spaced cylinders, such as slender buildings or chimneys, leading to the phenomenon of interference galloping;
- divergence and flutter, which are torsional or multi-mode instabilities that can occur at flexible structures with cross sections of an elongated shape, such as signboards or suspension-bridge decks,
- oscillations of slender cylinders induced by vortex-shedding.

All considered phenomena with the exception of divergence, which is a static phenomenon, depend on the mass damping ratio, also called Scruton number (*Sc*), which includes structural damping and the ratio of the oscillating or equivalent mass and the mass of the flow mechanically displaced air. The treatment of the considered aeroelastic phenomena is closely linked to the dynamic characteristics of structures – eigenfrequencies, eigenforms and structural damping - as regulated in the technically amended annex I.

Due to the strongly nonlinear nature of the considered aeroelastic phenomena, Section 10 and annex H indicate rather procedures for verification of structural integrity instead of load assumptions. For this purpose, critical velocities are determined involving both structural as well as aerodynamic parameters and are contrasted and compared with extreme, site specific mean wind velocities v_m at specific heights of application. Divergence, galloping and flutter are phenomena for which the critical velocities v_{crit} are determined as threshold values at which an aeroelastic phenomenon is expected to have its point of onset. The proof of structural resistance is provided by including a safety margins of mostly 25% when comparing site values of extreme mean velocities and critical velocities. This margin should also take into account gust sustained long enough to build up an oscillation, correlated over a major proportion of the structural span.

In this context, vortex excitation must be seen as a specific type of phenomenon which comprises a quasi-static, a resonant and finally an aeroelastic behaviour. Vortex-shedding occurs when separated shear layers are shed alternately from opposite sides of a slender structure. This gives rise to a harmonically fluctuating load perpendicular to the wind direction, and vortex-induced vibrations may occur when the load frequency coincides with the natural frequency of the structure. The frequency of vortex shedding is a function of the Strouhal number, see Section 10 for an approximate value which however is not always accurate, and find more detailed information in annex H. The critical velocity of vortex shedding is calculated from the Strouhal number as the flow velocity at which the vortex shedding frequency meets the eigenfrequency of the structure. As a boundary condition, $Sc < 40$ has been introduced in the first draft as a general, upper limit of Scruton numbers at which the possible occurrence of vortex-induced oscillations larger than a few percent of the cross-wind dimension shall be considered. At smaller values, larger structural oscillation in crosswind directions may occur when the ratio of the critical velocity and the extreme mean wind speed at the heights of oscillation remains smaller than 1.25. For a real aeroelastic behaviour of vortex-induced vibrations reciprocal interaction of structural displacements and the nearfield surrounding flow becomes active. The vortices are triggered to shed in the eigenfrequency although slightly different shedding frequencies are predicted via the Strouhal number. This is inherently adopted to the approaches in Section 10.3.4 (novel closed formula for the inertia force per unit length) and Annex H (calculations of the cross wind amplitudes). The number of load cycles follows the approach of the current code. As a novelty, different safety margins are specified in Section 10 for vortex-induced oscillations interacting with galloping.

In order to improve ease of use, guidance is offered with respect to identify cases where turbulence-induced resonance generates crucial structural effects instead of vortex-induced resonance. Two new figures (Fig. 4a and b) are introduced where domains are shown in which the base bending moments for square buildings (a) and chimneys (b) due to vortex shedding and due to wind buffeting are compared

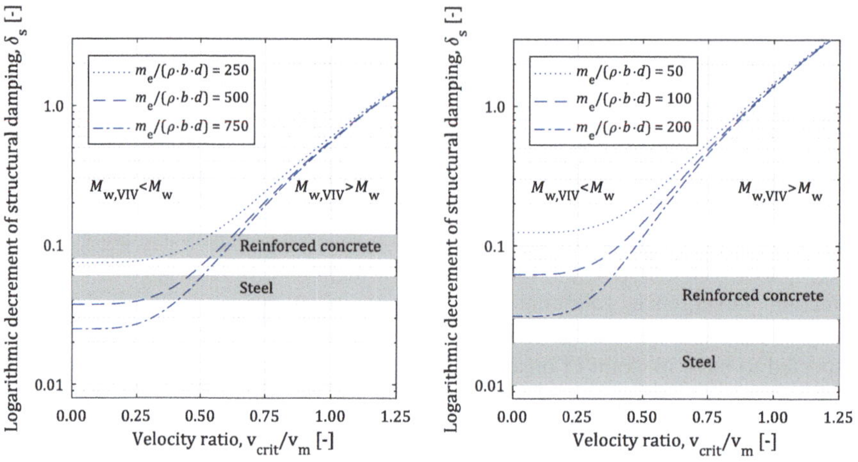

(a) Base moment for square buildings (b) Base moment for chimneys

Fig. 4. Characteristic diagrams: Assessment of structural effects due to wind buffeting and due to vortex shedding for the identification of the main load case for the proof of structural integrity

respectively. The abscissa and the ordinate of the characteristic diagrams are the velocity ratio v_{crit}/v_m and the structural damping.

It summary, it can be said that the following technical amendments have been incorporated into the draft of EN1991-1-4 for the assessment of aeroelastic phenomena:

- an approximate value of 0.2 as a general Strouhal number for standard problems of vortex shedding;
- a general threshold of $Sc < 40$ as an upper limit of Scruton numbers at which the possibility of vortex-induced oscillations larger than a few percent of the cross-wind dimension shall be considered. Smaller vibrations may become relevant for some structures, e.g. high-rise buildings;
- a novel closed formula for the inertia force per unit length due to vortex shedding;
- overworked approaches for the calculation of cross wind amplitudes due to vortex shedding, especially regarding the inclusion of wind turbulence and close-to-reality models for aerodynamic damping (see annex H);
- an amendment of other types of vortex induced oscillations than crosswind oscillations (see annex H);
- adapted safety margins for the interaction of vortex shedding and galloping (see Section 10 and annex H);
- overworked criteria for the identification of flutter resistant structural configurations (see annex H and compare to E.4.2 in the current code);
- an update of structural damping values compared to those tabulated in the current version (see annex I).

7 Conclusions

The first draft of the revised Eurocode on wind actions was completed in April 2018. The structure of the draft follows the Alan G. Davenport wind loading chain, and the extensive specifications of pressure coefficients, net pressure coefficients and force coefficients have been moved to 3 annexes. This format makes the relevant parts of the EN needed for a certain type of structure easy to locate, and achieves one of the main goals of the ongoing revision.

The draft specifies aerodynamic coefficients for many structures not covered by the current version. Furthermore, the current 1 m^2 and 10 m^2 pressure coefficients have been supplemented with a set of simplified global pressure coefficients useful for the determination of global forces on a structure, e.g. foundation loads.

The extension of structural heights from 200 m in the current version of up to 300 m in the draft has been made possible by including the Deaves & Harris wind model in the specifications, and sections with buffeting-induced vibrations in the cross-wind directions will be added in the next draft scheduled for April 2019. This addition is important because the dimensioning load on many high-rise structures are buffeting-induced cross-wind vibrations.

With respect to the assessment of aeroelastic phenomena it can be said that technical amendments have been incorporated into the draft of EN1991-1-4, Section 10 and annex H, which improve the ease of use through closed-form solutions for standard cases of vortex shedding and vortex-induced loading as well as for the identification of crucial load cases.

References

British Standards Institution, BSi (1986) Lattice towers and masts, Part 2, Guide to the background and use of Part 1: 'Code of Practice for loading'. BS8100: Part 2: 1986

Comité Européen de Normalisation, CEN (1995) Eurocode 1: Basis of design and actions on structures - Part 2-4: Actions on structures - Wind actions, EN1991-2-4

Comité Européen de Normalisation, CEN (2002) Eurocode 0: Basis of design, EN1990

Comité Européen de Normalisation, CEN (2005) Eurocode 1: Actions on structures - Part 1-4: General actions - Wind actions, EN1991-1-4

Comité Européen de Normalisation, CEN (2006) Eurocode 3: Design of steel structures Part 3-1: Towers, masts and chimneys – Towers and masts, EN1993-3-1

Consiglio Nazionale delle Ricerche, CNR (2008) Guide for the assessment of wind actions and effects on structures. CNR-DT 207/2008

Cook NJ (1986) The designer's guide to wind loading of building structures. Part 1 Background, damage survey, wind data and structural classification, Butterworths London UK [for] Building Research Establishment

Cook NJ (1990) The designer's guide to wind loading of building structures. Part 2 Static structures, Butterworths London UK [for] Building Research Establishment

Deaves DM (1981) Terrain-dependence of longitudinal r.m.s. velocities in the neutral atmosphere. J Wind Eng Ind Aerodyn 8:259–274

Deaves DM, Harris RL (1978) A mathematical model of the structure of strong winds. Construct Ind Res Inf Assoc, Report 76

Dyrbye C, Hansen SO (1997) Wind loads on structures. Wiley, Hoboken
ECCS (1987) Recommendations for calculating the effects of wind on constructions. European Convention for Constructional Steelwork - Report 58
Lumley JL, Panofsky HA (1964) The structure of atmospheric turbulence. Wiley/Interscience, New York
Tennekes H (1973) The logarithmic wind profile. J Atmos Sci 30:234–238
Vickery BJ (1994) Internal pressures and interaction with the building envelope. J Wind Eng Ind Aerodyn 53:125–144

Extreme Wind Prediction – The Australian Experience

J. D. Holmes(✉)

JDH Consulting, PO Box 269, Menton, VIC 3194, Australia
john.holmes@jdhconsult.com

Abstract. The extreme wind climate of Australia is discussed with a description of the available historical data recorded by the Bureau of Meteorology. The necessity of correction of gusts for gust duration, and for terrain, and for separating the data into storm type is outlined. The 'bootstrapping' technique for assessing sampling errors and confidence limits is also discussed. As examples of the methods discussed, recent data from three locations in South Australia were processed. These illustrated the mixed climate near the southern coastline, and the dominance of non-synoptic downdraft winds at a northern station far from the coast.

Keywords: Bootstrapping · Extreme · Non-synoptic · Probability · Wind speed

1 Introduction

The greatest source of uncertainty in the prediction of wind loads on structures is undoubtedly the estimation of extreme wind speeds, at the high recurrence intervals associated with low risk of exceedance of wind loads. In the loading of quasi-static structures, the wind speed is squared, so that errors in wind speed estimation are effectively doubled. For structures with significant resonant dynamic response, for which the response relation increases with wind speed to a higher power than two, the magnification of the errors in wind speed is even greater.

The emergence of limit-states structural design from the 1980s onwards has led to some countries adopting very high return periods/average recurrence intervals for wind speeds for ultimate limit states. For example, the national wind loading standard in Australia, AS 1170.2, adopted a return period of 1000 years for most structures in the 1989 edition (Standards Australia 1989). The wind load factor for wind loads calculated in this way was (and still is) 1.0. In 2002, a system was introduced in which the average recurrence interval, *R*, varies with the importance of the structure; for example, for farm buildings, *R* would be taken as 100 years, and for post-disaster buildings, such as hospitals, as 2000 years. For most buildings, *R* is taken as 500 years, giving an annual risk of exceedance of about 10% in a nominal building life of 50 years.

To predict extreme wind speeds, wind engineers must rely on data collected by bodies such as national meteorological agencies – in Australia it is the Bureau of Meteorology. Over periods of 40–50 years, anemometer locations often will be changed several times, the instrument itself may be changed, as may be the method of

F. Ricciardelli and A. M. Avossa (Eds.): INVENTO 2018, LNCE 27, pp. 365–375, 2019.
https://doi.org/10.1007/978-3-030-12815-9_29

acquiring data (for example, with the introduction of automatic weather stations (AWS) with digital sampling from the 1990s). The surrounding terrain often changes significantly over time, particularly at airports in, or near, urban areas. In the case of Australia where a peak gust speed is used as a basic wind speed, the effective gust duration is another factor that may require correction. The introduction of the 3-s moving average gust, as mandated by the World Meteorological Organization in the 1980s, means that corrections to the shorter duration gust used in Australian standards are required.

Assuming all the above corrections have been made and procedures have been followed, there remains the problem of extrapolating data from a station recorded over, say, 50 years, to make predictions to return periods of 500 to 1000 years for ultimate limit states design. Although E.J. Gumbel's classic text (Gumbel 1958) on extreme value theory discusses sampling errors and confidence limits for long-return-period predictions using the Type I Extreme Value Distribution, assessment of statistical sampling errors is often lacking in published analyses of extreme wind speeds in recent decades. This paper discusses the sampling errors and confidence limits for predictions using the distributions adopted for the main wind region covering the majority of Australia not affected by tropical cyclones – known as 'Region A' in AS/NZS 1170.2 (Standards Australia 1989, 2002, 2011). Prediction of extreme winds for regions affected by tropical cyclones requires alternative techniques and will not be discussed in this paper.

2 Extreme Wind Types

In a large continental country like Australia with several climate zones, there are several mechanisms that can generate high wind speeds: tropical cyclones, extratropical depressions, cold fronts, severe thunderstorm (convective) downdrafts, and occasionally tornados.

Figure 1 shows a map of Australia showing parts of the country where:

I. Synoptic winds are dominant. These include tropical cyclones, which are the dominant extreme wind event, at high annual recurrence intervals, for coastal strips north of about 25° South degrees latitude,
II. Convective downdrafts are dominant. This is the inland area or 'outback' more than about 200 km from the coastline.
III. The coastal region in the southeast of the country (i.e. the green zone in Fig. 1) where synoptic and convective (non-synoptic) strong wind events both play a significant role.

Since strong winds generated by different mechanisms generally will follow different probability distributions, it is necessary to separate high gusts into defined storm types before extreme-value analysis.

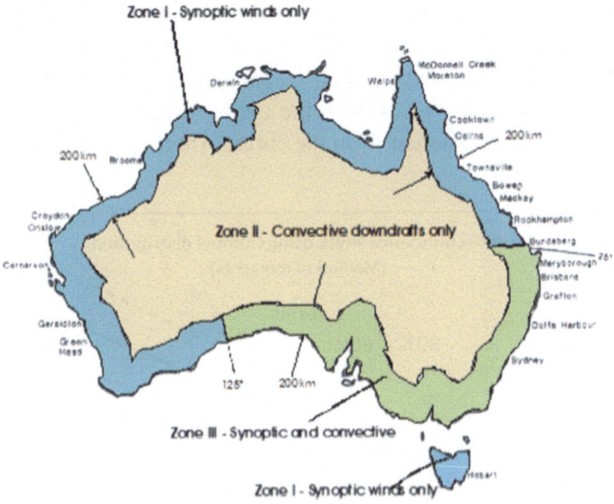

Fig. 1. Zones in Australia showing dominant extreme wind types.

3 Extreme Value Probability Distributions

Between 1989 and 2002, the wind gust speed versus return period for the region incorporating most of the populated part of the country, not affected by tropical cyclones, (known as 'Region A') was given in AS 1170.2-1989 by Eq. (1):

$$V_R = 29.2 + 3.0 \log_e R \,(\mathrm{m/s}) \tag{1}$$

This is the high-return period approximation to the Type I, or Gumbel, Extreme Value Distribution with a mode of 29.2 m/s and scale, or slope, of 3.0 m/s. As a special case of the Generalized Extreme Value (GEV) Distribution, the shape factor (k) is 0.

Since 2002, the GEV with a shape factor, k, fixed at +0.1, has been used for Region A in the Australian/New Zealand standard, giving the regional gust speed with an average recurrence interval of R years:

$$V_R = 67 - 41\,R^{-0.1}\,(\mathrm{m/s}) \tag{2}$$

4 Bootstrapping and Confidence Limits

As far back as the 1950s, Gumbel (1958) emphasized the importance of calculating confidence limits when making long return period predictions from limited data sets. A convenient way of doing this is by use of a simulation or 'bootstrapping' technique, in which a large number of samples of random data are generated using random numbers representative of the cumulative distribution function, with parameters determined from the actual data (e.g. Naess and Clausen 2001).

Figure 2 shows the 80% confidence limits obtained by simulating 50 samples of 50 years of annual maxima (2500 in total), and fitting each sample to the Gumbel Distribution of Eq. (1), using the Method of Moments. It can be seen that the predictions of V_{1000} can be up to 4 m/s from the correct value of 50 m/s (from Eq. (1)). Even the V_{50} predictions can depart from the correct value of 41 m/s by up to 2.5 m/s.

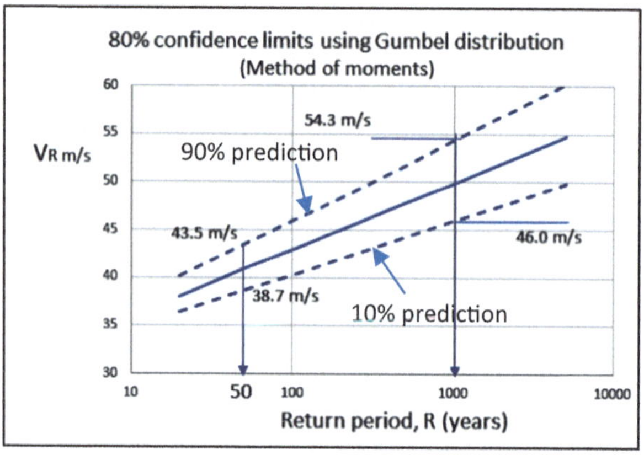

Fig. 2. 90% & 10% confidence limits for predictions using the Gumbel Distribution using 50-year samples

A similar 'bootstrapping' approach, in which the excesses above a threshold of 25 m/s were simulated with the Generalized Pareto Distribution corresponding to Eq. (2), produced Fig. 3. Again 50 groups of 50 samples each were simulated. The 80% confidence limits for V_{500} are 2.3–3.7 m/s. Figures 2 and 3 illustrate that even relatively high sample sizes of 50 years will result in wide confidence limits, for the return periods required for structural design for ultimate limit states.

5 Examples (from South Australia)

The examples discussed are from recent extensive analyses of extreme gusts from the state of South Australia, funded in part by the company which owns and operates the high-voltage transmission line network in that state.

The stations used for this paper are some of the oldest anemometer stations operated by the Bureau of Meteorology in South Australia, and hence have the longest record lengths. The former Dines pressure-tube anemometers were replaced in the 1990s by automatic weather stations (AWS) with cup anemometers and digital processing. The latter incorporates a 3-s moving-average digital filter, and a correction of the peak gust data to the nominal 0.2 s gust used in Australian/New Zealand Standard AS/NZS 1170.2 (Holmes and Ginger 2012) is required.

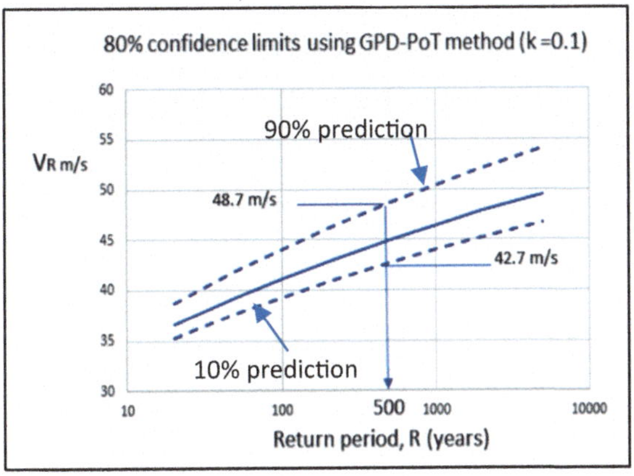

Fig. 3. 90% & 10% confidence limits for predictions using the Generalized Extreme Value/Generalized Pareto distributions.

5.1 Corrections and Storm-Type Separation

High gusts in South Australia are produced by a number of meteorological mechanisms, including large depressions (classified as 'synoptic'), severe local downdrafts or 'downbursts', and those associated with the passage of cold fronts. There are also cases which are a combination of these mechanisms. For the present analysis, wind gusts were classified as 'synoptic' or 'non-synoptic'.

For the AWS data the following method was adopted:

- Identify the occurrence time of the peak gust, V_M, of the event.
- Compute the average V_B of recorded wind gusts in the two-hour time period before the occurrence of the peak gust.
- Compute the average V_A of recorded wind gusts in the two-hour time period after the occurrence of the peak gust.
- Compute the wind gust ratios, $r_B = V_M/V_B$ and $r_A = V_M/V_A$.
- If $r_B < 2.0$ and $r_A < 2.0$, the event was classified as 'synoptic'
- If the preceding condition was not satisfied, the event was classified as 'non-synoptic'.

Figure 4 shows a plot of the recorded wind speeds from one station (Oodnadatta) that was classified as 'non-synoptic'.

The classification method described above is similar, but not identical, to that used by De Gaetano *et al.* for separation and classification of wind events in coastal Northern Italy (De Gaetano *et al.* 2014)

Corrections for terrain and gust duration were applied after separation of the gusts into synoptic and non-synoptic types. Corrections for gust duration are required for the AWS data, as the 3-s moving average gusts recorded by the Bureau of Meteorology are lower than the nominal 0.2-s gusts used in AS/NZS 1170.2, and those recorded by the Dines anemometers. The corrections are different for synoptic and non-synoptic gusts, and the values used were derived from Holmes and Ginger (2012).

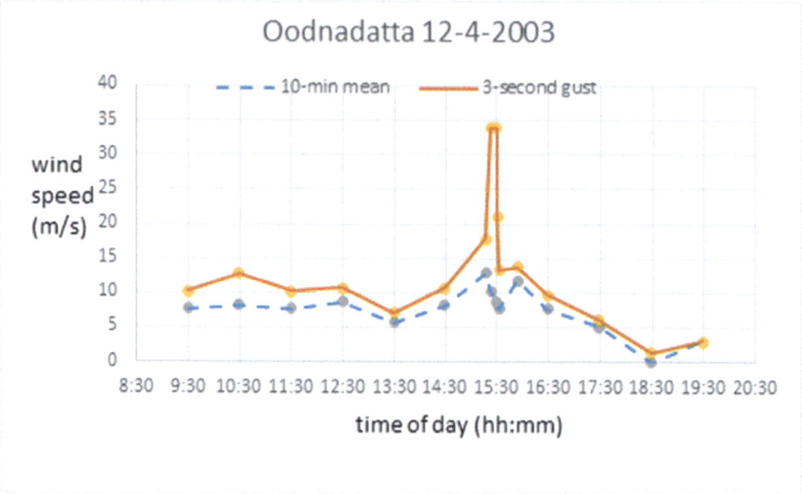

Fig. 4. Time history of a 'non-synoptic' event (m/s).

5.2 Extreme Value Distribution and Fitting Method

Equation (3) gives the general form of the relationship between regional wind speed, V_R, and average recurrence interval, R, for the distributions used in AS/NZS 1170.2. Distributions with the form of Eq. (3) were used to fit the combined Dines/AWS South Australian data for the present study.

$$V_R = C - D.R^{-k} \tag{3}$$

It may be noted that when the shape factor, k, approaches 0.0, Eq. (3) takes the form of Eq. (4):

$$V_R = A + B\,log_e.R \tag{4}$$

However, Eq. (3) has an upper limit (equal to C), unlike Eq. (4) which predicts unlimited values of wind speed with increasing average recurrence interval, R – this is physically unrealistic.

There are several methods available for fitting the GEV including the method of moments and the maximum likelihood method. For the present work, a simple method with a fixed shape factor, based on peaks-over-threshold (POT) data was used. The use of POT data allows use of all data above a defined threshold, instead of just the annual maxima, like traditional methods (see also: Holmes and Moriarty 1999; Naess and Clausen 2001).

The method used here is summarized as follows:

1. Set a threshold V_0 – e.g. 25 m/s,
2. Calculate average rate of exceedences of V_0, λ,
3. Calculate average of all positive excesses above the threshold, E,

4. Choose a shape factor, k; in this case k was fixed at 0.1,
5. Then calculate predictions for high average recurrence intervals, (Davison and Smith 1990):

$$V_R = V_0 + \left(\frac{1+k}{k}\right) E\left[1 - (\lambda R)^{-k}\right] \tag{5}$$

5.3 Results

Figures 5, 6 and 7 show the predictions of extreme wind speeds for both synoptic and non-synoptic winds for three locations. In Fig. 5, the data from three Adelaide area stations (Adelaide Airport, Edinburgh and Parafield) have been combined and analysed as a group (i.e. as a 'super-station'). For Adelaide, the total number of station-years processed was 78 for synoptic winds (AWS data only), and 144 for non-synoptic winds (Dines plus AWS data).

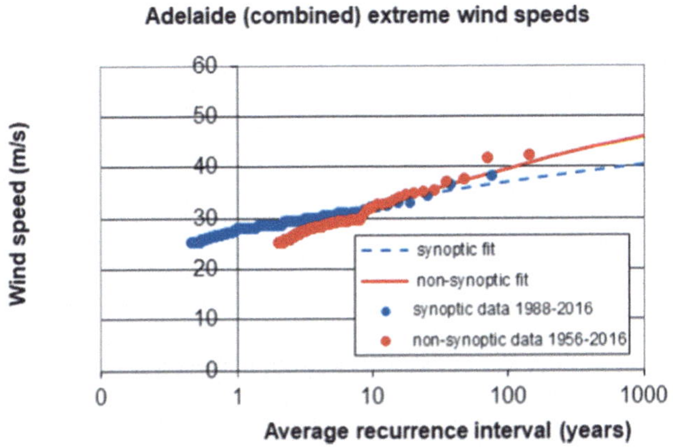

Fig. 5. Extreme wind predictions for the Adelaide area (data from 3 stations combined)

For average recurrence intervals of 100 years and greater, it is clear from Figs. 5, 6 and 7 that non-synoptic winds dominate at all locations, and at Oodnadatta, the most northern station, synoptic wind gusts are insignificant for all average recurrence intervals.

For the two southern coastal locations (Adelaide and Mount Gambier), synoptic and non-synoptic winds both contribute to the combined probability of exceedance of extremes at all recurrence intervals, whereas at the northern 'outback' station – Oodnadatta, located about 600 km from the coastline, the extreme winds are completely dominated by non-synoptic wind events. These are local convective downdrafts, and this extreme wind climate is typical for all locations in Australia more than about 200 km from the coast (see Fig. 1).

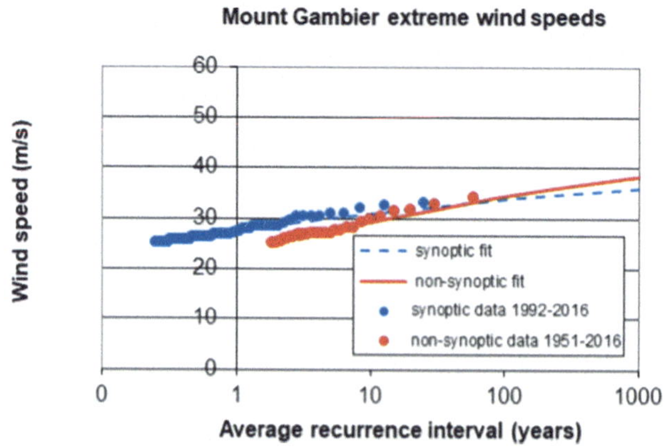

Fig. 6. Extreme wind predictions for Mount Gambier (southernmost station, near the coast)

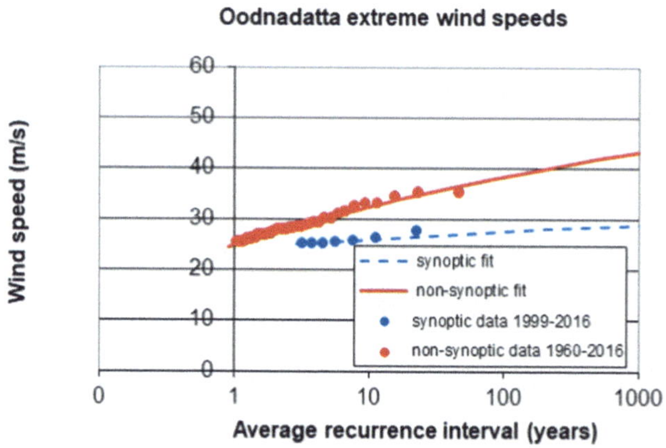

Fig. 7. Extreme wind predictions for Oodnadatta (northernmost station, in the 'outback')

5.4 Errors and Confidence Limits

Using the bootstrapping technique described in Sect. 4, the standard deviation, 10 and 90 percentile values of V_{500}, were calculated and are tabulated in Table 1. Table 1 shows standard deviations in the estimates of V_{500} between 5 and 6% of the nominal value, and 80% confidence limits departing by up to ±4 m/s from the nominal value.

Table 1. Sampling errors and 80% confidence limits for non-synoptic wind gusts speeds with 500-year A.R.I.

Location	V_{500} (m/s)	Std. devn.	10%ile	90%ile
Adelaide	44.3	2.09	41.7	47.1
Mount Gambier	37.4	2.11	34.4	40.2
Oodnadatta	41.9	2.16	39.5	45.2

6 Return Period and Average Recurrence Interval

Gumbel (1958) apparently was the first to use the term 'return period' as the reciprocal of the probability of exceedance of an extreme in a certain time period or 'epoch', such as an annual maximum wind speed. However, the term 'average recurrence interval' (A.R.I.) denotes the expected time interval between exceedances of a defined gust speed, and hence differs from return period. Denoting here the return period by 'R_P' and the average recurrence interval by 'R_I', Wang and Holmes (2018) showed that the relationship between the two is given by:

$$\frac{1}{R_P} = 1 - \exp\left(-\frac{1}{R_I}\right) \tag{6}$$

noting that Wang and Holmes (2018) also used the term 'mean inter-arrival time' for R_I.

Figure 8 shows Eq. (6) plotted, from which it is clear that the return period, R_P, and the average recurrence interval become virtually identical for values of both greater than about 10. However, R_P cannot take a value less than 1, whereas this restriction does not apply to the A.R.I. With this in mind it is unfortunate that some codes and standards occasionally refer to a '1-year return period' (e.g. ISO 10137:2007, Annex D); this clearly contravenes Gumbel's original definition of R_P.

The method based on the peaks-over-threshold approach discussed in Sect. 5.2, using Eq. (5), can be shown to give a distribution function for wind speed as a function of average recurrence interval, R_I, with the form of Eq. (3), in which:

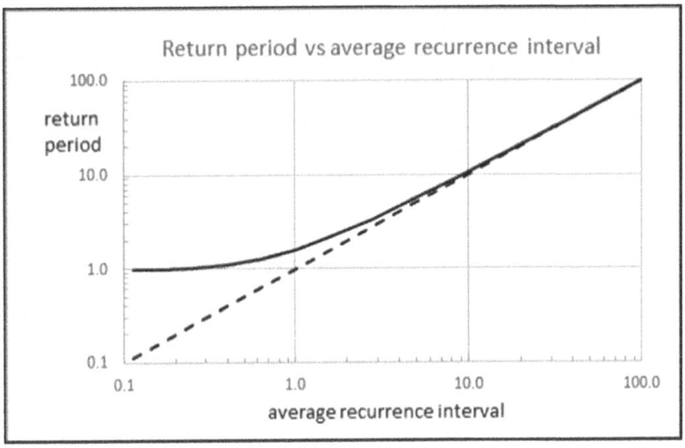

Fig. 8. Relationship between 'return period' and 'average recurrence interval' (from Wang and Holmes 2018)

$$C = V_0 + \frac{\sigma}{k} = V_0 + \left(\frac{1+k}{k}\right)E, \quad \text{and} \quad D = \left(\frac{1+k}{k}\right)E.\lambda^{-k} \tag{7}$$

However, direct fitting of the GEV will give the return period, R_p. in the form of:

$$\frac{1}{R_P} = 1 - \exp\left\{-\left[\frac{C - V_R}{D}\right]^{1/k}\right\} \tag{8}$$

It can be noted that Eq. (8) approaches the inverse of Eq. (3) at high values of R_P.

$$\text{i.e.} \quad \frac{1}{R_P} \rightarrow \left[\frac{C - V_R}{D}\right]^{1/k}$$

The difference in the two methods is illustrated by Fig. 9, which shows the distributions with C = 67 m/s, D = 41 m/s, and k = 0.1, (i.e. the values in Eq. (2)) plotted together.

The GPD/PoT approach, which allows average recurrence intervals less than 1 year was used to fit the wind data from the South Australian stations described in Sect. 5.3.

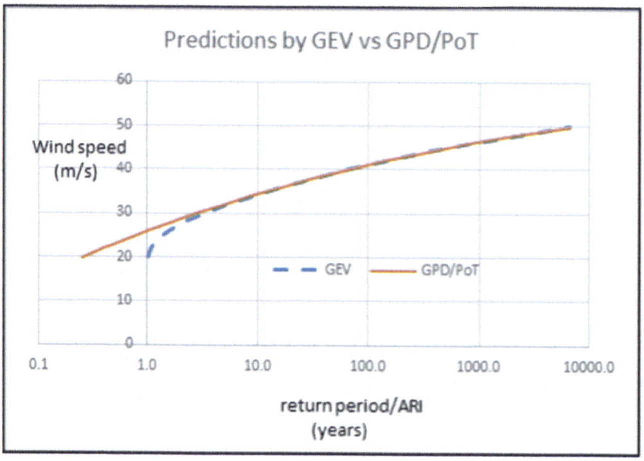

Fig. 9. Comparison of extreme wind distributions fitted by two methods

7 Conclusions

In the present paper, the extreme wind climate of Australia is discussed with a description of the available historical data recorded by the Bureau of Meteorology. The necessity of correction of gusts for gust duration, and for terrain, and for separating the data into storm type is outlined. The 'bootstrapping' technique for assessing sampling errors and confidence limits is also discussed.

The Generalized Extreme Value (GEV) and peaks-over-threshold method based on the General Pareto Distributions (GPD/PoT) are outlined, with a simple fitting method for the latter when the shape factor is fixed, were described. The difference between 'return period' and 'average recurrence interval' is also explained.

As examples of the methods discussed, recent data from three locations in South Australia were processed. These illustrate the mixed climate near the southern coastline, and the dominance of non-synoptic downdraft winds at a northern station far from the coast.

Acknowledgements. Some of the work described in this paper was partly funded by ElectraNet Pty. Ltd, Adelaide, South Australia. This work was carried out in collaboration with Stephen Oliver of Weather Solutions Pty. Ltd. and Chi-Hsiang Wang of CSIRO, who contributed several of the ideas outlined in this paper.

References

Davison AC, Smith RL (1990) Models for exceedances over high thresholds. J Roy Stat Soc 52:339–442

De Gaetano P, Repetto MP, Repetto T, Solari G (2014) Separation and classification of extreme wind events from anemometric records. J Wind Eng Ind Aerodyn 126:132–143

Gumbel EJ (1958) Statistics of extremes. Columbia University Press, New York

Holmes JD, Ginger JD (2012) The gust wind speed duration in AS/NZS 1170.2. Aust J Struct Eng 13:207–217

Holmes JD, Moriarty WW (1999) Application of the generalized Pareto distribution to extreme value analysis in wind engineering. J Wind Eng Ind Aerodyn 83:1–10

International Organisation for Standardization, ISO (2007) Bases for design of structures – serviceability of buildings and walkways against vibrations. International Standard ISO 10137:2007, ISO, Geneva, Switzerland

Naess A, Clausen PH (2001) Combination of peaks-over-threshold and bootstrapping methods for extreme value prediction. Struct Saf 23:315–330

Standards Australia (1989) SAA loading code. Part 2: wind loads. Australian Standard AS 1170.2-1989, Standards Australia, Homebush, New South Wales

Standards Australia (2002) Structural design actions. Part 2: wind actions. Australian/ New Zealand Standard AS/NZS 1170.2:2002, Standards Australia, Sydney, New South Wales

Standards Australia (2011) Structural design actions. Part 2: wind actions. Australian/ New Zealand Standard AS/NZS 1170.2:2011, Standards Australia, Sydney, New South Wales

Wang C-H, Holmes JD (2018) The equivalence of generalized extreme value distribution and generalized Pareto distribution for extreme gust analysis. In: 19th Australasian Wind Engineering Society Workshop, 4–6 April, Torquay, Victoria, Australia

Life-Cycle Cost-Based Wind Design of Tall Buildings

L. Ierimonti[1(✉)], I. Venanzi[1], L. Caracoglia[2], and A. L. Materazzi[1]

[1] Department of Civil and Environmental Engineering,
University of Perugia, Perugia, Italy
{laura.ierimonti,ilaria.venanzi,materazzi}@unipg.it,
ierimonti@strutture.unipg.it

[2] Department of Civil and Environmental Engineering,
Northeastern University, Boston, MA, USA
lucac@coe.neu.edu

Abstract. An automated and computationally cost-effective procedure, designated Life-Cycle Cost Wind Design (LCCWD) and recently proposed by the authors, is presented for the design of wind-excited tall buildings. The LCCWD is based on the evaluation of the life-cycle cost associated with wind-induced damage through the PEER equation. The "best" design configuration is obtained by comparing the lifetime cost of different design alternatives. All the main peculiarities involved in the design of high-rise buildings can be included: uncertainties in the wind load characterization and in the damage estimation, wind directionality effects, modeling of the response (modal superposition accounting for power-law function mode shapes, torsional effects, etc.), comfort issues. Moreover, the long-term monetary benefits associated with the installation of a control system can be considered. The present work is aimed at enhancing the investigation of the potentialities of the LCCWD procedure by including direct and indirect losses, different fragility/cost models and various design strategies.

Keywords: Life-cycle cost · Tall buildings · Non-structural damage · Business downtime

1 Introduction

Life Cycle Cost Analysis (LCCA) is devoted to quantifying the monetary losses expected in structures during their lifetime (Wen and Kang 2001; Cui and Caracoglia 2015; Venanzi et al. 2018). In the framework of the LCCA, the LCCWD (Life Cycle Cost Wind Design) procedure has been recently proposed by the authors (Ierimonti et al. 2017a, b, 2018), which is specifically devoted to wind-excited tall buildings. By exploiting the PEER integral equation considering suitable fragility models, it relates the probability of exceeding specific damage states to the repair costs needed to restore the structure to its original functions after a strong wind event. In non-hurricane prone regions, the main monetary losses can be related to the failure of drift-dependent and

F. Ricciardelli and A. M. Avossa (Eds.): INVENTO 2018, LNCE 27, pp. 376–386, 2019.
https://doi.org/10.1007/978-3-030-12815-9_30

acceleration-dependent nonstructural elements and to the temporary business downtime due to occupants' discomfort.

2 The LCCWD Procedure

The LCCWD accounts for various design aspects:

I. The structural system, configuration of the structural elements, the materials and the geometry of the building.
II. The building's orientation, considering wind exposure and geographical location.
III. The structural control system, which is often employed in tall buildings to reduce wind-induced c.
IV. The typology, number and location of nonstructural elements, which can be considered as integrated elements of the design process as they significantly influence the life-cycle costs.

The outline of the LCCWD procedure (Fig. 1), developed for each design alternative of steps I–IV, can be summarized as follows:

1. Select the structural model and, if applicable, the control system (*cs*) model (e.g. tuned-mass damper - TMD, etc.).
2. Identify the wind hazard model: wind tunnel tests, mean-wind speed annual maxima and direction-dependent joint probability distribution $f(V_{ref,}\theta + \delta)$, where wind speed V_{ref} is evaluated at a reference height (building top floor); θ is the mean-wind incidence angle and δ is the building orientation.
3. Evaluate structural response through engineering demand parameters (***EDP***) inducing damage.
4. Select the damage states related to: direct losses associated with nonstructural components; indirect losses related to occupant's discomfort (business downtime). Hence, evaluate the corresponding complementary cumulative distribution function (CCDF), i.e. the fragility function $P[\boldsymbol{DS}_j\,|\boldsymbol{EDP}(V_{ref},\theta,cs)]$, also accounting for the presence of the control system, where $\boldsymbol{DS}_j$ represents the jth selected damage state.
5. Compute the annual damage probability in a generic direction of the building ($h = x,\ y$) as a function of the building orientation δ and floor level (height z).

$$P^h_{j,cs}(\delta, z) = \int\int\int P\left[\boldsymbol{DS}_j|\boldsymbol{EDP}\left(V_{ref}, \theta, cs\right)\right] f\left[\boldsymbol{EDP}\left(V_{ref}, \theta, cs\right)\right] f\left(V_{ref}, \theta + \delta\right) \mathrm{d}\boldsymbol{EDP}\mathrm{d}V_{ref}\mathrm{d}\theta \quad (1)$$

6. Calculate the expected life-cycle cost ($E[\bullet]$) by considering: the loading occurrence (index l); the total number of loading occurrences (L) between time 0 and time t (in

years); each jth damage state; the deterministic cost C_j associated with the jth damage state; the yearly discount rate (λ):

$$E[C_{cs}(t)] = C_{0,s} + C_{0,cs} + E\left[\sum_{i=1}^{n_H}\sum_{h=x,y}\sum_{l=1}^{L}\sum_{j=1}^{K}\tilde{n}_j^h(z_i)C_j(z_i)e^{-\lambda t_l}P_{j,cs}^h(\delta, z_i)\right] \quad (2)$$

where $\tilde{n}$ is the number of damage-sensitive elements at the generic floor (height z); i is the reference index of each floor, i.e. first floor z_1 = 3 m and roof-top z_{nH} = 180 m; $C_{0,s}$ and $C_{0,cs}$ are the initial construction cost and the installation cost of the control system, respectively. By normalizing with respect to $C_{0,s}$, Eq. (2) becomes:

$$c_{cs}(t,\delta) = c_{o,cs} + E\left[\sum_{i=1}^{n_H}\sum_{h=x,y}\sum_{l=1}^{L}\sum_{j=1}^{K}\tilde{n}_j^h(z_i)c_j(z_i)e^{-\lambda t_l}\left[-\frac{1}{vt}\log\left(1 - P_{tj,cs}^h(\delta, z_i)\right)\right]\right] \quad (3)$$

7. Repeat steps 1–6 to compare different design solutions.

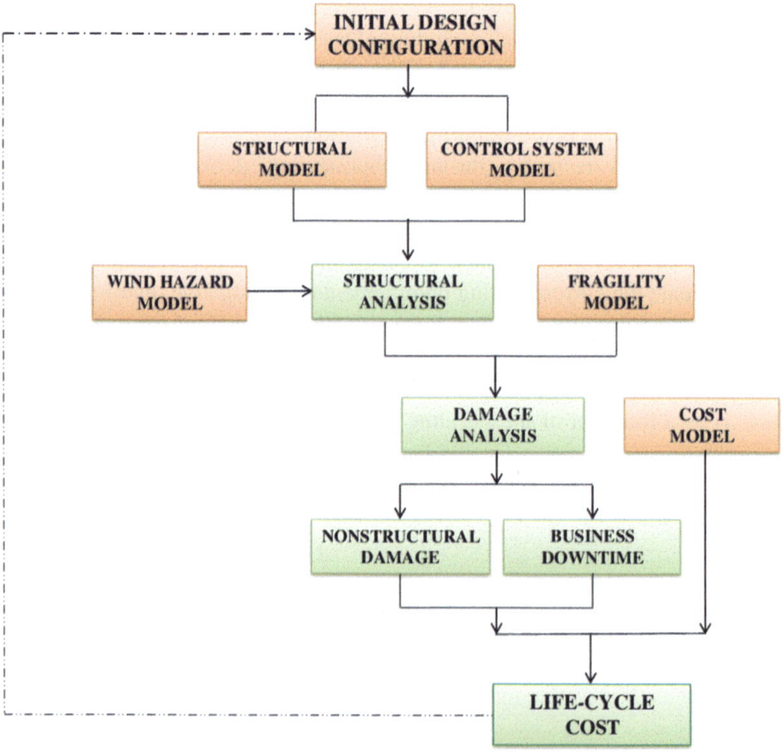

Fig. 1. Outline of the LCCWD procedure.

3 The Case Study

The efficiency of the LCCWD approach is demonstrated by making use of a case study, i.e. a 180-m high tall steel building, for which wind tunnel load data are available. Wind speed and direction data for a hypothetical example site location (Boston, Massachusetts, USA) are derived from full-scale records and the literature. The structural analysis is carried out in the frequency domain and considers modal superposition with power-law function mode shapes and torsional response effects (Ierimonti et al. 2017a, b, 2018).

A passive control system is installed. It is composed of two unidirectional TMDs, located at the elastic center (also coincident with the mass center) of the building roof top. The parameters of the TMDs are tuned to control the response of the first two lateral modes according to Warburton equations (Warburton 1982).

With the main objective of comparing different design alternatives, various distributions of indoor spaces are considered, in relation to their designated use. Four working groups, characterized by a specific use, are examined (Fig. 2): office area with a uniform distribution of partition walls (G0), sport area or relax area (G1) with separate rooms (changing rooms), meeting/conference area (open space) (G2) with a minimal number of private rooms, management office area (G3) developed with head offices distributed along the perimeter and employees offices clustered in the building core.

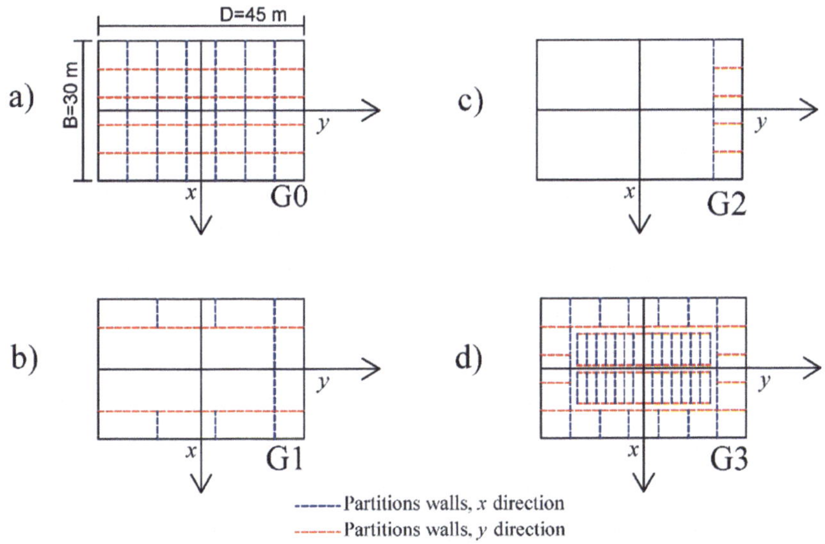

Fig. 2. Plan view with selected locations of the partition walls related to the G0 group (a); G1 group (b); G2 group (c); G3 group (d).

The following design scenarios are analyzed and compared: CI, where G0 configuration is considered on each floor; CA, with G1 group on floors 9, 19, 29, 39, 49,

59, G2 group on floors 10, 20, 30, 40, 50, 60 and G3 group is considered on the remaining floors; CB, where G1 group is considered on floors 20 to 22, 55 to 57, G2 group on floors 23 to 25, 58 to 60 and G3 group on the remaining floors.

3.1 Fragility Model

For nonstructural damage estimation, damage to partition walls (drift-sensitive) is considered, as a simulation example; the corresponding empirical fragility curve (Fig. 3b) is reconstructed from literature (FEMA-P-58 2012). Without loss of generality, other nonstructural components, limit-state and damage scenarios could be included in the analysis.

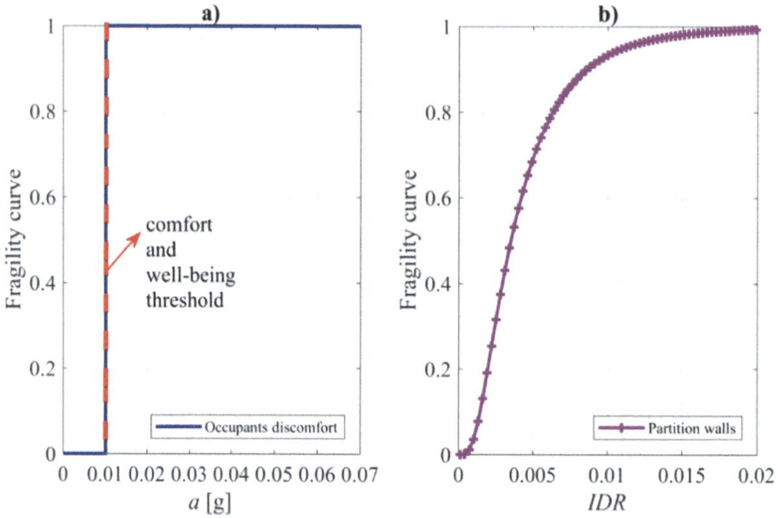

Fig. 3. Fragility curves for: (a) occupants discomfort as a function of peak acceleration *a*; (b) partition walls as a function of *IDR (inter-story drift ratio).*

The LCCWD approach also allows to quantify the annual damage probability associated with a revenue loss due to business downtime through Eq. (1), with the introduction of an appropriate fragility function. In the absence of more detailed information, a Heaviside step fragility function is chosen (Fig. 3a), centered on the common accepted average value of the perceptibility threshold, i.e. 10 milli-*g* (Burton et al. 2015; Griffis 1993; Chang 1973) and based on peak acceleration *a*: "damage" i.e. business downtime is experienced when the threshold *a* is exceeded.

Since partition walls fragility curve refers to standard unit elements, the number of units in the *x* and *y* directions are presented in Table 1 for each working group.

Table 1. Nonstructural components – number of reference-element units

Group	$\tilde{n}_j^x(z)$	$\tilde{n}_j^y(z)$
G0	3	3
G1	1.8	3
G2	1	0.9
G3	10	7.8

3.2 Cost Model

According to (Ierimonti et al. 2017a, b, 2018), the initial construction cost $C_{0,s}$ is assumed as a percentage of the structural steel mass. Similarly, the initial installation cost of the TMD system $C_{0,cs}$ is assumed as a fraction of the initial construction cost $C_{0,s}$. Since the real cost of the TMDs is not known a priori, the percentage value used as the initial TMD cost ($c_{0,cs} = C_{0,cs}/C_{0,s}$) is assumed as 0.01 through a preliminary sensitivity analysis, by imposing that the initial installation cost of the TMD is recovered about 10 years after the initial construction time (at $t = 0$).

Indirect losses (cost values) associated with business downtime are evaluated according to information provided in the "Information Technology Intelligence Consulting" (ITIC 2016), which is an online research database with information on a wide variety of technological topics, including unplanned downtime and their impact on business operation.

According to Eq. (3) the normalized costs needed for repairing the partition walls and for one hour of business downtime are reported in Table 2 for each group. The latter values conform to the hypothesis that the arrival time of a windstorm causes one hour of business interruption. Without loss of generality, since the perceptibility threshold can vary on the basis of the designated use of each floor (at height z), different values are assumed per working group unit.

Table 2. Normalized downtime costs c_j (per working group unit)

	c_j [10^{-4}]	
	Partition walls	Business downtime
G0	2.3	2
G1	2.3	1
G2	2.3	1
G3	2.3	2

4 Numerical Results

This section reports the numerical results obtained from the application of LCCWD procedure to the case study presented in Sect. 3.

Figure 4 shows the annual damage probabilities for partition walls along the x principal horizontal direction and for business downtime, as a function of the height above ground z (i.e. the floors) for systems with and without TMDs. The building orientation is set to $\theta + \delta = 0°$ (wind that blows from the North). Reductions of about 50% are noticeable at the top floor for damage to partition walls and of about 80% for acceleration-induced business downtime. Figure 5 illustrates the total normalized lifetime expected cost as a function of the floor number for scenarios CI, CA and CB. The life-cycle cost is separately evaluated for the systems with and without TMDs. As expected the presence of the TMDs reduces considerably the costs and minor differences along the height of the building can be noted for the various working groups.

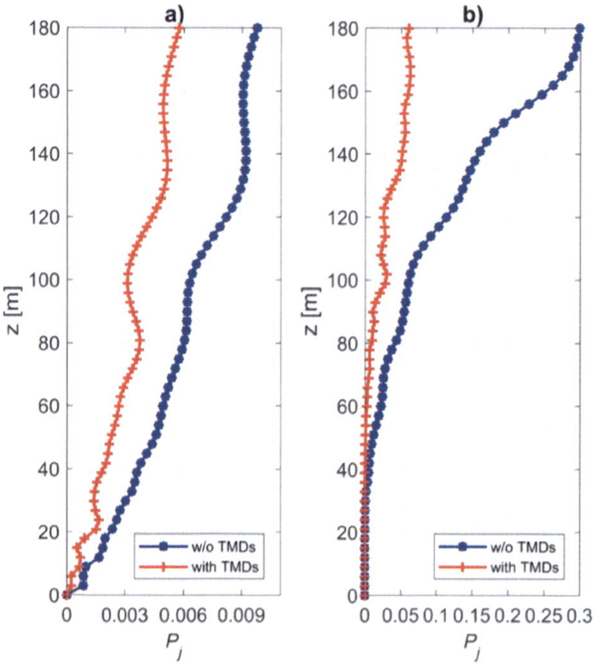

Fig. 4. Annual damage probability as a function of the floor height (z varying between first floor z_1 = 3 m and roof-top z_{nH}= 180 m): (a) partition walls (x direction); (b) business downtime.

Subsequently, the expected normalized life-cycle cost is evaluated over time t and results are reported in Fig. 6. The scenarios CI, CA and CB are analyzed and compared for the configurations with and without TMDs. The point of intersection between the controlled and the uncontrolled system is designated as Break Even Time (BET), defined as the time after which the initial investment costs associated to the installation of the TMDs are completely recovered (Ierimonti et al. 2018). From the inspection of the figure and consistently with the results obtained in Fig. 5, it can be noted that, despite a more marked difference between the three configurations (CI, CA, CB) in the uncontrolled case, the presence of the TMD reduces the gap between the design configurations. On the one hand, this observation confirms that the correct choice of damage-sensitive nonstructural elements, such as partition walls, and of the designated

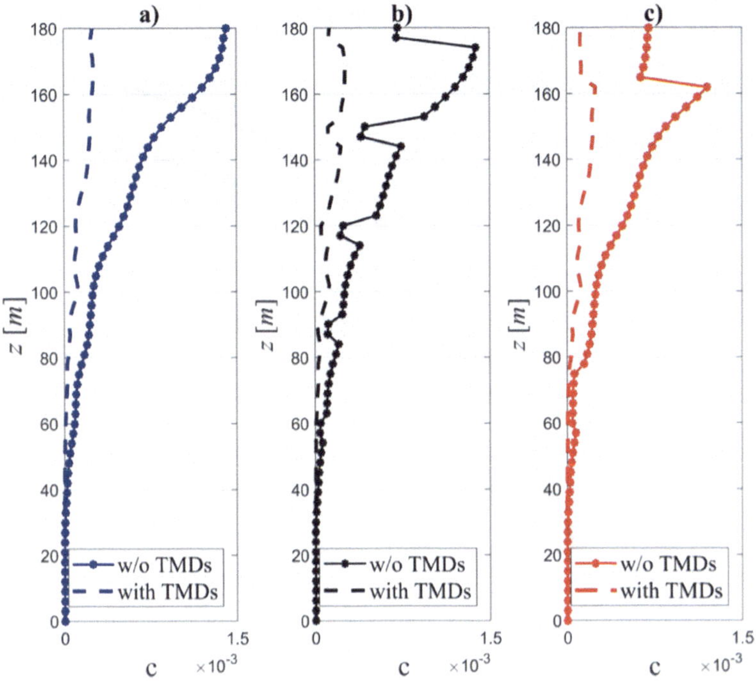

Fig. 5. Total normalized lifetime expected intervention costs (t = 100 years) as a function of the floor height (z varying between first floor z_1 = 3 m and roof-top z_{nH}= 180 m), examining design scenarios CI (a), CA (b) and CB (c) for controlled configuration (with TMDs) in comparison with the controlled configuration (w/o TMDs).

use of each floor can significantly influence the final design for the uncontrolled case. On the other hand, the presence of the TMD, reduces the differences between the design scenarios (CI, CA, CB).

Moreover, Fig. 5 also reveals lower values of the cost over time for configuration CB since areas (G1, G2), less sensitive to acceleration, are located at the top floor. By comparing the solutions with and without TMD installation, a decrease of the cost over time corresponds to a slight increase of the "break-even time" (BET; Ierimonti et al. 2018). This result confirms the fact that the cost of the TMD is an important aspect in the total life-cycle cost assessment; therefore, it should be considered at the early stages of the design.

4.1 Parametric Study on the Fragility Model for Business Downtime

As introduced in Sect. 3.1, the fragility curve representing the business downtime due to occupants' discomfort is assumed as a Heaviside step function as a first approximation. In order to relax this hypothesis and study the influence of different fragility models, a possibly more general lognormal distribution model is introduced with median value centered at a = 0.01 g and standard deviation σ parametrically varying

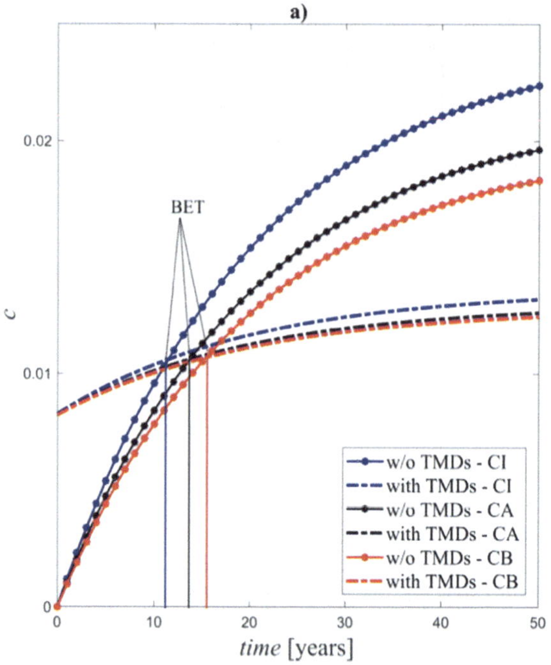

Fig. 6. Comparison between the expected values of the total investment cost for CI, CA and CB and indication of BET (break-even time).

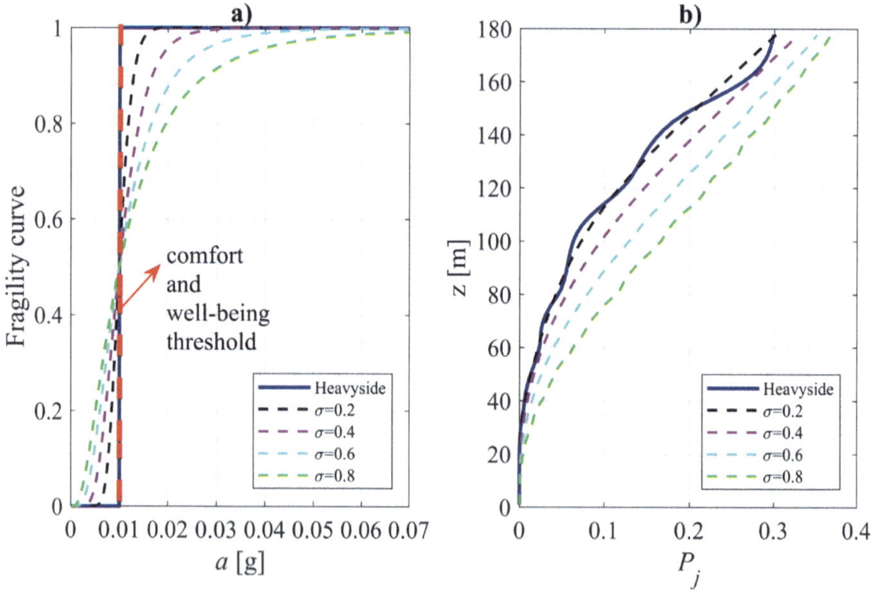

Fig. 7. (a) Business downtime fragility curves as a function of σ; (b) Annual damage probability due to occupants' discomfort as a function of σ and floor height (z varying between first floor z_1 = 3 m and $z = z_{nH}$ = 180 m) - CI configuration.

., Abdelrazaq A (2015) Wind-induced motion of tall buildings: designing for
rt. Int J High-Rise Build 4(1):1–8
Human response to motions in tall buildings. J Struct Div 499(6):1259–1272
L (2015) Simulation and analysis of intervention costs due to wind-induced
building. Eng Struct 87:183–197
Performance assessment calculation tool, vol 1. http://www.fema.gov/media-
cuments/90380
rviceability limit states under wind load. Eng J/Am Inst Steel Constr 30:1–16
glia L, Venanzi I, Materazzi AL (2017a) Investigation on life-cycle damage
ited tall buildings considering directionality effects. J Wind Eng Ind Aerodyn

oglia L, Venanzi I (2017b) Life-cycle loss estimation in tall buildings
wind directionality effects. In: 7th European and African conference on wind
CWE 2017
nzi I, Caracoglia L (2018) Life-cycle damage-based cost analysis of tall
ped with tuned mass dampers. J Wind Eng Ind Aerodyn 176:54–64
of Hourly Downtime Soars: 81% of Enterprises Say it Exceeds $ 300K on
tic-corp.com/blog/2016/08/cost-of-hourly-downtime-soars-81-of-enterprises-
00k-on-average/
), Ierimonti L, Fabrizi S (2018) Multi-hazard loss analysis of tall buildings
seismic loads, Structure and Infrastructure Engineering (in press)
2) Optimum absorber parameters for various combination of response and
eters. Earthquake Eng Struct Dynam 79:381–401
(2001) Minimum building life-cycle cost design criteria. I: methodology.
CE 127:330–337

from 0.2 g to 0.8 g with discrete increments
presents the annual damage probability du
configuration CI, calculated for each fragility
observed that higher values of σ, i.e. larger un
probabilities. This effect is more marked for
hypothesis of low uncertainty, the step functi

5 Conclusions

In this paper an automated life-cycle cost-
specific for wind-excited tall buildings, is pr
direct losses, like those associated with w
indirect losses, like those associated with bu
comfort. Alternative design configurations ar
and compared for the solutions with and w
building, equipped with two unidirectional T
the LCCWD approach. The main novel contr
that it is possible to independently examine
non-structural (or structural) configurations b
costs due to direct and indirect sources of l
lution over time confirms the importance
components in terms of element strength
number, and of the location on the floor, espe
results showed that the presence of a TMD
ment costs, reduces the dependence on th
numerical applications confirm that the LC
"best" configuration among different scenari
approach and can be employed for different
for designers and stakeholders.

Acknowledgements. This research and the stud
University in 2016 have been supported by the Uni
of the International PhD program between the
Braunschweig. Luca Caracoglia would like to ack
Foundation (NSF) of the United States, CAREER A
partial support of NSF Award CMMI-1434880. Lu
of the University of Perugia, mobility program f
torale D.R. 2244, 2014). Any opinions, findings a
of the authors and do not necessarily reflect the vi

References

Burton M, Kwok
occupant com
Chang F-K (1973
Cui W, Caracogli
damage on tal
FEMA-P-58 (201
library/assets/d
Griffis L (1993) S
Ierimonti L, Cara
cost of wind-e
171:207–2018
Ierimonti L, Car
accounting for
engineering, E
Ierimonti L, Ven
buildings equi
ITIC (2016) Cost
Average. http:
say-it-exceeds-
Venanzi I, Lavan
under wind an
Warburton G (19
excitation para
Wen Y, Kang Y
J Struct Eng A

Motion-Dependent Forces on Streamlined Bridge Girders and Their Influencing Parameters – Observations from Wind Tunnel Buffeting Response Data

J. B. Jakobsen(✉)

Department of Structural and Mechanical Engineering and Material Sciences, University of Stavanger, 4036 Stavanger, Norway
jasna.b.jakobsen@uis.no

Abstract. Aerodynamic stability of long-span bridges is most readily examined in the frequency domain, with self-excited forces modelled as linearized functions of the bridge velocities and displacements. The present paper briefly recalls the experimental approaches to characterize the motion-dependent forces, and further explores the validity of superposition of the linearized load components, in presence of various influencing parameters. Flutter derivatives obtained with section models of two streamlined bridge girders in ambient vibrations are revisited. Emphasis is placed on the influence of the twisting to vertical frequency ratio and the type of the participating degrees of freedom on the identified wind-structure interaction. A novel aspect of the analysis of the buffeting response data is also attempted, in order to isolate the significance of the twisting amplitude on the self-excited wind forces in ambient vibrations.

Keywords: Flutter derivatives · Wind tunnel investigation · Section model · System identification · Response amplitude · Frequency ratio · Buffeting response

1 Introduction

1.1 Formulation of the Motion-Induced Wind Forces

Aerodynamic stability is one of the integral aspects of the design of long-span bridges. It is commonly evaluated by modelling the self-excited wind forces in terms of the flutter derivatives as a link between the bridge deck motion and the resulting wind loading. Motion-dependent lift and overturning moment, $L_{ae}(t)$ and $M_{ae}(t)$, are e.g. formulated as functions of the vertical and torsional displacements, $r_z(t)$ and $\varphi(t)$, and the associated velocities according to the following expressions:

$$L_{ae}(t) = \frac{1}{2}\rho U^2 B\left[kH_1^*\frac{\dot{r}_z(t)}{U} + kH_2^*\frac{B\dot{\varphi}(t)}{U} + k^2 H_3^*\varphi(t) + k^2 H_4^*\frac{r_z(t)}{B}\right] \tag{1}$$

F. Ricciardelli and A. M. Avossa (Eds.): INVENTO 2018, LNCE 27, pp. 387–401, 2019.
https://doi.org/10.1007/978-3-030-12815-9_31

$$M_{ae}(t) = \frac{1}{2}\rho U^2 B^2 \left[kA_1^* \frac{\dot{r}_z(t)}{U} + kA_2^* \frac{B\dot{\varphi}(t)}{U} + k^2 A_3^* \varphi(t) + k^2 A_4^* \frac{r_z(t)}{B} \right], \quad (2)$$

where the air density, the mean wind speed and the bridge deck width are denoted ρ, U and B, respectively. k represents the reduced oscillation frequency ω, $k = B/U = 2\pi/U_{red}$, directly proportional to the number of oscillation cycles during the flow passage over the bridge deck width B. Flutter derivatives H_i^* and A_i^* (i = 1, 2, 3, 4) are commonly presented as functions of the reduced mean wind speed, $U_{red} = U/(f{\cdot}B)$, which, similarly to the reduced frequency, captures the level of the flow unsteadiness.

The equations above are a slightly modified and extended version of the formulation of self-excited forces introduced by Scanlan and Sabzevari (1969). An augmented set of three equations, considering also the dynamic equilibrium in the horizontal direction and the related deck-flow interactions, has also been introduced, e.g. in Scanlan (1992). The present paper deals with the streamlined bridge girders, the stability of which is governed by the vertical and the twisting response, so the following analysis is concentrated on the force components defined in the above two equations.

1.2 Experimental Characterisation of Flutter Derivatives

Since the wind-structure interaction in case of bluff bodies such as bridge decks cannot be derived from the potential flow theory, the unsteady aeroelastic forces are quantified experimentally, with bridge deck section models. The overall division of the testing techniques can be made into (i) the forced vibration tests and (ii) the tests with the freely responding models. The first approach involves measurement of the loads imposed by the prescribed, controlled bridge deck motion. In its basic version, the tests are performed with separate, pure twisting and translational harmonic oscillations. In each test, the lift and the overturning moment in phase and out of the phase with respect to the controlling motion are determined. The conditions are different from the state of coupled flutter, in which a unified twisting-vertical motion at a single frequency becomes effectively undamped. The application of the loads from the separate tests to predict flutter then rests on the assumption of the superposition of the motion-induced force components, associated with the small amplitude motion. The forcing devices of different levels of complexity have been developed over the years, e.g. by Ukeguchi et al. (1966), Diana et al. (2004), Neuhaus et al. (2009) and Siedziako et al. (2017). Wind loads induced by a more general forced motion have also been investigated, providing more information on the non-linear motion-induced force components and their relative importance. In a study concerning a variable range of motion amplitudes and frequency components for a bridge deck with the width to height ratio of B/H = 11, Wang (2015) observed significant non-linear force components associated with the combined vertical and twisting motion at two different frequencies, when the twisting amplitude was as large as 10°. For the rotational amplitude of 5°, only limited non-linear effects were found. Surface pressure measurements recorded during the forced vibration tests have also played an important role in understanding of the flutter mechanism and various types of the non-linearities involved.

Starting from 1969, Scanlan (Scanlan and Subzevari (1969)) advocated the use of inexpensive tests with an elastically suspended section model in transient vibrations. Since only response and no excitation is recorded in such a test, the motional derivatives are determined by one or other type of the system identification method of the bridge-wind system. Scanlan's early wind tunnel tests with freely responding section model in smooth flow considered first the model in a single degree of freedom in-wind decaying motion, and thereafter the vertical and torsional response at a unified, single frequency, at the onset of the coupled flutter. Relevant range of the reduced wind velocity was then covered by varying the ratio between the twisting and vertical frequency. The investigation also concerned the flutter velocity as a function of the torsional amplitude.

Later on, the response data of a section model with a single, fixed ratio of twisting to vertical frequency were utilized for determination of the entire set of flutter derivatives, what necessitated a modal interpretation of the Eqs. (1) and (2), with reference to the two degrees of freedom involved. In such tests, both the vertical and torsional in-wind frequencies are used when calculating the reduced frequency and wind velocity, depending on which response component the flutter derivative in question is associated with. In Eqs. (1) and (2), this corresponds to reduced frequencies k_z and k_φ in front of the derivatives associated with the vertical and the torsional motion respectively. One example of such modelling of the section model aeroelastic system was developed by Poulsen et al. (1992), and used to compare the flutter derivatives obtained with different twisting to vertical frequency ratios in smooth flow. A robust system identification method in which all eight derivatives are determined simultaneously, also applicable to the transient response in turbulent flow, was developed by Sarkar et al. (1994), and later by Chowdhury et al. (2003).

The modeling of the motion-dependent forces accounting for two degrees of freedom, illustrates the extended application of the flutter derivatives, not only for the prediction of the aerodynamic stability, but also for an improved buffeting response estimation e.g. at the design wind speed. Flamand and de al Foye (2000) pointed out the complementary role of the forced and the free vibration methods, and Sarkar et al. (2009) investigated the consistency of the derivatives obtained by the two methods applied in a single wind tunnel as well as in two different laboratories.

The modal analysis of the section model transient response data in terms of flutter derivatives paved the way for the estimation of the flutter derivatives from an even more general type of response, such as buffeting response data of a section model. The response is forced by a sustained action of turbulence and controlled by the model dynamic characteristics and the wind-structure interaction. Similarly to the transient response data, the buffeting response data blend the effects of the (i) mean twisting angle, (ii) non-linear structural characteristics, (iii) vibration amplitude, (iii) frequency ratio between the modes studied (iv) turbulence intensity and length scales etc.

1.3 Scope of the Present Analysis

As mentioned above, several works have examined possible significance of different parameters relevant to the wind-structure interaction, using the forced vibration tests as well as the free decay tests. To the author's knowledge, buffeting response data which reflect the most complete exposure of a section model to the different influencing factors, have not been studied in this respect so far. In the following, flutter derivatives of the streamlined bridge girders from two wind tunnel studies with section models are compiled in order to assess the effect of the frequency ratio and the degrees of freedom involved on the wind-structure interaction. When doing so, the effect of the mean twisting angle is considered to be negligible, the non-linear structural characteristics are accounted for and the average response amplitudes are considered as small. Similarly, the three-dimensional turbulent flow is assumed to alter the motion-induced forces to a limited degree. The influence of the increasing ambient vibration amplitude on the motional derivatives at the onset of the flutter is examined separately, in a tailored analysis of the buffeting response data. Along with the presentation and discussion of the experimental data at hand, the method applied for the identification of flutter derivatives is briefly recalled.

2 Wind Tunnel Tests

2.1 Bridge Deck Cross-Sections

Cross-sections of the investigated two bridge girders are illustrated in Fig. 1. Section A represents the bridge deck of a network arch bridge with the main span of 220 m. The tie is formed as a 7.2 m wide concrete slab, with 0.4 m high edge beams accommodating prestressing cables. The second bridge deck (Section B) is a stiffening steel box girder of a suspension bridge with a main span of 1310 m. At the lower box corners, guide vanes are introduced to mitigate vortex shedding excitation, together with a spoiler at the center of the lower deck surface. The results presented in Sects. 3.2 and 3.3. concern the downstream orientation of the walkway.

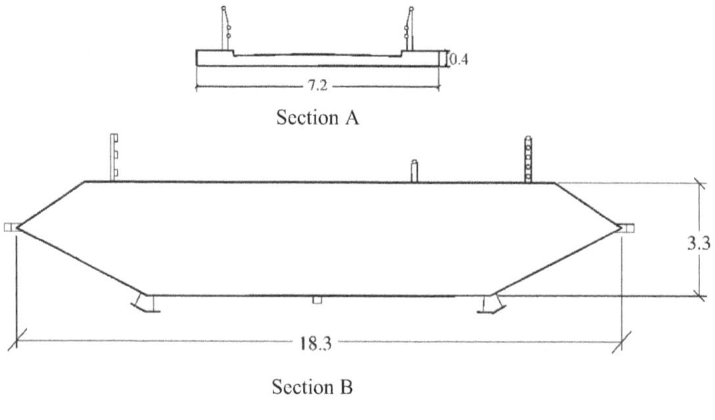

Fig. 1. Cross-section of two bridge girders investigated in the present paper. Full-scale dimensions, in meters.

The models were built in the 1:20 and 1:50 scales, for the Section A and B respectively. The static wind force coefficients determined in turbulent flow and their derivatives with respect to the angle of attack, at zero mean angle, are presented in Table 1. While the lift and the moment coefficients are expressed with reference to the deck width, the drag coefficient refers to the deck height.

Table 1. Static force coefficients for cross-sections in Fig. 1

Section	C_D	C_L	C_M	$dC_L/d\varphi$	$dC_M/d\varphi$
B	2.777	−0.190	0.021	5.27	0.66
A	0.704	−0.251	0.007	3.38	1.15

2.2 Dynamic Bridge Characteristics

The mass and the mass moment of inertia of the prototype bridges provided as basis for the wind tunnel tests are listed in Table 2, together with the degrees of freedom and the frequency ratios adopted in the wind tunnel test series. The model dynamic characteristics were arranged in a traditional section model setup, in which the length- and the wind velocity scaling governed the frequency scaling. Further details of the two investigations are provided in the wind tunnel test reports by Hansen et al. (2007, 2009).

Table 2. Bridge characteristics replicated in the section model setups for buffeting response and flutter investigations (ref Fig. 1)

Section	Degrees of freedom	Frequency ratio	Mass[kg/m]	Mass mom. of inertia [kgm^2/m]
A	V + T	2.11, 0.92	6600	29000
B	V + T	2.1, 2.5, 4.2	12900	427200
B	V + T + H	2.5, 4.7	12900	427200
B	T + H	1.5	12900	427200

For the narrow arch bridge (section A) the lowest twisting eigen-frequency ($f_{\varphi 0}$ = 0.53 Hz in full-scale) is smaller than the lowest vertical frequency (f_{z0} = 0.83 Hz). Since a coupled flutter instability cannot develop from the pair of the two modes, the main tests series was concentrated on the second twisting and the first vertical mode for which the frequency ratio was $f_{\varphi 02}/f_{z0} > 2.11$. Additionally, dynamic tests with a frequency ratio $f_{\varphi 0}/f_{z0} < 1$ were also carried out, to observe the torsional response with the lowest mode. Due to a limited relevance of the interaction between the two response components in this case, for which torsional divergence is a critical stability problem, no attempt was made to reproduce the exact ratio between the lowest torsional and the vertical modes, and the vertical response was in general considered of the secondary interest. Nonetheless the response data are very interesting in the present context, in which the superposition of the motion-induced wind load components can be examined in a case when the resulting aeroelastic response is different in nature from that developing into a coupled flutter.

For the suspension bridge deck (Section B), the initial test series was devoted to the lowest symmetric vertical and torsional modes which were fundamental for the (three-mode) coupled flutter, see e.g. Jakobsen and Hjorth-Hansen (2007). The ratio of the two eigen-frequencies was $f_{\varphi 0}/f_{z0} = 2.5$ ($f_{\varphi 0} = 0.359$ Hz and $f_{z0} = 0.142$ Hz). Several other frequency ratios were also investigated. The second round of tests also included flexibility in the horizontal directions, with an eigen-frequency 1.5 times lower than the torsional eigen-frequency. For practical reasons, the full-scale frequency ratio of seven was not reproduced at model scale, in the limited investigation involving the horizontal dynamic response.

2.3 Flow Conditions

The wind tunnel experiments with both bridge deck models were performed in the boundary layer wind tunnel (BLWT), with a 1.75 m wide and 1.55 high working cross-section, at Svend Ole Hansen Aps in Copenhagen. The along-wind turbulence intensity of 12–13% and 10% in the vertical direction were generated using three spires at the inlet of the cross-section together with a controlled, fluctuating fan velocity. The spires and the fan speed variations generated the high and the low frequency turbulence components respectively. The turbulence frequency content was biased towards the high frequencies compared to the targeted full-scale spectra, with less discrepancies between the two for the vertical turbulence.

For section B, motional derivatives estimated in the above described wind conditions are also compared to the values determined in a BLWT with a low turbulence intensity, at the Ruhr University of Bochum. The free flow along-wind turbulence intensity was about 3%, in the 1.8 m wide and 1.6 m high wind tunnel working section. Both wind tunnels are of the open circuit type.

2.4 Data Acquisition and the System Identification Analysis

The displacement data were deduced from the force signals transmitted by the load cells at the fixed end of the supporting springs. The time-series, acquired with the sampling frequency of 500 Hz, were first low-pass filtered and then resampled, so that the analyzed data included about forty records per twisting vibration cycle.

At high wind speeds, it was often found that slowly varying, quasi-static vertical response became significant, partly because of a high aerodynamic damping of the response at the vertical still-air frequency. A high-pass filter was then applied in order to isolate the resonant part of the buffeting response.

Structural damping varied in between the model setups and was of the order of 1% logarithmic damping. The dynamic model rig provided an amplitude dependent structural damping, what was accounted for when separating the motion-dependent forces from the pure structural dynamic characteristics. An equivalent amplitude equal to $\sqrt{2}\sigma_R$, where σ_R is the standard deviation of the (vertical or twisting) buffeting displacement response, was used to look up a structural damping ratio consistent with the magnitude of the buffeting response excited at each wind speed. Separate free decay tests in still-air were performed prior to the ambient vibration tests, in order to map the amplitude dependent structural damping. The decaying response time-histories were analysed in segments of ten or more oscillation cycles. The exponentially decaying envelopes were fitted to the local displacement maxima and the damping ratio then established as a function of the average local amplitude. The segment-wise analysis of the still-air free decay response also provided the amplitude dependent eigen-frequencies. The torsional eigen-frequency had a higher relative reduction at large amplitudes compared to the vertical one, but much smaller than the frequency changes caused by overturning moment induced by twisting. Both eigen-frequencies were thus treated as constant. The main step in the analysis of the displacement data is computation of the associated cross-covariance (coefficient) matrix for a selected number of time-lags. (Both are illustrated in Fig. 4.) For a matrix $\mathbf{y}$ containing the vertical and the twisting displacement data in two columns, the unbiased cross-covariance matrix is calculated as:

$$C_{yy}(j) = \frac{1}{N-j}\sum_{i=1}^{N-j} \mathbf{y}(i+j)\,\mathbf{y}^T(i),\ j = 0,\ 1,\ 2\ldots,\ l, \tag{3}$$

where j is number of sampling intervals for the discrete time delay $j\Delta t$, N is number of samples in the time-series, and i is counting index. The cross-covariance matrix reflects the in-wind structural properties in a manner similar to transient response data. The underlying aeroelastic system characteristics are identified from a singular value decomposition of the covariance block Hankel matrix, as described in Bogunovic Jakobsen and Hjorth-Hansen (1995).

3 Flutter Derivatives Estimated with Different Model Setups

3.1 Flutter Derivatives for Section a Model with Different Frequency Ratios

Motional derivatives for section model A estimated for two torsional to frequencies ratios are depicted in Fig. 2. The lower panels display the response standard deviations used to estimate the associated structural damping. For $f_{\varphi 0}/f_{z0} = 2.11$, the derivatives calculated from the two consecutive test series at a single mean wind velocity are presented with different markers. Each test series concerns an interval corresponding to 10 min period in full-scale. The difference between the two results is limited and reflects the random nature of the buffeting forces and the associated aeroelastic random response. The limited variations in the derivatives are plausible in the sense that smaller response magnitudes translate into slightly higher aerodynamic damping terms.

The model setup with the torsional frequency below the vertical eigen-frequency provides the flutter derivatives over an extended range of reduced wind velocities. The flutter derivatives in the present case follow the same trends as in the first case, however the two resulting aeroelastic responses have very different characteristics. With $f_{\varphi 0}/f_{z0} = 0.92$, a coupling between the torsional and the vertical mode does take the place, but the associated damping is in fact higher than that of the pure vertical mode. As already mentioned, the torsional divergence is the critical stability problem in this case. It can also be noted that the difference in structural damping in the two cases (about 2% logarithmic damping in the second case and half of that for $f_{\varphi0}/f_{z0} = 2.11$) also contributes to the difference in the dynamic response included in the lower part of the figure. However, the motion-induced forces are calculated by accounting for the difference.

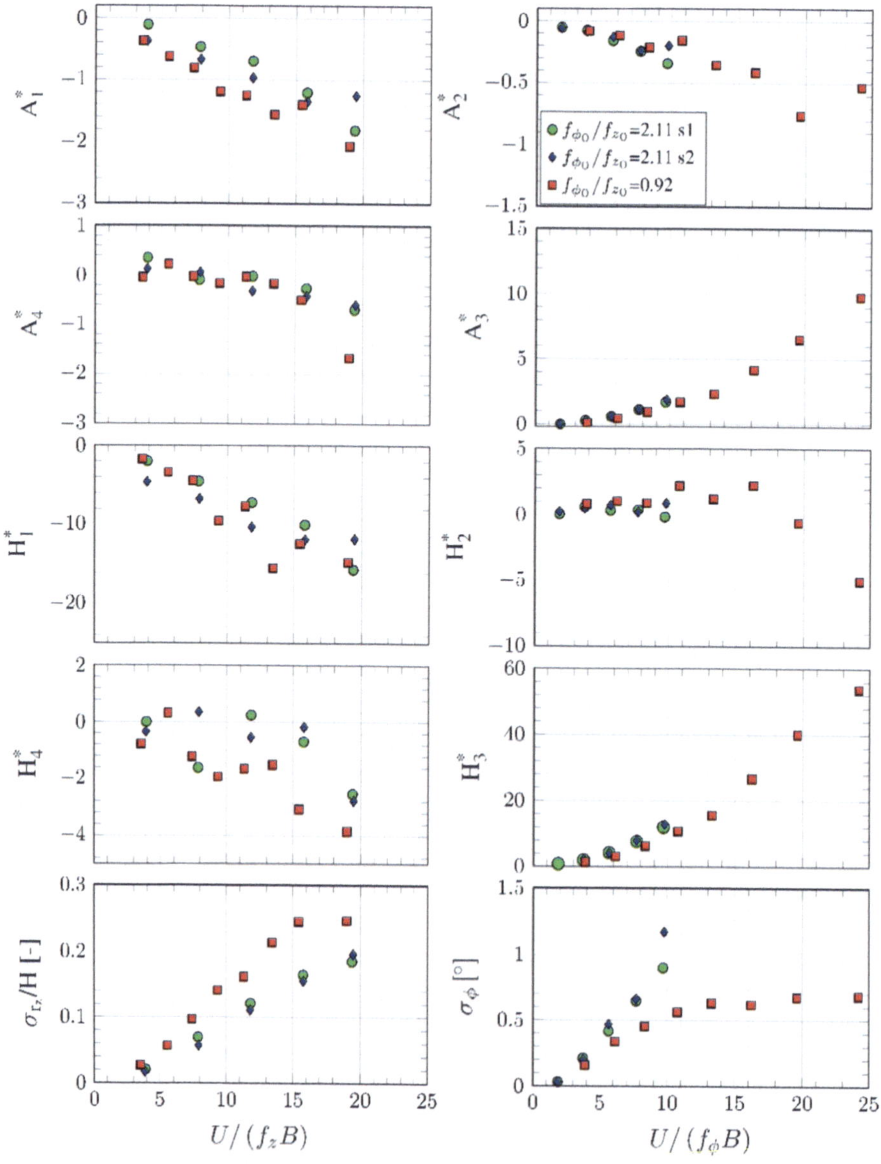

Fig. 2. Flutter derivatives for the bridge deck A for different frequency ratios.

To the author's knowledge, the motion-dependent forces "at work" in case of the twisting frequency lower than the vertical one have not been documented previously.

3.2 Flutter Derivatives for Section B Model with Different Frequency Ratios and Degrees of Freedom

In Fig. 3, blue squares represent the derivatives determined from the ambient vibration response of the model with the still-air frequency ratio $f_{\varphi 0}/f_{z0} = 2.5$. At a reduced frequency larger than $U/(f_{\varphi}B) = 10$, a large increase in torsional response shown in the lower part of the figure, indicates proximity to coupled flutter.

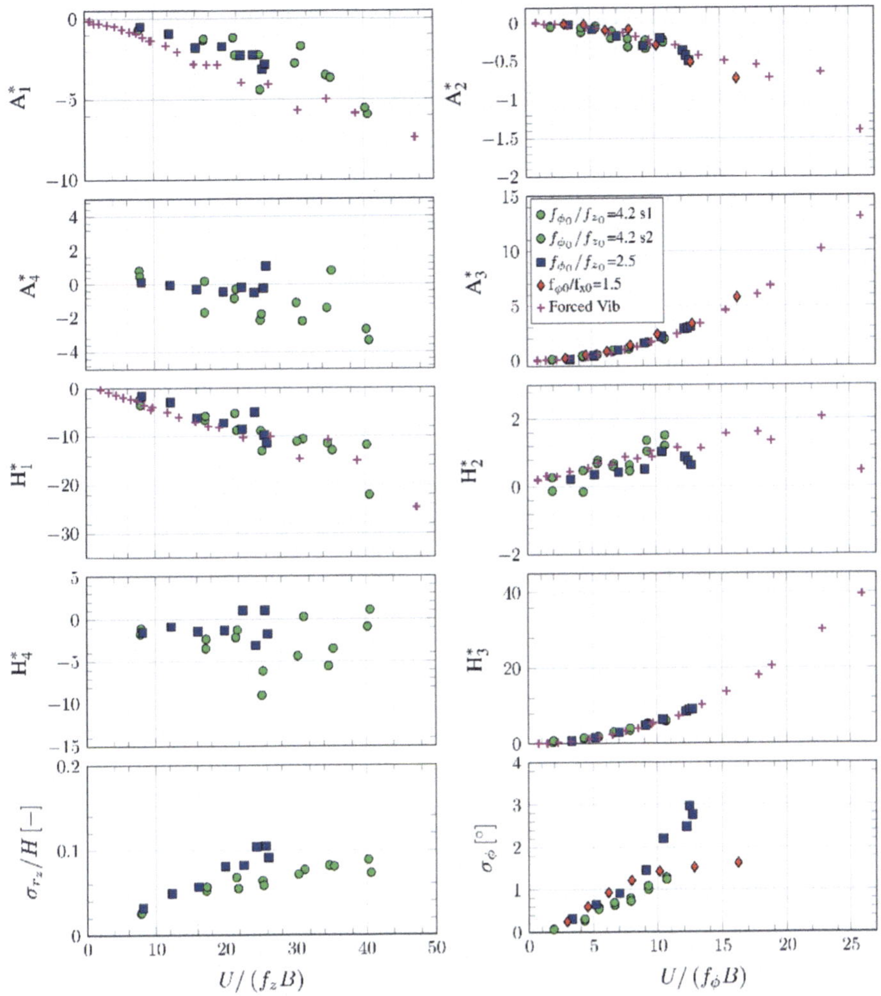

Fig. 3. Flutter derivatives for the bridge deck B for different frequency ratios, degrees of freedom and the test method

A consistent set of the motional derivatives, marked by green circles in the figure, is obtained with a higher frequency ratio of $f_{\varphi 0}/f_{z0} = 4.2$. The results calculated from the two consecutive runs at each wind speed are displayed using a single symbol, to limit the number of different markers in a rather busy figure.

The agreement between the derivatives determined with the two different frequency ratios is particularly good for those quantifying the most important motion-dependent load components, i.e. H_3^*, A_3^*, A_2^*, H_1^*, H_2^* and A_1^*.

The third case concerns the model response confined to the horizontal and torsional motion. The resulting derivatives associated with the rotational degree of freedom in these tests are included in Fig. 3 by red diamonds. Also this data align well with the aeroelastic interaction parameters identified in the two previous cases for Section B.

3.3 Comparison Between the Flutter Derivatives from the Buffeting Response and Forced Vibration Tests

The last set of flutter derivatives displayed in Fig. 3 stem from the forced vibration measurements undertaken at the Ruhr University of Bochum in collaboration with the University of Stavanger. The mechanism forcing the model into the harmonic oscillations was developed in Bochum, as described by Neuhaus et al. (2009). A new model was built for the forced vibration tests, also this time in the 1:50 scale. The main part of the tests concerned pure vertical, horisontal and twisting harmonic oscillations at a fixed frequency, with the vertical amplitude set to 6% of the bridge deck height the rotational one of 1°. The derivatives were calculated in a frequency domain analysis, from the real and the imaginary parts of the cross-spectra between the lift and moment and the underlying displacements, and corrected for the still-air "offsets". The entire set of 18 flutter derivatives is reported in Mikkelsen (2013) and the related rational functions discussed in e.g. Mikkelsen and Jakobsen (2017). A_4^* and H_4^* derivatives, which represent very small forces, had unexpected positive values from the forced vibration tests. The values most likely reflect practical difficulties in the estimation methodology rather than a genuine model-wind interaction and are omitted from the present comparison of the two load components of the secondary importance.

Considering the very different nature of the forced vibration and the buffeting vibration tests, the agreement between the motional derivatives estimated from the two test types can be characterised as striking, in particular concerning the A_3^* and H_3^* terms, which represent the dominant self-excited forces. Some discrepancy between the A_1^* and H_1^* derivatives can be observed. This is most likely due to the challenges in isolating the vertical resonant buffeting response from the significant background response, particularly with increase in wind speed. Possible differences in the self-excited loads in the low and the high turbulence intensity flows can neither be ruled out.

Lee and Kwon (2011) observed a large impact of the actively generated wind tunnel turbulence on the A_1^* derivative in case of a twin-box girder. In such a geometry, turbulence is expected to impact the wake of the upstream bridge box, in which the downstream box is immersed, and thus be more influential than in the present case.

In the present study, the vertical turbulence intensity is as high as 10% and the span-wise length scale $L_w \sim 0.5B$. Although occasionally and locally the model encounters large angles of attack due to turbulence, the span-wise integrated effect of the fluctuating flow field is expected to be more limited compared to e.g. a corresponding action of a monochromatic and/or fully correlated turbulence along the model span.

3.4 Wind-Structure Interaction as Function of the Buffeting Response Amplitudes

The results in Figs. 2 and 3 corroborate the superposition of the linearized force component in Eqs. (1) and (2) and the modal analysis of wind-induced bridge deck vibrations, on which the identification of the motion-dependent force components is founded. This indirectly substantiates a corresponding modal analysis of a prototype structure, which represents a simple and an efficient tool to assess the aerodynamic stability and the buffeting response. The consistency between the results based on the tests with different frequency ratios graphically illustrates the overriding role of the cross-sectional geometry on the unsteady aerodynamics of a structure in small motion.

Large twisting response amplitudes are known to significantly alter the flow-structure interaction and can in that sense also become a governing parameter. The influence of the torsional amplitude on the aerodynamic stability has been looked at in a series of studies, e.g. by Selberg (1961), Scanlan and Sabzevari (1969), Selberg and Hjorth-Hansen (1977), Falco et al. (1992), Noda et al. (2003), Diana et al. (2010) and Wang (2015). Noda et al. (2003) analysed surface pressure fluctuations on the flat plate like sections during a rotational oscillation cycle, similarly to the aerodynamic investigation of different generic bridge deck sections carried out by Matsumoto et al. (1994). For a rectangular cross-section with the width to depth ratio of B/H = 13, a sign change of the A_2^* derivative, from negative to positive, was detected at the reduced velocity of about $U/(f\varphi B) = 12.5$, for a twisting amplitude of 6°. For higher amplitudes and reduced velocities, a destabilizing overturning moment was shown to originate from a lift resultant occasionally crossing over to the downstream half width of the section, a feature of torsional flutter.

In the following, an attempt is made to isolate the shear effect of the vibration amplitude, by analysing amplitude dependent segments of the buffeting response time-histories. Figure 4 presents the response time-histories recorded with the section B, split into different segments, following a division of the local torsional maxima into three classes. The effective angles of attack associated with the largest vertical model velocities are considerably smaller, so the magnitude of the twisting amplitude is considered to control the flow-structure interaction and the division made accordingly.

For the pair of the vertical and torsional displacements in each different amplitude class, cross-covariance coefficients are calculated, with the remaining data values set equal to zero. Great care was taken to adapt the unbiased cross-covariance coefficient estimate to account for a single class at the time. In order to increase the number of samples in each of the classes, the three consecutive runs at a single mean wind speed, which were available for the upstream orientation of the walkway of Section B, are analysed as a single realization. In Fig. 4, the fluctuating part of the three runs, each lasting for 100.8 s after various data processing steps, are for the illustration purposes presented as a continuous record. The cross-covariance estimates are computed after

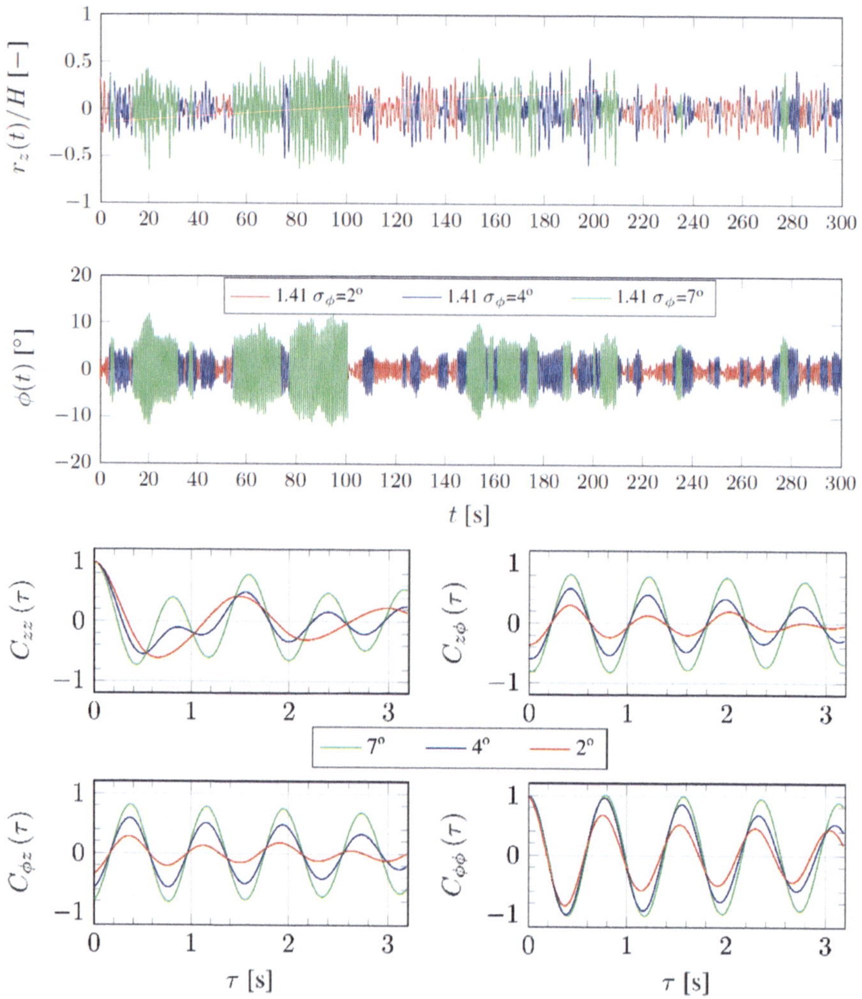

Fig. 4. Time-series of the buffeting displacement response of section model B with $f_{\varphi 0}$ = 1.590 Hz and f_{z0} = 0.637 Hz, i.e. frequency ratio $f_{\varphi 0}/f_{z0}$ = 2.5, at $U/(f_\varphi B)$ = 13.5 (two upper figure panels) and the associated cross-covariance coefficients (lowest two panel rows) separated for three different twisting amplitude levels.

inserting a sufficient number of data points with a zero value in between the test runs, in order to avoid erroneous correlation "overlap" of the non-synchronous records. The zero values (not visualized in Fig. 4) are added in a time-window equal to or longer than the largest time-lag for which the covariance function is computed, e.g. 3.2 s in the present example.

The three covariance estimates for time lags covering more than four twisting cycles are displayed in the lower part of Fig. 4. The three sets of functions indicate quite different average characteristics of the wind-structure system. A decreasing damping of the torsional response is clearly manifested with increasing amplitudes. It relates to a pure twisting damping corresponding to A_2^* values of 0.63, 0.52 and 0.38 for small to large amplitudes. The large amplitude response is also associated with a relatively higher participation of the coupled torsional-vertical mode, as visualized by the auto-covariance of the vertical displacement as well as the cross-correlation functions $C_{z\varphi}(\tau)$ and $C_{\varphi z}(\tau)$. A softening of the system in terms of the increasing twisting vibration periods for large amplitudes can be noted and is largely due to the pure structural non-linearity.

The analysis presented herein captures the essence of the motion-dependent forces at different amplitudes, but represents a preliminary result only. The approach needs to be further scrutinized and a larger data set used before a stringent set of flutter derivatives as a function of vibration amplitude can be presented.

4 Conclusions

Buffeting response data of two streamlined bridge deck section models are revisited with emphasis on the motion-dependent forces. Covariance based system identification method is used to determine the flutter derivatives associated with the model responses with different twisting to bending frequency ratios and degrees of freedom. The method builds on a state-space formulation of the equation of motions in the modal format. The focus is thus on the integrated, effective quantities rather than a precise description of forces, localized in space and time.

For the two bridge geometries, motional derivatives identified with the model setups with the twisting to bending eigen-frequency ratios different by a factor of two, are found to be consistent.

For the bridge deck with the width to depth ratio of B/H = 13, two cases of profoundly different aeroelastic responses were investigated – one with the twisting frequency higher than the vertical one, and another one with the opposite frequency ratio. Both types of the resulting vibrations were demonstrated to originate from the same motion-induced load parameters. The analysis provides a graphic example of the governing aerodynamic "signature" of the bridge deck cross-sectional geometry.

For the steel box girder with B/H = 5.5, a very good agreement is also found between the wind load components induced by twisting in two different model rigs, one enabling the vertical and the twisting response and the second one the horizontal and the twisting response.

For the same cross-section, the key derivatives determined from the ambient vibration tests and the separate forced vibration tests in the low turbulence flow were also in agreement, with slightly lower A_1^* and H_1^* magnitudes in high turbulent flow.

The above comparisons corroborate the validity of the adopted modal analysis of the two degree of freedom section model in ambient vibrations, and the complementary aerodynamic response analysis of a model and a prototype structure.

The superposition principle of the linearized load components holds for small response amplitudes, the assumption which is violated close to at the onset of the coupled flutter in turbulent flow. A preliminary analysis of the self-excited forces at different average twisting amplitudes in such conditions provides a new evidence of the expected destabilizing effects of large twisting motion in ambient vibrations.

Acknowledgements. The author is grateful to a number of colleagues who developed and conducted the wind tunnel studies revisited in this work: dr. Svend Ole Hansen and his wind tunnel team, Prof. Emeritus Erik Hjorth-Hansen from the Norwegian University of Science and Technology, The Norwegian Public Administration represented by dr. Bjørn Isaksen, Prof. Rudiger Höffer and dr. Christian Neuhaus from the Ruhr University of Bochum, and Ass. Prof. Ove Mikkelsen from the University of Stavanger. Assistance of dr. Etienne Cheynet from the University of Stavanger in the manuscript preparation is also greatly appreciated.

References

Bogunović Jakobsen J, Hjorth-Hansen E (1995) Determination of the aerodynamic derivatives by a system identification method. J Wind Eng Ind Aerodyn 57:295–305

Chowdhury AG, Sarkar PP (2003) A new technique for identification of eighteen flutter derivatives using a three-degree-of-freedom section model. Eng Struct 25:1763–1772

Diana G, Resta F, Zasso A, Belloli M, Rocchi D (2004) Forced motion and free motion aeroelastic tests on a new concept dynamometric section model of the Messina suspension bridge. J Wind Eng Ind Aerodyn 92:441–462

Diana G, Rocchi D, Argentini T, Muggiasca S (2010) Aerodynamic instability of a bridge deck section model: linear and nonlinear approach to force modeling. J Wind Eng Ind Aerodyn 98 (6–7):363–374

Falco M, Curami A, Zasso A (1992) Nonlinear effects in sectional model aeroelastic parameters identification. J Wind Eng Ind Aerodyn 42:1321–1332

Flamand O, de la Foye A (2000) Aerodynamic derivatives of three bridge decks measured by the forced oscillation technique. In: Proceedings of 3rd European & African conference on wind engineering, Eindhoven, 2–6 July, pp 605–610

Hansen SO, Lollesgaard M, Rex S, Jakobsen JB, Hjorth-Hansen E (2009) The Hardanger bridge: static and dynamic wind tunnel tests with a section model. Prepared for Norwegian Public Roads Administration, Revision 2, Svend Ole Hansen ApS (rev 0 in 2006)

Hansen SO, Lollesgaard M, Rex S, Jakobsen JB, Hjorth-Hansen E (2007). The Brandangersundet bridge: static and dynamic wind tunnel tests with a section model. Prepared for Norwegian Public Roads Administration, Revision 1, Svend Ole Hansen ApS

Jakobsen JB, Hjorth-Hansen E (2007) Arne Selberg's formula for flutter in light of multi-modal flutter analysis. In: Proceedings of 12th international conference on wind engineering, Cairns, Australia, 1–6 July, pp 135–142

Lee S, Kwon SD (2011) Effects of turbulence properties on bridge aerodynamics. In: Proceedings of the 13th international conference on wind engineering, Amsterdam, July 10–15

Matsumoto M, Kobayashi Y, Niihara Y, Shirato H (1994) Flutter mechanism and its stabilization of bluff bodies. In: Wind engineering: retrospect and prospect, proceedings of the ninth international conference on wind engineering, January 1995. John Wiley Eastern, New Delhi

Mikkelsen O (2013) Modelling of wind loads and wind-induced response of a long-span bridge in time-domain. PhD thesis, University of Stavanger

Mikkelsen O, Jakobsen JB (2017) Aeroelastic response from indicial functions with a finite element model of a suspension bridge. In: IOP conference series: materials science and engineering, vol 276, no 1, pp 1–18

Neuhaus CH, Roesler S, Höffer R, Hortmanns M, Zahlten W (2009) Identification of 18 flutter derivatives by forced vibration tests – a new experimental rig. In: Proceedings of the 5th european conference on wind engineering, 19–23 July. Firenze University Press, Firenze, pp 361–364
Noda M, Utsunomiya M, Nagao F, Kanda M, Shiraishi N (2003) Effects of oscillation amplitude on aerodynamic derivatives. J Wind Eng Ind Aerodyn 91(1–2):101–111
Sarkar PP, Jones NP, Scanlan RH (1994) Identification of aeroelastic parameters of flexible bridges. J Eng Mech 120(8):1719–1741
Sarkar PP, Caracoglia L, Haan FL, Sato H, Murakoshi J (2009) Comparative and sensitivity study of flutter derivatives of selected bridge deck sections, part 1: analysis of inter-laboratory experimental data. Eng Struct 31(1):158–169
Scanlan RH, Sabzevari A (1969) Experimental aerodynamic coefficients in the analytical study of suspension bridge flutter. J Mech Eng Sci 11(3):234–242
Scanlan RH (1992) Wind dynamics of long-span bridges. In: Larsen A (ed) Aerodynamics of large bridges: proceedings of the first international symposium on aerodynamics of large bridges, Copenhagen, 19–21 February. Balkema
Selberg A (1961) Oscillation and aerodynamic stability of suspension bridges. Acta Poytech Scand 13
Selberg A, Hjorth-Hansen E (1977) The fate of flat plate aerodynamics in the world of bridge decks. In: Proceedings of the theodorsen colloquim. Universitetsforlaget, Oslo
Siedziako B, Øiseth O, Rønnquist A (2017) An enhanced forced vibration rig for wind tunnel testing of bridge deck section models in arbitrary motion. J Wind Eng Ind Aerodyn 64: 152–163
Poulsen NK, Damsgaard A, Reinhold TA (1992) Determination of flutter derivatives for the great belt bridge. J Wind Eng Ind Aerodyn 41(1–3):153–164
Ukeguchi N, Sakata H, Nishitani H (1966) An investigation of aeroelastic stability of suspension bridges. In: Proceedings from the international symposium on suspension bridges, pp 79–100
Wang Q (2015) Study on nonlinear motion-induced aerodynamic force of streamline box girder under different harmonic motions using forced vibration test (2015). In: 14th international conference on wind engineering, Porto Alegre, June 21–26. Conference presentation and personal communication

Uncertainty Quantification for RANS Predictions of Wind Loads on Buildings

G. Lamberti(✉) and C. Gorlé

Department of Civil and Environmental Engineering, Stanford University,
473 Via Ortega, Stanford, CA 94305, USA
{giacomol,gorle}@stanford.edu

Abstract. Computational fluid dynamics simulations to calculate wind pressure loads on buildings can be strongly influenced by uncertainty in the inflow boundary conditions and the turbulence model. In the present work we investigate these uncertainties in Reynolds-averaged Navier-Stokes predictions for wind pressure coefficients of a high-rise building, and compare the results to wind tunnel measurements. The uncertainty in the inflow boundary condition is characterized using three uncertain parameters, the reference velocity, roughness length and model orientation, and propagated to the quantities of interest using a non-intrusive polynomial chaos expansion approach. The results indicate that the uncertainty in the inflow conditions is non negligible, but insufficient to explain the discrepancy with the wind tunnel data, in particular where flow separation occurs. The uncertainty related to the turbulence model is investigated by introducing perturbations in the Reynolds stress tensor. The results confirm that the turbulence model form uncertainty is dominant near the separation region that forms downstream of the windward building edge.

Keywords: Computational Fluid-Dynamics (CFD) · Atmospheric Boundary Layer (ABL) · Wind loading · Uncertainty Quantification (UQ)

1 Introduction

Computational fluid dynamics (CFD) simulations can provide a powerful tool to quantify pressure loads on buildings and assess wind hazards. However, CFD simulations of the atmospheric boundary layer (ABL) can be strongly influenced by uncertainty in the inflow conditions and turbulence model. Uncertainty quantification (UQ) methods can propagate these uncertainties to the quantities of interest to provide predictions for the wind pressure coefficients in terms of mean values and confidence intervals. This presents significantly more meaningful information to designers and engineers than the outcome of a single deterministic CFD simulation.

F. Ricciardelli and A. M. Avossa (Eds.): INVENTO 2018, LNCE 27, pp. 402–412, 2019.
https://doi.org/10.1007/978-3-030-12815-9_32

In the present work we quantify uncertainty in the time-averaged pressure coefficients predicted by RANS simulations that reproduce a wind tunnel experiment of a high-rise building. The experiment was originally performed at the wind tunnel of the Politecnico di Milano (PoliMi) [1], and subsequently repeated at the Wall of Wind (WoW) facility of Florida International University. The objective is to predict the distribution of the time-averaged pressure coefficient on the buildings lateral facade with 95% confidence intervals, and compare the results to the experimental data. In this paper, we focus on the PoliMi experiments, and consider uncertainty in the inflow boundary conditions and the turbulence model separately.

The uncertainty in the inflow condition is characterized using 3 uncertain parameters: the reference wind speed, the atmospheric boundary layer (ABL) roughness length, and the model orientation. Probabilities for these parameters are determined using information from the wind tunnel experiment, and propagated to the quantities of interest using a polynomial chaos expansion (PCE) approach [2]. The results of the inflow UQ study will be presented in terms of the predicted mean and 95% confidence intervals for the time-averaged pressure coefficients. To quantify turbulence model uncertainties, we employ a method that introduces perturbations in the eigenvalues of the Reynolds stress tensor [3,4]. The selected perturbations consistently increase or decrease the production of turbulence kinetic energy, and have been shown to predict an interval that plausibly contains the truth solution, i.e. the solution that would be obtained in absence of turbulence model uncertainty, in a variety of flow fields. The results for the 95% confidence intervals from the inflow UQ study and the plausible intervals from the turbulence model form UQ study will be compared to draw conclusions on the relative importance of these uncertainties when predicting mean pressure coefficients on buildings.

In the following sections we present more details on the test case and the wind tunnel measurements, the methodology used for the CFD UQ analysis, and the results. The final section presents the conclusions and future work.

2 Wind Tunnel Measurements

Several experiments on a high-rise building model were carried out in the ABL wind tunnel of the Polytechnic University of Milan [1]. The facility is a closed-circuit wind tunnel with a 35 m long, 14 m wide and 4 m high test section. The model was placed on a turntable with a radius of 6.5 m, located at 10 m from the inlet. The goal of the experiments was to measure wind pressure on the building's lateral facade, for an exposure representative of suburban terrain and different wind directions. In the following we briefly summarize the velocity measurements performed to characterize the wind tunnel ABL, and the wind pressure measurements on the building. For additional details regarding the experiments the reader is referred to [1].

2.1 ABL Characterization

3D hot-wire measurements were performed on a plane perpendicular to the flow direction at the center of the turntable, i.e. at the future location of the building model. In particular, 20 s time series of the three components of velocity were recorded at 5 spanwise locations and 56 vertical locations (Fig. 1) with a sampling frequency of 2000 Hz. The 5 spanwise locations are 0.6 m distant, while the resolution in the vertical direction is 43.7 mm below 0.75 m and 87.5 mm above.

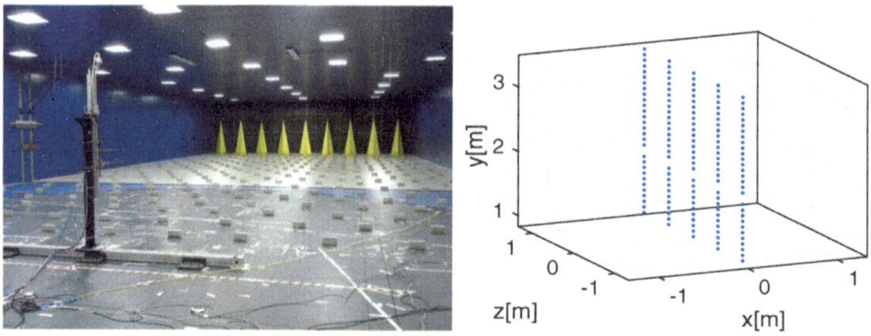

Fig. 1. Setup of the experiment to characterize the ABL in the wind tunnel (left); coordinates of the hot-wire measurements (right).

2.2 Pressure Measurements

The model of the high-rise building, shown in Fig. 2, is a 1 m long, 0.3 m wide and 2 m high rectangular box. The model was equipped with 4 aluminum tiles containing 223 pressure taps each, with a minimum spacing of 3 mm. The configuration was designed to allow a very detailed study of the pressure distribution in the regions of the building's lateral facade where the largest negative pressure peaks are expected, namely near the corners and edges. Figure 2 shows a detailed view of the pressure tap distribution on the tile positioned at the top corner of

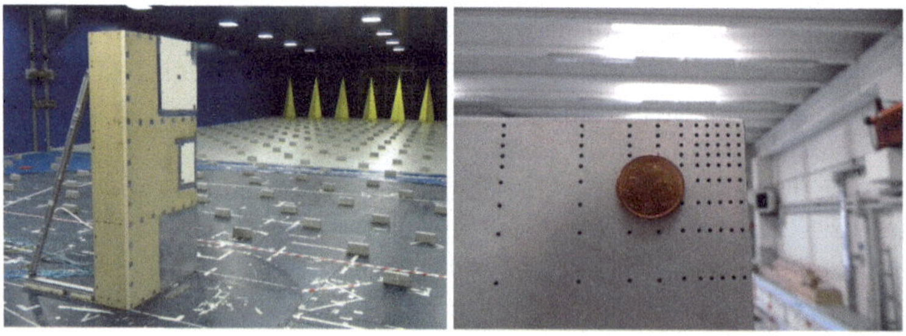

Fig. 2. Setup of the test on the high-rise building model (left); pressure tap distribution on the top-corner tile (right)

the building. The outcome of the test consists of 300 s time-series of pressure measurements in the 4 tiles (i.e. 892 pressure taps), sampled at a rate of 500 Hz.

3 Methods

In this section, we first describe the set-up of the RANS simulations. Subsequently, we introduce the methodology used to quantify the inflow uncertainty, focusing on the definition of the uncertain inflow parameters, and on the propagation method. Finally, we summarize the method used to quantify the turbulence model form uncertainty.

3.1 CFD Model Set-Up

We use the OpenFOAM v4.0 simpleFOAM solver to perform RANS simulations of the flow around the high-rise building model. The computational domain has the same dimensions as the test section of the wind tunnel. The mesh includes refinement regions close to the building model, resulting in a total number of 3.7 million hexahedral cells and an average y^+ of 500 on the building surface. We employ the RNG $k-\varepsilon$ model together with a standard log law wall function on the building surface. To properly represent a neutral ABL, we impose a logarithmic velocity profile at the inlet [5]:

$$U = \frac{u_\star}{\kappa} ln\Big(\frac{z+z_0}{z_0}\Big), \tag{1}$$

where U is the time-averaged velocity, $u_\star$ the friction velocity, κ the von Karman constant, z the height above the ground, and z_0 the roughness length. $u_\star$ and z_0 are found by fitting the logarithmic profile to the measurements. The corresponding appropriate boundary condition for the turbulence kinetic energy at the inlet is constant over height and the turbulence dissipation profile is defined to balance production of turbulence kinetic energy [5]:

$$k = \frac{u_\star^2}{\sqrt{C_\mu}}, \qquad \varepsilon = \frac{u_\star^3}{\kappa\,(z+z_0))}. \tag{2}$$

To ensure horizontal homogeneity of these profiles throughout the domain, we use a modified wall function on the ground that represents the effect of the roughness length of the terrain [5,6]. On the side walls we impose the inflow condition, while on the top of the wind tunnel we impose a slip boundary condition. The outlet is treated as a pressure-outlet with constant relative pressure equal to zero and a zero-gradient boundary condition for the other flow variables.

3.2 Inflow UQ Methodology

3.2.1 Characterization of the Inflow Uncertainty

As shown in Fig. 1, wind tunnel measurements of the ABL profiles were obtained in 5 spanwise locations. The variation in the mean velocity and turbulence kinetic energy profiles measured at these different locations is used to define the uncertainty in the inflow condition. We identify two uncertain parameters when fitting the logarithmic velocity profile from Eq. 1 to the experimental data: the roughness length z_0, and a reference velocity at 1 m height U_{ref}. Based on fitting the logarithmic profile to the measured velocity profiles at the 5 spanwise locations, we can define upper and lower limits for both uncertain parameters (see Table 1). In the present analysis, we assume the parameters are uniformly distributed on these supports. Figure 3 shows potential inflow profiles for the velocity and turbulence kinetic energy, considering all possible combinations of the minimum, mean, and maximum value for each uncertain parameter. The plots demonstrate that the resulting inflow profiles represent most of the spanwise variation of the experiment.

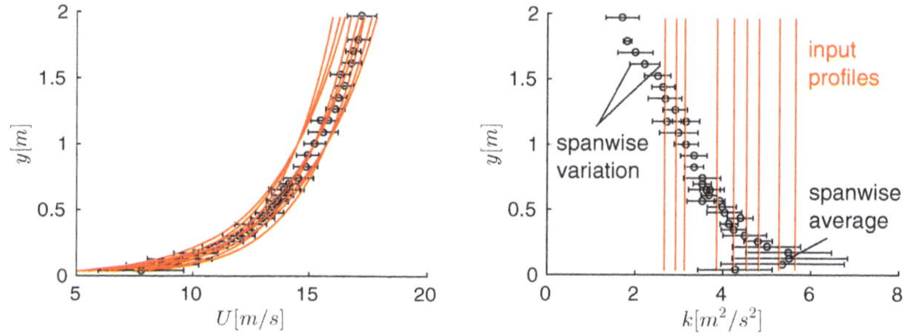

Fig. 3. Input profiles of the inflow UQ study: mean streamwise velocity (left) and turbulence kinetic energy (right).

In addition to defining uncertainty in the roughness length and reference velocity, we account for potential misalignment between the intended and actual model orientation in the wind tunnel. To represent this uncertainty, we introduce a third parameter representing the angle of the misalignment, α, with a uniform distribution between $-2°$ and $2°$. The resulting three uncertain parameters and their minimum, mean, and maximum values are summarized in Table 1.

Table 1. Inflow uncertain parameters.

	Min.	Mean	Max.
z_0 [mm]	1.3	2.9	7.1
U_{ref} [m/s]	9.9	10.3	10.7
α [°]	-2	0	2

3.2.2 Propagation Method

The open-source software Dakota v6.6 [7], developed at Sandia National Laboratories, is used to propagate the uncertainty in the input parameters to the output quantities of interest (QoIs) with a non-intrusive polynomial chaos expansion (PCE) approach. Given the assumed uniform distribution of the uncertain input parameters we use Legendre polynomial basis functions [2]. The coefficients are calculated using Galerkin projection on an anisotropic tensor grid, with three points for z_0 and U_{ref} and five points for α. This implies that a total number of 45 CFD model evaluations is performed.

Our QoIs are the time-averaged pressure coefficients at the 892 experimental pressure tap locations. The Fourier coefficients provide estimates for the mean and variance of the QoIs, while probability distributions of the QoIs can be obtained by sampling the PCE response surface. In the present study we compute mean and 95% confidence intervals from 10,000 realizations of the PCE.

3.3 Turbulence Model Form UQ Methodology

The turbulence model form UQ approach used in this study has been tested for different configurations [3,4,8,9]. The methodology consists of first performing a baseline RANS simulation, and subsequently introducing perturbations in the Reynolds stresses. The perturbations are defined in terms of a decomposition of the Reynolds stress tensor, i.e. based on the tensor magnitude and the eigenvalues and eigenvectors of the normalized anisotropy tensor. The perturbed Reynolds stresses R^*_{ij} can then be written in terms of the perturbed turbulence kinetic energy k^*, and perturbed eigenvectors v^*_{ik} and eigenvalues Λ^*_{kl}:

$$R^*_{ij} = 2k^* \left(\frac{1}{3}\delta_{ij} + v^*_{ik}\Lambda^*_{kl}v^*_{jl} \right). \tag{3}$$

In the present study we quantify the turbulence model form uncertainty by performing three additional simulations, introducing different perturbations to the eigenvalues Λ^*_{kl}. The perturbations are defined to introduce a bias in the Reynolds stress prediction, making the turbulence more one-, two-, or three-component. This is achieved by defining the perturbations as a change in the barycentric map coordinates of the original model prediction for the Reynolds stress tensor.

The barycentric map, shown in Fig. 4, provides a representation for all realizable Reynolds stress tensors. Its corners represent the different limiting states of turbulence anisotropy, e.g. isotropic turbulence corresponds to the three-component ($\mathbf{x}_{3C}$) corner. The map coordinates for any Reynolds stress tensor are a linear combination of the corner coordinates; the coefficients are a function of the anisotropy tensor eigenvalues, $\lambda_1 \geq \lambda_2 \geq \lambda_3$:

$$\mathbf{x} = (\lambda_1 - \lambda_2)\mathbf{x}_{1C} + (2\lambda_2 - 2\lambda_3)\mathbf{x}_{2C} + (3\lambda_3 + 1)\mathbf{x}_{3C}. \tag{4}$$

Hence, the intended bias can be introduced by first determining the barycentric map coordinates of the Reynolds stress computed by the baseline turbulence model ($\mathbf{x}$ in Fig. 4), and subsequently injecting uncertainty by moving towards one of the corners of the map. Because the extreme limiting states are not likely to occur, the bias introduced in the turbulent state is limited using a factor ΔB_i:

$$\mathbf{x_i^*} = \mathbf{x} + \Delta B_i(\mathbf{x_{ic}} - \mathbf{x}), \quad (5)$$

where i is the corner index, and $\Delta B_1 = 0.5$, $\Delta B_2 = 0.5$, and $\Delta B_3 = 0.1$. Once the new barycentric map coordinates are determined, the updated set of eigenvalues is evaluated and used to reconstruct the perturbed anisotropy tensor. In the final step, this perturbed anisotropy tensor is used to calculate the production of turbulence kinetic energy in its transport equation. It has been shown that perturbations towards $\mathbf{x_{1c}}$ will increase turbulence production, while those towards $\mathbf{x_{3c}}$ will decrease turbulence production [3,9].

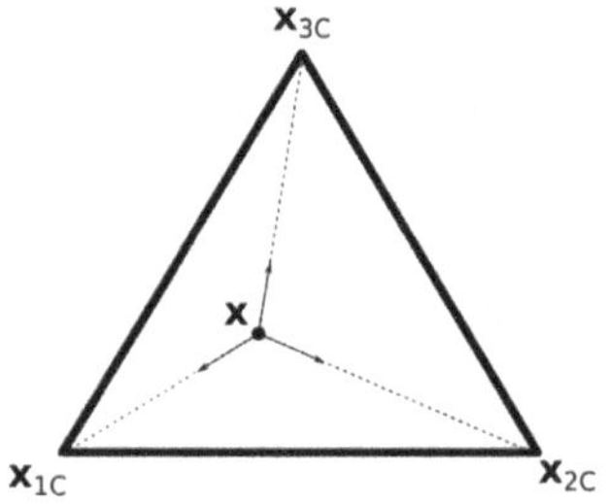

Fig. 4. Barycentric map.

4 Results

In the following, we first present results for the mean and 95% confidence intervals for the time-averaged pressure coefficients obtained from the inflow UQ study. Subsequently, we present the bounds obtained by the three perturbed simulations performed to quantify turbulence model uncertainty. In both cases, the results are compared to the experimental data obtained in the PoliMi wind tunnel.

4.1 Inflow UQ Results

Figure 5 compares the distribution of the time-averaged pressure coefficients measured on the building's lateral facade in the wind tunnel (left) to the mean values of the pressure coefficients computed from the RANS UQ study (right). Similar qualitative features can be appreciated: the flow separates at the windward edge of the building producing a region of strong suction (red region of Fig. 5) and subsequently reattaches towards the rear part of the model (yellow region of Fig. 5).

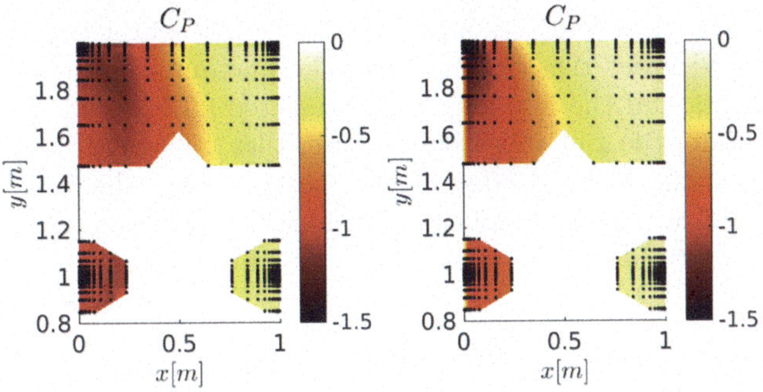

Fig. 5. Time-averaged pressure coefficient on building's lateral facade measured at PoliMi (left) and mean value computed from UQ (right).

To provide a more quantitative comparison, Fig. 6 show profiles of the time-averaged pressure coefficient along two rows of taps on the upper and lower tiles. Here, the mean and 95% confidence interval resulting from the UQ analysis are compared to the measurements. The 95% confidence intervals predicted by the

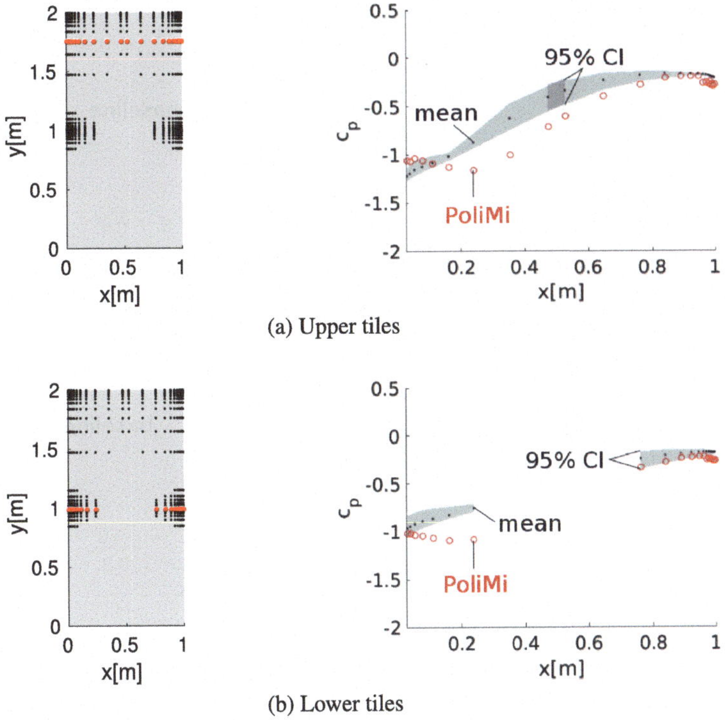

Fig. 6. Row of taps (left) and corresponding profiles of mean and 95% confidence interval (black dots and gray region respectively) of time-averaged pressure coefficient compared to the measurements (red circles).

inflow UQ framework encompass only ~21% and ~30% of the experimental data on the upper and lower tiles respectively. The maximum discrepancy between the measurements and the bounds of the predicted 95% confidence intervals is ~52% on the upper tile and ~28% on the lower tiles. In general, good agreement is obtained near the leeward edge of the model, where the flow is attached. The comparison deteriorates significantly inside the separation region, suggesting that the uncertainty in the turbulence model might be dominant in this region.

4.2 Turbulence Model UQ Results

As a first step towards assessing the turbulence model form uncertainty, we considered the mean values of the uncertain inflow parameters and performed three additional simulations, introducing different perturbations to the Reynolds stress tensor as discussed in Sect. 3.3. Figure 7 shows the resulting profiles of the time-averaged pressure coefficient along the two rows of taps on the upper and lower tiles; the result computed by the three perturbed simulations is compared to the wind tunnel data and the inflow UQ prediction. The perturbation towards maxi-

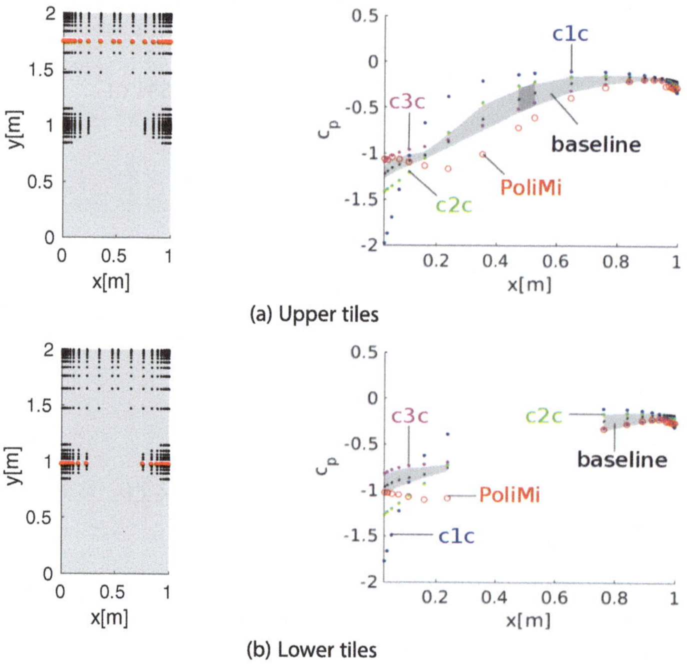

Fig. 7. Row of taps (left) and corresponding profiles of time-averaged pressure coefficient (right). Mean and 95% confidence interval (black dots and gray region respectively) computed from the inflow UQ are compared to the simulations with the one-component (blue dots), two-component (green dots) and three-component perturbations (magenta dots), and the measurements (red circles).

mum production of turbulence kinetic energy (at $\mathbf{x}_{1C}$) tends to minimize the size of separation regions and the building wake. This generates a stronger negative peak of pressure at the windward edge, but the recovery is faster (blue markers in Fig. 7). In contrast, the perturbation towards zero production (at $\mathbf{x}_{3C}$) maximizes the size of separation regions and the building wake, producing a slower recovery (magenta markers in Fig. 7). The result of the two-component perturbation falls between the one- and three-component perturbation, except around the streamwise location where the flow reattaches (green markers in Figs. 7). This is also the region of highest discrepancy with the measurements. The large difference between the results obtained with the different perturbations close to the windward edge indicates that the uncertainty related to the turbulence model is dominant when separation occurs; further downstream, where the flow is attached, the perturbed results all converge towards the baseline result.

5 Conclusions and Future Work

The current paper has presented methodologies to quantify inflow and turbulence model form uncertainty in RANS predictions of the mean pressure coefficient distribution on the lateral facade of a high-rise building. The results have been compared to available wind tunnel measurements to evaluate the predictive capabilities of the methods.

First, we quantified the uncertainty related to the inflow boundary condition using three uncertain parameters: the roughness length, the reference velocity and the model orientation. These uncertainties were propagated to the QoIs using a PCE approach to compute mean and 95% confidence intervals of the time-averaged pressure coefficient [2]. The resulting confidence intervals contain ~25% of the data; the highest discrepancy between the simulations and experiments is found to occur in the region of flow separation.

The second source of uncertainty investigated is the turbulence model. To quantify uncertainty related to the turbulence model form, we considered the mean values of the uncertain inflow parameters and performed three additional simulations, perturbing the eigenvalues of the Reynolds stress tensor of the baseline simulation [3,4]. The results show that introducing perturbations towards the one-component and three-component limiting states produce an upper and lower bound for the baseline turbulence model prediction in most locations on the lateral facade. The bounds are largest in the upstream region where flow separation occurs.

In summary, the results of the inflow and turbulence model UQ analysis show that uncertainty in the turbulence model is dominant in the region of flow separation, while the uncertainty in the inflow condition is more relevant where the flow is attached. Future work will focus on a more comprehensive integration of the inflow and turbulence model UQ methods to provide a single confidence interval that represents both types of uncertainty in the prediction for the time-averaged pressure coefficient. This will be achieved through a mixed aleatory-epistemic UQ analysis, where perturbed simulations will be performed for each of the inflow UQ simulations [10]. Subsequently, the overall framework will be extended to also include predictions with confidence intervals for the root mean square pressure coefficient, estimated from the RANS average flow field.

Acknowledgements. This material is based upon work supported by the National Science Foundation under Grant Number 1635137, and used the Extreme Science and Engineering Discovery Environment (XSEDE), which is supported by National Science Foundation grant number CI-1548562.

References

1. Amerio L (2018) Experimental high resolution analysis of the pressure peaks on a building scale model façades. PhD thesis, Politecnico di Milano
2. Smith RC (2013) Uncertainty quantification: theory, implementation, and applications, vol 12. SIAM
3. Gorlé C, Emory M, Larsson J, Iaccarino G (2012) Epistemic uncertainty quantification for RANS modeling of the flow over a wavy wall. In: Center for turbulence research, annual research briefs
4. Emory M, Larsson J, Iaccarino G (2013) Modeling of structural uncertainties in Reynolds-averaged Navier-Stokes closures. Phys Fluids 25(11):110822
5. Richards PJ, Hoxey RP (1993) Appropriate boundary conditions for computational wind engineering models using the $k-\varepsilon$ turbulence model. J Wind Eng Ind Aerodyn 46:145–153
6. Parente A, Gorle C, van Beeck J, Benocci C (2011) Improved kappa-epsilon model and wall function formulation for the RANS simulation of ABL flows. J Wind Eng Ind Aerodyn 99:267–278
7. Bohnhoff WJ, Dalbey KR, Eddy, JP, Frye JR, Hooper RW, Hu KT, Hough PD, Khalil M, Ridgeway EM, Rushdi A (2009) Dakota, a multilevel parallel object-oriented framework for design optimization, parameter estimation, uncertainty quantification, and sensitivity analysis: version 6.6 user's manual. Sandia National Laboratories, Technical report SAND2010-2183
8. Gorle C, Garcia-Sanchez C, Iaccarino G (2015) Quantifying inflow and RANS turbulence model form uncertainties for wind engineering flows. J Wind Eng Ind Aerodyn 144:202–212
9. Iaccarino G, Mishra AA, Ghili S (2017) Eigenspace perturbations for uncertainty estimation of single-point turbulence closures. Phys Rev Fluids 2:024605
10. Ng LWT, Eldred MS (2012) Multifidelity uncertainty quantification using non-intrusive polynomial chaos and stochastic collocation. In: 53rd AIAA/ASME/ASCE/AHS/ASC structures, structural dynamics and materials conference

A Model Extension for Vortex-Induced Vibrations

F. Lupi[1,2(✉)], H.-J. Niemann[1,2], and R. Höffer[1]

[1] Faculty of Civil and Environmental Engineering,
Department of Wind Engineering and Flow Mechanics,
Ruhr-Universität Bochum, Universitaetsstr. 150, 44801 Bochum, Germany
{francesca.lupi,hans-juergen.niemann,
ruediger.hoeffer}@rub.de

[2] Niemann & Partner Ingenieurgesellschaft,
Universitaetsstr. 142, 44799 Bochum, Germany

Abstract. The paper presents free-vibration wind tunnel tests performed at WIST Boundary Layer Wind Tunnel at Ruhr-Universität Bochum (Germany) on a 3D aeroelastic cylindrical model with circular cross-section. The aim of the tests is to validate a model extension to the original spectral method developed by Vickery & Basu, able to calculate the maximum oscillation of a structure subjected to vortex-induced vibration in the lock-in region. The peculiarity of the extension is the implementation of an experimental curve for the negative aerodynamic damping. It was previously developed by the authors through forced vibration wind tunnel tests. The model extension is based on a linear differential equation. In fact, linear – in case iterative – approaches are usually preferred for the design of structures. However, limitations due to linearization of an intrinsically non-linear phenomenon are unavoidable. Strengths and weaknesses of the linear approach are discussed in the paper.

Keywords: Circular cylinder · Free-vibration wind tunnel tests · Vortex-induced vibrations · Vortex shedding · Vortex resonance · Aerodynamic damping

1 Introduction

This work represents a further development of previous papers written by the authors (Lupi et al. 2017, 2018), dealing with modelling of vortex resonance. Firstly, wind tunnel experiments in forced vibrations performed at Ruhr-Universität Bochum (Germany) allowed the development of a new aerodynamic damping model (Lupi et al. 2018). The approach is based on a linear aerodynamic damping force and an amplitude dependent damping coefficient with positive curvature. The damping model was applied for the calculation of the largest cross-wind oscillation of full-scale chimneys in lock-in. An extension to the well-known spectral method - originally developed in Vickery and Basu (1983) - was proposed in (Lupi et al. 2017). The present paper elaborates and validates this model extension for vortex-induced vibrations (VIV) through free-vibration wind tunnel tests on a circular cylinder.

F. Ricciardelli and A. M. Avossa (Eds.): INVENTO 2018, LNCE 27, pp. 413–426, 2019.
https://doi.org/10.1007/978-3-030-12815-9_33

Many models were developed in the last decades to describe the fluid-structure interaction in vortex induced vibrations. They can be broadly classified in two categories. The first class corresponds to the one-degree-of-freedom models. The aeroelastic effect is included in the equation of motion of the structure, represented by 1-dof oscillator. Two main sub-classes of 1-dof models exist. In the first sub-class, the aeroelastic effect is represented by the energy transferred from the flow to the system, which acts as a negative damping. The concept goes back to Scruton (1963). Then, milestones in this sub-class are: Scanlan (1981), Vickery and Basu (1983), Simiu and Scanlan (1986), Ehsan and Scanlan (1990), Goswami et al. (1993), Larsen (1995), Hansen (2013), Sun et al. (2014). In the second sub-class, VIVs are modelled as forced-driven vibrations under a properly, response-dependent, modified load. Typical examples in this sub-class are the Ruscheweyh model (1986), which increases the correlation length of a purely harmonic load, the spectral load modifications in Flaga (1997) and Arunachalam and Lakshmanan (2015). Instead, the models of the second class describe the wake of the cylinder as a non-linear, self-excited and self-limited oscillator. The wake equation is coupled to the equation of motion of the structure through a forcing term. For these reasons, these two-equation models are named wake-oscillator models or coupled models. The concept was first proposed by Bishop and Hassan (1964) and then developed by Hartlen and Currie (1970), Skop and Griffin (1973, 1975), Iwan and Blevins (1974), Landl (1975), Tamura and Matsui (1980), Dowell (1981), Benaroya and Lepore (1983), Billah (1989), Blevins (1990), Krenk and Nielsen (1999), Plaschko (2000), Skop and Luo (2001), Facchinetti et al. (2004), Farshidianfar and Zangeneh (2010), Ogink and Metrikine (2010) and others.

Nevertheless, the long-lasting discussion about the most appropriate model to describe VIV and lock-in is not yet come to an end, especially concerning practical applications. Linear – in case iterative – approaches are usually preferred for the design of structures under cross-wind actions (see e.g. Eurocode 1, Method 1 and 2 in Annex E). However, limitations due to the linearization of an intrinsically non-linear phenomenon are unavoidable and they are sometimes responsible for inconsistencies among results of different methods.

The key role of the negative aerodynamic damping in the lock-in range and an extension to the original Vickery & Basu model are discussed and validated in this paper through free-vibration wind tunnel tests. This model extension goes back to a linear equation of structural motion with an amplitude-dependent damping force. It is not able to describe the whole phenomenon of VIV in all its aspects, but it focuses on the prediction of the largest cross-wind oscillation in lock-in. The range of application is the wind engineering field (not water) and structures with circular cross-section (investigations on different cross-sections are currently on-going at Ruhr-Universität Bochum). Within this range, predictions are realistic and reliable. Strengths and weaknesses of the linear approach are discussed in the paper. Attention is devoted to the concept of self-induced and self-limited oscillations and to the existence of a steady-state amplitude in lock-in. The paper cannot prove or answer the fundamental open question about the existence of a maximum possible amplitude for increasingly smaller mass and damping (Williamson and Govardhan (2008)), but the model extension takes a position in this regard and removes the existence of an upper limit.

2 Wind Tunnel Tests in Free-Vibrations

2.1 Boundary Layer Wind Tunnel and Experimental Set-up for Free-Vibration Testing

The wind tunnel tests are performed at WISt Boundary Layer Wind Tunnel at Ruhr-Universität Bochum. It is an open circuit wind tunnel with a test section of 1.8 m in width and 1.6 m in height (Fig. 1). The free-vibration wind tunnel tests are performed in uniform flow with turbulence intensity Iv = 4.5%.

Fig. 1. Wind tunnel model at Ruhr-Universität Bochum, Germany

The wind tunnel models for free-vibration tests are cantilever beams with circular cross-section (Fig. 1). Several cases with different values of the Scruton number Sc (ranging from 2.8 until 22.5) are investigated (see Table 1). The Scruton number is defined as follows:

$$Sc = \frac{4\pi \cdot \zeta_s \cdot m_{eq}}{\rho \cdot d^2} \tag{1}$$

where ζ_s is the structural damping ratio, m_{eq} is the equivalent mass in the first, cross-wind mode of vibration, ρ is the density of air (1.25 kg/m^3) and d is the diameter of the cylinder. Different values of the Scruton number are achieved by varying the diameter of the cross-section and by increasing the mass of the structure through a mass on top. In fact, the mass on top acts as a dead mass, which also affects the structural damping. In general, the structural damping slightly decreases as the dead mass on top increases. Therefore, the structural damping of each model configuration is measured through a free-decay wind tunnel test. Its value is reported in Table 1, together with all the other characteristics of the models.

The reactions at the base of the model (Fx, Fy, Fz and Mx, My, Mz, being x the wind direction, y the cross-wind direction and z positive upwards) are measured by a

Table 1. Configurations of wind tunnel models for free-vibration wind tunnel tests.

Configuration n.	1	2	3	4	5	6	7
Height h in mm	775	775	775	775	775	775	775
Diameter d in mm	0,050	0,050	0,050	0,050	0,032	0,032	0,032
Natural frequency f_1 in Hz	20,02	15,38	13,18	11,23	22,22	13,67	11,47
Equivalent mass m_{eq} in kg/m	0,232	0,44	0,615	0,848	0,232	0,615	0,848
Structural damping δ_s (log-dec.)	0,019	0,015	0,019	0,018	0,028	0,019	0,017
Scruton number Sc	2,81	4,22	7,48	9,77	10,11	17,78	22,53

6-component force balance. The sampling frequency is set to 2000 Hz. Every measurement lasts for 98 s.

The cross-wind oscillation of the model in the 1st mode shape, at the location z = h where the 1st mode shape has its maximum (equal to unity by normalization), can be calculated from the reactions measured at the base of the model. The mathematical relationship follows from the dynamic equation of motion of the system in resonance, being the inertial force in equilibrium with the elastic force.

$$\left(\int_0^h m(z)\cdot\Phi_{1,y}(z)\cdot dz\right)\cdot\omega_1^2\cdot y(t)\cdot = F_y(t) \rightarrow y(t) = \frac{F_y(t)}{\left(\int_0^h m(z)\cdot\Phi_{1,y}(z)dz\right)\cdot\omega_1^2} \tag{2}$$

The left-hand side of the Eq. (2) represents the resultant force at the base of the model produced by the distribution of the inertial force along the height in the 1st vibration mode $\Phi_{1,y}$. m(z) is the distributed mass of the model, ω_1 is its natural circular frequency in the 1st mode of vibration and y(t) is the time history of the displacement at the location h, where $\max(\Phi_{1,y}(z)) = 1$. Similarly, the bending moment at the base produced by the distribution of the inertial force along the height is calculated by the left-hand side of Eq. (3) and it is equal to M_x. Therefore, either the Eqs. (2) or (3) can be used to evaluate the top displacement in the first mode shape. In this work, the Eq. (3) is preferred because, in comparison to F_y, the time-history of the bending moment M_x is in general less affected by any disturbance.

$$\left(\int_0^h m(z)\cdot\Phi_{1,y}(z)\cdot z\cdot dz\right)\cdot\omega_1^2\cdot y(t) = -M_x(t) \rightarrow y(t) = -\frac{M_x(t)}{\left(\int_0^h m(z)\cdot\Phi_{1,y}(z)\cdot z\cdot dz\right)\cdot\omega_1^2} \tag{3}$$

2.2 Experimental Evidence

The phenomenon of vortex-induced vibration and lock-in is described in this section through some exemplary results of free-vibration experiments. Whereas in case of small

Scruton numbers the response at constant mean flow speed is predominantly harmonic, in case of larger Scruton numbers it is a mix of sinusoidal, small-band vibrations interrupted by stochastic broad band intervals. Among the available experiments listed in Table 1, let us consider more in detail the case of a relatively large Sc, which nevertheless shows an aeroelastic effect and a lock-in range: for instance, the case n.6 (Sc = 17.78).

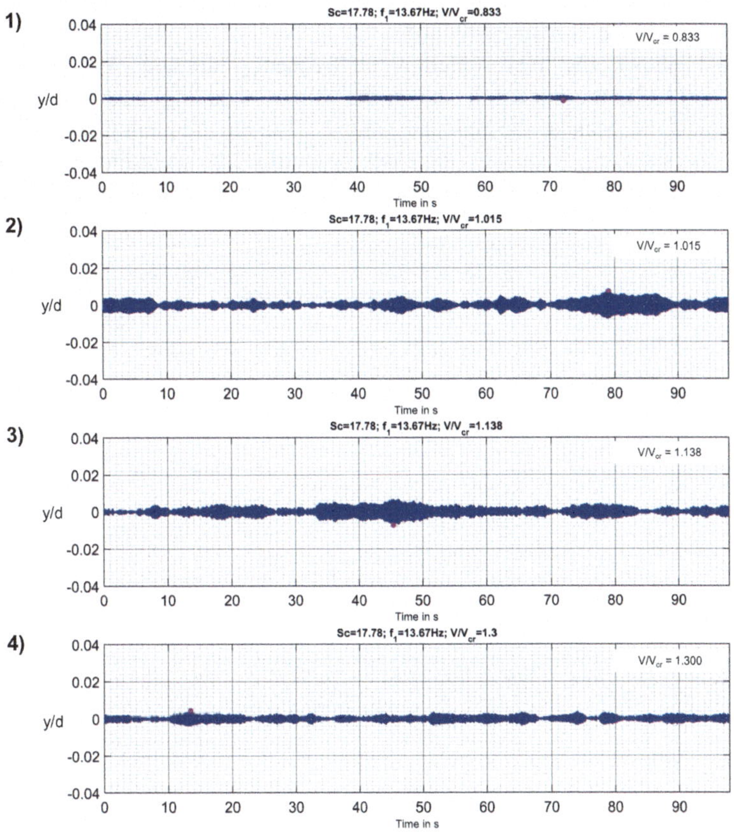

Fig. 2. Time histories of non-dimensional tip oscillation y(t)/d before (1. $V/V_{cr} = 0.833$), within (2. $V/V_{cr} = 1.015$ and 3. $V/V_{cr} = 1.138$) and after (4. $V/V_{cr} = 1.3$) the lock-in range. 98 s, Sc = 17.78.

Figure 2 shows typical time-histories (T = 98 s) of the non-dimensional vibration y (t)/d on top of the cantilever (Eq. (3)) at constant mean wind speed encompassing the lock-in range. The first time-history ($V/V_{cr} = 0.833$) precedes the lock-in range and therefore the amplitude is small. The second time-history occurs at a mean wind speed $V = 1.015V_{cr}$ and shows the largest amplitude measured in the lock-in range. However, because of the aeroelastic effect, a similar large oscillation also occurs at larger wind speed in the lock-in range, $V = 1.138V_{cr}$ (third time-history). The fourth time-history still presents some resonant grow-up events, but it is at the end of the lock-in range and the amplitude is smaller. The characterizing feature of these time-histories is the intermittency of the response. In fact, low-frequent fluctuations of the mean wind may cause lock-out from the lock-in range and eventually return into it. In such intervals, the vibration ebbs away and after return it starts building up again more or less rapidly, depending on damping (or Scruton number).

Figure 3 offers a better insight in the lock-in phenomenon through a selection of shorter time intervals (2 s) around the time instant where the largest amplitude occurs. The largest amplitude is marked by a dot.

Figure 4 shows the representation in frequency domain. Before resonance (1. $V/V_{cr} = 0.833$), the spectrum shows the vortex shedding peak at the Strouhal frequency and the peak at the reduced natural vibration frequency. As the velocity increases, the shedding frequency increases proportionally according to the Strouhal law, so that the Strouhal peak remains fixed. Vice versa, the peak of the reduced natural

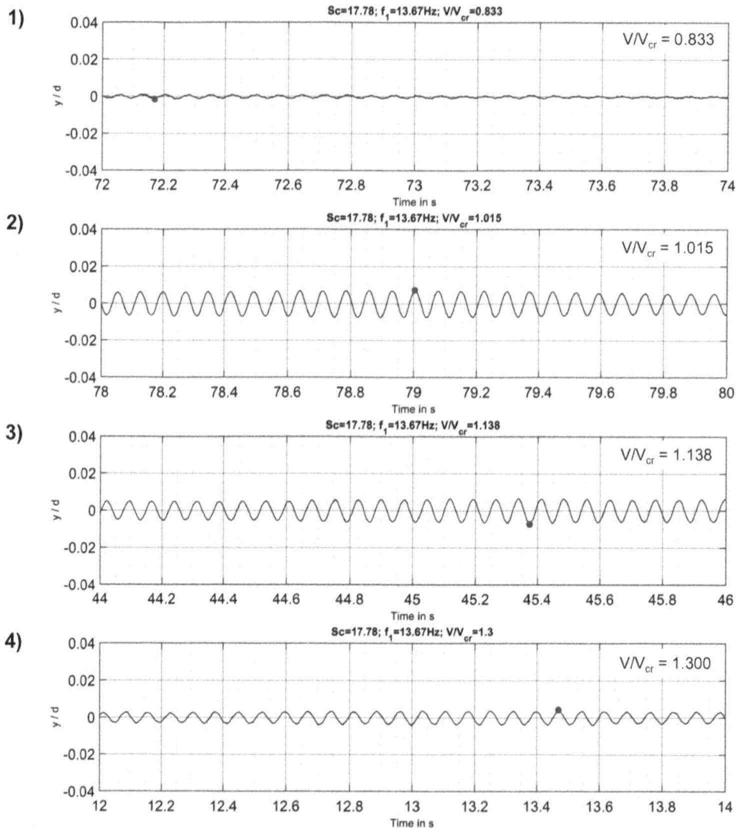

Fig. 3. Time histories of non-dimensional tip oscillation y(t)/d before (1. $V/V_{cr} = 0.833$), within (2. $V/V_{cr} = 1.015$ and 3. $V/V_{cr} = 1.138$) and after (4. $V/V_{cr} = 1.3$) the lock-in range. Zoom 2 s, Sc = 17.78.

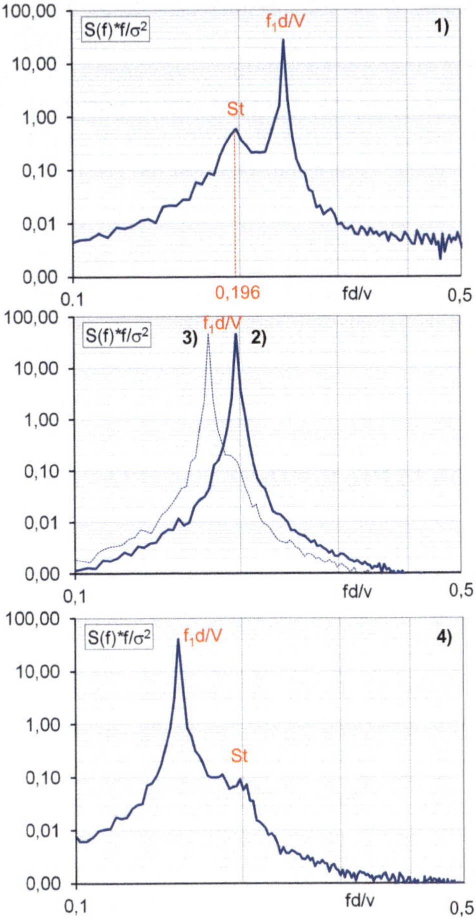

Fig. 4. Power spectral densities of base bending moment due to lift force before (1. $V/V_{cr} = 0.833$), within (2. $V/V_{cr} = 1.015$ and 3. $V/V_{cr} = 1.138$) and after (4. $V/V_{cr} = 1.3$) the lock-in range for Sc = 17.78.

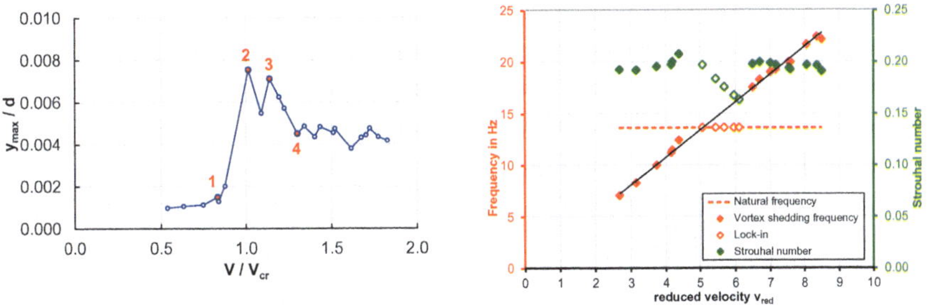

Fig. 5. Left: Lock-in range; right: Frequencies and Strouhal number (Sc = 17.78).

frequency moves leftwards as the velocity increases. In the lock-in range, the Strouhal peak cannot be distinguished anymore. As the velocity increases in the lock-in range (2. $V/V_{cr} = 1.015$ and 3. $V/V_{cr} = 1.138$), the resonance peak moves towards left, the Strouhal peak remains locked to the structural vibration and cannot be seen. Then, by further increase of the wind velocity out of the lock-in range (4. $V/V_{cr} = 1.3$), the Strouhal peak appears again, at its usual position, now on the right hand side of the structural vibration.

Figure 5 shows the lock-in range and summarizes the results. As the reduced velocity increases, the natural vibration frequency of the model remains constant. This is because non-linear effects like added mass of the air during the vibration are negligible. The vortex shedding frequency follows the Strouhal law before and after lock-in. Similarly, the Strouhal number is approximately constant before and after lock-in, whereas the Strouhal law is violated in lock-in.

3 Model Extension to Predict the Largest Oscillation in Lock-in

3.1 Innovative Aspects

The peculiar feature of the model extension proposed in this paper is the implementation of an experimental curve of the aerodynamic damping parameter K_a. It was derived in Lupi et al. (2017, 2018) through wind tunnel tests on a circular cylinder in forced-vibrations and it is described by the following equation:

$$K_a = 0,3475 \cdot \exp\left(-5,808\frac{\sigma_y}{d}\right) \cdot \left(\frac{\sigma_y}{d}\right)^{-0.3582} \tag{4}$$

The aerodynamic damping parameter K_a is defined as follows:

$$K_a = -\frac{\zeta_a \cdot m_{eq}}{\rho \cdot d^2} \tag{5}$$

where ζ a is the aerodynamic damping ratio. The Eq. (4) replaces the Eq. (6), which is applied in the Vickery & Basu model (1983) and also in the Eurocode Method 2:

$$K_a = K_{a0} \cdot \left(1 - \left(\frac{\sigma_y}{a_L \cdot d}\right)^2\right) \tag{6}$$

In Eq. (6), $K_{a0} > 0$ is a linear term, i.e. the negative aerodynamic damping for small oscillations. It represents the exciting mechanism, which introduces energy into the system. It depends on the Reynolds number and on the turbulence intensity of the incoming flow. The quadratic term into brackets represents the aeroelastic dissipation of energy. a_L is the non-dimensional rms value of the so-called limiting oscillation.

The two curves described by the Eqs. (4) and (6) are plotted in Fig. 6.

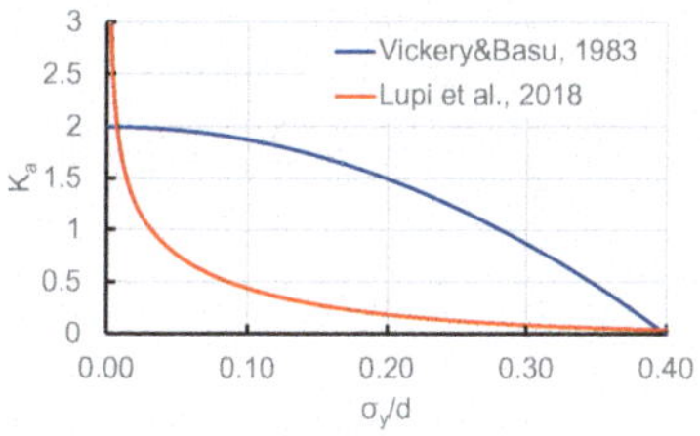

Fig. 6. Theoretical behaviour of the aerodynamic damping parameter (Vickery and Basu 1983) and experimental behaviour (Lupi et al. 2018).

The main difference between the two curves in Fig. 6 is their curvature. It is positive in the new experimental curve, whereas it is negative in the Vickery curve. Consequently, the Vickery curve becomes zero for $\sigma_y/d = a_L$ and therefore the model imposes an upper limit to the oscillation of structures with very low structural damping ($\zeta_s \rightarrow 0$), which is $y_{LIM} = \sqrt{2}\cdot a_L$. In fact, the existence of a maximum possible amplitude attainable for a cylinder undergoing VIV – in case of extremely small mass and damping – is still an open question (Williamson and Govardhan 2008). Furthermore, the Vickery curve crosses the y-axis for $K_a = K_{a0}$, being $K_{a0} = m_{eq}\cdot\zeta_{a0}/(\rho d^2)$. Because of that, the Vickery model distinguishes two regimes: the forced vibration regime in case $\zeta_s > \zeta_{a0}$ (or, equivalently $K_a > K_{a0}$) and the lock-in regime with aeroelastic effect in case $\zeta_s < \zeta_{a0}$. Instead, the experimental curve is asymptotic on both axes. It approaches zero for $\sigma_y/d \rightarrow \infty$, meaning that the deflection of structures with increasingly smaller mass and/or structural damping becomes larger and larger. However, the deflection cannot diverge to infinite because $m_{eq} \cdot \zeta_s$ must always be positive and cannot be zero. On the y-axis, the experimental curve approaches infinite asymptotically. This removes the arbitrary distinction between the forced vibration regime and the lock-in regime marked by the threshold K_{a0}. The aeroelastic effect does occur also for relatively large structural damping (as proved, e.g., by the experiments presented in Sect. 2), but it approaches zero as the mass and/or structural damping become larger and larger.

The expression of the aerodynamic damping described by the Eqs. (4) and (5) can be implemented in the following linear differential equation of motion, which describes the aeroelastic oscillation of a circular structure excited by a narrow-banded vortex shedding load F_L in the resonance zone of the first vibration mode.

$$\ddot{y}(t) + 2\omega_1 \cdot \left[\zeta_s + \zeta_a\left(\frac{\sigma_y}{d}\right)\right] \cdot \dot{y}(t) + \omega_1^2 \cdot y(t) = \frac{F_L(t)}{M_1} \quad (7)$$

This equation can be solved in frequency domain. The rms of the response is given by the expression (8), being S_L the spectral density of the vortex shedding load and K_1 the modal stiffness of the structure.

$$\frac{\sigma_y}{d} = \frac{1}{d}\sqrt{\frac{S_L(f_1)}{K_1^2} \cdot \frac{\omega_1}{8 \cdot \left[\zeta_s + \zeta_a\left(\frac{\sigma_y}{d}\right)\right]}} \tag{8}$$

The Eq. (8) corresponds to the Vickery & Basu model (1983) except for the expression used for the aerodynamic damping ζ_a. Because of that, the Eq. (8), together with (4) and (5), constitutes a model extension to the Vickery & Basu model. In this work, the spectral density of the modal load S_L is calculated through the original expression proposed by Vickery and Clark (1972) for the spectral density of the lift coefficient. The resonant contribution of the modal spectral load is given by Eq. (9) (see Lupi et al. 2018 for further details):

$$S_L(f_1) = 2 \cdot L_\lambda \cdot q_m^2 \cdot d^2 \cdot S_{cL}(f_1) = 2 \cdot L_\lambda \cdot \left(\frac{1}{2}\rho V^2\right)^2 \cdot d^2 \cdot \frac{\sigma_{cL}^2}{\sqrt{\pi} \cdot B \cdot f_1} \cdot \int_{z=0}^{h} \Phi_1^2(z) \cdot dz \tag{9}$$

L_λ is the correlation length of the vortex shedding load in lock-in (assumed equal to 1.5d in this work), Φ_1 is the mode shape and B is the bandwidth factor. V is the wind velocity for which is the largest amplitude is obtained. It generally corresponds to the critical velocity for larger Sc and it is higher than the critical velocity in case of smaller Sc. Relying on the experiments, it is assumed here: $V = V_{cr}$ for Sc $\leq$ 10 and $V = 1.25\ V_{cr}$ for Sc > 10.

Despite the amplitude-dependency of the aerodynamic damping, which requires an iterative or a graphic solution of (8), the model is linear, being (7) a linear differential equation. A linear model can experience neither self-induced vibrations nor limit cycles. In fact, the model reaches a steady-state amplitude under the external force F_L. Furthermore, a linear model is not able to reproduce the VIV phenomenon in all its aspects at once, being the phenomenon physically non-linear, but it focuses – in this case – on the prediction of the largest amplitude in lock-in.

Equation (8) provides the rms of the response. The amplitude of the response can be calculated by applying a peak factor. A typical feature of the free-vibration experiments, even in case of relatively large Scruton numbers, is that the time histories of the response in lock-in might be intermittent, but not stochastic (see Figs. 2 and 3). Therefore, a peak factor in the range 3–4, typical for stochastic vibrations and recommended for instance by the Eurocode for larger Sc, is considered here not appropriate. Instead, the investigation suggested an experimental value of 1.6 for the peak factor. The value is slightly above $\sqrt{2}$, which would correspond to a harmonic vibration.

$$\frac{y_{max}}{d} = k_p \cdot \frac{\sigma_y}{d} = 1,6 \cdot \frac{\sigma_y}{d} \tag{10}$$

3.2 Results

Figure 7 shows the model prediction (calculated through Eqs. (4), (5) and (8)–(10)) and the results of free-vibration wind tunnel tests in lock-in for selected cases (Sc = 2.81, 7.48, 17.78). The prediction matches the experimental results very well in case of smaller Sc, whereas it overestimates the oscillation in case of larger Sc (see for instance Sc = 17.78 in Fig. 7). However, the oscillation for Sc = 17.78 is small (y/d < 0.05), thus the overestimation is not significant for the design.

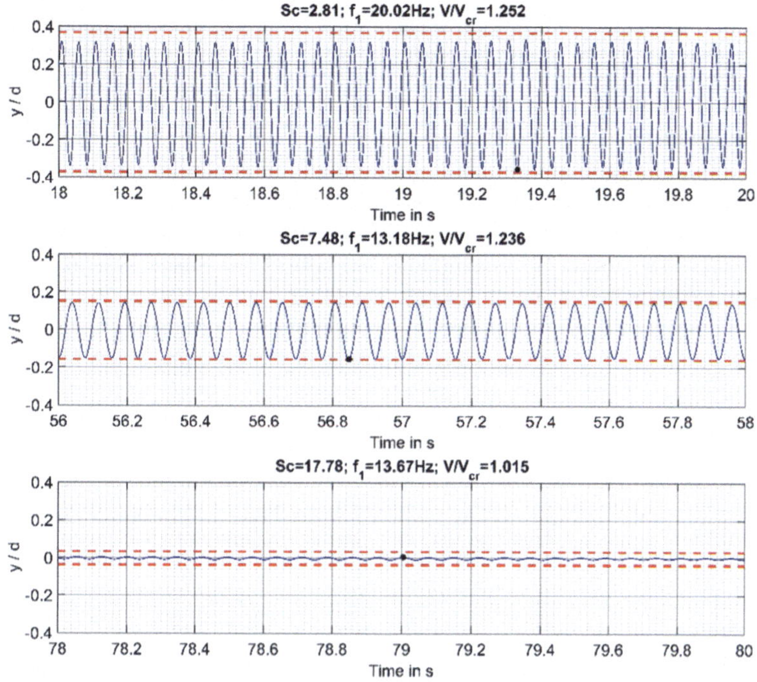

Fig. 7. Model predictions (red) vs. free-vibration wind tunnel tests in lock-in for selected cases (Sc = 2.81, 7.48, 17.78).

Figure 8 plots an overview of the largest amplitudes in the lock-in range as a function of the Scruton number. The model matches the experiments very well. However, it is apparent that the proposed model is not able to reproduce the whole lock-in phenomenon, but it focusses on the prediction of the largest amplitude. This limitation is imposed by having considered a linear equation (Eq. (7)) to describe a non-linear phenomenon. In particular, Eq. (7) is not able to model the "intermittent" behaviour of the response in the resonant zone (see Fig. 2). The intermittency is caused by the fluctuation of the turbulent wind speed around the mean, which causes a lock-in and lock-out effect in case of larger Sc. Capturing and modelling such a non-linear, quasi-stationary behaviour, a sort of detuning between the motion-induced force and the natural frequency of the system (see, for example, a similar approach in Náprstek and Fischer 2017) is a challenge that goes beyond the purpose of this paper.

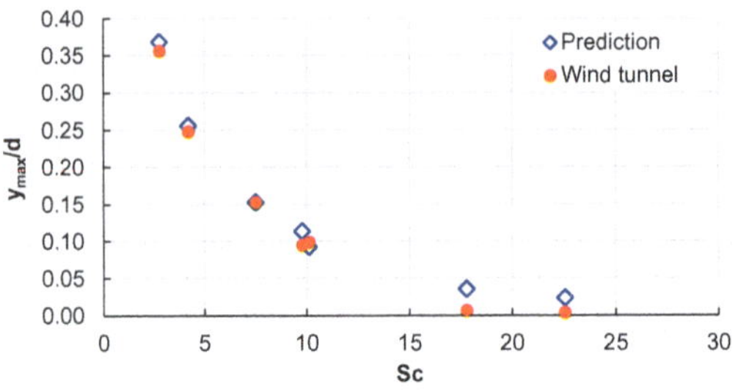

Fig. 8. Model predictions (red) vs. free-vibration wind tunnel tests in lock-in for the cases listed in Table 1. Overview of the largest amplitudes in lock-in as a function of the Scruton number.

4 Conclusion

The paper presents a model extension to predict the maximum amplitude of oscillation in the lock-in range, suitable for wind engineering applications, like chimneys and wind turbines. The following assumptions are considered: linear system, amplitude-dependent aerodynamic damping, vortex shedding load as narrow banded spectral force. The innovative aspect is the implementation of the experimental behaviour of the aerodynamic damping in the equation of motion of the structure.

Acknowledgements. The first author would like to acknowledge the Alexander von Humboldt Foundation (Germany) for the support to this research and the European Commission's Framework Program “Horizon 2020”, through the Marie Skłodowska-Curie Innovative Training Networks (ITN) “AEOLUS4FUTURE - Efficient harvesting of the wind energy” (H2020-MSCA-ITN-2014: Grant agreement no. 643167) for the possibility of cooperation.

References

Arunachalam S, Lakshmanan N (2015) Across-wind response of tall circular chimneys to vortex shedding. J Wind Eng Ind Aerodyn 145:187–195

Basu RI, Vickery BJ (1983) Across-wind vibrations of structures of circular cross-section. Part II: development of a mathematical model for full-scale application. J Wind Eng Ind Aerodyn 12:75–97

Benaroya H, Lepore JA (1983) Statistical flow-oscillator modeling of vortex-shedding. J Sound Vib 86(2):159–179

Billah KZR (1989) A study of vortex-induced vibrations. PhD-thesis, Princeton University

Bishop RED, Hassan AY (1964) The lift and drag forces on an oscillating cylinder. In: Proceedings of the royal society, vol 277

Blevins RD (1990) Flow-Induced Vibration. Van Nostrand Reinhold Co., Inc., New York

CICIND Model Code for Steel Chimneys, The CICIND Chimney Standard (2010)

CICIND Commentaries for Steel Chimney Code (2011)

Dowell EH (1981) Non-linear oscillator models in bluff body aero-elasticity. J Sound Vib 75 (2):251–264

Ehsan F, Scanlan RH (1990) Vortex-induced vibrations of flexible bridges. J Eng Mech 116 (6):1392–1411

Eurocode 1: Actions on Structures, Part 1–4: General actions - Wind actions (2010)
Facchinetti ML, De Langre E, Biolley F (2004) Coupling of structure and wake oscillators in vortex-induced vibrations. J Fluids Struct 19(2):123–140
Farshidianfar A, Zangeneh H (2010) A modified wake oscillator model for vortex-induced vibration of circular cylinders for a wide range of mass-damping ratio. J Fluids Struct 26(3):430–441
Flaga A (1997) Nonlinear amplitude dependent self-limiting model of lock-in phenomenon at vortex shedding. J Wind Eng Ind Aerodyn 69–71:331–340
Goswami I, Scanlan RH, Jones NP (1993) Vortex-induced vibration of circular cylinders. J Eng Mech 119(11):2270–2287
Griffin OM, Skop RA, Koopmann GH (1973) The vortex-excited resonant vibrations of circular cylinders. J Sound Vib 31(2):235–249
Hansen SO (2013) Vortex-induced vibrations – the Scruton number revisited. In: Proceedings of the institution of civil engineers – structures and buildings, vol 166, no 10, pp 560–571
Hartlen RT, Currie IG (1970) Lift-oscillator model of vortex-induced vibration. J Eng Mech Div ASCE 96:577–591
Iwan WD, Blevins RD (1974) A model for vortex induced oscillation of structures. J Appl Mech 41(3):581–586
Krenk S, Nielsen SRK (1999) Energy balanced double oscillator model for vortex-induced vibrations. J Eng Mech 125(3):263–271
Landl R (1975) A mathematical model for vortex-excited vibrations of bluff bodies. J Sound Vib 42(2):219–234
Larsen A (1995) A generalized model for assessment of vortex-induced vibrations of flexible structures. J Wind Eng Ind Aerodyn 57(2–3):281–294
Lupi F, Niemann H-J, Höffer R (2017) A novel spectral method for cross-wind vibrations: application to 27 full-scale chimneys. J Wind Eng Ind Aerodyn 171:353–365
Lupi F, Niemann H-J, Höffer R (2018) Aerodynamic damping model in vortex induced vibrations for wind engineering applications. J Wind Eng Ind Aerodyn 174:281–295
Náprstek J, Fischer C (2017) Analysis of the quasiperiodic response of a generalized van der Pol nonlinear system in the resonance zone. Comput Struct (in press). https://doi.org/10.1016/j.compstruc.2017.07.021
Ogink RHM, Metrikine AV (2010) A wake oscillator with frequency dependent coupling for the modelling of vortex-induced vibration. J Sound Vib 329:5452–5473
Plaschko P (2000) Global chaos in flow-induced oscillations of cylinders. J Fluids Struct 14 (6):883–893
Ruscheweyh H (1986) Ein verfeinertes, praxisnahes Berechnungsverfahren wirbelerregter Schwingungen von schlanken Baukonstruktionen im Wind. Beiträge zur Anwendung der Aeroelastik im Bauwesen, Heft 20. Innsbruck Lausanne (in German)
Scanlan RH (1981) On the state-of-the-art methods for calculation of flutter, vortex-induced and buffeting response of bridge structures. FHWA/RD-80/050, National Technical Information Service, Springfield, VA
Scruton C (1963) On the wind-excited oscillations of stacks, towers and masts. In: International Conference on Wind Effects on Buildings and Structures, Teddington, UK, 26–28 June 1963
Simiu E, Scanlan RH (1986) Wind effects on structures: an introduction to wind engineering, 2nd edn. Wiley, Hoboken
Skop RA, Griffin OM (1973) A model for the vortex-excited resonant response of bluff cylinders. J Sound Vib 27(2):225–233
Skop RA, Griffin OM (1975) On a theory for the vortex-excited oscillations of flexible cylindrical structures. J Sound Vib 41(3):263–274

Skop RA, Luo G (2001) An inverse-direct method for predicting the vortex-induced vibrations of cylinders in uniform and nonuniform flows. J Fluids Struct 15(6):867–884

Tamura Y, Matsui G (1980) Wake oscillator model of vortex-induced oscillation of circular cylinder. In: Proceedings of the fifth international conference on wind engineering, Fort Collins, vol 2, pp 1085–1094

Vickery BJ, Clark AW (1972) Lift or crosswind response of tapered stacks. J Struct Div ASCE 98(ST1):1–20

Vickery BJ, Basu RI (1983) Across-wind vibrations of structures of circular cross-section. Part I: development of a mathematical model for two-dimensional conditions. J Wind Eng Ind Aerodyn 12:49–73

Williamson CHK, Govardhan R (2008) A brief review of recent results in vortex-induced vibrations. J Wind Eng Ind Aerodyn 96:713–735

Sun Y, Li M, Liao H (2014) Nonlinear approach of vortex-induced vibration for line-like structures. J Wind Eng Ind Aerodyn 124:1–6

Aerodynamic Effect of Non-uniform Wind Profiles for Long-Span Bridges

T. M. Lystad[1,2(✉)], A. Fenerci[2], and O. Øiseth[2]

[1] Bridge Department, Norconsult AS, 1338 Sandvika, Norway
tor.martin.lystad@norconsult.com
[2] Department of Structural Engineering,
Norwegian University of Science and Technology, 7034 Trondheim, Norway

Abstract. Long-span bridges are often designed based on the assumption of wind field homogeneity. At the Hardanger Bridge, the wind field along the bridge span is monitored though 8 triaxial ultrasonic anemometers. Simultaneously recorded profiles for mean wind velocity and turbulence intensity along the span are used to investigate the effect of non-uniform wind profiles on the aerodynamic behaviour of the Hardanger Bridge. Extreme non-uniformity is considered using Monte Carlo simulations to generate extreme, but realistic wind profiles based on the variability of the measured wind field. When the buffeting response of the Hardanger Bridge is considered, significant effects on the behaviour is found.

Keywords: Long-span bridge · Non-uniform wind field · Field measurements · Complex terrain

1 Introduction

Long-span bridges located in complex terrain can be subjected to large wind field variations along the bridge span, as shown in (Lystad et al. 2018). Long-span bridges are often designed under the assumption of wind field homogeneity, however this assumption may not be valid for bridges located in complex terrain, such as the mountainous landscape along the west coast of Norway.

Since 2013, the Norwegian University of Science and Technology has been monitoring the wind field along the Hardanger Bridge girder through 8 triaxial ultrasonic anemometers. The Hardanger Bridge is a suspension bridge with a main span of 1 310 m, located in complex terrain. The measured wind field display large non-uniformity along the bridge span.

The effect of idealized non-uniform mean wind velocity profiles on the buffeting response and the aerodynamic stability of long-span bridges has been studied by (Arena et al. 2014; Zhang 2007), showing possible significant effects on the aerodynamic behavior. A non-uniform mean wind velocity profile measured from terrain model wind tunnel tests of the Stonecutters Bridge surroundings was also investigated in (Hu et al. 2017), without any significant impacts to the bridge behavior. However, to the authors knowledge, investigations of the effect of extreme, but realistic non-uniform wind profiles based on full-scale measurements has not been treated in the literature.

F. Ricciardelli and A. M. Avossa (Eds.): INVENTO 2018, LNCE 27, pp. 427–439, 2019.
https://doi.org/10.1007/978-3-030-12815-9_34

In this paper, non-uniform profiles of mean wind velocity and turbulence intensity measured in full-scale at the Hardanger Bridge site are investigated. A measure of the variability of the recorded non-uniformity is obtained by fitting the measured wind profiles to chosen shape functions using linear regression. Profiles of mean wind velocity and corresponding turbulence intensity are generated through Monte Carlo simulations from the joint probability distribution of the fitted shape function coefficients. Finally, the effect of some simulated extreme non-uniform wind profiles on the buffeting response of the Hardanger Bridge are shown.

2 Full-Scale Measurement Program

The Hardanger Bridge is a suspension bridge with a main span of 1 310 m, located along the west coast of Norway. The bridge is crossing the Hardanger fjord surrounded by tall and steep mountains, forming a very complex topography.

Since the opening of the bridge in 2013, a monitoring system has been recording the wind field along the bridge span as well as in the southern tower top. Also, acceleration response of the girder and towers has been recorded. An overview of the wind measurement system is given in Fig. 1 and the system is described in more detail in (Fenerci et al. 2017).

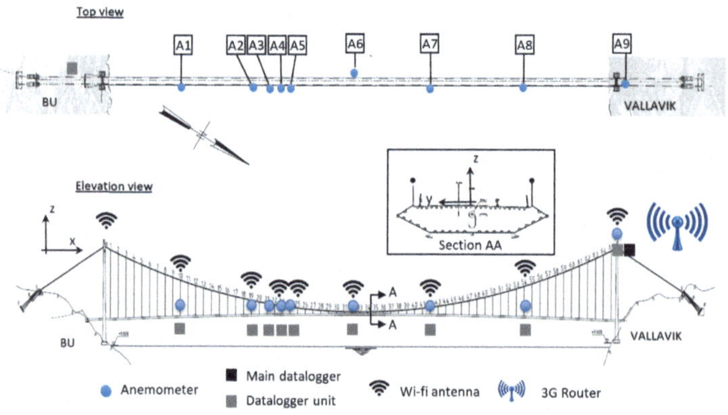

Fig. 1. Overview of the full-scale wind measurement system at the Hardanger Bridge

The Hardanger fjord is creating strong orographic channelling effects on the wind field subjected to the bridge structure, generating two very distinct incoming wind directions from the east and west. This behaviour can be observed in Fig. 2, where the midspan wind rose for strong winds with mean wind velocities above 15 m/s is shown. The wind rose shown in Fig. 2 is plotted on top of a map showing the surrounding topography, and it should be noted that the surrounding mountains reach up to above 1000 m. For a more thorough description of the surrounding topography, the reader is referred to (Fenerci et al. 2017; Fenerci and Øiseth 2017; Lystad et al. 2018).

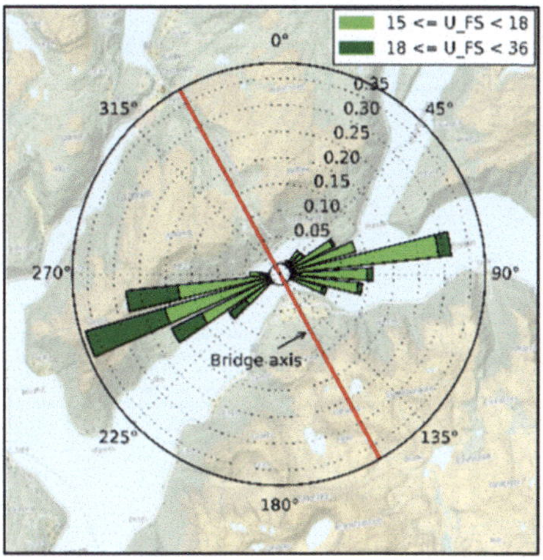

Fig. 2. Mean wind velocity rose for strong winds above 15 m/s, measured in the midspan. The wind rose is divided into two main wind directions, east and west, with each side adding up to 100% (base map with courtesy of Kartverket©)

3 Non-uniform Wind Profiles

3.1 Measurements

Along-span wind profiles for mean wind velocity and along-wind turbulence intensity for strong winds recorded at the Hardanger Bridge are shown in Fig. 3. Only measurements with a midspan 10-min time-averaged mean wind velocity above 15 m/s are considered, and extreme non-uniform measured profiles are highlighted.

By observing the general shape of the measured profiles, it is seen that the profiles may be described quite well by the combination of a linear variation and a cosine function. The following wind profile model is proposed:

$$U(x) = U_0 + a_1 x - a_2[1 - \cos(\frac{\pi}{L}x)] = U_0\{1 + \hat{a}_1 x - \hat{a}_2[1 - \cos(\frac{\pi}{L}x)]\} \tag{1}$$

$$I(x) = I_0 + b_1 x - b_2[1 - \cos(\frac{\pi}{L}x)] = I_0\{1 + \hat{b}_1 x - \hat{b}_2[1 - \cos(\frac{\pi}{L}x)]\} \tag{2}$$

where the a_1 and b_1 coefficients describe the linear variation and the a_2 and b_2 coefficients scale the cosine shape function. Another model for the non-uniformity of the mean wind velocity was shown in (Zhang 2007), but this model is deemed less suitable for the methodology described in the present study.

In the present study, only the non-uniformity in mean wind velocity and turbulence intensity is investigated. However, it should be noted that non-uniformity in other parameters, such as integral length scale, decay coefficients and yaw-angle may also be of importance. In (Lystad et al. 2018) it is shown how the yaw-angle is varying along the bridge span causing different upstream situations for the different parts of the span.

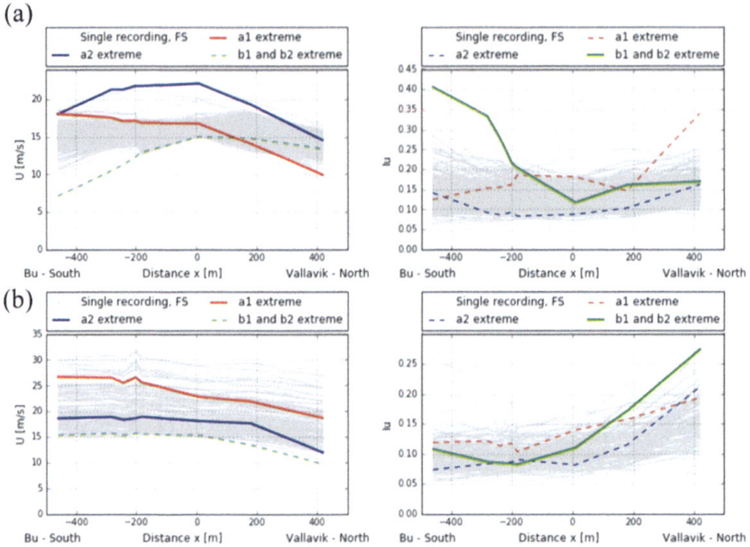

Fig. 3. Measured wind profiles and highlighted extreme non-uniformity, (a) The wind recordings of easterly winds, (b) recordings of westerly winds.

For comparison of the non-uniformity effect it is interesting to consider wind profiles with the same along span mean value. To achieve this the midspan value of the wind profile, U_0 and I_0, and thus the normalized non-uniformity coefficients $\hat{a}_i$ and $\hat{b}_i$ are scaled based on Eqs. (3) and (4), demanding the same along span mean values, U_{eq} and I_{eq} for all profiles.

$$U_{eq} = \frac{1}{L}\int_{-L/2}^{L/2} U(x)dx = U_0[1 - \hat{a}_2 + \frac{2}{\pi}\hat{a}_2] \tag{3}$$

$$I_{eq} = \frac{1}{L}\int_{-L/2}^{L/2} I(x)dx = I_0[1 - \hat{b}_2 + \frac{2}{\pi}\hat{b}_2] \tag{4}$$

Using linear regression, the non-uniform wind profile models described in Eqs. (1)–(4) are fitted to all measured wind profiles shown in Fig. 3. By doing so, the distribution of the normalized fitted coefficients, $\hat{a}_i$ and $\hat{b}_i$, can be obtained and a probability distribution can be adopted. The mean wind velocity profiles and the corresponding along-wind turbulence intensity profiles are used, so the joint probability density function (PDF) of all the normalized coefficients, $\hat{a}_1$, $\hat{a}_2$, $\hat{b}_1$ and $\hat{b}_2$ can be fitted. A normal distribution is assumed in the present study for the simplicity of describing a joint PDF. The marginal fitted PDF's for the normalized coefficients are shown in Fig. 4.õ

It can be observed from Table 1 that the mean value of the linear variation described by $\hat{a}_1$ and $\hat{b}_1$ is larger for the westerly winds than the easterly winds,

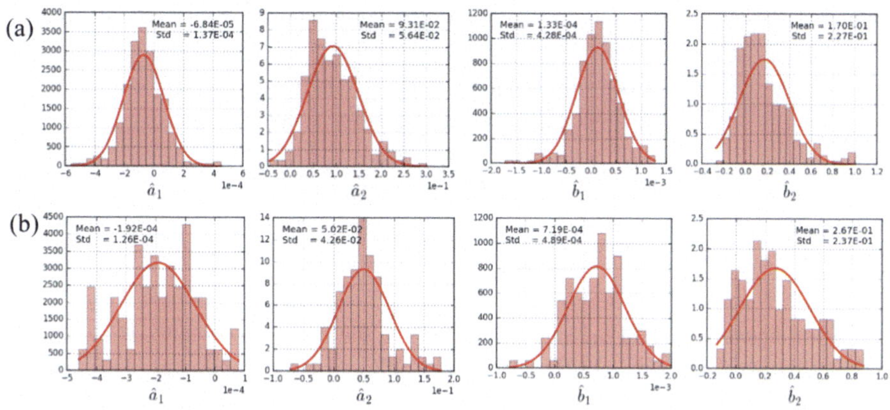

Fig. 4. Marginal PDF's for the normalized coefficients, (a) easterly winds, (b) westerly winds

confirming the qualitative observation made from Fig. 3. However, considering the cosine shape function coefficients $\hat{a}_2$ and $\hat{b}_2$ the easterly winds are displaying the largest mean value for the normalized non-uniformity coefficient for the mean wind velocity profiles, but the westerly winds are showing generally larger non-uniformity in the turbulence intensity profile.

Table 1. Key statistics for the fitted normalized non-uniformity coefficients

	Easterly winds		Westerly winds	
Coeff	Mean	Std	Mean	Std
$\hat{a}_1$	−6.84E−05	1.37E−04	−1.92E−04	1.26E−04
$\hat{a}_2$	9.31E−02	5.64E−02	5.02E−02	4.26E−02
$\hat{b}_1$	1.33E−04	4.28E−04	7.19E−04	4.89E−04
$\hat{b}_2$	−1.70E−01	2.27E−01	−2.67E−01	2.37E−01

Table 2. Correlation coefficients for the normalized non-uniformity coefficients for easterly winds

Coeff	$\hat{a}_1$	$\hat{a}_2$	$\hat{b}_1$	$\hat{b}_2$
$\hat{a}_1$	1.000	0.136	−0.784	−0.151
$\hat{a}_2$	0.136	1.000	−0.208	−0.704
$\hat{b}_1$	−0.784	−0.208	1.000	0.111
$\hat{b}_2$	−0.151	−0.704	0.111	1.000

A high correlation between $\hat{a}_i$ and $\hat{b}_i$ in mean wind velocity and turbulence intensity can be expected, as the turbulence intensity is, by definition, inversely proportional to the mean wind velocity. By studying Table 2 this expected effect is observed as the

Table 3. Correlation coefficients for the normalized non-uniformity coefficients for westerly winds

Coeff	$\hat{a}_1$	$\hat{a}_2$	$\hat{b}_1$	$\hat{b}_2$
$\hat{a}_1$	1.000	−0.390	−0.504	0.215
$\hat{a}_2$	−0.390	1.000	0.465	−0.502
$\hat{b}_1$	−0.504	0.465	1.000	−0.657
$\hat{b}_2$	0.215	−0.502	−0.657	1.000

correlation between $\hat{a}_1$ and $\hat{b}_1$, and $\hat{a}_2$ and $\hat{b}_2$ is in the range of 70–80%. However, by studying the westerly winds in Fig. 4(b) and Table 3 a larger randomness is observed in the normalized non-uniformity coefficients as well as in the correlation between them.

3.2 Simulations

Based on the fitted joint PDF of the normalized non-uniformity coefficients, Monte Carlo simulations can be applied to generate a large number of non-uniform wind profiles. The simulations are based on a chosen equivalent mean wind velocity of 40 m/s and an equivalent along-wind turbulence intensity of 15%. The fitted coefficients are generally based on lower measured wind speeds than the chosen equivalent mean wind velocity, but utilizing the normalized format of the non-uniformity coefficients it is interesting to investigate the effects of the observed non-uniformity at design wind speed levels. Extreme, but realistic, simulated profiles may be picked based on a chosen criterion. In this study the simulations are sorted in ascending order based on the absolute value of the normalized non-uniformity profiles, $\hat{a}_i$ and $\hat{b}_i$. To obtain an extreme non-uniform situation from the simulated profiles the nearest-rank method may be applied to the sorted lists of simulations for each normalized non-uniformity coefficient. Let, N, be the number of simulations and, n, be the simulation number in the sorted list, the simulated profile with a chosen percentile may be calculated as

$$n = \left\lceil \frac{P}{100} N \right\rceil \tag{5}$$

In Fig. 5, $N = 1000$ simulated profiles for each main wind direction, westerly and easterly winds, are shown. Extreme profiles picked according to the criterion described above are indicated as highlighted solid lines. The simulated wind profile of the other variable corresponding to the ranked variable extreme profile are dotted in the same colour in the figure. The normalized non-uniformity coefficients for the picked extreme profiles are shown in Table 4. Significant non-uniformity in the extreme simulated wind profiles both for mean wind velocity and along-wind turbulence intensity can be observed.

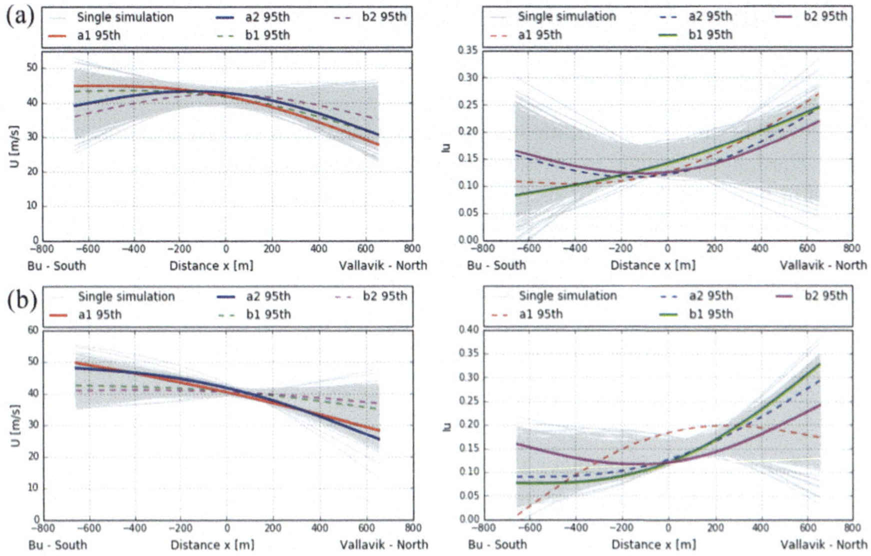

Fig. 5. Monte Carlo simulated wind profiles for, (a) easterly winds and, (b) westerly winds. Extreme simulations are highlighted along with the simulation of the corresponding variable (dotted).

Table 4. Model coefficients for the highlighted extreme non-uniformity profiles

Wind dir	Extreme type	U_0	$\hat{a}_1$	$\hat{a}_2$	I_0	$\hat{b}_1$	$\hat{b}_2$
East	$\hat{a}_{1,95th}$	42.02	−3.10E−04	1.32E−01	0.13	9.72E−04	−4.94E−01
	$\boldsymbol{\hat{a}_{2,95th}}$	**42.87**	**−1.50E−04**	**1.84E−01**	**0.12**	**5.48E−04**	**−6.51E−01**
	$\hat{b}_{1,95th}$	41.69	−2.28E−04	1.12E−01	0.14	8.75E−04	−1.58E−01
	$\hat{b}_{2,95th}$	42.49	−1.64E−05	1.61E−01	0.13	3.35E−04	−5.31E−01
West	$\hat{a}_{1,95th}$	40.53	−4.04E−04	3.61E−02	0.18	6.82E−04	5.01E−01
	$\hat{a}_{2,95th}$	41.85	−4.12E−04	1.22E−01	0.13	1.23E−03	−5.18E−01
	$\boldsymbol{\hat{b}_{1,95th}}$	**40.69**	**−1.41E−04**	**4.69E−02**	**0.12**	**1.59E−03**	**−6.83E−01**
	$\hat{b}_{2,95th}$	40.60	−7.74E−05	4.04E−02	0.12	5.25E−04	−6.61E−01

It should be noted that the described criterion is not a statistically complete procedure to identify a situation corresponding to an extreme load event, but it can provide valuable information for illustration purposes.

4 Case Study: The Hardanger Bridge

4.1 Theoretical Background

The effect of the simulated extreme non-uniform wind profiles on the dynamic response of The Hardanger Bridge is investigated. The Hardanger Bridge is a symmetrical suspension bridge with a main span of 1 310 m and no side spans.

The response calculations are performed in the frequency domain using the multimode theory described in (Chen et al. 2001; Jain et al. 1996a, b). The equation of motion of the dynamic system can be described in the frequency domain as

$$\begin{aligned}&\tilde{\mathbf{M}}_S\mathbf{G}_{\ddot{\eta}}(\omega)+[\tilde{\mathbf{C}}_S-\tilde{\mathbf{C}}_{ae}(U,\omega)]\mathbf{G}_{\dot{\eta}}(\omega)\\&+[\tilde{\mathbf{K}}_S-\tilde{\mathbf{K}}_{ae}(U,\omega)]\mathbf{G}_{\eta}(\omega)=\mathbf{G}_{\tilde{\mathbf{Q}}_{load}}(U,\omega)\end{aligned} \tag{6}$$

where ω is the angular frequency, $\tilde{\mathbf{M}}_S$, $\tilde{\mathbf{C}}_S$ and $\tilde{\mathbf{K}}_S$, are the structural mass-, damping - and stiffness matrices, respectively, in modal coordinates. $\tilde{\mathbf{C}}_{ae}$ and $\tilde{\mathbf{K}}_{ae}$, are the aeroelastic damping and stiffness matrices respectively, representing the motion induced forces. $\mathbf{G}_{\eta}$, $\mathbf{G}_{\dot{\eta}}$, $\mathbf{G}_{\ddot{\eta}}$ and $\mathbf{G}_{\mathbf{Q}_{load}}$ are the Fourier transforms of the displacement-, velocity-, acceleration response and the load process, respectively. In practice, the structural part of the equation has been calculated based on the finite element method, with ABAQUS (Dassault Systèmes, Waltham n.d.), and the aeroelastic motion induced forces are described based on wind tunnel experiments (Siedziako et al. 2017).

The wind field is described as a stationary stochastic process through the cross-spectral density matrix

$$\mathbf{S}_V(\Delta x,U,\omega)=\begin{bmatrix}S_{uu}(\Delta x,U,\omega) & S_{uw}(\Delta x,U,\omega)\\ S_{uw}(\Delta x,U,\omega) & S_{ww}(\Delta x,U,\omega)\end{bmatrix} \tag{7}$$

where S_{nm} are the cross spectral densities for the n and m components of the turbulence between two points separated in space by the distance Δx. In the current study, the off-diagonal terms of the cross-spectral density matrix are assumed to be negligible. The cross-spectral density for a single turbulence component can be described through the auto-spectral density function and the normalized co-spectra:

$$S_{nn}(\Delta x,U,\omega)=S_n(U,I_n,\omega)C_n(\Delta x,U,\omega) \tag{8}$$

$$C_n(\Delta x,\omega)=\exp(-K_n\frac{\omega\Delta x}{2\pi U}) \tag{9}$$

where S_n is the auto-spectral density function, C_n is the normalized co-spectra, and K_n is the decay coefficient. Having established the cross-spectral density matrix for the wind field process, the spectral matrix of the buffeting force on the structure in modal coordinates can be defined as

$$\mathbf{S}_{\tilde{\mathbf{Q}}_{load}} = \iint_L \Phi^T(x_1)\mathbf{B}_q(U,\omega)\mathbf{S}_V(\Delta x,\omega)\mathbf{B}_q^H(U,\omega)\Phi(x_2)dx_1dx_2 \tag{10}$$

where $\Phi(x_i)$ is the mode shape matrix and $\mathbf{B}_q$ is the load transfer matrix for the buffeting load on the structure. The load spectral density matrix is calculated by considering two points at a time, x_1 and x_2. Having a non-uniform wind field, the mean wind velocity and turbulence intensity will differ at the two points considered, so an adjustment to the original buffeting theory first described by (Davenport 1962; Scanlan and Tomko 1971) has to be made. In the present study, a simple adjustment is done by defining the cross-spectral density for the considered turbulence component as the product of the square root of the auto-spectral density in each point and a normalized co-spectra based on the average of the mean wind velocity from the two points considered. Hence, the following adjustment can be made to Eq. (8):

$$\begin{aligned} &S_{nn}[\Delta x, U(x_1), U(x_2), \omega] = \\ &\sqrt{S_n[U(x_1), I_n(x_1), \omega]}\sqrt{S_n[U(x_2), I_n(x_2), \omega]}C_n(\Delta x, U_{avrg}, \omega) \end{aligned} \tag{11}$$

where $U_{avrg} = [U(x_1) + U(x_2)]/2$.

The buffeting load transfer matrix is also a function of the mean wind velocity, and to account for the non-uniformity of the buffeting matrix corresponding to each of the points considered in Eq. (10), it is adjusted using the actual mean wind velocity in that point. Thus, the cross-spectral density matrix for the buffeting force on the structure can be redefined as

$$\mathbf{S}_{\tilde{\mathbf{Q}}_{load}} = \iint_L \begin{matrix} \{\Phi^T(x_1)\mathbf{B}_q[U(x_1),\omega]\mathbf{S}_V[\Delta x, U(x_1), U(x_2),\omega] \\ \mathbf{B}_q^H[U(x_2),\omega]\Phi(x_2)\}dx_1dx_2 \end{matrix} \tag{12}$$

The motion induced forces are integrated using the full non-uniform mean wind velocity profile as follows:

$$\tilde{\mathbf{K}}_{ae}(U(x),\omega) = \int_L \Phi^T(x)\mathbf{K}_{ae}(U(x),\omega)\Phi(x)dx \tag{13}$$

$$\tilde{\mathbf{C}}_{ae}(U(x),\omega) = \int_L \Phi^T(x)\mathbf{C}_{ae}(U(x),\omega)\Phi(x)dx \tag{14}$$

where $\mathbf{K}_{ae}$, and $\mathbf{C}_{ae}$ are described by the dimensionless aerodynamic derivatives.

4.2 Modal Analysis

Contributions from the 90 first still air modes was included in the modal analysis, resulting in a cut off frequency of 1.6 Hz. In Fig. 6 the two first lateral, vertical and torsional vibration mode shapes are shown, and in Table 5 the first 15 modes are summarized.

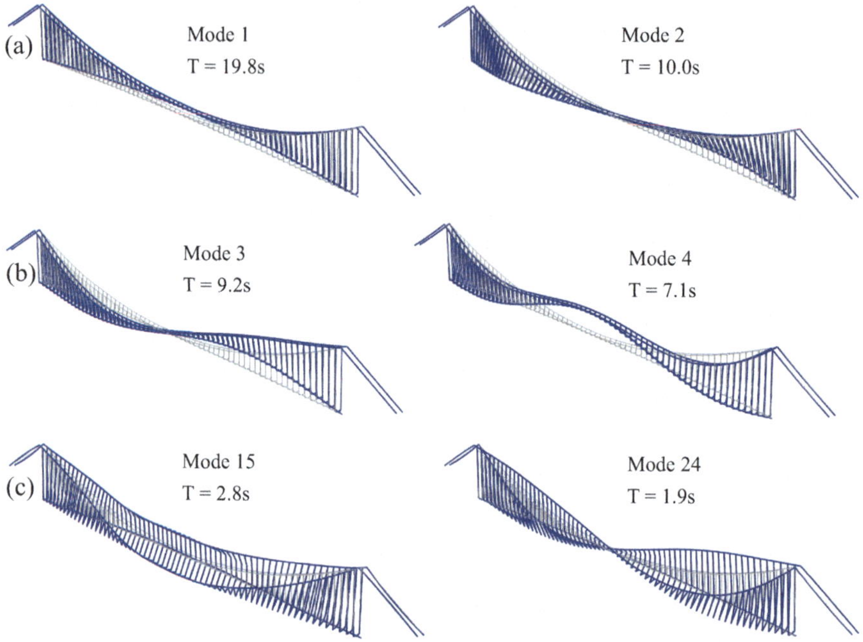

Fig. 6. Mode shapes of the Hardanger Bridge, (a) Two first lateral vibrations modes, (b) Two first vertical vibration modes, (c) Two first torsional vibration modes

Table 5. The frequency, period and shape of the first 15 modes of the Hardanger Bridge

Mode	Frequency	Period	Shape description
1	0.051	19.77	Lateral sym.
2	0.100	10.00	Lateral asym.
3	0.109	9.18	Vertical asym.
4	0.141	7.08	Vertical sym.
5	0.174	5.75	Lateral sym.
6	0.200	4.99	Vertical sym.
7	0.211	4.74	Vertical asym.
8	0.222	4.51	Cabel vibration
9	0.230	4.34	Cabel vibration
10	0.235	4.26	Cabel vibration
11	0.245	4.08	Cabel vibration
12	0.273	3.66	Vertical sym.
13	0.302	3.31	Lateral asym.
14	0.329	3.04	Vertical asym.
15	0.356	2.81	Torsional sym.

4.3 Buffeting Response

Using the methodology described above, the buffeting short-term extreme value for the horizontal, vertical and torsional displacement responses as well as the bending - and torsional cross-sectional moments has been calculated for the Hardanger Bridge girder. The tested extreme non-uniformity profiles are the cosine shaped extreme for the mean wind velocity from the easterly direction and the linearly varying extreme along-wind turbulence intensity profile for the westerly winds, highlighted in Table 4 and shown again in Fig. 7. Only the along-wind turbulence intensity is based on the full-scale measurements in this study, and the vertical component is assumed to be 50% of the along-wind component, adopting the same along span profile shape.

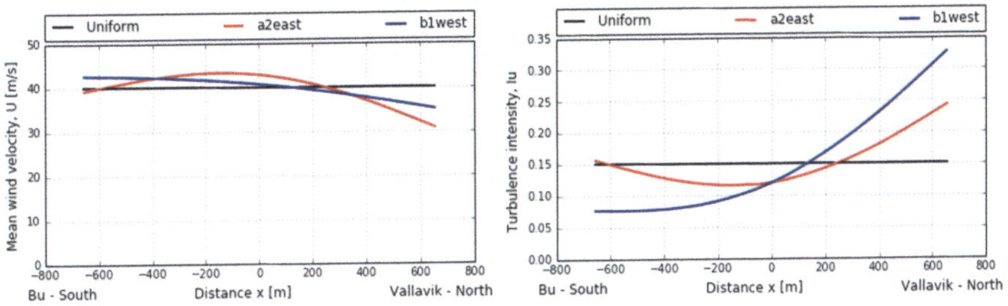

Fig. 7. Extreme non-uniform wind profiles used in buffeting calculations

Significant effects on the response can be observed in Fig. 8, for all response quantities considered. For the *a2east* profile all buffeting response quantities are lower than for the uniform situation. The reason for this is the opposite trends of the mean wind velocity and the turbulence intensity, cancelling each other out when considering buffeting action. A combination like this is natural, since the turbulence intensity is inverse proportional to the mean wind velocity. This observed effect highlights the limitations of the simple sorting criterion used in the present study where only extreme non-uniformity for each variable, either mean wind velocity or turbulence intensity, is sorted, and the corresponding profile has no influence on the sorting criterion. Adopting such an approach would not be suitable to identify the combined extreme non-uniformity situation for the considered bridge response, but merely to identify extreme profiles for the purpose of illustration. Also, different sorting criterions would be appropriate for the consideration of different response processes. For instance, an unfavorable combination of mean wind velocity and turbulence intensity would be critical for the buffeting response, but considering flutter instability the mean wind velocity is the driving process and an extreme mean wind velocity profile should be chosen with less emphasis on the combination with turbulence intensity. Possible sorting criterions for the simulated non-uniformity profiles are not investigated in the present study and could be subject for further work on this topic.

Although a cancelling out effect is observed for the considered *a2east* profiles, the *b1west* simulation show unfavorable conditions for the buffeting response of the Hardanger Bridge. The very high turbulence intensity towards the north end of the bridge is

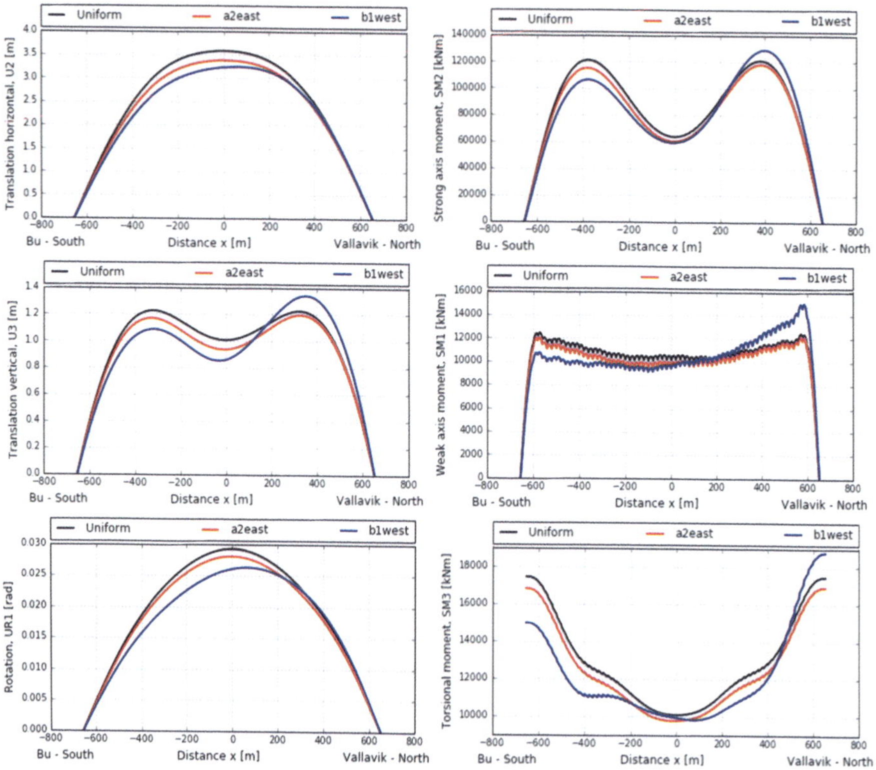

Fig. 8. Along span buffeting response of the Hardanger Bridge subjected to different wind profiles

generating large responses for all sectional moments shown in Fig. 8. Especially the weak axis moment, *SM1*, shows an underestimation of about 25% using the uniformity assumption compared with the simulated extreme non-uniformity profile.

5 Conclusions

A proposed method for investigating extreme non-uniformity effects for long-span bridges has been presented in this study. Measured full-scale wind profiles from the complex terrain of the Hardanger Bridge site are used as basis for simulation of extreme, but realistic non-uniform profiles of mean wind velocity and along-wind turbulence intensity and the following conclusions are drawn:

- The measured wind field profiles for westerly winds show a linearly varying trend. The easterly winds display a more typical effect where mean wind velocity is larger in the midspan than towards the bridge ends, and the opposite is the case for the turbulence intensity, due to terrain roughness changes along the span.

- A larger randomness in the fitted normalized non-uniformity coefficients was observed for the westerly winds than the easterly winds, indicating a dependence between the observed non-uniformity and the upstream terrain complexity.
- In complex terrain, extreme non-uniformity in the wind field can be expected, to the extent that should not be neglected in design of long-span bridges.
- The inverse nature of the mean wind velocity and the corresponding turbulence intensity will to some extent cancel out the non-uniformity effects on the buffeting response. So, to investigate extreme effects due to non-uniformity, an unfavourable combination of mean wind speed and turbulence intensity should be considered.

Acknowledgements. The research presented in this study has been financed by the Norwegian Public Roads Administration (NPRA), the Norwegian Research Council and Norconsult AS. The authors appreciate all these contributions.

References

Arena A, Lacarbonara W, Valentine DT, Marzocca P (2014) Aeroelastic behavior of long-span suspension bridges under arbitrary wind profiles. J Fluids Struct 50:105–119. https://doi.org/10.1016/j.jfluidstructs.2014.06.018

Chen X, Kareem A, Matsumoto M (2001) Multimode coupled flutter and buffeting analysis of long span bridges. J Wind Eng Ind Aerodyn 89:649–664. https://doi.org/10.1016/S0167-6105(01)00064-2

Dassault Systèmes, Waltham (n.d.) ABAQUS

Davenport AG (1962) Buffeting of a suspension bridge by storm winds. J Struct Div 88:233–270

Fenerci A, Øiseth O (2017) Measured buffeting response of a long-span suspension bridge compared with numerical predictions based on design wind spectra. J Struct Eng 143. https://doi.org/10.1061/(ASCE)ST.1943-541X.0001873

Fenerci A, Øiseth O, Rönnquist A (2017) Long-term monitoring of wind field characteristics and dynamic response of a long-span suspension bridge in complex terrain. Eng Struct 147:269–284. https://doi.org/10.1016/j.engstruct.2017.05.070

Hu L, Xu Y, Zhu Q, Guo A, Kareem A (2017) Tropical storm – induced buffeting response of long-span bridges: enhanced nonstationary buffeting force model. J Struct Eng 143:04017027. https://doi.org/10.1061/(ASCE)ST.1943-541X.0001745

Jain A, Jones NP, Scanlan RH (1996a) Coupled aeroelastic and aerodynamic response analysis of long-span bridges. J Wind Eng Ind Aerodyn 60:69–80. https://doi.org/10.1016/0167-6105(96)00024-4

Jain A, Jones NP, Scanlan RH (1996b) Coupled flutter and buffeting analysis. J Struct Eng 122:716–725

Lystad TM, Fenerci A, Øiseth O (2018) Evaluation of mast measurements and wind tunnel terrain models to describe spatially variable wind field characteristics for long-span bridge design. J Wind Eng Ind Aerodyn 179:558–573. https://doi.org/10.1016/J.JWEIA.2018.06.021

Scanlan RH, Tomko JJ (1971) Airfoil and bridge deck flutter derivatives. J Eng Mech Div 97:1717–1737

Siedziako B, Øiseth O, Rønnquist A (2017) An enhanced forced vibration rig for wind tunnel testing of bridge deck section models in arbitrary motion. J Wind Eng Ind Aerodyn 164:152–163. https://doi.org/10.1016/J.JWEIA.2017.02.011

Zhang X (2007) Influence of some factors on the aerodynamic behavior of long-span suspension bridges. J Wind Eng Ind Aerodyn 95:149–164. https://doi.org/10.1016/J.JWEIA.2006.08.003

A Comparative Analysis of Construction Costs of Onshore and Shallow- and Deep-Water Offshore Wind Farms

C. Maienza[1(✉)], A. M. Avossa[1], F. Ricciardelli[1], F. Scherillo[2], and C. T. Georgakis[3]

[1] Department of Engineering (DI), Università della Campania "Luigi Vanvitelli", Via Roma 9, 81031 Aversa, Italy
{carmela.maienza, albertomaria.avossa, friccia}@unicampania.it
[2] Seapower scrl, Naples, Italy
ferdinando.scherillo@seapowerscrl.com
[3] Department of Engineering, Aarhus University, Aarhus, Denmark
cg@eng.au.dk

Abstract. Recent years have seen renewed effort in the technological development of wind energy industry, where alongside onshore and shallow water offshore turbines, deep water offshore turbines have made their appearance, which represent the next frontier. One of the main aspects while developing a wind energy project is the overall cost and its breakdown between the main components. A comparative analysis of the costs associated with the different solution, the assessment of the key parameters of the wind economy, is therefore crucial when evaluating the possible investments. The aim of this work is to compare typical costs of onshore, shallow- and deep-water offshore wind farms, through an analysis of the aspects that characterize the wind economy, such as the cost of capital and operation and maintenance, and of their subheadings.

Keywords: Onshore wind turbine · Shallow water wind turbine · Deep water wind turbine · Capital cost · Operation & maintenance costs

1 Introduction

Over the last few decades, the increase in energy demand, the prevailing use of unsustainable resources and environmental degradation have become critical threats that require quick action. In this scenario, the use of renewable energy sources is an opportunity that thanks to technological progress is becoming competitive compared to traditional sources based on fuels. Among renewable sources, wind is one of the most secure, reliable and cost-effective. Recent years have seen a renewed effort towards the development of wind turbine technology, where alongside onshore and shallow water offshore turbines, deep water offshore turbines have made their appearance, which represent the next challenge in the wind power industry. Accumulated global installed capacity of wind energy has increased from approximately 48 GW in 2004 to more than 321 GW at the end of 2013, with an annual growth of more than 20%. Installed

F. Ricciardelli and A. M. Avossa (Eds.): INVENTO 2018, LNCE 27, pp. 440–453, 2019.
https://doi.org/10.1007/978-3-030-12815-9_35

wind capacity across all the EU member states is expected to quintuple, to more than 200 GW, between 2005 and 2020 (Beurskens 2011).

A similar trend, can be observed for offshore wind power that is expected to increase, between 2005 and 2020, from about 2% to 20% of the total installed capacity (Larsen and Sonderberg Petersen 2014). These figures reflect the expected decrease in the costs of offshore wind farms, and of the possible shortage of suitable onshore areas for wind generation, especially for densely populated countries, or areas prone to high environmental risk (Avossa et al. 2017a, b, c). This brought to the vision of deep water offshore wind energy generation, which has now entered into a prototyping stage. The advantages of deep water wind generation are the availability of an almost unlimited resource, together with lower visual and acoustic impact. Moreover, in the deep water environment a larger producibility can be achieved, due to higher winds with lower turbulence, the latter being also beneficial for fatigue loads (Bilgili et al. 2011).

On the other hand, offshore wind turbines have the disadvantage of higher complexity and of higher installation, maintenance and decommissioning costs; this is due to restricted site access because of harsh weather conditions, expensive installation procedures and grid connections.

One of the main aspects while developing a wind farm project is the overall cost and its breakdown. In recent years, studies have been carried out with the aim of optimizing the costs of wind energy projects, in particular, offshore ones. With reference to wind farms located in South Korea, Sajid et al. (2017) evaluated each wind turbine according to techno-economic parameters and considering, among the different values, also that of the Levelized Cost of Energy (LCOE).

In the case of deep water offshore farms, the planning, installation and operation process is becoming more complex due to the tendency in the offshore wind energy to implement projects more and more distant from the coast and in deeper waters.

Since the sector is under development and due to the absence of well-established tools, LCOE parameter proves to be useful, with which is possible to reduce the costs of wind turbine farms, by analyzing key aspects already during the planning and pre-design phase (Ebenhoch et al. 2015).

Within this topic, the paper aims at highlighting the typical costs of onshore, shallow- and deep-water offshore wind farms, through a specific analysis of the aspects that characterize the wind power economy. Moreover, a comparative analysis of the costs of the various typologies of wind turbines and the main economic key parameters will be carried out to evaluate the best possible field of investment. This is the counterpart of the wind resource available, that can be drawn from mean wind maps. Comparison of resource and installation costs in the three different situations considered gives a framework of the possible evolution of the wind energy industry for the next years. In particular, an effective economic analysis of the deep water offshore market is necessary, which to date has never been carried out. This economic analysis is fundamental to underline the greater producibility of offshore wind turbines in relation to sustainability and therefore, allows to highlight the great value to support the technical values of the next and innovative wind farms.

2 Current Costs of Wind Energy Farms

The key parameters of wind power economics are the capital cost and the operation and maintenance costs. They are often referred to as Capital expenditure (Capex) and Operational expenditure (Opex), respectively. Capex is the largest contributor to the life cycle costs of wind farms and covers all costs incurred before the commercial operation date. Opex instead, comprise all expenditures incurred after the commercial operation date, but prior to the decommissioning, that are necessary to operate the project and maintain turbine availability to produce energy.

2.1 Capital Cost

The capital cost can raise up to 80% of the total cost of the wind power project over its all lifetime (Blanco 2009). Capital costs are divided into: turbine costs, civil works, grid connection costs and other capital costs (e.g. construction of buildings, control systems).

Turbine costs include all the components of the wind generator, among which the main part consist of tower (the largest cost component), rotor, blades and nacelle.

Civil works include the tower foundations costs and the construction costs for site work. The largest cost components of foundations are the construction materials such as reinforced concrete and steel. The construction costs include transportation and installation of wind turbine (e.g. construction of access roads and other infrastructure required for the wind farm installation).

Grid connection includes transformers, substations and connection to the local transmission or distribution network. In fact, wind farms can be connected to electricity grids via the transmission network or distribution network. If the grid connection point is not far from the wind farm, the connection is typically a high voltage alternating current (HVAC) connection. The transmission capability of HVAC system reduces with distance because of dielectric losses. It is reasonable to use high voltage direct current (HDVC) in case of wind farms placed at a distance greater than 50 km from coast, because it has a distance independent power transmission capability (Brown et al. 2015).

2.2 Operation and Maintenance Costs

Kitzing and Monthorst (2015) states that the operations and maintenance costs (O&M) account around 25–30% of the total costs of a wind project. Moreover, this share might increase up to 30–35% towards the end of the project lifetime.

O&M costs are divided into fixed and variable. Fixed costs include the administration, insurance, grid access fees and service contracts for scheduled maintenance, while variable costs include scheduled and unscheduled maintenance not covered by fixed contracts, replacement parts and materials (IRENA 2012). Besides, O&M costs for offshore wind farms are significantly higher than onshore wind farms due to the higher costs involved in accessing and conducting maintenance on the wind turbines, towers and cabling and because of the harsh marine environment, tougher weather conditions and the higher expected failure rate for some components (Kitzing and Monthorst 2015).

3 Costs of Wind Turbines

The capital costs of a wind power project are split into several components. As reported by the International Renewable Energy Agency (IRENA 2012), the comparison of capital costs breakdown for typical onshore and offshore wind turbines are shown in Table 1.

Table 1. Comparison of capital costs breakdown for typical onshore and offshore, 2011

	Onshore	Offshore
Capital investment costs (USD/kW)	1750–2450	3300–5000
(a) Turbine	64–84%	30–50%
(b) Grid connection	9–14%	15–30%
(c) Construction	4–16%	15–25%
(d) Other financial costs	4–10%	8–30%

With reference to the various wind installations, the capital investment costs (USD/kW) and the percentages relating to the following parts are divided in four parts: turbine, grid connection, construction and other financial costs.

It can be observed that offshore turbines usually show double capital investment costs compared to onshore ones. Furthermore, there is a marked increase for all other cost, with the exception of the turbine.

In Table 2 detailed capital cost items of onshore, shallow- and deep-water offshore turbine are shown. It can be detected that for all types of wind installations, the turbine remains unchanged in terms of capital cost components. The other items differ mainly between the onshore and offshore types, in which the latter include additional capital cost components. Finally, shallow- and deep-water wind offshore turbines, differ in the type of foundation.

3.1 Onshore Wind Farm Costs

Regarding to a typical onshore wind farm, the main cost component is the turbine, whose most expensive parts are the tower, rotor blades and the gearbox which cost corresponds to 50–60% of the overall turbine cost. The generator, power converter, transformer and other miscellaneous costs (e.g. rotor hub, cabling) account for about 13% of the turbine costs (IRENA 2012).

The overall cost of the turbine also depends on the type of foundations (monopiles, jacket and other).

Grid connection is typically a high voltage alternating current (HVAC) because grid connection point is not far from the wind farm and there are low dieletric losses.

Table 2. Capital cost items for onshore, shallow- (SW) and deep-water (DW) offshore turbines

	Onshore	Offshore (SW)	Offshore (DW)
Turbine	Tower, Nacelle, Rotor, Blades	Tower, Nacelle, Rotor, Blades	Tower, Nacelle, Rotor, Blades
Grid connection	Onshore cables, Onshore substation	Onshore cable, Onshore & Offshore substations, Export cable, Interray cables	Onshore cable, Onshore & Offshore substations, Export cable, Interray cables
Construction	Land-based tower, Site access & Staging	Fixed-bottom tower, Port, Staging & Vessels	Floater, Mooring lines, Anchors, Port, Staging & Vessels
Other financial costs	Contingency, Construction finance	Contingency, Construction finance, Insurance, Decommissioning	Contingency, Construction finance, Insurance, Decommissioning

The total cost of grid connection varies from the place where the turbine is located (e.g. ridges of the mountains, hills) and the distance of the national electrical connection point to that of the wind farm.

An example of the cost data can be achieved analysing onshore, shallow- and deep-water offshore wind turbine projects from of National Renewable Energy Laboratory (NREL 2015). In particular, a comparison of the costs split that characterize the economy of different typologies of wind turbine installation can be developed. In the onshore case, the percentage costs are reported, based on a hypothetical turbine having the average of the characteristics among those of 68 wind projects installed in 2015 in the United States. The reference project is made up of several hundreds 2 MW turbines. Figure 1 shows the percentages of capital expenditures, with four groups of colors (red, green, blue and yellow) corresponding to the four main cost category reported in Table 1. Similar diagrams will be presented in the next Figs. 2 and 4 for the shallow- and deep-water offshore, respectively.

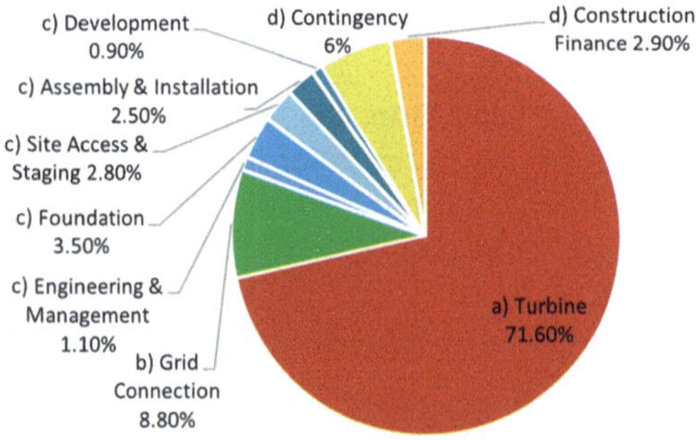

Fig. 1. Capital expenditures for onshore turbine reference wind plant project

3.2 Shallow Water Offshore Wind Farm Costs

Offshore wind farms are characterized by a higher producibility than onshore wind farms, main due to the presence of a higher wind speed and lower roughness condition.

For a typical offshore wind farm, the largest cost component is still the wind turbine.

The capital cost of shallow water offshore wind turbines is about two times higher than onshore wind energy projects. This higher cost is due to the expensive foundation constructions at sea, equipment, transporting materials, additional protection against corrosion and to the higher costs deriving from laying cables on seabed (IRENA 2012).

Grid connection is typically a high voltage alternating current (HVAC) connection. The distance to shore and water depth influence the total costs of a shallow offshore wind farm. In particular, grid connection and installation costs are the most affected.

Grid connection cost decreases as wind farm size increases. The distance to shore influences especially the cable installation costs within the total electrical installation costs. Installation cost almost double when the distance to the onshore grid connection point goes from zero to 60 km (EEA 2009). Besides the export cable, which connects the wind farm to an onshore connection point, is affected by distance from shore. In particular, the factors influencing cable costs are the cable size, sea bed conditions and possible presence of transformer stations.

In addition to grid connection, installation costs are also much effected. At larger distances, installation costs increase due to the larger travelling time required from the holding port to the offshore site.

Foundation cost increases due to the raising of water depth. For offshore wind turbines, foundation costs are estimated to increase of about 10% for a depth of the fixed tower going from 8 m to 16 m (EEA 2009). Furthermore, foundation supply costs can triple as the depth increases from 15 m to 40 m, using monopiles (EEA 2009). Scale factors associated with the reference case aim at deriving offshore investment costs, as a function of both distance from the coast and water depth, are shown in Table 3.

Operation and maintenance costs for shallow-water offshore wind farms are significantly higher than for onshore wind farms due to the larger costs involved in accessing and conducting maintenance on the wind turbines, towers and cabling and because of the harsh marine environment, tougher weather conditions and the higher expected failure rate for some components (Kitzing and Monthorst 2015).

Weather conditions can make installation more difficult leading to downtime which can affect about 20–30% on the overall installation time. Downtime is a factor that reflects the percentage of the time in which the weather conditions do not allow the vessel to operate safely. Therefore, it is an additional factor acquainting the real time needed to install offshore. For the case of shallow water systems, results from the project of the National Renewable Energy Laboratory are taken into account as reference (NREL 2015). The project was created by conducting several parallel assessments on 14 offshore wind projects installed in 2015 (e.g. considering global data analysis, cost reports, collaboration with industry). The average turbine capacity that were installed in 2015 was 4.14 MW and the corresponding percentages of capital expenditures are shown in Fig. 2.

Table 3. Scale factors for cost increases as a function of distance from coast and water depth

		Distance from coast (km)							
		0–10	10–20	20–30	30–40	40–50	50–100	100–200	>200
Depth (m)	10–20	1	1.022	1.043	1.065	1.086	1.183	1.408	1.598
	20–30	1.067	1.090	1.113	1.136	1.159	1.262	1.501	1.705
	30–40	1.237	1.264	1.290	1.317	1.344	1.464	1.741	1.977
	40–50	1.396	1.427	1.457	1.487	1.517	1.653	1.966	2.232

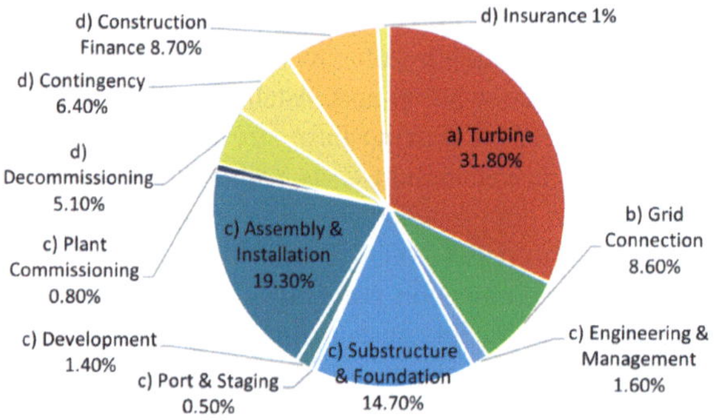

Fig. 2. Capital expenditures for shallow offshore turbine reference wind plant project

3.3 Deep Water Offshore Wind Farm Costs

The capital cost of wind turbines in deep water is about 3 times larger than onshore and about 1.5 times than for shallow water wind generators. This higher cost is mainly due to the laying of cables at sea, to expensive foundations at sea, and to shipping and construction site setup costs. As for turbines in shallow water, deep water offshore wind turbines require additional protection against the harsh marine environment and to corrosion, such to contain maintenance costs.

Wind turbine does not result the largest cost component such as in the other previous cases, but the substructure and foundation became the major cost component.

Grid connection are High Voltage Alterning Current (HVAC) or High Voltage Direct Current (HVDC) connections. The latter type of connection is typically used in deep water (for distances larger than 50 km), as the transmission capability of HVAC system reduces with distance because of dielectric losses, so it is reasonable to use HVDC because transmission capability is independent of distance. Therefore, direct current often provides the best technical and economic solution for transmission of power to the shore, without being prohibitively expensive in terms of Capex.

Figure 3 shows a comparison between the costs of AC and DC submarine transmission cables technologies as a function of distance from the coast. For a case study of 1,500 MW of transmission capacity, a break-even distance of 190 km is reached before the active power transmission capability of the AC system is achieved. Therefore, the DC option is not economical for a higher distance from the coasts. The break-even distances between HVAC and HVDC transmission depends on different factors such as transmission distances, capacities, and voltage levels (NREL 2016).

Installation cost increase both due to increasing distance to the coast and water depth.

Consecutively, in deep water offshore installations, distance and therefore travelling time from holding port to site affect the final cost, as also does the use of purpose-built vessels and cranes. In addition, deep water offshore installations turn out to be more expensive than shallow-water ones due to the high cost of construction and installation of the floater, of the mooring lines and of the anchoring system.

BVG Associates (2014) estimate the average cost of decommissioning of a wind farm to be 65% of the installation cost. Decommissioning operations are completed over time scales that range from a few days to several weeks. Once ashore, material is cut and transported to a scrap or disposal site. Disposal cost is calculated on a per ton basis and is estimated for each wind farm component. If the material can be sold, the operation receives income, otherwise if it is disposed of, the operation records a cost (Kaiser and Snyder 2010).

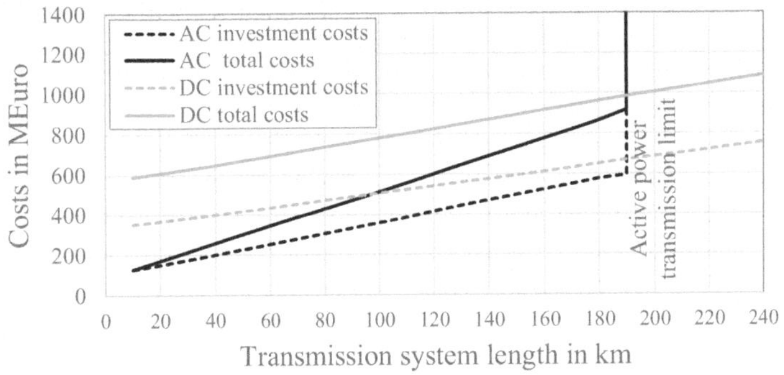

Fig. 3. Relation between AC and DC transmission cables

In deep water, strongest drivers of O&M are: distance from shore, accessibility limits related to local metocean conditions (e.g., wave height), and turbine rating (i.e., fewer, larger turbines suggest lower O&M costs per megawatt).

Refers to the analysis of deep water installations from NREL (2015), capital expenditures were formulated on the basis of a combination of empirical data and scalability ratios (Fig. 4).

3.4 Prediction of Deep Water Offshore Wind Farm Costs

In this paper, a prediction of a deep water offshore wind farm costs is attempted, and for the purpose the characteristics of the Hywind Scotland Park (HSP) pilot project was taken as a reference for making economic calculations and estimates. This wind farm is an optimised version of a prototype of wind turbine that has been operating off the coast of Norway since 2009. HSP project is the first multiunit floating wind farm and is realized by Statoil to create cost-efficient and low risk solutions for commercial-scale wind farms. In particular, this new installation aim at developing the following tasks:

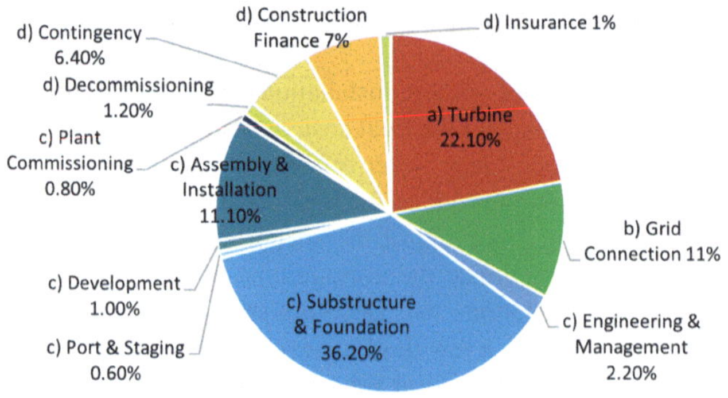

Fig. 4. Capital expenditures for deep water offshore turbine reference wind plant project

(a) test multiple units in farm configuration; (b) optimize assembly and installation; (c) verify up-scaled design; (d) evaluate reliability and availability of optimized multi-turbine concept; (e) mobilize supply chain. The HSP system includes five 6 MW Siemens wind turbines mounted on spar buoy platforms. Each turbine has a total height of 258 m, 178 m are below sea level and 80 m are above.

In Table 4 the main characteristics of the HSP project are shown (Statoil 2015). Before showing the results here obtained, it is useful to define the main cost criteria affecting the overall project costs of a deep-water wind farm, namely grid connection, assembly and installation, vessels typologies and, substructures and foundation.

Grid connection typically has three components: the array cable system, the export system, and the onshore transmission infrastructure. The array cables collect power from wind turbines and delivers it to the offshore substation transformer or directly to the export cable (typical rating of 33 kV). In this case a direct connect to the onshore power infrastructure is considered. This includes onshore cable and onshore substation to connect to existing power grid on land. Thus, based on the chosen site the grid connection consists of 5 interray cables each 3 km long, a 25–35 km long export cable, a 2–3 km long onshore cable and an onshore substation.

There are different methods of assembly and installation for the turbine. Each type of installation involves different costs based on a series of variables, such as the number of lifts, crane capacity, the type and movements carried out by the vessel and the type of assembly of turbine, in particular whether carried out partly or totally at the port or in the open sea (Fig. 5). Moreover, the installation time must be considered as an important variable affecting the final cost. The components of turbine are traditionally at least seven, namely two tower sections, nacelle, hub and three rotor blades and these parts, can be assembled with different methods.

In economic terms, the optimal installation method consists of turbine components lifted onto floater and tower configuration near shore, while the least convenient method is to lift every single component of the turbine offshore.

Furthermore, the time spent for the turbine installation, methods 1 and 2 take about 2 days to install, while methods 3, 4, 5 take an average of more 5 days to install and finally, method 6 is the fastest, taking only a day to install (Ahn et al. 2016).

Table 4. Characteristics of the Hywind Scotland Park project

Hywind project features	
Place	North Sea Scotland (Peterhead)
Area (sea level)	4 km^2
Number of turbines	5
Power	6 MW
Total power	30 MW
Height	258 m
Platform	Spar buoy
Depth range	95–120 m
Distance from coast	25 km
Average wind speed	10.1 m/s
Mean waves, Hs	1.8 m
Transmission voltage	33 kV
Installation period	21 months (2016-17)

Regarding the vessels, different types of them participate in the installation of an offshore wind farm, and are used for each specific phase of the life cycle of wind turbines. Vessels are distinguished into two main categories, such as installation and maintenance vessels, the latter being usually smaller than the formers. Installation vessels, such as crane, cable-laying and anchor handling vessels are designed for installation and decommissioning activities and large maintenance operations. Instead, maintenance vessels supply installation, operation and decommissioning phases through cargo and personnel transportation and tugging operations.

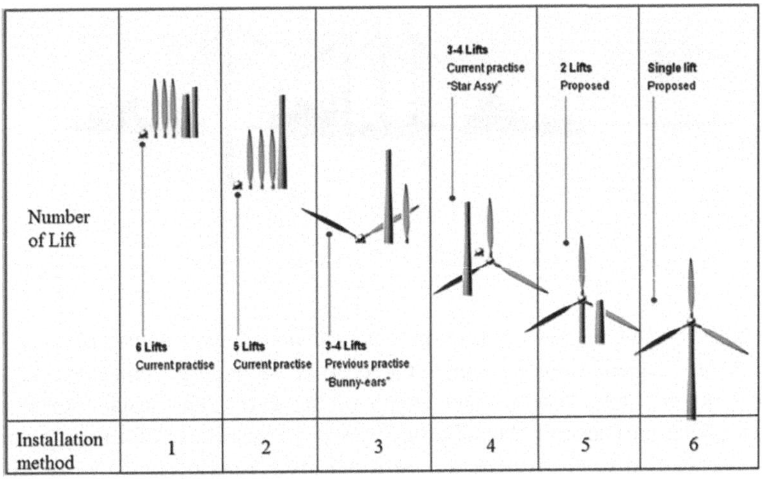

Fig. 5. Offshore wind turbine installation method

The main typologies of installation vessels, here considered for the cost estimation, are the Anchor Handling Tug Supply vessels (AHTSs), Barge tugs, Large vessels, Crane vessels and Platform Supply Vessels (PSVs).

Moreover, at Mather vessel is used for maintenance. AHTS is the multi-purpose vessel considered to assist in mooring, tugging and transportation of equipment (Fig. 6a). A barge tug was used to install interray cables and, generally has a lay rate of 0.3 km/day with a range of 0.15 to 0.60 km/day. A large vessel was used to install export cable and, empirically derived export cable installation rates range from 0.2 to 1.4 km/day with an expected value of 0.7 km/day. PSV was considered to support offshore activities and to transport cargo, such as turbine components or personnel (Fig. 6b). A crane vessel was used to lift massive loads such as substructures and complete turbines using a deck-mounted crane (Fig. 6c). Finally, a mother vessel is used as an offshore base for technicians during O&M phase. Table 5 shows the daily or annual cost of the different vessels typologies (Kaiser and Snyder 2010; Bjerkseter et al. 2013).

Substructures and foundations include the underside of the wind turbine and all the structures placed below sea level (floater, mooring lines and anchors). In this case there are in total 15 anchors and 15 mooring catenary lines each 800 m long.

Capital expenditures of the project were calculated also considering data costs collected by Bjerkseter et al. (2013). In particular, a thorough review of these data has been developed, that have been assembled and reorganized according to the characteristic features of the HSP project, to estimate its total capital expenditures.

In particular, the total cost of the turbine was calculated considering the values of each component, and the strategy based on three turbine components lifted onto floater offshore is taken onto account.

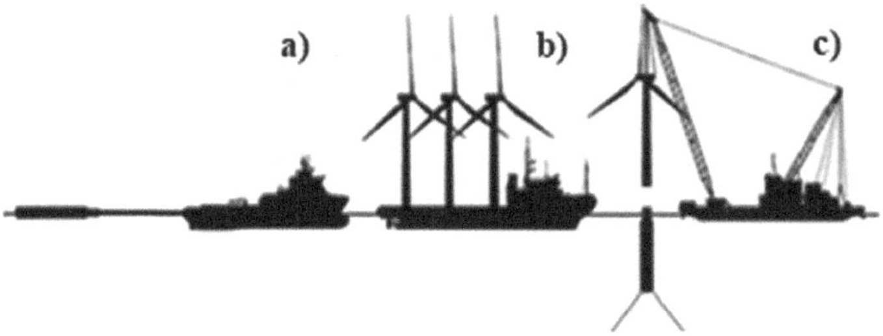

Fig. 6. Typologies of installation vessels

The costs related to grid connection were calculated considering the cost of the 33 kV HVAC transmission, to which the cost of secondary connection is added, depending on the characteristics and length of the latter. The installation cost was calculated taking into account the daily rate by type of vessel used for their positioning and their installation time. A barge tug operating over a period of 10 to 15 days was considered for interray cables installation. A large vessel operating for 15 to 20 days

Table 5. Rate per day/annual of installation and maintenance vessels

Installation vessels	Day rate $
AHTS	70,000
Barge tug	50,000
Large vessel	125,000
Crane vessel	45,000
PSV	33,000
Maintenance vessel	Annual rate $
Mother vessel	13,000,000

was used for laying the export cable. For the onshore cable and onshore substation costs estimation, the length and total cost were considered, respectively.

A similar calculation procedure was carried out for the substructure elements. In particular, materials and manufacturing costs for the floater wew taken into account. The total number and the cost of anchors and mooring lines were considered. The use of a AHTS vessel operating for a period of 15 to 20 days for anchoring, to which 5 days are added for laying of the mooring lines, have respectively been summed up for the installation. Finally, the results of the other cost categories (development, plant commissioning, engineering and management, contingency, insurance, construction finance, and port and staging) were derived from the corresponding unit cost per MW.

Figure 7 shows the results in terms of percentages capital expenditures for the overall project here obtained. It can be observed that these results are in good agreement with the capital expenditures calculated by NREL and reported in Fig. 4. As in the case of NREL offshore deep water wind plants, the turbine appears to be not the most expensive component of the HSP project (22.68% of the total costs).

Substructure and Foundation costs turn out to be the largest capital expenditure, 25.70% of total costs, being around 10% lower than the NREL analysis. This can be explained by the higher cost associated to other types of deep water offshore foundation, considered in the NREL economic analyses. In fact, an estimated average cost value, including more expensive floaters typologies (e.g. Spar buoy, Tension Leg Platform, Semi-submersible platform) in terms of materials and dimensions, is computed.

Assembly and installation is slightly higher than the NREL analysis, as it applies the installation method 4, of three turbine components lifted onto floater offshore, which is not the least expensive strategy. This cost is influenced by the number of lifts and by a pre-assembly of all or some components of the wind turbine at the port. Decommissioning has a much higher value than the one in Fig. 4. It has not been taken into account the gain from scrap steel that would decrease costs for decommissioning.

The results achieved out show that the turbine and the installation costs are in good agreement with the values estimated by NREL. Some differences are observed for the substructures and foundations, due to the different typologies taken into account for these calculations. Finally, it should be mentioned there may be margins of variable inaccuracy for some cost items, such as Assembly and installation and Contingency. These are mainly due to the downtime caused by accessibility restrictions, as severe weather conditions offshore and to their unpredictability.

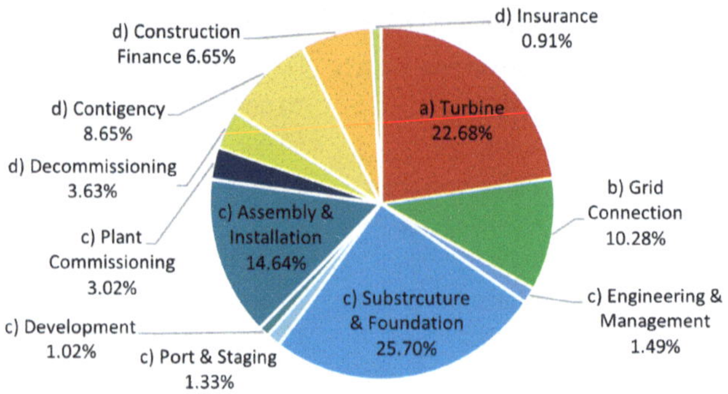

Fig. 7. Capital expenditures for deep water offshore turbine reference Hywind project

4 Conclusions

In this study, a comparative analysis of the overall construction, operational and maintenance costs of onshore and shallow- and deep-water offshore wind farms has been carried out. First, an estimation of the individual cost differences between wind installation typologies and an assessment of deep water offshore wind farm costs, are developed. Then, a comparative analysis of the overall costs affecting the various typologies of wind turbines and the main economic key parameters has been carried out. Consequently, a prediction of deep water offshore wind turbines overall costs is derived, borrowed from shallow water wind projects that now represent a well-established wind industry application. Thus, the paper also permit to carry out a cost assessment for deep water turbines, where at the moment only research prototypes exist and no detailed studies are available on their possible investment cost. In conclusion, the basic elements of the overall life cycle of the different types of wind farms has been set. These results could be useful to the future evaluation of the LCOE (Levelized Cost of Energy) expressing the key performance economic parameter and then, to move towards further studies for the definition of a business plan for deep water offshore wind farms.

References

Ahn D, Shin S, Kim S, Kharoufi H, Kim H (2016) Comparative evaluation of different offshore wind turbine installation vessels for Korean west-south wind farm. Int J Nav Arch Ocean 9:45–54

Avossa AM, Demartino C, Contestabile P, Ricciardelli F, Vicinanza D (2017a) Some results on the vulnerability assessment of HAWTs subjected to wind and seismic actions. Sustainability 9:1525

Avossa AM, Demartino C, Ricciardelli F (2017a) Peak response of hawts to wind and seismic actions. In: Proceedings – 7th European and African conference on wind engineering, EACWE, Liège, 4–7 July 2017

Avossa AM, Demartino C, Ricciardelli F (2017c) Assessment of the peak response of a 5 MW HAWT under combined wind and seismic induced loads. Open Constr Build Technol J. 11:441–457

Beurskens J (2011) Converting offshore wind into electricity: the Netherlands' contribution to offshore wind energy knowledge, We@Sea Research Programme 2004–2010

Bilgili M, Yasar A, Simsek E (2011) Offshore wind power development in Europe and its comparison with onshore counterpart. Renew Sust Energ Rev 15:905–915

Bjerkseter C, Agotnes A (2013) Levelised costs of energy for offshore floating wind turbine concepts. Master thesis, Norwegian University of Life Sciences

Blanco MI (2009) The economics of wind energy. Renew Sust Energ Rev 13:1372–1382

Brown C, Poudineh R, Foley B (2015) Achieving a cost-competitive offshore wind power industry: what is the most effective policy framework? Oxford Institute for Energy Studies, EL 15

BVG Associates (2014) Future renewable energy costs: onshore wind. How technology innovation is anticipated to reduce the cost of energy from European onshore wind farms, Renewable Energies

Ebenhoch R, Matha D, Marathe S, Cortes Munez P, Molins C (2015) Comparative levelized cost of energy analysis. In: 12th deep sea offshore wind R&D conference, EERA DeepWind'2015. Energy Procedia 80:108–122

European Environment Agency, EEA (2009) Europe's onshore and offshore wind energy potential. An assessment of environmental and economic constraints. Technical report

International Renewable Energy Agency, IRENA (2012) Renewable Energy technologies: cost analysis series. Volume 1: Power sector, Issue 5.5. Wind Power

Larsen HH, Sonderberg Petersen, L (2014) Wind energy - drivers and barriers for higher shares of wind in the global power generation mix, Denmark Technical University - International Energy Report

Kaiser MJ, Snyder B (2010) Offshore Wind Energy Installation and Decommissioning Cost Estimation in the U.S. Outer Continental Shelf, Energy Research Group

Kitzing L, Monthorst PE (2015) Trends in offshore wind economics - the past and the future. In: Proceedings - 14th wind integration workshop, Technical University of Denmark

National Renewable Energy Laboratory (NREL) (2015) 2015 Cost of Wind Energy Review

National Renewable Energy Laboratory (NREL) (2016) Offshore Wind Technologies Market Report

Sajid A, Sang-Moon L, Choon-Man J (2017) Techno-economic assessment of wind energy potential at three locations in South Korea using long-term measured wind data. Energies 10:1442

Statoil (2015) Hywind Scotland Pilot Park. Environmental Statement

Some Topics About Wind Engineering that Curtain Walling Design Might Be Longing for in Standards

G. Manara(✉)

Department of Structural Analysis, Permasteelisa S.p.A, Vittorio Veneto, Italy
g.manara@permasteelisagroup.com

Abstract. Curtain walling is assuming more and more a leading role in tall building architectural design.

From a structural standpoint, there is no doubt that the wind action is the main loading condition, governing all the dimensioning of any curtain wall single component. On the other hand, the wind action is sensitive to any curtain wall local shape discontinuity, or composition peculiarity, which may modify the load intensity on a micro-scale, but still important for structural safety.

On such respect, the structural engineer may have to cope with some recurrent issues:

- to define wind loads during an initial tender phase, when no wind tunnel test has been developed yet;
- to define wind loads for the cladding of a building that will never have the aid of a wind tunnel test;
- to define wind loads on components that have not been simulated in the wind tunnel test, because their size was too small when scaled to the wind tunnel test model.

Published standards are the main support to this activity, but this paper will show some frequent cases that are still missing, or just partially treated, in the international standards, like:

- Façade projections, i.e. canopies, balconies, sun-shadings, etc.;
- Buildings with curved surfaces but non-circular shape;
- Cladding elements with more than one skin.

Examples from the everyday design work will be shown.

1 Introduction

Façade engineering is strictly related to wind engineering. As a matter of fact, wind represents the most important source of actions on the façades and the definition of the local c_p values is always a tough task.

F. Ricciardelli and A. M. Avossa (Eds.): INVENTO 2018, LNCE 27, pp. 454–482, 2019.
https://doi.org/10.1007/978-3-030-12815-9_36

The standards are the basis of any assessment of wind loads, but they still ignore, or partially deal with, some common architectural issues that the structural engineer has to cope with quite frequently, like:

- Façade projections, i.e. canopies, balconies, sun-shadings, etc.;
- Buildings with curved surfaces but non-circular shape.
- Cladding elements with more than one skin.

This paper will show some frequent cases where a guideline of international standards is sought, but still missing or inadequate, and it will try to suggest possible improvements and sources to support it, whenever possible.

2 Façade Projections

Façade projections are frequent elements in curtain walls.

The most common projecting elements are:

- canopies;
- sun-shading elements, horizontal or vertical ones, like louvres, blades, fins, wings etc.;
- balconies.

All of them are subjected to wind loads; balconies are comparable to canopies, as far as wind load is concerned, so we will concentrate on the latters.

Canopies and sun-shading elements are frequently added to façades, to shelter the building entrance, or to shield the façade from the sun irradiation, but they are ignored by international standards, like Eurocode 1 (EN 1991-1-4) in Europe and ASCE 7 in the USA.

Wind load is, generally, the most important loading condition for façade projections and also the most complicated to ascertain. Nevertheless, some rules can be found in the literature and they could be considered as the basis for including this topic in the international standards.

We will go now in detail into the load assessment on these components.

The first step of such load evaluation is related to the same analysis that is needed by the façade global wind load study, up to determine the peak velocity pressure, $q_p(z_e)$.

The next step is related to assessing a proper local pressure coefficient (c_p), so to define the wind pressure on the element:

$$w_e = q_p(z_e) \cdot c_{pe} \tag{1}$$

where:

$q_p(z_e)$ is the peak velocity pressure
z_e is the reference height for the external pressure
c_{pe} is the pressure coefficient for the external pressure.

The assessment of the c_{pe} value is surely the most complicated and uncertain matter.

As a general remark, it should be noticed that projecting elements may be subjected both to direct wind flow, impinging the building surface, and to air flow that is parallel to the façade (ascending/descending flow, side flow), due to wind deviated by the building itself (see Fig. 1):

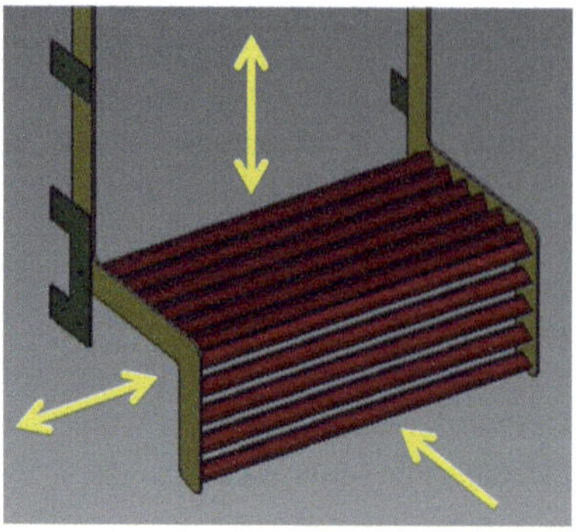

Fig. 1. Possible directions of wind acting on façade projections

Furthermore, they can be affected by different conditions according to their position along the height of the building, to their proximity to building corners, to their pattern if aligned in a row, to their porosity, and to the space which may (or may not) separate the projecting element from the façade surface.

We will go now to discuss this topic item by item.

Canopies

They can be placed at different heights along the building surface, with different orientations as far as the wind direction is concerned, and the façade itself influences the wind action on canopies.

According to Cook [1], Pt.2 §20.8.2, p. 472, canopies attached higher than halfway the height of the building ($H/h < 2$) can be calculated according to the classic rules of freestanding canopies, fully blocked at one edge. The relevant c_f could be derived, for instance, from EN 1991-1-4:2005 [2], tab. 7.6, in case of monopitch freestanding canopies fully blocked at one edge. Their values are about –1.4.

The $H/h < 2$ position is typical for low-rise buildings, but no reference or equivalence can be found in the international standards for canopies attached below half-way up the building height ($H/h \geq 2$), as normally occurs for high-rise buildings. For such position, the force coefficient values reported in the following Table 1 are proposed by Cook [1], once the wind peak velocity pressure is taken at the top of the building wall ($\theta = 0°$ canopy on the wind ward wall, $\theta = 90°$ canopy on the side wall, see Fig. 2):

Table 1. Pseudo-steady force coefficients, $c_f(\theta)$ (from Cook [1]).

q_{ref} as height of the building, $z_{ref} = H$								
H/h =	2	3	6	12	18	24	30	36
$\theta = 0°$	+0.30	+0.40	+0.69	+0.87	+0.92	+0.93	+.93	+0.91
$\theta = 90°$	–0.24	–0.70	–0.95	–1.04	–1.16	–1.30	–1.29	–1.16

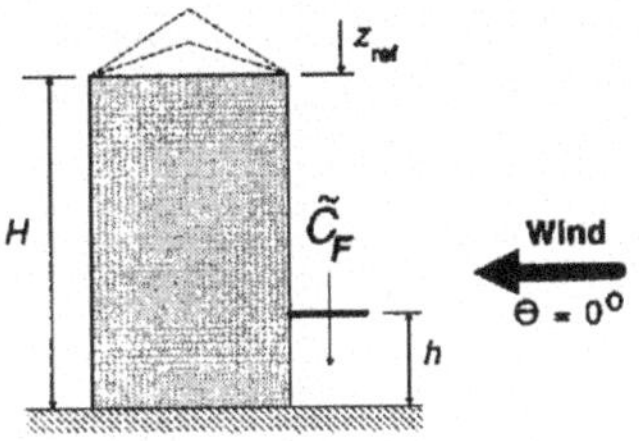

Fig. 2. Canopy position parameters and wind direction (from Cook [1])

Also Jancauskas and Holmes [3] presented the results of an investigation of the wind loads on canopies attached to buildings, including different building-wall/canopy-height (Fig. 3). For ratios between 1 and 2 ($1 \leq H/h < 2$), they demonstrated that the dominant net loads (highest peak) for the windward canopy were directed upwards and happened for a wind direction perpendicular to the windward façade which the canopy was attached to, as shown in Fig. 3(a). However, for building-height/canopy-height ratios greater than 2 ($H/h \geq 2$), the dominant net loads (highest peak) for the windward canopy were directed downwards for the same wind direction, as shown in Fig. 3(b).

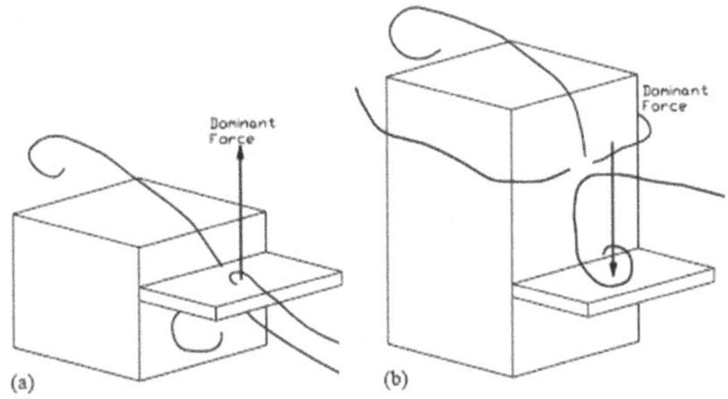

Fig. 3. Dominant wind force on a canopy according to its position (from Jancauskas and Holmes [3])

Moreover, for the case of building-height/canopy-height ratios greater than 2 ($H/h \geq 2$), the highest peak upward net load was proved to occur for a wind direction parallel ($\theta = 90°$) to the façade to which the canopy was attached to. As a matter of fact, when the canopy is on the side face, the accelerated flow around the building produces high upward loads. All this is consistent with Cook's [1] values here above. Finally, when the canopy is on the leeward face, the loads are much smaller.

Roh and Kim [4] also studied a special case of canopy in a corner of a building, for different sizes, heights and wind directions, as shown in Figs. 4 and 5:

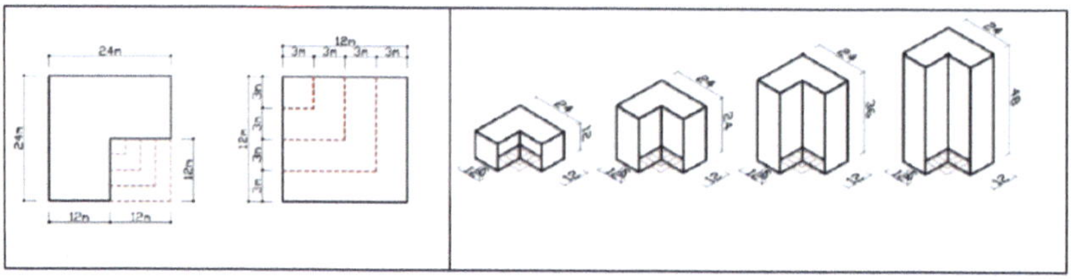

Fig. 4. Canopies of different size at the corner of buildings having different heights (from Roh and Kim [4])

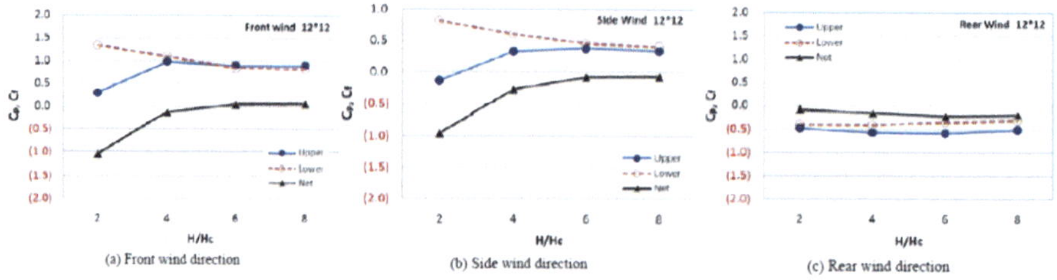

Fig. 5. c_p and c_f values for a 12 × 12 m canopy at the building corner (from Roh and Kim [4])

Some reference for canopies can be found in some national standards too.

The Suisse standard SIA 261 [5] offers quite detailed rules about canopy c_p and c_f values, proposing both a dependency on the height and on the slope, and a "zonage" of the canopy surface (Fig. 6):

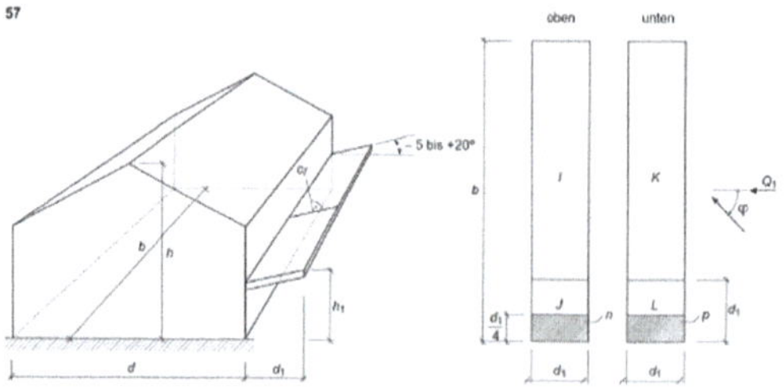

Fig. 6. Sketch identifying the basic parameters to calculate c_p and c_f (from SIA 261:2014 [5])

Nevertheless, the applicable height ratio h_1/h appears very limited.

The DIN EN 1991-1-4/NA:2010 [6], German National Annex to the Eurocode 1, provides $c_{p,net}$ values for canopies (Fig. 7), accounting both for the height ratio (from 0.1 up to the full building height) and for the canopy slope, considering also two zones, A and B, on its surface:

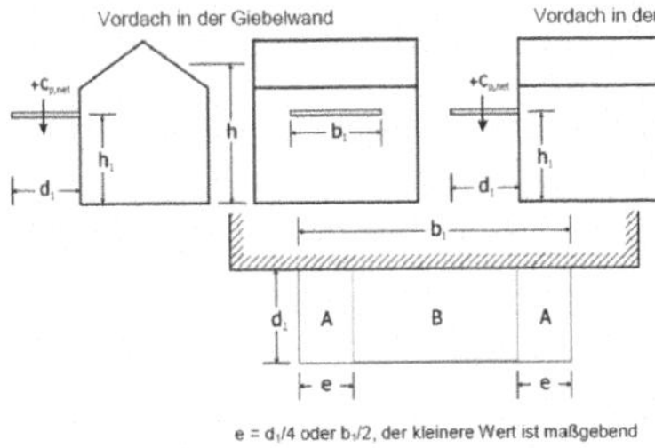

Fig. 7. Sketch identifying the basic parameters to calculate the $c_{p,net}$ (from DIN EN 1991-1-4/NA:2010 [6])

The $c_{p,net}$ values proposed by DIN are always both positive and negative, at every height ratio.

Also the Australian/New Zealand Code AS/NZS 1170.2:2002 gives recommendations regarding $c_{p,net}$ values of canopies attached to sloping roof low-rise buildings, with a roof slope of 10° or less (see Fig. 8). The code recommends that the canopies should be designed for both downward (positive) and upward (negative) net wind pressures. All pressure coefficients shall be used with the value of wind speed applying at average roof height (h). The standard provides values for wind direction normal to the attached wall. For wind parallel to the wall, it recommends that a canopy should be considered as a free roof and the net pressure coefficients should be obtained accordingly.

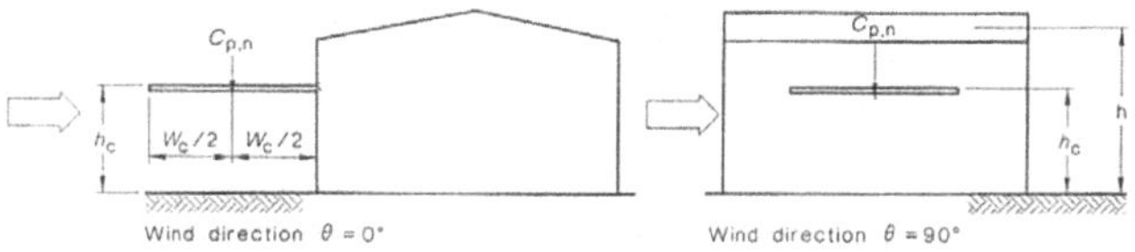

Fig. 8. Sketch identifying the basic parameters to calculate the $c_{p,net}$ (from AS/NZS 1170.2:2002 [7])

In conclusion, it appears that there are data about canopies, in literature and national codes, to be collected and included in the international standards. An endorsement by international standards is what is needed to allow an effective application of these rules to the daily design work for canopies, considering that most of Curtain Wall manufacturers operate in various national markets.

Sun-Shading Elements

Current architecture offers a large variety of sun-shading elements, which are becoming an essential complement of façades, to fulfil the current energy saving laws issued in most of countries.

Unfortunately, sun-shading elements (louvres, blades, fins, etc.) find even less attention than canopies into standards and literature, which currently provide no design guidelines on such items.

A possible rough calculation approach could be to consider them as follows:

- when separated from the façade by a distance z_g greater than *h/4* (see Fig. 9), they could be treated with the same approach as "signboards". E.g., in this case, EN 1991-1-4:2005, §7.4.3 [2], recommends for signboards to apply a force coefficient $c_f = \pm 1.80$. Such value is also applicable where z_g is less than *h/4* if $b/h \leq 1$; nevertheless, such geometrical condition can seldom occur in façades.
- when separated from the façade by a distance z_g less than *h/4* and $b/h > 1$ (see Fig. 9), they could be treated with the same approach as "free-standing walls", applying the calculation method proposed by the standards for this case (e.g.: EN 1991-1-4:2005, §7.4 and §7.4.1 [2])

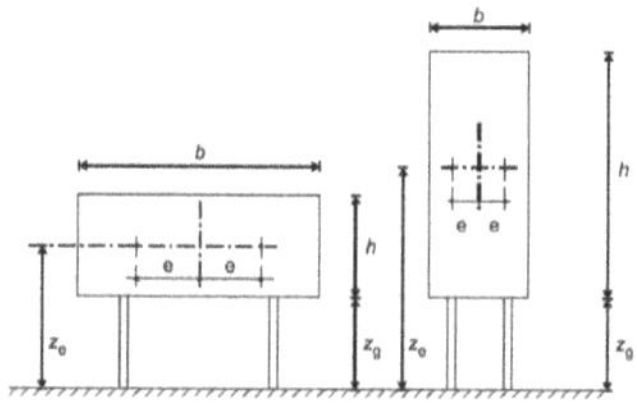

Fig. 9. Basic parameters to calculate the c_f for signboards (from EN 1991-1-4:2005 [2])

Considering that the sunshade surface is sometimes made of perforated metal sheet, the proposed approach has also the advantage of allowing the application of the related coefficient for porosity.

For instance, EN 1991-1-4:2005, §7.4 [2] recommends a solidity factor $\phi = 1$ for fully solid elements, and $\phi = 0.8$ for elements which are 80% solid (i.e. have 20% openings). Porous elements with a solidity ratio $\phi \leq 0.8$ could be treated as plane lattices (e.g. according to EN 1991-1-4:2005, §7.11 [2]).

Nevertheless, all these calculation procedures do not account for the presence of the building, so they should be applied with careful consideration.

Furthermore, sunshades are frequently installed in regular patterns (rows etc.). So, when considering wind flow parallel to the façade, one element could partially shelter the following one and a "sheltering factor" could be applied. The value of the sheltering factor ψ_s depends on the ratio *x/h* (spacing *x* between the elements divided their height *h*, see Fig. 10), and on the solidity factor ϕ of the upwind (sheltering) element, as suggested in EN 1991-1-4:2005, §7.4.2 [2].

It should be remarked that the sheltering factor should never be applied in the end zones, within a distance of at least h measured from the free end of the sunshade element.

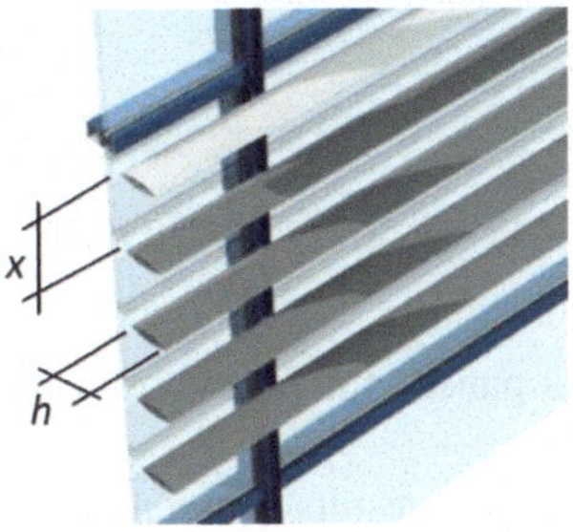

Fig. 10. Sun-shading spacing of single elements and height

When developing hand calculations, the decision if a sheltering factor should be applied or not, which value and how much extensively, is always a troubled one and, frequently, for a common engineer, it is not easy to understand which elements could be sheltered and which could not, without a guideline.

On such respect, it may be worth to show the results of a numerical analysis of the forces applied by the wind to a sun-shading panel made of aluminium RHS elements (called "blades" in the following), shown in the following rendering picture (Fig. 11). The CFD analysis automatically accounts for sheltering effects on some blades; it can be interesting to carry out a comparison with the calculation by formula.

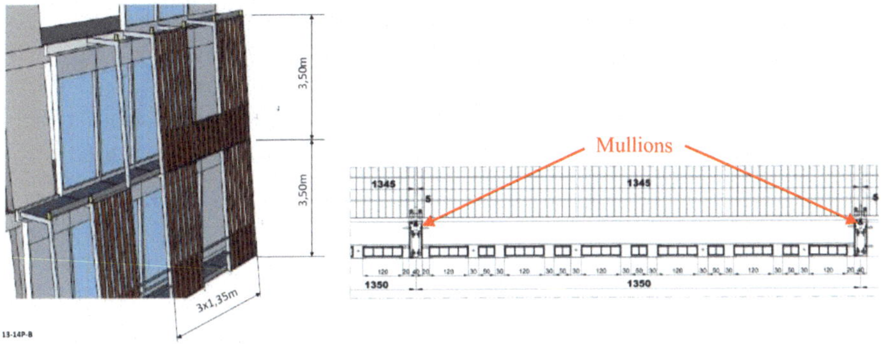

Fig. 11. Sun-shading panels (1.35×3.5 m^2) – Left: rendering [m]. Right: horizontal section [mm].

We can notice that the sun-shading panel is supported by two mullions. As for the blades, the smaller tubular sections stop at parapet level, while the bigger ones continue for the full height of the panel.

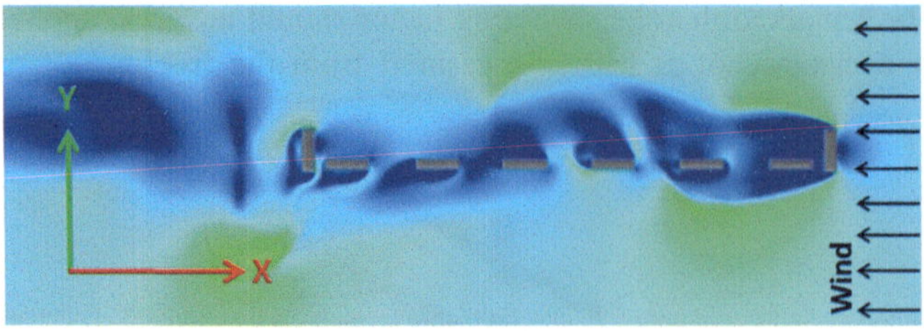

Fig. 12. Sun-shading panel: CFD analysis for wind direction parallel to the panel (0°)

The figures above show flat panels; the façade project included both flat and curved ones.

The following Fig. 12 shows the results of the numerical 2D analysis of one panel, carried out by means of the "*Autodesk® CFD*" software, for wind direction parallel to the façade.

This picture shows quite clearly that the first and the second blades, on the right hand, are in the wake of the right mullion, what can be appreciated also in the following Table 2, showing the drag force applied per unit length to each element, at progressive integration iterations.

Table 2. Drag force per unit length applied to each element in X direction F_x [N/mm].

Iteration	5250	5300	5350	5400	5450	5500	5550	5600
Mullion 1	–0.126	–0.129	–0.130	–0.128	–0.125	–0.126	–0.127	–0.127
Blade 1	0.003	0.001	0.004	0.003	0.002	0.003	0.006	0.001
Blade 2	0.020	0.024	–0.002	–0.016	0.007	0.011	0.013	0.024
Blade 3	–0.015	–0.005	0.002	–0.005	–0.035	–0.003	0.003	–0.009
Blade 4	–0.013	–0.005	–0.017	–0.008	–0.004	–0.015	–0.032	0.009
Blade 5	–0.024	–0.027	0.000	–0.021	–0.010	–0.007	–0.004	–0.036
Blade 6	–0.013	–0.018	–0.025	0.003	–0.018	–0.026	–0.022	–0.008
Mullion 2	–0.072	0.001	–0.030	–0.046	–0.100	–0.078	–0.091	–0.056
Panel Tot	**–0.241**	**–0.158**	**–0.198**	**–0.218**	**–0.283**	**–0.242**	**–0.256**	**–0.202**

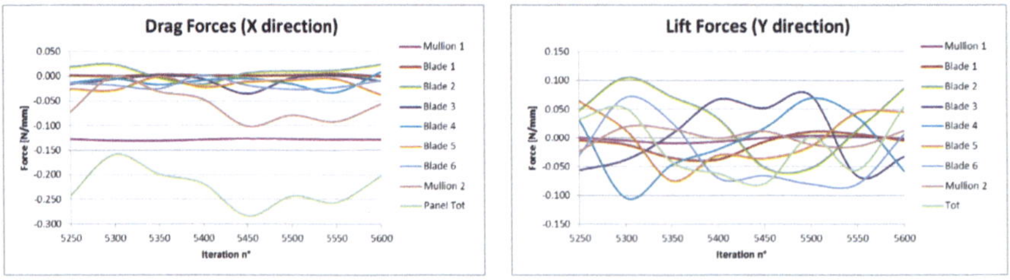

(a) (b)

Fig. 13. Graphs of the forces applied on each element by wind blowing parallel to the sun-shading panel (a) drag forces – (b) lift forces

The subsequent blades, from the 3rd to the 6th, show to partially shelter each other and also the leeward mullion appears to be partially shielded, if compared to the upwind mullion.

The results of this analysis are shown also in the following graphs (Fig. 13), where not only the drag (Fig. 13a) but also the lift forces (Fig. 13b), orthogonal to the panel (90°), are calculated thanks to the numerical analysis, what could have been quite hard to do by hand calculation.

Being the drag forces calculated per unit length and knowing that the panel is 3.5 m tall, it is possible to compare the total drag force applied to one panel according to the CFD analysis and that one calculated according to the §7.6 of the EN 1991-1-4. This was calculated neglecting the sheltering effects on the blades, but accounting for sheltering of the leeward mullion, which could be supposed as the most influenced, as reported in Table 3.

Table 3. Total drag force applied to one panel 3.5 m high in X direction F_x [N].

Aver. CFD [N/mm]		Hand calc. [N/mm]
Mullion 1	–0.127	–0.160
Blade 1	0.003	–0.010
Blade 2	0.010	–0.010
Blade 3	–0.008	–0.010
Blade 4	–0.011	–0.010
Blade 5	–0.016	–0.010
Blade 6	–0.016	–0.010
Mullion 2	–0.059	–0.115
TOT. FORCE	**–788 N**	**–1173 N**

The consequence of the sheltering effect is quite evident, decreasing the total force to the 67%.

However, we can consider now the same panel lying on a plane that is slightly inclined with respect to the wind direction, e.g. 15°. We can immediately remark in the following Fig. 14 that the sheltering effect applied by the upwind mullion is substantially lower than before and that its wake involves just the first blade from the right hand.

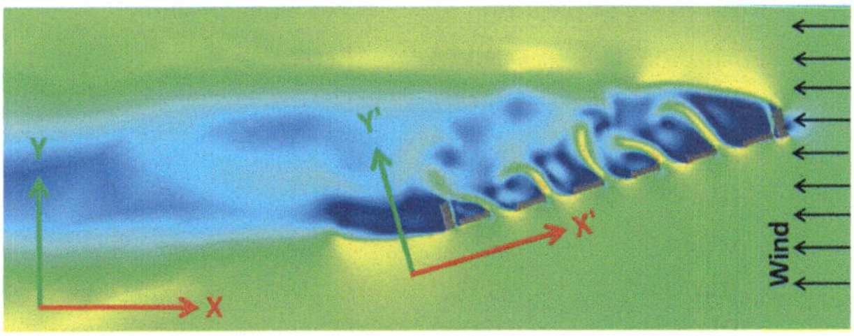

Fig. 14. Sun-shading panel: CFD analysis for wind direction inclined to the panel (15°)

On the contrary, we can notice that the 6th blade offers an efficient sheltering to the leeward mullion, even better than for 0°, what can be observed also reading the values reported in the following Table 4.

Table 4. Drag force per unit length applied to each element in X′ direction $F_{x'}$ [N/mm].

Iteration	5950	6000	6050	6100	6150	6200	6250	6275
Mullion 1	–0.132	–0.136	–0.135	–0.134	–0.128	–0.129	–0.135	–0.137
Blade 1	–0.002	–0.002	–0.002	–0.001	0.000	–0.001	–0.002	–0.002
Blade 2	–0.033	–0.030	–0.031	–0.034	–0.031	–0.028	–0.032	–0.034
Blade 3	–0.032	–0.036	–0.030	–0.026	–0.037	–0.040	–0.028	–0.026
Blade 4	–0.021	–0.028	–0.036	–0.030	–0.019	–0.030	–0.036	–0.032
Blade 5	–0.029	–0.025	–0.027	–0.035	–0.031	–0.020	–0.026	–0.032
Blade 6	–0.036	–0.033	–0.035	–0.032	–0.038	–0.034	–0.026	–0.026
Mullion 2	0.002	–0.005	0.006	–0.007	0.009	–0.004	–0.023	–0.023
Panel Tot	**–0.283**	**–0.295**	**–0.290**	**–0.298**	**–0.275**	**–0.286**	**–0.308**	**–0.311**

Again, we can look at the relevant drag (Fig. 15a) and lift (Fig. 15b) force graphs, here below.

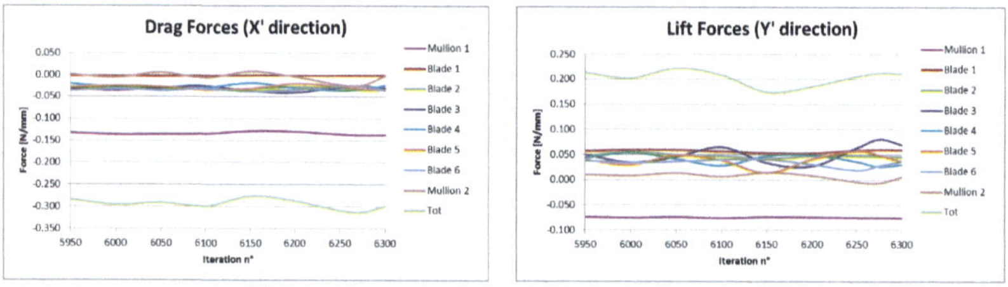

(a) (b)

Fig. 15. Graphs of the forces applied on each element by wind blowing at 15° to the sun-shading panel

Now we can repeat the same comparison of the total drag force applied to one panel, as before (Table 5):

Table 5. Total drag force applied to one panel 3.5 m high in X′ direction $F_{x'}$ [N].

Aver. CFD [N/mm]		Hand calc. [N/mm]
Mullion 1	–0.134	–0.160
Blade 1	–0.001	–0.010
Blade 2	–0.032	–0.010
Blade 3	–0.032	–0.010
Blade 4	–0.029	–0.010
Blade 5	–0.029	–0.010
Blade 6	–0.033	–0.010
Mullion 2	–0.005	–0.115
TOT. FORCE	**-1029 N**	**-1173 N**

As expected, the sheltering effect has strongly decreased; nevertheless, some influence is still there, reducing the total force to the 88% of that calculated according to a conservative normative approach.

Though, all what has been discussed here above does not account for the façade influence.

Considering that, in standard wind tunnel tests of a building, applying the most common scale factors, there is no possibility to take into consideration the small sun-shading elements, our suggestion would be to carry out devoted tests and studies aimed at defining a multiplying factor K_s which may allow to ascertain the sun-shading element $c_{p,net}$ simply starting from the façade measured c_{pe}[1]:

$$\textit{Sun-shading } c_{p,net} = K_s \cdot \textit{Façade } c_{p,e} \tag{2}$$

Considering the infinite variety of possible conditions, our proposal would be to concentrate the attention on the following "master cases" as reported in Fig. 16 here below.

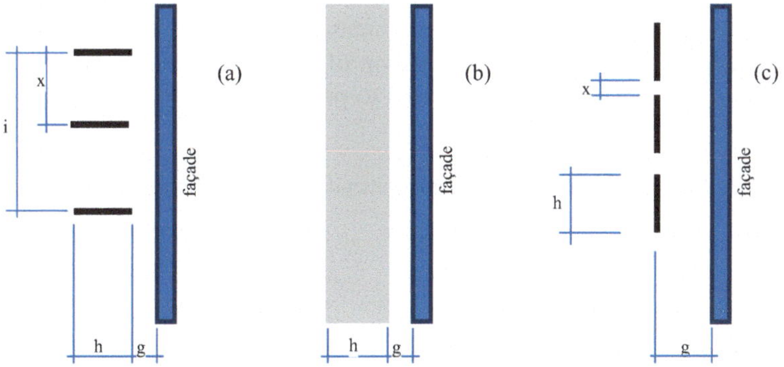

Fig. 16. Horizontal sun-shading (a). Vertical sun-shading (b). Screen-like sun-shading (c).

Where the following basic parameters are proposed:

g: gap between the façade and the sun-shading element, *g* = 0 mm, 200 mm, 600 mm.
h: width of the sun-shading element, *h* = 200 mm, 400 mm, 600 mm.
x: gap between subsequent sun-shading elements, x = 200 mm, 500 mm, 1500 mm
i: height of a group of *n* sun-shadings ($i = (n - 1) \cdot x$).

[1] A similar approach has been followed by *Windtech Consultants*® in some of their wind tunnel test reports.

Then, for each combination of the basic values above, a table like the following one (Table 6) could be suggested to collect the measured K_s values (e.g.):

Table 6. Example: horizontal sun-shadings, table for collected K_s values.

	Horizontal sun-shadings – h=400mm, g=0mm, x=500mm					
	Maximum peak c_p			Minimum peak c_p		
Location along the façade	Façade c_{pe}	Sun-shade $c_{p,net}$	Sun-shade Multipler K_s	Façade c_{pe}	Sun-shade $c_{p,net}$	Sun-shade Multipler K_s
Top Region						
Centre						
Edges						
Mid Region						
Centre						
Edges						
Low Region						
Centre						
Edges						

Finally, averaging the values in central areas in one group, and those at building edges in another group, for each combination of geometrical values, a table like the following one (Table 7) could be suggested to present the recommended K_s values:

Table 7. Example of final recommended K_s value table format.

Horizontal sun-shades		h = 200 mm				h = 400 mm				h = 600 mm			
		K_s at central areas		K_s at building edges		K_s at central areas		K_s at building edges		K_s at central areas		K_s at building edges	
x [mm]	g [mm]	When $+c_{pe,1}$	When $-c_{pe,1}$	When $+c_{pe,1}$	When $-c_{pe,1}$	When $+c_{pe,1}$	When $-c_{pe,1}$	When $+c_{pe,1}$	When $-c_{pe,1}$	When $+c_{pe,1}$	When $-c_{pe,1}$	When $+c_{pe,1}$	When $-c_{pe,1}$
200	0												
	200												
	600												
500	0												
	200												
	600												
1500	0												
	200												
	600												

At the end, nothing of more complicated than what we can find now in the current Eurocode about (e.g.) the different types of roofs and their "zonage".

These values, together with other similar tables, would allow the designer to interpolate or extrapolate the recommended values to most of sun-shade design conditions, just starting from the façade c_{pe} measured or calculated values.

3 Non-circular Buildings with Curved Surfaces

The wind action analysis for cylindrical buildings having a circular cross section is supported by clear rules in the Eurocode 1, based on Reynolds' number. Instead, if we are dealing with non-circular buildings having locally rounded vertical surfaces, no rule can be found in the standards.

For instance, a plan view of a non-circular building, featuring rounded surfaces, is shown in the following Fig. 17.

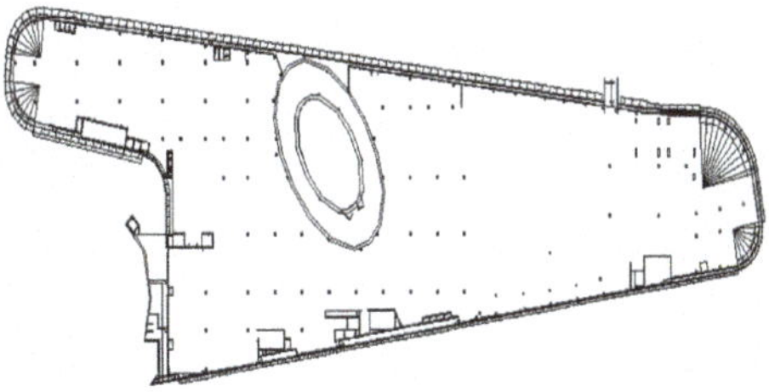

Fig. 17. Example of non-circular building having rounded corners.

No true circular cylindrical shape can be defined in this case, but experience has taught that certain curved areas may give rise to strong effects of wind suction, very similar to those of circular cylindrical buildings, whereas others can produce opposite results, smoothing the effects.

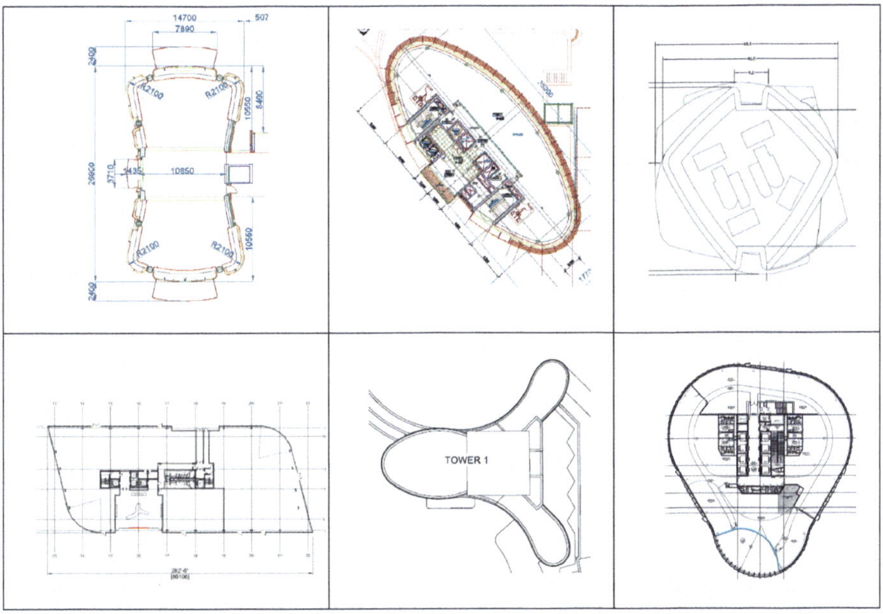

Fig. 18. Other examples of plane views of non-circular buildings having curved corners or wall surfaces.

Curved wall surfaces of any shape (circular, elliptical, free-form, ect.) are very frequent in modern architecture, but no hint can be found in the standards about how to deal with these shapes (Fig. 18).

Here follows an example showing the negative pressure values on two faces of the same building, featuring a curved face and a flat face.

The central part of the curved face of the building envelope shows distributed negative pressure values of –2.3 kPa (Fig. 19(a)), whereas the opposite flat surface of the same building, at its central part, shows distributed negative pressure values around –1.0 kPa (Fig. 19(b)). In addition, the intense negative values shown in Fig. 19(a) tend to remember the strong effects of separation typical of cylindrical buildings.

(a) (b)

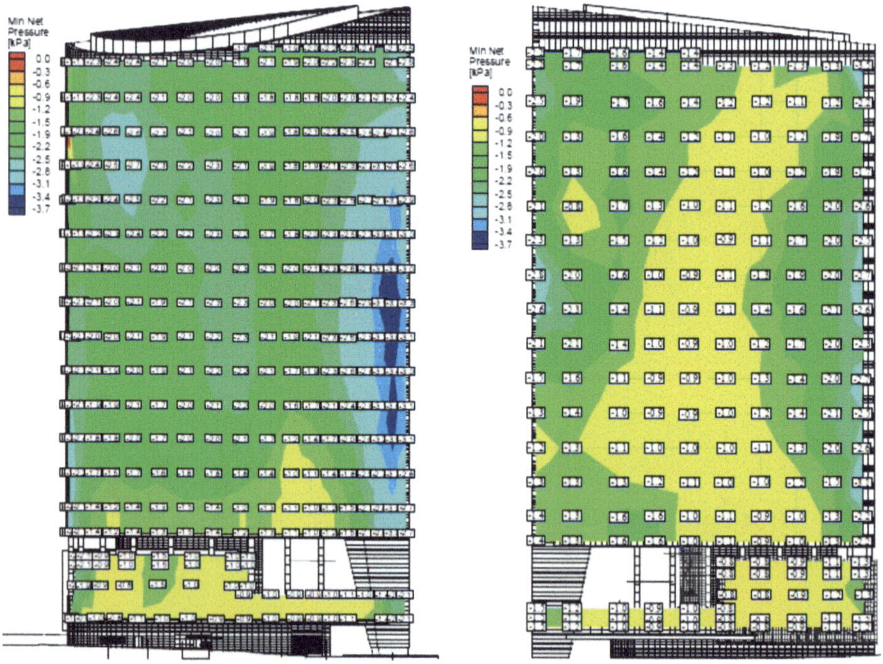

Fig. 19. Left (a) curved face of the building – Right (b) flat face of the building. (*Credits: BMT*®)

All values are the envelope of the minimal values in all the wind directions.

The only justification we can imagine of such a difference between the two faces is based on the effect of the curved surface, although the whole building is not a circular cylinder.

This kind of situation is so frequent that the façade engineers would really need to find some guideline in the standards about that.

4 Pressure on Cladding Elements with More Than One Skin

Wind load sharing in double skin cladding elements is a complex topic. This involves not only double skin façades having two glazed surfaces, but also ventilated claddings, shadow box elements, etc., whenever any ventilation hole has been opened.

Eurocode EN 1991-1-4 gives some guidance, at clause §7.2.10, but this frequently proves to be quite limited, compared to the actual possible configurations and it would require some development.

As a matter of fact, for instance, some in-house tests on shadow box panels of a project in USA have shown unexpected results, if compared to expectations according to the Eurocode, although the test procedure should be carefully questioned. We are going to report test procedure and results here below.

A shadow box generally consists of:

- Clear glazing,
- A cavity behind the glazing,
- An insulated panel/tray, internally closed by a sealed metallic back pane.

One of the key features of a shadow box is the cavity behind the external glazing, as shown in Fig. 20 here below.

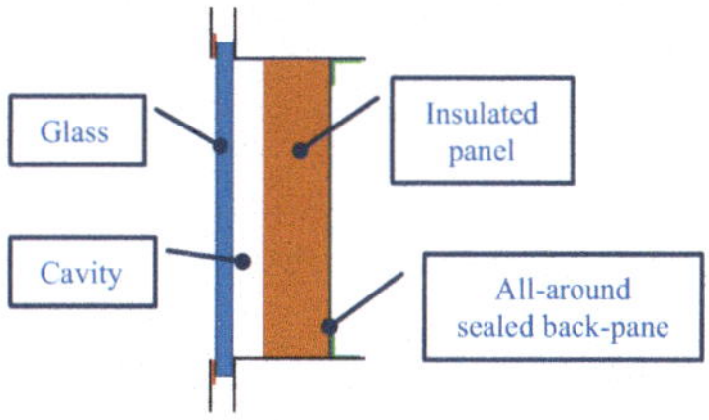

Fig. 20. Conceptual scheme of a shadow box.

Shadow boxes are used for the particular appearance they give. The cavity behind the glazing adds depth to the appearance, creating greater visual interest than what can be achieved with a more typical glazed spandrel. In addition, they allow the use of the same glass as that used in the vision areas, thus giving visual continuity between the different zones of the façade.

The shadow box we examine here featured a 1380 × 4394 mm^2 size (Fig. 21), with external double glazed pane 12/14/8 mm, an internal cavity 33 mm deep and an aluminium back pane 1.2 mm thick, silicone sealed.

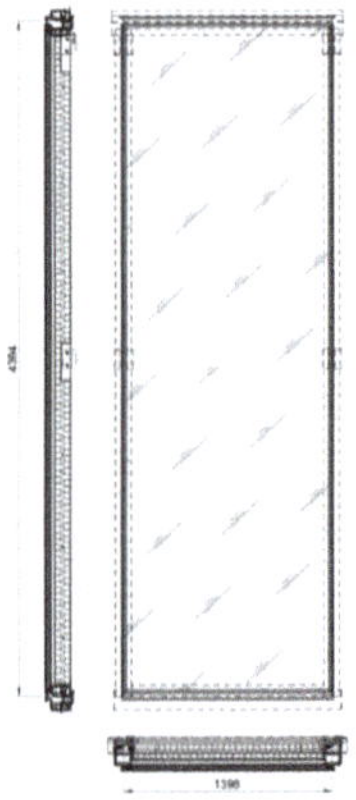

Fig. 21. Full scale shadow box (1380 × 4394 mm^2)

The lower transom showed two small holes, 14 mm diameter, aimed just at helping drying any possible condensation in the cavity. These holes were protected by filters allowing the 60% of free surface (Figs. 22 and 23).

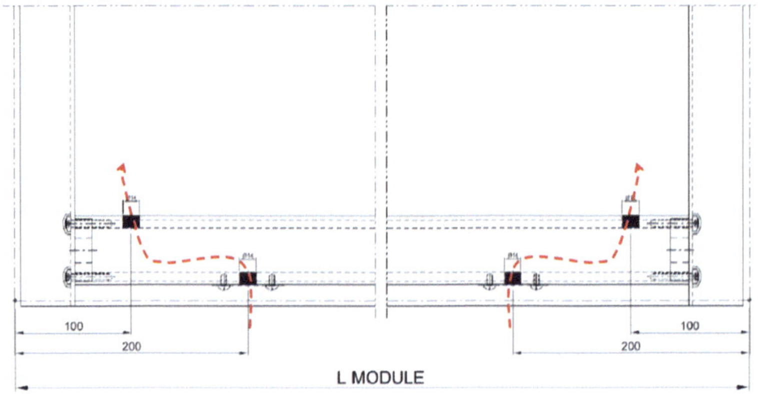

Fig. 22. Lower transom vertical section, frontal view; holes for drying the condensation.

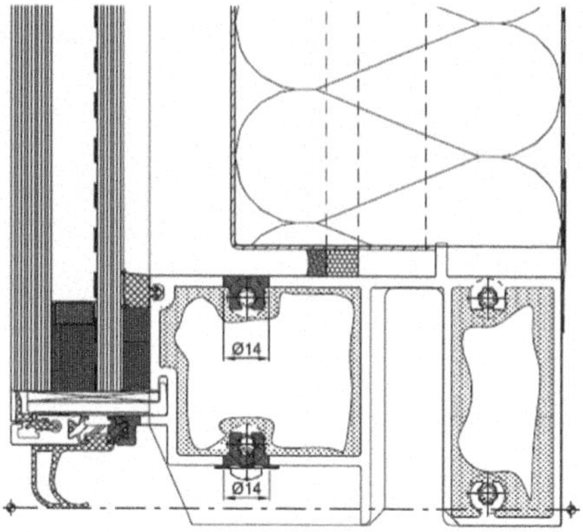

Fig. 23. Lower transom vertical section, side view.

Considering the definition of permeability given by EN 1991-1-4 §7.2.10 (2), we can calculate:

$$A_{op}/A_s = (2 \times 154) \times 0.6/6063720 = 0.000003 = 0.0003\% < 0.1\% \quad (3)$$

concluding that this shadow box surface could be considered as impermeable.

According to EN 1991-1-4 §7.2.10 (3), for walls and roofs with an impermeable inside skin and an impermeable, more rigid, outside skin, the wind force on the outside skin may be calculated from $c_{p,net} = c_{pe} - c_{pi}$, whereas the internal skin will support c_{pi} only. No doubt that the external 12/14/8 glass is more rigid than the internal 1.2 mm thick aluminium sheet panel, hence this shadow box unit was expected to comply with these rules.

Our test aimed at checking if these criteria could apply to our shadow box and specifically at verifying the wind load supported by the internal back pane.

So, we tried to reproduce, as far as possible, the on-site wind pressure loadings on a shadow box unit in full scale (1380 × 4394 mm^2).

Our testing equipment (Fig. 24) was composed of a sealed box, made of a SHS steel frame and thick plywood walls, featuring a closed chamber in front of the external glass pane, whereas the aluminium back pane was in open air.

A fan producing the requested air pressure was connected by a big pipe to the chamber of the sealed box, together with a shutter device, so to allow a quick opening and closing of the air inlet (Fig. 25).

The shadow box was fitted with several displacement transducers applied on the back pane, on the glass and on the frame, and with a pressure transducer inside the front sealed chamber (Fig. 26).

Considering that, in our test box, the back pane was in direct contact with the external environment, $c_{pi} = 0$; hence, on the glass skin $c_{p,net} = c_{pe} - c_{pi}, = c_{pe}$. On consequence, the over or under pressure acting on the back pane was expected to be approximately zero and that one acting on the front glass equal to the applied pressure.

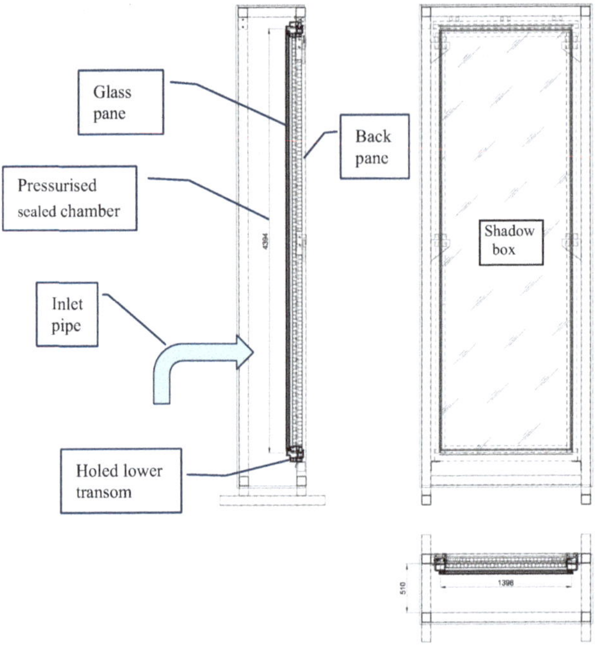

Fig. 24. Testing box.

Seeing that the design wind load values on the façade of this project were provided by a wind tunnel test, so the peak values of short duration were known too, we wanted to simulate somehow also the transient effect of gusts, in our test.

Fig. 25. Air supply pipe and shutter device.

As a matter of fact, considering the dynamic development of the gust pressure, we wanted to check whether there was any delay between the applied peak of pressure and its effects on the shadow box internal and external skins.

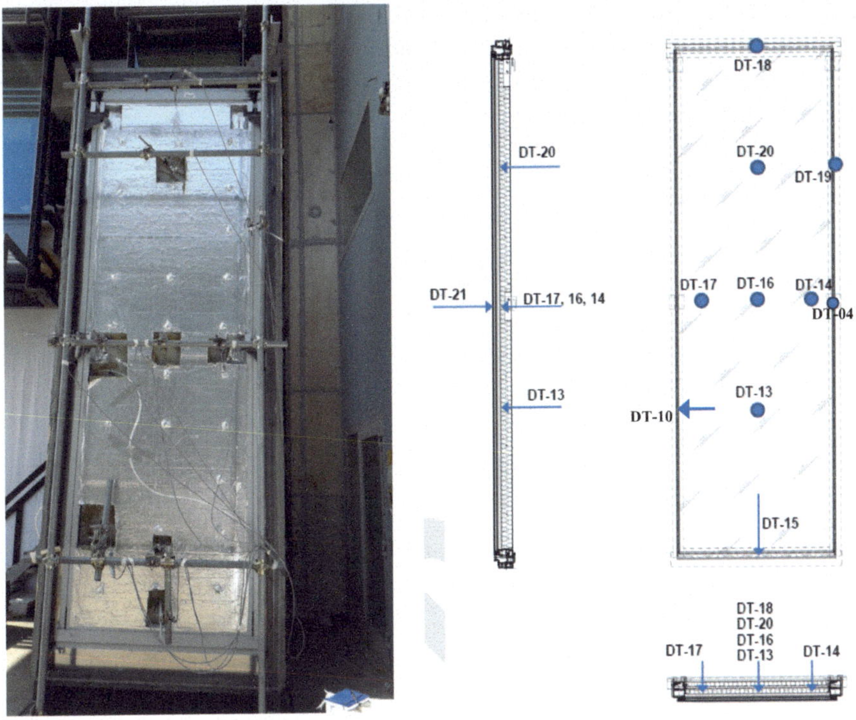

Fig. 26. Left: picture of the back pane and its transducers – Right: displacement transducer map.

Given that the gusts do not develop from steady air, but from a wind flow blowing at a certain average speed, we wanted to reproduce both the pressure effects of such average speed flow and of sudden gust peaks on top of it. We considered the 10-min average speed pressure as average wind speed pressure in our test.

Last but not least, considering that we tested a big shadow box at full size, it should be remarked that we could apply just pressure conditions to the unit, not an air flow with a certain speed in a boundary layer.

The first problem was how to ascertain the average speed, considering that the wind tunnel test was developed according to ASCE 7 standard and ASCE7-basic-velocity, defined at 10 m of height, already includes the 3-s gust effect: the peak wind velocity was 43.8 m/s (98 mph) in the examined project. Last but not least, the wind tunnel test output, shown in its report, was presented as a map of pressure values, which depend on the wind speed at different heights and on the local c_p values, which were not provided.

Hence, the process has been developed in the following steps:

(1) Once taken into account the 3-s gust speed, we selected suitable c_p values such to reproduce the most severe wind tunnel pressure values, –2873 Pa (–60 psf) and +1915 Pa (40 psf), as on the building surface portion cladded by the shadow boxes concerned by this study;

(2) We calculated the 10-min average speed corresponding to the known 3-s gust speed. This was based on Durst's diagram from ASCE7 (Fig. 27);

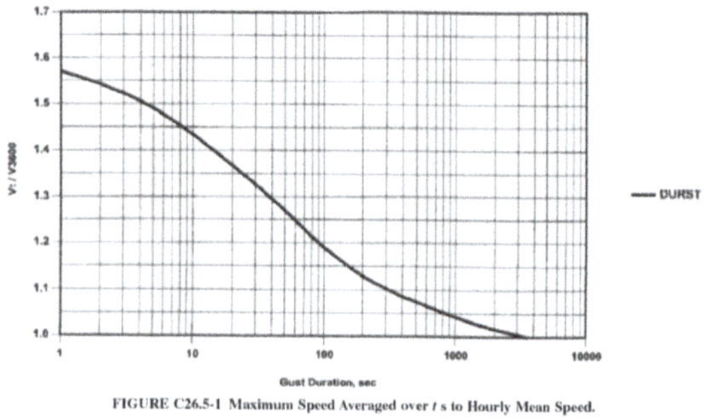

Fig. 27. Durst's diagram (from ASCE 7).

(3) We calculated the pressure values (positive and negative) corresponding to the 10-min average speed, by means of the same c_p values as calculated at bullet 1. Such pressure values resulted: –1006 Pa (–21 psf) and +670 Pa (14 psf).

This process and our intention to simulate both the average pressure and the peak pressure, produced pressure diagrams like the one shown here in Fig. 28.

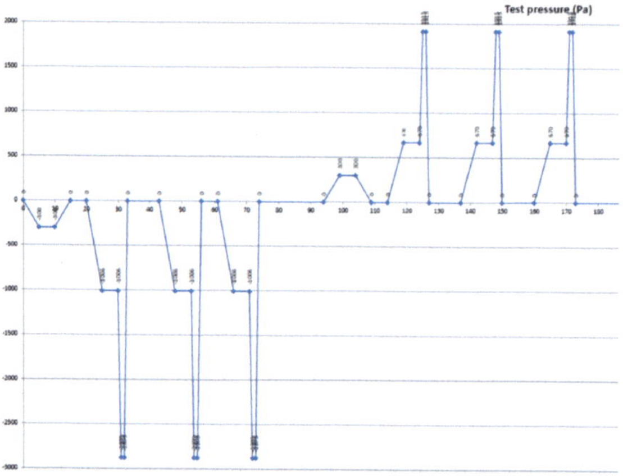

Fig. 28. Ideal applied pressure diagrams.

The first –300 Pa/+300 Pa peak is applied just to settle the prototype and zero its free movements; after that, there is the increase to the "10-min average pressure", –1006 Pa/+670 Pa, maintained for a while, and then the sudden increase to the gust peak value, –2873 Pa/+1915 Pa, which develops during 3-s. Each type of cycle is repeated 3 times, to collect comparable values and check their effect distribution. We were compelled to compress the "sustained" wind flow step duration down to 10 s, for practical reasons, but, considering we directly applied pressure values, this appeared to work properly.

The pressure diagrams shown in Fig. 28 are the "ideal" ones; inertias and friction loss produced, in reality, diagrams like those shown here below in Fig. 29.

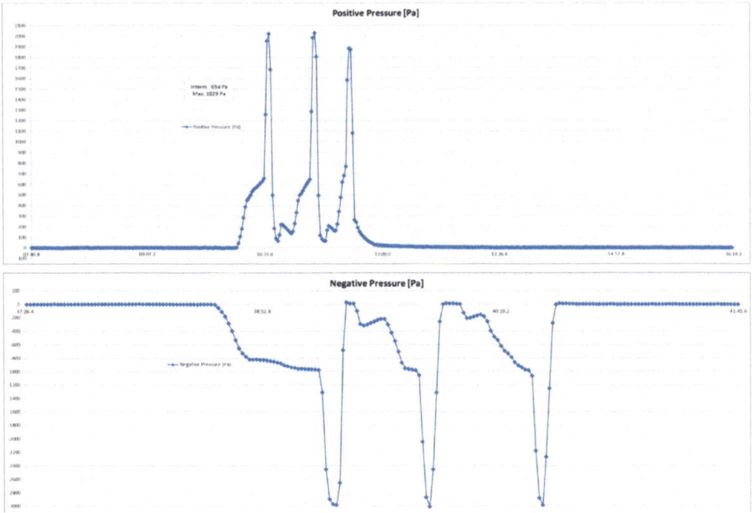

Fig. 29. Actual applied pressure diagrams.

We started our tests with the negative pressure cycles, but they proved to be too intense for our measuring equipment. Some displacement transducers lost their contact and some others exceeded the measuring range. Although unable to take a proper measure, this behaviour was showing effects that were beyond our expectations.

Hence, we went on with the positive pressure tests only.

Here follow some displacement plots measured at different positions. It should be noted that, to get the displacement either of the back pane sheet or of the glass pane, their readings had to be cleared of the supporting bracket free displacements and of the supporting frame deformation (Fig. 30).

These diagrams show that the glass pane centre has a maximum displacement of approximately 17 mm (averaged), whereas the back pane centre has a maximum displacement of approximately 25 mm. This is not a negligible displacement for an element that was supposed to carry nearly no load.

On the other hand, the back pane has very little inertia, so even a little load could cause significant deflections.

Hence, the question was: how much is the load magnitude causing a deflection of 25 mm at the back pane centre?

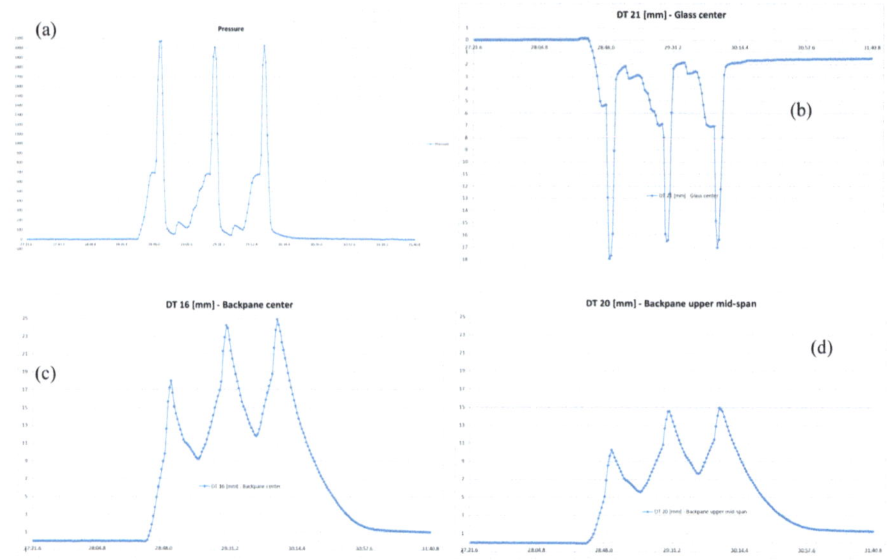

Fig. 30. Applied pressure (a) – Displacement at glass pane centre (b). Displacement at back pane centre (c) – Displacement at back pane upper mid-span (d).

To answer to this question, we developed a simple finite element model of the back pane, by means of shell elements, and in a geometrically non-linear analysis we increased the load gradually until we reached 25 mm of maximum displacement at the centre of the panel, as shown in Fig. 31.

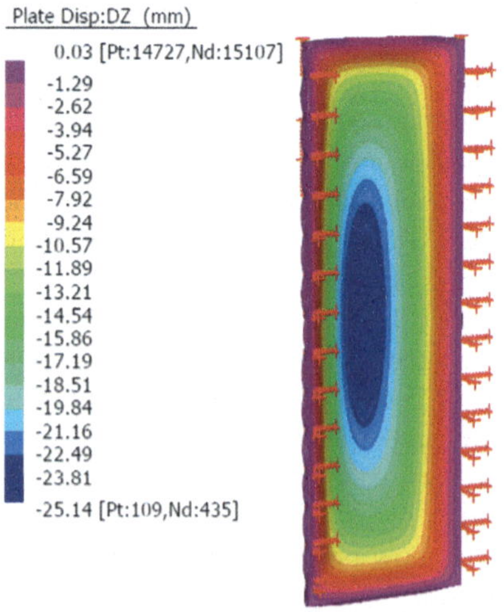

Fig. 31. Finite element simulation of the back pane deflection under distributed load.

This maximum deflection, 25 mm, corresponded to a distributed load of 105 Pa.

If the load would have been merely shared proportionally to the stiffness of each component, considering that the stiffness ratio between the back pane and the glass is in the range of 10^{-5}, we could have expected approximately 0.02 Pa acting on the back pane. On the contrary, it is quite evident that it is supporting a higher load.

Another interesting diagram is shown here below, in Fig. 32, where the time paces of the applied pressure and of the displacement peaks are compared, at different locations of the back pane: centre, lower mid-span, and upper mid-span.

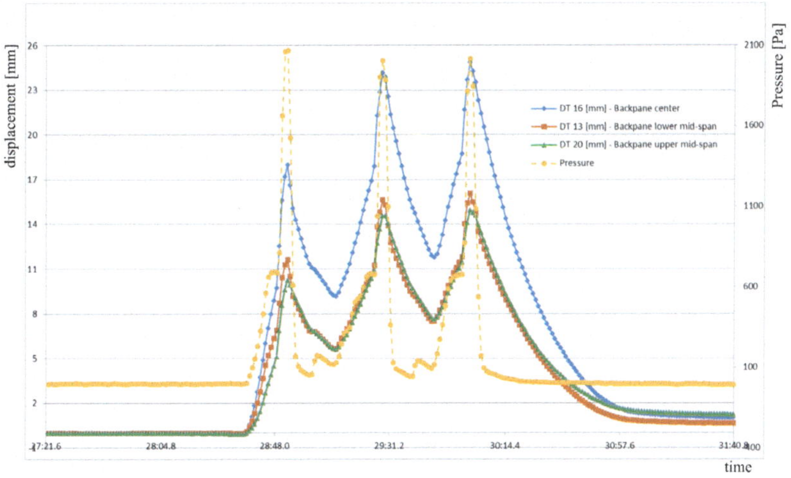

Fig. 32. Two open holes - Pressure and back pane displacements vs. time.

We can notice that there is no perceivable delay; the peak of applied pressure appears to be simultaneously transmitted inside the whole shadow box cavity, causing the deflections measured and shown here above.

Then, we decided to check what could happen having just one open hole and finally without any hole.

As far as the one-open-hole condition is concerned, we can notice in Fig. 33 here below that the maximum displacement at back pane centre has decreased to 19.7 mm (averaged) and that, once again, there is no delay between the applied pressure and the consequent displacement of the back pane.

Here follows the graph of the corresponding glass deflection, which reaches a maximum of approximately 17.0 mm (averaged), as shown in Fig. 34.

Last but not least, here follows in Fig. 35 the graph of displacement at the back pane centre and of the applied pressure, vs. time, in the condition with no open holes.

In this case the magnitude of displacement has dramatically decreased to 3.8 mm.

Here follows also the graph of the corresponding glass deflection, which reaches a maximum of approximately 18.3 mm (averaged), as shown in Fig. 36.

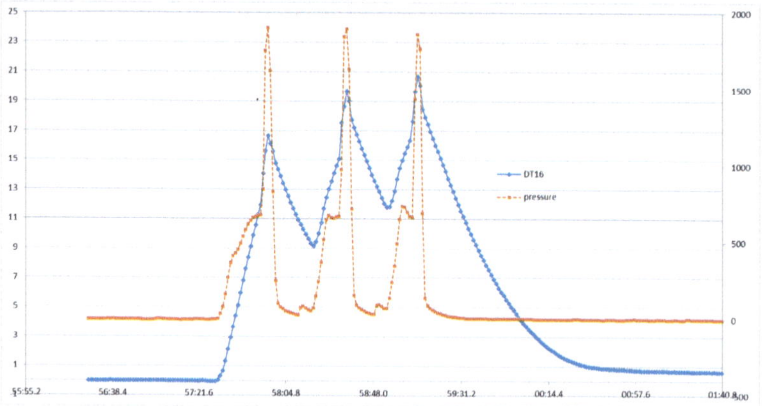

Fig. 33. One open hole - Pressure and displacements at the back pane centre, vs. time.

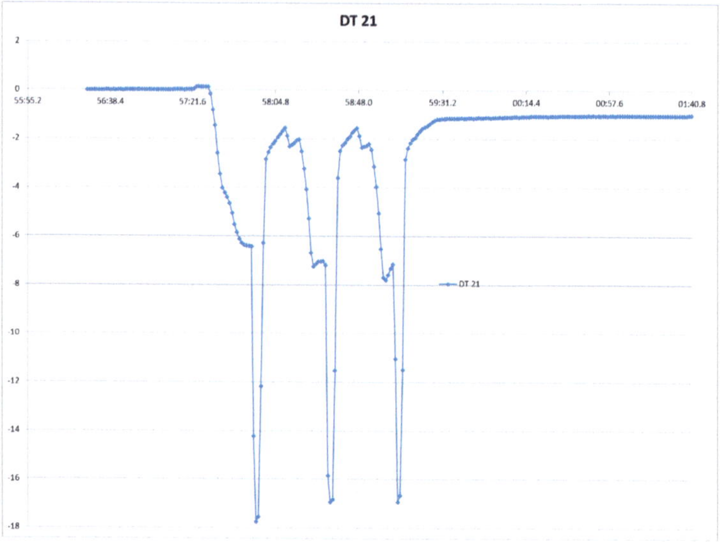

Fig. 34. One open hole - Displacements at the glass pane centre.

Finally, we can resume in the following Fig. 37 a comparison of the displacement graphs at back pane centre in the three different conditions: two open holes, one open hole, zero open holes.

As done before, comparing the measured displacement with that calculated by a finite element model, we can derive the load that is affecting the back pane in the three different conditions, as shown in the following Table 8.

Similarly, we can also resume a comparison of glass and back pane displacements in Table 9.

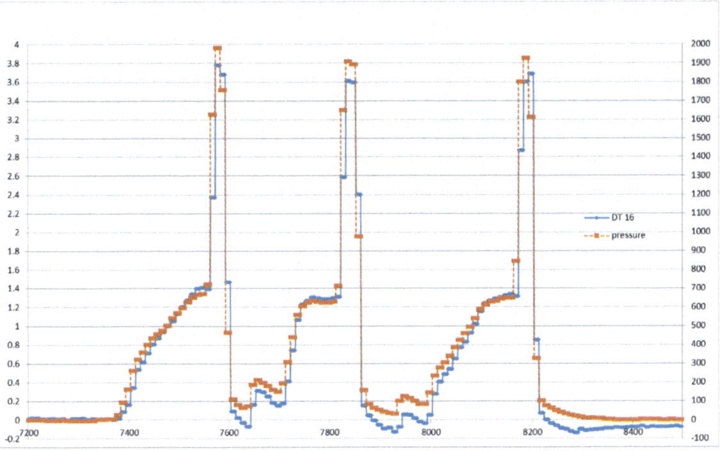

Fig. 35. Zero open holes - Pressure and displacements at the back pane centre, vs. time

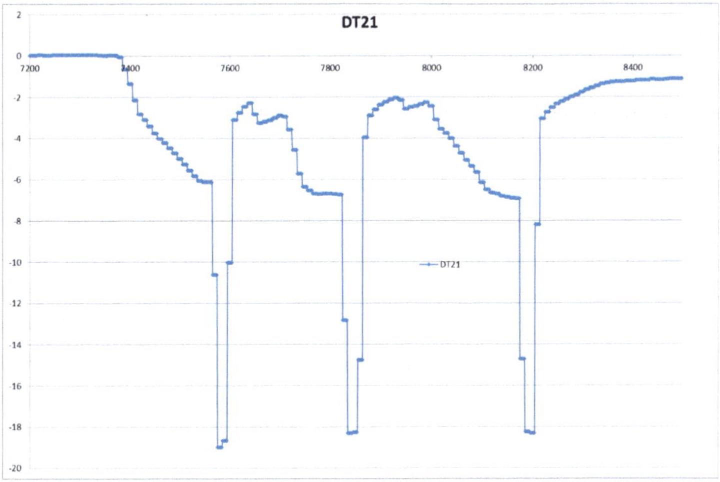

Fig. 36. Zero open holes - Displacements at the glass pane centre.

As noted before, the load values producing the measured displacements do not appear to be simply and only related to the stiffness ratio between glass pane and back pane.

It looks like that the glass pane didn't significantly "feel" the load sharing with the back pane, due to the huge inertia difference (10^{-5} ratio) and it just felt some effect of pressure equalisation in the shadow box cavity, at least until one hole was open.

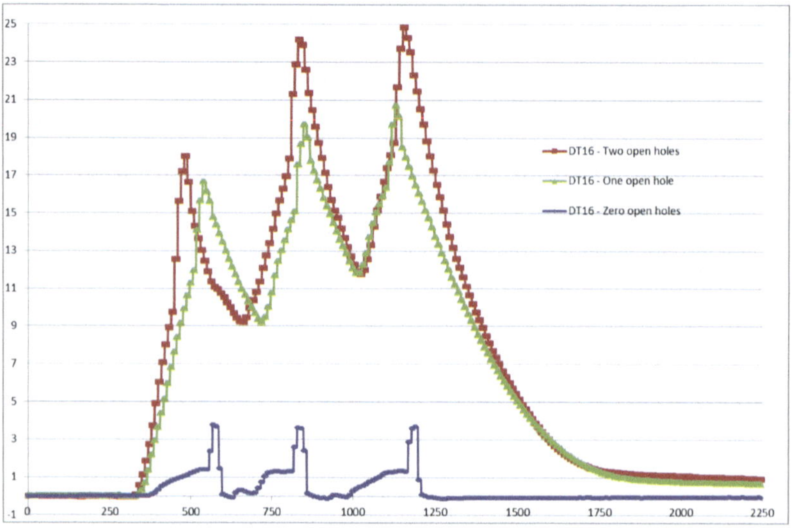

Fig. 37. Measured displacements at the back pane centre vs. time in the three different conditions: two open holes, one open hole, zero open holes.

Table 8. Displacement vs. acting load at the back pane centre.

Open holes	Displacement [mm]	Load [Pa]
2	24.9	120
1	20.6	90
0	3.8	8

Table 9. Displacements at the back pane centre vs. those at glass pane centre.

Open holes	Back pane displacement [mm]	Glass pane displacement [mm]
2	24.9	17.0
1	20.6	17.0
0	3.8	18.3

The back pane, on its side, appeared to have been strongly affected by the pressure entering the cavity, again until any hole was open. Then, with no open holes, it seems it was deflected by the load transmitted by the "gas spring" effect due to the air in the sealed cavity volume, compressed by the front glass deformation.

Here follow some possible considerations derived from this analysis:

- The limit value of permeability, as defined in the Eurocode to identify an impermeable volume against a permeable one, appears to have questionable applicability, particularly to a single panel as the shadow box that we tested.

- Supposing the shadow box surface as impermeable, its skins should have shared the applied load according to their stiffness, but this did not appear to be the only law driving their behaviour, which departed from what we were expecting according to the Eurocode. The glass pane showed to feel some pressure equalisation effect and the metallic back pane strongly felt the incoming pressure, as long as any opening was present, whatever its size.
- The propagation of pressure inside the panel cavity proved to be nearly instantaneous, despite the very small surface of the openings and the applied filters; the friction loss did not appear to affect the velocity of propagation.
- It should be remarked that we were compelled to apply a pressure in a confined volume, the sealed front chamber, communicating, through the two holes, with the only internal shadow box cavity, sealed as well. Such sealed front chamber represented the "external environment" subjected to the wind action and that was the only way for us to apply a pressure load, due to the tested sample size. Actually, in a façade on a building, the situation is quite different, because the wind action is not confined in a volume and the gusts do not have a perfect timing correlation on the whole surface of one shadow box like this one (approximately 6 m^2).
- Our measuring equipment had one pressure transducer, reading the applied pressure in the front chamber, but it was missing a direct reading of pressure inside the shadow box cavity; on consequence, the propagation of the pressure impulse was derived from its effects (i.e. from the displacement readings) instead to be directly measured. It is our intention to improve our measuring equipment and, possibly, repeat these tests.
- The pressure load was applied as an impulse and its action involved the dynamic response of the shadow box skins (glass and back pane), as like as the fluid dynamic response of the air inside the shadow box cavity. In other words, the phenomenon is dynamic and it should be studied in a fully dynamic domain. It might be reasonable supposing that some "transfer function" may rule the behaviour of the full system and the propagation of the pressure inside the shadow box cavity, and that such transfer function might change according to the cavity characteristics. This could be input for future studies and developments.
- The behaviour of double skin elements needs still some studies. The relevant section in the Eurocode deserves development.

So, concluding about this topic of double skin claddings, the Eurocode has the credit to be the only international standard dealing with this subject, but a substantial development of its section would be of paramount importance for the curtain walling designers.

5 Conclusions

Some issues that belong to the daily work of façade engineers, but still miss an official guideline like the international standards, were addresses in this paper to the kind attention of researchers and standard writers in wind engineering, for their consideration.

We also hope to have proved that, at least for some of these items, there are the bases for encouraging such progress into the international standards, and to have stimulated interest in studying and improving those topics that still have a limited development.

References

1. Cook NJ (1990) The designer's guide to wind loading of building structures. Part 2: static structures. Building Research Establishment Report, Butterworths
2. EN 1991-1-4 (2005) Eurocode 1: actions on structures - Part 1–4: general actions - wind actions
3. Jancauskas E, Holmes J (1985) Wind loads on attached canopies. In: US national conference on wind engineering, proceedings, Texas Tech University, Lubbock
4. Roh HW, Kim HR (2011) Wind pressure distribution on canopies attached to tall buildings. J Mech Sci Technol 25(7):1767–1774
5. SIA 261 (2014) Actions on structures
6. DIN EN 1991-1-4/NA (2010) National annex – nationally determined parameters – Eurocode 1: actions on structures – Part 1–4: general actions - wind actions
7. AS/NZS 1170.2:2002. Australian/New Zealand standards, structural design actions- Part-2, wind actions, Jointly published by Standards Australia International Ltd., Sydney and Standards New Zealand, Wellington

Flow Separation Delay and Drag Reduction Through Contoured Transverse Grooves

A. Mariotti(✉), G. Buresti, and M. V. Salvetti

Dipartimento di Ingegneria Civile e Industriale, Università di Pisa, Via G. Caruso 8, 56122 Pisa, Italy
alessandro.mariotti@unipi.it, {g.buresti,mv.salvetti}@ing.unipi.it

Abstract. The results of several investigations aimed at assessing the performance of contoured transverse grooves as a method to delay flow separation are described. The physical mechanism at the basis of this passive technique is the local relaxation of the no-slip boundary condition, with a consequent reduction of the viscous losses and an increase of the downstream near-wall momentum. Numerical simulations and experiments showed that the application of one groove transverse to the flow direction may delay boundary layer separation both in laminar and turbulent conditions. As a consequence, significant increases of the pressure recovery in plane diffusers and decreases of the drag of boat-tailed axisymmetric and two-dimensional bluff bodies were obtained. It is shown that robust configurations may be devised provided the shape and dimension of the grooves are suitably chosen in order to assure the formation of steady and stable local flow recirculations.

1 Introduction and Description of the Idea

Flow control is an extremely important topic both for applications and for basic fluid dynamic research. In particular, the delay of boundary layer separation over curved solid surfaces immersed in internal or external flows is of great importance in many engineering applications in order to attain desired characteristics of the aerodynamic loads, such as the reduction of the aerodynamic drag of vehicles, the increase of the maximum lift of airfoils or the increase of the pressure recovery in diverging ducts. More in general, the desired objective may be obtaining configurations providing optimum values of certain load components or of suitable design-connected parameters, such as the ratio between lift and drag.

Different active or passive techniques may be used to achieve these objectives (see e.g. the review in [1,2] and the references herein) and the present one belongs to the class of passive control methods. The idea is to introduce small and suitably-shaped grooves transverse to the flow (i.e. *contoured transverse grooves, CTGs*) to generate local steady and stable flow recirculations, producing on the outer flow an effect that is analogous to a local relaxation of

F. Ricciardelli and A. M. Avossa (Eds.): INVENTO 2018, LNCE 27, pp. 483–495, 2019.
https://doi.org/10.1007/978-3-030-12815-9_37

the no-slip boundary condition. The use of small local recirculations as a passive method for the control of boundary layer separation was already proposed in [3], where, in spite of the preliminary nature of the numerical analysis, small contoured steps (of the same order of the incoming boundary layer) were shown to be able to control separation and to improve the performance of boat-tails. The use of cavities for flow control is also at the basis of the so-called "trapped vortex" concept, first introduced by [4], in which a large vortical structure with almost constant vorticity creates a recirculation region separated from the outer flow by a thin and stronger shear layer. This flow configuration may lead to a downstream boundary layer with higher momentum, with a consequent separation delay; however, in that case the cavities are typically much larger than the incoming boundary layer thickness and it is extremely improbable that stable vortices may form inside them. Therefore, an active control, for instance through adequate blowing and suction, is practically always necessary (see, e.g. [5]). Conversely, our objective was to achieve this result without the necessity of any active stabilization procedure. In practice, this implies that the grooves considered in the present method must have a depth that is smaller than the thickness of the upstream boundary layer and a suitable shape contributing to the steadiness and stabilization of the flow. The shape and width of the groove are fundamental to avoid the presence of self-sustained cavity oscillations, which may be caused by shear layer instability, pressure wave propagation and acoustic resonance (see, e.g., [6,7]). Indeed, the occurrence and magnitude of these oscillations are closely related with the geometry of the grooves and the characteristics of the incoming boundary layer. In particular, a rear sharp edge must be avoided and the boundary layer thickness should be at least of the same order as the cavity depth [6].

The effectiveness of the proposed technique in delaying separation was verified both in internal and external flows. First, it was applied to the improvement of the pressure recovery in plane diffusers and the results of those investigations showed that significant increases in the efficiency of the diffusers could be obtained both in laminar and in turbulent flow conditions (see [8–10]). Then, in [11] the same technique was applied to different axisymmetric boat-tailed bluff bodies, with flow separations occurring over the boat tail lateral surface and in [12] to a two-dimensional boat-tailed bluff body with a cross-section profile similar to one of the axisymmetric bodies tested in [11]. Boat tailing is a method to reduce the drag of bluff bodies through a gradual reduction of the body cross-section before a sharp-edged base, which causes a pressure recovery over the final portion of the lateral surface and over the base (see e.g. [13,14]). The drag reducing performance increases with decreasing ratio between the final and initial cross-flow dimensions of the boat tail; however, given a certain length of the boat tail, when this ratio is too small boundary layer separation may occur before the base, thus limiting the pressure recovery. Through a synergic use of experiments and large eddy simulations, the introduction of contoured grooves was shown to delay boundary layer separation, leading to a higher effectiveness of boat tailing.

In the present work the main results obtained previously by using CTGs are summarized and the physical mechanisms responsible for the separation delay effect are described. The main analogies and differences in the various applications are pointed out, in order to highlight general trends and conclusions. The results regarding the flow around two-dimensional and axisymmetric bodies with a similar cross-section profile are presented and discussed together. Compared to the axisymmetric cases, the main difference is that the two-dimensional body is characterized by a strong pressure gradient along the boat tail lateral surface and by high pressure and velocity fluctuations caused by the alternate vortex shedding in the wake. Thus, this test case is a very challenging one for the proposed flow control device.

2 Test Cases and Investigation Methodologies

The different numerical and experimental method and set-ups used in the above described investigations are herein briefly recalled. The diffuser geometry investigated in laminar flow in [8, 10] is shown in Fig. 1(a). Three different values of the divergence half-angle α were considered, viz. $\alpha = 2°$, $\alpha = 3.5°$ and $\alpha = 5°$, while the diffuser area ratio, AR, i.e. the ratio between the diffuser outlet and inlet cross-areas, was kept constant to the value $AR = 2$. Two-dimensional incompressible simulations were carried out at a Reynolds number $Re = hu/\nu = 500$, where u is the x-velocity on the diffuser axis at the inlet section and ν is the kinematic viscosity. The inlet velocity field is specified by using a Blasius boundary layer profile with thickness $\delta_0 = 0.10h$ on the two walls. The simulations were carried out by using a commercial code, based on the finite-volume discretization method, with unsteady time advancing and a second-order implicit scheme. A second-order upwind scheme was used for the space discretization. The computational grid was unstructured and was made of triangular elements (see [8, 10]) for further details on the simulations).

As for the diffuser in turbulent regime, the considered geometry was aimed at reproducing the experimental conditions in [15] and is shown Fig. 1(b)). The diffuser area ratio is $AR = k/h = 4.7$ and the diverging angle is $\alpha = 10°$. The upstream constant-width channel is sufficiently long to obtain a fully-developed turbulent channel flow at the inlet of the diffuser diverging section. The simulations were carried out at $Re = hu/\nu = 2 \times 10^4$, where u is the x-velocity on the centreline at an upstream reference cross-section where the flow has already become fully-developed turbulent. Two-dimensional simulations of the flow inside the diffuser were carried out by solving the RANS equations using the same commercial code as in the laminar case. The Shear-Stress Transport $k - \omega$ turbulence model was chosen, no wall functions were used and a suitable grid refinement was adopted in order to have $y^+ \leq 1$ at the wall (see [9] for further details).

The axisymmetric and two-dimensional boat-tailed bluff bodies have the same streamwise section, consisting in a 3:1 elliptical forebody and a rectangular main part followed by a circular-arc boat tail (see Fig. 1(c)). The cross-flow sections are sketched in Fig. 1(d) and (e) for the axisymmetric and the two-dimensional cases,

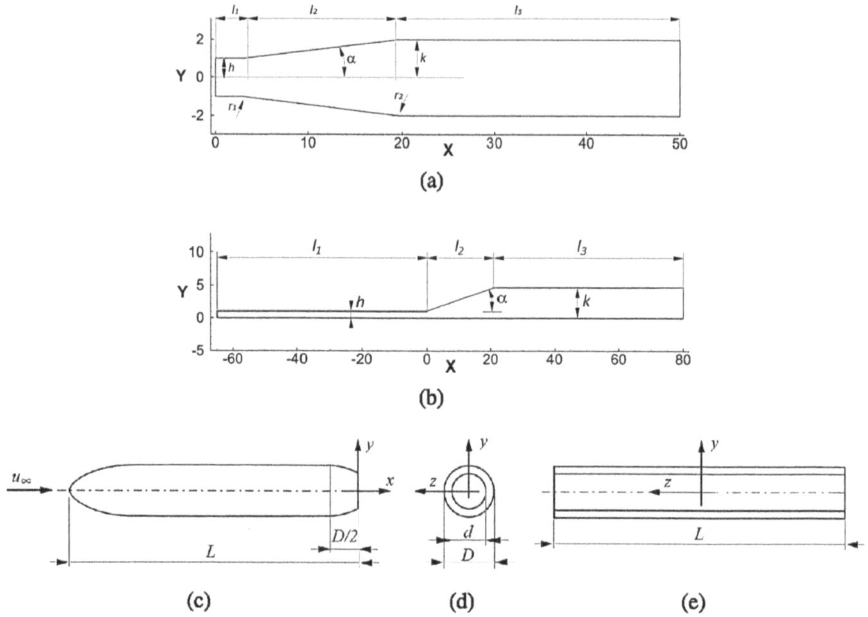

Fig. 1. Sketch of the considered geometries for: (a) diffuser in laminar flow regime, (b) diffuser in turbulent flow regime, (c) streamwise section of the axisymmetric and of the two-dimensional boat-tailed bluff bodies, (d) cross-flow section of the axisymmetric boat-tailed bluff body, (e) cross-flow section of the two-dimensional boat-tailed bluff body.

respectively. The ratio between the main body cross-flow dimension, D, and the overall length, L, is $D/L = 0.175$. The boat-tail length is $D/2$ and different ratios between its final dimension, d, and that of the main body are considered, ranging from $d/D = 0.632$ to $d/D = 0.886$. When no groove is present, at the considered Reynolds number equal to $Re = Du_\infty/\nu = 9.6 \times 10^4$ the boat-tail geometries imply different degrees of flow separation along their lateral surface. The effectiveness of the CTGs was assessed through a synergic use of experiments and large eddy simulations in the axisymmetric case and through numerical simulations in the two-dimensional one. Simulations have been carried out by using AERO, an in-house developed compressible flow solver. The code had already been successfully used for the simulation of bluff-body flows (see, e.g. [16,17]). The numerical discretization is based on a linearized implicit time advancing and on a mixed finite-volume/finite-element method, applicable to unstructured grids. The accuracy of the numerical method is second order both in space and in time and the Smagorinsky subgrid scale model is used to close the VMS-LES equations. Simulations are carried out for laminar freestream conditions and the freestream Mach number is equal to 0.1 in order to make the compressibility effects negligible. For the axisymmetric case, the computational domain is cylindrical, with a diameter of 15D and a length of 50D (see [11] for further details). Conversely, in the two-dimensional case the computational domain is rectangular, with a cross section of $5.71D \times 150D$ (in the z and y directions, respectively) and an overall length of $50D$. Periodic boundary conditions are imposed in the spanwise direction, characteristic-based boundary conditions [16] are used at the inflow, outflow and upper and lower surfaces of the computational domain, while no slip is imposed at the body surface. The rectangular computational domain is discretized by an unstructured grid with approximately 2×10^6 nodes, which is particularly refined near the body surface and in the near wake, with a wall $y^+ \leq 1$ (see [12] for further details).

The experiments were carried out in the closed-return subsonic wind tunnel of the University of Pisa, having a circular open test section 1.1 m in diameter and 1.42 m in length, with a free-stream turbulence level of 0.9%. The aluminum-alloy model has a diameter D = 70 mm and an overall length L = 400 mm. The final $0.5D$ long portion of the model, which is removable, may be substituted with four ABS polymer elements, produced with a 3D printer, comprising $0.5D$ long boat tails with $d/D = 0.791$ and $d/D = 0.632$, both plain and with the groove. The models were supported above a flat plate by means of a faired strut. The pressures were measured using two miniature electronic pressure scanners, directly positioned inside the model. The measurement points were 288 over the lateral surface and 97 over the base and were sufficient to permit to obtain, by integration, reliable values of the contributions of the lateral surface and of the base to the pressure drag. The velocity measurements for the evaluation of the boundary layer profiles and the wake field were carried out through hot-wire anemometry. Single-wire probes were moved in the x, y and z directions with an accuracy of 0.1 mm by means of a computer-controlled traversing system. A sampling frequency of 2 kHz, which is significantly higher than the highest analysed fluid-dynamic frequency, and a time length of approximately 65 s were chosen for the pressure and hot-wire measurements (see [11,18,19] for further details on the experimental set-up).

3 Geometry of a Contoured Groove and Physical Mechanisms

The geometry of a typical contoured transverse groove, shown in Fig. 2, is characterized by an initial sharp edge and semi-elliptical contour – in order to fix and promote the start of the desired local recirculation region – and by a final spline tangential to the basic surface. The chosen shape avoids the presence of sources of local perturbations, such as downstream sharp edges; however, to assure the steadiness of the flow and the absence of cavity oscillations, the depth of the groove is chosen to be definitely smaller than the thickness of the incoming boundary layer. Referring to Fig. 2, the parameters defining the geometry of a groove are then the starting point, s, the total length, t, the depth h, and the length of the ellipse axis in the streamwise direction, a. The groove starting point s is measured from the beginning of the diffuser divergent part or from the beginning of the boat tail to the initial sharp edge of the groove.

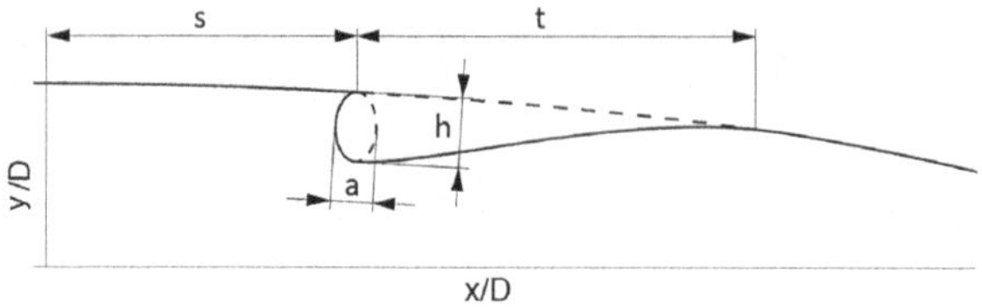

Fig. 2. Geometry of the contoured transverse groove

In all the investigated flows, local steady flow recirculations are generated when the suitably-shaped grooves are introduced transverse to the flow in a curved solid wall (see e.g. the flow streamlines for the axisymmetric boat tail $d/D = 0.791$ without and with groove shown in Fig. 3(a) together with the azimuthal vorticity magnitude). The physical mechanism responsible for the success of this flow control strategy is essentially the effective relaxation of the no-slip condition along the outer boundary of the recirculation region inside the groove; in effect, slip velocities up to $0.4u_\infty$ are found over the groove at the vertical coordinate corresponding to the surface of the boat tail without groove. This reduces the momentum losses near the wall and thus leads to a boundary layer with higher-momentum downstream of the groove, allowing separation to be delayed (see e.g. the velocity profiles downstream of the groove in Fig. 3(b)). As can also be seen in Fig. 3(a), outside the groove region – i.e. in the free shear layer bounding the recirculation – the magnitude of the vorticity is significantly smaller than that of the vorticity present near the surface of the plain boat tail, leading to a reduction of the enstrophy in that region. Conversely, the thickness of the boundary layer just before and after the groove is smaller for the grooved boat tail and this increases the near-wall vorticity in those regions. In the boat tail with groove a non-negligible vorticity of opposite sign is also present at the bottom of the groove due to the reverse velocity in the recirculation region. Considering the effect of these variations on the enstrophy integral (around the groove and also in the wake), when the groove is introduced a decrease of the global dissipation is found, thus witnessing a reduction of the viscous losses.

The steadiness of the recirculation region inside the groove has been experimentally and numerically assessed by considering the reduction of the velocity and pressure fluctuations in the groove region and in the boundary layer downstream of the reattachment point. In particular, a reduction of the velocity fluctuations within the boundary layer in the groove region is found when the groove is present (see Fig. 4(a) and (b), respectively for experiments and simulations on the axisymmetric boat tails). Moreover, the distribution of the pressure fluctuations over the lateral surface of the two-dimensional boat tail $d/D = 0.791$, shown in Fig. 4(c), confirms the remarkable reduction of the pressure fluctuations inside the groove region compared to those present over the plain boat tail. Finally, no peak is present in the spectra of the velocity signals in the boundary layer over the above-mentioned positions, further substantiating that no cavity resonance phenomena are present. In summary, all the available evidence strongly corroborates the inference that the small recirculation regions present within the grooves are effectively steady.

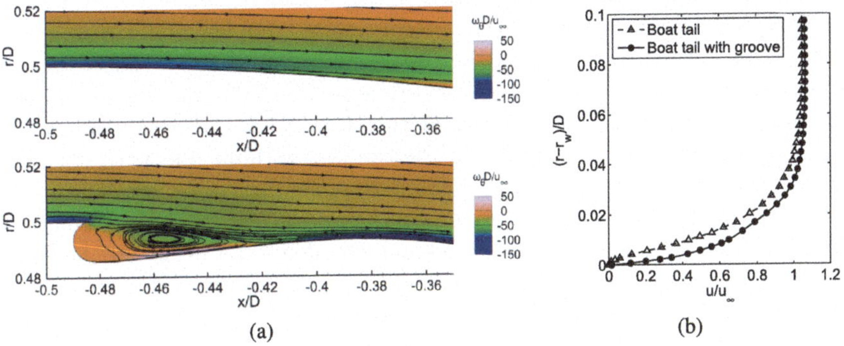

Fig. 3. Results for the axisymmetric boat-tailed bluff body $d/D = 0.791$: (a) Azimuthal vorticity field, averaged in time and in azimuthal direction for the boat tail without groove (top) and with groove (bottom); (b) numerical boundary layer x-velocity profiles at $x/D = -0.35$, averaged in the azimuthal direction.

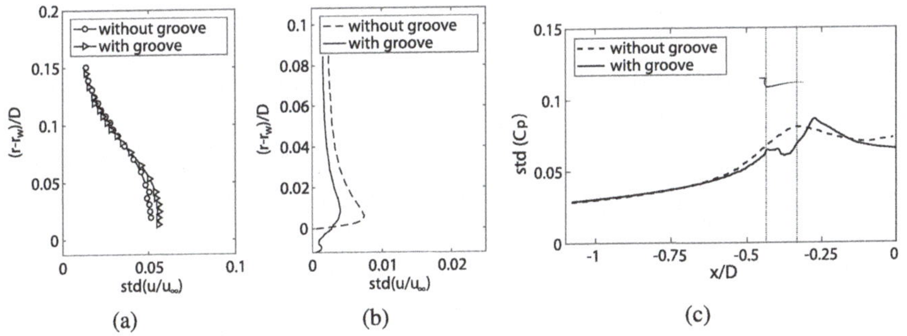

Fig. 4. (a) Experimental boundary layer x-velocity fluctuations at $x/d = -0.37$ for the axisymmetric boat tail with $d/D = 0.632$. (b) Numerical boundary layer x-velocity fluctuations at $x/d = -0.46$ for the axisymmetric boat tail with $d/D = 0.791$. (c) Numerical standard deviation of the pressure coefficient over the lateral surface of the two-dimensional boat tail with $d/D = 0.791$, averaged in the spanwise direction.

4 Groove Effect in Practical Applications

In this Section some results from the previous investigations [7–10] are briefly recalled and compared with the new ones regarding the flow around a two-dimensional boat-tailed body. The main analogies and differences between the various applications are pointed out. For plane diffusers, the introduction of CTGs was found to significantly reduce flow separation and to increase the pressure recovery both in laminar [8,10] and in turbulent conditions [9]. The grooves were applied to laminar diffuser configurations with the same area ratio and different diverging angles, characterized by significantly different flow features (from completely attached flow to the presence of a large asymmetric separation region). Considering, for instance, the results for the diffuser having divergence half-angle $\alpha = 3.5°$, a pressure recovery increase of 13.0% was found by introducing a couple of symmetric cavities at the start of the diverging walls. From the visualization of the streamlines in the diffuser without and with optimized grooves, sketched in Fig. 5(a), it is evident that the flow separates at the cavity edge but reattaches immediately downstream, forming a small recirculation region. Furthermore, the subsequent asymmetric flow separation is delayed and its extent is strongly reduced. The behaviour of the mean pressure coefficient $\overline{C_{p_x}}$ along the length of the diffuser, shown in Fig. 5(b), evidences the larger increase in pressure at the beginning of the diffuser diverging walls produced by the grooves. More downstream, the pressure losses are lower than in the reference configuration due to a reduction of the separated region extent. Modifications of the groove geometry have also been tested, by allowing the presence of a bulge or a depression after the groove as well as by introducing multiple couples of cavities. The use of multiple grooves further delays flow separation in the diffuser and hence enhances the pressure recovery performance.

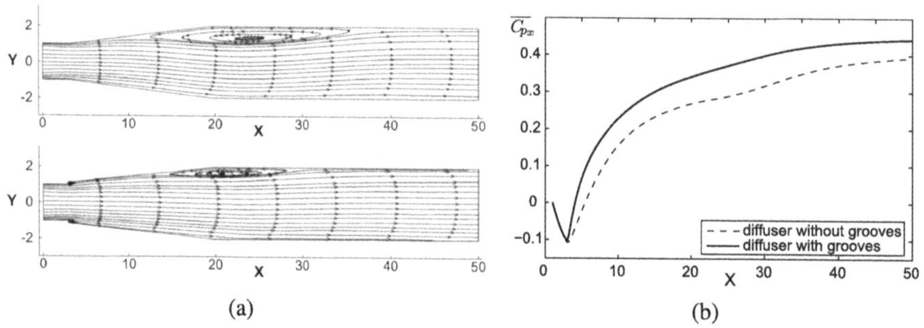

Fig. 5. Results for the laminar diffuser having $AR = 2$ and divergence half-angle $\alpha = 3.5°$: (a) flow streamlines in the reference diffuser (top panel) and in the diffuser with contoured grooves (bottom panel); (b) mean pressure coefficient at different X sections.

The proposed control strategy is effective also for turbulent flows at higher Reynolds numbers. Indeed, the introduction of one optimized groove in the diffuser of Fig. 1(b) leads to an increase in pressure recovery of the order of 6.9% and to a significant reduction of the separation extent. A further improvement (9.6%) is obtained by introducing two subsequent grooves in the divergent wall.

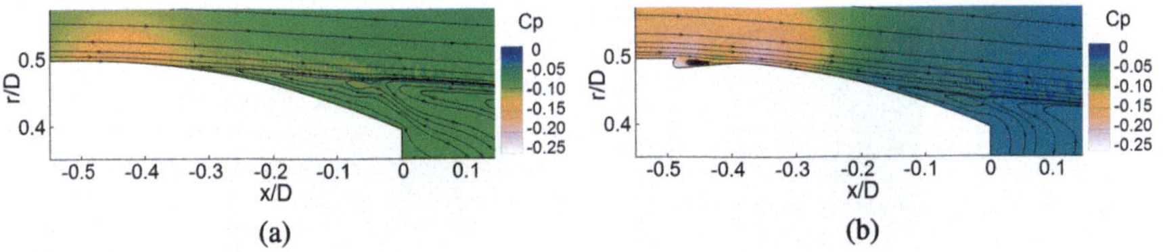

Fig. 6. Mean velocity streamlines and pressure coefficient field for an axisymmetric boat-tailed bluff body: reference boat tail (a) and boat tail with groove (b).

As for external flows, the method is found to be effective to reduce the drag of axisymmetric boat-tailed bluff bodies (see [11]). When no groove is present, the boat-tail geometries imply flow separation along their lateral surface. As demonstrated by wind tunnel tests and large eddy simulations, the introduction of one suitably contoured transverse groove on the lateral wall produces a significant boat-tail drag reduction (up to 29%, see [11]) due to a significant delay of the flow separation and a consequent increase of the base pressure. Indeed, as can be observed in Fig. 6, the delay of separation causes a larger pressure recovery in the last part of the body, leading to a higher base pressure. In particular, the presence of the groove induces an earlier and enhanced decrease of the pressure before and at the beginning of the boat tail. However, a rapid increase in pressure occurs around the reattachment region after the recirculation, followed by another decrease due to the subsequent acceleration of the flow. Finally, the pressure recovery corresponding to the flow deceleration is much steeper when the groove is present, leading to a significantly higher pressure at separation, which remains practically constant over the whole base.

In the experiments, no measurements allowing a direct detection of the separation point were available; however, the different positions of the boundary layer separation may be inferred from the values of the pressure standard deviation along the lateral surface, which are shown in Fig. 7. Indeed, it is reasonable to suppose that the region of separation be connected with an increase of the pressure fluctuations. The plain boat tail shows a significant increase of the fluctuations over the rear part of the lateral surface, indicating a flow separation occurring well before the base. The introduction of the groove has a notable effect and the data seem to indicate that separation is now practically avoided, as witnessed by the remarkable decrease of the pressure fluctuations over the last portion of the lateral surface and over the base. Moreover, the separation line, which is not axisymmetric on the reference body due to the presence of the model support, is noticeably regularized by the introduction of the groove.

The flow control performance of the grooves is also confirmed by the recent results on two-dimensional boat-tailed bluff bodies [12]. Indeed, the introduction of one couple of suitably contoured transverse grooves on the lateral walls was found to produce a boat-tail drag reduction of the order of 9.7%, as a result of a significant delay of flow separation and of the consequent increase of the base pressure. As expected, the flow within the grooves is found to be steady and the

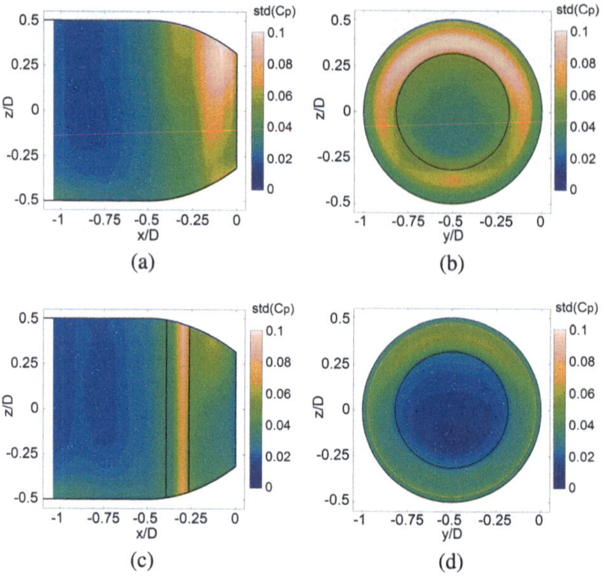

Fig. 7. Standard deviation of the pressure coefficient for the boat tail $d/D = 0.632$: (a) lateral surface of the plain boat tail; (b) base of the plain boat tail; (c) lateral surface of the boat tail with groove; (d) base of the boat tail with groove.

separation-delay effect to be connected with the enhanced near-wall momentum present in the boundary layers immediately downstream of the grooves. The wake flow is characterized by the periodic alternate shedding of vortices, which is evident from the analysis of the instantaneous vortex indicator λ_2 [20], illustrated for the plane $z/D = 0$ in Fig. 8 and with three-dimensional isosurfaces in Fig. 9. From the analysis of these two figures, the main effects of the introduction of the groove seem to be a reduction of the wake width and an increase of the spanwise correlation of the near-wake vortical structures. The narrowing of the wake produces an increase of the related Strouhal number, defined as $St = fD/u_\infty$. Indeed, it increases from $St = 0.304$ for the body with the plain boat tail to $St = 0.319$ for the one with groove, in inverse proportion to the wake width reduction. The increased regularity of the vortex shedding in the spanwise direction may be assessed through the evaluation of the correlation coefficient in the $z-$direction of the x-velocity component in the near wake and of the pressure field on the boat tail. The correlation coefficient in the $z-$direction is defined as $\rho(q) = \overline{q(x,y,z)q(x,y,z+\Delta z)}/(\sigma_{q(x,y,z)}\sigma_{q(x,y,z+\Delta z)})$, where Δz is the distance from the reference symmetry plane $z/D = 0$, $\overline{q(x,y,z)q(x,y,z+\Delta z)}$ is the covariance of the quantity of interest q, $\sigma_{q(x,y,z)}$ and $\sigma_{q(x,y,z+\Delta z)}$ are the related standard deviations. For the evaluation of the correlation coefficient we consider a line in the spanwise direction at a $y-$coordinate $0.07D$ outside the wake borders at the cross section $x/D = 2$ and its values are shown in Fig. 9(c). As can be seen, a significant increase in the spanwise correlation of the $x-$velocity component is found for the boat tail with groove. This increased correlation is probably connected with the straighter flow separation line in the spanwise direction induced by the presence of the grooves.

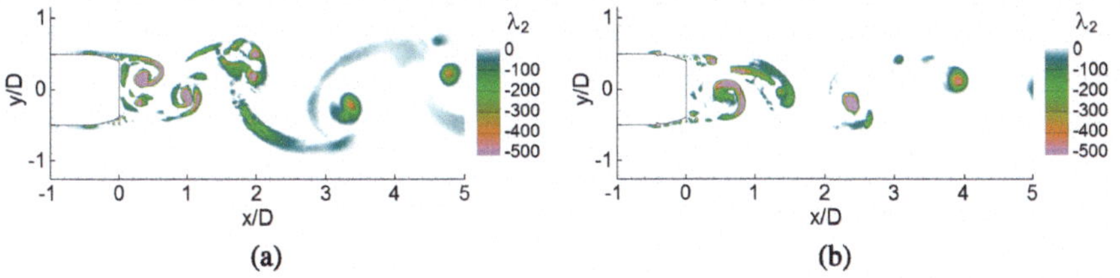

Fig. 8. Values of the λ_2 vortex indicator on the plane $z/D = 0$ downstream of a two-dimensional boat-tailed bluff body: (a) reference boat tail and (b) boat tail with groove.

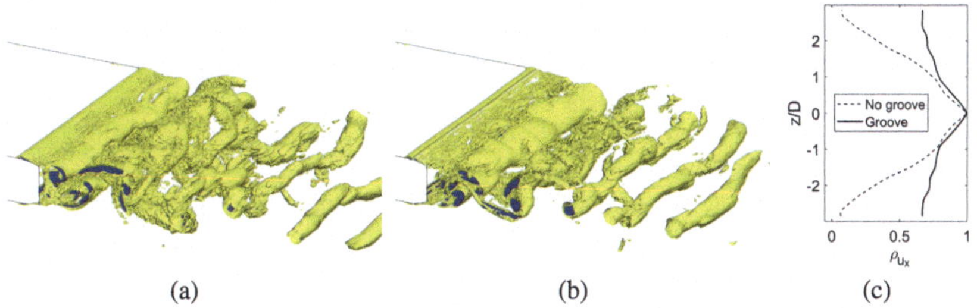

Fig. 9. Iso-surfaces of λ_2 for the two-dimensional boat-tailed bluff body (a) without the groove and (b) with the groove. (c) cross correlation of the x–velocity component evaluated $0.07D$ outside the wake boundary at $x/D = 2$.

5 Conclusions and Future Developments

Numerical simulations and experiments showed that the application of one groove transverse to the flow direction may delay boundary layer separation both in laminar and turbulent conditions. As a consequence, significant increases of the pressure recovery in plane diffusers and decreases of the drag of boat-tailed axisymmetric and two-dimensional bluff bodies may be obtained. In addition, in all the considered applications, the flow control device has been proved to be robust to small variations of its geometrical parameters and of the flow conditions. The only requirement to obtain this effect is that the position and length of the groove be chosen so that the flow may reattach downstream of the localized recirculation in the groove region. On the other hand, it is found that an increase of the groove depth may lead to an increase of the effectiveness of the groove in moving the separation point downstream. However, since an excessive depth of the groove might imply the appearance of phenomena of self-sustained cavity oscillations, which should be avoided to assure an adequate operation of the groove as an effectively passive flow-control device, only grooves whose depth is lower than a suitable percentage of the boundary layer thickness should be chosen. Possible objectives for future activities may be the optimization of the groove geometrical parameters for different applications and more basic investigations to understand the connection of those parameters with the physical mechanisms at the basis of the effect of the grooves. Furthermore, the possibility of reducing the friction drag through the introduction of multiple consecutive grooves might be analysed.

References

1. Gad-el-Hak M (2000) Flow control: passive, active, and reactive flow management. Cambridge University Press, Cambridge
2. Demartino C, Ricciardelli F (2017) Aerodynamics of nominally circular cylinders: a review of experimental results for civil engineering applications. Eng Struct 137:76–114
3. Buresti G, Tondi D (2007) Stepped boat-tails: a proposal for the control of the aerodynamic loads on bluff bodies. In: Atti del Dipartimento di Ingegneria Aerospaziale, ADIA 2007-7, ETS Editrice, Pisa
4. Ringleb F (1961) Separation control by trapped vortices. In: Lachmann GV (ed) Boundary layer and flow control, vol 1. Pergamon Press, Oxford, pp 265–294
5. Iollo A, Zannetti L (2001) Trapped vortex optimal control by suction and blowing at the wall. Eur J Mech (B/Fluids) 20:724
6. Rossiter JE (1964) Wind-tunnel experiments on the flow over rectangular cavities at subsonic and transonic speeds. Aeronautical Research Council, R & M, No. 3438 (1964)
7. Rockwell D, Naudascher E (1978) Review - self-sustaining oscillations of flow past cavities. J Fluid Eng 100:152–165
8. Mariotti A, Grozescu AN, Buresti G, Salvetti MV (2013) Separation control and efficiency improvement in a 2D diffuser by means of contoured cavities. Eur J Mech B Fluid 41:138–149
9. Mariotti A, Buresti G, Salvetti MV (2014) Control of the turbulent flow in a plane diffuser through optimized contoured cavities. Eur J Mech B Fluid 48:254–265
10. Mariotti A, Buresti G, Salvetti MV (2015) Use of multiple local recirculations to increase the efficiency in diffusers. Eur J Mech B Fluid 50:27–37
11. Mariotti A, Buresti G, Gaggini G, Salvetti MV (2017) Separation control and drag reduction for boat-tailed axisymmetric bodies through contoured transverse grooves. J Fluid Mech 832:514–549
12. Mariotti A, Buresti G, Salvetti MV (2019) Separation delay through contoured transverse grooves on a 2D boat-tailed bluff body: effects on drag reduction and wake flow features. Eur J Mech B Fluid 74:351–362
13. Maull DJ, Hoole BJ (1967) The effect of boat-tailing on the flow round a two-dimensional blunt-based aerofoil at zero incidence. J R Aeronaut Soc 71:854–858
14. Mair WA (1969) Reduction of base drag by boat-tailed afterbodies in low-speed flow. Aeronaut Q 20:307–320
15. Buice CU, Eaton JK (2000) Experimental investigation of flow through an asymmetric plane diffuser. Trans ASME J Fluids Eng 122:433–435

16. Camarri S, Salvetti MV, Koobus B, Dervieux A (2004) A low-diffusion MUSCL scheme for LES on unstructured grids. Comput Fluids 33:1101–1129
17. Mariotti A, Buresti G, Salvetti MV (2015) Connection between base drag, separating boundary layer characteristics and wake mean recirculation length of an axisymmetric blunt-based body. J Fluid Struct 55:191–203
18. Mariotti A, Buresti G (2013) Experimental investigation on the influence of boundary layer thickness on the base pressure and near-wake flow features of an axisymmetric blunt-based body. Exp Fluids 54:1612
19. Mariotti A (2018) Axisymmetric bodies with fixed and free separation: base-pressure and near-wake fluctuations. J Wind Eng Ind Aerodyn 176:21–31
20. Jeong J, Hussain F (1995) On the identification of a vortex. J Fluid Mech 285:69–94

Experimental Study on the Effect of Secondary Vortices at the Trailing Edge on Motion-Induced Vortex Vibration

K. Matsuda(✉), K. Kato, N. Cao, and R. Higashimura

Department of Civil Engineering and Architecture,
Kyushu Institute of Technology, 1-1 Sensui-cho, Tobata-ku, Kitakyushu,
Fukuoka 804-8550, Japan
{matsuda, kato}@civil.kyutech.ac.jp,
sounatoku@gmail.com, o105061r@mail.kyutech.jp

Abstract. According to the results of conventional wind tunnel tests on rectangular cross sections with side ratios of *B/D* = 2–8 (*B*: along-wind length (m), *D*: cross-wind length (m)), motion-induced vortex vibration was confirmed. The generation of motion-induced vortex vibration is considered to be caused by the unification of separated vortices from the leading edge and secondary vortices at the trailing edge. Spring-supported tests and smoke flow visualization tests for *B/D* = 0.62, 0.75, 1.0, 1.18 and 1.50 were conducted in a wind tunnel at Kyushu Institute of Technology. As a result, it was considered that the secondary vortices at the trailing edge are not always essential for the generation of the motion-induced vortex vibration in heaving motion of each cross section. In this study, smoke flow visualizations were performed for side ratios of *B/D* = 2.0, 4.0, 6.0 in order to elucidate the role of secondary vortices at the trailing edge in motion-induced vortex vibration in heaving motion. Spring-supported tests were also carried out in order to obtain the response characteristics of the models.

Keywords: Motion-induced vortex vibration · Smoke flow visualizations · Spring-supported tests · Rectangular cross sections · Side ratios

1 Introduction

According to Shiraishi and Matsumoto (1983), the vortices separated from rectangular cross sections in wind tunnel tests are broadly classified into Kármán vortices and motion-induced vortices (vortices separated from the leading edge). Kármán vortices are accompanied by interference of the two separated shear layers at both the top and bottom surfaces of the structures. Motion-induced vortices are shed separately from the leading edges of the top and bottom surfaces, because the separated shear layers at the top and bottom surfaces are excited alternately due to the vibration of the rectangular cross section. The vibration caused by the latter vortices was found in past wind tunnel tests by Novak (1971) and Otsuki et al. (1971). The vibration is known as either motion-induced vortex vibration (Shiraishi and Matsumoto (1983), Komatsu and Kobayashi (1980)) or impinging-shear-layer instability (Rockwell and Naudascher

F. Ricciardelli and A. M. Avossa (Eds.): INVENTO 2018, LNCE 27, pp. 496–506, 2019.
https://doi.org/10.1007/978-3-030-12815-9_38

(1978); Nakamura et al. (1991); Ohya et al. (1992); Naudascher and Wang (1993)). The onset wind speed of this vibration depends on the side ratio of the rectangular cross section, and this relation has been schematically clarified by Shiraishi and Matsumoto (1983), Naudascher and Wang (1993). The generation of motion-induced vortex vibration is considered to be caused by the unification of separated vortices from the leading edge and secondary vortices at the trailing edge. Motion-induced vortex vibration was confirmed by conventional wind tunnel tests on rectangular cross sections with side ratios of *B/D* = 2–8. However, in order to clarify the relation between the generation of motion-induced vortex in heaving motion and lower side ratios *B/D*, a visualization test for the smoke flows around the cross section was performed by providing a forced heaving oscillation on models of the rectangular cross sections of *B/D* = 0.62, 0.75, 1.0, 1.18 and 1.50 by Matsuda et al. (2016) and Matsuda et al. (2017). As a result, it was considered that the secondary vortices at the trailing edge are not always essential for the generation of the motion-induced vortex vibration in heaving motion of each cross section.

The purpose of this study is to find out the role of the secondary vortex at trailing edge in heaving motion-induced vortex vibration of rectangular cross sections with side ratios of *B/D* = 2.0, 4.0 and 6.0 by carried out smoke flow visualizations and spring-supported tests. Moreover, based on the results, this research aimed to verify whether or not the secondary vortices at the trailing edge are essential for the generation of motion-induced vortex vibration in heaving motion.

2 Experimental Setup

Smoke flow visualizations around the model during oscillating condition were conducted in a small-sized wind tunnel (0.4 m high × 0.4 m wide × 2.0 m long) at Kyushu Institute of Technology. Figures 1 and 2 show the schematic illustration of experimental setup for smoke flow visualization and the forced oscillation system, respectively. It was presumed that wind speeds of $V = 0.6$–1.0 m/s in the wind tunnel were good for visualization, so eventually $V = 0.8$ m/s was selected as the onset wind speeds of motion-induced vortex vibration. Accordingly, the experimental Reynolds number was $Re = VD/v = 1.1 \times 10^3$, where *V*: wind speed (m/s), *D*: cross-wind length (m), *v*: kinematic viscosity (m^2/s). For the forced-oscillating amplitude η, the non-dimensional double amplitudes were set as $2\eta/D = 0.02$–0.20 (every 0.01 intervals). Because the onset reduced wind speeds of motion-induced vortex vibration and Kármán vortex-induced vibration come close to each other on the cross section at roughly *B/D* = 2.8–6.0, according to Matsumoto et al. (1993a), smoke flow visualizations were conducted using a splitter plate (hereinafter referred as "S. P.") installed at the downstream side of the rectangular cross section model with the gap length *g* for the purpose of weakening the Kármán vortex, as shown in Fig. 3. Table 1 shows the experimental conditions for the smoke flow visualization tests. Spring-supported tests were also performed in order to get the response characteristics of the models. Table 2 shows the experimental conditions for the spring-supported tests. Laser displacement detectors were used for measuring an amplitude of the models. The spring-supported tests were conducted in a closed circuit wind tunnel (cross section: 1.8 m high × 0.9 m wide) at Kyushu Institute of Technology. All wind tunnel tests were carried out in a smooth flow.

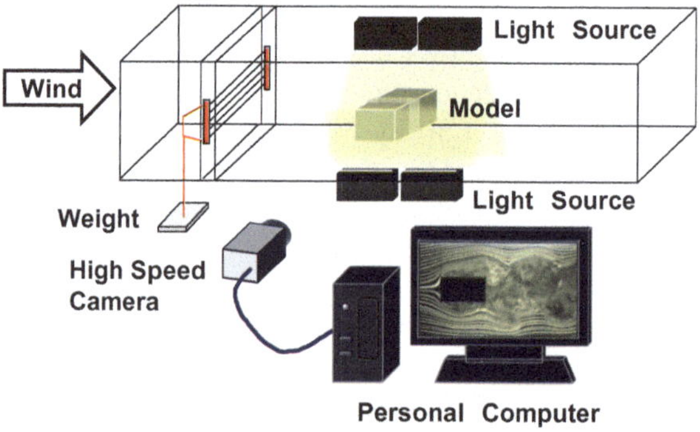

Fig. 1. Schematic illustration of experimental setup for smoke flow visualization

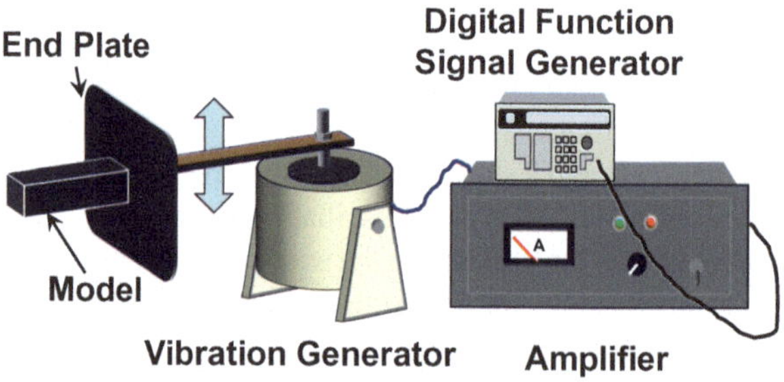

Fig. 2. Forced oscillation system for smoke flow visualization

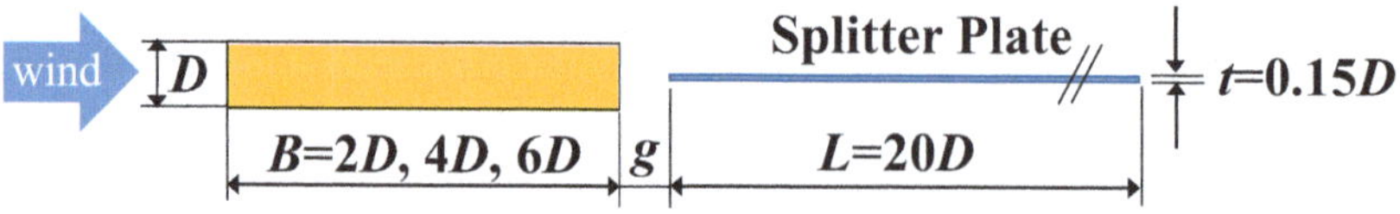

Fig. 3. Rectangular cross section model and splitter plate

Table 1. Experimental conditions for smoke flow visualization

B/D	B (m)	D (m)	Splitter plate (S.P.)	Gap length between model and splitter plate g		Wind speed V (m/s)	Reynolds number $Re = VD/\nu$	Forced-oscillating frequency f (Hz)	Reduced wind speed $Vr = V/fD$	Non-dimensional forced-oscillating double amplitude 2η/D
2	0.04	0.02	w/o	–	–	0.8	1.1×10^3	12.0	3.33 = 1.67*B/D*	0.02–0.20
			w/o	–	–					
			w/	0.20*D*	0.10*B*			7.34	5.45	0.20
			w/	0.50*D*	0.25*B*					
4	0.08	0.02	w/o	–	–	0.8	1.1×10^3	12.0	3.33 = 0.83*B/D*	0.02–0.18
			w/o	–	–					
			w/	0.40*D*	0.10*B*			9.32	4.29	0.20
			w/	1.25*D*	0.31*B*					
			w/o	–	–	0.8	1.1×10^3	6.00	6.67 = 1.67*B/D*	0.02–0.34
			w/o	–	–					
			w/	0.40*D*	0.10*B*			4.8	8.31	0.20
			w/	1.25*D*	0.31*B*					
6	0.12	0.02	w/o	–	–	0.8	1.1×10^3	8.00	5.0 = 0.83*B/D*	0.02–0.18
			w/o	–	–					
			w/	0.60*D*	0.10*B*			7.43	5.38	0.20
			w/	2.0*D*	0.33*B*					
			w/o	–	–			4.00	10.0 = 1.67*B/D*	0.02–0.46

Table 2. Experimental conditions for spring-supported tests

B/D	B (m)	D (m)	Splitter plate (S.P.)	Gap length between model and splitter plate g		Air density ρ (kg/m³)	Mass per unit length m (kg/m)	Structural damping (logarithmic decrement) δ	Natural frequency f (Hz)	$Sc = 2m/\rho D^2$
2	0.09	0.18	w/o	–	–	1.23	5.49	0.1143	5.22	4.7
			w/	0.20*D*	0.10*B*	1.24	5.35	0.0044	5.15	4.7
			w/	0.50*D*	0.25*B*	1.24	5.52	0.0048	5.20	5.2
4	0.09	0.36	w/o	–	–	1.25	3.82	0.0133	6.25	10.1
			w/	0.40*D*	0.10*B*	1.24	3.69	0.0135	6.20	9.9
			w/	1.25*D*	0.31*B*	1.23	3.85	0.0130	6.23	10.1
6	0.09	0.54	w/o	–	–	1.22	4.55	0.0156	5.73	14.4
			w/	0.60*D*	0.10*B*	1.23	4.44	0.0247	5.66	21.9
			w/	2.0*D*	0.33*B*	1.22	4.41	0.0162	5.67	14.4

3 Experimental Results and Discussion

3.1 B/D = 2.0 Rectangular Cross Section

The results of the spring-supported test of *B/D* = 2.0 are shown in Fig. 4. The vortex-induced vibrations occurring at approximately *Vr* = 3 were confirmed. These vortex-induced vibrations are considered to be motion-induced vortex vibrations because they are occurring at around $V_{cr,n=1} = (1/n)1.67 \times B/D = 1.67 \times 2.0 = 3.33$ and there was no change in onset reduced wind speed due to S.P. being installed. The result shows that the nearer the S.P. was installed at the downstream side of the model, the lower the response amplitude became. Furthermore, the occurrence of Kármán vortex-induced vibration and galloping could be confirmed in the high wind speed region. Because the vibration occurred at the higher wind speed region in the two cases with S.P., rather than that without S.P., it was confirmed that Kármán vortices were weakened by S.P.

The result of the experiment of flow visualization under the condition that reduced wind speed was $Vr = 5.45$, where the response amplitude becomes the maximum and the forced-oscillating heaving amplitude set as $2\eta/D = 0.2$ was compared with the results of three cases where no S.P. was installed and cases where it was installed at $g = 0.10B$ and $g = 0.25B$. The results are shown in Fig. 5. It was confirmed that the development of the integrated vortex of the separated vortex from the leading edge with the secondary vortex at the trailing edge was inhibited by the existence of S.P. It is considered that this is the reason why the response amplitude of motion-induced vortex vibration decreased when S.P. was installed.

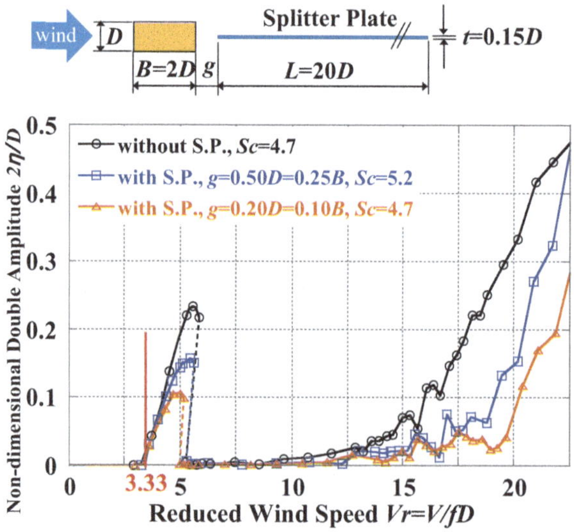

Fig. 4. Spring-supported test result of $B/D = 2.0$

(a) without S.P. (b) with S.P. of g=0.10B (c) with S.P. of g=0.25B

Fig. 5. Flow visualization test results of forced-oscillating model of $B/D = 2.0$ at the bottom displacement in $2\eta/D = 0.20$ at $Vr = 5.45$

3.2 B/D = 4.0 Rectangular Cross Section

The result of the spring-supported test of the case of *B/D* = 4.0 is shown in Fig. 6. The vortex-induced vibrations occurring at roughly *Vr* = 3 and *Vr* = 7 were confirmed. They are considered to be motion-induced vortex vibrations because these vortex-induced vibrations were occurring at $V_{cr,n=2} = (1/n)1.67 \times B/D = (1/2)$ $1.67 \times 4.0 = 3.33$ and $V_{cr,n=1} = (1/n)1.67 \times B/D = 1.67 \times 4.0 = 6.67$, and there was no change in onset reduced wind speed when S.P. was installed. The result obtained was that when S.P. was installed at *g* = 0.10*B*, the response decreased. This result corresponds to those of previous research of Matsumoto et al. (2008). Meanwhile, the result obtained was that when S.P. was installed at *g* = 0.31*B*, the response increased. This result also corresponds to the results of previous research of Matsumoto et al. (1993b).

The results of the flow visualization experiment under the condition that reduced wind speeds were *Vr* = 4.29 and *Vr* = 8.31, where response becomes the maximum, and the forced-oscillating heaving amplitude was set as $2\eta/D = 0.2$ were compared with the results of three cases where no S.P. was installed and cases where it was installed at *g* = 0.10*B* or *g* = 0.31*B*. The results are shown in Figs. 7 and 8, respectively. In either case of reduced wind speed of *Vr* = 4.29 or *Vr* = 8.31, when no S.P. was installed, it can be confirmed that the secondary vortex at the trailing edge generated almost constantly separating from the trailing edge. From the fact that when S.P. was installed at *g* = 0.10*B*, the generation of the integrated vortex of the separated vortex from the leading edge with the secondary vortex at the trailing edge was inhibited by S.P., it is supposed that the exciting force acting on the model decreased and as a result, the response of motion-induced vortex vibration decreased. It was confirmed when S.P. was installed at *g* = 0.31*B*, the integrated vortex can enter the gap strongly between S.P. and the model. In other words, it is considered that the Kármán vortex was not only

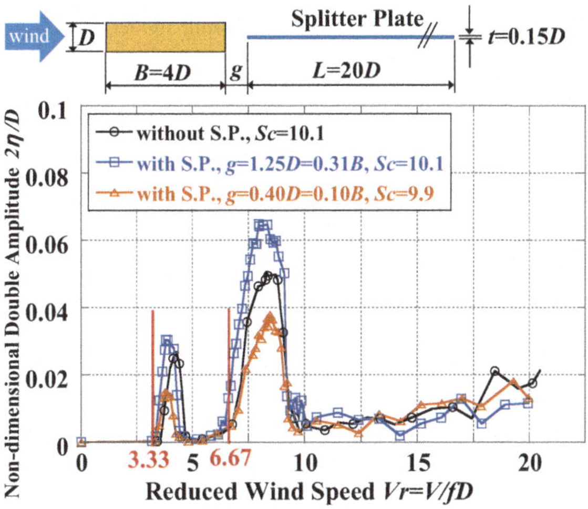

Fig. 6. Spring-supported test result of *B/D* = 4.0

weakened by the installation of S.P. and separated vortices from leading edge were elicited, but the entrainment of the integrated vortex into the gap between the model and S.P. had a tendency to become strong. As the result, the exciting force acting on the model became large, and the response also became large.

(a) without S.P. (b) with S.P. of g=0.10B (c) with S.P. of g= 0.31B

Fig. 7. Flow visualization test results of forced-oscillating model of B/D = 4.0 at the bottom displacement in $2\eta/D$ = 0.20 at Vr = 4.29

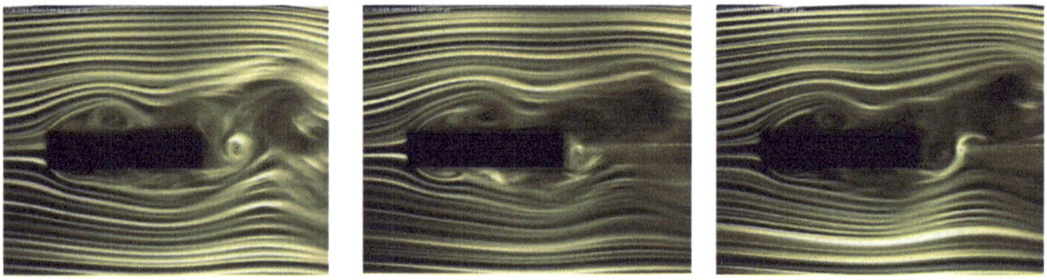

(a) without S.P. (b) with S.P. of g=0.10B (c) with S.P. of g=0.31B

Fig. 8. Flow visualization test results of forced-oscillating model of B/D = 4.0 at the bottom displacement in $2\eta/D$ = 0.20 at Vr = 8.31

3.3 B/D = 6.0 Rectangular Cross Section

The result of the spring-supported test in the case of B/D = 6 is shown in Fig. 9. The vortex-induced vibrations occurring from approximately Vr = 5 were confirmed. These vortex-induced vibrations are considered to be motion-induced vortex vibrations because they were occurring at $V_{cr,n=2} = (1/n)1.67 \times B/D = (1/2)1.67 \times 6.0 = 5.0$ and there was no change in onset reduced wind speed when S.P. was installed. The result was obtained that although the case was the same as that of B/D = 4.0, when S.P. was installed at g = 0.10B, the response of motion-induced vortex vibration decreased; however, the response increased when S.P. was installed at g = 0.33B. Meanwhile, motion-induced vortex vibration was not confirmed, which is considered to occur from approximately $V_{cr,n=1} = (1/n)1.67 \times B/D = (1/1)1.67 \times B/D = 1.67 \times 6.0 = 10.0$.

The result of the experiment of flow visualization under the condition that reduced wind speed was $Vr = 5.38$, where response amplitude becomes the maximum, and the forced-oscillating heaving amplitude was set as $2\eta/D = 0.2$ was compared with the results of three cases when no S.P. was installed and cases where it was installed at $g = 0.10B$ and $g = 0.33B$. The results are shown in Fig. 10. It can be seen that the generation patterns of secondary vortex at trailing edge in all three cases have the same tendency as those in the case of $B/D = 4.0$. Therefore, it is considered that the response of motion-induced vortex decreased when S.P. was installed at $g = 0.10B$ and it increased when S.P. was installed at $g = 0.33B$, compared with the response without S.P., just as the same as in the case of $B/D = 4.0$.

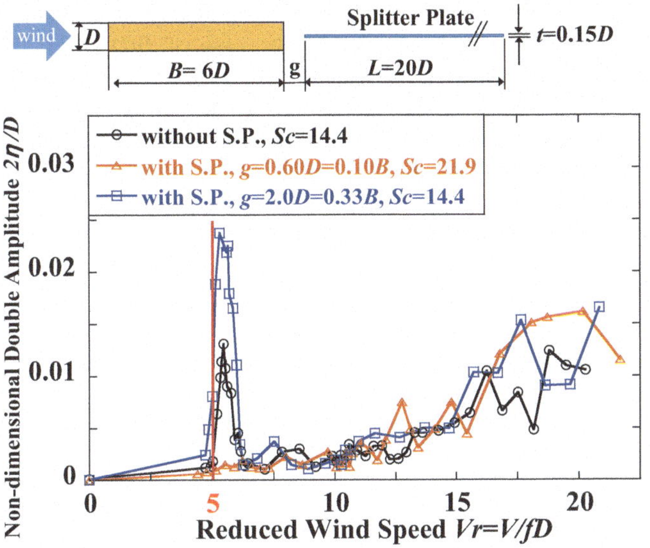

Fig. 9. Spring-supported test result of $B/D = 6.0$

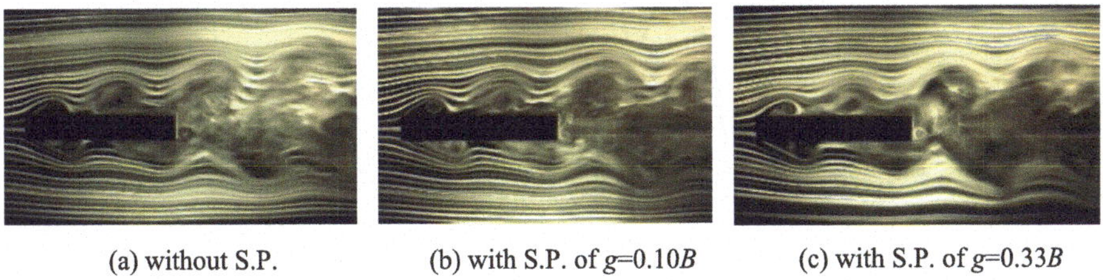

Fig. 10. Flow visualization test results of forced-oscillating model of $B/D = 6.0$ at the bottom displacement in $2\eta/D = 0.20$ at $Vr = 5.38$

3.4 Generation Probability for Secondary Vortex at the Trailing Edge

When the model is photographed for 10 cycles while it is forced-oscillated at reduced wind speed $V_{cr,n=2} = (1/n)1.67 \times B/D = (1/2)1.67 \times B/D = 0.83 \times B/D$ or $V_{cr,n=1} = (1/n)1.67 \times B/D = (1/1)1.67 \times B/D = 1.67 \times B/D$, 20 secondary vortices at the trailing edge are generated at maximum. The generation probability was defined as (the number of generation times/20) × 100 (%), and it was used as an index showing the ratio of contribution of secondary vortex at trailing edge to motion-induced vortex vibration. The results are shown in Fig. 11. The generation probability of the secondary vortices at the trailing edge tends to become higher as the forced-oscillating heaving amplitude $2\eta/D$ increases as the same as the case of $B/D = 0.62$–1.50 shown in Fig. 12 of Matsuda et al. (2017). However, the forced-oscillating amplitude becomes larger for the generation probability to reach 100% in the case of $B/D = 4.0$ and 6.0 at $V_{cr,n=1} = (1/n)1.67 \times B/D = 1.67 \times B/D$ than that in the case of $B/D \leqq 2.0$. This cause is considered to be the interference between motion-induced vortices and Kármán vortices due to the fact that the onset reduced wind speeds of motion-induced vortex vibration and Kármán vortex-induced vibration come close to each other on the cross section at roughly $B/D = 2.8$–6.0. As a result of the comparisons of Fig. 11 with Figs. 4, 6 and 9, which are the results of spring-supported tests of the rectangular cross sections of $B/D = 2.0$, 4.0 and 6.0, respectively, motion-induced vortex vibration is occurring with the amplitudes smaller than the forced-oscillating amplitude with which the secondary vortex at the trailing edge can fully generate; therefore, it is considered that the secondary vortex at trailing edge does not fully contribute to the onset of motion-induced vortex vibration. Therefore, it is suggested that the secondary vortex at the trailing edge is not always necessary for the onset of motion-induced vortex vibration of the rectangular cross sections of $B/D = 2$, 4 and 6, as the same as the case of cross section of $B/D = 0.62$–1.50, which are the results of previous research of Matsuda et al. (2017).

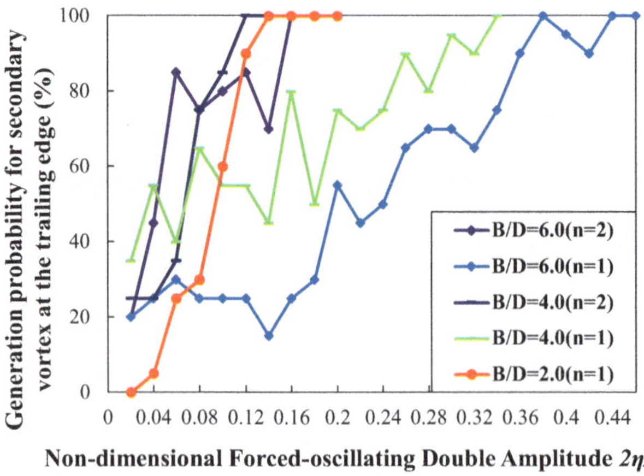

Fig. 11. Generation probability for secondary vortex at the trailing edge ($B/D = 2.0$, 4.0, 6.0)

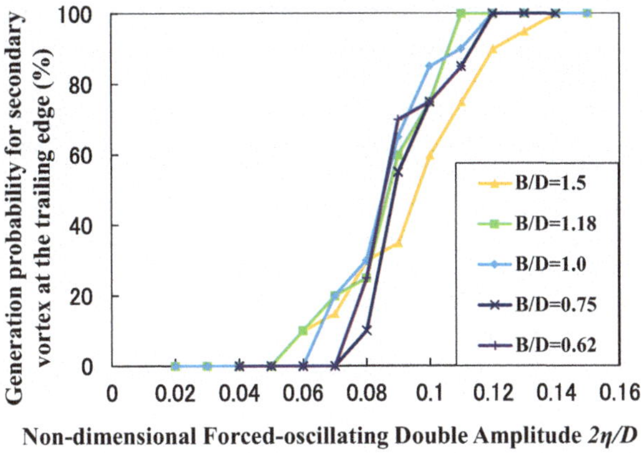

Fig. 12. Generation probability for secondary vortex at the trailing edge (*B/D* = 0.62–1.5) (Matsuda et al. 2017)

4 Conclusions

Smoke flow visualizations and spring-supported tests were performed for side ratios of *B/D* = 2.0, 4.0, 6.0 in order to elucidate the role of secondary vortices at the trailing edge in motion-induced vortex vibration. The findings obtained from this research are as follows:

1. The secondary vortex at the trailing edge is not always necessary for the onset of motion-induced vortex vibration in heaving motion with *B/D* = 2.0, 4.0 and 6.0 as the same as the results of our previous research targeting of rectangular cross section with the side ratio of *B/D* = 0.62–1.50.
2. It was suggested that the extent of the exciting force of motion-induced vortex vibration in heaving motion of the cross section with the side ratio of *B/D* = 2.0, 4.0 and 6.0 is dependent on the existence of a secondary vortex at the trailing edge.

References

Komatsu S, Kobayashi H (1980) Vortex-induced oscillation of bluff cylinders. J Wind Eng Ind Aerodyn 6:335–362

Matsuda K, Kato Tamai Y, Suda K (2016) Experimental study on aerodynamic vibrations of rectangular cross sections having low side ratios. In: Proceedings of 8th international colloquium on bluff body aerodynamics and applications, Northeastern University, Boston, Massachusetts, USA

Matsuda K, Kato K, Arise K, Ishii H (2017) Study on the relation between side ratios of rectangular cross sections and secondary vortices at trailing edge in motion-induced vortex vibration. In: Proceedings of the ASME 2017 pressure vessels & piping division conference, PVP2017, 65565

Matsumoto M, Shiraishi N, Shirato H, Stoyanoff S, Yagi T (1993a) Mechanism of, and turbulence effect on vortex-induced oscillations for bridge box girders. J Wind Eng Ind Aerodyn 49(1–3):467–476

Matsumoto M, Shiraishi N, Shirato H, Yagi T, Tanaka M, Yamagishi M (1993b) On mechanism of vortex-induced oscillations for rectangular cylinder. J JSCE Struct Eng 39A:891–896 (in Japanese)

Matsumoto M, Yagi T, Tamaki H, Tsubota T (2008) Vortex-induced vibration and its effect on torsional flutter instability in the case of B/D = 4 rectangular cylinder. J Wind Eng Ind Aerodyn 96(6–7):971–983

Nakamura Y, Ohya Y, Tsuruta H (1991) Experiments on vortex shedding from flat plates with square leading and trailing edge. J Fluid Mech 222:437–447

Naudascher E, Wang Y (1993) Flow-induced vibrations of prismatic bodies and grids of prisms. J Fluids Struct 7:341–373

Novak M (1971) Galloping and vortex induced oscillations of structures. In: Proceedings of the third international conference on wind effects on buildings and structures, Tokyo, Japan, pp 799–809

Ohya Y, Nakamura Y, Ozono S, Tsuruta H, Nakayama R (1992) A numerical study of vortex shedding from flat plates with square leading and trailing edges. J Fluid Mech 236:445–460

Otsuki Y, Washizu K, Tomizawa H, Ohya A, Fujii K (1971) Experiments on the aeroelastic instability of prismatic bars with rectangular sections. In: Proceedings of the third international conference on wind effects on buildings and structures, Tokyo, Japan, pp 891–898

Rockwell D, Naudascher E (1978) Review - self-sustaining oscillations of flow past cavities. Trans ASME J Fluids Eng 100:152–165

Shiraishi N, Matsumoto M (1983) On classification of vortex-induced oscillation and its application for bridge structures. J Wind Eng Ind Aerodyn 14(1–3):419–430

Wind–Wave Loading and Response of OWT Monopiles in Rough Seas

A. Mockute[1(✉)], C. Borri[1], E. Marino[1], and C. Lugni[2]

[1] Department of Civil and Environmental Engineering (DICEA), Università di Firenze, Via Santa Marta 3, 50139 Florence, Italy
{amockute, cborri, enzo.marino}@dicea.unifi.it

[2] National Research Council – Maritime Research Institute (CNR-INSEAN), Via di Vallerano, 00128 Rome, Italy
claudio.lugni@cnr.it

Abstract. Monopile-supported OWTs are prone to ringing – a dangerous nonlinear resonant phenomenon which is still not fully understood. Both wind and waves have a major influence on it, wind with its aerodynamic damping in the case of an operating wind turbine, and waves due to the fact that only in the case of fully nonlinear wave kinematics such phenomenon has been observed numerically. This study is interested in both of these influences, especially in the case of misaligned wind and waves – when the nonlinearities of waves are coming from a direction of reduced aerodynamic damping. However, for the influence of wave nonlinearities to be fully understood and the most accurate loading model to be used in the study, an intermediate investigation is conducted comparing the loading from combinations of six models for wave kinematics in increasing nonlinearity and three hydrodynamic loading models – Morison equation, slender-body theory in its two used formulations, and FNV perturbation theory on a fixed cylinder representative of an OWT monopile. It has been found that wave kinematics from fifth order could already be sufficient to capture the nonlinear loading in such steep waves as long as a suitable more sophisticated loading model is used. Moreover, none of the considered combinations in this two-dimensional study manage to capture the secondary load cycle, which used to be directly linked to ringing, therefore in the next phase a moving cylinder is to be considered to analyse the potential of capturing ringing in the most efficient manner before moving to the full dynamic OWT in wind and waves.

Keywords: Offshore wind turbines · Numerical wave modelling · Ringing · Highly nonlinear waves

1 Introduction

Wind is a prominent renewable energy resource which can help cater for the ever-growing energy demand while the offshore sites allow for much larger wind turbines which have significantly higher capacity (GWEC 2017). However, the technology is still relatively expensive, the support structure being one of the largest contributors of the cost. Therefore a more efficient design of the support structures could not only

F. Ricciardelli and A. M. Avossa (Eds.): INVENTO 2018, LNCE 27, pp. 507–518, 2019.
https://doi.org/10.1007/978-3-030-12815-9_39

increase the safety of these gigantic structures, but also contribute towards the reduction in the total price.

A major catalyst in the improvement of the numerical modelling of offshore wind turbines has been the Offshore Code Comparison Collaboration OC3 (Jonkman and Musial 2010) where multinational participants modelled three support structures for a baseline offshore wind turbine, both fixed and floating; its Continuation OC4 (Jonkman et al. 2012) which included additional two support structures, and the most recent OC5 project – Offshore Code Comparison Collaboration Continuation, with Correlation (Robertson et al. 2016), which included a comparison to an experimental study as well. However, one of the crucial findings of OC4 has showed that major discrepancies have been seen when it comes to hydrodynamic loading (Robertson et al. 2014), and a further detailed hydrodynamic study in the Phase I of OC5 has showed that while the first order loading is captured relatively well by the participants, the majority of them still strongly underpredict or completely omit the higher order loading components which are crucial for nonlinear resonant phenomena such as springing and ringing (Robertson et al. 2015).

Springing and ringing, illustrated in Fig. 1, are highly nonlinear phenomena that cause strong amplifications in the amplitude of loading, therefore can lead to serious damage to the structure (NPD 1992; Grue et al. 1994; Huseby and Grue 2000). Springing is more continuous, persistent, associated with 2nd order wave loading components and has been successfully modelled numerically since the 1990s. Ringing, on the other hand, is more burst-like and associated with 3rd or higher order loading, or with secondary load cycle in the case of a fixed rigid cylinder, and has been a research topic of interest since then.

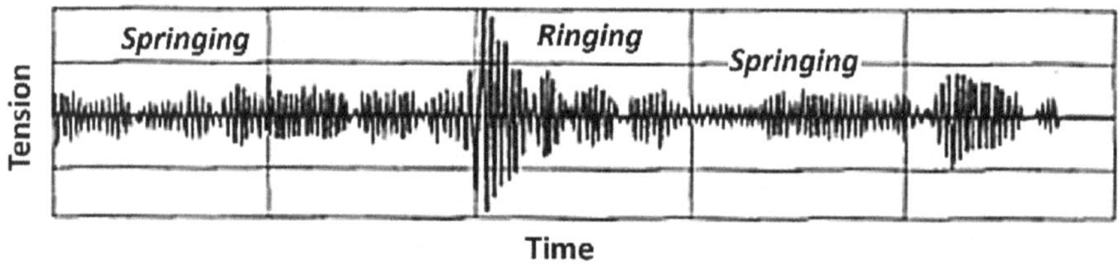

Fig. 1. Illustration of springing and ringing in tension response of an offshore structure over time. Adapted from NPD (1992).

These nonlinear responses have been previously observed in the numerical studies on monopile-supported offshore wind turbines, and fall into two main categories of studies: the influence of wind and associated aerodynamic damping, especially in the cases of parked versus operating wind turbine, and the importance of nonlinearities in wave kinematics.

First of all, wind has a crucial influence on the dynamic response of the monopile-supported offshore wind turbine. All of the cases of ringing observed on the offshore wind turbines have been reported on a parked wind turbine, when it is behaving as a relatively static structure. However, when the turbine is operating such resonant amplifications have been seen to be diminished, associating the absence with the aerodynamic damping in the direction of waves (Marino et al. 2013b, Schløer et al. 2016, Marino et al. 2017). Interest of this study is not only on the damping influence of wind and turbulence on an operating wind turbine in relation to the nonlinear resonant responses in general, but also more specifically in the case of wind-wave misalignment, which is associated with rough seas.

A past study on the influence of wind-wave misalignment on the fatigue of a monopile-supported offshore wind turbine has showed increased fatigue loads in the case of misaligned wind and waves, presumably due to the reduced aerodynamic damping (van der Meulen et al. 2012). Similarly, when the nonlinear wave loading is coming from a different direction than the main wind loading and its associated damping, the sensitivity of the structure to the nonlinear resonant effects may be increased. A preliminary study has found that even though the influence of the nonlinearities in wave loading is still most critical in the case of co-aligned wind and waves on a parked wind turbine, the consideration of wind-wave misalignment in the case of an operating wind turbine shows an increased influence of wave nonlinearities (Mockute et al. 2017). Nonetheless, wave nonlinearities are an additional topic of complexity and interest by itself which would contribute to an improved modelling of OWT loading.

The resonant amplifications have only been observed when fully nonlinear wave kinematics have been used, but have been numerously reported to be omitted if linear wave kinematics were used (Marino et al. 2013a, 2014; Schløer et al. 2016). Marino et al. (2014) have also investigated the potential of second order wave kinematics in recreating such nonlinear resonant phenomena, and have assessed that just as linear wave kinematics second order Stokes waves did not manage to capture ringing. Moreover, in most of the previous studies only Morison equation was employed as the hydrodynamic loading model, which is the commonly used model for hydrodynamic loading on slender monopiles. However, it is often described as underperforming in very steep waves or shallower depths (Grue et al. 1994; Malenica and Molin 1995; Chaplin et al. 1997; Huseby and Grue 2000; Paulsen et al. 2014). To accommodate for that, during the period of interest in ringing more sophisticated hydrodynamic loading models have been derived, such as slender-body or perturbation theories which are anticipated to perform better in the conditions where ringing could be expected (Faltinsen et al. 1995; Malenica and Molin 1995; Rainey 1995). Both the wave kinematics and the hydrodynamic loading models have nonlinearities stemming from them and the importance of each with regards to ringing is not yet fully understood and the behaviour of the offshore structure with a combination of more nonlinear wave theories and a more sophisticated hydrodynamic loading model remains an unanswered question.

Therefore in this first phase, before moving to the full monopile-supported offshore wind turbine, wave loading is assessed on a fixed surface-piercing cylinder in all combinations of wave kinematics from linear to fully nonlinear including Stokes 2nd, 3rd and 5th order waves, in combination with three hydrodynamic loading models: Morison equation, slender-body theory and Faltinsen-Newman-Vinje (FNV) perturbation theory.

2 Methodology

The wind-wave loading is modelled on NREL 5-MW baseline offshore wind turbine (Jonkman et al. 2009) to get a realistic response of an actual offshore wind turbine. To model the combined loading of wind and waves coupling will be done between an open-source aero-hydro-servo-elastic FAST by NREL (Jonkman and Buhl 2005) and the separately modelled wave loading. The need for coupling arises from the fact that the hydrodynamic loading in FAST is computed in the HydroDyn time-domain module (Jonkman et al. 2014) which, just as the majority of Offshore Code Comparison Collaboration participants, by default has up to second order wave kinematics and Morison equation for the hydrodynamic loading model.

The wave kinematics employed in this study are of a gradually increasing non-linearity. Firstly for linear wave kinematics the simplest Airy theory is employed, also known as small-amplitude theory. Weakly nonlinear waves which provide sharper crests and shallower troughs are modelled with Stokes 2nd, 3rd and 5th Order theories, as described in numerous textbooks, e.g. Chakrabarti (1987). Finally, the fully non-linear waves which do not have depth or amplitude restrictions up to a breaking wave are modelled with Rienecker-Fenton theory (Rienecker and Fenton 1981) and a two-dimensional higher-order Boundary Element Method (BEM) model, initiated with the same Rienecker-Fenton theory. The BEM model, developed by Marino (2010), solves Laplace equation in a domain bounded by impermeable bottom, periodic vertical walls and a free water surface, with quadratic special elements and fourth order time-stepping.

Three hydrodynamic loading models are analysed in this study: full Morison equation which considers both inertia and drag, the slender-body theory which suggests corrections to the Morison inertia term, and Faltinsen-Newman-Vinje (FNV) third order perturbation theory, which takes flow nonlinearities and diffraction into account; each discussed in detail in the following paragraphs.

The first hydrodynamic loading model is the commonly implemented Morison equation (Morison et al. 1950). The equation consists of a linear inertia term M and quadratic drag d, which are both integrated from the bottom to the instantaneous free surface, as described in Eq. 1 and illustrated in Fig. 2. It is a semi-empirical loading model, where the drag and inertia coefficients (C_d and C_m consequently) have to be tuned for the specific conditions by experimental data or best practice guidelines. In this study the C_m coefficient is fixed to the value of 1 since most of the analyzed cases are in the inertia-dominated regime, and the C_d coefficient is dependent on the local Keulegan-Carpenter number (= $|v_x|*T/(R/2)$, where v_x is the horizontal water particle velocity, T – wave period, and R – cylinder radius).

$$F_{Mor} = \underbrace{\int_{-h}^{\eta} C_m \rho \pi R^2 \frac{\partial v_x}{\partial t} dz}_{M} + \underbrace{\int_{-h}^{\eta} C_d \rho R |v_x| v_x dz}_{d} \tag{1}$$

Where η – instantaneous free surface elevation; ρ – water density; g – gravitational constant; h – water depth.

The slender-body theory was introduced by Rainey (1989) and then extended to include the surface distortion due to the presence of the cylinder in Rainey (1995). It suggests to separate the Morison inertia term and correct it by the slender-body terms. The terms which are applicable to a fixed circular monopile are summarized in Chaplin et al. (1997) as axial divergence term *A* (Eq. 2), integrated from the bottom to the instantaneous free water surface just as the Morison terms, and surface intersection term *I* (Eq. 3) and surface distortion term *D* (Eq. 4), both applied as point loads at the free surface intersection with the cylinder, as illustrated in Fig. 2.

$$A = \int_{-h}^{\eta} \rho\pi R^2 \frac{\partial v_z}{\partial z} v_x dz \tag{2}$$

$$I = -(\rho\pi R^2/2) v_x^2 \partial\eta/\partial x \tag{3}$$

$$D = \left(\frac{7\rho\pi R^2}{2g}\right) v_x^2 \frac{\partial v_x}{\partial t} \tag{4}$$

Where v_z – horizontal water particle velocity.

The axial distortion term *A* and the surface intersection load *I* both correct the linear inertia term *M* to second order (Rainey 1989), and then the third-order error is corrected by the surface distortion term *D* introduced by Rainey (1995). However, since the derivation of the surface distortion term *D* is resting on the Stokes small amplitude assumption its validity in very steep waves has been doubted and therefore it is sometimes preferred to apply only the *A* and *I* corrections (e.g. Chaplin et al. 1997; Schløer et al. 2016). Therefore in this study both *M+A+I* and *M+A+I+D* combinations are considered.

The final hydrodynamic loading method is the FNV perturbation theory, introduced by Faltinsen et al. (1995) for infinite depth regular waves and only recently generalised to finite depth by Kristiansen and Faltinsen (2017). This formulation is applied in this study and it consists of an integrated term *F'* up to the instantaneous free surface (Eq. 5) and a point load F^{ψ} (Eq. 6), applied at the still water level in accordance with the original theory as illustrated in Fig. 2. The integrated term *F'* not only considers the flow nonlinearities in the first term, but also diffraction by integrating the scattering potential on horizontal 2D planes over the depth in the second term, where a_{11} is the 2D added mass coefficient. The point load F^{ψ} considers the local free surface non-linearities to the third order.

$$F' = \int_{-h}^{\eta} \left(\rho\pi R^2 \left(\frac{\partial v_x}{\partial t} + v_x \frac{\partial v_x}{\partial x} + v_z \frac{\partial v_x}{\partial z} \right) + a_{11} \left(\frac{\partial v_x}{\partial t} + v_z \frac{\partial v_x}{\partial z} \right) \right) dz \tag{5}$$

$$F^{\psi} = \left(\frac{4\rho\pi R^2}{g}\right) v_x^2 \frac{\partial v_x}{\partial t} \tag{6}$$

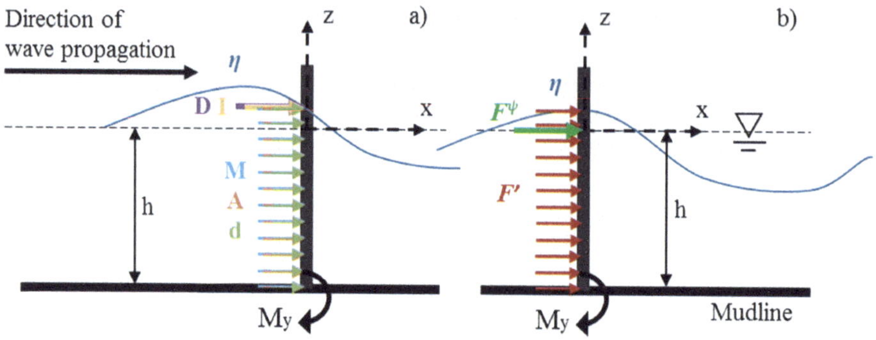

Fig. 2. Coordinate system and forcing components of (a) Morison equation and slender-body theory; (b) FNV theory.

3 Results and Discussion

For the behavior of the wave loading models in very steep waves, where the structures are susceptible to highly nonlinear events and both the wave theories and loading models tend to lose validity, an experiment on fixed rigid cylinder by Stansberg (1997), cited in Robertson et al. (2015), was recreated numerically and selected results in horizontal force over time are displayed in Figs. 3, 4, 5 and 6. The experiment was conducted in waves with period $T = 1.53$ s, wave height $H = 0.33$ m, in water depth of $h = 10$ m and cylinder radius $R = 0.1$ m, leading to very high steepness $ka = 0.282$. The wave height and period were tuned by the least square method (lowest root-mean-square-error) on the wave elevation, just as some of the participants of OC5 project in Robertson et al. (2015).

The performance of Morison equation in such steep waves is shown in Fig. 3. It can be seen at the peak of the loading that not even with the most nonlinear wave kinematics Morison equation manages to capture the highly nonlinear shape of the crest, underestimating the higher order loading components which are apparent in this case. To compare, the M+A+I+D model, shown in Fig. 4, can be seen to capture the trend in loading perfectly.

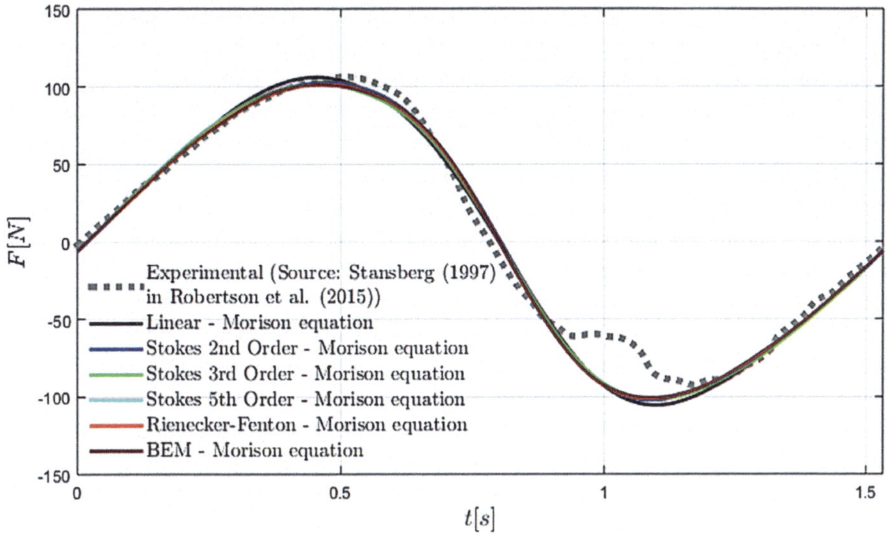

Fig. 3. Normalised horizontal loading over a normalised period on a fixed cylinder in the steepness of $ka = 0.282$; experimental measurements by Stansberg (1997, cited in Robertson et al. 2015), all wave kinematics with Morison equation.

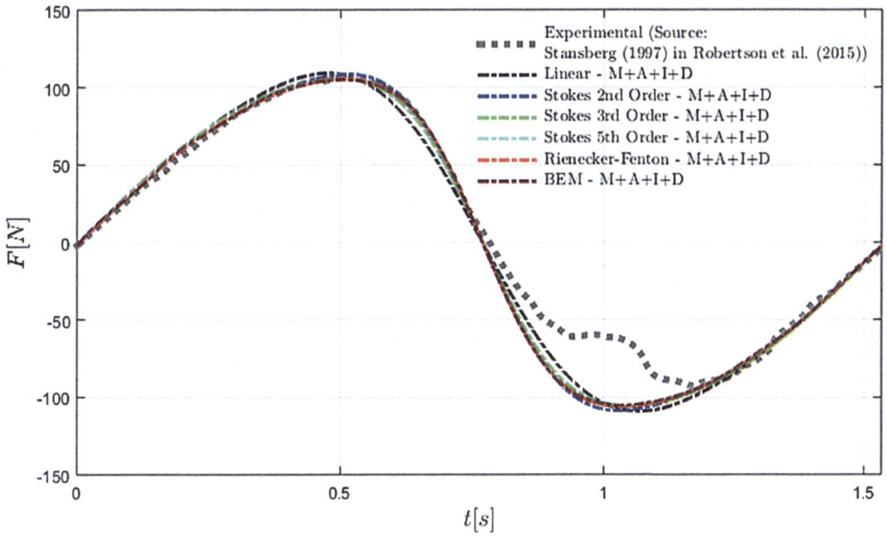

Fig. 4. Normalised horizontal loading over a normalised period on a fixed cylinder in the steepness of $ka = 0.282$; experimental measurements by Stansberg (1997, cited in Robertson et al. 2015), all wave kinematics with M+A+I+D loading model.

The influence of wave kinematics is shown with the M+A+I+D loading model in Fig. 4. Firstly, the loading already from linear wave kinematics is not completely sinusoidal due to the nonlinearities in the hydrodynamic loading model M+A+I+D. Nonetheless, the linear wave kinematics are not capturing the shape of the peak of loading, just as Morison equation seen in Fig. 3. The peak loading is not ideally captured neither by the second or third order wave kinematics with second order overestimating, and third order underestimating the maximum. Stokes 5th order, Rienecker-Fenton and BEM kinematics all show the best behaviour, therefore fifth order kinematics are sufficient in these wave conditions when they are used with a sophisticated loading model.

To analyse the suitability of the Stokes 5th order wave kinematics all hydrodynamic loading models are shown based on them in Fig. 5. It can be seen that the only loading model recreating the main trend in the loading with the fifth order kinematics is the M+A+I+D model. Nonetheless, with fully nonlinear Rienecker-Fenton wave kinematics, as shown in Fig. 6, the peak of loading is successfully captured by both M+A+I +D and FNV while M+A+I and Morison equation are underestimating the sharp crest of the peak loading. Such difference by the addition of the surface distortion term D implies that the third order loading becomes of strong importance in such steep waves, therefore a hydrodynamic loading model of at least third order is required to avoid underestimation of the loading.

Nonetheless, the main discrepancy is seen at time $t \approx [0.85\text{–}1.25]$ s, where secondary load cycle occurs. Secondary load cycle is a second peak observed experimentally in the wave loading on a fixed rigid cylinder in the same wave conditions as ringing, therefore it has been directly linked to ringing by many researchers (Grue et al. 1994; Chaplin et al. 1997; Grue and Huseby 2002). None of the analysed combinations for wave loading managed to capture the secondary loading cycle numerically, as shown in Figs. 3, 4, 5 and 6. Morison equation is not expected to capture the secondary load cycle as it consists of linear inertia and quadratic drag, therefore it is of second order, just as M+A+I, as discussed in Sect. 2. Nevertheless, both M+A+I+D and FNV are including the third order loading, in addition to FNV considering diffraction.

The fact that not even the most sophisticated models with fully nonlinear wave kinematics (both Rienecker-Fenton and BEM) are incapable of recreating the secondary load cycle in this two-dimensional study implies that secondary load cycle is caused by either three-dimensional effects or the effects of higher order than third. This is in accordance with the findings of the 3D-CFD study by Paulsen et al. (2014) which showed that the secondary load cycle is caused by diffraction – a diffracted wave forming a vortex at the back of the cylinder and the interaction between the opposing flows in the vortex versus the wave still propagating in the same direction are creating a pressure reduction that is observed as secondary loading. The same study also displayed the harmonic analysis of secondary load cycle which revealed that not even the first six harmonics were sufficient to recreate the secondary load cycle, therefore suggesting that it should be disassociated from ringing which is normally linked to the third harmonic. Such suggestion has been confirmed experimentally by Suja-Thauvin et al. (2017) by showing that parallel fixed and moving cylinders were not always experiencing secondary load cycle and ringing at the same events.

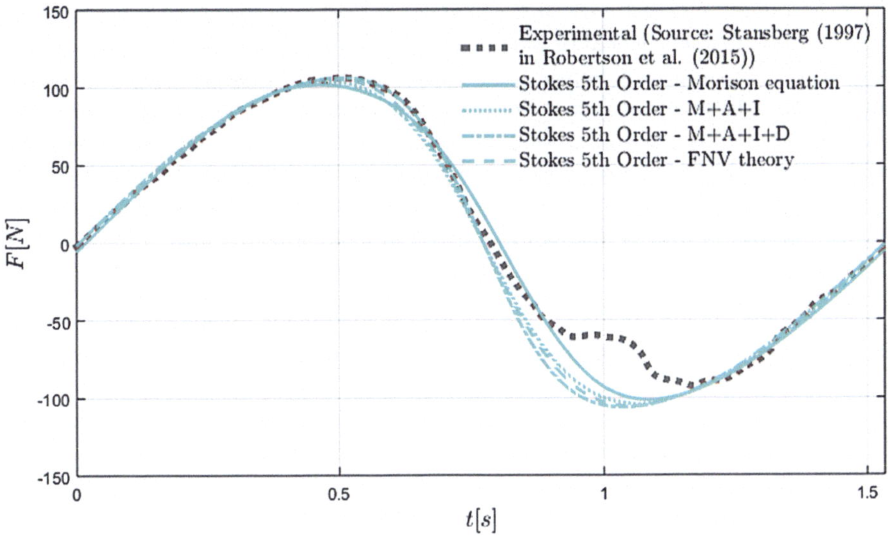

Fig. 5. Normalised horizontal loading over a normalised period on a fixed cylinder in the steepness of ka = 0.282; experimental measurements by Stansberg (1997, cited in Robertson et al. 2015), all loading models with Stokes 5th order wave kinematics.

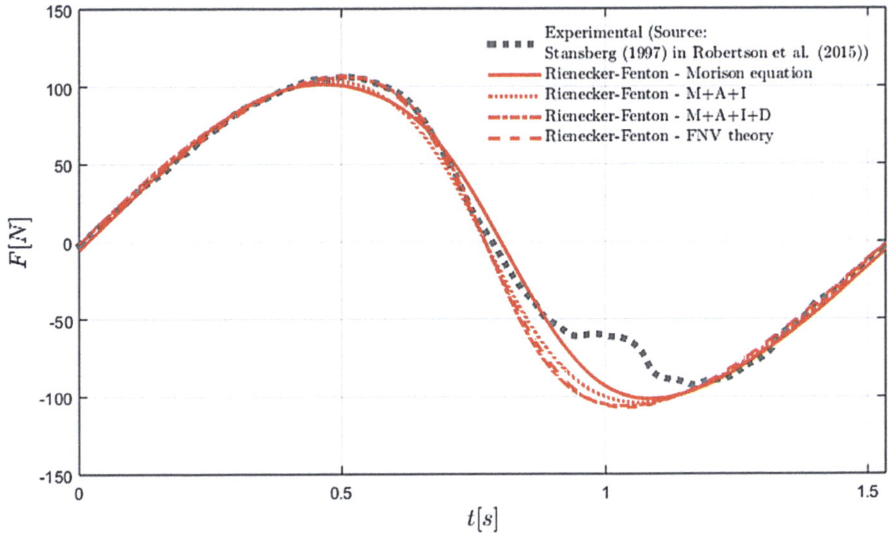

Fig. 6. Normalised horizontal loading over a normalised period on a fixed cylinder in the steepness of ka = 0.282; experimental measurements by Stansberg (1997, cited in Robertson et al. 2015), all loading models with Rienecker-Fenton nonlinear wave kinematics.

Therefore the fact that none of the models are capturing the secondary load cycle does not mean that they would be omitting ringing too. In fact, resonant responses have been observed with fully nonlinear wave kinematics and Morison equation in the past, but omitted with linear and second order wave kinematics, as discussed in the introduction, therefore it is crucial to continue this study on a moving cylinder in the next phase to assess whether an intermediate combination of wave kinematics in addition to a more sophisticated hydrodynamic loading model could capture ringing in a more efficient manner before moving to the complex case of combined wind-wave action on a full offshore wind turbine.

4 Conclusions and Outlook

In the present study a numerical analysis has been carried out to evaluate three hydrodynamic loading models in combination with six models for wave kinematics in order to assess their appropriateness for very steep waves and to understand the associated nonlinear phenomena in order to efficiently analyse the wind-wave loading on OWTs in rough seas. It has been found that in the very steep wave conditions associated with ringing the main loading is best recreated already with Stokes fifth order wave theory as long as a more sophisticated loading model is used, namely the full slender-body theory, or with fully nonlinear Rienecker-Fenton wave kinematics and FNV or slender-body theories. The commonly used Morison equation has been seen to underpredict the highly nonlinear loading. Finally, in this two-dimensional study none of the analysed combinations manage to capture the secondary load cycle; nonetheless, the secondary load cycle has recently been disassociated from ringing, therefore it is crucial to continue this study to consider the motion of the cylinder. With the findings of this intermediate study it is expected to continue to the main study of a full offshore wind turbine in open seas including the wind-wave misalignment with the most accurate and efficient wave loading modelling.

Acknowledgements. The activity presented in the paper is part of the European Commission's Framework Program "Horizon 2020", through the Marie Skłodowska-Curie Innovative Training Network (ITN) "AEOLUS4FUTURE - Efficient harvesting of the wind energy" (H2020-MSCA-ITN-2014: Grant agreement no. 643167). The COST TU1304 action WINERCOST is also gratefully acknowledged.

References

Chakrabarti SK (1987) Hydrodynamics of offshore structures. 1. Offshore structures - Hydrodynamics I. Springer, Berlin

Chaplin JR, Rainey RCT, Yemm RW (1997) Ringing of a vertical cylinder in waves. J Fluid Mech 350:119–147

Faltinsen OM, Newman JN, Vinje T (1995) Nonlinear wave loads on a slender vertical cylinder. J Fluid Mech 289:179–198

Grue J, Huseby M (2002) Higher harmonic wave forces and ringing of vertical cylinders. Appl Ocean Res 24(4):203–214

Grue J, Bjorshol G, Strand Ø (1994) Nonlinear wave loads which may generate 'ringing' responses of offshore structures. In: 9th international workshop on water waves and floating bodies, 17–20 April, Kuju, Fukuoka, Japan

GWEC (2017) Global wind report: annual market update 2017. Global Wind Energy Council

Huseby M, Grue J (2000) An experimental investigation of higher-harmonic wave forces on a vertical cylinder. J Fluid Mech 414:75–103

Jonkman J, Buhl M (2005) FAST user's guide, Technical report. National Renewable Energy Laboratory, Golden, CO

Jonkman J, Musial W (2010) Offshore code comparison collaboration (OC3) for IEA task 23 offshore wind technology and deployment, Technical report. National Renewable Energy Laboratory, Golden, CO

Jonkman JM, Butterfield S, Musial W, Scott G (2009) Definition of a 5-MW reference wind turbine for offshore system development, Technical report. National Renewable Energy Laboratory, Golden, CO

Jonkman JM, Robertson AN, Hayman GJ (2014) HydroDyn user's guide and theory manual, Technical report. National Renewable Energy Laboratory, Golden, CO

Jonkman J, Robertson A, Popko W et al (2012) Offshore code comparison collaboration continuation (OC4), phase i – results of coupled simulations of an offshore wind turbine with jacket support structure. In: 22nd international conference on ocean, offshore and arctic engineering, 17–22 June, Rhodes, Greece

Kristiansen T, Faltinsen OM (2017) Higher harmonic wave loads on a vertical cylinder in finite water depth. J Fluid Mech 833:773–805

Malenica S, Molin B (1995) Third-harmonic wave diffraction by a vertical cylinder. J Fluid Mechanics 302:203–229

Marino E, Lugni C, Borri C (2013a) The role of the nonlinear wave kinematics on the global responses of an OWT in parked and operating conditions. J Wind Eng Ind Aerodyn 123:363–376

Marino E (2010) An integrated nonlinear wind-waves model for offshore wind turbines. Firenze University Press

Marino E, Lugni C, Borri C (2013b) A novel numerical strategy for the simulation of irregular nonlinear waves and their effects on the dynamic response of offshore wind turbines. Comput Methods Appl Mech Eng 255:275–288

Marino E, Lugni C, Stabile G, Borri C (2014) Coupled dynamic simulations of offshore wind turbines using linear, weakly and fully nonlinear wave models: the limitations of the second-order wave theory. In: 9th international conference on structural dynamics EURODYN, June 30–July 2, Porto, Portugal

Marino E, Giusti A, Manuel L (2017) Offshore wind turbine fatigue loads: the influence of alternative wave modeling for different turbulent and mean winds. Renewable Energy 102:157–169

Mockute A, Marino E, Borri C (2017) Effects of wind-wave misalignment on the dynamic response of an offshore wind turbine in highly nonlinear waves. In: 7th European and African conference on wind engineering, 3–6 July, Liege, Belgium

Morison JR, O'Brien MP, Johnson JW, Schaaf SA (1950) The force exerted by surface waves on piles. Pet. Trans. AIME 189:149–154

NPD (1992) Annual report, 1992. Norwegian Petroleum Directorate, Stavanger

Paulsen BT, Bredmose H, Bingham HB, Jacobsen NG (2014) Forcing of a bottom-mounted circular cylinder by steep regular water waves at finite depth. J Fluid Mech 755:1–34

Rainey RCT (1989) A new equation for calculating wave loads on offshore structures. J Fluid Mech 204:295–324

Rainey RCT (1995) Slender-body expressions for the wave load on offshore structures. Proc Math Phys Sci 450(1939):391–416

Rienecker MM, Fenton JD (1981) A Fourier approximation method for steady water waves. J Fluid Mech 104:119–137

Robertson AN, Jonkman JM, Vorpahl F (2014) Offshore code comparison collaboration continuation within IEA wind task 30: phase II results regarding a floating semisubmersible wind system. In: 33rd international conference on ocean, offshore and arctic engineering, 8–13 June, San Francisco, California, USA

Robertson AN, Wendt F, Jonkman JM et al (2015) OC5 project phase i: validation of hydrodynamic loading on a fixed cylinder. In: Twenty-fifth international ocean polar engineering conference, 21–26 June, Kona, Hawaii

Robertson AN, Wendt F, Jonkman JM et al (2016) OC5 project phase Ib: validation of hydrodynamic loading on a fixed, flexible cylinder for offshore wind applications. Energy Procedia 94:82–101

Schløer S, Bredmose H, Bingham HB (2016) The influence of fully nonlinear wave forces on aero-hydro-elastic calculations of monopile wind turbines. Mar. Struct. 50:162–188

Stansberg CT (1997) Comparing ringing loads from experiments with cylinder of different diameters - an experimental study. In: BOSS 1997 Behaviour of offshore structures 2, pp 95–109

Suja-Thauvin L, Krokstad JR, Bachynski EE, de Ridder E-J (2017) Experimental results of a multimode monopile offshore wind turbine support structure subjected to steep and breaking irregular waves. Ocean Eng 146:339–351

Van der Meulen MB, Ashuri T, van Bussel GJW, Molenaar DP (2012) Influence of nonlinear irregular waves on the fatigue loads of an offshore wind turbine. In: The science of making torque from wind, 9–11 October, Oldenburg, Germany

Transitional Shear Layers on Rectangular Sections

D. M. Moore[1(✉)], C. W. Letchford[2], and M. Amitay[1]

[1] Department of Mechanical, Aerospace and Nuclear Engineering, Rensselaer Polytechnic Institute, Troy, NY 12180, USA
{moored7, amitam}@rpi.edu
[2] Department of Civil and Environmental Engineering, Rensselaer Polytechnic Institute, Troy, NY 12180, USA
letchc@rpi.edu

Abstract. The present work implements 2D Particle Image Velocimetry measurements in an effort to study the transition process of the separated shear layer on rectangular sections. Inspection of the average and fluctuating vector fields enable quantitative definitions of important length scales such as transition lengths and reattachment lengths. The effect of free stream Reynolds number is examined in detail which, among other points, show clear trends of increasingly rapid transitions lengths over the wind speeds investigated. Additionally, transitional shear layer dynamics are discussed for sections exposed to inclined winds. It is shown that under small angles of attack, relatively large changes in shear layer trajectory result in significant alterations to the global loads felt by the body.

Keywords: Bluff body · Shear layer · Transition · Turbulence

1 Introduction

Introduced in 2008, the Benchmark on Aerodynamics of Rectangular Cylinders (BARC) aimed to address the fundamental fluid physics that govern the aerodynamic performance of the 5:1 rectangular cylinder. While many research teams contributed to this benchmark study, the collection of data pertaining to the unsteady loading on the BARC varied significantly. The study appears to have two open areas yet to be completely understood. First, it can be shown that proper prediction of the true center of the recirculating region alongside the body is not easily calculated (Bruno et al. 2014). The second issue pertains to the variation of the lift coefficient for the BARC under inclined winds where the lift coefficient has been observed to decrease as the Reynolds number climbs (Schewe 2013). Bluff bodies with salient edges are traditionally trusted to behave independently of the Reynolds number, however these works suggest otherwise.

The edges themselves subject the oncoming flow to accelerations which are unattainable without infinite pressure gradients. Thus, separation of the flow at the leading and/or trailing edge corners becomes a defining feature of a bluff body (Roshko 1954). For rectangular sections, the edge is typically sharp enough to fix the separation

F. Ricciardelli and A. M. Avossa (Eds.): INVENTO 2018, LNCE 27, pp. 519–533, 2019.
https://doi.org/10.1007/978-3-030-12815-9_40

point, as opposed to a circular sections where the separation point moves according to the nature of the local pressure gradient and boundary layer behavior at the surface (Achenbach 1971). The separated flow, initially in alignment with the surface on the front face, experiences significant shearing as it bends into the direction of the free stream flow. Mean velocity profiles drawn across the shear layer near the leading edge corner reveal extreme velocity gradients and an inflection point consistent with features listed in (Drazin 2002) as necessary conditions for the Kelvin-Helmholtz (KH) instability. It follows that the turbulence within separated shear layer demands the acknowledgment of fluctuations in the fluidic loading of the body and should therefore be included in any subsequent analysis. Turbulence within shear layers has been observed in two major capacities. For initially laminar flows, the planar mixing layer received significant attention perhaps most famously by (Brown and Roshko 1974). The applied problem of a shear layer interacting with a bluff body is given an intriguing discussion in (Unal and Rockwell 1988) where the interactions of shear layers and wakes are shown to be closely coupled. The second approach is more directly applicable to the practicing wind engineer, which is the study of Free Stream Turbulence (FST) in the flow. (Gartshore 1973) showed that the shear layer is sensitive to fluctuations injected on to the stagnation streamline. The increase in fluctuations and added vorticity foster global changes in the growth rates and trajectories, affecting the bulk force coefficients for a rectangular section. The variation of spectral content within FST has also been shown to have consequences in the loads felt by the body. (Mooneghi et al. 2015; Saatoff and Melbourne 1997) have shown that perhaps only a set of specific scales within the FST spectrum contribute to the pressures felt by the body. These examples demonstrate the immense depth of the FST problem which the community has only just begun to unpack. Numerical investigations have also made efforts to resolve the nature of the turbulence within the shear layer. In the mean sense, there appears to be good agreement between experiments and simulation. However, the fluctuating surface pressures and resulting loads on the body are difficult to resolve using CFD (Mannini et al. 2010; Shimada and Ishihara 2001).

Ultimately it is the FST contribution to the shear layer behavior at high Reynolds numbers that is sought after by the wind engineering community. Full scale structures rarely experience turbulence intensities that are comparable to smooth flow conditions in a wind tunnel laboratory. However, benchmark studies such as the BARC shed light on the fact that the self-generated turbulence within the bluff body shear layer is still an area which has yet to be completely understood. Once fully learned, the laminar flow instance will inform FST studies by exposing the exchanges of energy and length scales most preferred by turbulence and most critical to structural design. In pursuit of that goal, an effort is presented here to further the community's knowledge of the inner workings of the bluff body shear layer considering both fluctuations as well as Reynolds number effects.

2 Experiments

A modular two dimensional rectangular prism was constructed out of 5 aluminium square tubes with side dimensions of $h = d = 50.8$ mm and $w = 508$ mm. Each module was fabricated such that they were easily bolted together from the downstream side. The final product formed a rectangle with various aspect ratios including d/h =1:1, 3:1 and 5:1, see Fig. 1. However, the span remained constant for all tests. The surfaces of each module were faced with a milling machine to maintain accurate dimensionality, and more importantly, to ensure sharpness of the corners. The final assembly was then mounted on a test rig with end plates in the Large Wind Tunnel at the Center for Flow Physics and Control (CeFPaC) at Rensselaer Polytechnic Institute. The end plates extended 14 h upstream and 20 h downstream of the model from its rotational axis located at the geometric center of each shape. Tests were then carried out spanning a range of Reynolds numbers between 1.3×10^4 to 1.2×10^5 based on the model thickness and the free stream wind speed. 2C-2D Particle Image Velocimetry (PIV) was employed to document the flow field near the prisms' leading edge and near flow domains. The wind tunnel itself is a blow down design with two centrifugal blowers turned by a 100 hp motor. A 9:1 contraction ratio divides the settling chamber from the test section which is 0.8 m × 0.8 m × 5 m in width, height, and length, respectively. Turbulence intensities within the test section were measured to be approximately 0.3% with a maximum available speed of 50 m/s. With the test rig installed, the blockage ratio reached 7.2% based on the area between the endplates at zero degrees angle of attack. No corrections were made to the freestream velocity to account for the blockage.

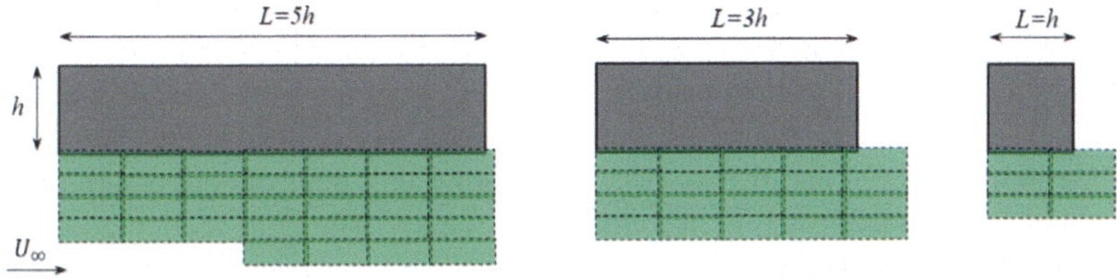

Fig. 1. Schematic of the three models tested and their respective PIV interrogation windows

PIV was used to investigate each of the Fields of View (FOV) marked by the small green rectangles in Fig. 1. By focusing the cameras on a small region, the resulting spatial resolution is higher than an equivalent setup focused on a large domain. This strategy was implemented in this study to adequately resolve the width of the shear layer at separation. 2C-2Cam PIV was used in each FOV to document the flow. The pair of cameras were LaVision Imager LX with 2MP each. Each of their windows corresponded to 1608 px × 1208 px with 10% of their image borders overlapping. Both cameras were calibrated simultaneously by using the same target. An articulating optical arm allowed the laser sheet to mirror each move performed by the camera setup. The laser was a New Wave Solo PIV 120 mJ/pulse per frame Nd: YAG (532 μm) dual-head laser. The system captured image pairs at a rate of 15 Hz and in each FOV 1000 images were captured to ensure the convergence appropriate turbulence statistics and reduce uncertainties of unbiased measurement errors. A multi-pass cross-correlation algorithm was implemented to calculate the vector fields. The first pass utilized a 64 px × 64 px interrogation window. Secondary and tertiary passed used a 32 px × 32 px window each having a 50% overlap. Upon gaining vector fields for each FOV post processing procedures then allowed the time-averaged quantities to be stitched together to reconstruct the global flow fields presented in this study. Figure 1 shows a varying number of FOVs needed for each of the three bodies investigated. The reconstructed flow fields presented here include between 6 and 43 FOV's. Resulting spatial resolution in all cases is 5.45 vectors per millimeter or 279 vectors per body height, h. Time-averaged pressures were recorded along the centerline of each body using an array of Scanivalve model DSA 3217 pressure scanners with 16 channels each. The scanners were sampled at 125 Hz and averaged over 30 s. Three of the six scanners had a range of 2500 Pa and three had a maximum pressure of ±6800 Pa. The accuracy of scanners was 0.2% full scale. Taps were 1.59 mm in diameter and tap locations were evenly distributed around each module for a spatial resolution of 8.4 mm between ports. The total number of parameters explored for the current study using PIV include the following: 6 Reynolds numbers, 2 angles and 3 geometries amount to 36 different datasets. Pressure measurements covered a similar parameter map with intermediate angles between 0 and 5°.

3 Results

Preliminary results demonstrate the fine level of detail within the flow captured by PIV. Figure 2 illustrates the reconstruction of the time-averaged quantities taken from each of the FOV's described in the preceding section. In the upper half of the image, streamlines are presented to demonstrate the various regions of recirculation occurring alongside the body. In the figure, as the wind moves from left to right, there are two visible recirculation regions. It may be worth noting that their directions are opposite with the larger of the two rotating in a clockwise manner. The large recirculation region is contained within the bounds of the leading edge and the reattachment point, x_r, at approximately $x_r/h = 4.4$ where two streamlines are observed to diverge upon approaching the surface. This reattachment is in reasonable agreement with the ensemble statistics in Table 10 of (Bruno et al. 2014) whose averaged reattachment length is 4.7. Furthermore, the coordinates of the recirculation bubble also fit well the ensemble statistics with the rotational centre of the large recirculation region occurring at streamwise and vertical coordinates of $x_c/h = 2.51$ and $y_c/h = 0.37$ respectively. There also appears to be a secondary opposite recirculation region immediately downstream of the leading edge and contained between the corner and $x/h = 0.4$. These two regions appear only in the mean flow and are not detectable as coherent structures in the instantaneous fields. Nevertheless, the mean topology of the flow around the BARC has proven to be a metric for comparing datasets between investigators and as such one more contribution is made here. Furthermore, the time-averaged vorticity is also shown in the lower half of the Fig. 2. Upon comparing the streamlines from the upper half with the vorticity contours on the lower half it can be seen that the recirculation regions exist in regions of oppositely signed vorticity which explains their rotational directions. While the concentration of positive vorticity accumulates on the front face due to an extreme pressure gradient and is subsequently diffused along the body, a faint region of negative vorticity is visible close to the surface. This is a necessary condition for vortices in the shear layer developing due to the Kelvin Helmholtz instability. Once the primary vortex is formed, an image vortex will also form and induce flow in the opposing direction. One such primary vortex is shown in the inset of the Fig. 2 as a way to get a glimpse of the flow patterns that comprise the time-averaged result. The inset is taken from FOV 1, which is closest to the leading edge corner. In the image a very clear band of vorticity is seen to undergo an undulation which is abruptly followed by the roll up in what can be argued to be a coherent vortex (Haller 2005). Noting that past work on sharp-edged bodies has suggested the front face boundary layer is likely to remain laminar for Reynolds numbers upward of $O(10^9)$ (Sigurdson 1986) it seems as though transition from laminar to turbulence may be initiated here by the formation of this vortex just downstream of the leading edge. This vortex is considered to be a shear layer vortex, or a Kelvin Helmoltz (KH) vortex. It is the spatio-temporal evolution and lifecycle of these structures which is thought to contribute towards the unsteady loads felt by the body and therefore the transition from laminar to vortical and eventually turbulent flow requires careful examination.

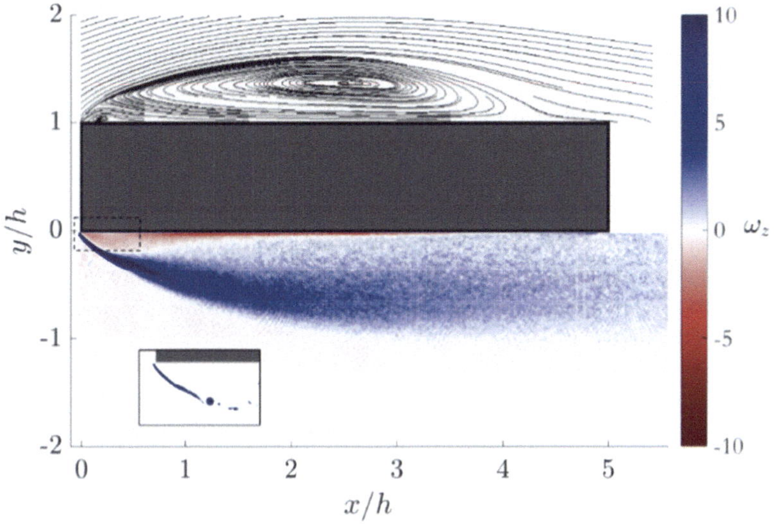

Fig. 2. Upper surface: streamlines. Lower surface: Time-averaged spanwise vorticity. Inset: Instantaneous spanwise vorticity. Reynolds number is $Re_h = 1.3 \times 10^4$. 5:1 body

However, what is also noticeable in Fig. 2 is how rapid the spatial diffusion of vorticity is. The ratio of vorticity magnitude to the noise floor set by differentiating discrete data approaches unity in regions near the trailing edge making any quantitative analysis uncertain. Therefore, it may be more instructive to choose a different parameter in place of vorticity that demonstrates a more favorable signal to noise ratio.

In recognition of this, the Reynolds stresses and Turbulent Kinetic Energy (TKE) are considered. While vorticity is a conservation quantity describing the rotational nature of the fluid elements, the TKE and contributing Reynolds stresses are produced anywhere there are fluctuations in the presence of a mean velocity gradient. This suggests that the integrated TKE and stress quantities are non-zero and increasing along the length of the body. Following a standard Reynolds decomposition, the two dimensional TKE parameter, $\tilde{E}$, is defined as,

$$\tilde{E} = \frac{1}{2}\left(\overline{u'u' + v'v'}\right) \tag{1}$$

where $u'(x,t) = u(x,t) - \overline{U}(x)$. Although vorticity and TKE are fundamentally different from one another, certain flow features are detectable using either parameter with the later of the two being preferable in the event of a discretized dataset. We submit that laminar-turbulent transition is well defined by inspection of the TKE distributions.

3.1 Separation and Transition

Figure 3 shows a series of profiles of TKE on the 5:1 body demonstrating the increasing nature as the flow interacts with the body. Along the vertical direction, each of the profiles exhibit a distinct peak along with a distribution that is relatively smooth. On the freestream side, each profile eventually decays to the freestream value which is close to zero for smooth flow conditions. Considering profiles changing along the streamwise coordinate, the area under each profile appears to increase as the width of the distributions grows. The peaks appear to grow in amplitude faster near the leading edge but it is difficult to discern the location of a peak in the latter half of the body. In effect, the profiles in Fig. 3 demonstrate the increasing fluctuating energy contained in the shear layer, bordered on one side by freestream fluid and the body surface on the other. Upon integrating these profiles Fig. 4 is realized. To underscore the universality of this trend to all rectangular sections, the integrated TKE profiles are also included for the 3:1, and 1:1 bodies.

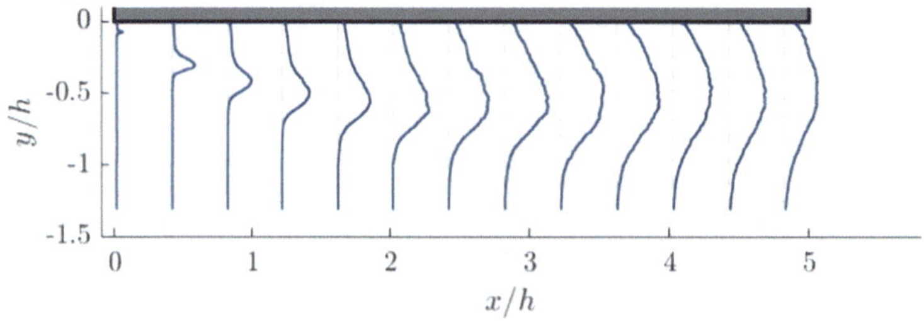

Fig. 3. Profiles of TKE along the 5:1 body. Reynolds number is $Re_h = 4.1 \times 10^4$

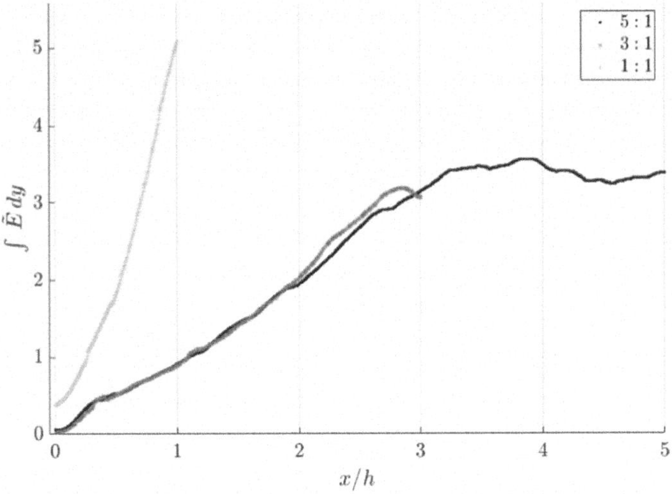

Fig. 4. Integrated TKE along the length of body Reynolds number is $Re_h = 4.1 \times 10^4$

The trends exhibited by Fig. 4 show several interesting features. First, it is noted that of the three bodies investigated, only the 5:1 and 3:1 experience mean reattachment. This notion is confirmed by the increasing nature of $\tilde{E}$ until the location where mean flow reattaches, as seen in the upper half of Fig. 2 for the 5:1 body. At the point of reattachment, the integrated TKE saturates and remains constant in the latter parts of the body length. Similarly, the distribution for the 3:1 body indicates that the flow attaches is the last 10% of the body. The aggressive rate of increase for the 1:1 may be attributed to activity in the wake where the Von Kármán vortex street is formed in close proximity to the leading edge of the body, merely a single body dimension apart. In contrast the 3:1 and 5:1 necessarily displace the activity in the wake farther from the leading edge allowing for a slower development of fluctuating energy. In fact, one could adopt a scaling for the horizontal axis where the normalization parameter is the distance between leading and trailing edges. Upon doing that the values at each coordinate may be more comparable between bodies. However, the height, h, is maintained here as the normalization parameter for all figures. As such, it may simply be concluded that the longer bodies allow for a closer monitoring of the shear layers' evolution without being overwhelmed by instabilities present in the wake. A final feature of Fig. 4 is a short, increasingly rapid rise in TKE levels, common between the three body types existing immediately downstream from the leading edge corner. This exponential increase in fluctuating energy is reminiscent of qualities described by linear stability theory and is given more detail in Fig. 5.

Linear stability analyses of laminar flows, including the planar mixing layer has undergone substantial scrutiny from a wide range of investigators including (Huerre and Monkewitz 1985; Michalke 1964; Monkewitz 1982). The theory states that a small sinusoidal velocity perturbation superimposed onto the mean velocity profile of a shear layer is subject to exponential growth in space or time, or both. The mathematical treatment stops at the evaluation of any unstable frequencies and wavenumbers existing in the flow. Of course, a full description would follow the nonlinear interactions of waves as they grow in amplitude, become nonlinear and roll up into vortices. However, such a description has not yet been offered to the authors' knowledge. Since the problem is based on periodic disturbances, the TKE parameter should also see a rise in amplitude as the perturbation amplitude grows. Specifically, the linear analysis describes the amplitude of a velocity perturbation as a complex exponential whose exponents serve as the growth rates. This suggests that evidence of the linear portion of this problem ought to be verified by tracking an exponential increases in the amplitude of a fluctuating energy. Figure 5 shows an example of such an event. In the figure, 3 sample curves are displayed for the 5:1 body. The curves are $\tilde{E}_{\max}$, which is defined by the maximum value from each streamwise profiles in Fig. 3 and then repeated for each Reynolds number tested. For the lowest Reynolds number, shown with dark black symbols, the trend is clear where $\tilde{E}_{\max}$ increases exponentially from near zero at the leading edge corner where the front face boundary layer becomes a separated shear layer. The increase continues a short distance, extending up to the first half of the body dimension at which point the distribution saturates and decreases in a shallow manner. This corroborates with the behavior of the profiles in Fig. 3 whose amplitudes increased initially and then saturate only to grow in width beyond $x/h = 1$. As the Reynolds number is increased, the initial curvatures become steeper indicating more aggressive exponential growth rates and the saturation points migrate upstream and downward along the two axes, respectively. Downstream, all curves appear to converge. Appealing to physical insight from linear theory, it is expected that the perturbation amplitude grows exponentially until it is sufficiently large enough to invalidate the assumption of an infinitesimally small disturbance. In other words, the wave amplitude grows until the waves themselves are no longer linear at which point the energy budget cannot be accurately predicted. It is at this point, which has been labelled above as the saturation points, where one might be interested to see if the most amplified wavelength has matured enough to roll up into a vortex.

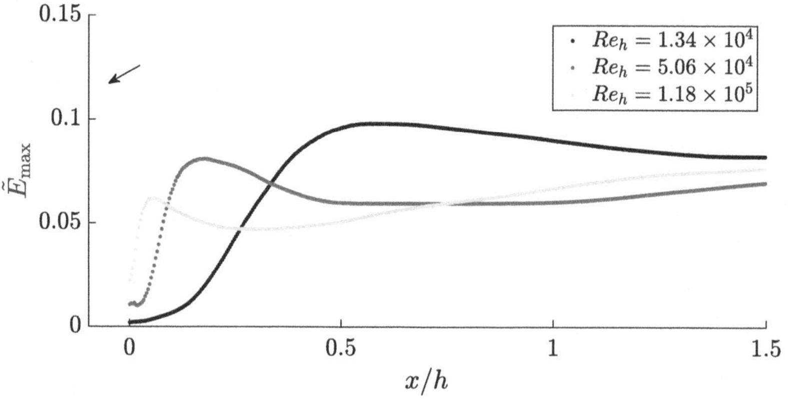

Fig. 5. Examples of maximum distributions of TKE plotted over the first body dimension for the 5:1 body. Arrow is in the direction of increasing Reynolds number.

These questions as to the flow's instantaneous behaviour informed by the TKE can be answered by inspection of Fig. 6 which shows the same dataset from Fig. 5 aligned with two example instantaneous PIV snapshots. Figure 6a shows the spanwise vorticity distribution expanded from Fig. 1 which is a single instant in time where the Reynolds number is $\mathrm{Re}_h = 1.34 \times 10^4$. Figure 6b also shows spanwise vorticity but for a Reynolds number an order of magnitude higher, $\mathrm{Re}_h = 1.18 \times 10^5$. The two distributions are similar in that there is a band of concentrated vorticity emanating from the leading edge that rolls up into a KH vortex. For the uppermost row, the lowest Reynolds number, this roll up event takes place at approximately $x/h \approx 0.4$. For the middle row at the higher Reynolds number, the same event takes a fraction of the distance and the roll up occurs almost immediately after separation, $x/h \approx 0.05$. Also, the resulting KH vortex has a reduced diameter and relative spacing. What is most impressive perhaps is the ability of the $\tilde{E}_{\max}$ distributions to predict the location of these vortex roll up events. For example, the roll up in Fig. 6a lines up well with the location of the saturation point in Fig. 6c for the same Reynolds number (the black symbols). Similarly, the roll up in Fig. 6b coincides with the saturation point of the higher Reynolds number curve in Fig. 6c marked by the grey symbols. This pattern of TKE trajectories lining up with appearances of the first vortex as seen by the PIV is common among all three bodies and both angles tested although the same multi-window figures are not presented for those instances in the interest of brevity.

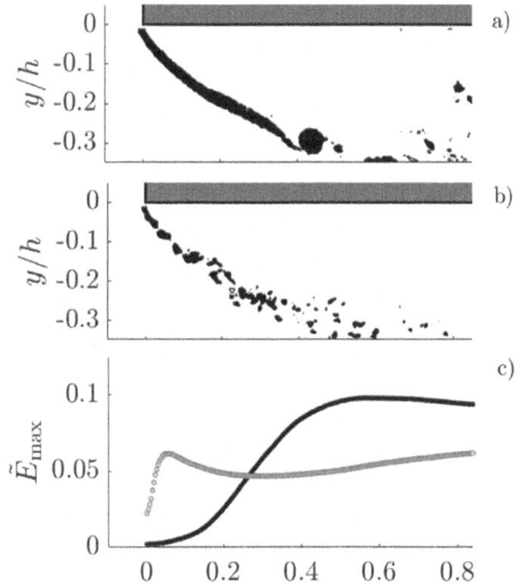

Fig. 6. PIV comparison. (a) Instantaneous image of normalized spanwise vorticity, $Re_h = 1.34 \times 10^4$. (b) Instantaneous image of spanwise vorticity, $Re_h = 1.18 \times 10^5$. and (c) distributions of maximum TKE; black circles $Re_h = 1.34 \times 10^4$, grey squares $Re_h = 1.18 \times 10^5$

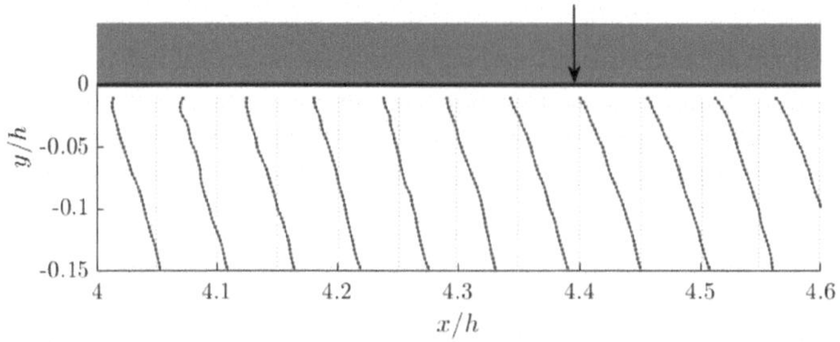

Fig. 7. Time-averaged profiles of the mean streamwise velocity component zoomed in near the location of reattachment. The black arrow marks the point of reattachment for $Re_h = 1.34 \times 10^4$ at $\frac{x_r}{h} = 4.39$.

Seeing an exponential rise in TKE levels that is followed by saturation and then punctuated by the appearance of a vortex provides a convincing series of clues that the shear layer, which is born laminar (Sigurdson 1986), is undergoing turbulent transition in the regions very near the leading edge corner. Moreover, that the transition process appears to be Reynolds number dependent shows that there are implications for proper scaling of bluff bodies in wind tunnel testing.

3.2 Reattachment

Farther along the body, the separated shear layer grows in width via turbulent entrainment developing larger fluctuations. For the longer two bodies tested, the 5:1 and 3:1, the shear layer was observed to reattach to the surface. This was detected by monitoring the distribution of velocity gradient very close to the wall surface. Due to reflections of the wall surface from the PIV laser, the surface itself is not resolved. However, Fig. 7 shows that there is enough data to resolve the location where the slope of the velocity profiles changes sign, approximately ½ millimeter away from the wall. The location where the change of slope takes place indicates where the recirculation bubble ends, displayed with x_r.

In previous investigations, authors have used the maximum mean pressure coefficient along the side face of the body to define the edge of the recirculation bubble (Mannini et al. 2017); however, here the velocity gradient is used in part because its spatial resolution is much finer than that of the pressure taps. Recent work from Schewe (2013) provides a phenomenological argument for the behavior of the reattaching shear layer on the 5:1 body. In particular, he mentions that the dependency of the mean lift coefficient is a function of an asymmetrical reattachment of shear layers on opposite sides of the body, the extreme scenario being when the angle of attack is sufficient to completely separate the flow on one side (suction side) and force early reattachment on the other (pressure side). Data from the current study agree with the characteristic lift curves presented in Schewe (2013) at moderate Reynolds numbers, (see Fig. 8a) however the migration of the reattachment points defined above appear to be weak functions of Reynolds number, contrary to his findings. That the initial development of the shear layer is Reynolds number dependent and yet the reattachment dependency is not, raises interesting questions about the fluid dynamic interactions in between.

Figure 8a displays the measured lift curves for all bodies and Reynolds numbers tested in the current study. It is immediately apparent that the 3:1 and 1:1 bodies have negative lift slopes indicating higher suctions on the face that is traditionally referred to as the *pressure side*. The magnitude of the lift slopes, or the sensitivity to angle of attack, also appears to scale with the length of the body indicating that longer bodies are more sensitive to small angles of attack. Recognizing that all three bodies increase linearly only roughly between 0 and 3°, the PIV data taken at 5° represent instances where the suction side is fully separated for all three bodies. As such, any further Reynolds number dependencies are attributed to a single shear layer, rather than a combined effort from opposing sides of the body. Lastly, the tight clustering of all data at each angle suggests a mild dependence of the lift coefficient on Reynolds number. That relative invariance may be explained in more detail by Fig. 8b which shows the measured reattachment lengths with respect to Reynolds number. Data taken from the 5:1 and 3:1 bodies shows the same mostly constant behavior for Reynolds numbers above $Re_h \approx 5 \times 10^4$. However, there are distinct shifts in the reattachment lengths based on the length of the body (e.g. 5:1 compared to 3:1 for orthogonal winds). It follows to explore the parameter map that categorizes the movement of the reattachment position and then compare driving mechanisms behind the shear layer, both at the leading edge and at reattachment.

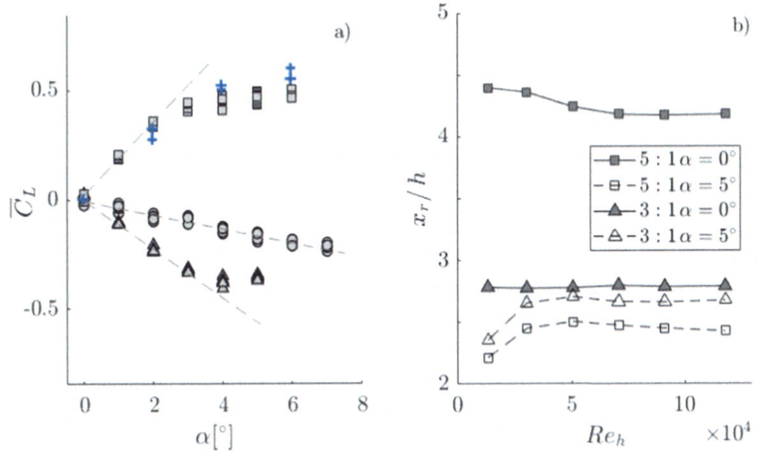

Fig. 8. (a) Lift curves for all bodies and all Reynolds numbers tested. (■) 5:1. (▲) 3:1. (●) 1:1. '+', Schewe 2013. (b) Reattachment lengths for the 5:1 and 3:1 bodies. For positive angles of attack the reattachment length is measured on the pressure side of the body. Symbols are listed in the legend.

4 Discussion

Upon shortening the body the reattachment lengths apparently decrease up to a certain point, close to $x_r/h \approx 2.7$. Bodies shorter than that have no reattachment at all for (i.e. the 1:1 square prism). This decrease in reattachment length increases the curvature of the shear layer trajectory accelerating the flow, thus, enhancing local suctions near the leading edge. Changing the angle of attack also moves the reattachment point, although sensitivity is significantly higher for longer bodies such as the 5:1 and minimal for shorter bodies like the 3:1. This is reflected in the lift measurements. At small angles of attack, the movement of the reattachment points helps to explain the lift curves. As Fig. 9 shows, when the body's suction side is completely separated, then the unsteady loads on the body coming from shear layers are mainly due to a single shear layer on the pressure side, rather than contributions from both sides.

It has also been shown that the presence of freestream turbulence shortens the lengths of the reattachment (Gartshore 1973). In those cases, both enhanced vorticity and turbulent entrainment are found to be the mechanisms for the shear layers behavior. Here, it appears that either the lack of an afterbody or an increased angle of attack does the same work of turbulence although it remains an open question as to what further reductions in the recirculation lengths could be observed by adding fluctuations in the freestream.

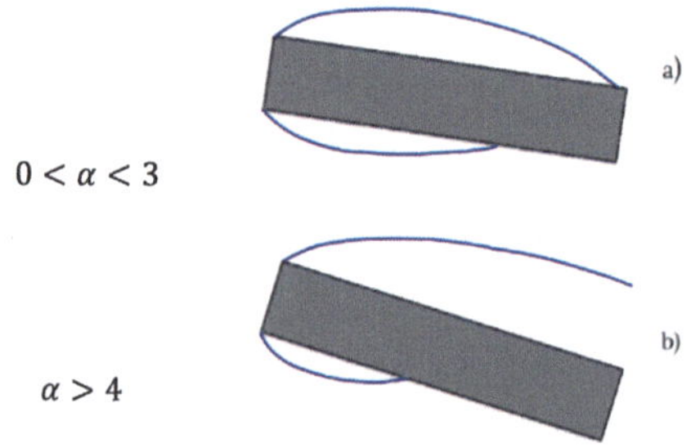

Fig. 9. Schematic showing the migration of the reattachment points under slight angles of attack. Blue lines indicate the approximate position of the time-averaged shear layer position. Wind is left to right.

For angles of attack larger than 4°, the influence of a single transitioning shear layer remains open. It is shown here that the shear layer is born laminar and quickly turns toward turbulence with transition lengths being on the order of 5–40% of a body height. Treating the KH instability as a convective instability it may be expected that the instability's spectral information travels along the shear layer in the direction of the mean flow. Alternatively, shedding in the near wake of the body has been shown to be an absolute instability (Betchov and Criminale 1966). Absolute instabilities are defined by localized perturbations growing both upstream and downstream (Schmid and Henningson 2001). One consequence of this is that the wake shedding frequency can be detected everywhere in the flow surrounding the body. This suggests that the interval between shear layer transition and reattachment contains spectral content from both vortex shedding mechanisms. The two frequencies are orders of magnitude apart at the birth of the KH instability according to Lander et al. (2018). Farther downstream at reattachment, the elevated high frequency energy associated with the KH frequency redistributes itself as the KH vortices break down into smaller and smaller eddies following the turbulence cascade. At the same time, Fig. 4 shows that the net amount of TKE increases along the body and so it is unlikely that dissipation will relaminarize the shear layer. Towards the reattachment point and ultimately at trailing edge of the body, the dominating scales in the flow are attributed to those in the wake, and the smallest scales being the remains of the dissipated KH vortices. This range of active length scales and influx of fluctuating energy is likely what drives the turbulence machine for the shear layer on a rectangular section. The natural tendency of turbulence to erode organized energy into a turbulence cascade of diminishing length scales reduces the coherence of the KH vortex behavior, neutralizing Reynolds number effects farther downstream. Of course, the reattachment length is not assumed to be completely independent of the Reynolds number. If the Reynolds number of a turbulent shear layer is representative of the range between the smallest and largest scales, a larger disparity between scales ought to be possible. But given the limited range of wind tunnel facilities in size and speed the additional scales provided by changing the Reynolds number are secondary to the natural behavior of the flow and thus its effect on the reattachment length is overshadowed by the self-generated turbulence within the separated shear layer. The spatial domain containing this self-generated turbulent energy exchange is seemingly short. Initial distributions of maximum TKE for the 5:1 body in Fig. 5 converge on one another after approximately 1.5 body heights indicating a relative similarity for different Reynolds numbers well beyond the leading edge. It is likely that at least some of the fluctuations of the KH exit the measurement plane as the instability waves become nonlinear and more three dimensional. However, that the TKE parameter tends toward self similarity is an early clue as to why the reattachment lengths are not well organized by Reynolds number.

As a final note, the normalized lengths mentioned in the preceding sections are shown in Fig. 10 as a concluding remark about the Reynolds number dependency of the bluff body shear layer. Figure 10a shows the transition lengths defined by the TKE curves in Fig. 5 normalized by the initially observed value. The transition lengths for all geometries and angles tested are included and appear to fit the phenomenological scaling offered in (Lander et al. 2018) with an exponential decay proportional to $Re_h^{-0.6}$. For shear layers experiencing a range of pressure gradients and on different bodies it is remarkable that the fit line describes their upstream migration in a reliable manner. It would seem as though the shear layer, prior to transition, is perhaps more associated with phenomena on the front face than with interactions along the side faces. The reattachment lengths in Fig. 10b are presented a similar in normalized plot, although they paint a more obscured picture as to any clear trend. There are some interesting patterns appearing when the pressure side of the inclined bodies are compared to the orthogonal wind case. For those situations, the reattachment length is relatively longer although the changes are on the order of 10%, conforming to the ideas presented above.

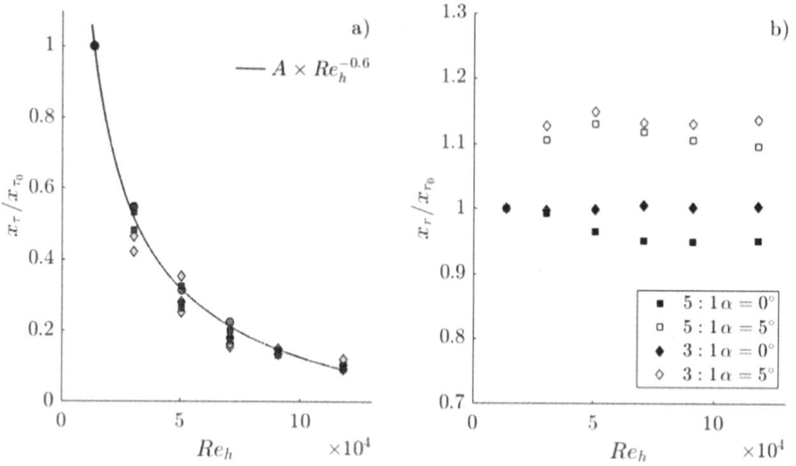

Fig. 10. Normalized transition, x_τ, and reattachment lengths, x_r, for all rectangular bodies tested. (■) 5:1. (◆) 3:1. (●) 1:1. A trend line is added with a scaling constant extracted from Lander et al. (2018). (b) Reattachment lengths for the 5:1 and 3:1 bodies. For positive angles of attack the reattachment length is measured on the pressure side of the body. Symbols are listed in the legend.

5 Conclusions

Wind tunnel experiments using 2D PIV have been carried out capturing the transition process of the separated shear layer for rectangular sections and highlighting the Reynolds number dependence of the transition length. Particular attention is devoted toward the 5:1 body. Comparisons to relevant metrics of the BARC are made showing good agreement of mean flow topology under orthogonal winds, however, the variation of lift slopes are found to be weak functions of Reynolds number contrary to available literature (Schewe 2013). The noteworthy disappearance of Reynolds number effects between the leading edge and reattachment points along the length of the transitioning shear layer remains an open area of interest. It is further proposed that the small length scales governing the dynamics in areas near the upstream corner of the section are transient and slowly migrate toward larger scales, on the order of the body's thickness, by the time fluid particles reach the aft portions of the body. Future measurements will be made in pursuit of a more complete description of these active scales residing within the separated shear layer.

Acknowledgements. Portions of this work are gratefully supported through the National Science Foundation under grant number 1727401.

References

Achenbach E (1971) Influence of surface roughness on the cross-flow around a circular cylinder. J Fluid Mech 46(2):321–335

Betchov R, Criminale WO (1966) Spatial instability of the inviscid jet and wake. Phys Fluids 9 (2):359–362

Brown G, Roshko A (1974) On density effects and large structure in turbulent mixing layers. J Fluid Mech 64:775–816

Bruno L, Salvetti MV, Ricciardelli F (2014) Benchmark on the aerodynamics of a rectangular 5:1 cylinder: an overview after the first four years of activity. J Wind Eng Ind Aerodyn 126:87–106

Drazin PG (2002) Introduction to Hydrodynamic Stability. Cambridge University Press, Cambridge

Gartshore I (1973) Gartshore 1973 BLWTL report-4-73. CAN, London

Haller G (2005) An objective definition of a vortex. J Fluid Mech 525:1–26

Huerre P, Monkewitz PA (1985) Absolute and convective instabilities in free shear layers. J Fluid Mech 159:151–168

Lander DC, Moore D, Letchford CW, Amitay M (2018) Scaling of square prism shear layers. J Fluid Mech 849:1096–1119

Mannini C, Marra AM, Pigolotti L, Bartoli G (2017) The effects of free-stream turbulence and angle of attack on the aerodynamics of a cylinder with rectangular 5:1 cross section. J Wind Eng Ind Aerodyn 161:42–58

Mannini C, Šoda A, Schewe G (2010) Computers & fluids unsteady RANS modelling of flow past a rectangular cylinder: investigation of reynolds number effects. Comput Fluids 39 (9):1609–1624

Michalke A (1964) On the inviscid instability of the hyperbolic tangent velocity profile. J Fluid Mech 19(04):543–556

Monkewitz PA (1982) Influence of the velocity ratio n the spatial instability of mixing layers. Phys Fluids 25(7):1137–1143

Mooneghi MA, Irwin P, Chowdhury AG, Technology A, Francisco S (2015) Partial turbulence simulation method for small structures, June, pp 1–11

Roshko A (1954) On the drag and shedding frequency of two-dimensional bluff bodies. National Advisory Committee for Aeronautics, Washington, D.C.

Saatoff PJ, Melbourne WH (1997) Effects of free-stream turbulence on surface pressure fluctuations in a separation bubble. J Fluid Mech 337(1997):1–24

Schewe G (2013) Reynolds-number-effects in flow around a rectangular cylinder with aspect ratio 1:5. J Fluids Struct 39:15–26

Schmid P, Henningson D (2001) Stability and Transition in Shear Flows. Springer-Verlag New York Inc., USA

Shimada K, Ishihara T (2001) Application of a modified k-e model to the prediction of aerodynamic characteristics of rectangular cross-section cylinders. J Fluids Struct 16 (15):399–413

Sigurdson LW (1986) The structure and control of a turbulent reattaching flow. California Institute of Technology

Unal MF, Rockwell D (1988) On vortex formation from a cylinder. Part 1. the initial instability. J Fluid Mech 190:491–512

An Immersed Boundary CFD Code for Building and Environment Applications

M. Mulas(✉) and R. Zucca

Vento CFD, Cagliari, Italy
marco.mulas@vento-cfd.com

Abstract. We present a novel Computational Fluid Dynamics (CFD) software developed for the Building and Environment applications. The technical noveltries are represented by the use of the Immersed Boundary method which allows a fully automated and extremely quick Cartesian Grid generation process. The result is an easy work-flow that takes less than 10 min to prepare the application, no matter of the geometrical complexity of the object of the simulation.

The software focuses on 4 outdoor wind applications: the evaluation of the wind load on constructions, the so called pedestrian comfort analysis on urban districts, the natural ventilation in buildings and the transport and diffusion of contaminants on cities and environmental sites, both in steady-state and transient modes.

The equation for the contaminant is decoupled from the solution of the 5 thermo-fluid dynamic equations (conservation of mass, equation of motion and conservation of energy), and it can be solved on an available, previously computed solution, in frozen mode.

A fully implicit solution algorithm drives the iterative method to the steady-state solution. In case a transient analysis is of interest, the same solution algorithm is used to follow the time evolution time-step after time-step.

The software has been validated against available wind tunnel data.

1 The Equations of Fluid Mechanics

The coupled system of the unsteady Navier-Stokes equations, which represents the equations governing all fluid flows, are here written

$$\begin{aligned}
\frac{\partial \rho}{\partial t} + \frac{\partial}{\partial x_j}\left(\rho u_j\right) &= S_{mass} \\
\frac{\partial \rho u_i}{\partial t} + \frac{\partial}{\partial x_j}\left(\rho u_i u_j + \delta_{ij} p - \tau_{ij}\right) &= S_{momentum_i} \; [i = 1, 2, 3] \\
\frac{\partial \rho E}{\partial t} + \frac{\partial}{\partial x_j}\left(\rho H u_j + \; - u_i \tau_{ij} + q_i\right) &= S_{energy}
\end{aligned}$$

The beauty of the equations of Fluid Mechanics is that they represents conservation laws. From top to bottom: the conservation of mass, the conservation of momentum in the 3 coordinate directions x, y and z (or Newton's second law for fluids) and the conservation of energy (the first principle of thermodynamics).

F. Ricciardelli and A. M. Avossa (Eds.): INVENTO 2018, LNCE 27, pp. 534–547, 2019.
https://doi.org/10.1007/978-3-030-12815-9_41

The dependent variables that are to be solved are: the mass per unit volume (density), the momentum per unit volume (the density multiplied by the velocity components, where the subscript i = 1, 2, 3 stands for the three cartesian velocity components u, v and w), and the total energy per unit volume (the total energy E is the sum of the kinetic and internal energy, multiplied by the density).

The equations are written in tensor notation: where the same subscript j is repeated in the same term, that implies a summation over 3 terms for j = 1, 2 and 3. The following example clarifies the use of the notation

$$\begin{aligned}\frac{\partial}{\partial x_j}(\rho u_j) &= \frac{\partial}{\partial x_1}(\rho u_1) + \frac{\partial}{\partial x_2}(\rho u_2) + \frac{\partial}{\partial x_3}(\rho u_3) \\ &\equiv \frac{\partial}{\partial x}(\rho u) + \frac{\partial}{\partial y}(\rho v) + \frac{\partial}{\partial z}(\rho w)\end{aligned}$$

The structure of the five equations is the same: an unsteady term and a divergence term. When integrated over a volume, the unsteady term represents the time rate of increase/decrease of mass, momentum and energy in the volume, and the divergence term represents the sum of the fluxes through the volume boundaries. Essentially, the mass in the volume (or the momentum, or the energy) increases or decreases depending on the net flux of mass entering/exiting the volume through the boundaries.

At the right hand side of all equations, there can be a source term representing a volume addition/extraction of mass, momentum or energy per unit time. Mass source terms may represent a source of contaminant, momentum sources can be used for modelling a fan, for indoor ventilation, and energy sources can model the heat power of radiators, for instance. All source terms can be made active only in limited, user defined, portions of the computational domain.

In order to "close" the system of equations, the following constitutive relations are needed:

$$\rho = \rho(p, T)$$

$$\Upsilon_i = D\frac{\partial Y}{\partial x_i} \quad \tau_{ij} = \mu\frac{\partial u_i}{\partial x_j} \quad q_i = \kappa\frac{\partial T}{\partial x_i}$$

These relations represent: the equation of state on top, in this case the ideal gas equation of state, the relation of mass mixing coefficient, the stress-strain relation and the law of heat conduction. It has to be noticed that in Fluid Mechanics, differently than in Solid Mechanics, the stresses are linearly related to the rate of change of the deformations (i.e. the fluid velocities). These relations also introduce the physical properties for the diffusivities of mass, momentum (the coefficient of viscosity) and energy (the heat conductivity).

It is worth noting how the 3 conservation laws for mass, momentum and energy, keep their simmetry and simplicity. Their corresponding mixing are related to the gradients of mass, momentum and temperature, which represents the internal energy. Whenever there is a gradient of one of the dependent variables, there will be physical (microscopic) mixing going on.

In order to model the turbulence, an equation for a turbulence model has to be added to the main 5 equations. All turbulence models aim to compute a so-called eddy viscosity, or turbulent viscosity, which is to be added to the physical coefficient of viscosity

$$D = D_p + D_t \quad \mu = \mu_p + \mu_t \quad \kappa = \kappa_p + \kappa_t$$

The 3 mixing coefficients are related to each other, for any given fluid, so knowing one of them suffices. The subscript p is usually dropped.

The role of the turbulent viscosity is to provide sufficient and "correct" mixing due to the turbulence in regions where shear flows and wakes are present. Moreover, it should model the boundary layer, the very thin fluid regions near solid boundaries.

Two turbulence models are available in VENTO AEC, the software used in this study: a 1-equation model that solves directly for the turbulence viscosity (the Spalart-Allmaras model) and the 2-equation Kappa-Epsilon model, where the turbulent viscosity is calculated from two addition variables, the turbulence kinetic energy and the turbulence dissipation. The turbulence models are described in all details in the Ref. [1].

2 The Numerical Method

The equations are solved in a Cartesian grid, i.e. all cells are hexahedra with all 12 edges aligned with one of the Cartesian axes. The system of the equations are written in compact form as follows:

$$\frac{\partial Q}{\partial t} = \frac{\partial F(Q)}{\partial x_j} + S = RES$$

where Q is an array of the 5 dependent variables, and $F(Q)$ is the array of their corresponding fluxes. The steady state sought solution is $Q(t, x, y, z)$ for which $RES(Q) = 0$. The above system is discretised implicitly, i.e. by evaluating the right hand side at the next, yet unknown, iterative step $N + 1$, and then linearized:

$$Q^{N+1} = Q^N - \Delta t\, RES^{N+1}$$

$$\left[\frac{1}{\Delta t} + \frac{\partial}{\partial Q}(RES^N)\right](Q^{N+1} - Q^N) = RES^N$$

The resulting system of algebraic equations (5 algebraic equations for each computational cell) are moved toward the steady state solution by a Gauss-Seidel relaxation method, starting from a given initial solution. The initial solution can be the wind and atmospheric conditions for outdoor applications. The iterations are stopped when the Residual term (the sum of the net flux through the cell boundaries and the source term over all of the cells in the computational domain) are zeroed, i.e. they fall below a certain, arbitrarily assigned low value.

It is worth noticing that the time dependency is lost because of the linearization step, and the independent variable t only represents a sort of iterative counter. In case the time accuracy has to be recovered, for simulating real transients, an additional term must be introduced:

$$\frac{\partial Q}{\partial t} = \frac{\partial Q}{\partial \tau} - \frac{\partial F(Q)}{\partial x_j} + S = -\frac{\partial Q}{\partial \tau} + RES \equiv RES^*$$

$$\frac{\partial Q}{\partial \tau} = \frac{3Q - 4Q^N + Q^{N-1}}{2\Delta\tau}$$

The new term is the "true" time rate of change of the dependent variables, and the greek letter *tau* is now the real time. The same relaxation iterative procedure is now applied to drive to zero the new Residual (*RES**) that includes the real time discretization term.

All details of the equations as well as of the numerical method can be found in reference [1].

3 The Contaminant Equation

An equation for simulating the time accurate transport and the diffusion of a contaminant can be added to the main system of 5 equations, plus the turbulence model equation.

$$\frac{\partial \rho Y}{\partial t} + \frac{\partial}{\partial x_j}\left[\rho u_j Y - (D + D_t)\frac{\partial \rho Y}{\partial x_j}\right] = \rho \dot{Y}$$

Here Y represent the mass fraction of a gaseous contaminant. Once multiplied by the density, it represents a concentration [kg of contaminant per cubic meter].

The structure of the contaminant equation is the same as that of the 5 main equations: a conservation equation with an unsteady term, a flux term and a possible source term. The solution algorithm is also the same, both for steady-state and transient situations.

An interesting feature is that the solution of a contaminant equation does depend on the solution of the main equations, namely the momentum equation, because the velocity field is responsible for transporting the contaminant in time and space. However, the viceversa doesn't hold: the evolution of the contaminant does not influence back the solution of the momentum equation.

For this reason the contaminant is called a passive scalar, and its equation con be solved once the solution of the main 5 equations is available, i.e. in "frozen mode". It is an efficient way to analyse both indoor and outdoor scenarios, changing the contaminant source location and looking at how and where the contaminant is trasported.

4 The Immersed Boundary Method

The Immersed Boundary (IB) method is a technique that uses Cartesian grids which are non conformal with the boundaries of the objects under numerical investigation. Standard CFD techniques for generating the grids follow the bodies' boundaries, so that grid lines and bodies' contours coincide. This grids are called conformal, or body-fitted, and are necessarily curvilinear. Moreover, the grids may become highly stretched and skewed, particularly close to the solid boundaries, which in turn degrades the accuracy of the numerical methods where it is mostly needed.

The R&D on the Immersed Boundary method started at the beginning of 2000, and produced so far hundreds of journal and conference papers and research works. However, no commercial CFD product has yet offered such a technology.

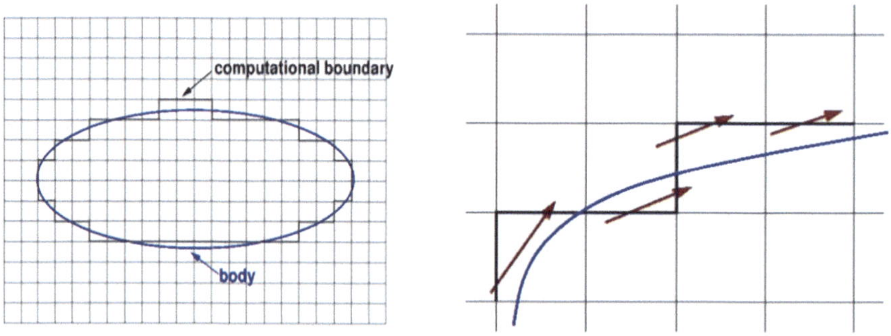

By using the IB technique, the cells intersecting the bodies are kept entire, neither cut nor adapted to the body's contour. That makes the boundaries of the computational domain look like a staircase. Since IB is not staircase computation, the key point, which may be seen as a definition of the IB method, is that mass, momentum and energy fluxes are allowed through the computational boundaries, and not allowed through the bodies' contours, as it is schematically shown in the figure above.

Essentially, the only difference between IB and standard CFD techniques are in the boundary condition treatment. Everywhere else, the two approaches are identical: Finite Volume, cell-centered RANS CFD.

There are plenty of literature about the IB methods, for instance in Refs. [2–4]. The method we developed and used in this study is different and yet unpublished.

The priceless benefit coming from the use of the IB technique is that the process of grid generation can be made completely automatic and extremely quick, no matter of the complexity of the geometrical definition of the objects under investigation. Moreover, the grid generation process remains free of potential geometrical errors because the cells are neither cut nor adapted.

The automatic procedure shrinks the whole preparation phase, between importing the geometry and starting the simulation, to a few minutes compared to many hours, and sometimes days, using standard techniques. Moreover, accuracy will benefit by using Cartesian grids, which carry the least numerical truncation error, compared to curvilinear (and necessarily distorted) grids.

5 Applications

In this paragraph we present 4 test cases in the field of Building and Environment. Wind tunnel measurements are available for all cases but one, together with CFD results obtained with other commercial CFD products.

1. Aerodynamic Load on Building and Structures
A scaled 1:400 model of a building 183 m high, with a cross section of 30 m by 46 m, has been measured in a wind tunnel. The reference [5] shows the experimental results together with CFD results obtained with ANSYS Fluent CFD product, the most popular CFD package.

We used both available turbulence models available: the 1-equation Spalart-Allmaras model, and the 2-equation Kappa-Epsilon model. The Fig. 1 shows the convergence of the two simulations to the steady-state solution:

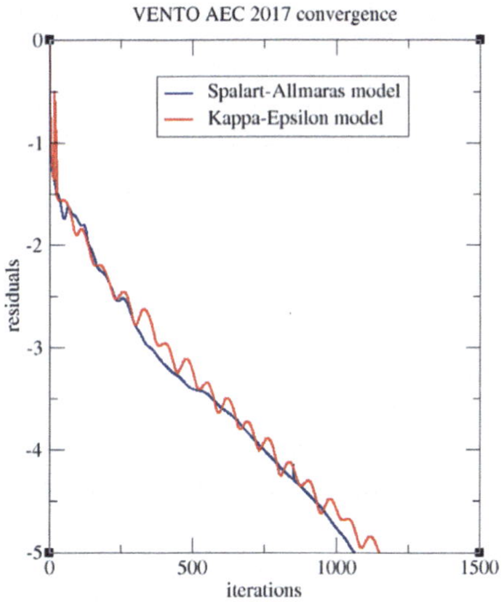

Fig. 1. Convergence of the wind pressure test case

The simulations were stopped after the Residuals dropped by a factor 100.000 (the Fig. 1 shows a log scale on the ordinate axis, versus the iterations on the abscissa axis). The computations took some 70–80 min on a 4-core desktop using a grid of little less than 1M cells.

The Fig. 2 shows the map of the pressure coefficient on the windward section.

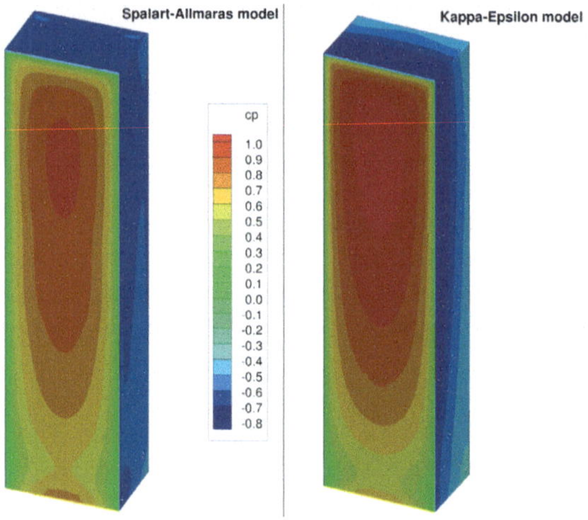

Fig. 2. Colour map of the computed pressure coefficient colour map.

The next Fig. 3 shows VENTO AEC Cp distributions along the perimeter of the building at two thirds height. The results obtained with the two turbulence models are both shown:

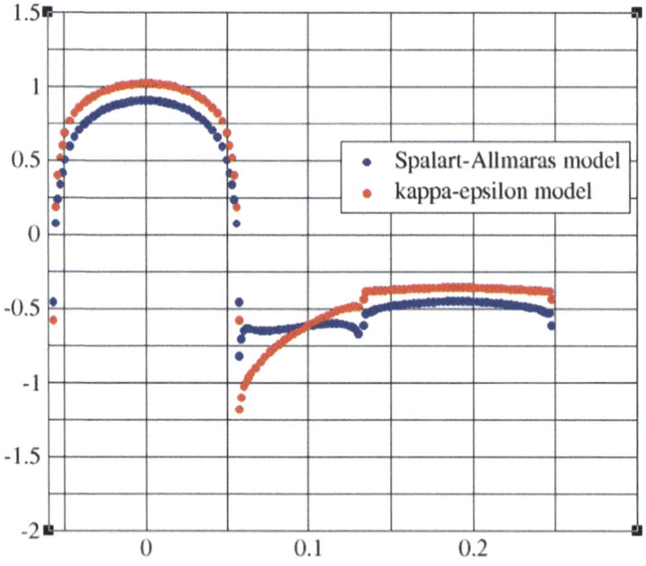

Fig. 3. Cp distributions on the building perimeter. VENTO AEC results

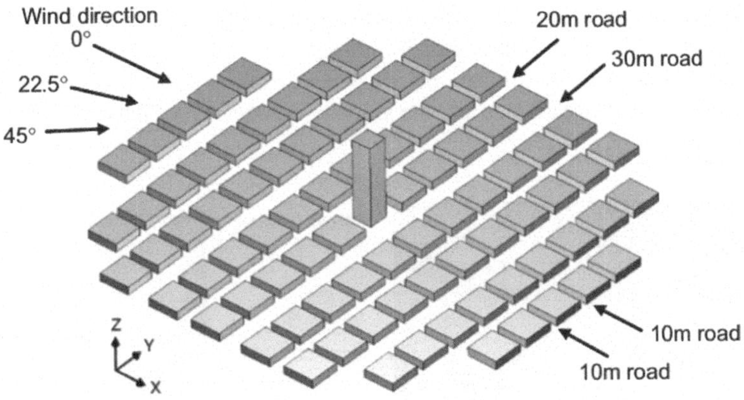

Fig. 4. Model of a urban area for pedestrian comfort analysis

The results obtained with the Spalart-Allmaras model of VENTO AEC code compare well with wind tunnel data and some of the CFD results of Ref. [5] on all sides (windward, sidewall and leeward). It is interesting noticing that the Kappa-Epsilon results over predict the Cp on the windward side (stagnation pressure) and shows a larger depression when the wind turns to the sidewall. The result reported in the Ref. [5] obtained by using the Kappa-Epsilon model shows a very similar behaviour, which illustrates that this is a model shortcoming.

2. Pedestrian Comfort

The Fig. 4 shows a scaled 1:400 wind tunnel model of a high rise building in a simple urban area. The high rise building is 100 m high (25 cm the model). This test case is taken from Ref. [6].

The Ref. [6] reports the wind tunnel measurements taken by a number of probes located near the ground at 5 mm height, which would correspond to 2 m in the real model.

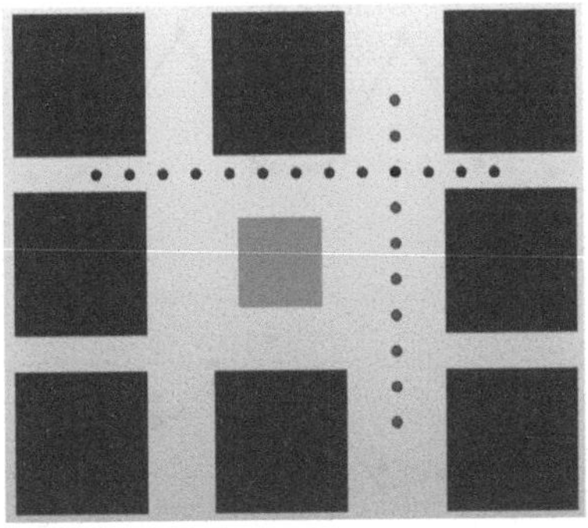

Fig. 5. Location of the probes: orange row and blue row. The wind direction is from the left side

The Fig. 5 shows the location of two such rows of probes, at 2 m height from the ground (5 mm in the wind tunnel model).

We carried out the CFD simulation for a wind direction of 0° (shown in the Fig. 4). The Fig. 6 shows the streamlines and the velocity colour map of the computed steady-state solution.

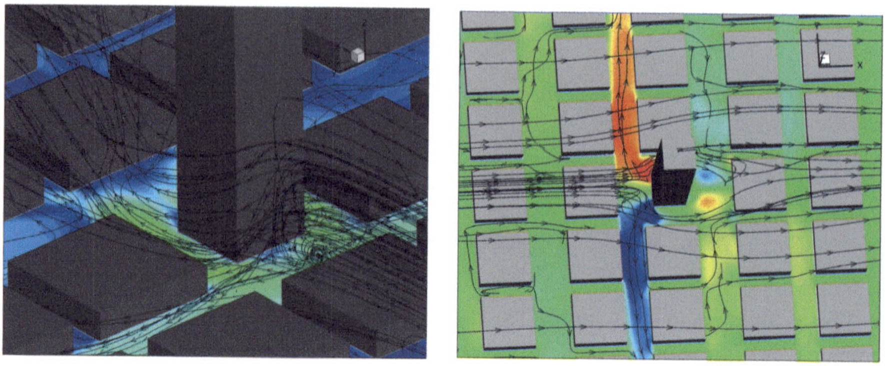

Fig. 6. Streamlines of the wind field near the high rise building. Velocity magnitude colour map on the left, and Y component of the velocity on the right (red is positive and blue is negative, going in the negative Y direction)

The large mass of wind that run onto the high obstacle, is forced to turn toward the two sides and accelerates in the streamwise direction (X direction) as well as along the two streets perpendicular to the wind direction (the Y direction).

The Fig. 7 shows the comparison of the measured and computed velocity in the locations corresponding to the two rows of probes (orange and blue probes, shown in the Fig. 5).

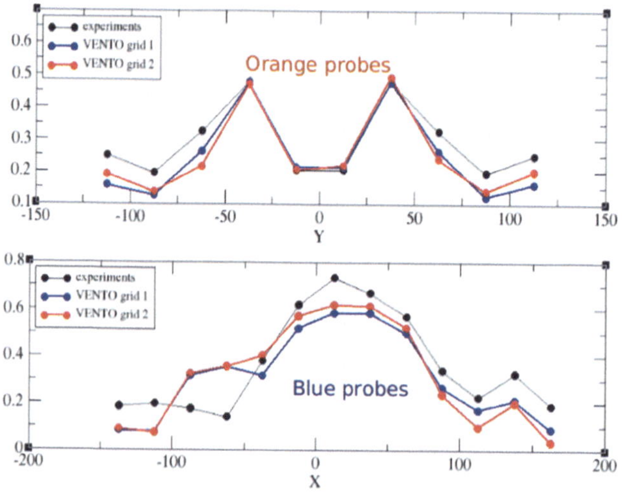

Fig. 7. Comparison of wind tunnel data versus VENTO AEC results

Computed results compare well on the orange probes, downstream of the high rise building and perpendicular to the wind direction. There are some discrepancies on the blue probes, near the intersection of the street perpendicular to the wind direction and upstream of the high rise building. The computation shows a greater acceleration (the red area shown in the Fig. 6, on the right) of the wind that enters the street with 90° turn.

3. Natural Ventilation

The Fig. 8 shows an isolated building with two asymmetric openings, on the windward and the leeward sides. The stationary solution was measured in a wind tunnel [7].

Fig. 8. Wind tunnel model of an isolated building with two asymmetric openings

The Fig. 9 shows the computed results versus the wind tunnel measurements and other CFD results obtained with ANSYS Fluent CFD software.

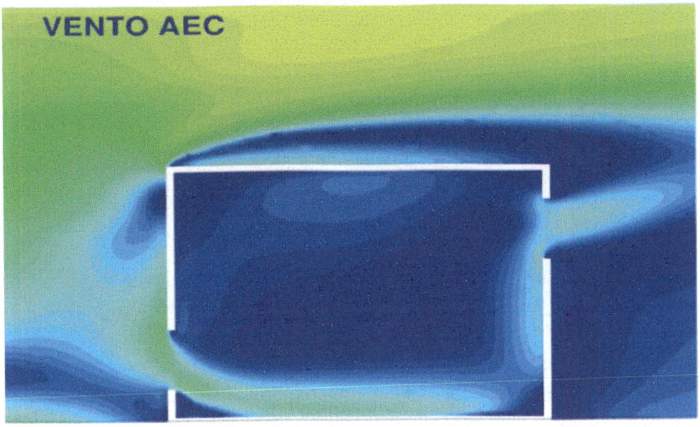

Fig. 9. Flow visualization obtained with VENTO AEC

The size of the recirculation region, on the building roof, is overestimated by ANSYS Fluent [7]. Wind tunnel data [7] and VENTO AEC produce a much similar picture. The Fig. 10, shows the streamwise normalized wind speed along an horizontal line inside the building.

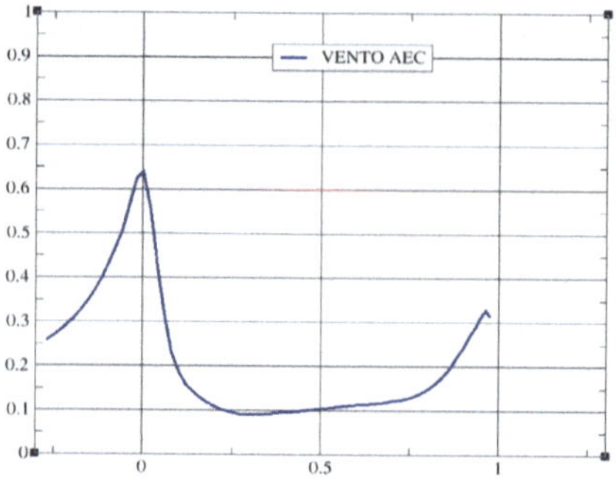

Fig. 10. Streamwise normalized velocity on an horizontal line through the windward entrance, obtained with VENTO AEC

Upstream of the windward opening ANSYS Fluent [7] overestimates the velocity, VENTO AEC underestimates it. VENTO AEC shows a steeper negative gradient, right after the entrance, more similar to the measured data [7], and matches the last value better than ANSYS Fluent.

4. Transport and Diffusion of a Contaminant in a Milan

The last test case we present concerns the time accurate simulation of the transport and diffusion of a contaminant in the city of Milan. We don't have data to compare with, for this last application.

The wind speed is 20 m/s, coming from nord. The Fig. 11 shows the map of the wind velocity magnitude at 1 m height above the ground level, together with the streamlines.

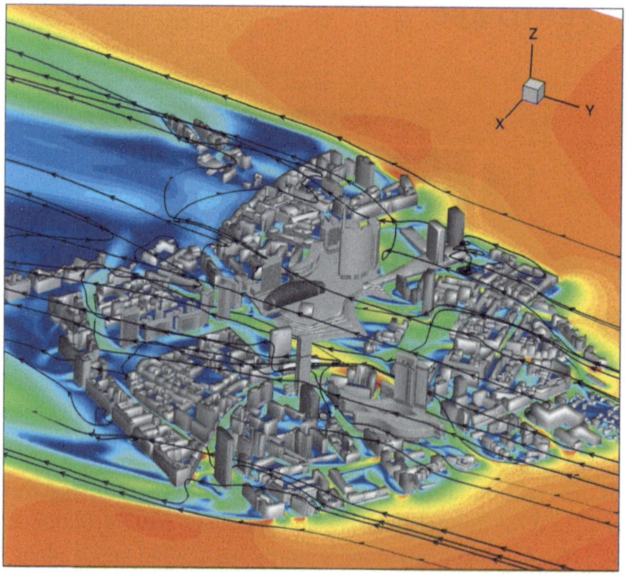

Fig. 11. Wind velocity map of Milan. Wind speed = 20 m/s, wind direction Nord.

The computed flow field has been used to determine the transport of a contaminant release over time ("frozen flow" simulation). The contaminant release was set to 1 kg/s, and the simulation was carried out for 3 min, with 180 steps of 1 min. The following 5 Figs. 12, 13, 14 and 15 show the concentrations after 30, 60, 90 and 180 s, when the stationary state was reached.

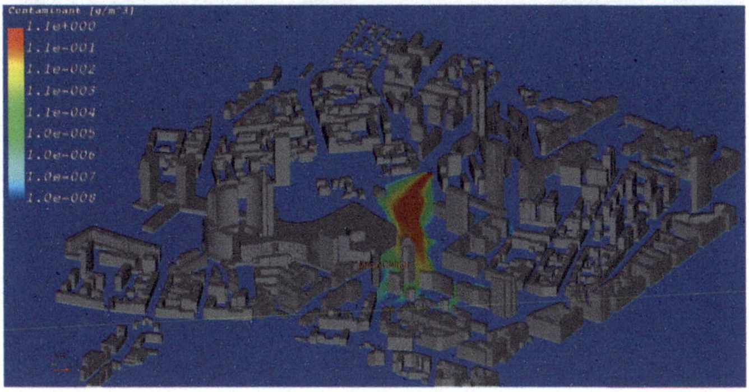

Fig. 12. Map of contaminant after 30 s

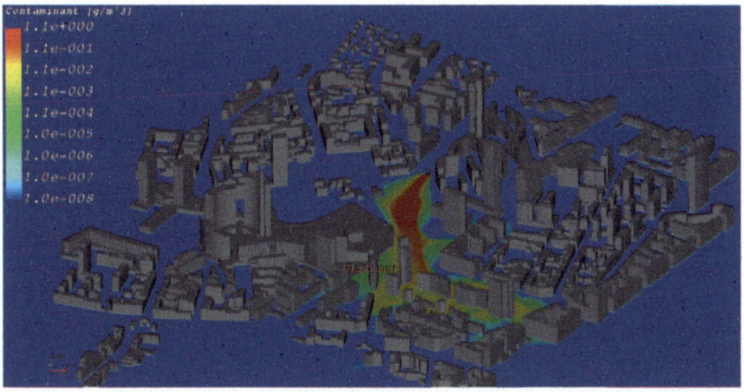

Fig. 13. Map of contaminant after 60 s

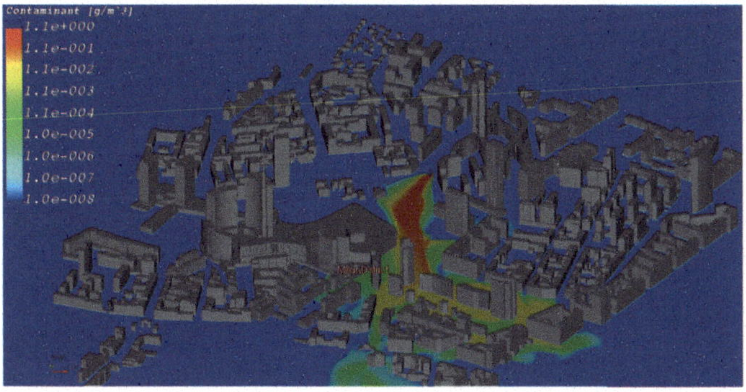

Fig. 14. Map of contaminant after 90 s

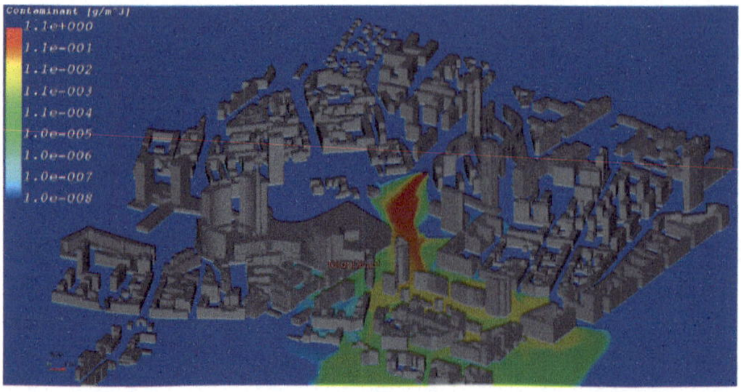

Fig. 15. Map of contaminant after 3 min (stationary conditions)

The last Fig. 16 shows a 3D iso-surface of the contaminant, once the stationary conditions were reached. The iso-surface is at 1 microgram per cubic meter.

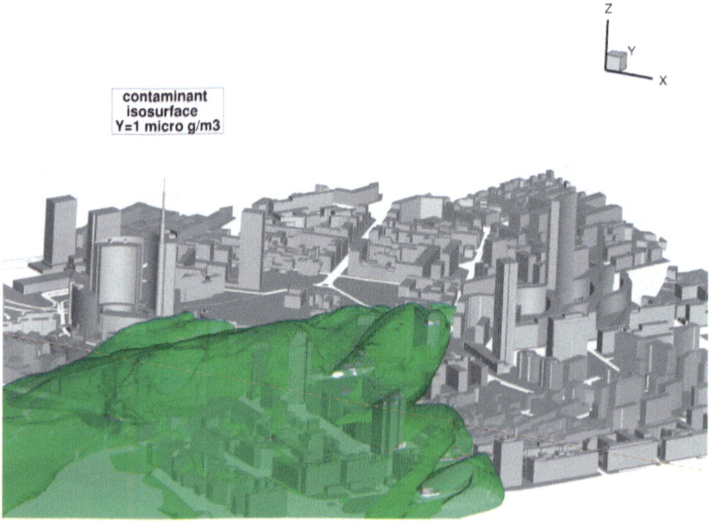

Fig. 16. 3D isosurface of the concentration cloud

6 Conclusions

In this paper we presented a novel CFD software, VENTO AEC, based on the innovative Immersed Boundary method. The IB technology allows to shrink the pre-processing phase, due to a fully automatic grid generation process, to only few minutes, regardless of the complexity of the geometrical object under analysis. Standard CFD softwares may take from an hour or two, with "easy" geometrical design, to a few days.

We compared the results with wind tunnel data and other CFD simulations.

The use of Cartesian grids, and an efficient fully implicit iterative process, returns quick and accurate results.

References

1. Mulas M et al (2002) Efficient parallel computation of flows of arbitrary fluids for all regimes of Reynolds, Mach and Grashof numbers. Int J Heat Fluid Flow 12(6):637–657
2. Iaccarino G, Verzicco R (2003) Immersed boundary technique for turbulent flow simulation. Appl Mech Rev ASME 56:331–347
3. Kang S, Iaccarino G, Moin P (2009) Accurate immersed boundary reconstructions for viscous flow simulations. AIAA J 47(7):1750–1760
4. Mittal R, Iaccarino G (2005) Immersed boundary method. Ann Rev Fluid Mech 37:239–261
5. Dagnew AK, Bitsuamalk GT, Merrick R (2009) Computational Evaluation of wind pressure on tall buildings. In: 11th American conference on wind engineering, San Juab, Puerto Rico, 22–26 June 2009
6. Yoshie R et al (2007) Cooperative project for CFD prediction of pedestrian wind environment in the Architectural Institute of Japan. J Wind Eng Ind Aerodyn 95:1551–1578
7. Peren JI et al (2015) CFD analysis of cross-ventilation on a generic isolated building with asymmetric opening positions. Build Environ 85:263–276

Simultaneous Vibration Suppression and Energy Harvesting in Wind Excited Tall Buildings Equipped with the Tuned Mass Damper Inerter (TMDI)

F. Petrini[1(✉)], Z. Wang[2], and A. Giaralis[2]

[1] Department of Structural and Geotechnical Engineering, Sapienza University of Rome, Via Eudossiana 18, 00184 Rome, Italy
francesco.petrini@uniromal.it

[2] Department of Civil Engineering, City, University of London, Northampton Square, London EC1V0HB, UK
{zixiao.wang, agathoklis.giaralis.1}@city.ac.uk

Abstract. This paper investigates the potential of tuned mass dampers (TMDs) coupled with inerter devices in different tuned mass dampers inerter (TMDI) topologies to dissipate oscillations in tall buildings due to vortex shedding in the across wind direction while generating electric energy. The TMDI is first optimized for minimizing peak accelerations for serviceability purposes in a 74 storey benchmark steel building under different wind intensity levels. It is seen that TMDI stiffness and damping optimal parameters are robust to design/reference wind velocity and, therefore, to potential climate change effects, while achieving same level of performance using significantly smaller attached mass compared to the classical TMD. Then, a regenerative electromagnetic motor (EM) is added to the TMDI allowing for varying the TMDI damping property as well as transforming part of the dissipated kinetic energy to electricity. It is shown that by increasing TMDI damping above the optimal value for vibration suppression and/or by reducing the inerter property increases the available energy for harvesting at the expense of larger floor accelerations. Therefore, it is concluded that by relaxing serviceability limit state requirements associated with occupancy considerations renders possible an increase in energy generation in wind-excited tall buildings.

Keywords: Tuned mass damper inerter · Energy harvesting · Tall buildings

1 Introduction

Modern tall buildings can be particularly susceptible to oscillations in the across-wind direction due to vortex shedding generated at their edges; this is especially true for structures with rectangular floor-plan and high height-to-width aspect ratios (Ierimonti et al. 2017). In such buildings, ensuring that floor accelerations remain below a certain threshold becomes the dominant structural design requirement to achieve acceptable performance for occupants comfort and serviceability. In this context, over the past two decades, tuned mass dampers (TMDs), among other devices and configurations for

F. Ricciardelli and A. M. Avossa (Eds.): INVENTO 2018, LNCE 27, pp. 548–561, 2019.
https://doi.org/10.1007/978-3-030-12815-9_42

supplemental damping, have been widely used in practice for vibration mitigation in wind-excited tall buildings to meet serviceability criteria prescribed by the relevant building codes and guidelines (Ierimonti et al. 2018). In its simplest form, the classical linear passive TMD comprises a mass attached towards the building top via linear stiffeners, or hangers in case of pendulum-like TMD implementations, and energy dissipation devices (dampers). Stiffness and damping properties of the TMD are optimally designed/tuned to control solely the first (fundamental) building vibration mode for a pre-specified attached mass (e.g. Rana and Soong 1998). Nevertheless, as the building aspect ratio increases, peak floor accelerations are more and more influenced by higher modes of vibrations and, therefore the classical TMD becomes less efficient to meet serviceability. Further, TMD effectiveness for vibration control depends heavily on the attached mass. However, the latter is limited in practice by the need to accommodate its weight and volume.

Recently, Giaralis and Petrini (2017) addressed the above drawbacks of the TMD to suppress wind-induced vibrations in the across wind direction of tall buildings by incorporating an inerter device in the so-called tuned mass-damper-inerter (TMDI) configuration originally developed for suppressing vibrations in base-excited structures (Marian and Giaralis 2014). The inerter is a two-terminal device with negligible mass/weight which develops a resisting force proportional to the relative acceleration of its terminals by a scalar coefficient termed "inertance" (Smith 2002). Giaralis and Petrini (2017) demonstrated that appreciable gains in reducing peak top floor accelerations in typical tall/slender benchmark buildings are achieved by TMDIs compared to TMDs for the same attached mass. These gains are attributed partly to the mass-amplification effect and partly to higher-modes-damping effect endowed to the TMD by the inerter.

In this contribution, the potential benefits of adding a regenerative electromagnetic motor (EM) to TMDIs with varying inertance in wind-excited tall buildings is numerically explored. Note that the consideration of the EM allows for varying the TMDI damping property while enables transforming part of the kinetic energy into electricity. As shown by Marian and Giaralis (2017) for the case of harmonically excited single degree-of-freedom structures, the varying inertance leverages the TMDI vibration control and energy harvesting capabilities. Herein, performance of TMDI with varying inertance and damping properties for concurrent suppression of floor acceleration and generation of energy in wind-excited tall buildings is investigated. To this aim, a planar dynamical model of a 74-storey TMDI-equipped building with square floor plan is considered. Wind action is represented by an analytical spectral density matrix modelling correlated across-wind induced forces accounting for vortex shedding and the structural analysis step is undertaken in the frequency domain for efficiency. Starting from an optimally designed TMDI defined via a standard numerical optimization search to minimize peak top floor acceleration, the peak top floor acceleration and the available energy for harvesting are quantified as both TMDI damping and inertance properties vary. Numerical results are reported for different TMDI topologies and pre-specified attached masses.

2 Benchmark Building Description and Modelling

The same tall building examined in Giaralis and Petrini (2017) is herein considered as a case-study. The structure is a 74-storey steel frame with a 50m-by-50m footprint and of 305.9 m total height (13.0 m for the ground/1st floor, 4.9 m for the last/74th floor, 4.0 m for the rest).

The building comprises two spatial steel frames, one inner including 12 columns, and one outer formed by 28 columns, the two frames are connected by three outriggers located at 100 m, 200 m, and 300 m in elevation. All columns have hollow square sections, with varying outer dimensions and thickness along the building height ranging in between 1.20 m to 0.50 m, and 0.06 m to 0.025 m, respectively. Beams are of various standard double-T steel section profiles and all beam-to-column joints are taken as rigid. The outriggers are braces consisted of double-T beams and hollow-square diagonal struts. The first three natural frequencies of these modes and the corresponding modal participating mass ratios in parentheses are 0.185 Hz (0.6233), 0.563 Hz (0.1900), and 1.052 Hz (0.0745). Frequency-dependent modal damping ratios are assumed in the analysis 2% for j = 1, 2, 3; 4% for 4, 5, 6; 6% for j = 7, 8, 9, 10; 9% for j = 11, 12, ..., 20; 12% for j = 21, 22, ..., 40; 15% for j = 41, 42, ..., 60; and 18% for j = 61, 62, ..., 74.

2.1 TMDI-Equipped Tall Buildings

Conceptually defined by Smith (2002), the ideal inerter is a linear massless two-terminal mechanical element resisting relative acceleration at its terminals through the so-called inertance coefficient, b, measured in mass units. In this regard, the inerter element develops a force $F = b(\ddot{u}_1 - \ddot{u}_2)$, where u_1 and u_2 are terminal displacements and a dot over a symbol signifies differentiation with respect to time. The inerter can be realized in different ways. A common inerter type is the so-called gearbox-flywheel inerter (Smith 2002). It comprises a rotating flywheel linked to a rack and pinion mechanism through a gearbox. In this manner the horizontal motion of the rack (rigidly connected to the TMDI mass) is transformed to rotational motion of the flywheel. For this kind of device, the inertance b is given by

$$b = m_f \frac{\gamma_f^2}{\gamma_{pr}^2} \left(\prod_{i=1}^{n} \frac{r_i^2}{pr_i^2} \right), \tag{1}$$

where m_f and γ_f are the mass and radius of the gyration of the flywheel respectively, γ_{pf} is the radius of gyration of the flywheel's pinion, $r_i/(pr_i)$ is the gearing ratio on the i-th stage of the gearbox (with a total of "n" stages). Clearly, through changes to the gearing ratios, the same device can provide different inertance values (Lazarek et al. 2018).

In the TMDI configuration, originally proposed by Marian and Giaralis (2014), an inerter element links the attached mass to a different floor from the one that the spring and damping elements are connected to, as shown in Fig. 1. In this manner, the inerter contributes additional inertia to the TMDI without increasing its weight. In this context, consider a planar linear n-storey frame structure modelled as an n-DOF dynamical system with mass m_k (k = 1, 2, …, n) lumped at the k-th floor as shown in Fig. 1. The TMDI comprises a mass m_{TMDI} attached to the top floor via a linear spring of stiffness k_{TMDI} and a linear dashpot of damping coefficient c_{TMDI}, and linked to the bottom floors by an ideal inerter of inertance b. The mass **M**, the damping **C**, and the stiffness **K** matrices characterizing a n-storey linear frame equipped with a "-p" TMDI topology (where "p" represents the number of floors spanned by the inerter in connecting the m_{TMDI} with one of the floor diaphragm under the top floor) can be written as

$$\begin{aligned} \mathbf{M} &= \mathbf{M}_{\mathbf{s}}^{\mathbf{n}+1} + (m_{TMDI}+b)\mathbf{1}_{\mathbf{n}+1}\mathbf{1}_{\mathbf{n}+1}^{T} + b\mathbf{1}_{\mathbf{n-p}}\mathbf{1}_{\mathbf{n-p}}^{T} - b\left(\mathbf{1}_{\mathbf{n}+1}\mathbf{1}_{\mathbf{n-p}}^{T} + \mathbf{1}_{\mathbf{n-p}}\mathbf{1}_{\mathbf{n}+1}^{T}\right) \\ \mathbf{C} &= \mathbf{C}_{\mathbf{s}}^{\mathbf{n}+1} + c_{TMDI}\left(\mathbf{1}_{\mathbf{n}+1}\mathbf{1}_{\mathbf{n}+1}^{T} + \mathbf{1}_{\mathbf{n}}\mathbf{1}_{\mathbf{n}}^{T} - \mathbf{1}_{\mathbf{n}+1}\mathbf{1}_{\mathbf{n}}^{T} - \mathbf{1}_{\mathbf{n}}\mathbf{1}_{\mathbf{n}+1}^{T}\right) \\ \mathbf{K} &= \mathbf{K}_{\mathbf{s}}^{\mathbf{n}+1} + k_{TMDI}\left(\mathbf{1}_{\mathbf{n}+1}\mathbf{1}_{\mathbf{n}+1}^{T} + \mathbf{1}_{\mathbf{n}}\mathbf{1}_{\mathbf{n}}^{T} - \mathbf{1}_{\mathbf{n}+1}\mathbf{1}_{\mathbf{n}}^{T} - \mathbf{1}_{\mathbf{n}}\mathbf{1}_{\mathbf{n}+1}^{T}\right) \end{aligned} . \tag{2}$$

where n – p is the floor number where the second terminal of the inerter is attached to. In the above expressions, $\mathbf{M}_{\mathbf{s}}^{\mathbf{n}+1} \in \mathbb{R}^{(n+1)\times(n+1)}$, $\mathbf{C}_{\mathbf{s}}^{\mathbf{n}+1} \in \mathbb{R}^{(n+1)\times(n+1)}$, and $\mathbf{K}_{\mathbf{s}}^{\mathbf{n}+1} \in \mathbb{R}^{(n+1)\times(n+1)}$ are the primary structure mass, $\mathbf{M}_{\mathbf{s}} \in \mathbb{R}^{n\times n}$, damping, $\mathbf{C}_{\mathbf{s}} \in \mathbb{R}^{n\times n}$, and stiffness, $\mathbf{K}_{\mathbf{s}} \in \mathbb{R}^{n\times n}$, matrices, respectively, augmented by one last (bottom) row with zero entries and by one last (rightmost) column with zero entries. Further, the vector $\mathbf{1}_{\mathbf{u}} \in \mathbb{R}^{(n+1)\times 1}$ has zero entries except from the u-th entry which is equal to one, and the superscript "T" denotes matrix transposition. Note that the inclusion of the inerter device influences only the mass matrix **M** of the controlled structure (i.e., the matrices **C** and **K** are the same for the TMD and for the TMDI). Moreover, note that for $b = 0$ the mass matrix of a frame building with a TMD attached to the top floor is retrieved (i.e., the TMD is a special case of the TMDI). Furthermore, for $b \neq 0$ the mass matrix **M** in Eq. (2) is not diagonal since the inerter introduces cross-terms equal to $-b$ that couples the DOF of the attached mass, numbered as $n + 1$, with the DOF of the (n – p)-th floor. Three TMDI connectivity (topologies) are investigated as shown in Fig. 1. In Giaralis and Petrini (2017) it is shown that, non-optimally tuned TMDIs achieve broadband vibration suppression and mitigate vibration contributions associated to higher vibration modes of the building structure.

Lastly, the fact that the effective inertia corresponding to the DOF of the attached mass is equal to (m_{TMDI} + b) (mass amplification effect of the inerter) motivates the definition of the following frequency ratio υ_{TMDI} and damping ratio ξ_{TMDI}.

$$\upsilon_{TMDI} = \frac{\sqrt{\frac{k_{TMDI}}{(m_{TMDI}+b)}}}{\omega_1}, \quad \xi_{TMDI} = \frac{c_{TMDI}}{2\sqrt{(m_{TMDI}+b)k_{TMDI}}}. \tag{3}$$

to characterize the dynamics of the TMDI given an attached mass m_{TDMI} and inertance b.

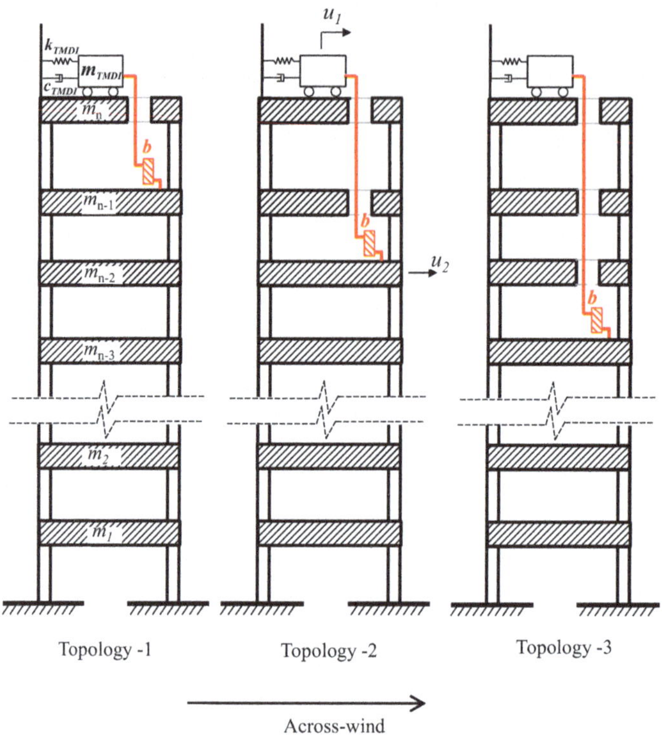

Fig. 1. Connection topologies for lumped-mass linear model of a wind excited n-storey frame building equipped with a two-terminals gearbox type TMDI

2.2 Frequency-Domain Across-Wind Response Analysis

Wind forces acting at the slab heights of the building structure in the across wind direction are modeled as a zero-mean Gaussian ergodic spatially correlated random field represented in the frequency domain by a $\mathbf{S}_{\mathbf{FF}}^{\mathbf{74}}(\omega) \in R^{74\times 74}$ power spectral density (PSD) matrix. The stochastic wind field model of Liang et al. (2002) is herein adopted to define the above force PSD matrix. The response displacement, velocity and acceleration PSD matrices of the TMDI-equipped frame structure defined by the **M**, **C**, and **K** matrices in Eq. (1) due to the $\mathbf{S}_{\mathbf{FF}}^{\mathbf{74}}$ matrix are obtained using the frequency domain input-output relationships of random vibrations

$$\begin{aligned} &\mathbf{S}_{\mathbf{xx}}(\omega) = \mathbf{B}(\omega)^*\mathbf{S}_{\mathbf{FF}}(\omega)\mathbf{B}(\omega); \\ &\mathbf{S}_{\dot{\mathbf{x}}\dot{\mathbf{x}}}(\omega) = \omega^2\mathbf{S}_{\mathbf{xx}}(\omega) \quad \text{and} \quad \mathbf{S}_{\ddot{\mathbf{x}}\ddot{\mathbf{x}}}(\omega) = \omega^4\mathbf{S}_{\mathbf{xx}}(\omega) \end{aligned} \tag{4}$$

respectively. In Eq. (4), $\mathbf{S}_{\mathbf{FF}}$ is the PSD wind force matrix $\mathbf{S}_{\mathbf{FF}}^{\mathbf{74}}$ augmented by a zero row and a zero column corresponding to the DOF of the TMDI which is not subjected to any wind load (internally housed). Further, the "*" superscript denotes complex matrix conjugation, and the transfer matrix **B** is given as

$$\mathbf{B}(\omega) = \left[\mathbf{K} - \omega^2\mathbf{M} + i\omega\mathbf{C}\right]^{-1}. \tag{5}$$

where, $i = \sqrt{-1}$. Next, the response displacement, velocity and acceleration variances of the k-th floor for the TMDI-equipped primary structure are obtained as

$$\begin{aligned} \sigma_{x_k}^2 &= \int_0^{\omega_{\max}} S_{x_k x_k}(\omega)d\omega \\ \sigma_{\dot{x}_k}^2 &= \int_0^{\omega_{\max}} S_{\dot{x}_k \dot{x}_k}(\omega)d\omega \, . \\ \sigma_{\ddot{x}_k}^2 &= \int_0^{\omega_{\max}} S_{\ddot{x}_k \ddot{x}_k}(\omega)d\omega \end{aligned} \tag{6}$$

respectively. That is, by integrating the response auto-spectra populating the main diagonal elements of the response PSDs in Eq. (4), on the frequency axis up to a maximum (cut-off) frequency ω_{max} above which the energy of the underlying processes is negligible. Furthermore, the variance of a relative response (e.g. acceleration) between two different floors (or DOFs) k and q is obtained by

$$\sigma_{\ddot{x}_{kq}}^2 = \sigma_{\ddot{x}_k}^2 + \sigma_{\ddot{x}_q}^2 - 2\int_0^{\omega_{\max}} S_{\ddot{x}_k \ddot{x}_q}(\omega)d\omega. \tag{7}$$

where the integrand is the acceleration response cross-spectrum corresponding to the k and q DOFs. Finally, k-th peak floor displacements, velocity and accelerations and peak relative velocity and acceleration between k and q floors, are estimated by the expressions

$$\begin{aligned} &peak\{x_k\} = g\sqrt{\sigma_{x_k}^2}, \quad peak\{\dot{x}_k\} = g\sqrt{\sigma_{\dot{x}_k}^2}, \quad peak\{\ddot{x}_k\} = g\sqrt{\sigma_{\ddot{x}_k}^2}, \\ &\text{and} \\ &peak\{\dot{x}_{kq}\} = g\sqrt{\sigma_{\dot{x}_{kq}}^2}, \quad peak\{\ddot{x}_{kq}\} = g\sqrt{\sigma_{\ddot{x}_{kq}}^2}, \end{aligned} \tag{8}$$

respectively. In the above expressions, g is the peak factor estimated by the widely used empirical formula due to Davenport (1964): $g = \sqrt{2ln(\eta T_{wind})} + 0.577/\sqrt{2ln(\eta T_{wind})}$, where $\eta = 2\pi/\omega$ is the effective structural response frequency in Hz (e.g., can be taken equal to the fundamental natural frequency of the primary structure), and T_{wind} is an assumed time duration of exposure to the wind action during which the peak response quantities in Eq. (8) are evaluated. The latter consideration implies that the underlying stochastic input/output processes are quasi-stationary (i.e., stationary/ergodic time-limited processes) and narrow-band. Regarding this point, well known other works (e.g. Huang et al. 2012) have shown that the response process can be non-Gaussian, and then imply the need for other equations for the peak factor.

3 Performance-Based Optimization of the TMDI for Vibration Suppression

The performance of the building in term of occupants comfort is evaluated by comparing the hourly peak accelerations ($T_{wind} = 3600$ s in Davenport's peak factor) of the floors as experimented under different design wind speeds V_{ref} (gradient wind velocity of the atmospheric boundary layer taken at an height of 810 m in urban congested area), with code-prescribed threshold values depending on the first natural frequency of the building in across-wind direction (equal to 102.9 m/s for the considered building). Then, considering the configurations in Fig. 1, optimization of the TMDI parameters (Design Variables - DVs) υ_{TMDI} and ξ_{TMDI} in Eq. (3) is conducted for fixed values of the TMD mass ratio $\mu = m_{TMDI}/M_{building}$ and inertance ratio $\beta = b/M_{building}$, treated as secondary design parameters, where $M_{building}$ is the total mass of the building. This is achieved by using the pattern search algorithm (Charles and Dennis Jr. 2003). Since the maximum peak acceleration is always attained at the top floor of the building, the goal of the optimization problem is to minimize the hourly peak top floor acceleration peak$\{\ddot{x}_{74}\}$ induced by the considered wind loads. The range of values of both secondary parameters and DVs has been chosen to ensure practically feasible configurations. To this end, the upper bounds of υ_{TMDI} and μ are set equal to 1.2 and 0.9%, respectively, on the basis of a literature review focusing on real-word TMD installations in high-rise buildings with tuning of the devices to the first mode of the building. Further, ξ_{TMDI} and β are limited to maximum values of 0.8 to avoid unrealistically high viscous damping coefficients and inertance for the considered structure (see also Giaralis and Petrini 2017 for discussion on feasibility of large-scale inerters). The details of the structural optimization problem are summarized in Table 1.

Table 1. Parameters' values in the TMDI optimization

Design variables	min	Max
υ_{TMDI}	0.2	1.2
ξ_{TMDI}	10^{-5}	0.8
Secondary design parameters	Values	
μ	0.1%–0.9%	
β	0.001–0.8	

From a computational viewpoint, strong convex behavior of the objective function peak$\{\ddot{x}_{74}\}$ in the υ_{TMDI}, ξ_{TMDI} plane is noted with a single global optimal design point being observed for all TMDI optimization cases considered.

The values ξ_{opt}, and υ_{opt} of the DVs resulting from the optimization are shown in Fig. 2 for the case of $V_{ref} = 30$ m/s and for the three considered topologies in Fig. 1. The y-axis of the charts shown in Fig. 2, represents the classical TMD configurations ($\beta = 0$), while the x-axis represents configurations with very low (almost-zero) physical mass. The latter configuration is termed tuned inerter damper (TID) in the literature (Lazar et al. 2014). It can be seen that for fixed β the trend of υ_{opt} is linear (decreasing) with μ, both for the TMD and for the TMDI for any β value. Increasing β and/or topology (increasing the number of floors spanned by the inerter terminals) makes the deviation of the υ_{opt} values from those of the TMD larger. Note that this trend is partly due to definition of υ_{TMDI}, which has the inertance at the denominator. Regarding ξ_{opt}, it is seen that its value is not monotonically varying on the μ-β plane, as already shown in Giaralis and Petrini (2017), while significantly different values are observed for different TMDI topologies.

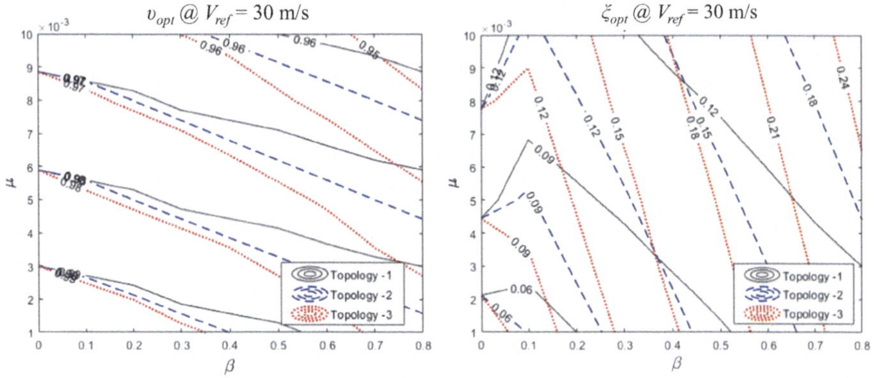

Fig. 2. Optimal values of the DVs in the secondary design parameters plane

The change of the optimal DV values with the reference wind speed is shown in Fig. 3, for the three TMDI topologies and for β = 0.1; 0.8 and μ = 0.5%. The consideration of different V_{ref} values aims to verify the sensitivity/robustness of optimal DV values respect to the different location and/or exposures of the building as well as to climate change effects which are expected to increase V_{ref} at a given location in the near future (e.g. Steenbergen et al. 2012). It is seen that the optimal DVs obtained for each topology are clustered together, with bounds represented by trends obtained at lower and higher considered β and shown in Fig. 3. Notably, optimal DV values are at large insensitive to the variation of reference wind speed. In particular, ξ_{opt} values remain constant at different V_{ref} values. On the other hand, υ_{opt} values vary only slightly with V_{ref}, especially for the large β value considered. Furthermore it can be observed that by switching from connectivity -1 to -3, there is larger spread of optimal values as inertance changes: υ_{opt} values increases in variability and shift to lower frequencies, while ξ_{opt} values increases in variability and shifts towards higher damping values. Optimal peak top floor acceleration for V_{ref} = 30 m/s and for the three considered topologies are shown in Fig. 4, as a function of the attached mass and for various inertance values, including the special case of β = 0 corresponding to the classical TMD. As previously reported in Giaralis and Petrini (2017) for non-optimized TMDI configurations, by increasing the inertance and/or by increasing the number of floors spanned by the inerter improves vibration suppression performance of the TMDI across the board. Figure 4 further suggests that the TMDI performs better than TMD as long as μ remains below some value which depends strongly on the inertance and the topology. For example, for -1 topology a TMD with 0.5% attached mass ratio performs better from TMDIs with $\beta < 0.3$ and same mass ratio. Nevertheless, a TMDI with $\beta = 0.1$ and 0.5% mass ratio in topology -3 performs better than a TMD with 0.5% mass ratio. Overall, it is found that TMDIs with relatively small mass ratio and inerters spanning more than one floor are most efficient for lightweight floor acceleration suppression.

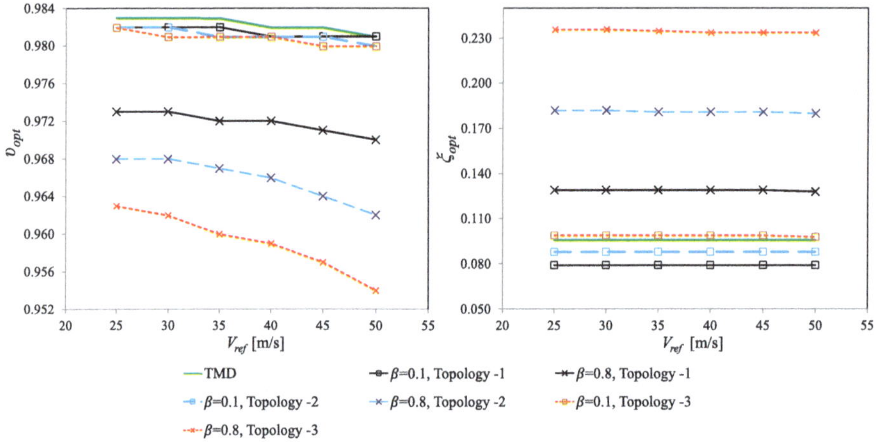

Fig. 3. Sensitivity of the optimal values of the DVs to the variation of the wind speed. Case μ = 0.5%

4 Energy Harvesting Potential for Optimal TMDI with Varying Inertance and Damping

Having established the benefits of the TMDI for suppressing floor accelerations in wind-excited tall buildings, this section explores its potential for harvesting energy from building oscillations. To this aim, a linear translational electromagnetic motor (EM) shunted by a purely resistive load is added in parallel to the dashpot TMDI element as shown in Fig. 5. Compared to the standard TMD-based energy harvesters proposed in the literature for electric generation from low-frequency large-amplitude oscillations (see e.g., Tang and Zuo 2012), the herein considered energy harvesting-enabled TMDI (EH-TMDI) differs by the inclusion of the inerter.

A standard linear EM is assumed such as one comprising a moving magnet within a coil coupled with a rectifying circuit to ensure unidirectional electric current flow irrespective of the magnet direction of motion (see e.g. Gonzalez-Buelga et al. 2014). By neglecting the mechanical parasitic damping leading to energy losses, the EM resists the relative motion between the primary structure and the attached mass by developing a damping force F_{EM} in the mechanical domain written as

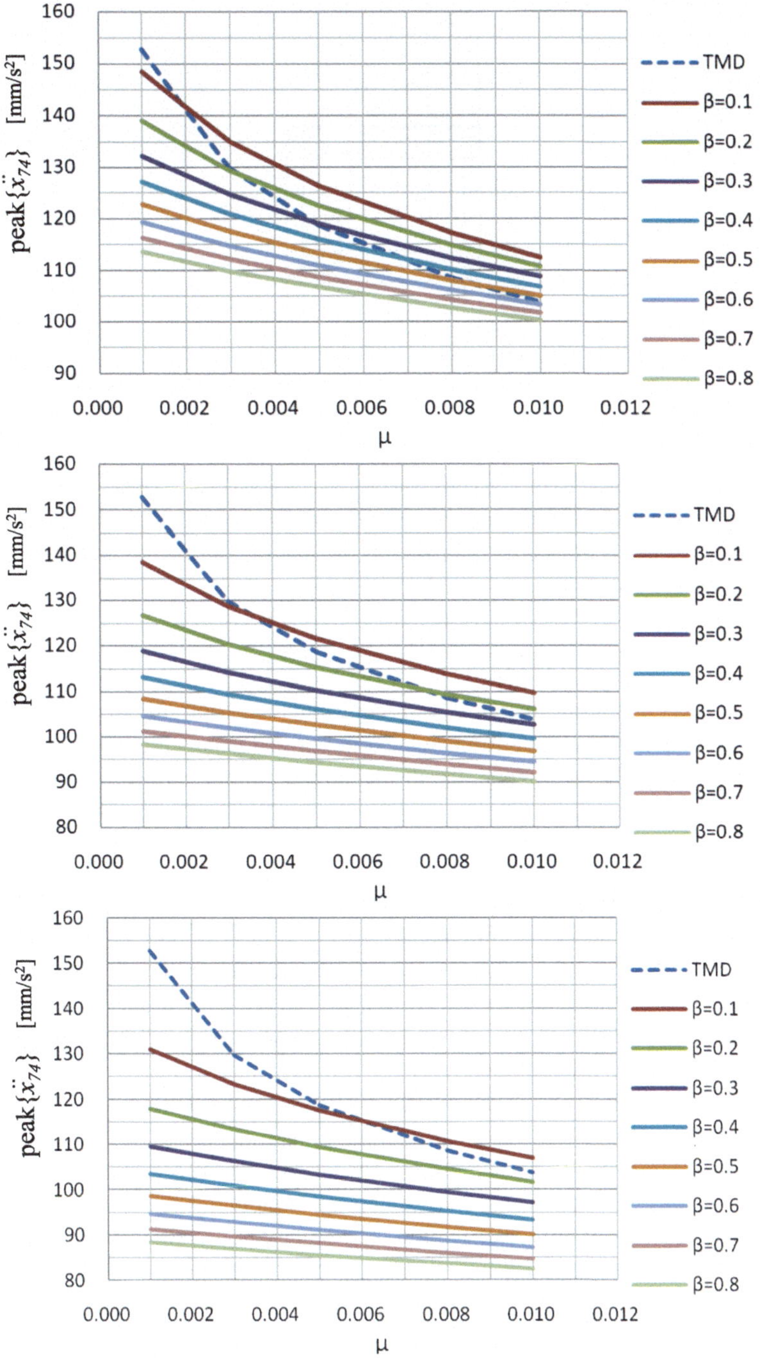

Fig. 4. Performances with optimally designed TMDI, V_{ref} = 30 m/s. Topologies: -1 (upper), -2 (middle), -3 (bottom)

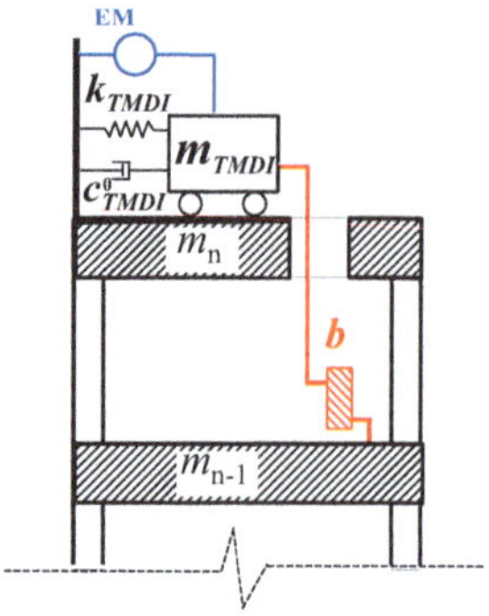

Fig. 5. Connection of the EH-TMDI system in presence of an EM

$$|F_{EM}| = c_{EM}|(\dot{x}_{74} - \dot{x}_{TMDI})| = c_{EM}|(\dot{x}_{74} - \dot{x}_{75})|. \tag{9}$$

where c_{EM} is the electromechanical damping coefficient of the EM. The equations of motion of previous sections are applicable to the herein considered building equipped with the EH-TMDI by setting

$$c_{TMDI} = c^0_{TMDI} + c_{EM}. \tag{10}$$

Herein, it is assumed, that the EM contributes to half of the total damping ($c^0_{TMDI} = c_{EM}$) as this consideration is related to optimal harvesting criteria (Gonzalez-Buelga et al. 2014). In this regime, the available energy for harvesting is quantified as

$$EHpot = c_{EM} \cdot \text{peak}|(\dot{x}_{74} - \dot{x}_{75})|^2. \tag{11}$$

The *EHpot* is plotted in Fig. 6 as function of c_{EM} for the case of μ = 0.5%, V_{ref} = 30 m/s, topology -1, and three different β (=0.1, 0.2 or 0.3) values. Note that c_{EM} = 100% denotes that $c_{TMDI} = c^0_{TMDI} + c_{EM}$ is taken equal to the optimal value for vibration suppression obtained as shown in previous sections. In the same figure, the peak square value of the relative velocities at the EM terminals (peak$|(\dot{x}_{74} - \dot{x}_{75})|^2$), and the peak$\{\ddot{x}_{74}\}$ are also shown.

It is seen that *EHpot* increases initially with c_{EM} until it reaches a plateau due to reduction of peak$|(\dot{x}_{74} - \dot{x}_{75})|^2$ with c_{EM}. At the same time, peak top floor acceleration increases as c_{EM} deviates from its optimal value. This result suggests that it is possible to increase energy generation in wind-excited tall buildings by relaxing performance requirements in terms of floor acceleration. The latter consideration may be applicable to buildings with significantly different occupancy between day and night hours as is the case of office buildings: during off-hours minimal occupancy suggests that through varying the damping coefficient of the EM, larger amounts of energy can be harnessed since the serviceability limit state in terms of occupants comfort (i.e., peak floor acceleration threshold) is alleviated. Moreover, similar effects are seen in Fig. 6 for the

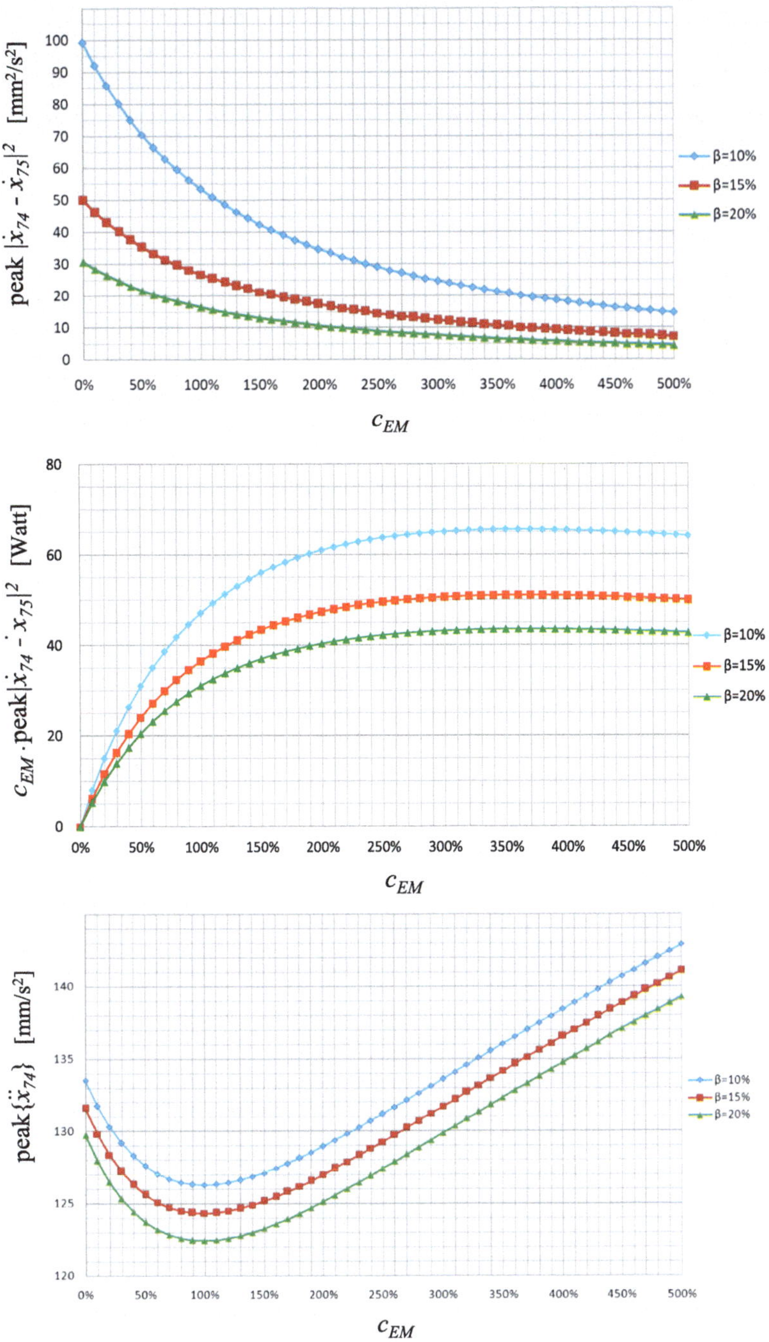

Fig. 6. EH potential and performances of the EH-TMDI. μ = 0.5%, V_{ref} = 30 m/s. Topology: -1. peak relative velocity (upper), EH potential (middle), peak acceleration at top floor (bottom)

case of varying the inertance: for a fixed, possibly non-optimal c_{EM}, a reduction of the inertance (e.g., through reducing the gearing ratio in flywheel-based inerters driven by geared rack-and-pinion mechanisms as discussed in Marian and Giaralis 2017) leads to increased available energy for harvesting at the expense of increased peak top floor acceleration.

5 Conclusions

In this paper, optimal TMDI tuning for suppressing floor accelerations along the across wind direction of wind-excited tall buildings for different wind intensities has been first pursued. It was found that optimal stiffness and damping properties do not change significantly with the reference wind intensity used in design and assessment for occupants' comfort and serviceability. This finding suggests that the TMDI is robust to changes of building wind exposure and to climate change effects. Moreover the TMDI has been coupled with a regenerative electromagnetic motor and TMDI potential for simultaneous energy harvesting and vibration suppression was quantified allowing for changes to the optimal TMDI damping property and/or inertance. It was reported that changes to the considered properties effectively leverages the conflicting objectives of maximizing the available energy to be turned into electricity by the motor and of minimizing the peak top floor acceleration. It is therefore concluded that by relaxing occupants' comfort requirements (i.e., serviceability limit state thresholds for floor accelerations) in wind-excited tall buildings based on occupancy considerations (e.g., off-hours in an office building) renders possible an increase in energy generation to be used to support power sensors and actuators required used for climate control or health monitoring in modern structures.

References

Charles A, Dennis JE Jr (2003) Analysis of generalized pattern searches. SIAM J Optim 13 (3):889–903

Davenport AG (1964) Note on the distribution of the largest value of a random function with application to gust loading. Proc Inst Civil Eng 28:187–196

Giaralis A, Petrini F (2017) Wind-induced vibration mitigation in tall buildings using the tuned mass-damper-inerter. J Struct Eng ASCE 143(9). https://doi.org/10.1061/(asce)st.1943-541x.0001863. Article Number 04017127

Gonzalez-Buelga A, Clare LR, Cammarano A, Neild SA, Burrow SG, Inman DJ (2014) An optimised tuned mass damper/harvester device. Struct Control Health Monit 21(8):1154–1169

Huang MF, Chan CM, Lou WJ, Kwok KCS (2012) Statistical extremes and peak factors in wind-induced vibration of tall buildings. J Zhejiang Univ Sci A 13(1):18–32

Ierimonti L, Caracoglia L, Venanzi I, Materazzi AL (2017) Investigation on life-cycle damage cost of wind-excited tall buildings considering directionality effects. J Wind Eng Ind Aerodyn 171:207–218

Ierimonti L, Venanzi I, Caracoglia L (2018) Life-cycle damage-based cost analysis of tall buildings equipped with tuned mass dampers. J Wind Eng Ind Aerodyn 176:54–64

Liang S, Liu S, Li QS, Zhang L, Gu M (2002) Mathematical model of across-wind dynamic loads on rectangular tall buildings. J Wind Eng Ind Aerodyn 90:201–251

Lazar IF, Neild SA, Wagg DJ (2014) Using an inerter-based device for structural vibration suppression. Earthquake Eng Struct Dynam 43(8):1129–1147

Lazarek M, Brzeski P, Perlikowsi P (2018) Design and identification of parameters of tuned mass damper with inerter which enables changes of inertance. Mech Mach Theor 119:161–173

Marian L, Giaralis A (2014) Optimal design of a novel tuned mass-damper–inerter (TMDI) passive vibration control configuration for stochastically support-excited structural systems. Probab Eng Mech 38:156–164

Marian L, Giaralis A (2017) The tuned mass-damper-inerter for harmonic vibrations suppression, attached mass reduction, and energy harvesting. Smart Struct Syst 19(6):665–678

Rana R, Soong TT (1998) Parametric study and simplified design of tuned mass dampers. Eng Struct 20(3):193–204

Smith MC (2002) Synthesis of mechanical networks: the inerter. IEEE Trans Autom Control 47 (10):1648–1662

Steenbergen RDJM, Koster T, Geurts CPW (2012) The effect of climate change and natural variability on wind loading values for buildings. Build Environ 55:178–186

Tang X, Zuo L (2012) Simultaneous energy harvesting and vibration control of structures with tuned mass dampers. J Intell Mater Syst Struct 23(18):2117–2127

Probabilistic Assessment of Windblown Sand Accumulation Around Railways

L. Raffaele[1,2(✉)] and L. Bruno[1,2]

[1] Department of Architecture and Design (DAD), Politecnico di Torino, Viale Mattioli 39, 10125 Turin, Italy
{lorenzo.raffaele,luca.bruno}@polito.it
[2] Windblown Sand Modeling and Mitigation Joint Research Group, Turin, Italy

Abstract. New ultra-long transnational railway megaprojects are currently being planned or under construction. Along their route, they increasingly cross arid and desert regions. In these regions, railways are vulnerable to windblown sand. Several failure cases recently occurred, e.g. along the Linhai-Ceke railway in China, or the Aus-Lüderitz railway in Namibia. We consider windblown sand as an environmental key factor, analogously to wind or snow drift. We categorize its effects into Sand Ultimate Limit States and Sand Serviceability Limit States. In the design perspective, the quantitative prediction of the windblown sand sedimented along the railway is mandatory. We propose a probabilistic approach to sedimented windblown sand modelling because of the inborn variability of the phenomenon. The proposed method allows evaluating the design value of the accumulated windblown sand volume for a given site.

Keywords: Windblown sand · Sand limit states · Railway · Monte Carlo

1 Introduction

Railway megaprojects traditionally includes long and ultra-long railway projects that are characterized by long-lasting impact on the economy, the environment, and society. Some historic examples are the Pacific Railroad (3.100 km, 1863–1869) in North America, and the Trans-Siberian Railway (9.200 km, 1891–1916) in Asia. New ultra-long transnational railway megaprojects are currently being planned (Fig. 1). The Arab League Countries have recently conceived the Arab Network Railway, a 30.000 Km long, high-speed/high-capacity railway network across Middle East and North Africa. The Countries of the Gulf Co-operation Council are building the Gulf Railway, a 2.217 Km long railway network. The Trans-Asian Railway Southern Corridor, within the broader Iron Silk Road plan, is a 8.000 Km long transcontinental network between Europe and Southeast Asia. The corresponding investments are significant: for instance, the Middle

F. Ricciardelli and A. M. Avossa (Eds.): INVENTO 2018, LNCE 27, pp. 562–572, 2019.
https://doi.org/10.1007/978-3-030-12815-9_43

East Countries have allocated about USD 260 billion to build 40.000 Km of railway tracks up to 2030. Given the extension of their route, railway megaprojects cross different environments, and are subjected to related specific environmental actions. The sand blown by the wind is the most important specific key environmental factor that affects the design of railways crossing arid and desert regions.

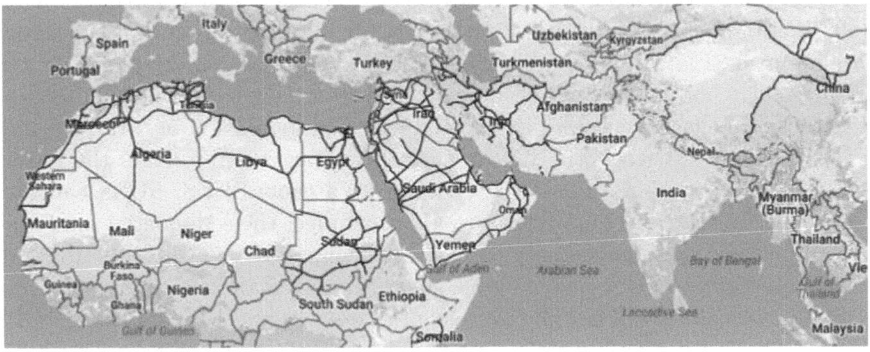

Fig. 1. Built and planned railway lines in arid and desert regions

Windblown sand induced effects on railways range from construction delays to service suspension and train derailment. They result in expensive track maintenance and sedimented sand removal. In order to mitigate windblown sand effects, a number of windblown Sand Mitigation Measures (SMMs) have been proposed in the last decade. SMMs generally aim at avoiding sand sedimentation on the railway track. In particular, the so-called Path-SMMs [1] promote sand sedimentation away from the railway. Path SMMs mostly translate into barriers located between the sand source and the railway

Despite pioneering ad-hoc studies for specific projects, a systematic and comprehensive problem setting is still missing. As a consequence, the design, analysis and performance assessment of SMMs remain in the realm of qualitative empiricism. In this study, windblown sand effects are categorized by the resulting level of performance of the railway, and by corresponding sand limit states. Then, we propose a probabilistic approach to windblown sand sedimented around railways because of the inborn variability of both wind and sand characteristics. The proposed method is applied to a case study in the Arabic peninsula.

2 Windblown Sand Limit States

Windblown Sand Limit States (SLSs) are defined as threshold performance levels, beyond which the infrastructure no longer fulfils its design criteria. SLSs are separated into Sand Ultimate Limit States (SULSs) and Sand Serviceability Limit States (SSLSs), analogously to safety formats widespread in Civil

Engineering [2]. Attaining SULS involves service interruption and/or passengers unsafe conditions. Attaining SSLS involves railway partial loss of capacity and/or passenger discomfort. SLS may be attained by one or more components of the railway infrastructure, e.g. civil works, track superstructure, signalling system, rolling stock. Some SLSs are described in the following. They are fully reviewed in [3].

SULSs are mainly induced by civil works, when the railway embankment or cutting are buried by sedimented sand. This schematically occurs under two different environmental conditions. First, when a field of migrating dunes overruns the alignment (Fig. 2a). Secondly, when the line crosses a sandy plane [4]: the overall railway body acts as an obstacle to the incoming wind flow; it induces the deceleration of the flow around it, and the sand sedimentation in turn (Fig. 2b). SULSs are attained when the full or partial coverage compromises the infrastructure safety or operation. Once civil works attains SULS, SULS can also be attained by the track superstructure. For instance, the sedimented sand may jam railroad switches, analogously to snow and ice in cold environments (Fig. 2c). Service interruption necessarily follows, since switches malfunction leads to train derailment or collision. The full coverage of current segments induces the SULS of rolling stock, i.e. running train derailment [5] (Fig. 2d).

Fig. 2. Sand Ultimate Limit States: full coverage by encroaching sand dune (a, explicit publishing permission from the owner of the photo Giles Wiggs), partial coverage in a sandy plain (b), jammed turnout (c), train derailment (d, reprinted from [6] with the permission from the editor).

Under SSLSs, windblown sand affects only a single component of the railway. However, this can reverberate on the overall railway system performances, i.e. its speed [7]. The sole SSLS attained by civil works is the partial obstruction of culverts by sedimented sand (Fig. 3e). Among track superstructure SSLS, the most common one is ballast contamination (Fig. 2a–c). It is generally quantified by referring to a permitted level of contamination, quantitatively expressed by the Percentage Void Contamination (PVC). In particular, an allowable limit of PVC is set to 30% for a concrete sleeper track with a 250 mm thick ballast layer [8]. Ballast contamination leads to a series of side effects, e.g. increasing of the stiffness and decreasing of the damping of ballast bed and rail support modulus [9], accumulation of permanent deformation, rail corrugation (Fig. 3d). Other examples of track superstructure SSLSs are corrosion and degradation of fastening systems (Fig. 3f), r.c. sleeper and rail (Fig. 3g) due to the salt content of the sand [10]. Furthermore the sand induces abrasive wear inside switches components (Fig. 3h), while the thin layer of sedimented sand on the wheel-rail contract surface leads to rail grinding [11] (Fig. 3i), and rolling stock wheel profiling (Fig. 3j).

Fig. 3. Sand Serviceability Limit States: different levels of ballast contamination (a–c), rail corrugation (d, reprinted from [12] with the permission from Elsevier), culverts obstruction (e), contamination of fasteners (f), rail head corrosion (g, courtesy of Astaldi), contamination of turnout components (h), downwind rail head (i, courtesy of Astaldi), wheel profiling (j, reprinted from [13] with the permission from Voestalpine).

The metric describing windblown sand effects and resistance can vary depending on the SLS. However, for most SLSs, the local windblown sand is defined as the volume of sedimented sand around an SMM or a railway component.

3 Modelling Windblown Sand Around Railways

In the following, we propose a probabilistic model to assess the windblown sand sedimented around railways by analogy with environmental actions in Civil Engineering [14], e.g. wind, windblown snow or wind-induced ice accretion.

3.1 Incoming Windblown Sand

Windblown sand results from the interaction of two physical subsystems, i.e. the wind and the sand. Among all transport mechanisms, saltation mainly contributes to the total transported sand mass [15]. In particular, windblown sand saltation occurs when the wind shear velocity u_* is higher than the aeolian threshold shear velocity u_{*t} (Fig. 4).

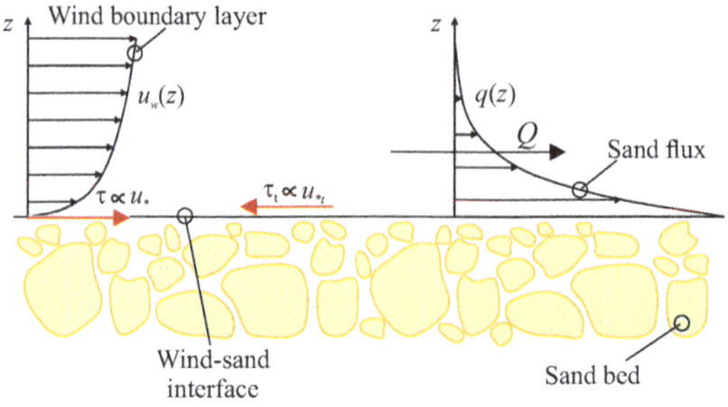

Fig. 4. Wind-sand interaction during saltation

The incoming windblown sand is defined as the amount of sand carried by the incoming wind undisturbed by any obstacle, in analogy to the incoming mean wind velocity in wind engineering practice. The resulting sand transport rate Q is defined as the integration of the sand flux q along the vertical direction. However, Q is usually estimated by means of semi-empirical models [15]. Such models are macroscopic constitutive laws that do not explicitly account for the variability of both wind and sand subsystems. The effects of the variability of the sand subsystem on u_{*t} have been investigated in a number of papers, reviewed in [16]. The evaluation of Q in a fully probabilistic approach is firstly introduced by [17], where both the variability of wind subsystem, i.e. speed and direction, and sand subsystem, i.e. aeolian threshold velocity, are taken into account. In this study, the following semi-empirical model is adopted [18]:

$$Q = 6.7\sqrt{\frac{d}{d_r}}\frac{\rho_a}{g}u_*^3\left(1-\frac{u_{*t}}{u_*}\right) \quad \text{if} \quad u_* > u_{*t} \tag{1}$$

$$Q = 0 \quad \text{if} \quad u_* \leq u_{*t} \tag{2}$$

where $d_r = 0.25$ mm is the reference grain diameter, ρ_a is the air density, g is the acceleration of gravity.

3.2 Local Windblown Sand Accumulation

The local sand accumulation is expressed by the time-cumulated sedimentation volume V. It is related to the incoming windblown sand by the sedimentation coefficient C_s. It depends on both the windblown sand yaw angle and the geometry of the obstacle, analogously to the aerodynamic coefficients in wind engineering. However, it is worth stressing a substantial difference: the obstacle is now described by a virtual geometry that varies over time and depends on the sedimentation volume. Given the lack of a closed form of C_s, its semi-empirical expression can be obtained by wind tunnel tests and/or computational simulations. Figure 5a shows the conceptual scheme of the modelling framework. The incoming windblown sand Q_{in} splits into the sedimentation rate $Q_s = C_s Q_{in}$

and the outgoing transport rate $Q_{out} = (1 - C_s)Q_{in}$. The local accumulated sand volume V results from the time-cumulation of Q_s.

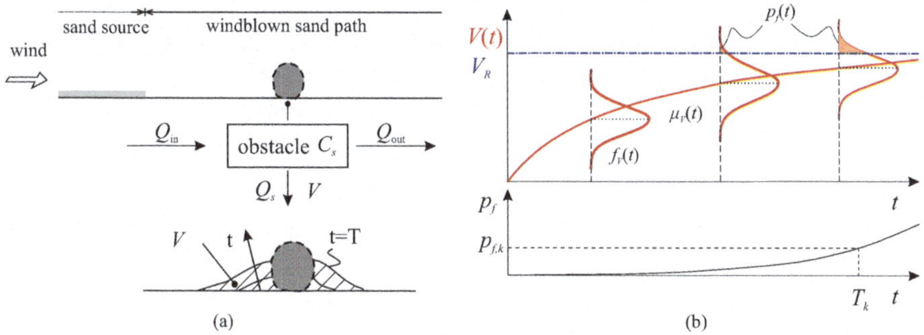

Fig. 5. Local windblown sand accumulation: conceptual scheme (a), time-variant sand volume and probability of failure (b)

By analogy with common wind engineering practice, the incoming windblown sand direction is assumed in the worst case scenario, i.e. perpendicular to the obstacle. The generic obstacle splits the plane into two sides. As a result, the Probability Density Function (PDF) of Q_{in} derives from the side-PDF of the incoming wind speed, i.e. the non-directional PDF of the incoming wind speed from the left or the right side. The PDFs of Q_s and Q_{out} can therefore be obtained as

$$f(Q_s) = C_s(V)f(Q_{in}) \tag{3}$$

$$f(Q_{out}) = [1 - C_s(V)]\, f(Q_{in}) \tag{4}$$

Hence, the PDF of the accumulated sand volume can be obtained as the mixture of convolutions:

$$g(V) = \sum_{n=1}^{\infty} (f_1 * f_i * \ldots * f_n)\, Q_s P\,[N = n] \tag{5}$$

with $f_i = f$ for $i = 1, \ldots, N$, where $f(Q_s)$ is the probability density function of Q_s and N is the number of occurrences in which the wind blows in that direction over the reference time t. In order to assess the variables above, a Monte Carlo approach based on bootstrapping techniques from a dataset of observed wind values is adopted.

For a generic obstacle, a resistant sand volume can be defined according to a given SLS, e.g. the sand volume that leads to the overturning of a vertical wall for lateral earth pressure. In general, such a volume can be described by its nominal value V_R whatever the nature of the obstacle is. Indeed, its degree of uncertainty is much lower than the one of the sand volume. Hence, the basic condition for a satisfactory state [19] is:

$$V(t) < V_R \tag{6}$$

Figure 5b shows the time-variant sand volume through its probability density function $f_V(t)$ and its mean value $\mu_V(t)$.

Given the nominal value V_R, the probability of failure $p_f(t)$ is equal to:

$$p_f(t) = P\left[V(t) > V_R\right] = 1 - F_V\left(V_R, t\right) \tag{7}$$

where F_V is the cumulative distribution of V. The characteristic failure time T_k can be defined as the time during which the condition (6) is violated only with a given probability $p_{f,k}$, i.e. $p_f(T_k) = p_{f,k}$. Assuming the monotonous function $V(t)$, p_f is also monotonous and its inverse function can be defined. Hence, the failure time can be derived as

$$T_k = p_f^{-1}\left(p_{f,k}\right) \tag{8}$$

4 Explorative Application

In the following an effort is accomplished to put the proposed model at work aimed at demonstrating the technical feasibility of the approach in an engineering perspective.

The selected site is in the Arabian peninsula, along the North-South railway line. The railway alignment around the site develops along the West-East direction: it follows that the incoming sand transport rate Q_{in} attacks the railway from the North side and South side. The mean sand grain diameter is equal to $d = 0.35$ mm. The aerodynamic roughness z_0 is equal to $z_0 = 4e{-}3$ m. The wind velocity in situ measurements refer to 6 years from Jan 2007 to Dec 2012. The 10-min average wind velocity U_{10} is recorded with a sampling yaw angle $\Delta\theta = 10°$

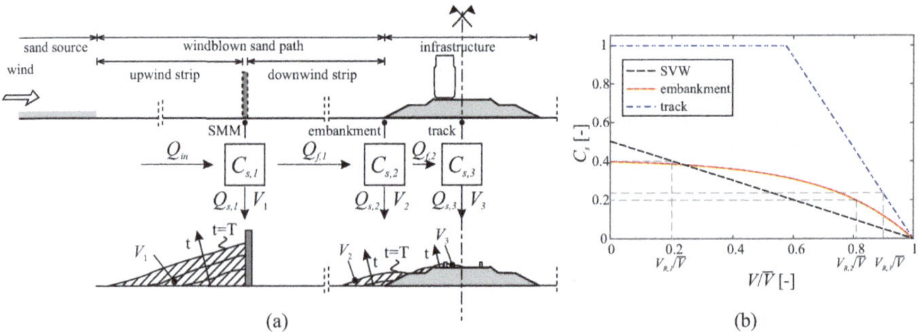

Fig. 6. Scheme of the setup geometry and related state variables (a), sedimentation coefficients versus filling ratios (b).

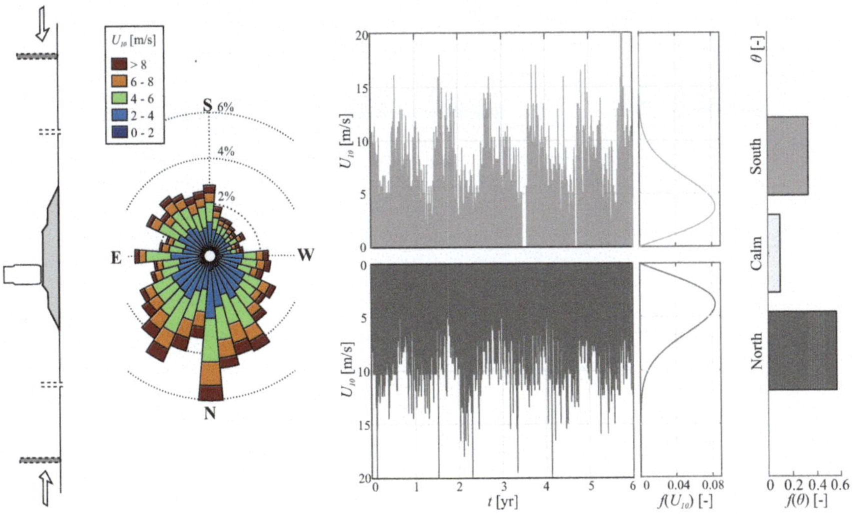

Fig. 7. Incoming wind statistics: setup geometry (a), wind rose (b), wind speed time histories (c), non-directional wind speed probability density functions (d), wind direction discrete probability density function (e).

and a sampling time interval $\Delta t = 1$ h. Figure 6a shows the scheme of the setup geometry, and related state variables for a single side of the railway corridor. Such a scheme results directly from the general one in Fig. 5, by putting in series three obstacles, i.e. an SMM, the railway embankment, and the railway track. The same scheme holds for the opposite side of the railway. The height of the SMM and of the embankment are set equal to 4 m and 2.5 m, respectively. A double-track railway is considered, with a 0.25 m-width ballast bed. The considered SMM is a common Straight Vertical Wall (SVW) [1]. The sedimentation coefficients C_s as a function of the dimensionless sedimented sand volume are plotted in Fig. 6b. They are obtained from [20]. Due to the lack of data on the railway track, its C_s is conjectured constant up to the filling of ballast voids (PVC = 100%), and then linearly decreasing up to the maximum sedimented sand volume $\bar{V}$. The sand sedimented around the SMM and the embankment is planned to be removed when their sedimentation coefficients are lowered by 20% and 50%, respectively.

Both U_{10} and u_{*t} are random variables according to [17]. Monte Carlo simulation account for about 22e+6 realizations for each obstacle. The results discussed in the following refer to the attainment of SULS for full covering of the track under North winds only.

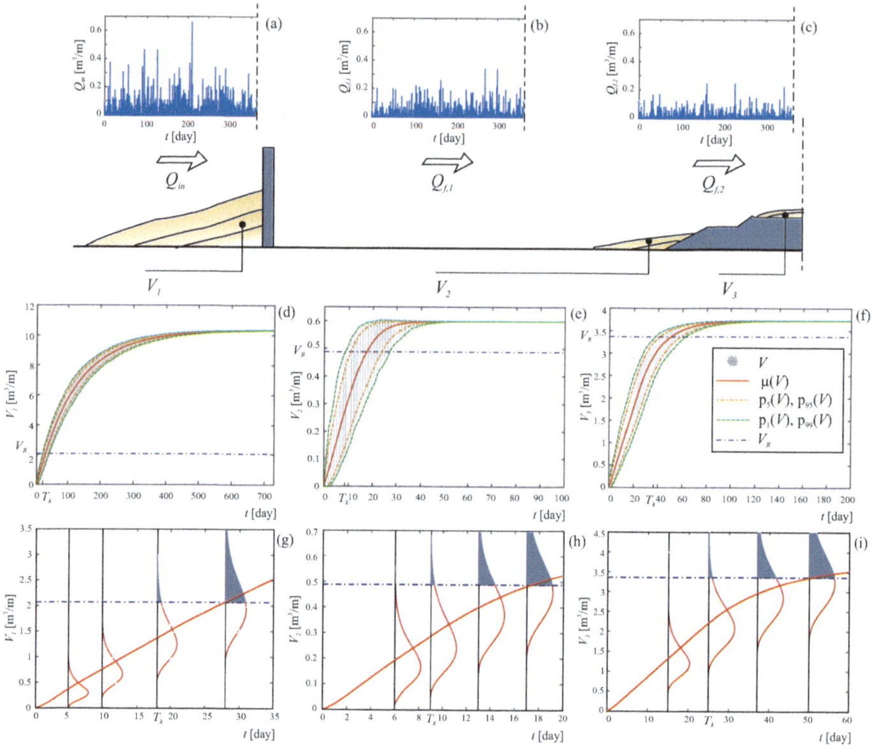

Fig. 8. Sedimented sand volume on SVW, embankment and track under North wind. Sand transport rate time histories (a–c) and time-variant accumulated volume (d–i).

Figure 7 shows the full wind rose (b), the time history of the non-directional wind speed incoming from both North and South railway sides (c), and the resulting statistics of both the wind speed U_{10} (d) and direction θ (e). $f(U_{10})$ is classically Weibull distributed. Figure 8 gives an example of the framework at work, with regard to SVW and incoming wind from North side. For each obstacle: (a–c) plot 1-year long time histories of the incoming and filtered sand transport rate; (d–f) collect the realizations of the time-varying sedimented volume and related statistics, i.e. mean values μ and i-th percentiles p_{th}; (g–i) provide a close-up view of the PDF of the sedimented volume, the nominal sand resistance, the time varying resulting probability of failure, and the characteristic sand removal and failure times T_k corresponding to $p_{f,k} = 1\%$. The sand transport rate is progressively reduced by in series obstacle filtering. The sedimented sand volume statistics for all obstacles share a horizontal asymptote at $\bar{V}_i$, while their trends depend on the sedimentation coefficient of each obstacle. The design value of the sand volume V_d corresponding to a target failure probability, e.g. $P_F \approx 10^{-6}$ for SULS, can be evaluated by the same proposed fully probabilistic approach, provided a large enough number of realizations are obtained. The corresponding sand removal and failure design times derives from Eqs. 7 and 8, i.e. $T_d = p_f^{-1}(P_F)$. The SULS partial safety factor γ can be derived a posteriori as $\gamma = V_d/V_k$ in a semi-probabilistic perspective.

5 Conclusions

We proposed the innovative probabilistic evaluation of the windblown sand accumulated volume on railways. The proposed framework allows evaluating sand removal and failure design times, and associated partial safety factors. The resulting safety format paves the way to the rationale design of sand mitigation measures and sand removal planning from both the sand mitigation measures and railway tracks.

Acknowledgements. The study has been developed in the framework of the Windblown Sand Modeling and Mitigation (WSMM) joint research, development and consulting group established between Politecnico di Torino and Optiflow Company. The authors wish to thank the other members of the WSMM group for the helpful discussions about the topics of the paper.

References

1. Bruno L, Fransos D, Lo Giudice A (2018) Solid barriers for windblown sand mitigation: aerodynamic behavior and conceptual design guidelines. J Wind Eng Ind Aerodyn 173:79–90
2. EN 1990 (2002) Eurocode - basis of structural design
3. Bruno L, Horvat M, Raffaele L (2018) Windblown sand along railway infrastructures: a review of challenges and mitigation measures. J Wind Eng Ind Aerodyn 177:340–365
4. Cheng J-j, Jiang F-q, Xue C-x, Xin G-w, Li K-c, Yang Y-h (2015) Characteristics of the disastrous wind-sand environment along railways in the Gobi area of Xinjiang, China. Atm Env 102:344–354
5. Nathawat V, Sharda A (2005) Challenges of track maintenance in desert area - problems and remedies. Permanent Way Bull 32:1–8
6. Davel Wallis M (2014) Freight Train Derails. Namib Times
7. Zakeri JA, Forghani M (2012) Railway route design in desert areas. Am J Env Eng 2:13–18
8. Indraratna B, Su L, Rujikiatkamjorn C (2011) A new parameter for classification and evaluation of railway ballast fouling. Can Geotech J 48:322–326
9. Ionescu D, Fedele D, Trounce M, Petrolito J (2016) Deformation and degradation characteristics of sand-contaminated railway ballast. In: Pombo J (ed) Proceedings of the third international conference on railway technology: research, development and maintenance. Civil-Comp Press, Stirlingshire
10. Carrascal I, Casado J, Diego S, Polanco J (2016) Dynamic behaviour of high-speed rail fastenings in the presence of desert sand. Constr Build Mat 117:220–228
11. Faccoli M, Petrogalli C, Lancini M, Ghidini A, Mazzù A (2018) Effect of desert sand on wear and rolling contact fatigue behaviour of various railway wheel steels. Wear 396–397:146–161

12. Tyfour WR (2008) Predicting the effect of grinding corrugated rail surface on the wear behavior of pearlitic rail steel. Tribol Lett 29:229–234
13. Kóllmann J (2013) Railway operations under harsh environmental conditions sand, dust & humidity problems and technical solutions/mitigation measures. In: AHK workshop be a partner of Qatar Rail, Berlin
14. EN 1991 (2003) Eurocode 1: actions on structures
15. Kok JF, Parteli EJR, Michaels TI, Karam DB (2012) The physics of wind-blown sand and dust. Rep Prog Phys 75:106901
16. Raffaele L, Bruno L, Pellerey F, Preziosi L (2016) Windblown sand saltation: a statistical approach to fluid threshold shear velocity. Aeolian Res 23:79–91
17. Raffaele L, Bruno L, Fransos D, Pellerey F (2017) Incoming windblown sand drift to civil infrastructures: a probabilistic evaluation. J Wind Eng Ind Aerodyn 166:37–47
18. Lettau K, Lettau H (1987) Experimental and micro-meteorological field studies of dune migration. In: Exploring the world's driest climate (IES Report), vol 101, pp 110–147
19. Melcher ER, Beck AT (2018) Structural reliability analysis and prediction. Wiley, Chichester
20. Hotta S, Horikawa K (1990) Function of sand fence placed in front of embankment. In: Coastal engineering. American Society of Civil Engineers, New York

Wind-Induced Fatigue Verification Standard Methods

M. P. Repetto and M. Damele(✉)

Department of Civil, Chemical and Environmental Engineering (DICCA), University of Genoa, Via Montallegro 1, 16145 Genoa, Italy
repetto@dicca.unige.it, michela.damele@edu.unige.it

Abstract. Wind-induced fatigue is a crucial topic in the design of slender structures. Based on a closed-form solution proposed by one of the authors, this paper reports an engineering approach developed to evaluate the buffeting-induced fatigue life of structures and structural elements. Two classes of formulas, referred to as *detailed* and *simplified calculations*, are shown. The *detailed calculation* provides refined approximations of the reference target solution, while the *simplified calculation* provides easy solutions on the safe side. The novelty of this paper is that it completes the formulation with analytical assessment of input parameters, fully coherent with standards format. Furthermore, the method has been applied to a turbulence-sensitive structural type of sign support structures. Large-scale inspections show that many of these structures suffer for fatigue damages. The comparison between the inspection outcomes and the proposed method results shows a very good agreement, while the European standard assessment provides unreliable results of fatigue life of this kind of structures.

Keywords: Design codes · Fatigue damage · Highway signal support · Stochastic analysis · Variable amplitude fatigue

1 Introduction

Fatigue of metallic materials is a critical, local and progressive phenomenon of damage accumulation which can lead to structural failure under cyclic actions of amplitude lower than design loads, without appreciable plastic deformations. Thus, fatigue may determine weak structural failure without exceeding ultimate strength. Steel elements often experience this process in the welded or bolded joints.

Assessment and design procedures involves the knowledge of resistance and actions. Considering the conventional nominal stress approach, the fatigue strength of a structural component is obtained by S-N experimental curves provided by codes. Fatigue international standards and recommendations give information on structural fatigue resistance for steel structures and for aluminium ones (Eurocode 3, 2005; Eurocode 9, 1998; IIW Recommendations 2016).

In order to identify the number of cycles that stresses the structures, cycle counting methods have been defined, by which a cycle histogram can be obtained from the stress time-history at the detail. The total cumulative damage is expressed by means of a

F. Ricciardelli and A. M. Avossa (Eds.): INVENTO 2018, LNCE 27, pp. 573–586, 2019.
https://doi.org/10.1007/978-3-030-12815-9_44

suitable cumulative damage law. Among all the proposed procedures, the linear accumulation law is still adopted in many standards. The critical total damage value, corresponding to failure, is conventionally defined as equal to one. The fatigue life is defined as the time in which total damage reaches the unit.

Slender structures exposed to wind may experience large vibrations and repeated stress cycles, so that they can be strongly affected by fatigue. This phenomenon may occur at moderate and frequent wind velocities and can be related to the vortex shedding phenomenon or to the dynamic response of structure to turbulence actions. Several collapses due to wind loading are related to fatigue (Holmes 2002; Repetto and Solari 2010). Typical kinds of structures sensitive to wind induced fatigue are traffic signal structures, poles and towers, that are strongly affected by dynamic response due to turbulence action; chimneys or cylindrical part of structures that suffer for vortex shedding induced-vibrations; other complex structures such as wind turbines and cranes, which can be damaged by the non-linear combination of wind and other variable loading fatigue. Despite a wide research literature addressed to this problem, standards and codes related to wind-induced fatigue are still fragmentary. Eurocode 1 (2005) provides an expression for cycle histogram in alongwind fatigue assessment based on very simplified assumptions (Kemper and Feldmann 2011). This code also provides reliable methods for evaluating the amplitude and number of load cycles due to vortex shedding. Other international standards are addressed to special kinds of structures, such as chimneys (CICIND 1999), poles (AASHTO, 2018) and wind turbine (IEA 1990), adopting different approaches.

A wide research project aimed at formulating and calibrating a general procedure for determining wind-induced fatigue has carried out between 2001 and 2012 at University of Genova (Repetto and Solari 2001, 2002, 2004; Repetto 2006; Repetto and Solari 2006, 2009, 2010, 2012). A refined closed-form solution of the alongwind-induced fatigue has been obtained, from which a simplified procedure suitable for engineering evaluation and code provisions has been derived (Repetto and Solari 2012).

This paper further discusses the input parameters of this procedure, defining their expressions in complete accordance with Eurocode standards format for wind induced Ultimate Limit State (ULS) analysis. The proposed method is then applied to a particular type of turbulence-sensitive structure, namely traffic signals support structures, for which a huge inspection campaign has been carried out, reporting fatigue damage in welded and bolted joints due to buffeting-induced loads. The proposed method and the Eurocode 1 procedure have been applied to one critical inspected structure, comparing the fatigue evaluations with the in-situ inspection report. The comparison with the proposed method shows a very satisfactory agreement, on the contrary, the Eurocode 1 appears unreliable in the present case. Analyses also show the effect of uncertainties on the local climate parameters, discussing the great variability of the results on changing the site of the installation. The perspectives of the research aims at furnishing a general and complete procedure at standard level for wind-induced fatigue assessment.

2 Wind-Induced Fatigue Life and Damage Evaluation

The wind induced fatigue assessment procedure already presented in (Repetto and Solari 2012) is considered. This approach takes into account the alongwind turbulence response and it considers, for safety's sake, that the incoming wind has always the worst direction for the structural response (unidirectional analysis) (Repetto and Solari 2004).

The proposed formulation considers the joint effect of the wind-induced variable loadings, of the wind-induced static actions and other permanent loads on the structure; it does not consider the joint effect due to other variable loadings on the structure, such as mobile loads, vehicular traffic, waves. As the fatigue damage phenomenon is strongly non-linear, superposition effect does not apply; therefore, when different types of variable actions affect the structure, the current procedure can't be used for structural verification.

The developed formulation gives two different fatigue life assessment levels (Fig. 1): a *simplified calculation*, that provides preventive values, and a *detailed calculation*, that provides reliable values.

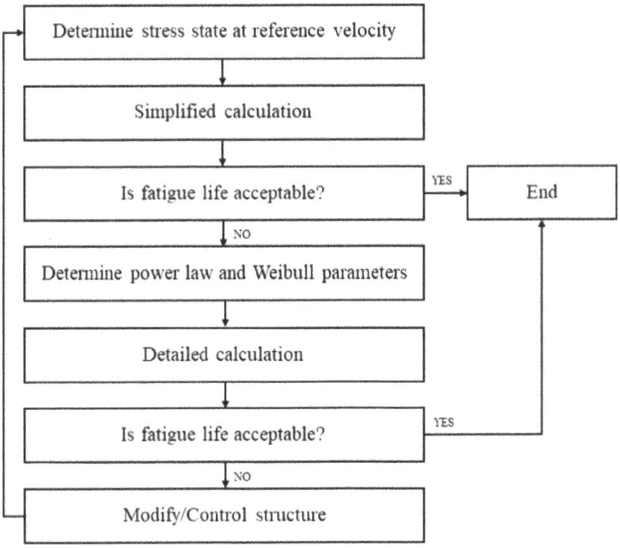

Fig. 1. Standard procedure for wind-induced fatigue life assessment

The *simplified calculation* is based on the structural response evaluated at the reference mean velocity, $\bar{u}_{ref}$, adopted for ULS analysis. Coherently with the Eurocode 1 (2005), $\bar{u}_{ref}$ is the wind velocity with 50 years return period in the site, at height z_e, where z_e is a fixed reference height for evaluating external wind action on a structure in the wind flow. The resulting fatigue life obtained with the *simplified method* is underestimated; consequently, if the obtained fatigue life is higher than the nominal life of the structure, the wind-induced fatigue verification is met, otherwise the *detailed calculation* should be applied.

The *detailed calculation* is based on the structural response evaluated at the reference mean velocity, $\bar{u}_{ref}$, and at the fatigue velocity, $\bar{u}_{fat}$, in the area where the structure is located. The fatigue velocity $\bar{u}_{fat}$ is the wind velocity corresponding to the maximum wind-induced fatigue damage conditions. The resulting fatigue life obtained with the *detailed method* is reliable; consequently, if the obtained fatigue life is higher than the nominal life of the structure, the wind-induced fatigue verification is met, otherwise the structure isn't safe and structural design has to be modified. Engineers may choose to carry out the *detailed calculation* directly.

Fatigue life due to alongwind turbulence-induced actions, expressed in years, is provided by:

$$T_F = \frac{1}{\bar{D}(1)} \tag{1}$$

where the annual mean damage $\bar{D}(1)$ is provided by (Repetto and Solari 2009, 2012):

$$\bar{D}(1) = \bar{D}_0(1) C_{BM} C_M C_{SN} \tag{2}$$

where $\bar{D}_0(1)$ is referred to as the 0 level solution, C_{BM} is the bi-modal factor, C_M is the mean stress factor, C_{SN} is the fatigue curve factor.

The 0 level solution, $\bar{D}_0(1)$, is inspired by a similar method proposed by Holmes (2002), and adopts three classic simplifications traditionally used in the technical literature: (1) the stress process is narrow band; (2) the stress process is zero mean; (3) the fatigue curve is a straight line on a log (stress range) – log (number of cycles that cause the collapse) diagram, with constant slope equal to 3.

The bi-modal factor, C_{BM}, corrects the 0 level solution, removing the first simplifying hypothesis and taking into account the actual bi-modal spectrum of the dynamic response process at turbulent wind actions, with quasi-static and resonant contents. It satisfies the condition $C_{BM} \leq 1$.

The mean stress factor, C_M, corrects the 0 level solution, removing the second simplifying hypothesis and taking into account the non-null value of the mean stress due to static wind actions and to other static loadings on the structure. It satisfies the condition $C_M \geq 1$.

The fatigue curve factor, C_{SN}, corrects the 0 level solution, removing the third simplifying hypothesis and taking into account the actual tri-linear S-N fatigue curve in a bi-logarithmic diagram for steel, in particular the cut-off limit of this curves. It satisfies the condition $C_{SN} \leq 1$.

For the particular case of damage due to normal stresses for steel, the 0 level solution and corrective factors are expressed by:

$$\bar{D}_0(1) = 15.77 \frac{\sigma_{sD}^3}{\Delta\sigma_C^3} \nu_{sD} \cdot A_0 \tag{3}$$

$$C_{BM} = \frac{\nu_{uD}}{\nu_{sD}} + \frac{n_D}{\nu_{sD}} \lambda_{RD}^{3/2} \cdot A_{BM} \tag{4}$$

$$C_M = \left[\frac{f_{tk}}{(f_{tk} - s_P)}\right]^3 \left[1 + \frac{\bar{s}_D}{(f_{tk} - s_P)} \cdot A_M\right] \tag{5}$$

$$C_{SN} = 1 - \exp\left\{-A_{SN}\left(\frac{\sigma_{sD}}{\Delta\sigma_C}\right)^{B_{SN}}\right\} \tag{6}$$

where $\Delta\sigma_C$ is the reference value of fatigue normal resistances of the analysed structural detail for 2 million cycles (Eurocode 3, 2005); $\bar{s}_D$, σ_{sD}, ν_{sD} and λ_{RD} are the mean value, standard deviation, expected frequency and normalized variance of the resonant part of the alongwind induced stress in the critical section at the reference velocity, respectively; n_D is the first mode of vibration frequency in the wind direction; ν_{uD} is the expected frequency of the alongwind component of turbulence; f_{tk} is the steel failure characteristic stress; s_p is the permanent loadings-induced stress in the examined section (calculated at serviceability).

The input parameters have been obtained coherently with ULS wind response evaluation proposed by Eurocodes. In particular, $\bar{s}_D$ is obtained applying the static aerodynamic force $\bar{f}_D(z)$ induced by reference mean velocity at the structure; σ_{sD} is given by:

$$\sigma_{sD} = \gamma_F \cdot \bar{s}_D \cdot 2I_u(z_e) \cdot \sqrt{B_D^2 + R_D^2} \tag{7}$$

where γ_F is the safety factor for fatigue analysis (Eurocode 3, 2005); $I_u(z_e)$ is the longitudinal turbulence intensity evaluated at reference height z_e; B_D and R_D are quasi-static and resonant response factors, respectively, the first one considers the non-perfect correlation of the pressure acting on the structure, the second one considers the resonance between the turbulent excitement and the first mode of vibration of the structure (Eurocode 1, 2005); ν_{sD} is given by:

$$\nu_{sD} = n_D \cdot \sqrt{\frac{R_D^2}{B_D^2 + R_D^2}} \geq 0.08\,\text{Hz} \tag{8}$$

λ_{RD} is given by:

$$\lambda_{RD} = \frac{R_D^2}{B_D^2 + R_D^2} \tag{9}$$

and ν_{uD} is given by:

$$\nu_{uD} = 0.179 \frac{\bar{u}_{ref}}{L_u(z_e)} \left[\frac{L_u(z_e)}{\tau \cdot \bar{u}_{ref}}\right]^{0.72} \tag{10}$$

where $\bar{u}_{ref}$ is the reference mean wind velocity; $L_u(z_e)$ is the longitudinal turbulent length scale evaluated at reference height z_e and τ is the averaging time for the peak wind velocity.

A_0, A_{BM}, A_M, A_{SN} and B_{SN} are parameters that depend on structural response at wind velocities lower than or equal to reference velocity and on the probability distribution of the wind velocity in situ. These parameters may be evaluated by *simplified* or *detailed calculations*.

In the *detailed calculation*, these parameters are provided by the following expressions:

$$A_0 = 24\pi(0.2k - 0.12)^{3\alpha_{\sigma D}} \left(\frac{3\alpha_{\sigma D}}{k} + 1\right)^{\left(\frac{3\alpha_{\sigma D}}{k} + 0.5\right)} \left(\frac{1}{e}\right)^{\left(\frac{3\alpha_{\sigma D}}{k} + 1\right)} \tag{11}$$

$$A_{BM} = (0.2k - 0.12)^{\frac{3\alpha_{\lambda D}}{2}} \left(\frac{3\alpha_{\sigma D}}{k} + 1\right)^{\left(\frac{3\alpha_{\lambda D}}{2k}\right)} \tag{12}$$

$$A_M = 3(0.2k - 0.12)^2 \left(\frac{3\alpha_{\sigma D}}{k} + 1\right)^{\left(\frac{2}{k}\right)} \tag{13}$$

$$A_{SN} = 342.25(0.2k - 0.12)^{2k} \tag{14}$$

$$B_{SN} = \frac{2k}{\alpha_{\sigma D}} \tag{15}$$

where k is the shape parameter of the Weibull probability distribution of the current values of the wind velocity in situ. This parameter has been studied by Pagnini and Solari (2016) for Italian territory, but knowledge has to be enhanced in all the countries. $\alpha_{\sigma D}$ is the exponent of the power law expressing the fluctuating stress variation, σ_{sD}, when mean wind velocity changes (Repetto and Solari 2012):

$$\alpha_{\sigma D} = 2 - 0.721 \cdot \ln\left[\frac{B_D^2 + R_{Dfat}^2}{B_D^2 + R_D^2}\right] \tag{16}$$

$\alpha_{\lambda D}$ is the exponent of the power law expressing the normalized variance variation of the stress resonant part, λ_{RD}, when wind mean velocity changes (Repetto and Solari 2012):

$$\alpha_{\lambda D} = -1.443 \cdot \ln\left[\frac{B_D^2 + R_D^2}{B_D^2 + R_{Dfat}^2} \cdot \frac{R_{Dfat}^2}{R_D^2}\right] \tag{17}$$

where B_D, R_D are response factors evaluated at reference wind velocity, $\bar{u}_{ref}$, and R_{Dfat} is the resonant response factor evaluated at the fatigue velocity, $\bar{u}_{fat}$. In the Italian territory, the fatigue velocity has been evaluated (Repetto and Solari 2012) as:

$$\bar{u}_{fat}(z_e) = 0.5 \cdot \bar{u}_{ref}(z_e) \tag{18}$$

In the *simplified calculation*, the parameters A_0, A_{BM}, A_M, A_{SN} and B_{SN} are provided by assuming the input parameters in Eqs. (11)–(15) on the safe site. In the case of alongwind induced fatigue evaluation of steel structures, $A_0 = 0.085$; $A_{BM} = 0.9$; $A_M = 1.4$; $A_{SN} = 2.2$; $B_{SN} = 0.8$ (Repetto and Solari 2012).

3 Sign Support Structures

Traffic-signal-support structures with cantilevered mast arms are known to exhibit large-amplitude vibrations under wind actions. Many damages have been reported in literature (Fig. 2) and many studies have been addressed to such kind of structures (Wieghaus *et al.* 2017). Based on full-scale measurement of a traffic-signal-support structure, Zuo and Letchford (2010) demonstrated that two types of excitation mechanisms, vortex-shedding and buffeting actions, lead to main vibration effects.

Fig. 2. An example of highway sign support structure collapsed in Italy in 2011

A wide inspection campaigns on sign support steel structures have been recently carried out by IIS (Italian Institute of Welding), monitoring the presence and the characteristics of damage. On the whole, 67 structures have been inspected, with different structural schemes, ages and locations. Despite all the structures have been verified at ULS, inspections indicated presence of fatigue damage in many welded or bolted joints; the risk has been classified as low for 0% of structures, intermediate for 48%, high for 28% and very high for 24%. Many of them have been repaired, some have been definitely removed.

One of the "flag-frame" structures has been evaluated adopting the procedure presented in Sect. 2. The cantilever structure is composed by a vertical element linked to a horizontal "arm" at 7.49 m height. Structural elements are made of steel S355 with square cross-section; the section dimension is variable with the height in the vertical structural component and it is constant in the horizontal one. The connection of the post to the ground is typically end plate jointed; the connection of the post to the horizontal "arm" is a welded joint; the horizontal "arm" is composed by three trunks, connected by end plate joints. The structure supports two traffic signals; the central one is 2 m high and 5 m wide and the extreme one is 1.8 m high and 3 m wide. The installation of this structure is quite recent, since it was less than 10 years ago. Figure 3 reports a schematic representation of the structure, putting in evidence four joints.

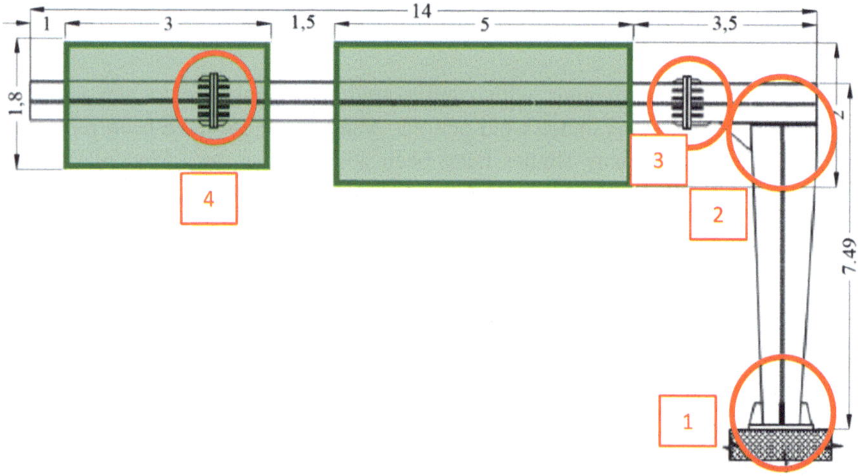

Fig. 3. Schematic representation of the considered "flag-frame" sign support structure

Despite ULS verifications were met, inspections have found out critical damage in some joints, actually. In particular, the most critical detail is Joints 2 (Fig. 4).

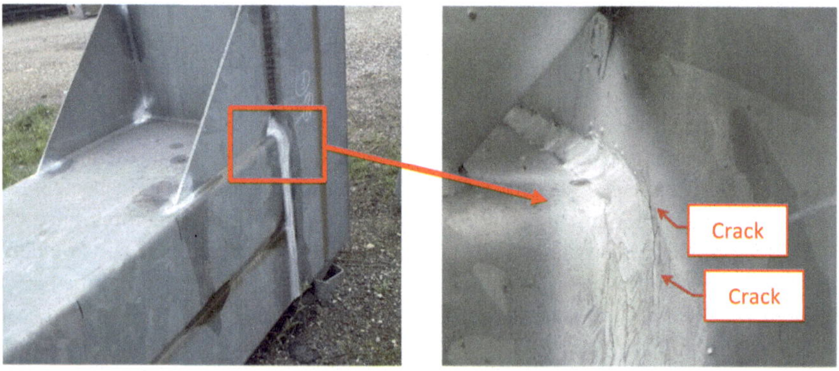

Fig. 4. Damages inspections on Joints 2

Adopting the procedure previously explained, the stress state and the fatigue damage associated with the wind loading conditions is examined in this critical joint.

A modal analysis of the structure has been carried out and alongwind dynamic response has been evaluated in order to obtain nominal stresses in the critical section. Table 1 shows the characteristics of the first five vibration modes.

Table 1. First five vibration modes, periods, natural frequencies and brief description

Modes	T [s]	n [s]	Description
First	0.571	1.752	Longitudinal (horizontal element)
Second	0.531	1.869	Transversal (vertical and horizontal element)
Third	0.170	5.889	Transversal (vertical and horizontal element)
Fourth	0.115	8.681	Longitudinal (horizontal element)
Fifth	0.059	16.979	Transversal (horizontal element)

The probability of occurrence of the loading conditions is associated only with the mean wind velocity with 50 years of return period, $\bar{u}_{ref}$. The site is located in conventionally called zone 3 (Italian standard CNR, 2018), therefore $\bar{u}_{ref} = 27$ m/s; $k = 1.20$ (Pagnini and Solari 2016); the fatigue velocity, $\bar{u}_{fat}$, is assumed equal to 13.5 m/s, according to Eq. (18). Wind loading and dynamic analyses are based on the standard prescriptions, obtaining values reported in Table 2. The structure is verified according to ULS, vortex shedding and galloping verifications.

Table 2. Alongwind wind loading and structural response parameters

Parameters	Units	Joint 2
n_D	Hz	1.752
f_{tk}	N/mm^2	510
$\bar{u}_{ref}$	m/s	27
$\bar{u}_{fat}$	m/s	13.5
k	-	1.20
γ_F	-	1.25
ν_{sD}	Hz	1.377
ν_{uD}	Hz	0.068
λ_{RD}	-	0.617
$\alpha_{\sigma D}$	-	2.395
$\alpha_{\lambda D}$	-	0.868
B_D^2	-	0.750
R_D^2	-	1.209
R_{Dfat}^2	-	0.383
s_P	N/mm^2	33.27
$\bar{s}_D$	N/mm^2	44.62
σ_{sD}	N/mm^2	31.93

Stress status in Joint 2 is characterized by shear stress, due to the wind-induced torsional effect on the structure. Due to the limited information about fatigue resistance for shear stress in this kind of joints, in the present example the ideal normal stress in the joint has been considered (reported in Table 2).

For this critical joint, the category detail is selected, considering Eurocode 3 (Fig. 5). The choice of the most suitable S-N curve is a crucial point in the procedure.

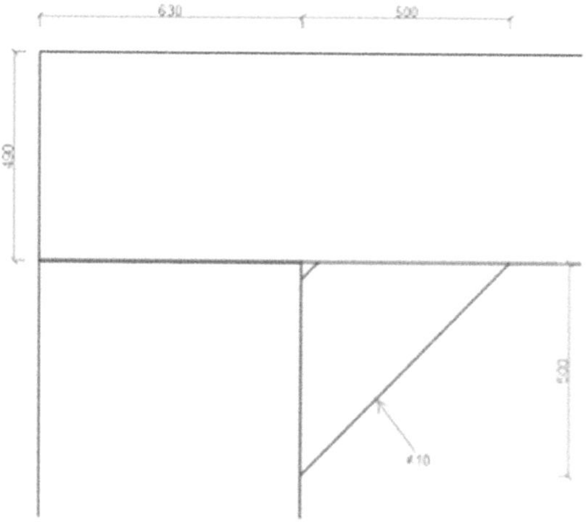

Fig. 5. Joint 2 detail

The geometric detail of the actual joint is not provided in Eurocode classification, therefore detail category 36 is assumed on the safe side.

The *simplified calculation* of the proposed procedure has been firstly applied to Joint 2, showing a resulting preliminary fatigue life value lower than nominal life of the structure, assumed as 20 years. Thus, the *detailed calculation* has been applied (Table 3).

Table 3. Resulting 0 level solution, correction factors, annual mean damage and fatigue life for structural Joint 2 obtained with *simplified* and *detailed calculation*

	Simplified calculation	Detailed calculation
$\bar{D}_0(1)$	0.296	0.150
C_{BM}	0.790	0.371
C_M	1.889	1.351
C_{SN}	1.000	0.904
$\bar{D}(1)$	0.441	0.068
T_F	2 years	15 years

The fatigue life of Joints 2 obtained with *simplified calculation* is hugely lower than the 20 years nominal life of the structure, therefore *detailed calculation* is required. A reliable fatigue life of 15 years is obtained, a value that is still lower than design life, showing that the structure is not verified according to the proposed procedure, as expected.

The 0 level solution of damage is strongly sensitive to stress value. Bi modal factor reduces hugely this damage estimation, mean stress factor ups this value by around 30%; fatigue curve factor reduces slightly damage estimate because stress ranges are high enough to be far from cut-off limits in S-N curves. Due to the strong non-linearity of the fatigue damage accumulation, the propagation of uncertainties is significant. Despite this, the result obtained with the proposed procedure is well in line with the inspection, which had pointed out large damages in this joint.

This is not equally true for the method proposed in actual European standard codes. Applying the Eurocode 1 (part 1–4) calculation for alongwind-induced fatigue evaluation it is possible to figure out cycle counting N_g to reach or exceed an effect value ΔS due to wind loading in 50 years of return period. ΔS is expressed as a rate of S_k, which is assumed equal to the maximum normal stress range in the considered structural section. Figure 6 shows $\Delta S/S_k$ respect to N_g in percentage terms for Joint 2.

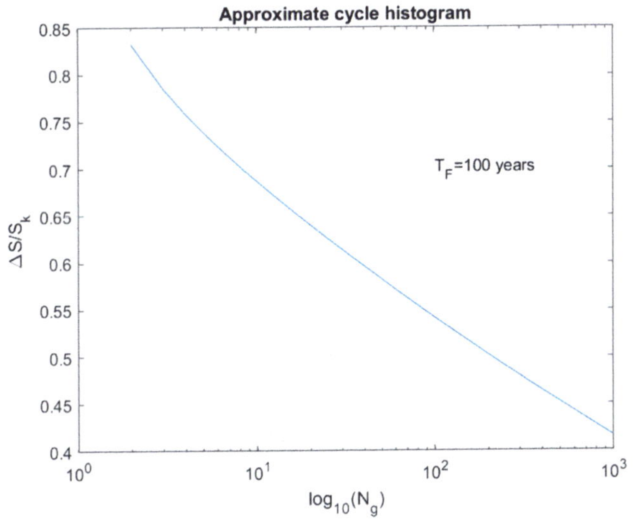

Fig. 6. Alongwind-induced fatigue life evaluated by Eurocode 1 (2005) procedure for Joint 2.

The equation provided by Eurocode allows to obtain N_g respect to ΔS variation. Knowing the detail category and its fatigue resistance S-N curve, it is also possible to know N respect to ΔS variation, which is the number of cycles that leads the structure to collapse. Total damage in 50 years is the summation, or the integral, of damage fractions $N_g(\Delta S)/N(\Delta S)$. Applying the standard procedure for the case in analysis, a fatigue life of T_F = 100 years is obtained. It is therefore clear that this counting cycle equation isn't reliable in the present case, since the obtained outcome is completely not coherent with inspection report and it is critically not on the safe side.

Finally, a comparison of the results obtained with different local climate conditions has been carried out, pretending to locate the sign support structure in every Italian zone defined in CNR code (Fig. 7a). As expected, fatigue life is heavily dependent on reference wind velocity and Weibull shape parameter k, varying from about 50 years to about 2 years of residual life (Table 4; Fig. 7b). Since this type of structure is widespread in the territory, considering a fatigue verification procedure is so much important, for design of new structures and for planning inspections campaign.

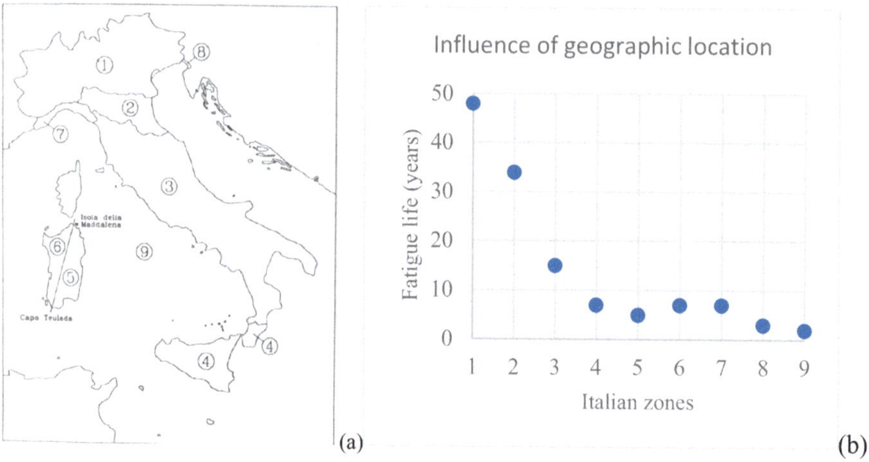

Fig. 7. Italian zones defined in CNR code (a) and fatigue life values of the structure obtained referred to different Italian geographical zones (b)

Table 4. Fatigue life values of the structure obtained referred to different Italian geographical zones

Zones	u_{ref} (m/s)	k	T_F (years)
1	25	1.00	48
2	25	1.15	34
3	27	1.20	15
4	28	1.35	7
5	28	1.50	5
6	28	1.35	7
7	28	1.35	7
8	30	1.50	3
9	31	1.15	2

4 Conclusions and Perspectives

In the present study, a simple analytical approach aimed to predict wind-induced fatigue life of structures sensitive to wind dynamic actions is presented. This procedure allows to obtain reliable results using a reasonably simplified analytical formulation.

The current paper provides the formulation of a *detailed calculation*, which may be reasonably adopted in engineering applications to evaluate reliable values of fatigue life of wind-induced fatigue sensitive structures, and the formulation of a *simplified calculation*, which may provide easy results on the safe side. The input parameters equations of this procedure have been completely defined in the present work, in accordance with standards format.

The second part of the paper is addressed to the traffic-sign support structures, which have been proved to be turbulence-sensitive structural types. A wide inspection campaign has been presented, showing that many inspected structures were affected by fatigue damages. The proposed method for fatigue assessment has been applied to one of the inspected structure. The comparison between the inspection and the outcome of the proposed method shows a very good agreement, while the European standard assessment provides unreliable and unsafe results of fatigue life of this kind of structures. It means that probabilistic approach can have the potentiality of catching the main features of the wind-induced fatigue phenomenon, giving reliable results with simplified procedures.

It is worth noting that the choice of the S-N curve greatly affects the results and constitute a crucial point in the analysis of such a simple structural type, thus extensive fatigue tests should be carried out in order to create a catalogue of S-N curves ad hoc for the details usually adopted by this structural typology. This is particularly critical for shear stress induced fatigue, for which the standard fatigue resistance curves are only related to circular section and furnish unrealistic resistance values in civil structural details.

Finally, a strong dependency on climatological input parameters of fatigue life outcomes has been pointed out, showing the importance of more accurate methods to define these parameters in different areas, in particular Weibull parameter k.

Future perspectives of this research are addressed to generalize the proposed procedure considering every type of S-N fatigue resistance curves, and also generalising the method for alongwind and crosswind dynamic response.

Acknowledgements. The authors would like to acknowledge IIS (Italian Institute of Welding) for fruitful advice and discussion.

References

AASHTO, American Association of State Highway and Transportation Officials (2018 Edition) Standard Specifications for Structural Supports for Highway Signs, Luminaires, and Traffic Signals

CICIND: International Committee on Industrial Chimneys (1999, Revision 1, Amendment A). Model Code for Steel Chimneys

CNR–DT 207, Advisory Committee on Technical Recommendations for Construction (2008, review 2017–2018) Guide for the assessment of wind actions and effects on structures

Eurocode 1–UNI EN 1991-1-4 (2005) Actions on structures - Part 1–4: General actions-Wind action

Eurocode 3–EN 1993-1-9 (2005) Design of steel structures–Part 1-9: Fatigue

Eurocode 9–ENV 1999-2 (1998) Design of aluminium structures – Part 2: Structures susceptible to fatigue

Holmes JD (2002) Fatigue under along-wind loading–closed-form solutions. Eng Struct 24:109–114

IEA, International Energy Agency (1990) Recommended practices for wind turbine testing and evaluation–Part 3–Fatigue

IIW, International Institute of Welding (2016 edition) Recommendations for fatigue design of welding joints and components, document XIII/XV revision

Kemper FH, Feldmann M (2011) Fatigue life prognosis for structural elements under stochastic wind loading based on spectral methods, Part I: Linear structures. In: Proceedings of the 8th International Conference on Structural Dynamics, EURODYN 2011, July 4–6, Leuven, Belgium

Letchford C, Cruzado H, Zuo, D (2008) Risk assessment model for wind-induced fatigue failure of cantilever traffic signal structures, Report No. 4586-4, Texas Department of Transportation (TxDOT), Austin, TX

Pagnini LC, Solari G (2016) Joint modeling of the parent population and extreme value distributions of the mean wind velocity. J Struct Eng (ASCE) 142(2):1–10

Repetto MP (2005) Cycles counting methods for bi-modal stationary Gaussian processes. Probab Eng Mech 20(3):229–238

Repetto MP, Solari G (2001) Dynamic alongwind fatigue of slender structures. Eng Struct 23 (12):1622–1633

Repetto MP, Solari G (2002) Dynamic crosswind fatigue of slender structures. Wind Struct 5 (6):527–542

Repetto MP, Solari G (2004) Directional wind-induced fatigue of slender vertical structures. J Struct Eng 130(7):1032–1040

Repetto MP, Solari G (2006) Bimodal alongwind fatigue of structures. J Struct Eng 132(6):899–908

Repetto MP, Solari G (2009) Closed form solution of the alongwind-induced fatigue damage to structures. Eng Struct 31(10):2414–2425

Repetto MP, Solari G (2010) Wind-induced fatigue collapse of real slender structures. Eng Struct 32(12):3888–3898

Repetto MP, Solari G (2012) Closed-form prediction of the alongwind-induced fatigue of structures. J Struct Eng (ASCE) 138:1149–1160

Wieghaus KT, Mander JB, Hurlebaus S (2017) Damage avoidance solution to mitigate wind-induced fatigue in steel traffic support structures. J Constr Steel Res 138:298–307

Zuo D, Letchford CW (2010) Wind-induced vibration of a traffic-signal-support structure with cantilevered tapered circular mast arm. Eng Struct 32:3171–3179

Experimental and Computational Analysis of Microscale Wind Conditions in the Port of Amsterdam

A. Ricci[1,2(✉)] and B. Blocken[1,3]

[1] Department of the Built Environment, Eindhoven University of Technology, P.O. Box 513, 5600 MB Eindhoven, The Netherlands
{a.ricci,b.j.e.blocken}@tue.nl

[2] Department of Civil, Chemical and Environmental Engineering (DICCA), University of Genoa, via Montallegro 1, 16145 Genoa, Italy
alessio.ricci@unige.it

[3] Department of Civil Engineering, KU Leuven, Kasteelpark Arenberg 40 - bus 2447, 3001 Louvain, Belgium
bert.blocken@bwk.kuleuven.be

Abstract. The present paper deals with on-site measurements and 3D steady RANS simulations performed on the *"IJmuiden vault"* which falls under the Port Authority of Amsterdam. The numerical results were validated, in terms of amplification factor and local wind direction, using on-site measurements carried out by four 2D ultrasonic anemometers for a period of nine months. To quantify the deviation between measured and simulated data, the metric *FAC1.3* was used. In that respect, 90% of simulated data (in terms of amplification factor) was found to be within 30% of deviation from the measured data. In terms of local wind direction, 86% of simulated data were found within ±30° and only the 6% of the whole database showed a large deviation equal and greater than ±60°. Finally, a software application was developed to convert the macroscale wind conditions to the local wind conditions near and in the newly built vault.

Keywords: Computational fluid dynamics (CFD) · On-site measurements · Microscale wind conditions · Urban environment · Validation

1 Introduction

The macroscale wind conditions by standard weather stations show considerable differences with the local microscale wind conditions in harbor basins (Blocken et al. 2015). This is often caused by the local-scale forcing effects (e.g. buildings) and the influence of terrain roughness on the local wind flow (Ricci et al. 2017a,b). Knowledge of microscale port wind conditions is important for maneuvering and mooring of ships and also for optimizing the harbor design (Janssen et al. 2017). In particular, the increasing ship sizes cause a larger ship inertia which renders the navigation through the harbor and especially through vaults by the tugboat pilots more difficult (Burlando et al. 2014). For this purpose, wind-tunnel (WT) testing and Computational Fluid

F. Ricciardelli and A. M. Avossa (Eds.): INVENTO 2018, LNCE 27, pp. 587–598, 2019.
https://doi.org/10.1007/978-3-030-12815-9_45

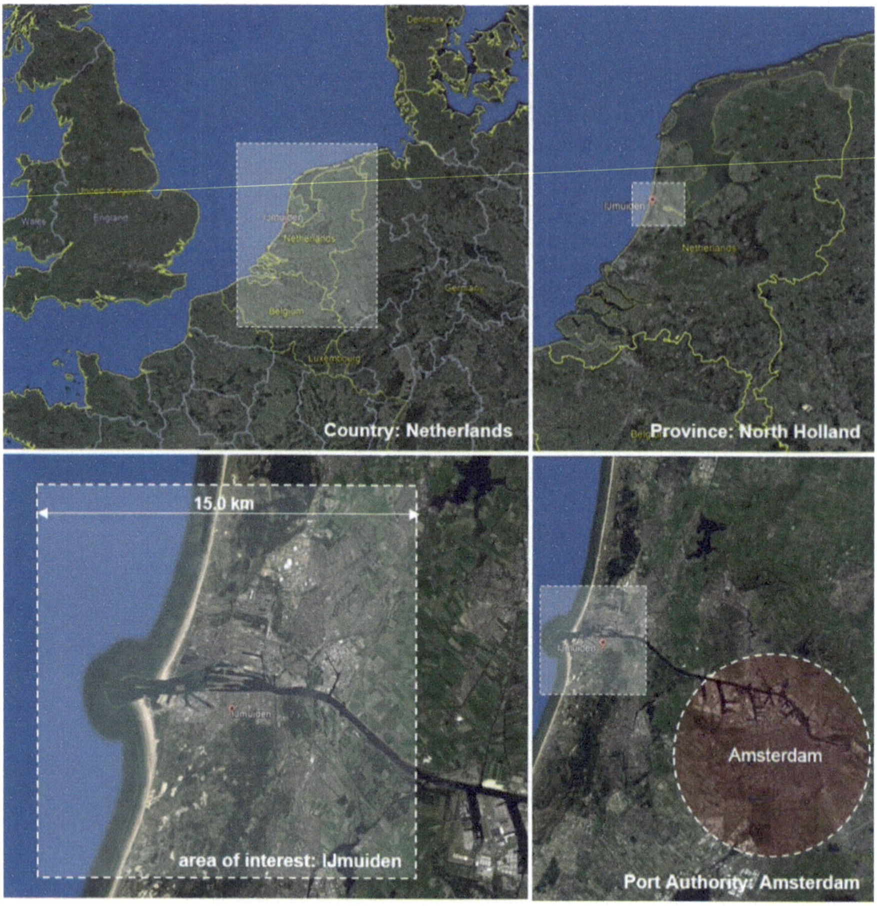

Fig. 1. Geographical localization of the area of interest (photo credit to Google Maps).

Dynamics (CFD) techniques are often employed to properly model wind flows, under neutral atmospheric conditions over complex urban environments (Murakami 1997; Stathopoulos 1997; Baker 2007; Solari 2007). However, both WT tests and CFD simulations need an accurate validation commonly carried out by on-site measurements (e.g. 2D and 3D anemometers and LiDAR) (Blocken 2014).

For this purpose, in order to develop an accurate transformation tool by which convert the macroscale wind conditions to the local wind conditions, on-site measurements (first stage) and CFD simulations (second stage) were performed on *"IJmuiden vault"* which falls under the Port Authority of Amsterdam, in North Holland (Netherlands). (Figure 1). In a third stage, not discussed in this paper, the results of this transformation were used in two different ways: first to provide realistic simulations to train tugboat pilots, second as input for a real-time software application that provides the local wind conditions based on standard weather station recordings to help the tugboat pilots to guide the ship more precisely through the vault.

In Sect. 2 the on-site measurements will be described. In Sect. 3 the computational settings will be described. In Sect. 4 the validation of CFD results will be discussed. Section 5 concludes the paper with discussion and conclusions.

2 On-Site Measurements

As first stage, on-site measurements of wind speed with 2D ultrasonic anemometers were performed at five different positions, termed *IJm1, IJm2, IJm3, IJm4* and *KNMI* from 5th of July to 5th of April 2018 with some interruption in between (Fig. 2). The stations, termed *IJm1, IJm2, IJm3* and *IJm4*, were installed in the vault at a height of about 10 m above the ground level by the Building and Physics research unit of the Eindhoven University of Technology. The measurements of these four stations were used to validate the CFD results performed in the second stage of this work. Conversely, the *KNMI* station, which belongs to Koninklijk Nederlands Meteorologisch Instituut (KNMI), were used in a third stage (not discussed in this paper) to convert the macroscale wind conditions obtained by the numerical simulations to the local wind conditions near and in the newly built vault. While the *KNMI* station was already present in the IJmuiden vault since a long time, the four positions of the stations *IJm1, IJm2, IJm3, IJm4* were accurately chosen in order to (i) properly represent the wind conditions around the vault and (ii) to avoid as much as possible the local disturbance effects caused by small-scale terrain and building features not included in the CFD geometrical model.

In order to validate the CFD results, the gathered on-site measurements of stations *IJm1, IJm2, IJm3* and *IJm4* were 1-min data and successively averaged into 10-min data. The database composed of 10-min mean wind speed values was then furtherly analyzed in order to select those values associated with a possible neutral atmospheric boundary layer (ABL) conditions (Stull 1988). Since all the CFD simulations were performed for neutral atmospheric conditions and ignoring any thermal effects, in accordance with the Pasquill (1961) and Venkatram (1996) stability classes and in absence of temperature data, only samples (i.e. the 10-min mean wind speed values) greater than 6 m/s were considered for the validation. This is important to exclude thermal effects from the comparison. Around every of the 24 wind directions for which CFD simulations were performed ($\theta = 0°, 15°, \ldots, 345°$), a relatively narrow 15° (i.e. ±7.5°) wind direction sector was defined. Therefore, all 10-min data values that had mean 10-min wind directions at the reference station *IJm1* within this sector, were averaged (U) and standard deviations (σ) were obtained. Afterwards, the so-called *amplification factor* (K_{exp}) was calculated at the positions *IJm2, IJm3* and *IJm4* and for every 15°. The amplification factor (K_{exp}) is defined as the ratio between the mean wind speed (U) value at the measurement position (e.g. at *IJm2, IJm3* and *IJm4*) and the mean wind speed (U_{ref}) value at the reference station *IJm1*. In Sect. 4 the amplification factors obtained by the on-site measurements (K_{exp}) will be compared to the equivalent ones (K_{num}) calculated using the CFD results in order to quantify the deviations between simulated and measured data.

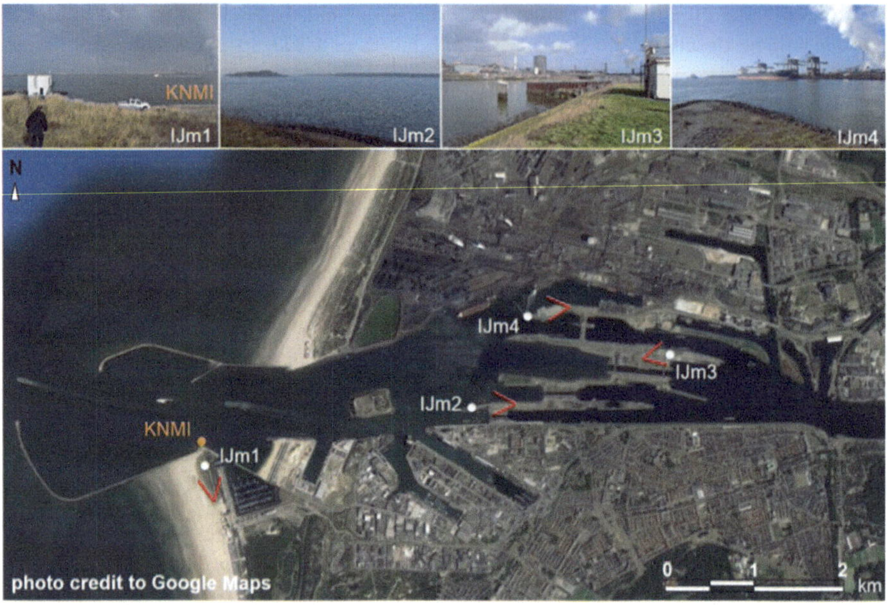

Fig. 2. Positions of the 2D ultrasonic anemometers installed in the IJmuiden vault.

3 CFD Simulations

3.1 Computational Geometry, Domain and Grid

CFD simulations were performed for 24 reference wind directions ($\theta = 0°$, 15°, …, 345°) with an interval of 15°. A high-resolution computational grid with an area of 225.0 km^2 (i.e. $L \times L \times H = 15.0 \times 15.0 \times 500.0$ km^3) and approximately 74 million control volumes was built using the software Gambit 2.4.6 (Figs. 3 and 4). In order to achieve a high-level of control over the docks (where the measurement stations are located), the walls of buildings and inside the canals and streets, the grid generation technique presented by van Hooff and Blocken (2010) was used. Furthermore, to avoid possible convergence problems and maximize numerical accuracy, only hexahedral and prismatic cells were constructed throughout the whole computational domain (Fig. 4). An expansion rate (although different throughout the domain) below or equal to 1.3 was adopted during the grid generation process in adherence to the best practice guidelines (Franke et al. 2007; Tominaga et al. 2008).

As indicated in Fig. 3, three different levels of refinement were defined during the construction both for the geometry and grid (e.g. buildings, bridges and docks) in the area of interest. More detailed geometries and finest grids were constructed in the zone termed *"vault"* of approximately 7.2 km^2 (i.e. 6.0 *x* 1.2 km^2) (Fig. 3a). In the *vault*, the buildings, bridges and docks were constructed with their real ground plan and heights, but replacing pitched roofs with flat roofs for the buildings. A maximum cell size and an expansion rate of respectively 10 m and 1.1 were adopted. Overall, the computational grid of the *vault* counted about 27 million control volumes. In the second zone, termed *"surrounding"*, an intermediate level of resolution both for the geometries and grids was defined (Fig. 3). In this zone of about 50.8 km^2, the buildings were constructed with their real ground plan and heights, but replacing pitched roofs with flat ones. A coarser grid with respect to the one constructed in the *vault* zone was built here. A maximum cell size and an expansion rate of respectively 30 m and 1.2 were employed. The *surrounding* zone counted about 24 million control volumes. In the third zone, termed *"district"* and of about 167 km^2, the lowest resolution (compared to the aforementioned zones) both for geometries and grids was adopted (Fig. 3). Due to the significant dimensions of the computational domain, a further simplification both of geometries and grids was required in this zone to reduce the computational costs. Group of buildings corresponding to small urban surroundings (e.g. with a surface of 0.06–0.50 km^2) were approximated to bluff bodies for which an average height of buildings group was calculated. An expansion ratio of the grid equal to 1.3 and a maximum cell size equal to 100 m were here used. This zone counted about 23 million control volumes.

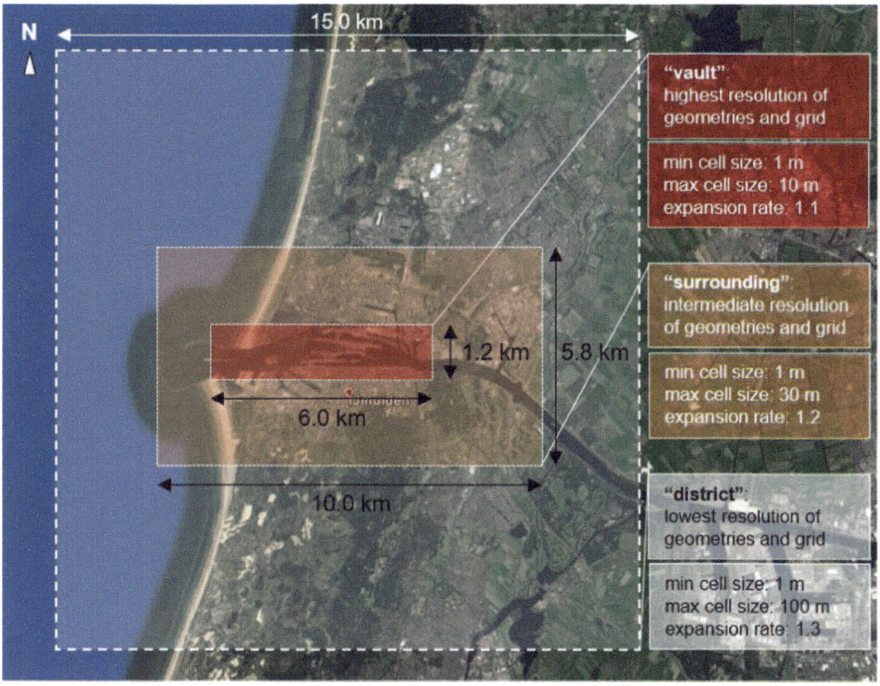

Fig. 3. Overview of the area of interest reproduced by the computational domain with an indication of the three zones for which different levels of refinement of geometries and grids have been defined (Photo credit to Google Maps).

Fig. 4. Comparison between an aerial photograph (photo credit to Google Maps) and the computational grid of the Port of Amsterdam

3.2 Boundary Conditions and Solver Settings

In accordance to some previous publications dealing with wind flows modeling in urban environments (e.g. Blocken et al. 2012; Blocken 2015) a distinction between two types of roughness was made: the roughness included by different patches on the ground of the computational domain and the roughness not directly included in the computational domain but imposed on the inlet profiles as function of the terrains surrounding the area of interest and situated outside the computational domain. In that respect, in accordance with the roughness classification proposed by Davenport and updated by Wieringa (1992), five aerodynamic roughness lengths (z_O) values were employed to determine the inflow profiles blowing from different sectors. The mean wind speed (U), turbulent kinetic energy (k) and turbulence dissipation rate (ε) profiles, applied to the inlet faces of the computational domain, were calculated using the equations by Richards and Hoxey (1993):

$$U(z) = \frac{u*}{\kappa} ln\left(\frac{z + z_0}{z_0}\right) \quad (1)$$

$$k(z) = \frac{u*^2}{\sqrt{C_\mu}} \quad (2)$$

$$\varepsilon(z) = \frac{u*^3}{\kappa(z + z_0)} \quad (3)$$

where $u*$ is the friction velocity, z_O is the aerodynamic roughness length, κ indicates the von Karman constant (equal to 0.41) and C_μ is a constant (equal to 0.09).

As reported in Fig. 5, five roughness patches, each one representative of a different aerodynamic roughness length (z_O) value, were defined on the ground of the computational domain (Wieringa 1992): *"sea"* (z_O = 0.0002 m), *"smooth"* (z_O = 0.005 m), *"open"* (z_O = 0.03 m), *"roughly open"* (z_O = 0.10 m) and *"rough"* (z_O = 0.25 m). Afterwards, the standard wall functions by Launder and Spalding (1974) with roughness modification by Cecebi and Bradshaw (1977) were used to convert the aerodynamic length (z_O) into the corresponding equivalent sand-grain roughness height (k_S) and the roughness constant (C_S). For every roughness patch, different equivalent sand-grain roughness height (k_s) values were calculated by the equation k_s = 9.793 z_O/C_s in accordance with Blocken et al. (2007). Hence, different roughness constant (C_S) values were chosen in order to comply with the necessary condition $y_p > k_s$, where y_p is the distance of the centroid of the first cell from the wall.

At the sides (only for the reference wind directions θ = 0°, θ = 90°, θ = 180° and θ = 270°) and top of the computational domain slip walls conditions were imposed. At the outlet, zero static gauge pressure was set.

CFD simulations were performed for 24 reference wind directions (θ) using the 3D steady RANS approach and the realizable k-ε turbulence model (Shih et al. 1995). The second-order discretization schemes for the governing equations and the SIMPLE algorithm for pressure-velocity coupling were used (Ferziger and Perić 2002). Each simulation was carried out by ANSYS Fluent 16.0 for approximately 72 h on 64 cores on the Dutch national supercomputer SURFSARA, Cartesius (www.surfsara.nl).

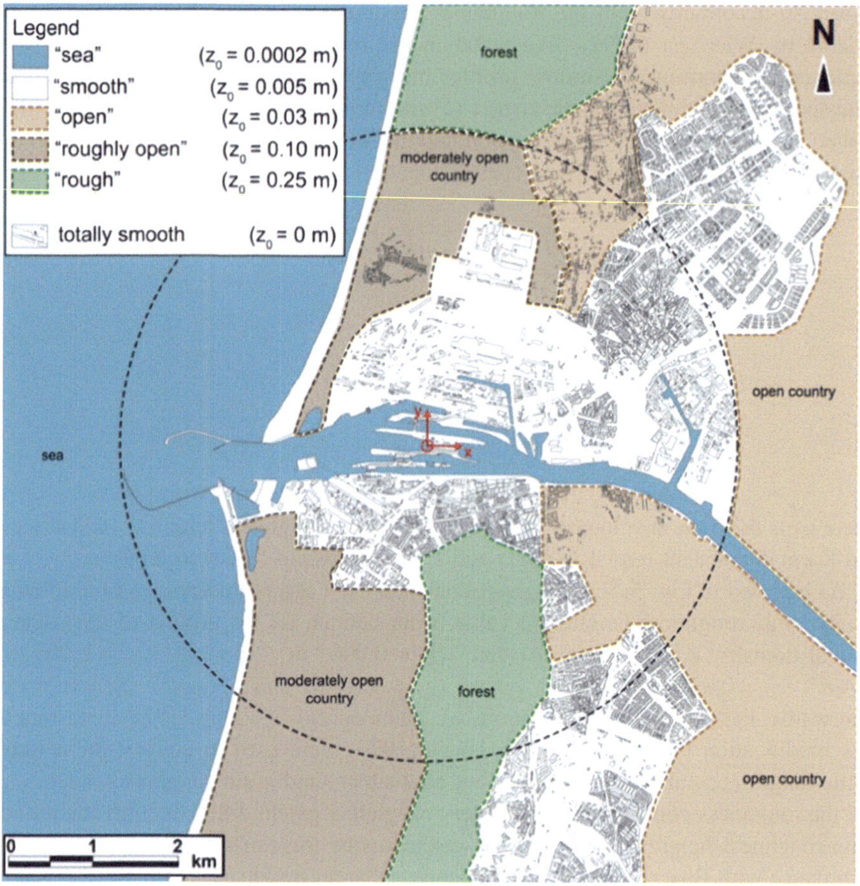

Fig. 5. Roughness map indicating the five roughness patches applied by wall functions on the ground of the computational domain.

4 Validation of the CFD Results

The on-site measurements (measured data) and CFD results (simulated data) were compared in terms of amplification factor (K), as previously introduced in Sect. 2, and local wind direction (ϕ) for all the 24 wind directions and 4 measurement positions (*IJm1*, *IJm2*, *IJm3* and *IJm4*). Furthermore, to quantify the discrepancies, in terms of amplification factor, between measured and simulated data, the fraction of data within a factor 1.3 (*FAC1.3*) was used (Schatzmann et al. 2010; Ricci et al. 2017). The ideal value of *FAC1.3*, corresponding to complete agreement between measured and simulated results, is equal to 1 and the results were graphically displayed by the scatter plot reported in Fig. 6a. In Fig. 6a the abscissa and the ordinate indicate the amplification factors (K) calculated for the measured (K_{exp}) and simulated data (K_{num}) respectively. In Fig. 6b the same axes indicate the local wind direction both for the mean measured (ϕ_{exp}) and simulated (ϕ_{num}) data.

Overall, Fig. 6a showed a satisfactory agreement between measured and simulated data in terms of amplification factor (K). Indeed, over a database of 72 samples (i.e. 24 for three stations), 90% of simulated data was found to be within 30% of deviation from the measured data. A large deviation between measured (K_{exp}) and simulated (K_{num}) data, referring to the stations *IJm3* and *IJm4*, was found only for the reference wind directions θ = 105° and θ = 120°. A satisfactory agreement, between simulated and measured data, was also found in terms of local wind directions (ϕ) as shown in Fig. 6b, where a tight distribution around the diagonal can be observed. In that respect, 86% of simulated data (over a database of 96 samples, i.e. 24 for each stations) were found within ±30° and only the 6% of samples showed a large deviation equal and greater than ±60°.

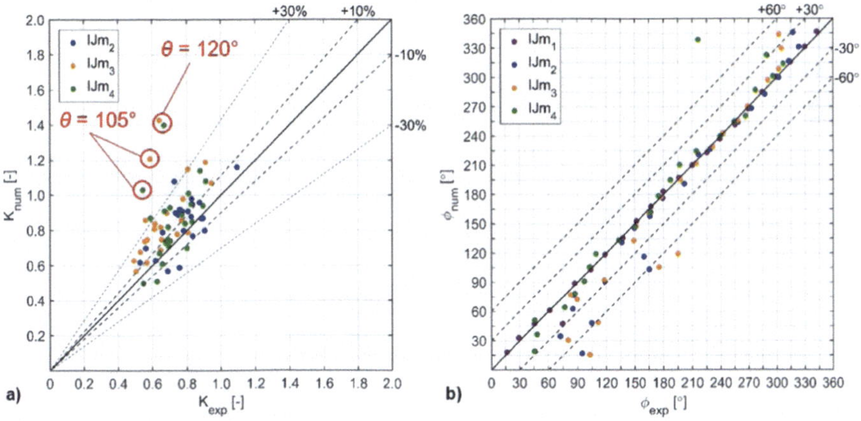

Fig. 6. Comparison of numerically simulated and experimentally measured speed ratio (K) at three measurement positions for 24 reference wind directions, θ, (a). Same for local wind directions, ϕ, (b).

5 Discussion and Conclusions

In order to better understand the deviation between simulated and measured data, contours of amplification factors (K_{num}) and of mean wind speed (U) calculated at 10 m above the ground (i.e. at the same height of the anemometric stations) were deeply analyzed for all the 24 reference wind directions. As an example, the contour of mean wind speed (U) of one out of two reference wind directions showing deviations greater than 30% in terms of amplification factor, between measured and simulated data is displayed in Fig. 7a. It is worth to note that only for these wind sectors (i.e. θ = 105° and θ = 120°), while the reference station *IJm1* is in an extensive leeward zone (with wind speed values below 6 m/s) caused by a building and the low hills located upstream the station *IJm1*, the stations *IJm3* and *IJm4* are in windward zones with mean wind speed values greater than 9 m/s (Fig. 7b,c). In that respect, it may be possible that the limitations of the 3D steady RANS approach in predicting the separation and zones nearby the station *IJm1* has been the cause of large deviations found

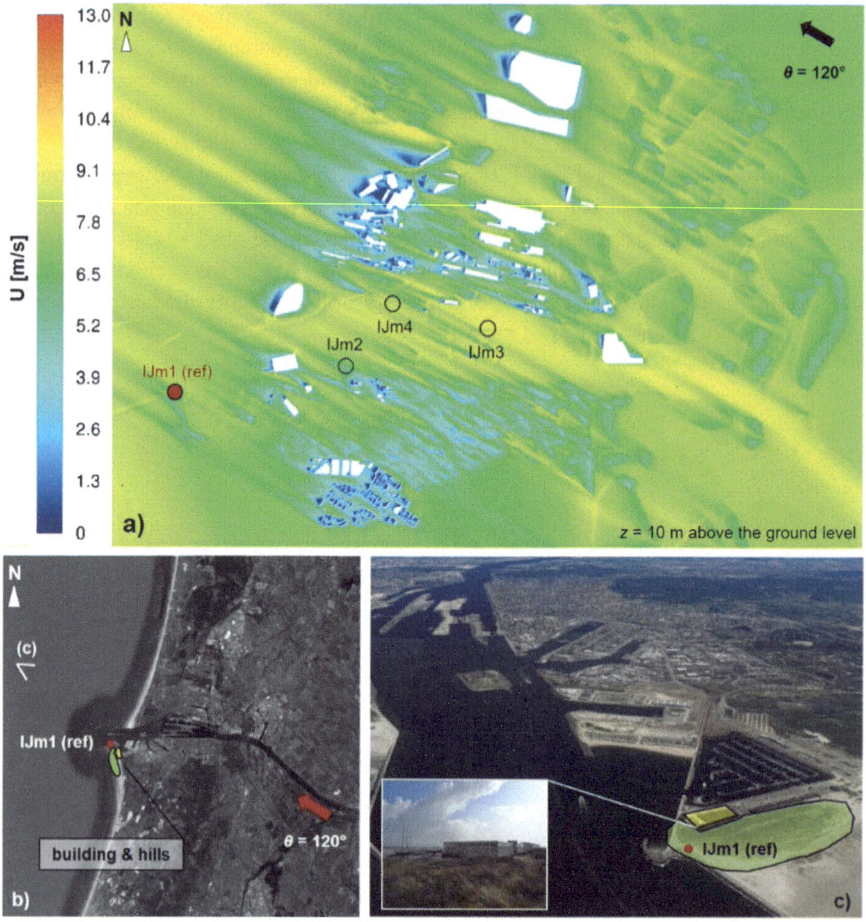

Fig. 7. Contour of mean wind speed (U) for the reference wind direction θ = 120° made at 10 m above the ground level (a). Sketch of the port of Amsterdam indicating the Eastern reference wind direction θ = 120° and the critical position of the station *IJm1* with respect to the surrounding (c,b) (Photo credit to Google Maps).

(in terms of amplification factor) between simulated and measured data for θ = 105° and θ = 120°. On the other hand, it is important to stress also that, for these two reference wind directions (i.e. θ = 105° and θ = 120°) a very limited number of *"samples"* (i.e. 10-min mean wind speed values greater than 6 m/s), were found along the measuring period of nine months.

In conclusion, the following observations are made:

- A satisfactory agreement in accordance with previous publications concerning similar studies was found between simulated and measured data, in terms of amplification factor (K) and local wind direction (ϕ).

- The wind flow effects caused by the complexity of the harbor and his surrounding (e.g. the presence of hills, harbor cranes and the whole shipbuilding) are hardly predictable by the 3D steady RANS approach for some wind directions.
- As a consequence of the aforementioned point, the use of more sophisticated numerical approach, as the large eddy simulation (LES), requires a much finer computational grid throughout almost the whole domain, which in turn makes CFD cases very time-consuming and not realizable in a reasonable time.
- The database of on-site measurements collected along nine months may be limited to validate CFD simulations performed under neutral atmospheric conditions for all 24 wind sectors.

Future steps will be targeted at improving the database of on-site measurements to fully understand the 3D steady RANS limitations when predicting microscale wind conditions.

Acknowledgements. This work was sponsored by NWO Exacte en Natuurwetenschappen (Physical Sciences) for the use of supercomputer facilities, with financial support from the Nederlandse Organisatie voor Wetenschappelijk Onderzoek (Netherlands Organization for Scientific Research, NWO).

References

Baker CJ (2007) Wind engineering - past, present and future. J Wind Eng Ind Aerodyn 95:843–870

Blocken B, van der Hout A, Dekker J, Weiler O (2015) CFD simulation of wind flow over natural complex terrain: Case study with validation by field measurements for Ria de Ferrol, Galicia, Spain. J Wind Eng Ind Aerodyn 147:43–57

Blocken B (2014) 50 years of computational wind engineering: past, present and future. J Wind Eng Ind Aerodyn 129:69–102

Blocken B, Janssen WD, van Hooff T (2012) CFD simulation for pedestrian wind comfort and wind safety in urban areas: General decision framework and case study for the Eindhoven University campus. Environ Model Softw 30:15–34

Blocken B, Stathopoulos T, Carmeliet J (2007) CFD simulation of the atmospheric boundary layer: wall function problems. Atmos Environ 41:238–252

Burlando M, Pizzo M, Repetto MP, Solari G, De Gaetano P, Tizzi M (2014) Short-term wind forecast for the safety management of complex areas during hazardous wind events. J Wind Eng Ind Aerodyn 135:170–181

Cebeci T, Bradshaw P (1977) Momentum Transfer in Boundary Layers. Hemisphere Publishing Corporation, New York

Ferziger JH, Perić M (2002) Computational Methods for Fluid Dynamics, 3rd edn. Springer, Heidelberg ISBN 978-3-540-42074-3

Franke J, Hellsten A, Schlünzen, H, Carissimo, B (2007) COST Action 732 quality assurance and improvement of microscale meteorological models

Franke J, Hellsten A, Schlünzen H, Carissimo B (2007) COST Action 732 quality assurance and improvement of microscale meteorological models. COST Action 732

Janssen WD, Blocken B, van Wijhe HJ (2017) CFD simulations of wind loads on a container ship: validation and impact of geometrical simplifications. J Wind Eng Ind Aerodyn 166: 106–116

Launder BE, Spalding DB (1974) The numerical computation of turbulent flows. Comput Methods Appl Mech Eng 3:269–289

Murakami S (1990) Computational wind engineering. J Wind Eng Ind Aerodyn 36:517–538

Pasquill F (1961) The estimation of the dispersion of windborne material. The Meteorological Magazine, vol 90, no 1063

Richards PJ, Hoxey RP (1993) Appropriate boundary conditions for computational wind engineering models using the k-ε turbulence model. J Wind Eng Ind Aerodyn 46–47:145–153

Ricci A, Burlando M, Freda A, Repetto MP (2017a) Wind tunnel measurements of the urban boundary layer development over a historical district in Italy. Build Environ 111:192–206

Ricci A, Kalkman I, Blocken B, Burlando M, Freda A, Repetto MP (2017b) Local-scale forcing effects on wind flows in an urban environment: impact of geometrical simplifications. J Wind Eng Ind Aerodyn 170:238–255

Schatzmann M, Olesen H, Franke J (2010) COST 732 model evaluation case studies: approach and results. COST Action 732

Shih TH, Liou WW, Shabbir A, Zhu J (1995) A new k-ε eddy-viscosity model for high Reynolds number turbulent flows e model development and validation. Comput Fluids 24:227–238

Solari G (2007) The international association for wind engineering (IAWE): progress and prospects. J Wind Eng Ind Aerodyn 95:813–842

Stathopoulos T (1997) Computational wind engineering: past achievements and future challenges. J Wind Eng Ind Aerodyn 67–68:509–532

Stull RB (1988) An Introduction to Boundary Layer Meteorology, 1st edn. Springer, Netherlands ISBN 978-94-009-3027-8

Tominaga Y, Mochida A, Yoshie R, Kataoka H, Nozu T, Yoshikawa M et al (2008) AIJ guidelines for practical applications of CFD to pedestrian wind environment around buildings. J Wind Eng Ind Aerodyn 96:1749–1761

van Hooff T, Blocken B (2010) Coupled urban wind flow and indoor natural ventilation modelling on a high-resolution grid: a case study for the Amsterdam ArenA stadium. Environ Model Softw 25:51–65

Venkatram A (1996) An examination of the Pasquill-Gifford-Turner dispersion scheme. Atmos Environ 30:1283–1290

Wieringa J (1992) Updating the Davenport roughness classification. J Wind Eng Ind Aerodyn 41–44:357–368

Experimental Tests on the Wave-Induced Response of a Tension Leg Platform Supporting a 5 MW Wind Turbine

L. Riefolo(✉), M. Vardaroglu, and A. M. Avossa

Department of Engineering (DI), Università della Campania "Luigi Vanvitelli", Via Roma 9, 81031 Aversa, Italy
{luigia.riefolo,mustafa.vardaroglu, albertomaria.avossa}@unicampania.it

Abstract. Floating offshore wind turbines are complex dynamic structures, and the analysis of their environmental loads requires experimental test investigations. This paper aims to provide the experience gained from wave basin experiments performed at the Danish Hydraulic Institute on a floating wind turbine Tension Leg Platform within the framework of the EU-Hydralab IV Integrated Infrastructure Initiative. Froude-scaled model was subjected to regular waves and steady wind loads. Measurements were taken of hydrodynamics, displacements and wave induced forces at the mooring lines. First, free vibration and hammer tests were performed to obtain the natural frequencies of the floating motions and tower elastic behaviour, respectively. Then, displacements, rotations, and forces were measured under regular waves and parked and rated wind conditions. Spectral analyses were carried out to investigate the dynamic response of TLP wind turbine. The results show that most of the dynamic response occurs at the wave frequency and natural frequencies.

Keywords: Floating wind turbine · TLP · Experimental tests · Free decay tests · Dynamic analyses

1 Introduction

Beginning from 2000's, there has been an increasing demand for offshore wind energy harvesting. The possible shortage of suitable onshore areas for wind generation, especially for densely populated countries or because areas prone to high environmental risk (Avossa et al. 2017a, b, c), together with stronger and more steady wind potential in off-shore areas and less environmental impact (Riefolo et al. 2016), leads the way to find solutions for deep-water regions. Offshore wind turbine systems are subjected to harsh environmental conditions. Statoil (2017) reported storms with wind speed of 44 m/s and maximum wave height of 19 m from the Hywind-Demo site.

The coupled nonlinear hydrodynamic and aerodynamic response makes the design of the floating wind turbines challenging. Design of the floating wind turbines originates from oil & gas platforms. Butterfield et al. (2007) classifies the floating wind turbines into three groups based on the physical principle or strategy that is used to achieve static stability. Spar-buoy platforms are an example of ballast-stabilized type

F. Ricciardelli and A. M. Avossa (Eds.): INVENTO 2018, LNCE 27, pp. 599–612, 2019.
https://doi.org/10.1007/978-3-030-12815-9_46

that uses ballast weights hung below a central buoyancy tank which creates a righting moment. Tension Leg Platform (TLP) is an example of mooring line stabilized platform that achieves stability through the use for mooring line tension. A barge-type platform is an example for buoyancy-stabilized platforms that achieve stability through the use distributed buoyancy, taking advantage of weighted water plane area for righting moment.

The complicated response of the floating wind turbines can be observed either by monitoring the prototype in the ocean or building model scales in the laboratories or conducting simulations with numerical methods. Jonkman and Buhl Jr. (2005) constructed "FAST" code (Fatigue, Aerodynamics, Structures, and Turbulence) for the numerical analysis of floating wind turbine systems. Jonkman et al. (2009) developed and proposed a standard 5 MW Horizontal Axis Wind Turbine (HAWT) system at National Renewable Energy Laboratory (NREL) as a standard and a base project for the future studies. Jonkman (2010) studied the design of a floater system for the Offshore Code Comparison Collaboration (OC3) project, Phase IV. Numerical analysis solutions need to be calibrated and verified by laboratory testing or prototype monitoring. Shin (2011) tested a scaled OC3-Hywind model in the laboratory. Browning et al. (2012) conducted an experimental study on a spar buoy, attached to a 1/50 scale model of the NREL 5 MW reference turbine. The resulting data were scaled using Froude scaling to full size and used for calibration and validation of the numerical model FAST. Goupee et al. (2012) conducted experiments on 1:50 scale models of NREL 5-MW reference turbine standing on spar buoy, TLP and a semi-submersible floater. In a similar study, Nihei et al. (2014) conducted an experimental campaign on a 5-MW turbine supported by a TLP, spar-buoy and semi-submersible floaters. According to the study, the TLP type showed the most favourable performance while the spar type demonstrated largest acceleration in almost all the range of environmental conditions. Pegalajar-Jurado et al. (2016) conducted tests on a DTU reference 10-MW TLP floating wind turbine with focusing on blade pitch control. Oguz et al. (2018) conducted a test campaign with a 5-MW NREL reference turbine on an Iberdrola TLP concept. Results from the test campaign were compared with FAST simulations. It was reported that in some cases FAST simulations were successful to predict the actual response, while, in other situations the numerical results did not match with the actual response.

Statoil, Norway has been monitoring the Hywind Demo located in North Sea near Norway since 2009. Hywind Demo consists of a single 2.3 MW horizontal axis wind turbine (HAWT) with a spar buoy floater (Skaare et al. 2015). Following the demo, Statoil has been generating data and energy from the Hywind Pilot Park with spar buoy supported HAWT's in Scotland since 2017. Principle Power from United States has been monitoring the WindFloat since 2011 in Portugal. WindFloat project consists of a single HAWT on a semi-submersible, three column stabilized offshore platform.

In 2014, an experimental test campaign was conducted on a 5 MW HAWT NREL standard turbine within Hydralab IV Integrated Infrastructure Initiative (Tomasicchio et al. 2014). The turbine response was monitored with both a spar buoy floater and a TLP platform. Tests were held in DHI Offshore Wave Basin in Hørsholm, Denmark.

Different wind and wave conditions were considered. Natural frequencies of the structure were measured by free vibration tests. Regular and irregular wave tests were performed under cut-in, rated and cut-out wind speed. A remarkable response was reported under wave frequency.

Experimental and numerical data from spar-buoy tests has been discussed recently by Tomasicchio et al. (2017, 2018) and Riefolo et al. (2018). In this study, TLP data are processed and discussed. Natural frequencies are extracted. Both parked and rated wind conditions are analysed. Mooring line forces and displacement data have been discussed in both time and frequency domains.

2 TLP Physical Model and Setup

The design of tension leg platform TLP physical model has been conducted with reference to the MIT/NREL prototype (Jonkman and Matha 2009; Matha 2009). The MIT/NREL floating wind turbine achieves its restoring essentially through its tension leg mooring system (Wayman et al. 2006). The prototype consists of a cylindrical platform, ballasted with concrete and moored by four pairs of vertical tendons in tension. Each pair of tendons attaches to a spoke that radiates horizontally from the bottom of the platform. The concrete ballast is used to ensure that the combined turbine-platform system remains stable during float-out, even without the tendons, in mild meteocean conditions (Tomasicchio et al. 2012). Table 1 summarizes the geometric and dynamic properties of the TLP prototype and model.

Table 1. Geometric characteristics of the TLP floating wind turbine. Length scale λ = 1:40.

TLP (MIT/NREL) wind turbine	Full scale	Unit	Scale factor	Model scale
Diameter of the platform	18	m	λ	0.450
Draft and Depth to fairleads	47.89	m	λ	1.197
Radius to fairleads	27	m	λ	0.675
CM location below still water level	40.61	m	λ	1.015
Tower height (hub level)	90	m	λ	2.250
Tower mass	347500	kg	λ^3	5.297
Mass (floating system) including ballast	8600000	kg	λ^3	131.1
Water displacement	12180	m^3	λ^3	0.190
Roll inertia about CM_z	5.72E + 08	kg*m^2	λ^5	5.446
Unstreached line length	151.7	m	λ	3.792
Line diameter	0.127	m	λ	0.003
Line mass density	116	kg/m	λ^2	0.071

2.1 Wave Basin and Instrumentation

The experimental tests have been performed at the DHI Offshore Wave Basin in Denmark. The wave basin is 20 m long, 30 m wide, it has a water depth of 3 m and a pit of 6 m.

The TLP wind turbine model was constructed out of plastic of density equal to 1200 kg/m^3, it was composed by four main components: floater, cover, tower and leg assembly (Fig. 1). The designed still water level (SWL) lies right at the edge of the larger cylinder cover. The mooring lines of TLP model have been anchored perpendicular to the table placed in the pit. With reference to the TLP MIT/NREL prototype, a 1:40 scale factor has been adopted. The draft of the considered TLP is 47.89 m in full scale. The immersed cross section area of the TLP is 0.54 m^2. In Fig. 2 a sketch of the TLP scaled model installed into the wave basin is shown.

The wave maker is equipped with 60 individually controlled flaps, able to generate regular and irregular waves. A parabolic wave absorber located opposite to the wave maker has minimized the reflection. In particular, incident and reflected waves were evaluated through a five wave-gauge array reflection analysis (Mansard and Funke 1980). The five wave-gauges were placed at the centre of the pit in order to carry out the wave calibration. During the model testing, the wave gauges were moved 3 m downstream the TLP wind turbine. Additionally, six wave gauges were located around the floating structure.

The velocity field was measured in the proximity of TLP through a Nortek Vectrino velocimeter. The acoustic doppler velocimeter ADV was located at a distance of 60 cm from the front size of the floating structure. A Qualisys Track System (www.qualisys.com) followed the rigid body motion of the TLP in terms of displacements (surge, sway, heave) and rotations (roll, pitch and yaw). The system is based on two cameras emitting infrared light. Five passive spherical markers, 40 mm in diameter, reflect the infrared light and were placed on a frame mounted at the tower base. Data acquired by the Qualisys Track Manager were directly transferred through an analog output to the main data acquisition system.

The DHI Wave Synthesizer synchronized all the sensors. Sampling took place at 40 Hz. The test duration for regular and irregular waves is equal to 3 and 10 min, respectively.

An overview of the instrumentation of the rotor and of the tower is given in Fig. 2. A six-component force gauge was mounted at the base of the tower, between the tower and the floater, measuring forces and moments in x, y and z direction. At the top of the tower, between the tower and the nacelle, a four-component force gauge was mounted, measuring forces and moments in x and y direction. Furthermore, three accelerometers were placed at different levels along the tower; in particular, two accelerometers were located underneath the nacelle, measuring the acceleration along y and z direction, and a third one at the bottom of the tower, measuring the acceleration along x direction. Two pressure gauges were mounted in the submerged part of structure measuring the pressures.

Fig. 1. TLP wind turbine model in the wave basin of DHI laboratory.

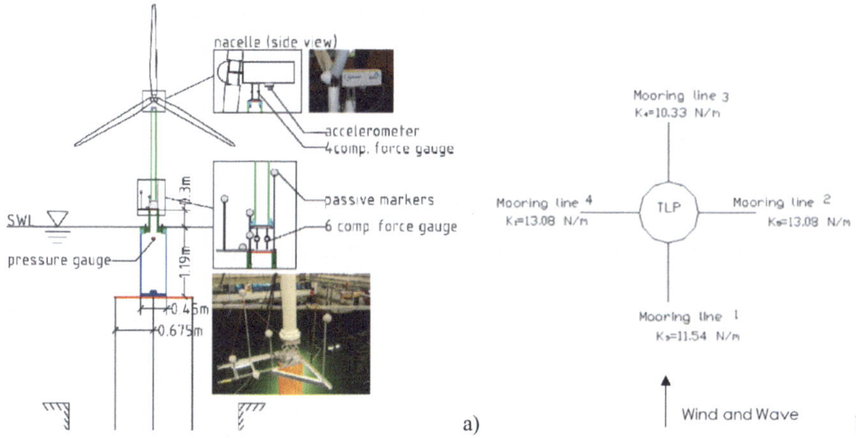

Fig. 2. (a) Overview of the TLP model with a part of the instruments. (b) Mooring lines stiffness and wind/wave direction with reference to the TLP.

2.2 Rotor, Blades and Wind Conditions

A motor inside the casing induced the rotation for the rotor. A potentiometer adjusted the rotational speed to 38 rpm, which corresponds to a rotational speed of 12.1 rpm full scale, allowing the gyroscopic effects.

The rotor blades were scaled geometrically and each have a length of 1.575 m. They have been constructed by using fiber glass and scaling the drawings from a real case (Fig. 3). The pitch of the blades was set to 30° in order to determine a thrust of 4 N at 38.1 rpm, model scale. Further tests to obtain a relationship between thrust and rotational speed were conducted with rotational speeds of 32 rpm and 42 rpm, model scale. Only static wind loads were reproduced, by applying the mean thrust force to the nacelle. In particular, a weightless line was connected to the nacelle, passing through a pulley and with a suspended mass. The full-scale thrust for the 5 MW NREL reference turbine was calculated by different researchers, for instance Sclavounos et al. (2008) found that the rotor thrust under an 11 m/s wind is around 800 kN, which corresponds to 12.5 N for the 1:40 scaled model. Therefore, 4 N came from the trust force generated by the rotor, and the difference was realized with a weight of 8 N.

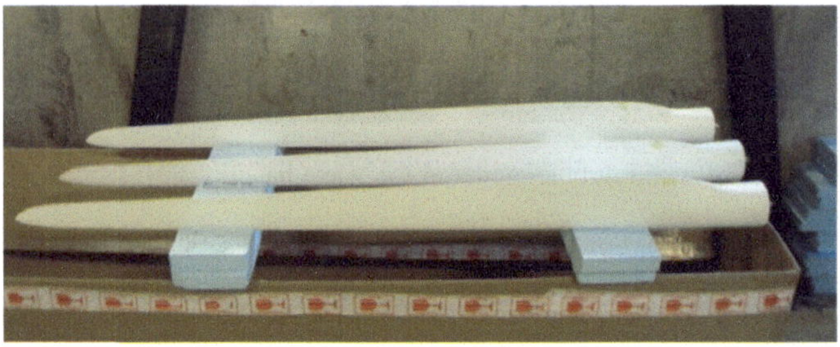

Fig. 3. Profile of the blades.

3 Test Program and Methodology

The three conditions of parked, rated and stalled were considered in the tests, according to the standards IEC 61400-1 and IEC 61400-3 (IEC 61400-1; IEC 61400-3). In particular, the parked condition refers to the case without rotation of the blades; the rated (operational) condition was simulated, combining mean thrust, rotating rotor and different sea states under regular and irregular waves; and the extreme wave condition was generated, with the rotor being stopped and mean thrust corresponding to the cut-out wind speed. Regular and irregular waves were heading in two different directions: orthogonal (0°) and oblique (20°) with reference to the TLP floating structure. The generated regular and irregular waves are shown in Table 2, in terms of H and T the regular wave height and wave period, respectively, and H_s and T_p the significant wave height and peak wave period, respectively.

On the basis of the time series of motions and mooring line tensions the spectral analysis has been performed to evaluate the power spectral density (PSD) induced by wave and wind action.

The hydrodynamic response of the TLP in the large wave flume under regular waves and a static wind load has been analysed.

In particular, to execute spectral analysis of each studied test, the Fast Fourier Transform (FFT) is applied through a specific function of the MATLAB™ software

(Wafo dat2spec) based on Welch's averaged periodogram method with no overlapping batches, has been used. Consequently, the peaks corresponding to low-frequencies, wave response and harmonics are detected from the spectra. The total power around the main frequencies is also examined in order to analyse the energy and the response of the TLP under parked and rated conditions.

Table 2. Test program of the TLP floating wind turbine.

Wind speed (rotor condition)	Waves	Prototype scale		Model scale	
		H or H_s [m]	T or T_p [s]	H or H_s [cm]	T or T_p [s]
0 m/s (parked) 11.4 m/s (rated)	Regular	1.00	10.1	2.5	1.6
		1.56	12.6	3.9	2.0
		1.80	15.2	4.5	2.4
		4	11.4	10	1.8
		6		15	
		8		20	
		6	12.6	15	2.0
			15.2		2.4
	Irregular	4	10.1	10	1.6
		6		15	

4 Results

The results are presented in full scale and only a part of the dataset is analysed in this paper. In particular, 8 tests under regular waves with different wave characteristics, H and T, and rotor conditions (parked/rated) were selected. Furthermore, the selected tests, the wave direction was orthogonal to the TLP (Table 3).

Table 3. Regular wave tests considered for time and frequency domain analyses.

H [m]	T [s]	Parked	Rated
1.8	15.2	1232	1247
6	11.4	1235	1249
6	12.6	1237	1251
6	15.2	1238	1252

4.1 Free Decay Tests

Free decay tests were carried out in order to determine the surge, sway and the tower natural frequencies. In particular, free oscillations induced by initial conditions of the motion consisting of displacements or rotations imposed for a certain degree of freedom, was analysed in time and frequency domains. Moreover, the system identification is based on the impact modal testing where the tower is excited by performing a hammer impact, at a fixed point during the test. The free oscillations in the other directions have not well determined because during the experimental tests the TLP has been excited in more than one direction, giving therefore not reliable results. In Fig. 4, the PSDs for surge and sway are shown. Natural frequencies of 0.0170 Hz for the surge motion and of 0.0178 Hz for the sway one, corresponding to the maximum peak amplitude, were found. Furthermore, the hammer test was carried out to evaluate the natural frequency of the tower. In Fig. 5 it is noted the PSD of the free decay of the tower. The natural frequency of the tower is equal to 1.58 Hz. This was calculated to be 0.6 Hz of the prototype structure (Matha 2009), and if Cauchy scaling were matched, it should have been the same on the model.

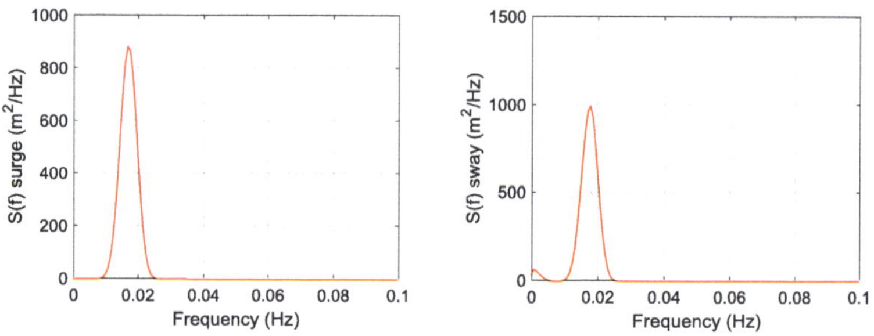

Fig. 4. PSDs from the free decay tests: surge (left) and sway (right).

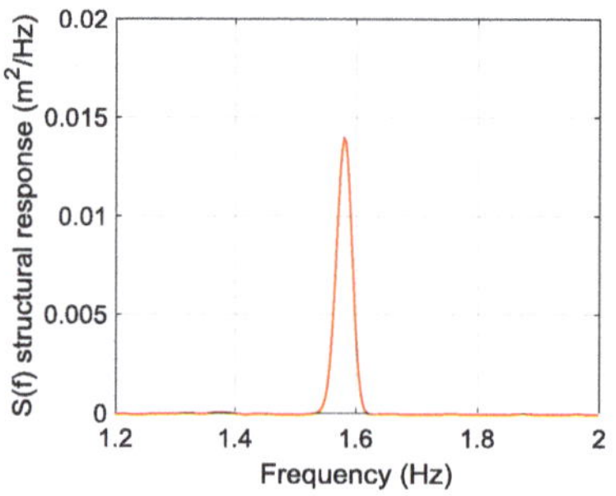

Fig. 5. PSD from the free decay test of the tower.

4.2 Dynamic Response to Regular Waves

The dynamic response of the TLP wind turbine is discussed in time and frequency domains. Figure 6 shows the Power Spectral Density Function (PSDF) of surge as measured in test #1232. It is noted that, in the calculation of PSDFs, initial portion of the measured data were eliminated to prevent an interference. PSDF calculations are handled with the software by WAFO-group (2000). In PSDF plots both axes are in logarithmic scale. Several peaks can be distinguished in Fig. 6. Wave frequency of 0.066 Hz and the first two harmonics of the wave frequency at 0.132 Hz and 0.198 Hz are clear. Figure 6 also shows a peak value at 0.0151 Hz that is close to the surge natural frequency value of 0.0170. The peak at 1.59 Hz corresponds to the tower first natural bending frequency value of 1.58 Hz. This value also coincides with the value reported by Tomasicchio et al. (2018). It is noted that same tower was tested in both studies.

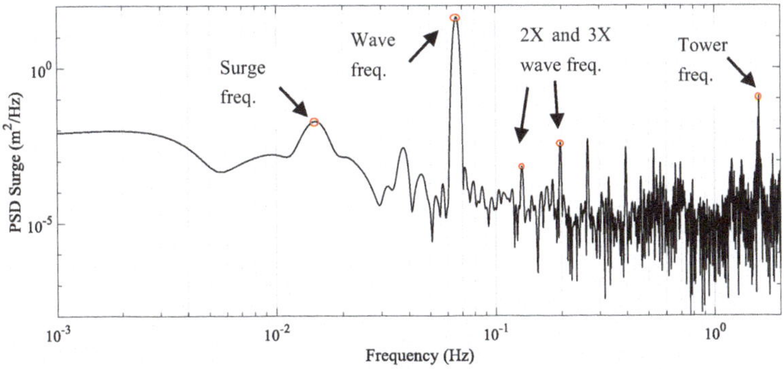

Fig. 6. PSDF of surge as measured in test #1232

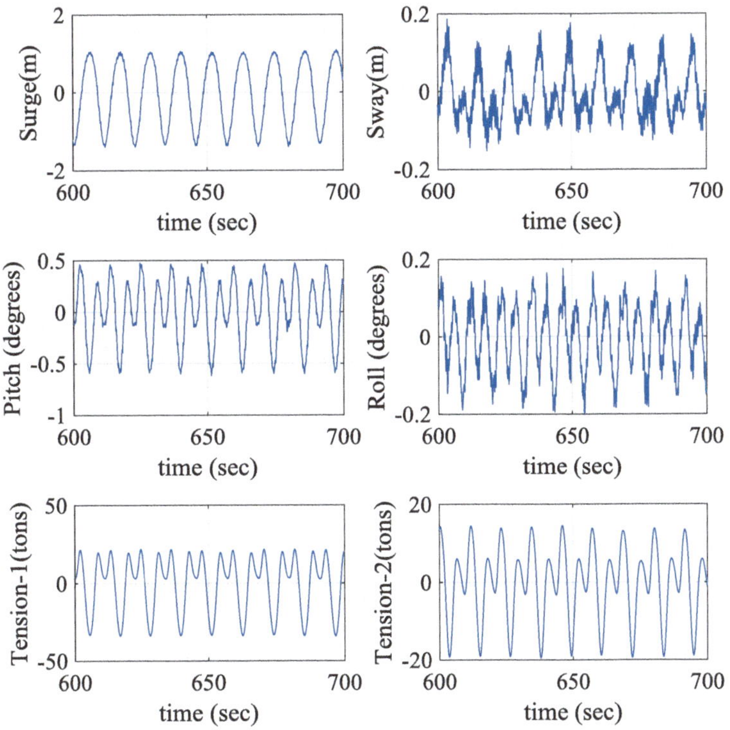

Fig. 7. Sample time histories of surge, sway, pitch, roll, mooring line tensions as measured in test #1235

Figure 7 shows sample time histories from test #1235. It can be observed that, surge and sway motion are governed by excitation frequency with a period of 11.4 s. Pitch and roll motion varies along the time with an amplitude about 0.5 and 0.2 degrees. Tensions along line 1 have less variability than those along line 2.

Time-history analysis results are discussed from the point of view of frequency domain analyses. Figure 8a shows the PSDFs of surge measured from all the tests presented in Table 3. In all the tests, response is dominated by the wave frequency. Figure 8b focuses on the peaks at the first and second harmonics of the fundamental wave frequency. Figure 9 confirms the results from Fig. 8 by means of pitch PSDF.

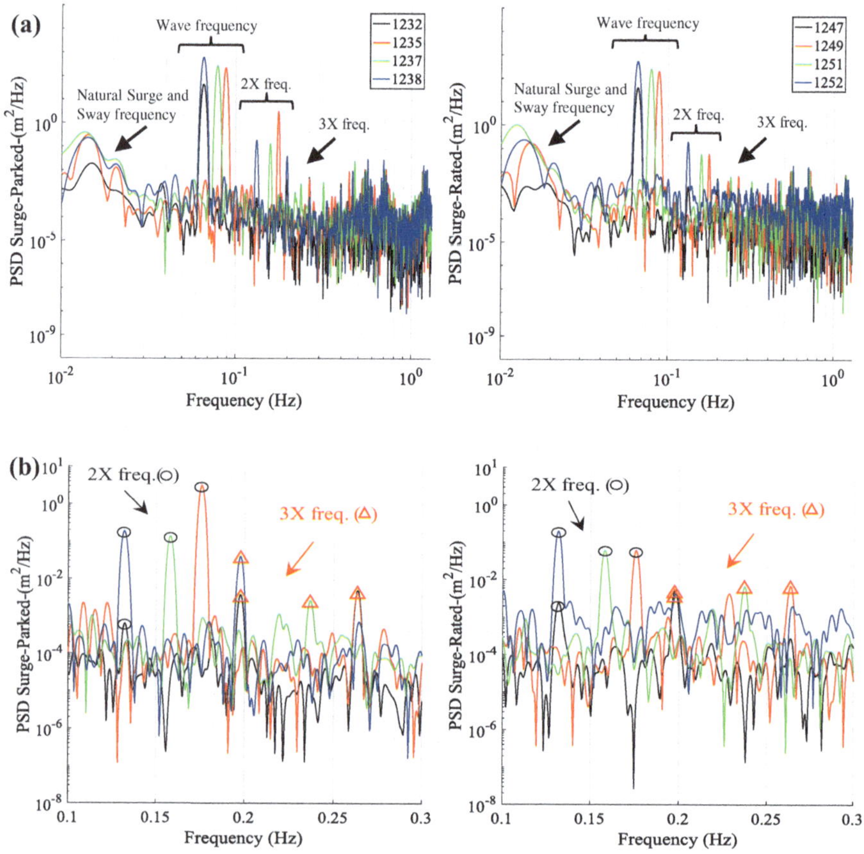

Fig. 8. (a) PSDFs of surge as measured in the different tests: parked conditions (left) and operational conditions (right). (b) Close-up view of the peaks at the first and second harmonic of the wave frequency.

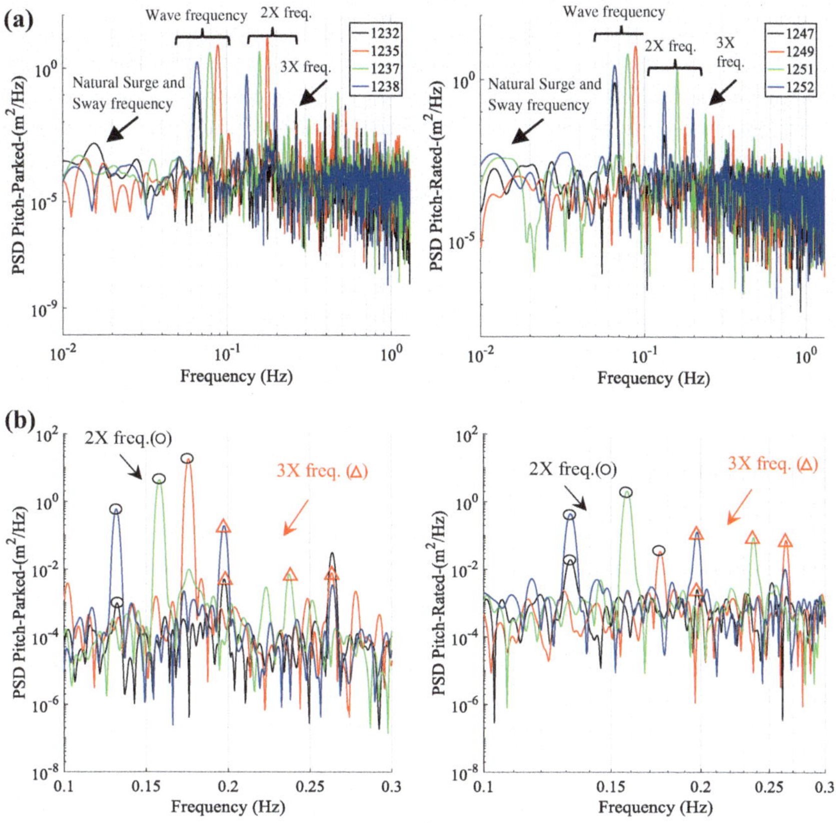

Fig. 9. (a) PSDFs of pitch as measured in the different tests: parked conditions (left column) and operational conditions (right column). (b) Close-up view of the peaks at the first and second harmonic of the wave frequency (second row).

Table 4. Surge narrow-band and total power (m^2)

Parked Test #	1232	1235	1237	1238
Sway/Surge frequency	7.23E–05	8.55E–04	1.60E–03	8.52E–04
Wave frequency	1.31E–01	6.98E–01	8.40E–01	1.88E+00
2X wave frequency	1.98E–06	9.40E–03	4.26E–04	5.78E–04
3X wave frequency	1.09E–05	1.29E–05	8.68E–06	1.17E–04
Total power	1.32E–01	7.11E–01	8.44E–01	1.89E+00
Rated Test #	1247	1249	1251	1252
Sway/Surge frequency	7.56E–06	6.26E–04	3.70E–03	8.88E–04
Wave frequency	1.25E–01	6.56E–01	8.15E–01	1.74E+00
2X wave frequency	9.26E–06	1.88E–04	2.02E–04	6.03E–04
3X wave frequency	1.70E–05	2.25E–05	2.41E–05	1.87E–05
Total power	1.26E–01	6.59E–01	8.21E–01	1.74E+00

Table 4 shows the power corresponding to surge narrow-band around the relevant frequencies, together with the total power, in order to analyse the contribution of different frequencies with respect to the total response. It is noted that the fundamental wave frequency is dominant for all the studied cases under parked and rated conditions and contributes to the total response from 94.08% to 99.94%. On the other hand, the oscillation frequency of surge/sway and the wave harmonics give a less contribution to the total response. The sum of the two components 2X and 3X the wave frequency is always lower than the contribution of the surge/sway frequency in comparison with the total response.

With regards of the rotor condition (rated or operational), the total response gives lower values than the parked condition due to the aerodynamic damping effect. This is not always confirmed when the contributions of the band power for surge/sway frequency and the harmonics are investigated.

5 Conclusions

In the present study, an experimental analysis has been carried out to evaluate the dynamic response of the TLP floating wind turbine. Different regular waves have been considered, together with two different wind conditions, parked and rated, respectively. Displacements, rotations and mooring line forces have been measured and analysed.

First, free decay tests were carried out to detect the natural periods. Furthermore, the hammer test was carried out to evaluate the first elastic bending frequency of the tower. The measured full-scale rigid body oscillation frequencies were found to be of 0.0170 Hz for the surge and of 0.0178 Hz for sway motion. The natural frequency of the tower was equal to 1.58 Hz; however, it was not properly scaled, as the Cauchy number was not considered in the TLP's design.

Then, analysis of the dynamic response in terms of displacements, rotations and mooring line forces reveals that this occurs mainly at the oncoming wave frequency, with smaller or larger components at its first and second harmonics. Wave frequency of 0.066 Hz and the first two harmonics of the wave frequency at 0.132 Hz and 0.198 Hz have been detected. Moreover, a peak value at 0.0151 Hz that is close to the surge natural frequency value of 0.0170 Hz was found. From PSDFs the peak at 1.59 Hz which refers to the tower first natural bending frequency was also detected.

The power corresponding to surge narrow-band around the relevant frequencies, together with the total power, was examined in order to analyse the contribution of different frequencies with respect to the total response. It was noted that the fundamental wave frequency is dominant for all the studied cases under parked and rated conditions. On the contrary, the oscillation frequency of surge/sway and the wave harmonics give a less contribution to the total response.

Concerning the rated condition, the total response gives lower values than parked condition due to the aerodynamic damping effect. This is not always confirmed when the contributions of band power for surge/sway frequency and harmonics are analysed.

It can be concluded that wave basin tests are a useful tool for investigating the dynamic response of TLP wind turbine when it is subjected to combined wind and wave loads. Further studies will be conducted on TLP under irregular waves and static wind.

References

Avossa AM, Demartino C, Contestabile P, Ricciardelli F, Vicinanza D (2017a) Some results on the vulnerability assessment of HAWTs subjected to wind and seismic actions. Sustainability 9:1525

Avossa AM, Demartino C, Ricciardelli F (2017b) Peak response of hawts to wind and seismic actions. In: Proceedings of the 7th European and African conference on wind engineering, EACWE 2017, Liège, 4–7 July 2017

Avossa AM, Demartino C, Ricciardelli F (2017c). Assessment of the peak response of a 5 MW HAWT under combined wind and seismic induced loads. Open Constr Build Technol J 11:441–457

Brodtkorb PA, Johannesson P, Lindgren G, Rychlik I, Rydén J, Sjö E (2000) WAFO - a MATLAB toolbox for analysis of random waves and loads. In: 10th international offshore and polar engineering conference, Seattle, USA, vol III, pp 343–350

Browning JR, Jonkman J, Robertson A, Goupee AJ (2012) Calibration and validation of a spar-type floating offshore wind turbine model using the FAST-dynamic simulation tool. J Phys Conf Ser 555(2014):012015

Butterfield S, Musial W, Jonkman J, Sclavounos P (2007) Engineering challenges for floating wind turbines. Report NREL/CP-500-38776, United States

Goupee AJ, Koo B, Kimball RW, Lambrakos KF, Dagher HJ (2012) Experimental comparison of three floating wind turbine concepts. In: 31st international conference on off-shore mechanics and arctic engineering, OMAE, Rio de Janeiro, Brazil, 1–6 July 2012

IEC 61400-1, Ed. 2, wind turbine generator systems, part 1: safety requirements (1999)

IEC 61400-3, Ed. 1, Wind turbines, Part 3: Design requirements for offshore wind turbines (2009) California at Berkeley, Fourth edn. Prentice Hall (2012)

Jonkman JM, Buhl ML Jr (2005) FAST user's guide, Technical report NREL, TP-500-38230

Jonkman J, Butterfield S, Musial W, Scott G (2009) Definition of a 5-MW reference wind turbine for offshore system development, NREL report, Golden, Colorado

Jonkman J, Matha D (2009) A quantitative comparison of the responses of three floating platform. In: Proceedings of European offshore wind 2009 conference and exhibition, NREL/CP-500-46726

Jonkman J (2010) Definition of the floating system for phase IV of OC3. NREL report, Golden, Colorado, May 2010

Mansard EPD, Funke ER (1980) The measurement of incident and reflected spectra using a least squares method. In: Proceedings of the international conference on coastal engineering, Sidney, Australia, pp 154–172

Matha D (2009) Model development and loads analysis of an offshore wind turbine on a tension leg platform, with a comparison to other floating turbine concepts. Technical report NREL/SR-500-45891, United States

Nihei Y, Iijima K, Murai M, Ikoma T (2014) A comparative study of motion performance of four different FOWT designs in combined wind and wave loads. In: 33rd international conference on offshore mechanics and arctic engineering, San Francisco, USA, 8–13 June 2014

Oguz E, Clelland D, Day AH, Incecik A, López JA, Sánchez G, Almeria GG (2018) Experimental and numerical analysis of a TLP floating offshore wind turbine. Ocean Eng 147:591–605

Pegalajar-Jurado A, Hansen AM, Laugesen R, Mikkelsen RF, Borg M, Kim T, Heilskov NF, Bredmose H (2016) Experimental and numerical study of a 10 MW TLP wind turbine in waves and wind. J Phys 753:092007

Riefolo L, Lanfredi C, Azzellino A, Vicinanza D, Tomasicchio GR, D'Alessandro F, Penchev V (2016) Offshore wind turbines: an overview of the effects on the marine environment. In: Proceedings of the international society of offshore and polar engineers (ISOPE), Rhodes (Rodos), Greece, pp 427–434, 26 June–2 July 2016

Riefolo L, del Jesus F, Guanche RG, Tomasicchio GR, Pantusa D (2018) Wind/wave misalignment effects on mooring line tensions for a spar buoy wind turbine. In: Proceedings of the international conference on offshore mechanics and arctic engineering, Madrid, Spain, 17–22 June, paper n° OMAE2018-77586 (in press)

Sclavounos PD, Tracy C, Lee S (2008) Floating off-shore wind turbines: responses in a sea state, Pareto optimal designs and economic assessment. In: Proceedings of the 27th international conference on offshore mechanics and arctic engineering, OMAE, Estoril, Portugal, 15–20 June 2008

Shin H (2011) Model test of the OC3-Hywind floating offshore wind turbine. In: Proceedings of the 21st international society of offshore and polar engineers (ISOPE), Hawaii, USA, 19–24 June 2011

Skaare B, Nielsen FG, Hanson TD, Yttervik R, Havmøller O, Rekdal A (2015) Analysis of measurements and simulations from the Hywind Demo floating wind turbine. Wind Energy 18:1105–1122

Statoil (2017). Hywind. Floating Offshore Wind 2017, Norwegian Energy Partners, 26 January, Bryggetorget, Oslo

Tomasicchio GR, Avossa AM, Riefolo L, Ricciardelli F, Musci E, D'Alessandro F, Vicinanza D (2017) Dynamic modelling of a spar buoy wind turbine. In: Proceedings of the international conference on offshore mechanics and arctic engineering, Trondheim, Norway, Paper N. OMAE2017-62246, pp V010T09A083–V010T09A093, 25–30 June 2017

Tomasicchio GR, Armenio E, D'Alessandro F, Fonseca N, Mavrakos SA, Penchev V, Schüttrumpf H, Voutsinas S, Kirkegaard, J, Jensen PM (2012) Design of a 3D physical and numerical experiment on floating off-shore wind turbines. In: 33rd Conference on Coastal Engineering, Santander, Spain, 1–6 July 2012

Tomasicchio G, D'Alessandro F, Avossa AM, Riefolo L, Musci E, Ricciardelli F, Vicinanza D (2018) Experimental modelling of the dynamic behaviour of a spar buoy wind turbine. Renew Energy 127:412–432

Tomasicchio GR, D'Alessandro F, Musci E, Fonseca N, Mavrakos SA, Kirkegaard J, Katsaounis GM, Penchev V, Schüttrumpf H, Wolbring J, Armenio E (2014) Physical model experiments on floating off-shore wind turbines. In: Hydralab IV joint user meeting, Lisbon, Portugal, 2–4 July 2014

Wayman EN, Sclavounos PD, Butterfield S, Jonkman J, Musial W (2006) Coupled dynamic modeling of floating wind turbine systems. In: The Proceedings of the Offshore Technology Conference, Houston, Texas, NREL/CP-500-39481, 1–4 May 2006

Mixture Model in High-Order Statistics for Peak Factor Estimation on Low-Rise Building

F. Rigo[1(✉)], T. Andrianne[1], and V. Denoël[2]

[1] Wind Tunnel Lab, Department of Aerospace and Mechanical Engineering, University of Liège, Liège, Belgium
{francois.rigo,t.andrianne}@uliege.be

[2] Stochastic and Structural Dynamics, Department of Urban and Environmental Engineering, University of Liège, Liège, Belgium
v.denoel@uliege.be

Abstract. To design reliable structures, extreme pressures and peak factors are required. In many applications of Wind Engineering, their statistical analysis has to be performed considering the non-Gaussianity of the wind pressures. With the increasing precision and sampling frequency of pressure sensors, short and local peak events of large amplitude are more usually captured. Their relevance is naturally questioned in the context of a structural design. Furthermore, the increasing computational power allows for accumulation and analysis of larger data sets revealing the detailed nature of wind flows around bluff bodies. In particular, in the shear layers and where local vortices form, it is commonly admitted that the Probability Density Function (PDF) of measured pressures might exhibit two or more significant components. These mixed flows can be modelled with mixture models (Cook 2016). Whenever several processes coexist, and when one of them is leading in the tail of the statistical distribution, as will be seen next in the context of corner vortices over a flat roof, it is natural to construct the extreme value model with this leading process and not with the mixed observed pressures. It is therefore important to separate the different processes that can be observed in the pressure histories. Once this is done, specific analytical formulations of non-Gaussian peak factors can be used to evaluate the statistics of extreme values (Kareem and Zhao 1994; Chen and Huang 2009). The separation of mixed processes is usually done by means of the PDF of the signals (Cook 2016). This information is of course essential to perform an accurate decomposition, but it might be facilitated by considering higher rank information like auto-correlations and higher correlations like the triple or quadruple correlation. Indeed, the two phenomena that need to be separated and identified might be characterized by significantly different timescales, which are not reflected in the PDF. In this paper, the large negative pressures measured on a flat roof are analyzed and decomposed into two elementary processes, namely, the flapping corner vortex and the turbulent flow detaching from the sharp upstream edge. This paper will finally show that an accurate decomposition of the recorded pressures into their underlying modes provides a more meaningful evaluation of the extreme pressures.

Keywords: Wind pressure · Peak factor · Non-Gaussianity · Mixture distribution · Wind tunnel tests · Low-rise building

F. Ricciardelli and A. M. Avossa (Eds.): INVENTO 2018, LNCE 27, pp. 613–629, 2019.
https://doi.org/10.1007/978-3-030-12815-9_47

1 Introduction and Motivation

This work focuses on the statistical analysis of wind pressures, involving strong non-Gaussianity, in the scope of building roof. The estimation of peak factor is necessary for the prediction of wind loads and thus the design of structures. If wind tunnel measurements are not enough precise, i.e. too short samples, too low sampling frequency or too coarse tap spatial distribution, some important information can be lost. With the increasing of pressure sensors precision in term of sampling frequency, local effects can be better captured but the question can be raised on the relevance of the study of very local effects, negligible in term of spatial and time scale compared to the entire structure.

Many research works have been carried about the non-Gaussian statistics analysis of wind pressure (Kareem and Zhao 1994; Sadek and Simiu 2002; Chen and Huang 2009; Kwon and Kareem 2011; Huang 2016). Their purpose is to estimate peak factors, based on the translation process (Grigoriu 1984). Peak factor can be estimated using the Hermite moment-based model (Winterstein 1988) applied to the Gaussian peak factor of the classical (Davenport 1964) model. This Hermite moment-based approach is obtained using a transformation of a Gaussian into a non-Gaussian softening process. For many wind pressures, there is a good correlation between the peak factor g and high-order statistics (skewness γ_3 and kurtosis γ_4). However, the Hermite moment-based model is less precise when pressures are significantly non-Gaussian. There exist also more advanced models, such as the modified Hermite model (Tognarelli et al. 1997) or the revised Hermite model (Winterstein et al. 1994) which are able to deal with stronger non-Gaussianity. The former is quite heavy to solve due to a non-linear system, the latter requires conditions on γ_3 and γ_4.

Wind pressure on a low-rise building roof can be highly non-gaussian and present different modes; the turbulent background and a component linked to the main flow behavior. By studying the probability density function (PDF), such modes can be highlighted, together with the analysis of higher rank properties such as the autocorrelation, the triple correlation. The attempt to separate the modes makes it possible to select the process of interest in the tail of the distribution and to restore the good agreement between g and (γ_3, γ_4). This is the option that is followed in our study.

Another kind of solution is a point-to-point distribution function mapping between Gaussian and non-Gaussian processes, assuming a Weibull or Gamma distribution (Sadek and Simiu 2002). This model allows to obtain an analytical evaluation of the peak factor but only in the case of a Weibull or Gamma distribution and is not enough general to handle multiple mode distribution.

The determination of an extreme wind load at a given point is not a final objective in itself. Many applications such as façade element design require the computation of the resultant of wind pressures on areas. Several researches have been carried out to make a link between time and space averaging. This tool allows to compute the average wind loads on a desired area by an equivalent time-averaging of the pressure at a location. The area-averaged quantity necessary for the dimensioning can be computed by multiplying pressure time histories by the aerodynamic admittance $\chi^2(n)$ (Holmes 1997). This function depends on the frequency of pressure fluctuation n and the coherence function γ defined in the area of interest $A(\bar{x}_1, \bar{x}_2)$. The main challenge is to set the expression of γ. Lawson (1980) proposes a solution to compute the area-averaged pressure by filtering the time pressure at a location using a moving average of window $T = KL/V$, known as the *TVL* equation. V is the freestream wind speed, L a characteristic length of the structure and K a constant that has to be computed. In the original *TVL* equation, K was estimated to 4.5 but (Holmes 1997) proposed to change it to 1 making the analogy to a square plate. In a more general case, (Amerio 2014, 2017) suggested to compute K by looking at its variability over a large number of peak locations. The *TVL* relation is set thanks to the definition of an equivalent width $w = \sqrt{A}$, with A the area where peak pressures occur. Nevertheless, no clear trend for K has yet been found. The main issue with the time-space equivalence is the fact that pressure is stationary (even ergodic) in time but not in space. For instance, over a low-rise building flat roof for example, the flow shows corner vortices, meaning that there is no spatial stationarity. This method has thus a limited range of applications.

The main contribution of the paper is to propose a new way to separate the different contributions in the wind flow and to estimate peak factors based on long time wind pressure measurements. High-order and ranks statistics are studied and the general methodology that is proposed to decompose the wind pressure can be applied to other practical cases of wind around buildings that require an estimate of wind loads on areas.

The developed methodology will be illustrated with the analysis of wind pressure measurement on a square low-rise building flat roof, with a 45° wind orientation. After understanding the kind of flow involved and basic pressure statistics, the parallel between autocorrelation and probability distribution decomposition is applied to decompose the pressure into simple components. Finally, it is shown that usual peak factor formulations are sufficiently accurate to model extreme values.

2 Case Study: Wind Tunnel Setup

The setup made by Blaise et al. (2017) consists of a square plan-form low-rise building (sides of 45 m and a height of 25 m), uniformly instrumented by 121 pressure taps in a quarter of its roof (see Figs. 1 and 2), sampled at $f_s = 500$ Hz. The model is placed at an incidence of 45° in the atmospheric boundary layer test section of the wind tunnel (WT) of the University of Liège. This direction is fixed to obtain a symmetry in the complex nature of the corner vortices developed at the edges (Kawai and Nishimura 1996).

The atmospheric boundary layer test section creates a turbulent wind of category III according to the Eurocode (1991), with $z_0 = 0.3$ m and $z_{min} = 5$ m. The mean velocity (Eq. (1)) and turbulence intensity (Eq. (2)) profiles are presented in Fig. 3.

Fig. 1. Model inside the WT of University of Liège

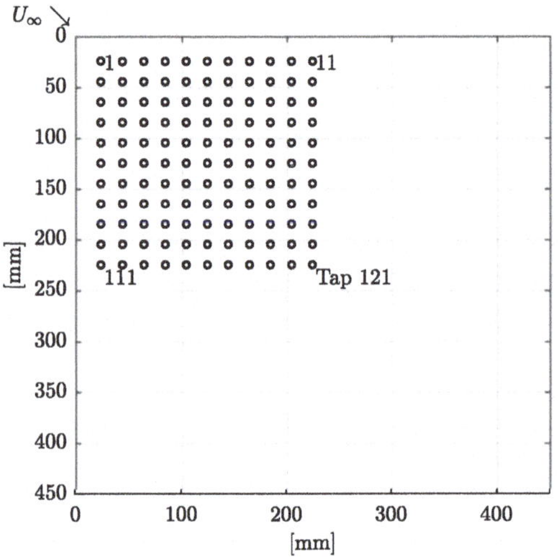

Fig. 2. Location of pressure taps on the model, the wind direction is fixed at 45°.

$$U_\infty(z) = U_{ref} k_r \ln(z/z_0), k_r = 0.19(z_0/0.05)^{0.07}, \quad (1)$$

$$I_u(z) = \frac{1}{\ln\left(\frac{z}{z_0}\right)}. \quad (2)$$

Pressure taps are linked to the scanner using pneumatic connection (vinyl tubes) of internal diameter of 1.32 mm and length 600 mm. The static pressure is well measured but unsteady components have to be dynamically corrected, thanks to the theoretical formulation of Bergh and Tijdeman (1965) and the experimental apparatus of Rigo (2017). Figure 4 presents the transfer function, in amplitude and phase, necessary to correct measurements, until the sampling frequency of 500 Hz.

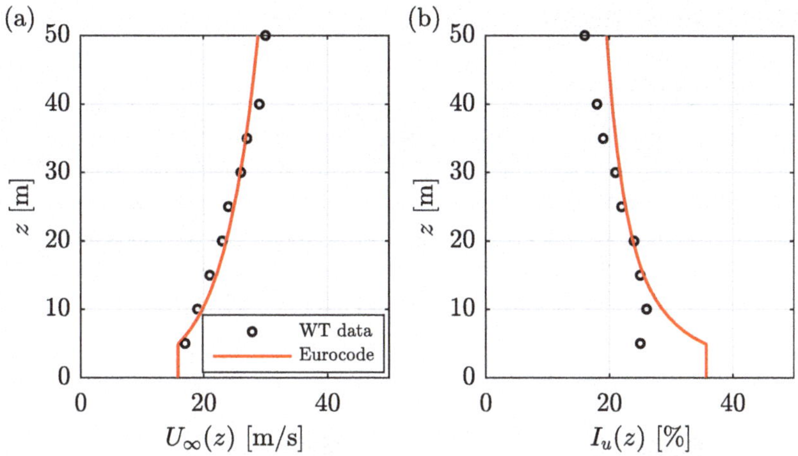

Fig. 3. (a) Mean velocity $U_\infty(z)$ and (b) turbulence intensity $I_u(z)$ profiles of the atmospheric boundary layer: measurements and comparison with suburban category III terrain, from Eurocode (1991), (Blaise et al. 2017).

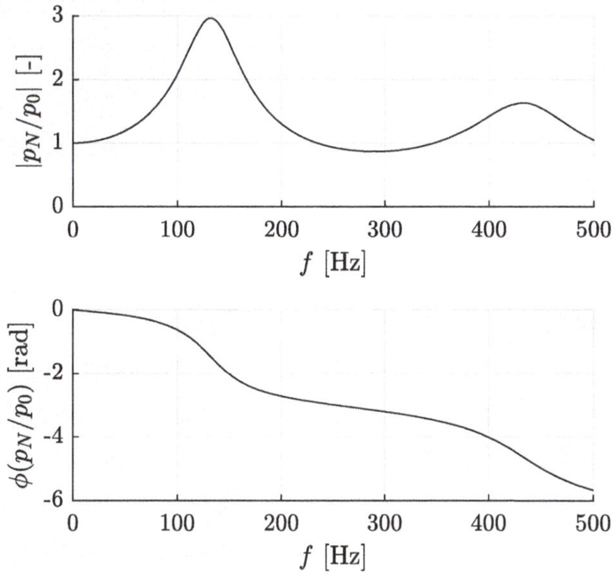

Fig. 4. Transfer function of the pressure tubes, in amplitude and phase of corrected pressure as a function of the frequency (Bergh and Tijdeman 1965; Rigo 2017).

3 Wind Tunnel Results

3.1 First Rank Statistics

The geometric scaling is $\lambda_L = 1/100$. The Reynolds condition would require a velocity scaling of $\lambda_U = \lambda_L^{-1} = 100$, which would impose WT speed of about 1000 m/s, impossible. The Reynolds dependency is more critical for smooth and mainly circular shapes such as cylinder or sphere. When the flow is completely separated and turbulent, the Strouhal scaling is preferred. Because of WT performances, the velocity scaling is $\lambda_U = 1/3.5$. The Strouhal condition imposes $\lambda_T = \frac{\lambda_L}{\lambda_U} = 1/28.6$ (time).

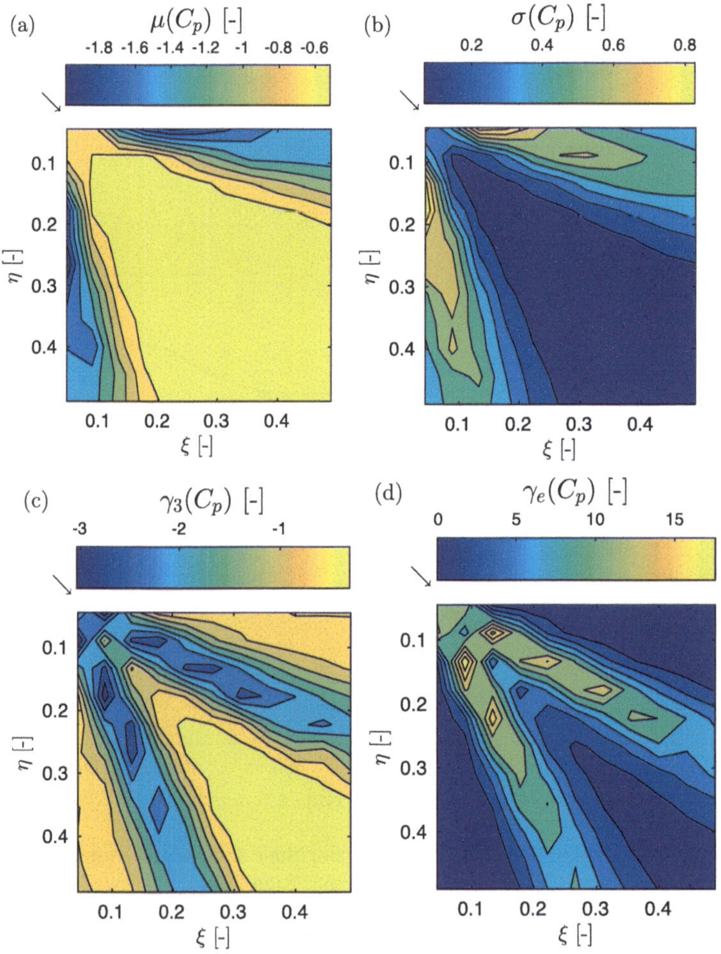

Fig. 5. Map of mean (a), standard deviation (b), skewness (c) and kurtosis excess (d).

Every measurement is converted in full scale. A total of 13 h of measurement has been taken in the WT, corresponding to 371.8 h full scale. This long information allows to compute precise PDF tail, necessary to study the extreme values. In the following, statistics are presented in a non-dimensional way: position $(\xi, \eta) = (x/L, y/L)$ and pressure coefficient with Eq. (3).

$$C_p = \frac{p - p_\infty}{\frac{1}{2}\rho U_\infty^2} \tag{3}$$

The Reynolds number of the WT model is 6.8×10^5. The flow around this low-rise building is characterized by corner vortices, that roll in a cone shape. The pressure is mainly negative because of the separated nature of the flow ($\mu(C_p)$ in Fig. 5(a)) and is the most negative where it starts to separate, just after the edges, on the corner. The two cones on both corners have a main axis, recognizable by the ridgelines of the standard deviation of the pressure coefficient $\sigma(C_p)$, in Fig. 5(b). The non-Gaussianity happens in the corner vortices boundaries, where $|\gamma_3(C_p)|$ and $|\gamma_e(C_p)|$ are the highest. For a Gaussian process, $\gamma_3 = 0$ and $\gamma_4 = 3$. The kurtosis excess $\gamma_e = \gamma_4 - 3 = 0$. On Fig. 5, a lower right triangle appears systematically, with low statistics values. This region is mainly Gaussian, with a low dispersion and mean pressure value. Physically, the flow in this region is the case of a simple turbulent flow on a flat plate. This region will not be studied, since it represents less interest compared to the complex flow in corner vortices.

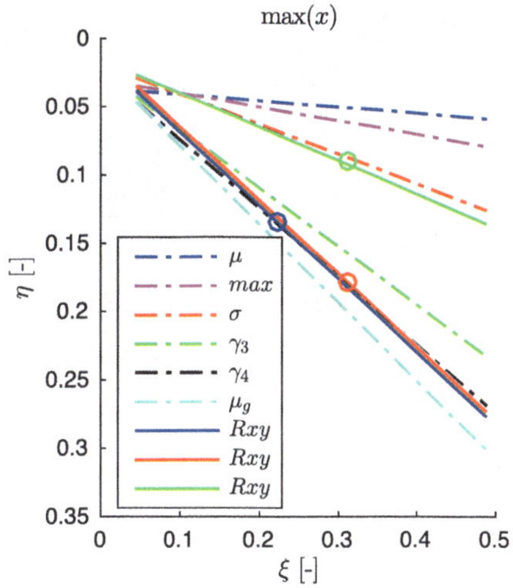

Fig. 6. Topology of corner vortices: ridgelines of pressure statistics.

The topology of the corner can be identified by comparing ridgelines of statistics, thanks to maximum statistics values in Fig. 6. The vortex starts where separation starts (highest $|\mu(C_p)|$), the main axis is located where $\sigma(C_p)$ is maximum and the vortex ends up by rolling down, where $|\gamma_3(C_p)|$ is the highest. For a point on the main vortex axis (green o), the cross-correlation is high on the axis. The same observation is valid for a point on the vortex roll down (in blue and red). Knowing this particular flow, pressure in the corner vortices has to be carefully analyzed.

As suggested by Kawai and Nishimura (1996), there is one main vortex at the corner, but a secondary small one develops just next to the edge, below the main vortex. Pressure taps in this setup were not enough close to the edge to capture well this secondary vortex. Nevertheless, in Fig. 5, $\gamma_3(C_p)$ increases just next to the edge, at $(\xi, \eta) = (0.05, 0.4)$, suggesting the presence of another vortex.

By looking at skewness and kurtosis values, each pressure distribution in the corner vortices has a negative skewness (negative extreme pressure events) and a positive excess kurtosis (called softening process, the PDF has a flatter shape compared to a Gaussian process). Based on time series, the peak factor is computed as $g = \max_T \left|\frac{C_p - \mu}{\sigma}\right|$, where T is the reference window size taken, conventionally 10-min. Figure 7(a) shows a map of the mean value of peak factor of the pressure coefficient $\mu_g(C_p)$, the highest values happening were the vortex cone touches the roof, similarly to γ_3 and γ_e.

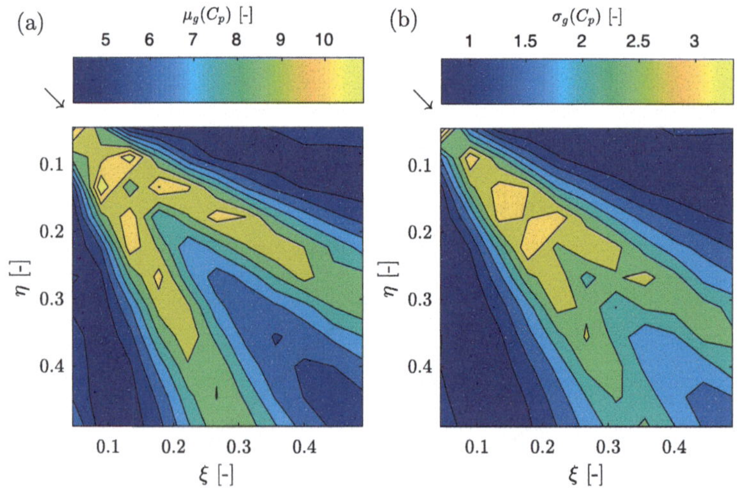

Fig. 7. Map of peak factor (a) mean and (b) standard deviation.

Figure 8 allows to have a quick qualitative view of the PDF. Each PDF is represented in log-scale at all 121 pressure taps location. The same scaling is used to represent them and have a quick comparative view of the dispersion and asymmetry of PDFs. Those in the corner vortices are highly non-Gaussian and two modes (two bumps) are clearly identifiable. This motivates the discussion of Sect. 4. Figure 9 shows a typical pressure coefficient signal (at tap 5): with local negative extreme pressures ($\gamma_3 < 0$) and wide range of distributed values, no concentrated values around the mean (softening, $\gamma_e > 0$). Moreover, the two modes are identifiable with two principal levels at around $C_p = -0.9$ and -2.1.

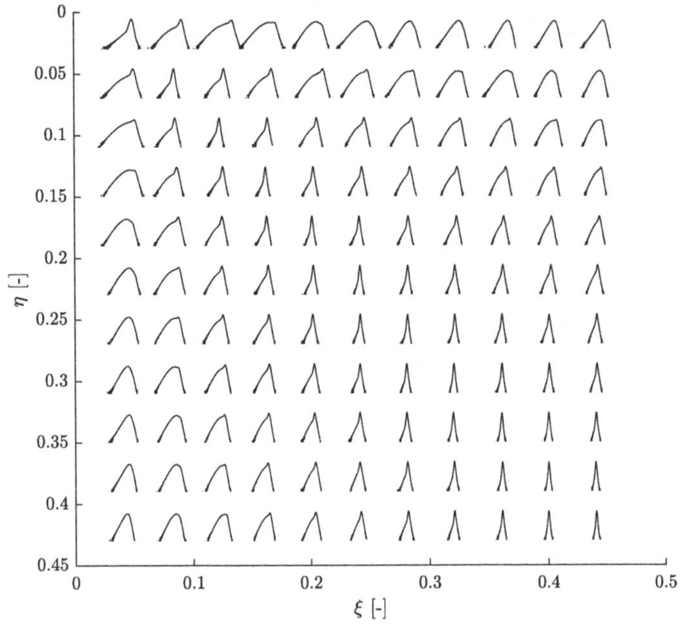

Fig. 8. Map of all PDFs.

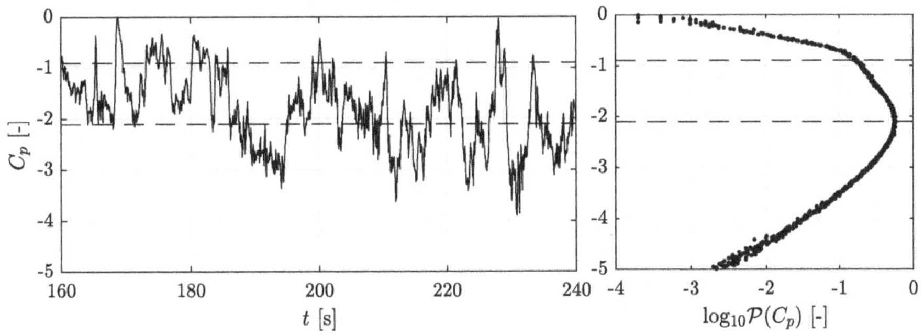

Fig. 9. Pressure signal at tap 5 and PDF (log-scale).

3.2 Peak Factor Estimation

As motivated, the purpose of this study is the peak factor estimation. A very widely used technique for non-Gaussian processes is the Hermite moment-based model (Kareem and Zhao 1994) (KZ):

$$\mu_{g_{Her}} = \kappa\left\{\left(\beta + \frac{\gamma}{\beta}\right) + h_3\left[\beta^2 + 2\gamma - 1 + \frac{1.98}{\beta^2}\right] + h_4\left[\beta^3 + 3\beta(\gamma - 1) + \frac{3}{\beta}\left(\frac{\pi^2}{6} - \gamma + \gamma^2\right) + \frac{5.44}{\beta^3}\right]\right\} \quad (4)$$

where h_3, h_4, κ are coefficients of moment-based Hermite model (Winterstein 1988), $\gamma \approx 0.5772$ is the Euler's constant, and

$$\beta = \sqrt{2\ln(\nu_0 T)}, \quad \nu_0 = \sqrt{\frac{m_2}{m_0}}, \quad m_i = \int_0^\infty n^i S_y(n)dn, \quad h_3 = \frac{\gamma_3}{4 + 2\sqrt{1 + 1.5\gamma_4}}, \quad h_4 = \frac{\sqrt{1 + 1.5\gamma_4} - 1}{18}, \quad \kappa = \frac{1}{\sqrt{1 + 2h_3^2 + 6h_4^2}} \quad (5)$$

where ν_0 is the mean zero upcrossing rate of process y (standardized non-Gaussian process obtained from a non-Gaussian process $x, y = (x - E[x])/\sigma_x$), T is the duration, m_i is the ith spectral moment of y, $S_y(n)$ is the one-sided power spectral density of y, n is the frequency in Hz. The validity range of the application of Hermite model is defined in a monotonic region, satisfying (Winterstein and MacKenzie 2012)

$$\gamma_e \geq (1.25\gamma_3)^2 \quad (6)$$

Clearly, γ_e is positive, valid for a softening process. Choi and Sweetman (2010) defined another practical boundary for γ_e vs γ_3 of a random process,

$$\gamma_e \geq \gamma_3^2 - 2 \quad (7)$$

Figure 10 shows γ_e vs γ_3 from WT data with boundaries of Eq. (6), which is not always satisfied, while Eq. (7) is always verified.

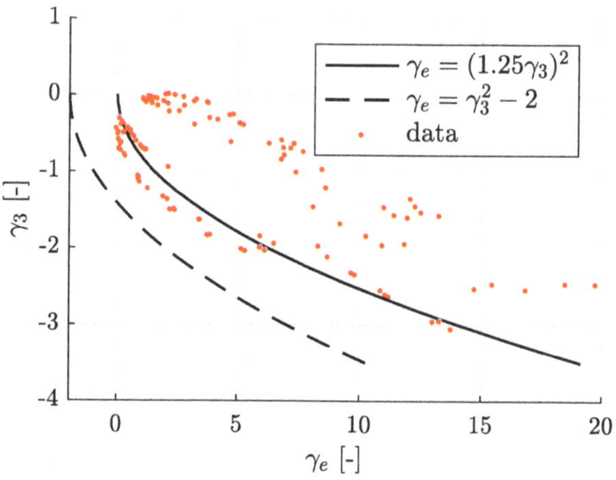

Fig. 10. Skewness γ_3 and kurtosis excess γ_e coefficients of all pressure taps with boundaries of Eqs. (6)–(7).

Figure 11 shows a comparison of the mean peak factor from data μ_g with the Hermite estimate $\mu_{g_{Her}}$. Most of estimates are well correlated, especially for processes with smaller peak factors, which tend to be just slightly non-Gaussian. Our aim is the apply to KZ formulation on a component only of the pressure (the one that matters in the tail) and also provide good correlation for larger peak factors. Peak factor estimates which do not verify Eq. (6) deviate from the bisector and have to be corrected.

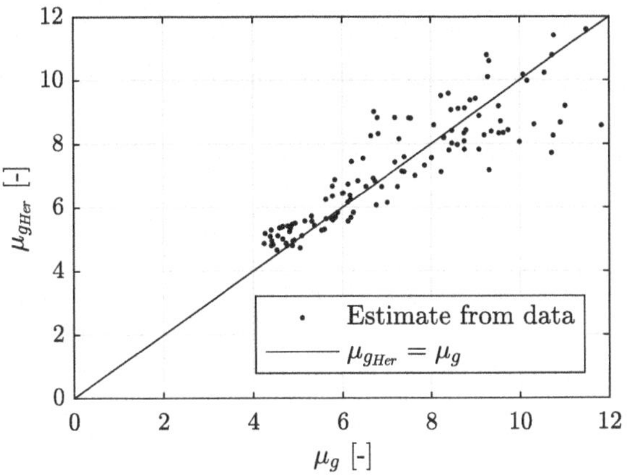

Fig. 11. Comparison of mean peak factor from direct statistical analysis μ_g and the Hermite estimated value $\mu_{g_{Her}}$.

3.3 Time Windowing Influence

As discussed in the motivation, the equivalent time-space averaging cannot be performed using the *TVL* equation because of the spatial non-stationarity inherent to corner vortices. Nevertheless, by choosing a sufficiently fine pressure tap distribution, wind loads can be computed over an appropriate area and the structure dimensioning can be carried out. When looking at the time resolution, the question can be raised on the necessary precision. With the increased pressure sensors sensitivity, sampling frequencies can be increased, at more precise events can be captured. If their characteristic time of apparition is very low, such extreme pressure events will not have a major impact on the whole structure. Starting from the recorded results, it is possible to see the change of statistics when the signal is filtered (considering typical 10 min windows). The influence of the windowing is studied using three kinds of filters:

- Moving average (MA): the pressure is filtered by taking its mean value over a window size of T_{MA} seconds.
- Low-pass filter (LP): this filter passes only frequencies lower than the cutoff frequency f_c.
- Decimation (Dec): reduction of the sampling frequency f_s of the signal. Its equivalent to downsample the LP filtered signal by keeping only every M^{th} sample.

Figure 12 shows the ratio of the filtered mean peak factor μ_g over the original one (using the unfiltered signal) depending on the type of filter and the value of the cutoff frequency or window size. The case of tap 24 (in the corner vortex, close to the edge) is presented. MA and LP give very close results, but Dec shows some discrepancies. This is due to the downsampling, that creates a distortion of the signal, called aliasing (Blachuta and Grygiel 2008). In the following, the LP filter is kept. Figure 13 compares the evolution of the peak factor μ_g, skewness γ_3, kurtosis γ_4 and the estimate $\mu_{g_{Her}}$ with the LP filtering frequency f_c. For pressure tap 24, there is a good correlation between μ_g and $\gamma_3 - \gamma_4$ and $\mu_{g_{Her}}$ is well estimated. However, at tap 111 the correlation is less evident, the Hermite model being outside its validity range, leading to a less precise estimation.

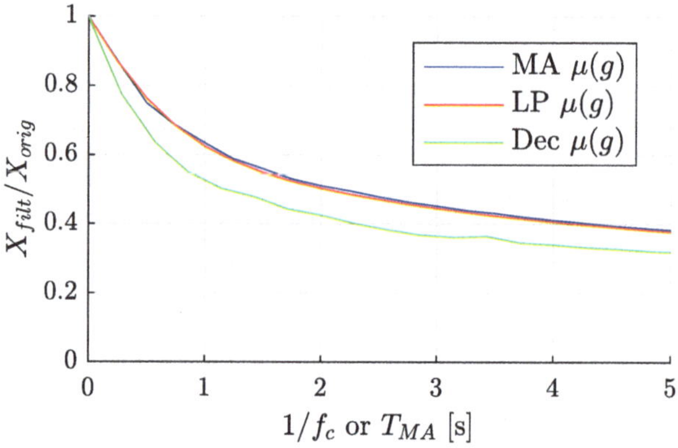

Fig. 12. Pressure tap 24: ratio filtered/unfiltered mean peak factor μ_g depending for three filters (MA, LP, Dec) as a function of the cutoff/sampling frequency $f_s = f_c$ and time window size T_{MA}.

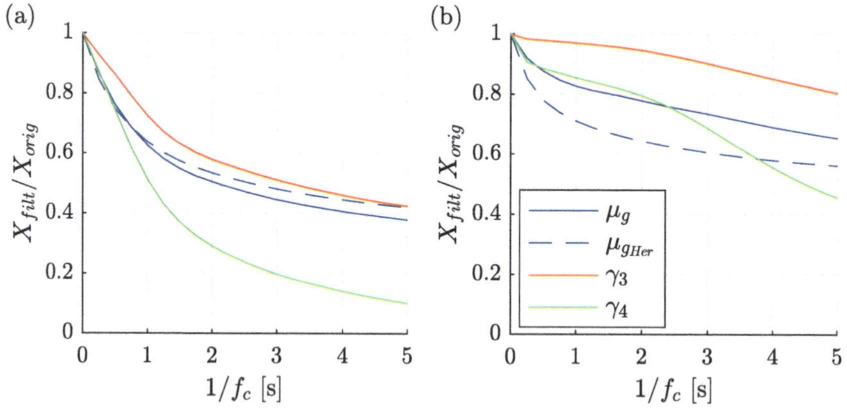

Fig. 13. Influence of LP filtering frequency f_c on peak factor μ_g, skewness γ_3, kurtosis γ_4 and comparison with the estimate $\mu_{g_{Her}}$. Pressure tap (a) 24 and (b) 111.

4 Decomposition and Mixture Model

By looking to the PDFs, it has been shown in Sect. 2 that several modes are present. Cook (2016) proposes a skew gaussian-exponential mixture model (SGEMM) based on a skew-gaussian and a skew hyperbolic secant distribution. The skew-gaussian (SG) is given by

$$p_{SG}(x) = 2\phi\left(\frac{x-\xi}{\omega}\right)\Phi\left(\alpha\frac{x-\xi}{\omega}\right) \tag{8}$$

where ϕ and Φ are the standard gaussian PDF and cumulative density distribution (CDF), ξ a location parameter, ω a scale parameter and α a skewness parameter. The skew-hyperbolic secant (SHS) distribution is

$$p_{SHS}(x) = \frac{\left[\frac{\alpha}{2}+\frac{1}{2\alpha}\right]\sin\left(\frac{\pi}{1+\alpha^2}\right)}{\exp\left(\frac{\alpha\pi}{2}\frac{(x-\xi)}{\omega}\right)+\exp\left(-\frac{\pi}{2\alpha}\frac{(x-\xi)}{\omega}\right)}. \tag{9}$$

The complete PDF can thus be decomposed into n_1 SG and n_2 SHS,

$$P(x) = \sum_{i=1}^{n_1} w_{SG,i}\, p_{SG,i}(x) + \sum_{i=1}^{n_2} w_{SHS,i}\, p_{SHS,i}(x) \tag{10}$$

where w_i are weighted parameters such that $\sum_{i=1}^{n_1} w_{SG,i} + \sum_{i=1}^{n_2} w_{SHS,i} = 1$. As each distribution requires 4 parameters, there is thus $4(n_1+n_2)-1$ free parameters to determine. By looking at PDF shapes, two SG are used for the main body of the PDF and one SHS for the tail. A fitting campaign is then carried out using a least square method. Firstly, by minimising $\log p(x)$ between the model and the data, to fit the tail. Then, tail parameters of SHS are fixed and SG parameters are re-fitted.

In the case of a bi-modal PDF, the mean and the variance of the total PDF can be linked to those of its components 1 and 2,

$$\mu = w_1\mu_1 + w_2\mu_2, \text{ with } w_1 + w_2 = 1 \tag{11}$$

$$\sigma^2 = w_1\sigma_1^2 + w_2\sigma_2^2 + w_1 w_2(\mu_1-\mu_2)^2 \tag{12}$$

When bumps are clearly identifiable, initial guesses for the location and scale parameters (ξ, ω) are easy to choose and the fitting converges quickly. Figure 14 shows the decomposition in modes of the PDF at pressure tap 34. The two modes are easy to identify by looking at the two levels in the time signal (Fig. 14(a)) and the two bumps in the PDF (Fig. 14(b)): (i) mode 1 corresponding to the turbulent background flow (as the one present on the roof region without vortex) and (ii) mode 2 to the vortices fluctuation (separated flow).

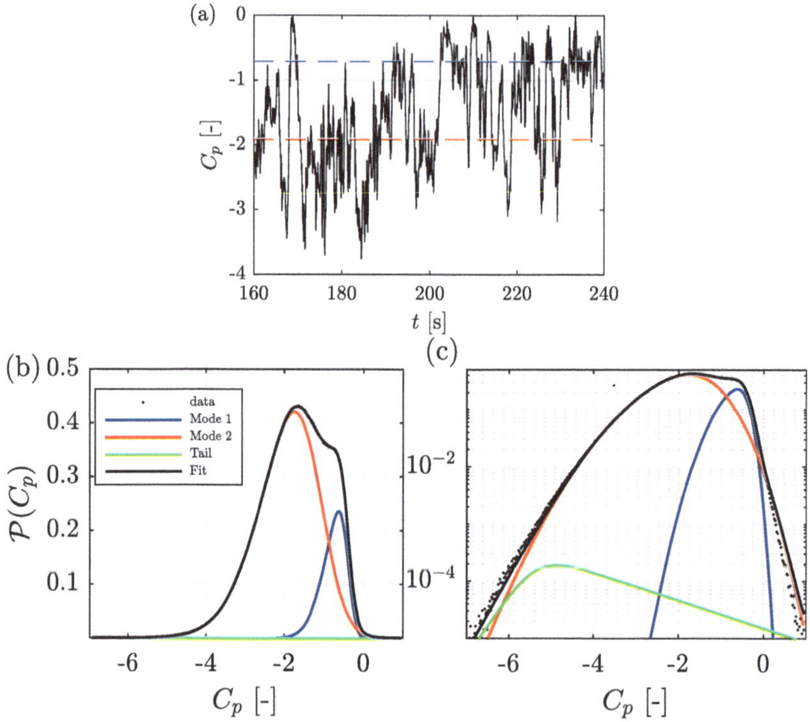

Fig. 14. Pressure tap 34: (a) time signal with modes mean and PDF decomposition in mode 1, 2, tail and resulting fit in (b) linear and (c) log scale.

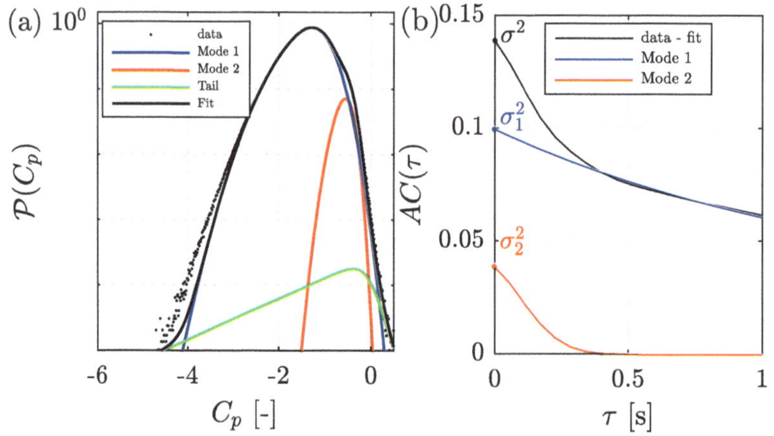

Fig. 15. Pressure tap 100: (a) PDF decomposition in mode 1, 2, tail and resulting fit in log scale and (b) autocorrelation decomposition in modes 1 and 2.

By going through the taps along the corner, the two bumps are getting closer and eventually merge, mode 2 becoming more and more important compared to mode 1. The method proposed by Cook (2016) is not robust anymore to identify initial values of the fitting algorithm. Nevertheless, by looking at the autocorrelation function in Figs. 15 and 16(b), the shape shows a break between two exponential decay functions, at around 0.4 s. This also suggests a decomposition in 2 modes, as before: (i) the turbulent background of mode 1 with a longer memory time than (ii) mode 2. The value of the autocorrelation at the origin is the variance. The autocorrelation can be often fitted using an exponential decay and is here decomposed in two components,

$$AC(\tau) = \int_{-\infty}^{+\infty} C_p(t)C_p(t-\tau)dt = w_1\sigma_1^2 e^{-A\tau} + w_2\sigma_2^2 e^{-B\tau} \tag{13}$$

The AC is composed of two exponential decays with a value at the origin corresponding to the weighted variance of the mode $w_i\sigma_i^2$. The value of AC at the origin is the variance σ^2. In the case of two close bumps, means of mode 1 and 2 are close and $w_1 w_2(\mu_1 - \mu_2)^2$ is negligible. The total variance is thus $\sigma^2 = \sigma_1^2 + \sigma_2^2$. Variances of both modes are identified and are used in the PDF decomposition algorithm, to make it more robust for the fitting convergence (see Figs. 15 and 16(a)).

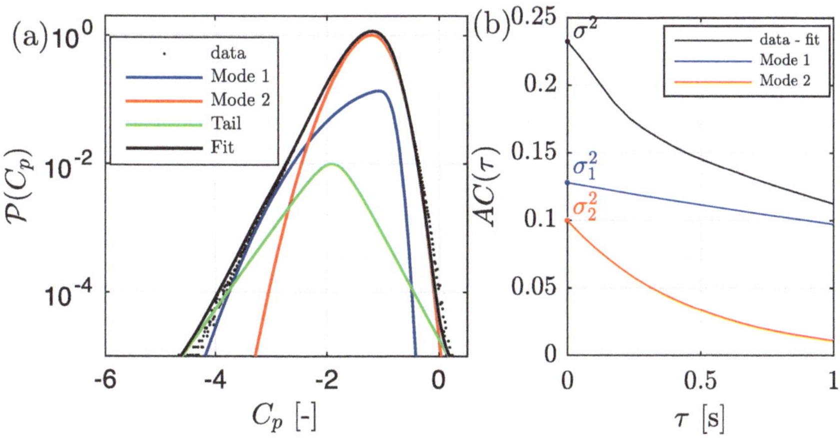

Fig. 16. Pressure tap 111: (a) PDF decomposition in mode 1, 2, tail and resulting fit in log scale and (b) autocorrelation decomposition in modes 1 and 2.

Figure 17(a) shows the map of the relative importance of variances of modes 1 and 2. Mode 1 is more present at the extremities of the vortex cone. Once PDFs are decomposed, one can access the statistical properties of a given mode. It allows to correct the value of skewness and kurtosis used in the extreme value model of Kareem and Zhao (1994). More precise estimation of peak factor is then obtained, compared to those obtained with the whole PDF, not robust in the model because of its strong non-Gaussianity. Figure 17(b) presents the new corrected peak factor estimations $\mu_{g_{Her}}$ more

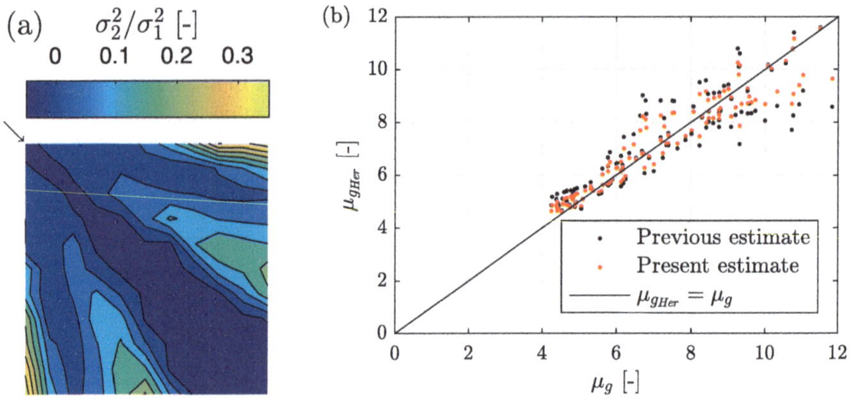

Fig. 17. (a) Map of relative importance of variance of mode 2 compared to 1 and (b) new peak factor estimate.

precise than the previous one. This was possible thanks to higher rank property, the autocorrelation.

The same procedure could be done using the triple correlation (with 2 delays) to decompose skewness of both modes.

5 Conclusions

In the present study, a statistical analysis has been carried out to estimate peak factor of wind pressure based on wind tunnel measurement on a low-rise building flat roof. High-order moments and ranks were used in the context of long time duration measurement. For most of the pressure taps, the trend of the peak factor is well correlated with skewness and kurtosis. When the validity range of Hermite moment-based model is respected, the estimated peak factor is precise. However, some pressure taps located on corner vortices boundary experienced strong non-Gaussianity and the Hermite estimate was less accurate. The physical analysis of the wind flow in parallel with the bi-modal nature of the PDFs in corner vortices suggested to decompose it and highlight the mode of interest. A methodology for the mixture model decomposition has been carried out. When the identification of the different modes was not clear on the PDF (first rank characteristic), the bimodal nature was highlighted thanks to a second rank characteristic: the autocorrelation, experiencing a kink. Variances of modes were estimated thanks to value at origin of the decomposed autocorrelation. They were then used as guesses for the PDF fitting. The peak factors were then re-estimated thanks to the main mode of interest. Analysis of higher-order and rank statistics allows to access new information that was hidden with basic and low-order properties. The triple correlation and the bispectrum could also be studied and used in this context.

Acknowledgements. The research described in this paper has been partially supported by the National Fund for Scientific Research.

References

Amerio L (2014) Numerical and experimental analysis of peak pressure loads on rectangular building. Master's dissertation, Politecnico di Milano

Amerio L (2017) Experimental high resolution analysis of the pressure peaks on a building scale model façades. PhD dissertation, Politecnico di Milano

Berg H, Tijdeman H (1965) Theoretical and experimental results for the dynamic response of pressure measuring systems. National Aero and Aeronautical Research Institute, NLR-TR F.238

Blachuta M, Grygiel R (2008) Sampling of noisy signals: spectral vs anti-aliasing filters. IFAC Proc Vol 41(2):7576–7581

Blaise N, Andrianne T, Denoël V (2017) Assessment of extreme value overestimations with equivalent static wind loads. J Wind Eng Ind Aerodyn 168:123–133

Chen X, Huang G (2009) Evaluation of peak resultant response for wind-excited tall buildings. Eng Struct 31:858–868

Choi M, Sweetman B (2010) The hermite moment model for highly skewed response with application to tension leg platforms. J Offshore Mech Arct 132:021602

Cook J (2016) Short communication: on the Gaussian-exponential mixture model for pressure coefficients. J Wind Eng Ind Aerodyn 153:71–77

Davenport AG (1964) Note on the distribution of the largest value of a random function with application to gust loading. J Inst Civ Eng 24:187–196

Eurocode (1991) EN 1991-1-3: Actions on Structures-General Actions-Wind Actions. European Committee for Standardization

Grigoriu M (1984) Crossings of non-Gaussian translation processes. J Eng Mech 110(4):610–620

Huang H, Huan S, Feng H, Lou W (2016) Non-gaussian time-dependent statistics of wind pressure processes on a roof structure. Wind Struct Int J 23:275–300

Holmes JD (1997) Equivalent time averaging in wind engineering. J Wind Eng Ind Aerodyn 72:411–419

Kareem A, Zhao J (1994) Analysis of non-Gaussian surge response of tension leg platforms under wind loads. J Offshore Mech Arct Eng 116:137–144

Kawai H, Nishimura G (1996) Characteristics of fluctuating suction and conical vortices on a flat roof in oblique flow. J Wind Eng Indus Aerodyn 60:211–225

Kwon D, Kareem A (2011) Peak factors for non-Gaussian load effects revisited. J Struct Eng 137 (12):1611–1619

Lawson TV (1980) Wind effect on buildings: design applications, vol 1. Applied Science Publishers, London

Rigo F (2017) Unsteady pressure measurement around aerodynamic bodies: development of a calibration apparatus and wind tunnel testing. Master's dissertation, University of Liège

Sadek F, Simiu E (2002) Peak non-Gaussian wind effects for database-assisted low-rise building design. J Eng Mech 128(5):530–539

Tognarelli MA, Zhao J, Kareem A (1997) Equivalent statistical cubicization for system and forcing nonlinearities. J Eng Mech 123(8):1772–1790

Winterstein SR (1988) Nonlinear vibration models for extremes and fatigue. J EngMech 114:1772–1790

Winterstein SR, MacKenzie CA (2012) Extremes of nonlinear vibration: models based on moments, L-moments, and maximum entropy. J Offshore Mech Arct 135(2):021602

Winterstein SR, Ude TC, Kleiven G (1994) Springing and slow-drift responses: predicted extremes and fatigue vs. simulation. In: Proceedings of BOSS-94, vol 3. MIT, pp 1–15

Vortex Induced Vibrations of Rectangular Cylinders Arranged on a Grid

F. Rigo[1(✉)], V. Denoël[2], and T. Andrianne[1]

[1] Wind Tunnel Lab, Department of Aerospace and Mechanical Engineering, University of Liège, Liège, Belgium
francois.rigo@uliege.be

[2] Stochastic and Structural Dynamics, Department of Urban and Environmental Engineering, University of Liège, Liège, Belgium

Abstract. A grid arrangement made of one hundred rectangular cylinders fixed to the facade of a house generates strong and disturbing monoharmonic noise. The cross-flow vibration of the rectangular cylinders is identified as the origin of the noise. The present article proposes a complete investigation of the Vortex Induced Vibration (VIV) combined with a grid effect. It is based on in situ measurements, numerical (finite elements and Computational Fluid Dynamics (CFD)) and extensive wind tunnel (WT) modelling. A comparison between unsteady pressure measurements and CFD results allows to understand the vortex shedding process and synchronization type depending on the wind incidence and spacing of the cylinders. On the basis of this multi- approach parametric investigation, a deep understanding of the VIV-grid phenomenon enables to propose two mitigation techniques. These techniques are tested and their effectiveness is reported in terms of vibration amplitude and acoustic intensity (Rigo et al. 2018).

Keywords: Vortex Induced Vibration · Unsteady · Pressure · Aerodynamics · Computational Fluid Dynamics · Bluff body · Wind tunnel · Finite element model

1 Introduction

A vertical louver structure installed on the facade of a house is composed of aluminium rectangular cylinders with a constant horizontal spacing (Fig. 1). This architectural structure is dedicated to isolate the occupants installed on the outdoor terrace from the neighbourhood. The grid was designed based on aesthetic and static loading constraints: it is clamped at four extreme points into the piers of the facade.

An intense acoustic emission was reported under specific wind conditions (fall and spring). This undesirable phenomenon creates a characteristic single frequency noise between 50 Hz and 100 Hz. Due to the large number of rectangular cylinders (100 in total), the phenomenon taking place is similar to a

F. Ricciardelli and A. M. Avossa (Eds.): INVENTO 2018, LNCE 27, pp. 630–651, 2019.
https://doi.org/10.1007/978-3-030-12815-9_48

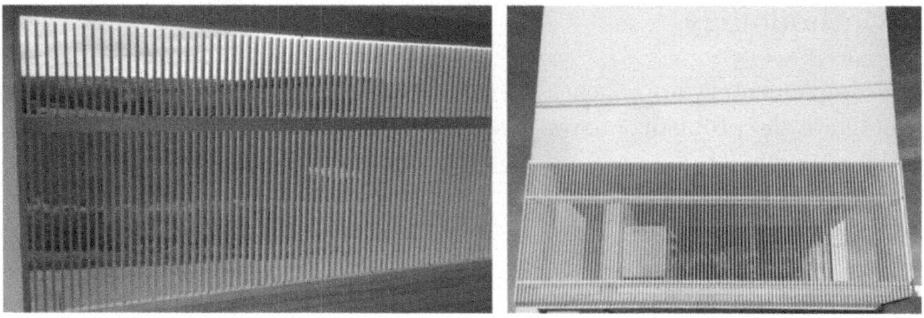

Fig. 1. Real grid on site, from inside (left) and outside (right)

mono-frequency wind instrument creating a strong and unacceptable noise. In addition, vibrations of the louver are transmitted to the structure of the building. On top of this comfort problem, the issue of the safety due to a potential accelerated fatigue degradation of the structure must be considered.

In this paper, a complete analysis of the behaviour of the grid under wind is performed. The objective is first to understand and characterize the aeroelastic instability and then to solve it using an adequate mitigation technique. The starting point is the study of the grid in its initial configuration, then some parameters will be explored, to understand their effects and how they can be changed to solve the problem. The results of the study can be extended to similar structures where the aesthetic objective leading to light and flexible designs must be consistent with external excitations such as atmospheric wind. For that purpose, *in situ* measurements are performed, to make the link between the vibration of cylinders and the wind conditions (amplitude and direction). Afterwards, a wind tunnel model is developed, based on the real grid, to reproduce real conditions and study other configurations through a parametric analysis. The controlled conditions in laboratory enable to investigate the interaction between wind conditions, vibration and acoustic levels. Finally, numerical methods are used to compare and validate results, by studying structural (modal) properties of the grid on one side (using FEM) and fluid dynamics around the grid on the other (using CFD, with `OpenFoam`).

From the initial observations of the structure installed on the house (single frequency, wind direction and amplitude dependencies), the aeroelastic effect responsible of these vibration is the vortex shedding process in the wake of each rectangular cylinder. Hence it consists of a Vortex Induced Vibration (VIV) phenomenon, which results from the interaction between fluid and structural forces, at a given frequency (Techet 2005; Blevins 1990). Indeed, observations show a phenomenon which is self-limited, directional and occurs for a certain velocity range (5 m/s to 8 m/s).

2 Methodology

Three complementary approaches are carried out to understand the phenomenon and mitigate the problem: *in situ* measurements, wind tunnel parametric study on a 1:1 scale model and comparison with numerical models (FE and CFD).

2.1 *In Situ* Measurement

Experimental measurements are performed on the grid during half a week in windy conditions leading to vibrations of the structure. Anticipating the mitigation technique, two rectangular cylinders were filled with sand. The instrumentation is shown in Fig. 2. It consists in

1. Three mono-axis accelerometers (PCB-Piezotronics 3713B112G) are placed at mid-height of 3 central cylinders of the grid. The two cylinders filled with sand are instrumented, plus another one empty. These three cylinders are next to each other.
2. An anemometer to measure the wind speed U and direction α. The wind components (u, v) are measured in the system of reference of the grid.
3. An acquisition system (computer and internet connection) to remotely acquire data at 640 Hz.

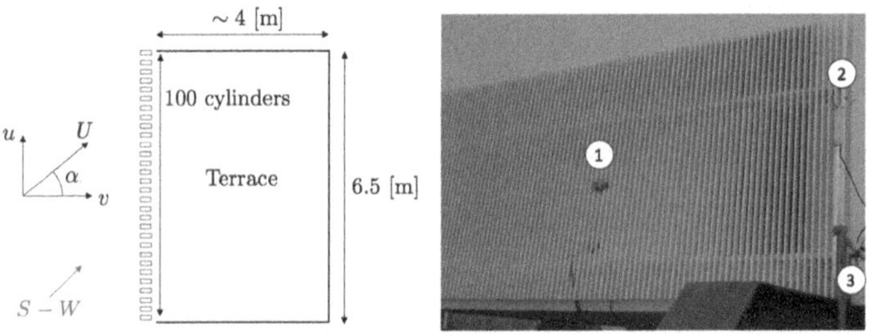

Fig. 2. Grid and terrace geometry, wind components (left) and *in situ* instrumentation: acclerometers (1), anemometer (2), data logger (3) (right)

The modal properties of the grid are identified experimentally. The grid is made with 100 cylinders of length 2.3 m, with a cross section of $B \times D = 5\,\text{cm} \times 2\,\text{cm}$, thickness of 2 mm and spaced by $T = 4\,\text{cm}$. The cylinders are welded on two horizontal supports, vertically spaced by 1.35 m. The welding is equivalent to a pinned support (picture in Fig. 2 and schematics in Fig. 3). Table 2 shows the eigenfrequencies and damping ratio of the three instrumented cylinders from ambient vibrations. Because of the higher mass of cylinders 1 and 2 filled with sand, the first eigenfrequency is lower than the one of the empty cylinder 3. In addition the sand in cylinders 1 and 2 increases their damping ratio. The Scruton numbers, calculated from the formula proposed in the Eurocode (EC) (introduced in Sect. 1), are presented in this table. The lower value of the third (empty) cylinder, $Sc = 36.5$, is an indicator that the structure might be affected by VIV (Hansen 2007).

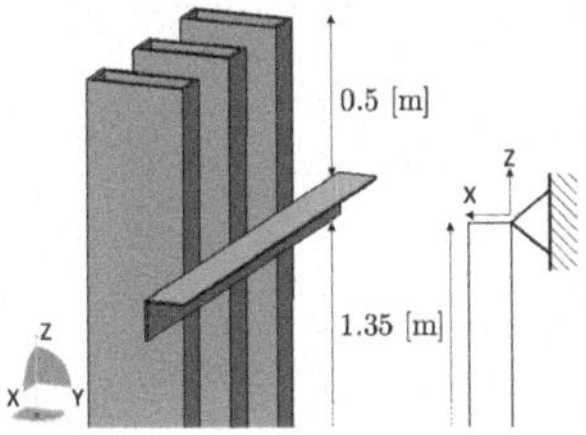

Fig. 3. 3D fixation and boundary condition simplification

Table 1. Modal parameters of the grid

	f_1 Hz	f_2 Hz
Experimental	59	176

Table 2. Modal characteristics from ambient vibrations of the 3 cylinders instrumented.

Cylinder	1	2	3
f_1 Hz	44	43	59
ς %	1.1	0.9	0.5
Sc from (EC)	196.7	160.9	36.5

2.2 Wind Tunnel Model

The wind tunnel model consists in a 1:1 scale reproduction of the real grid with 10 cylinders closely spaced and one isolated (Fig. 4). Due to height limitations of the test section (max 1.8 m), it is decided to limit the height of the cylinders to 1.35 m (equal to the spacing between the two supporting horizontal beams on site). The 11 rectangular cylinders are connected to a rigid frame with grooves by M3 screws. This allows the horizontal positioning of cylinders to study the spacing parameter T/D. To guarantee the same fixation type and strength, a setting torque tool is used to screw with a constant torque of 15 Nm. The ensemble (Fig. 4) is screwed on the turning table, to study the parameter angle of attack α.

The FEM model is used to justify the omission of the upper and lower parts of the cylinders and to retain only the central, bi-pinned part. The same aluminum tubes are used for the design. A comparison between the mode of interest of a single beam ($f_1 = 66$ Hz) and a FEM model of the real grid ($f_1 = 64$ Hz) shows very close results. The real grid shows other global mode associated to horizontal frame supports, which consist of simple bending at lower frequency ($f = 1.2$ Hz)

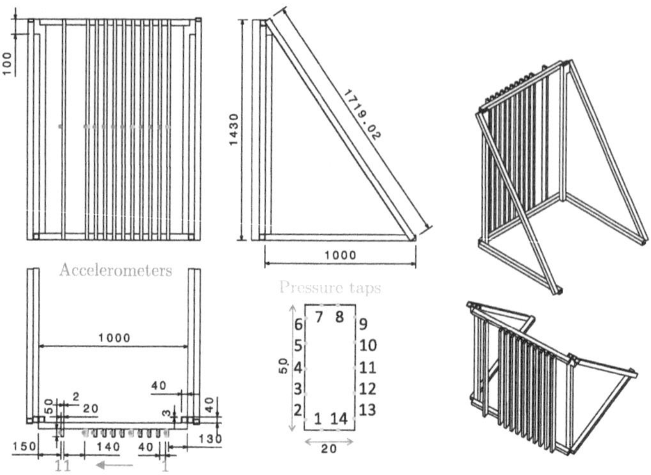

Fig. 4. Drawing of the grid model: top, side, front and axisymmetric views, locations of accelerometers, pressure taps and cylinders numbering, dimensions in mm

and it does not interact with the flexion mode of the cylinders. The analytic and FEM solutions show that using a beam of 1.35 m (between pin supports) with or without free ends have the same frequency. Modes associated with the free ends have higher frequencies.

Each cylinder is instrumented at mid-height by a mono-axis accelerometer, measuring the acceleration in the x-direction (see red points in Fig. 4). The complete instrumentation diagram together with the dimensions of the model installed in the test section are presented in Fig. 5.

Figures 6a and b show the Fast Fourier Transform (FFT) and damping ratios of the mode of interest measured experimentally on every cylinder of the wind tunnel model. The frequency is roughly the same for all cylinders (f = 61 Hz) and very close to the FE prediction (f_s^0 = 66 and 64 Hz in Table 1). Damping values are identified using logarithmic decrement. The damping was computed from period to period over the whole time history of the free decay: the mean and the standard deviation of results are reported for each cylinder in Fig. 6b. The damping ratio varies from one cylinder to the other despite the application of the same tightening torque. Note that this situation is probably the same on the real grid, where cylinders have been welded. This modal analysis validates the design of a grid composed of only 10 cylinders (instead of 100), because each cylinders behave structurally as a single beam (very close frequency, around $f_s^0 = 61 - 66$ Hz).

Unsteady pressure measurements are carried out on the surface of the cylinders. Five pressure taps are located along each cylinder chord B and two along cross flow dimension D, thus 14 in total (light points in Figs. 4 and 5). These taps are connected to a pressure scanner system (DPMS series by Turbulent Flow Instrumentation) by urethane tubes (L = 1.3 m and ID = 1.37 mm).

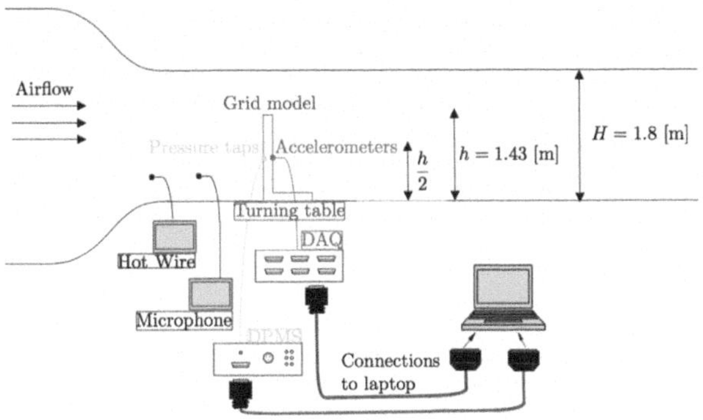

Fig. 5. Wind tunnel model and instrumentation

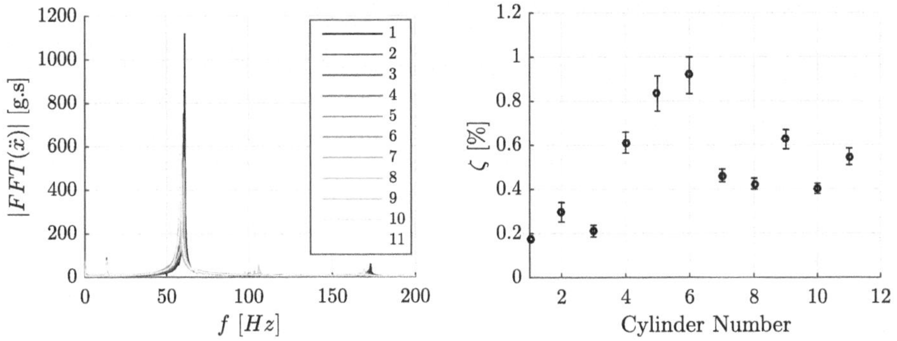

(a) FFT amplitude of cylinders acceleration $\ddot{x}$

(b) Damping ratio ζ of cylinders

Fig. 6. Experimental properties of wind tunnel model at wind-off conditions, free decay using hammer excitation

The objective is to study the vortex shedding generation and interaction (phase between vortex of two adjacent cylinders). Recorded pressure signals are dynamically calibrated thanks in the frequency domain using the theory by Bergh and Tijdeman (Bergh and Tijdeman 1967). For both fluid measurements, the sampling frequency is set to 500 Hz. The wind tunnel turbulence intensity is 0.2% and the blockage of the model is 9.5%.

2.3 Computational Fluid Dynamics

A CFD model of the grid is implemented in `OpenFoam`. An Unsteady Reynolds Averaged Navier-Stokes (URANS) turbulent flow is implemented using a $k - \varepsilon$ turbulence model and wall functions. The objective is to identify the unsteady characteristics of the flow (Strouhal number and force amplitudes) due to the vortex shedding process in grid configuration. The cylinders are static in the fluid domain. The size of the computational domain $L \times W = 48D \times 32.5D$. The boundary conditions of the fluid domain: at inlet (fixed constant velocity U, zero pressure gradient), at outlet (zero velocity gradient, fixed pressure $p = 0$), at cylinder wall (no slip: zero velocity and zero pressure gradient), up-down sides (slip: zero velocity-pressure gradient). Turbulent quantities are computed with `OpenFoam` functions, using the following initial conditions for inlet: turbulent intensity $I_u = \frac{U_1}{U} = 0.8\%$ and turbulent length scale $L/D = 0.0045$ (Ong 2013).

3 Results

3.1 *In Situ* Measurements

As stated previously, two cylinders were filled with sand to increase their mass and damping, hence increasing the Scruton number. The responses of the filled cylinders are compared to the one of an empty (non-modified) cylinder. Figure 7 shows the maximal acceleration for the three accelerometers together with the wind intensity U and direction α during the entire measurement campaign (78 h). It is observed that each vibration event (peak rms acceleration) happens when the wind speed is higher than 5 m/s, with a direction varying between 0° and 50°. The dominant wind direction in Belgium is the South-West and the average intensity is between 4 m/s and 5 m/s. Due to the orientation of the louver, the wind sector 0°–50° with a wind speed of 5 m/s is most likely to occur leading to frequent VIV event.

On the basis of these first observations, the aeroelastic phenomenon taking place is Vortex Induced Vibration (VIV). This single degree of freedom instability occurs at low wind speed in a given lock-in range. The critical velocity will be quantified in the following sections. The other single degree of freedom in cross-flow, namely transverse galloping, is rejected as responsible for these vibrations because the corresponding quasi-steady critical wind speed is 17.2 m/s. Note that this value corresponds to an estimation taking into account a structural damping of 0.5% (from Table 2) and a slope of the lift coefficient $A = \frac{dC_L}{d\alpha} + C_D = -5$ (calculated from the pressure measurements on the grid, see Fig. 13). This wind speed has never been reached during the in-situ measurement, as can be seen in Fig. 7.

It is proposed to study more precisely the strongest VIV event, taking place between t = 75–78 h. Figure 8 shows the vibration envelope $\ddot{x}_{rms}$ for each cylinders as a function of wind speed and associated wind direction. This figure summarizes the correlation between VIV events as a function the corresponding wind conditions $U - \alpha$, instead of time t only.

The maximum amplitude of filled cylinders is roughly half of the empty one. This observation is in opposition with the VIV models predicting maximum response amplitudes inversely proportional to the Scruton number: $y_{max} \sim 1/Sc$. On the basis of Table 2, a multiplicative factor of 5 or 6 between the amplitudes of the empty cylinder to the filled ones would be expected. Another important observation concerns the wind speed at which the VIV takes place: for the empty cylinder the calculated critical wind speed is around 5 m/s and it should be 3.7 m/s for the filled ones. Furthermore, it is observed from the upper plot of

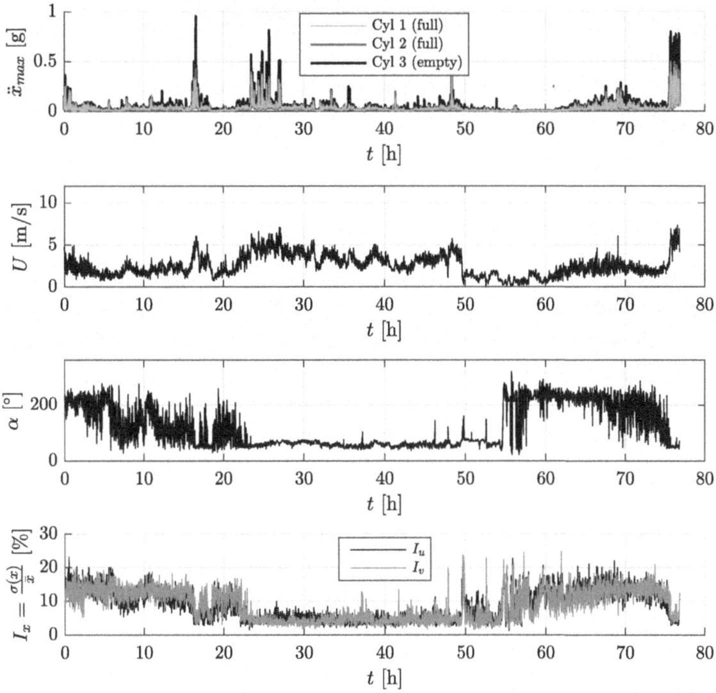

Fig. 7. *In situ* measurement: peak response during 7 half days, correlation event - $U - \alpha$

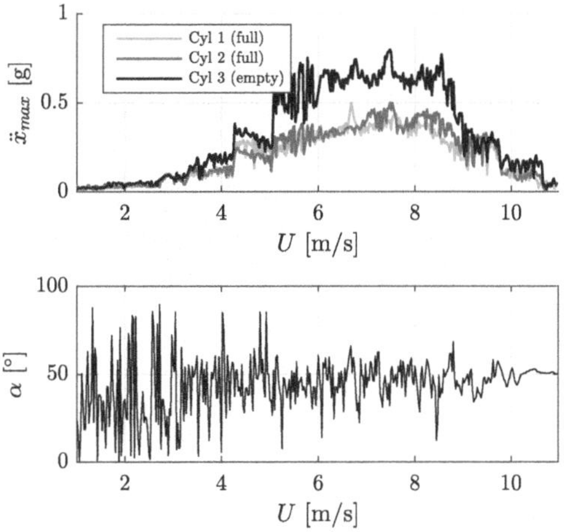

Fig. 8. *In situ* measurement: VIV curves $\ddot{x}_{rms}$ envelope for event at t = 77 [h] and corresponding α

Fig. 7 that the VIV take place simultaneously, i.e. for the same wind conditions, for the three instrumented cylinders. The explanation lies in the fact that the two filled cylinders are surrounded by 98 empty cylinders. If the wind conditions are met to have VIV on the latter, the resulting vibrations can be transmitted by the horizontal supporting beam and excite the filled cylinders in a seismic manner.

3.2 Initial Configuration in the Wind Tunnel

The wind tunnel model reproducing the geometry and dynamics feature of the grid is tested in controlled wind conditions. In this section we study the effect of the localisation of the cylinder in the grid and the wind direction. Figure 9 shows cylinders response as a function of airspeed for a wind direction $\alpha = 0°$. Similarly to the *in situ* measurement, the vibration starts around $U = 4-5$m/s for all cylinders, having the same frequency (see Fig. 6a). Unlike the site measurements, the acceleration reaches higher level (up to $6g$ here, vs. $1g$ on site). The comparison between site and laboratory measurements will be presented further. It is observed that the level of vibration is different from cylinder to cylinder. This can be explained by the different damping ratios (see Fig. 6b). An alternative manner to present the results is to multiply the rms acceleration by the damping ratio, a damping independent plot can be obtained (see right plot of Fig. 9). Cylinders at the center of the grid (4 to 6, numbering in Fig. 4) vibrate the most. They are not affected by end effects, while the amplitude vibration of cylinders 1, 2 and 9 10 at extremities is lower because they are influenced by a non-symmetric flow, even for $\alpha = 0°$. This observation is due to the grid effect, which is studied in more details in the following sections.

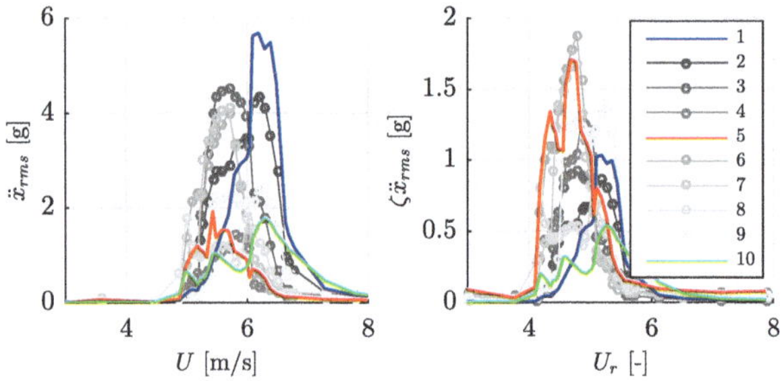

Fig. 9. Wind tunnel model: VIV curves at $\alpha = 0°$ and $T/D = 2$, $(\ddot{x}_{rms} - U)$ and $(\zeta\ddot{x}_{rms} - U_r)$

3.3 Parametric Studies

After identifying conditions that lead to VIV instability *in situ* and in the wind tunnel, a parametric study is performed to study the influence of wind direction and spacing of cylinders. Experimentally, the cylinders are free to vibrate and their acceleration, pressure distribution and wake behaviour are measured and analysed. The CFD results are compared to experimental data to obtain a complete knowledge of the flow around the cylinders. In the CFD model, the cylinders are held statically, so that the effect of the motion of the latters is neglected. This assumption is rather correct because of the small amplitude observed during *in situ* measurements (<1 mm). Hence, the CFD model is used to understand the flow, pressure and vortex shedding process.

3.3.1 Aeroelastic Responses

We focus first on the last two parameters: wind direction α and spacing T/D. Figure 10 shows the effect of α and T/D on cylinder response. The following observations are made: (i) increasing the spacing T/D increases the critical reduced velocity at which VIV occurs. (ii) The same tendency holds for the wind direction, hence these two parameters have an effect on the Strouhal number: increasing T/D or α decreases St. (iii) The cylinders located in the center of the grid reach the highest amplitude of vibration for zero and non zero wind direction. These cylinders are affected by a grid effect, amplifying the VIV phenomenon.

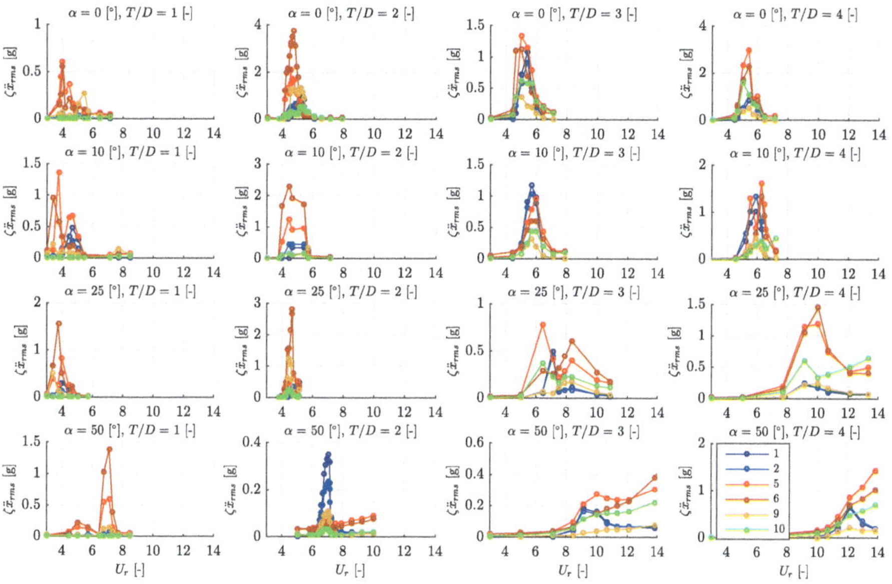

Fig. 10. Wind tunnel model: effect of α and T/D on the wind tunnel model

The acceleration of the flow through the grid alters the flow topology by changing the curvature of the separated shear layers and modifying the vortex shedding process. In the present parametric study, the change of flow topology is due to the grid effect or to the angle of attack. It is traduced by a variation of the Strouhal number. U_r^{crit} is lower and St appears higher for interfering cylinders than for a single one ($St = 0.22$ for $T/D = 2$ at $\alpha = 0°$ whereas $St = 0.17$ for a single cylinder, Fig. 12). This enhanced effect by the biased gap flow was previously suggested by Zdravkovich (Zdravkovich 1985). (iv) When wind direction is around 25° or 50°, last cylinders end (9–10) are subjected to end effect, lowering the vibration amplitude. Nevertheless, the first ones (1-2-3) in front of the flow are more sollicitated, according to the flow topology in Fig. 18.

In addition, an interaction between VIV and galloping is observed in the specific cases of $T/D = 3$ and 4 for an angle of attack of 50°. This interaction phenomenon is not studied in the scope of this work which focuses on the particular case of the grid with a spacing $T/D = 2$. The parametric analysis of the variation of the Strouhal number with α and T/D can be summarized graphically in Fig. 11, where St is identified from the VIV curves of Fig. 10, through $St = 1/U_r^{crit}$. St decreases when α and T/D increase. The effect of angle of attack is much more marked than spacing.

The grid effect is also presented in Fig. 12 where the Strouhal number vs α is plotted for an isolated body and different value of the spacing $T/D = 2$. It is observed that the Stouhal number monotonically decreases as a function of the angle of attack unlike the case of an isolated rectangle. In addition, in the case of an isolated cylinder, two values of St are reported in Fig. 12: a high value (0.17) and a low value (0.06). These values were measured by Knisely (Knisely 1990) who studied the St number of a rectangular cylinder characterized by different depth to chord ratios. Nakaguchi (Nakaguchi and Muto 1968) and Okajima (Okajima 1982) had shown by experiments that for cross sections with $2 < B/D < 2.8$, a double mode exists in the lift. Physically, it is because of an unsteady reattachment of the separated shear layer on side faces. There are discontinuities in the Strouhal number around $B/D \approx 2.5 - 2.8$. In the present work, only the high St value is observed experimentally in the case of a large spacing ($T/D = 8$, at $St = 0.18$, see Fig. 11). This value is identified because of the velocity range tested. This figure shows than the St approaches the one for a single cylinder when T/D increases.

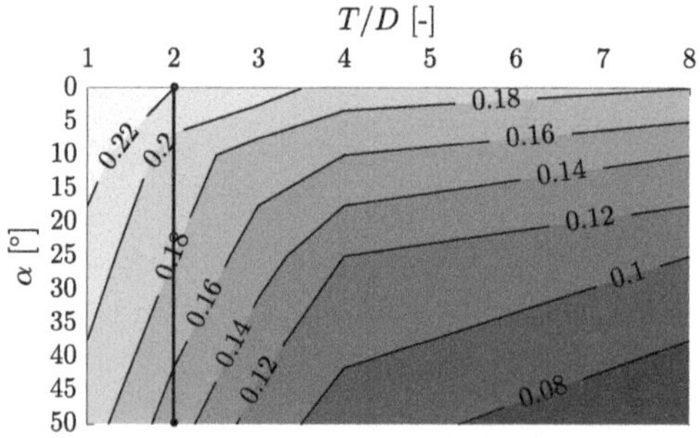

Fig. 11. Wind tunnel model: contour plot of Strouhal number (obtained from VIV response) vs. α and T/D, the black line corresponds to the studied grid.

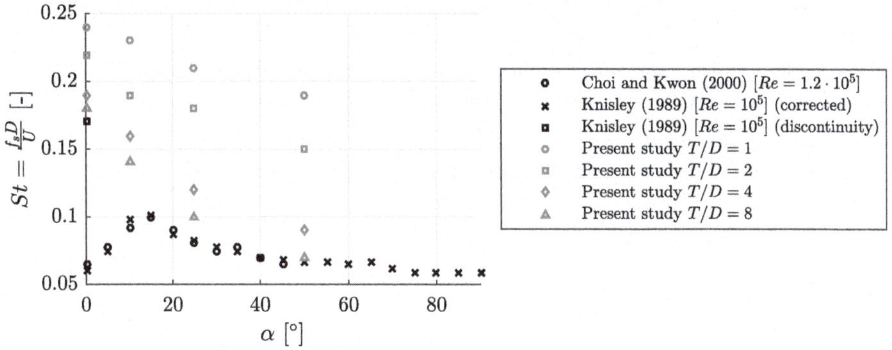

Fig. 12. Strouhal number of a single rectangular cylinder ($\frac{B}{D} = 2.5$), adapted from (Knisely 1990) and (Choi and Kwon 2000)

3.3.2 Pressure Distributions

Figure 13 show the mean of the pressure coefficients on cylinders 1, 5 and 10 for different values of the spacing parameter (T/D = 1, 2, 4 and 8) and wind direction ($\alpha = 0°$, 10° and 50°).

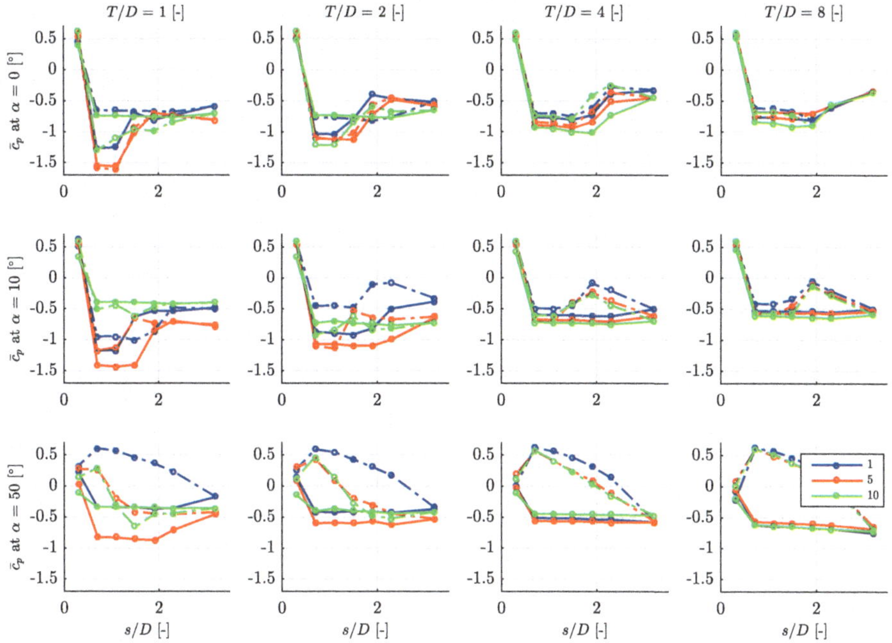

Fig. 13. Wind tunnel model: mean pressure distribution for $\alpha = 0°, 10°, 50°$ at $T/D = 1, 2, 4, 8$ for cylinders 1, 5, 10 (pressure coefficient $\bar{c_p}$ vs non-dimensional curvilign coordinate s/D), cylinder lower side in (-) and upper side in (-.)

Table 3. Strouhal St and lift standard deviation $\sigma(c_l)$ for cylinder 5 depending on α: comparison experiment (Wind tunnel model VIV response and lift coefficient) - CFD

	St			$\sigma(c_l)$	
α	Wind tunnel (VIV response)	Wind tunnel (lift coefficient)	CFD	Wind tunnel (lift coefficient)	CFD
0°	0.22	0.21	0.19	0.24	0.27
10°	0.19	0.20	0.18	0.20	0.26
50°	0.15	0.14	0.16	0.04	0.03

At $\alpha = 0°$, the mean pressure loading is symmetric on the upper and lower surfaces of cylinder 5 (central). It is also the case for the three cylinders but only when the spacing is high enough (T/D), so that they are considered as alone cylinders. It is observed that the mean pressure coefficients are more negative (up to $\bar{c_p} = -1.5$ for the case $T/D = 1$) on the central cylinder (numbered 5). The same remarks hold for the standard deviation of the pressure coefficient on this cylinder: the fluctuations are much higher than the external cylinders (numbered 1 and 10). This grid effect is important for small spacing ($T/D = 1$ or 2) and disappears for larger values. For all wind directions, the cylinders behave as single cylinders when the spacing is above 4. The plateau observed in Fig. 13 correspond to the flow separation region around the surface of the cylinder.

A quantitative comparison of the lift coefficient is presented in Table 3, where the capability of the numerical model to reproduce the flow is confirmed, except at large angle of attack. It is possible to make the link between this table and the summary Fig. 11: the values of St ≈0.22, ≈0.2 and ≈0.15 of this figure match the corresponding ones of the table for the three wind directions.

4 Discussions

This section discusses the main experimental and numerical findings presented above. The comparison of *in situ* and wind tunnel measurements is based on the level of turbulence of the incoming flow. A detailed analysis of the vortex shedding process of the cylinder in grid arrangement is presented. Finally, two mitigation techniques are proposed, tested and classified on the basis of their effectiveness to reduce vibration and acoustic emissions.

4.1 Comparison of the Approaches

Figure 14 gathers the VIV responses measured *in situ* and in the wind tunnel. The damping independent form is presented here to focus on the VIV-grid phenomenon only. The effect of direction has been studied in the previous section: a factor of 5 between 0° and 50° is observed in accordance with Fig. 10.

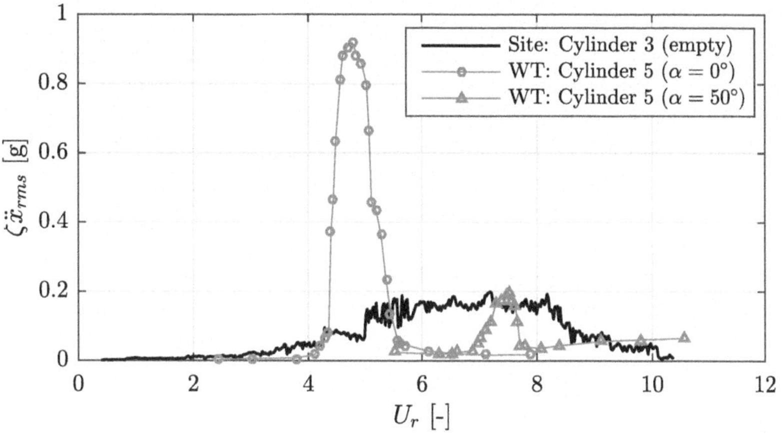

Fig. 14. VIV curve, damping independent acceleration amplitude $\zeta\ddot{x}_{rms}$, as a function of U_r: comparison between *in situ* (cylinder 3) and wind tunnel (cylinder 5 at $\alpha = 0°$ and 50°)

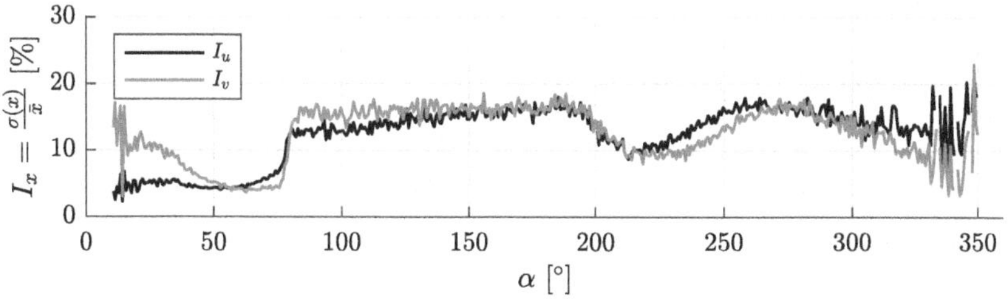

Fig. 15. *In situ* measurements: turbulence intensity (10 s windows) vs. wind direction

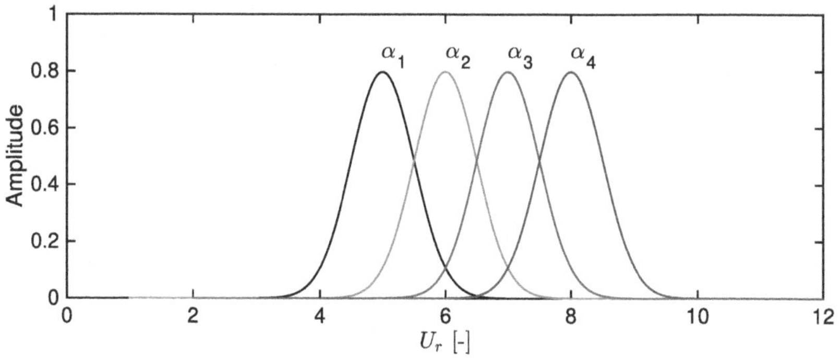

Fig. 16. Sketch of the VIV response on site as the enveloppe of multiple VIV responses

The VIV curves corresponding to *in situ* and wind tunnel at $\alpha = 50°$ are shown in Fig. 14. They reach the same maximum value of $\zeta\ddot{x}_{rms} \approx 0.2$. The correspondency between these two curves is justified by the wind direction measurement during the important VIV events on site, taking place when the wind direction reaches 50° (see Fig. 7). In addition, the analysis of the turbulent content of the wind as a function of its direction is presented in Fig. 15. It is observed that the wind at 50° has the lower turbulent intensity (I_u and $I_v \approx 5\%$), and hence is more prone to develop VIV than other wind directions for which I_u and $I_v > 10\%$ (Hansen 2007).

The difference in the size of the lock-in range measured in situ and in the wind tunnel is important. It cannot be explained by the value of the structural damping which is twice lower in situ than on the wind tunnel model. An alternative explanation is based on the change of wind direction taking place in situ. This fluctuation of α impacts the Strouhal number ($St(\alpha)$ see Fig. 11) and hence the critical wind speed at which VIV starts to occur. Assuming the same size of the lock-in range (roughly $\Delta U_r = 1.5$), the VIV response in situ corresponds to the enveloppe of several VIV curves corresponding to different wind direction. The situation is sketched schematically in Fig. 16. Note that the amplitude of the VIV response is not taken into account in the present discussion since it depends on the value of the lift coefficient, which is in turn a function of the wind direction.

4.2 Grid Effect on Vortex Shedding

By measuring simultaneously the pressure distributions on adjacent cylinders 4, 5, 6, the vortex shedding behaviour can be studied in term of phase and synchronization between cylinders. Figure 17 shows the time variation of the pressure c_p at corners of cylinders 4 to 6 for $\alpha = 0°$ and $T/D = 1, 2, 4$.

For small spacing ($T/D = 1$), a single vortex street region, such as commented by Zdravkovich (Zdravkovich 1985) is observed. The upper side of cylinder 5 (in blue) and the lower side of cylinder 6 (in magenta) are in phase. The lower side of cylinder 5 (in red) and the upper side of cylinder 4 (in green) are together in phase, but out of phase with the above mentioned blue and magenta curves. Cylinders are so close to each other and interfering strongly, that the symmetric sides of two adjacent cylinders are synchronized in term of vortex shedding (red-green or blue-magenta). The flow is separated everywhere but the pressure for the upstream corner (plain line in Fig. 17) is different from the one at the downstream corner (dashed line in Fig. 17). This corresponds to the step shape pressure of Fig. 13: different separated flow at $s/D = 0.5$ (upstream corner) and $s/D = 3$ (downstream corner).

For larger spacings ($T/D = 2$), a biased gap flow, or a coupled vortex street is observed (Zdravkovich 1985). Magenta and red curves are then in phase, in opposition to blue and green ones. Vortex shedding is no more synchronized in alternating channels. However, the upper sides of all cylinders are in phase, the lower sides are also in phase, but the upper-lower ejection alternates, because of the spacing of the cylinders, reducing their correlation and strength. The

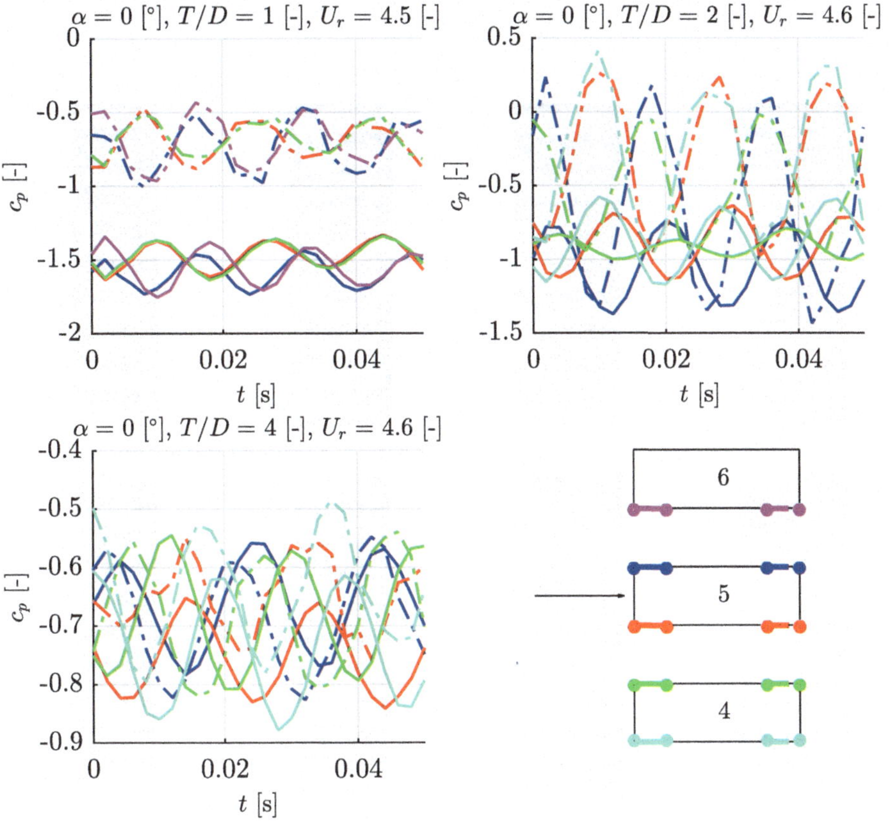

Fig. 17. Wind tunnel model: pressure $c_p(t)$ at corners of cylinders 4, 5, 6 at $\alpha = 0°$ and $T/D = 1, 2, 4$. Upstream = solid lines and the second point ($s/D = 0.5$) on the red line in the pressure distribution in Fig. 13). Downstream = dashed lines and the second last point ($s/D = 3$) on the red line in the pressure distribution in Fig. 13).

cylinders still influence each other because the pressure at the upstream and downstream corners is still different (but less than for $T/D = 1$). The same type of synchronization takes place for $T/D = 4$, but in that case the adjacent cylinders do not influence each other. The pressure at the upstream corners gets closer to the one at downstream ones. Vortices are alternatively ejected between the upper and lower cylinder sides.

Figure 18 shows instantaneous vorticity fields for a grid of 10 cylinders at $\alpha = 0°, 25°, 50°$ and $T/D = 1, 2, 3, 4$. The following observations are made about the spacing of cylinders from these snapshots: (i) for $T/D = 1$, the flow in the grid is governed by strong shear layers and the vortices are not clearly formed one chord downstream the cylinders. Strong interactions are observed in the wake of the cylinders. (ii) For $T/D = 2$ the formation of vortices is more noticeable than $T/D = 1$ and their interaction is important. (iii) For $T/D = 3$ and 4, the spacing between cylinders is such that the vortex shedding process is similar to the one of a single cylinder. Other observations are made about the wind direction: (i) for

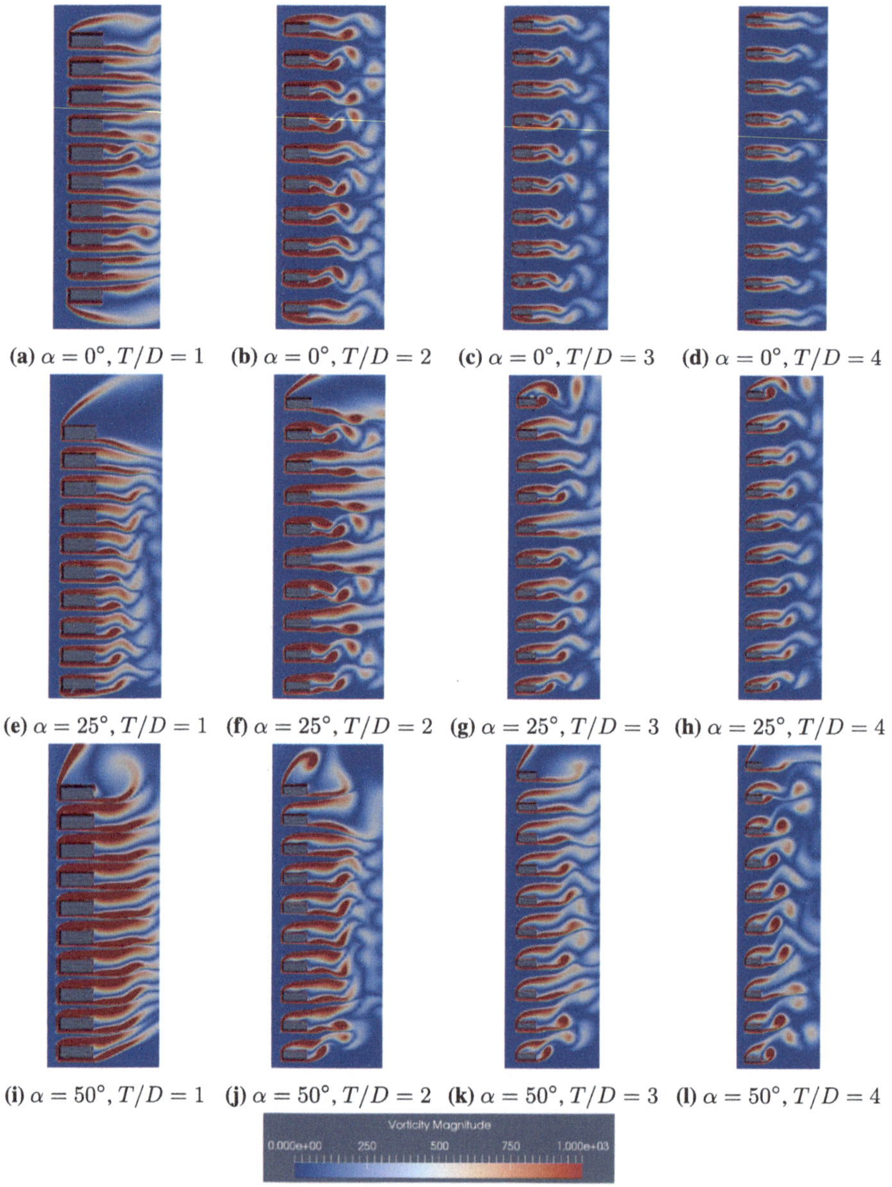

(a) $\alpha = 0°, T/D = 1$ (b) $\alpha = 0°, T/D = 2$ (c) $\alpha = 0°, T/D = 3$ (d) $\alpha = 0°, T/D = 4$

(e) $\alpha = 25°, T/D = 1$ (f) $\alpha = 25°, T/D = 2$ (g) $\alpha = 25°, T/D = 3$ (h) $\alpha = 25°, T/D = 4$

(i) $\alpha = 50°, T/D = 1$ (j) $\alpha = 50°, T/D = 2$ (k) $\alpha = 50°, T/D = 3$ (l) $\alpha = 50°, T/D = 4$

Fig. 18. CFD results - vorticity magnitude around the grid at $U = 5\,\mathrm{m/s}$: α and T/D effect

$\alpha = 25°$ the vortex shedding process is disorganised, except for the case of large spacing ($T/D = 4$). (ii) For $\alpha = 50°$, strong wake interaction is observed for $T/D = 1$ and 2. The end-effects are more pronounced for these spacings. A clear wake organisation is observed in the central part of the grid. (iii) For $T/D = 3$ and 4, the wake interaction decreases and typical vortex shedding phenomenon is observed.

4.3 Mitigation Techniques

On the basis of the fundamental understanding of the grid VIV phenomenon, it is possible to propose different solutions to mitigate the vibration and acoustic effects. The incidence α is not anymore a parameter but depends on the atmospheric wind, hence it is a stochastic variable. With the objective to keep the grid unchanged, the parameter T/D is also fixed equal to 2. Because the grid is geometrically unchanged, the variation of the Strouhal number with wind direction is unchanged and the only non-dimensional number to vary is the Scruton number Sc. The objective is to reduce the vibration of the phenomenon or to push it further for statistically rare wind velocities. The target is to avoid any vibration in the range 0–10 m/s, which represents the more probable wind speeds in Belgium.

Two solutions have been implemented and tested:

1. Solution 1: Installing an horizontal beam riveted to all cylinders;
2. Solution 2: Filling cylinders with sand.

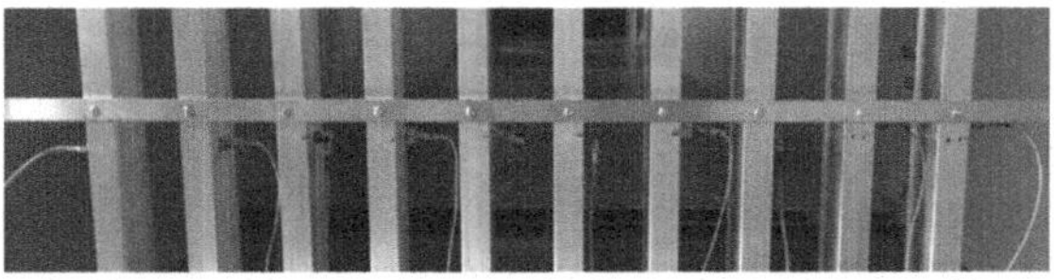

Fig. 19. Solution 1: small riveted horizontal beam (Aluminium)

The first solution has the advantage to increase slightly the frequency of the mode which is responsible for the grid-VIV instability (60 Hz to 66 Hz). In addition, damping is added to the system by the friction of the rivets on the cylinders (see picture in Fig. 19). Another effect of the beam is to distribute the energy of the excitation over several modes (at least 3), as can be seen from the 3 high peaks in the FFT of the bold black curve in Fig. 20. This spreading is interesting in order to avoid to concentrate all the energy on a single mode, as seen in the FFT's of the initial and solution 2. The horizontal beam has also a function of dashpot: as VIV does not take place simultaneously on all cylinders, the neighbour cylinders can act as a mass damper system limiting the excitation at the location where the vibration is maximal. The second solution takes advantage of the added mass of the structure (m_S) and added damping of the medium (sand). Both solutions increase the Scruton number. The former has the disadvantage of visual modification, while the latter increases strongly the total weight of the structure (+200%).

Figure 20 shows the modal properties of the wind tunnel model with the two solutions implemented, in comparison with the initial case. It is shown that the first solution doubles the structural damping, while second one increases it by a factor of 5. Taking into account the added mass of each solution, the Scruton number reach 45 and 225 for solution 1 and 2 respectively. The Scruton number of the initial configuration is equal to 18.

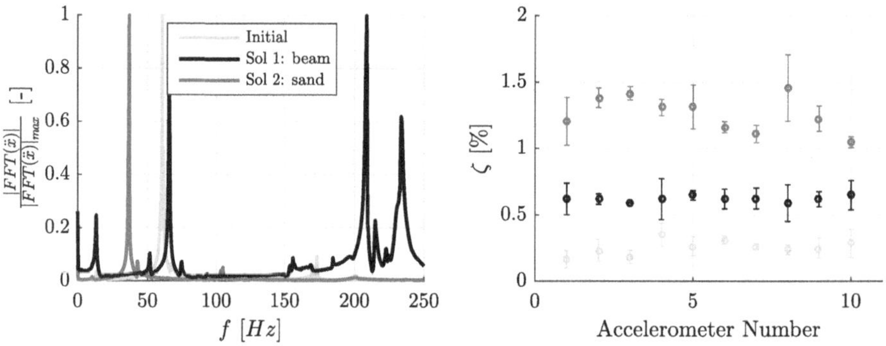

Fig. 20. Modal properties of the wind tunnel model ($U = 0$ m/s, cylinder 5): comparison of the initial structure with two solutions

The upper plots of Fig. 21 show the acceleration of the central cylinder for the three different configurations. A small VIV response is measured for solution 2 (cylinders filled with sand). The ratio between the highest acceleration of the initial and solution 2 is equal to 6, while the ratio of the corresponding Scruton numbers is equal to 12. This fact might be explained by the VIV grid effect, together with an unknown variation of the sand distribution in the cylinders. The wind speed at which VIV takes place is equal to 4 m/s, as expected from the lower frequency of the filled cylinders due to added mass.

The response observed at 15 m/s for the solution 1 has a typical clutch VIV shape. Nevertheless, it appears at a lower critical speed (ratio 2.5) than the one predicted by the ratio of frequency of 3.2. For this case, the prediction of critical galloping velocity is about 17.5 m/s, below the critical VIV speed of 19 m/s. A complex VIV-galloping interaction is triggered, known as a quenching (Mannini (Mannini et al. 2014)) with vibration beginning at a lower speed than the critical VIV prevision. Moreover, the first mode (at 66 Hz) is not excited. From our explanation relying the effect of horizontal beam, it creates a link between cylinder. The non-synchronous feature of VIV on cylinders is used so that the neighbour cylinder act as a damper. This is very effective on first bending mode because beam is placed at mid-height, where vibration and mitigation is maximal.

The lower plots of Fig. 21 show the acoustic intensity I_a measured by a microphone located in the flow field inside the test section of the wind tunnel. Absolute acoustic levels are not representative of reality because of the multiple sound wave reflections on the metallic walls of the test section. For this reason we focus here on the relative differences between the three configurations. As expected from the previous analysis of accelerations, the acoustic emission is much lower when mitigation solutions are installed: a decrease of 20 dB is reported for $\alpha = 0°$ and only 10 dB for $\alpha = 50°$. Note that the linear increase of the acoustic intensity is due to the turbulent flow signature of the grid, which is different from the mono-harmonic sound emission due to the VIV-grid phenomenon. It is interesting to note that the sound felt *in situ* was higher than the one measured in wind tunnel. This can be explained by the presence of the terrace downstream the grid, which plays a role. Indeed the frequency of vibration is $f = 60$ Hz, the speed of sound is $a_0 = 340$ m/s, hence the associated wavelength is $\lambda = a_0/f = 5.6$ m, very close of the dimensions of the terrace (4 m by 6.5 m in Fig. 2). It acts like a soundboard and amplifying the acoustic intensity, the noise experienced is felt stronger. Moreover, the terrace rear wall is made of glass, increasing this effect.

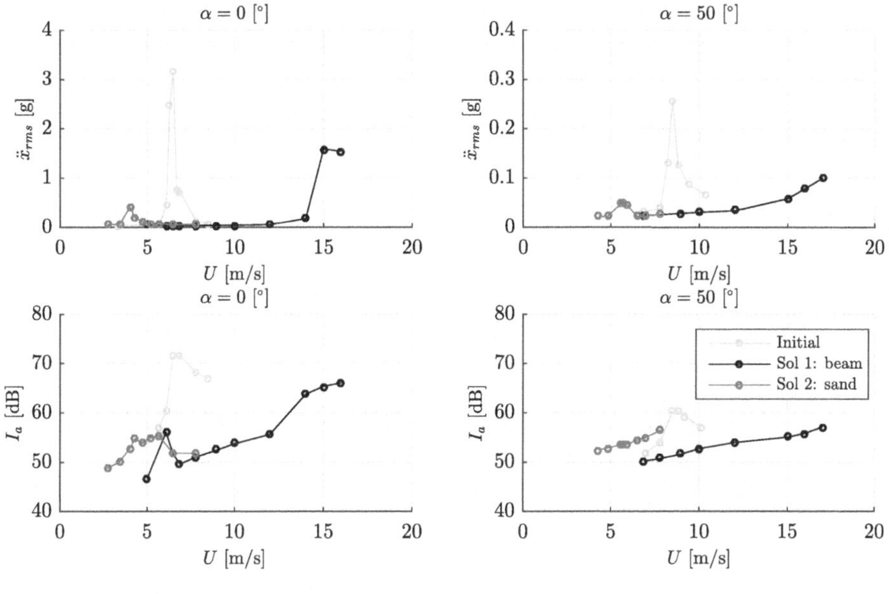

Fig. 21. Wind tunnel model: VIV curves $\ddot{x}_{rms}$ and acoustic intensity I_s vs U at $\alpha = 0° - 50°$ for the initial and solution models

On the basis of the analysis of the two mitigation techniques, it is concluded that the adjunction of the horizontal beam leads to better performances to cancel out the VIV grid phenomenon.

5 Conclusion

In this work, undesirable noise occurring on a grid made of rectangular cylinders motivated the study of the VIV-grid phenomenon. Based on *in situ* measurements, the wind conditions leading to this instability were studied: the dominant

South-West wind of Belgium at $U = 5\,\mathrm{m/s}$ produces high vibration amplitudes of the cylinders. A full scale wind tunnel model of the real grid was built to reproduce the real conditions and to study more configurations, not constrained by wind conditions on site. A parametric study on the wind incidence and the cylinder spacing was carried out, highlighting the effect on the flow around a grid arrangement. The grid arrangement has a tendency to decrease the critical velocity. The wind incidence has also a strong effect on the VIV grid phenomenon: the Strouhal number decreases when wind incidence increase. The experimental pressure analysis was useful to understand the vortex shedding process responsible of the VIV-grid phenomenon. The use of the CFD model allowed to study in depth the flow around the cylinders and the coupling/synchronization of vortices. The superposition of VIV curves in the wind tunnel pointed out the driving parameters of the VIV-grid phenomenon: coupling of wind incidence, magnitude and turbulence intensity. The combination of the three different approaches allowed to have a complete understanding of the phenomenon. Finally, two mitigation techniques were proposed and tested to cancel out the aeroelastic instability and corresponding acoustic emission within the range of most likely wind speeds.

Acknowledgments. The authors would like to thank the company V2i for providing technical support and devices necessary to make all modal and in-situ measurements.

References

Eurocode (1991) En 1991-1-4: Actions on structures-general actions-wind actions. European committee for standardization

Bergh H, Tijdeman H (1967) Theoretical and experimental results for the dynamic response of pressure measuring systems

Blevins R (1990) Flow induced vibrations. Krieger Publishing Co., Florida

Choi C, Kwon D (2000) Determination of the strouhal number based on the aerodynamic behavior of rectangular cylinders. Wind Struct 3(3):209–220

Hansen S (2007) Vortex-induced vibration of structures. In: Structural engineers world congress, Bangalore, India

Knisely C (1990) Strouhal numbers of rectangular cylinders at incidence: a review and new data. J Fluids Struct 4(4):371–393. https://doi.org/10.1016/0889-9746(90)90137-T ISSN 0889-9746

Mannini C, Marra A, Bartoli G (2014) Viv-galloping instability of rectangular cylinders: review and new experiments. J Wind Eng Ind Aerodyn 132:109–124. https://doi.org/10.1016/j.jweia.2014.06.021 ISSN 0167-6105

Nakaguchi H, Muto S (1968) An experimental study on aerodynamic drag of rectangular cylinders. J Japan Aeronaut Space Sci 16:1–5

Okajima A (1982) Strouhal numbers of rectangular cylinders. J Fluids Mech. 123:379–398

Ong M (2013) Unsteady RANS simulation of flow around a 5:1 rectangular cylinder at high reynolds numbers, vol 5

Rigo F, Denoel V, Andrianne T (2018) Vortex induced vibrations of rectangular cylinders arranged on a grid. J Wind Eng Ind Aerodyn 177:1–13. https://doi.org/10.1016/j.jweia.2018.04.011 ISSN 0889-9746

Techet A (2005) Vortex induced vibrations. MIT, Boston

Zdravkovich M (1985) Flow induced oscillations of two interfering circular cylinders. J Sound Vib 101:511–521

The Interplay Between Background Atmospheric Boundary Layer Winds and Downburst Outflows. A First Physical Experiment

D. Romanic(✉) and H. Hangan

Wind Engineering, Energy and Environment (WindEEE) Research Institute, Western University, 2535 Advanced Avenue, London, ON N6M 0E2, Canada
{dromanic,hmhangan}@uwo.ca

Abstract. This paper studies the interaction between downburst outflows and the background atmospheric boundary layer (ABL) winds close to the surface. Downburst is a buoyancy-driven downdraft of cold air that emerges from cumuliform clouds and results in a vigorous starburst outflows upon reaching the surface. Currently, there are neither satisfactory analytical models nor experimental results on the highly complex interaction between these two wind systems. One of the advanced modes of the WindEEE Dome operation, at Western University in Canada, enables the simultaneous generation of downbursts and ABL winds. In accordance with the WindEEE Dome capabilities, an experiment is designed to address this long-standing question on the relationship between ABL winds and downbursts. This paper shows the interaction between downburst and ABL winds for seven azimuthal positions in respect to the incoming ABL wind direction and six heights. The results show that the traditional approach of adding ABL winds to downburst outflow as either vector or algebraic sum is inaccurate for all heights and azimuth angles. A new empirical relationship between downbursts with and without ABL winds is presented herein.

Keywords: Downburst · Thunderstorm · Wind simulator · Flow interaction · WindEEE Dome

1 Introduction

A detail investigation of downbursts started in approximately the 1960s after several commercial airplane crashes that were thought to be caused by thunderstorm winds (Fujita 1981, 1985). Later, it has been demonstrated that downbursts are extremely damaging not only to airplanes, but also to the built environment on the earth's surface (e.g., Holmes et al. 2008; Solari et al. 2012, 2015; Elawady et al. 2017). Although a large progress has been made in understanding these severe windstorms since the 1960s, there is still a number of unknown facets associated with downbursts. One of these is the coupling between, in relative terms, the small scale downburst and the large scale synoptic winds. This topic is the main subject of the present paper.

F. Ricciardelli and A. M. Avossa (Eds.): INVENTO 2018, LNCE 27, pp. 652–664, 2019.
https://doi.org/10.1007/978-3-030-12815-9_49

Downbursts are created in the precipitation zone of thunderstorms, which is also the area where the evaporation and melting of hydrometeors takes place underneath the cloud base. The latent heats of evaporation and melting are taken from the surrounding air, which consequently results in temperature decrease of this air in respect to the ambient air around thunderstorm. Therefore, the air in the precipitation zone becomes denser (and thus heavier) than the ambient air, which further results in its descent towards the surface. Once the cool air reaches the surface, it spreads out radially creating vigorous winds, called downburst outflows. Although downbursts are buoyancy-created phenomena, the drag caused by the falling rain and/or hail and graupel additionally increases the strength of the descending current. Wet downburst are associated with precipitation that reached the surface while, in the case of dry downbursts, all hydrometeors evaporated during the descent. Another classification of downbursts is based on their spatial extent. In this approach, downbursts are categorized either as microburst or macroburst if the horizontal size of downburst is below or above 4 km, respectively.

The parent cumuliform cloud (typically cumulonimbus), as well as the developing downburst, are not an isolated system, but instead both are embedded into the background atmospheric boundary layer (ABL) winds. That is, the cloud-downburst system constantly interacts with the ABL winds, which, in turn, are characterized with the pronounced velocity shears during the atmospheric conditions favorable for development of thunderstorms. Note that the velocity shear indicates that both speed and direction of wind are changing with height. Therefore, downbursts are highly transient winds with distinct spatiotemporal variability. In addition, downburst outflows are highly three-dimensional with the vertical velocity component in many instances being larger than the horizontal wind vector.

Currently there are neither satisfactory analytical models nor experimental results on the nonlinear interaction between the two wind systems—i.e., downbursts and ABL winds. An attempt to combine the flows through a linear vector or algebraic summation is physically inaccurate, as recently demonstrated at the European Wind Engineering Conference in Liège, Belgium (Romanic et al. 2017). However, this simplification has often been invoked in literature (e.g., Holmes and Oliver 2000; Chay et al. 2006; Kim and Hangan 2007; Abd-Elaal et al. 2014).

This paper concentrates on experimental investigation of the interaction between downburst and straight flows. The experiments were carried out in the Wind Engineering, Energy and Environment (WindEEE) Dome (Hangan 2010) at Western University in Canada. Experimentally produced downbursts in the WindEEE Dome are analyzed in the presence of straight ABL winds as well as without them. The discrepancies between these two cases are investigated for different distances from downburst center and for different azimuth angles in respect to the incoming straight wind. Moreover, the interaction between downbursts and ABL winds as function of height is also analyzed.

The WindEEE Dome facility and the experiment setup are described in Sects. 2 and 3, respectively. The results and discussion are presented in Sect. 4. This paper only documents the results for one radial distance from downburst center, but the paper shows the results for multiple heights above surface as well as seven different azimuth angles in respect to the incoming ABL wind. Thus, an azimuth-dependent empirical model of downbursts in ABL winds is proposed herein. Since this study provides only a portion of the overall analysis that are going to be conducted in the future, the conclusions together with several prospects for further research are outlined in Sect. 5.

2 The WindEEE Dome

Since comprehensive description of the WindEEE Dome and the simulator capabilities are provided in Hangan et al. (2017), only a brief summary of this facility is presented in this paper. The WindEEE Dome (Hangan 2010; Hangan et al. 2017) is a three-dimensional and time dependent wind testing chamber of 25 m inner diameter and 40 m outer diameter, including the return circuit section. By using a system of 100 dynamic fans installed on the 6 peripheral walls coupled with 6 larger fans in the upper plenum, this wind simulator can produce different wind systems (Hangan et al. 2017; Refan and Hangan 2018; Jubayer and Hangan 2018), including downbursts of different intensities and downdraft dimensions. In addition, using a guillotine system and the bell mouth, downbursts can be translated at various translation velocities over a distance of 5 m through the testing chamber. A schematics of the WindEEE Dome downburst mode is depicted in Fig. 1. The pure downburst mode of the WindEEE Dome (i.e., no ABL winds) was recently investigated by Jubayer et al. (2016), (2017).

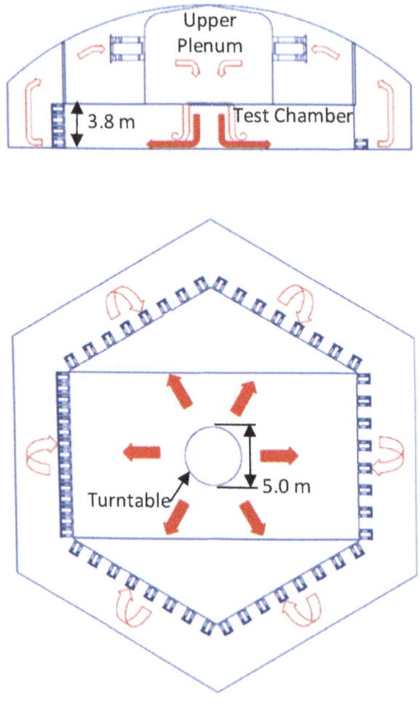

Fig. 1. Schematics of axisymmetric downburst mode of the WindEEE Dome

The ABL winds in the WindEEE Dome are created using the 60-fan wall as shown in Fig. 2. The profiles of ABL winds for different exposures as well as their spectral characteristics are discussed in more details in Hangan et al. (2017).

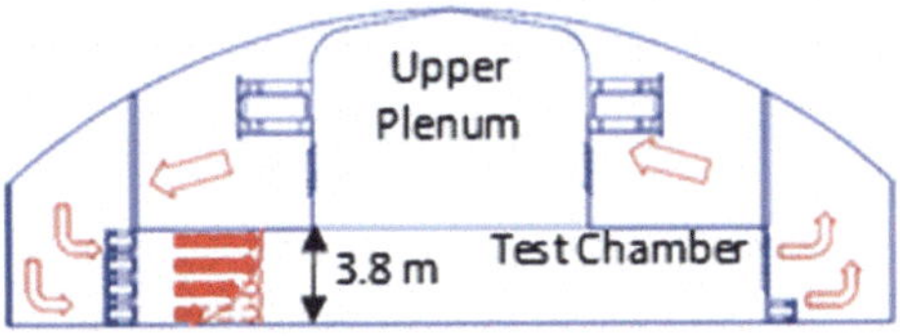

Fig. 2. Schematics of ABL winds in the WindEEE Dome

3 Experiment Setup and Data Processing

The experiment setup in the WindEEE Dome is depicted in Fig. 3. Six Cobra Probes (from Turbulent Flow Instrumentation Pty Ltd) are installed on a heavy mast in order to measure downburst outflow. Notice that these probes always point towards the centre of the turntable, which is also the centre of the downburst downdraft. On the other side, one Cobra Probe is used to measure ABL winds and this probe is in a symmetric position to the downburst probes in respect to the direction of incoming ABL winds (see Fig. 3). The ABL probe was always pointing towards the 60-fan wall.

Two sets of experiments are carried out. First, the downburst is physically simulated without the presence of ABL wind in order to estimate the undisturbed downburst outflow for different azimuth angles. Second, the downburst is physically created in the presence of ABL winds in such a way that the downdraft was released in already developed ABL winds.

The most convenient coordinate system to use in this case is the cylindrical system with the axis (r, θ, z), as schematically depicted in Fig. 3. Notice that the positive direction of the θ-axis is clockwise starting from the incoming direction of ABL winds (i.e., technically speaking, θ is measured from the 60-fan wall). The positive direction of the z-axis is upward. Due to the symmetry of the flow around $\theta = 0°$ direction of ABL winds, the measurements of downburst and ABL winds are performed only for a half of the circle (i.e., $0° \leq \theta \leq 180°$) with the azimuthal increment, $\Delta\theta$, of 30°. Due to the circular symmetry, the results can be mirrored for the other half of the circle (i.e., for $180° < \theta < 360°$). The measuring heights of downburst Cobra Probes were 0.04, 0.10, 0.15, 0.20, 0.27 and 0.42 m above floor. The Cobra Probe that is used to measure ABL winds was at 0.12 m above floor.

All results in this paper are presented in a non-dimensional form with the normalization procedure following the one used in the classical impinging jet studies of Gutmark et al. (1978) and Ho and Nosseir (1981). Namely, the reference velocity, U_0, used to normalize all velocities is:

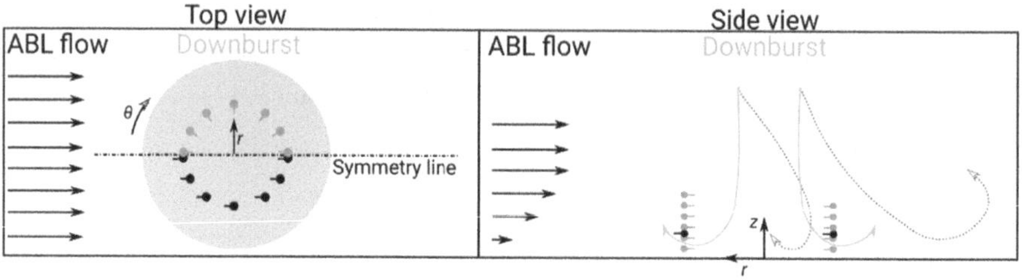

Fig. 3. Experiment setup from top and side views. The grey probes are used to measure downburst outflow while the black probe is used to measure ABL winds. Notice that the downburst probes always point towards the centre of the turntable. In the side view panel, the dotted downburst line is a schematics of a downburst distorted by the ABL winds

$$U_0 = U_{jet}\sqrt{\frac{D}{H}} \tag{1}$$

where $U_{jet} = 8.87\,\mathrm{ms}^{-1}$ is the velocity of the downburst downdraft (i.e., impinging jet) at the exit of bell mouth and measured in the centre of the downdraft, $D = 3.20\,\mathrm{m}$ is the diameter of the downdraft, and $H = 3.75\,\mathrm{m}$ is the vertical distance between the downdraft exit at bell mouth level and the impingement plane (i.e., testing chamber floor). Therefore, from Eq. (1) it follows that $U_0 = 8.19\,\mathrm{ms}^{-1}$. In a similar fashion, the non-dimensional time, t', is expressed as:

$$t' = t\frac{U_o}{D} \tag{2}$$

where t is the time in seconds. All geometric dimensions in this study are normalized with D, while all velocities are normalized with U_0 including the ABL winds denoted as U_c (the subscript "c" stands for "crossflow"). The total duration of measurements is $t' = 307.1$ ($t = 120\,\mathrm{s}$). Data acquisition always started prior to the downburst release and the downburst outflow was stopped by closing the bell mouth lovers prior to the end of measurements. That way both downburst ramp-up and downburst attenuation segments are measured. The sampling frequency of Cobra Probes was 2500 Hz and the velocity threshold for the radial velocity is 2 ms^{-1}. All results in this paper depict the radial velocity components captured by the Cobra Probes (i.e., lateral and vertical velocity components, as well as the overall velocity magnitude, are not analysed in this study).

The benchmark cases are the experiments conducted without ABL wind (i.e., $U_c/U_0 = 0$), while the case with the ABL wind corresponds to $U_c/U_0 \approx 0.43$. The approximate sign is used because U_c has a small variation for different values of θ (i.e., different distances from the 60-fan wall). The exact value of U_c/U_0 for each case is included in the presented graphs in the following section. It is important to emphasize that the downburst outflow and ABL winds are measured at the same radial distance from the centre of the turntable; $r/D = 1.0$ (see Fig. 3). However, as noted above, θ spans from 0° to 180°, with $\Delta\theta = 30°$, which effectively is a full circle due to the circular symmetry.

4 Results and Discussion

The first difference between downburst flows with and without background ABL winds is shown in Fig. 4. Even qualitatively, it still can be observed that the downburst supplemented with the ABL winds is weaker than the pure downburst outflow for $\theta = 90°$. The attenuation of the downburst outflow is observed at all six heights. Figure 4 further shows that the intensity of the first peak in the downburst time series with ABL winds is overall attenuated too. The first peak, which is also the largest peak in the time series, is associated with the passage of the primary vortex in the radially advancing downburst outflow (e.g., Hjelmfelt 1988; Hall and Ewing 2005; Kim and

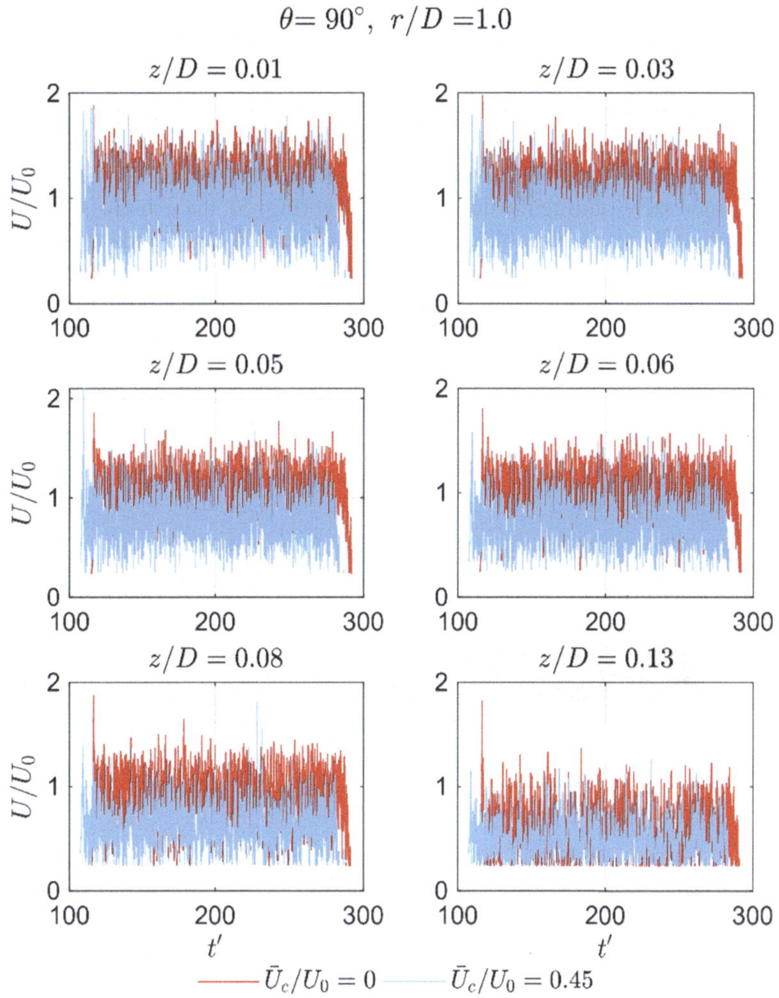

Fig. 4. An example of downburst time series without (red) and with (blue) ABL winds at $\theta = 90°$

Hangan 2007). The only exception from this rule is observed at the height $z/D = 0.0469$, where the peak of downburst outflow with ABL winds is larger than the ramp-up peak of the pure downburst. Interestingly, this peak is also the overall largest peak from all six heights ($U/U_0 = 2.1$ at $t' = 109.6$). At this point, it is not clear whether the peak is due to the random turbulent fluctuations in the flow field or it represents a distinct feature of the outflow. The former argument is more likely as the same signature is not observed in other cases in Fig. 4.

The time series of ABL winds for the same azimuth position ($\theta = 90°$) is shown in Fig. 5. The abrupt change in the time series of ABL winds at approximately $t' = 108$ coincides with the release of downburst. The mean wind speed of ABL wind at $z/D = 0.0406$ before downburst is 0.45 and the value drops down to 0.31 after the downburst. However, the fluctuations in the time series during the downburst outflow are much larger when compared to the undisturbed ABL winds. However, the most pronounced difference between the ABL and ABL plus downburst is observed when analysing kurtosis. Without downburst, ABL winds are profoundly Gaussian with the kurtosis equal to 3.12, while the kurtosis increases to 12.41 during the downburst event. Such a large value of kurtosis indicates the higher probability of extreme wind speed peaks in the time series, which is also qualitatively observed in Fig. 5. Although the data quality in the ABL time series during the downburst is low (Fig. 5), the non-Gaussian features of the flow are still quite noticeable. Lastly, the recovery of ABL winds after the end of the downburst at around $t' = 293$ is also observed in Fig. 5. However, notice that these "recovered" ABL winds after $t' = 293$ are weaker than the one prior to downburst. Similar finding is also observed in full scale events (Burlando et al. 2017).

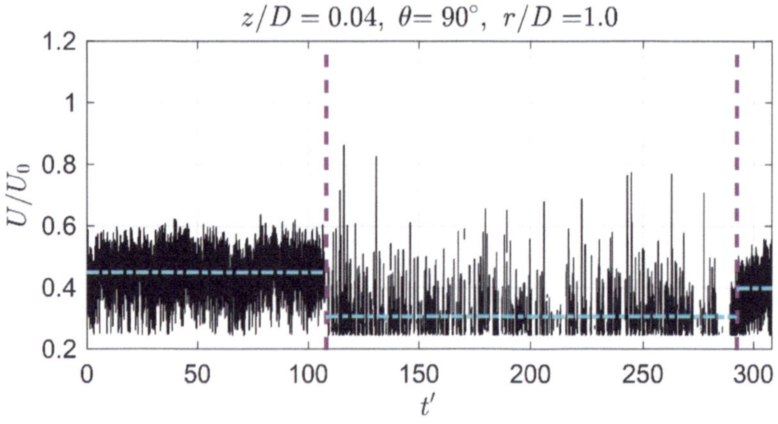

Fig. 5. An example of time series of ABL winds at $\theta = 90°$. The magenta lines show the beginning and end of the downburst while the cyan lines are the mean speeds of ABL winds prior to, during, and after the downburst

To complete the presentation of wind characteristics for $\theta = 90°$, Fig. 6 depicts the vertical profiles of downburst outflow with and without ABL wind. Similar to the qualitative observation previously made by inspecting Fig. 4, Fig. 6 quantitatively demonstrates that the instantaneous peaks in the pure downburst are higher than in the downburst flow supplemented with the ABL wind. The only exclusion from this rule is the height $z/D = 0.0469$.

More importantly, though, Fig. 6 (right panel) shows a clear difference between the vertical profiles of the mean wind speeds. The largest discrepancies between two flows are observed at lower heights. This finding is somewhat expected since the vertical extent of the downburst is smaller than the extent of ABL winds. In both cases, the nose-shape profile of downburst wind speed is preserved, but the nose-like roundness of the profile is more pronounced for the undisturbed downburst.

It is important to emphasize at this point that the presented case demonstrates inaccuracy of the common assumption that the background ABL winds can simply be added as an algebraic sum to the downburst wind in order to obtain the overall wind velocity. For example, at the height $z/D = 0.0313$, the mean downburst wind speed without ABL is 1.10. The velocity of ABL winds before the downburst is 0.45 (see Fig. 5 and legend in Fig. 6). Firstly, simply summing up these two values results in

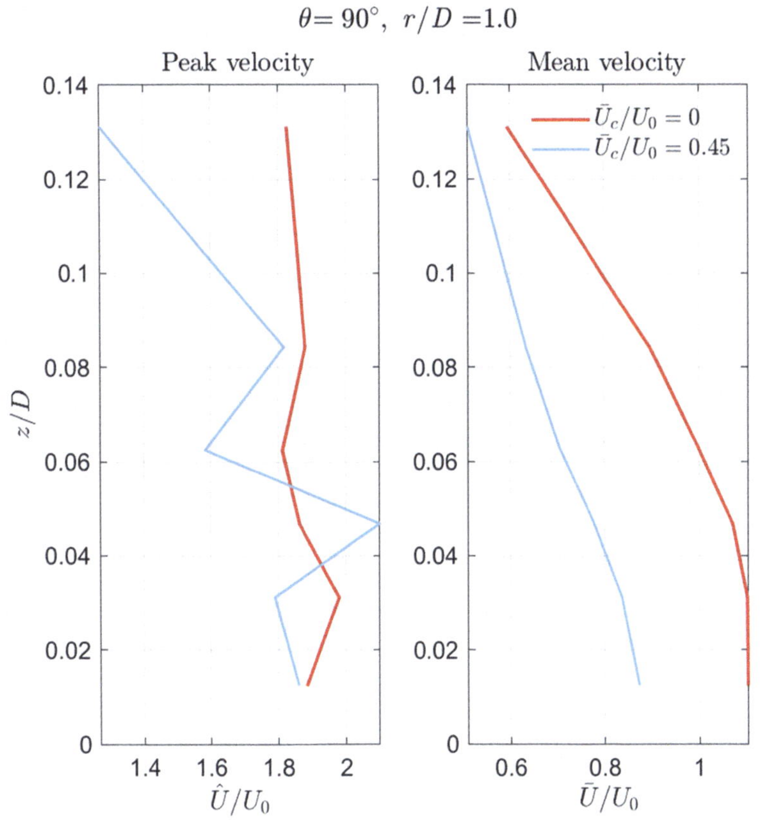

Fig. 6. Vertical profile of downburst winds without ABL (red) and with ABL (blue) for $\theta = 90°$. Left panel shows the instantaneous peak velocity while the right panel depicts the mean wind speed

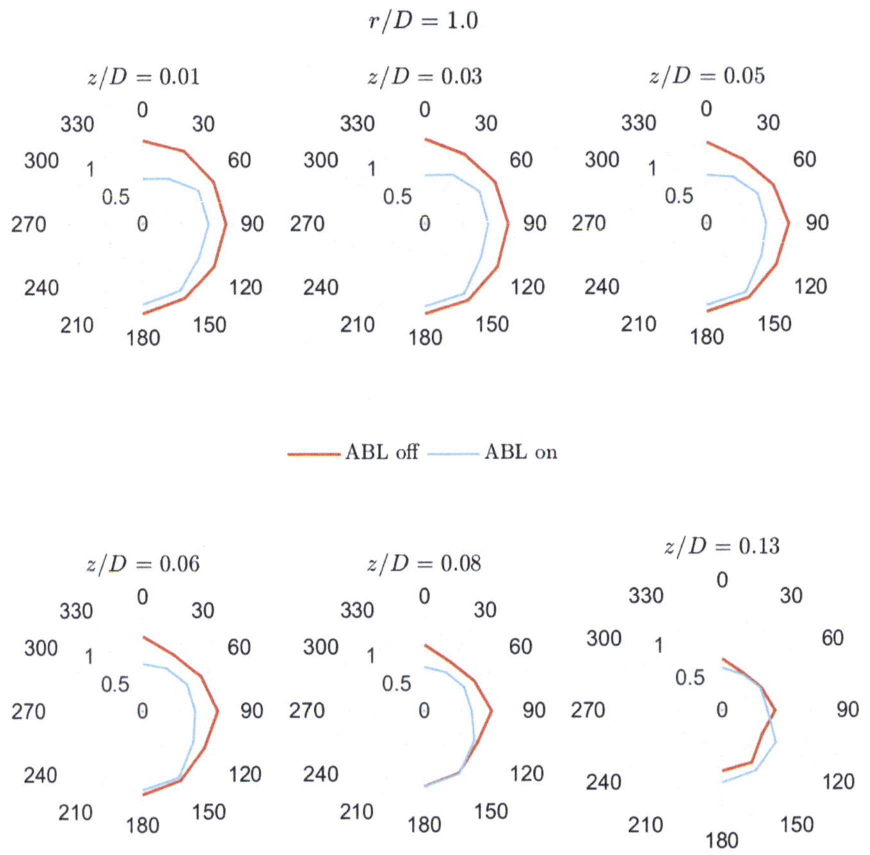

Fig. 7. Mean downburst wind speed as function of height (z/D) and azimuth angle (θ) with (blue) and without (red) ABL winds.

1.55 which is 1.87 times larger than the actual value observed in Fig. 6 (i.e., $\overline{U}/U_0 = 0.83$ with ABL winds). Secondly, and more importantly, the presence of ABL winds, in fact, diminishes the intensity of downburst, thus making this simple summation profoundly inaccurate. Thirdly, adding up the velocities as two vectors, $\sqrt{(1.10)^2 + (0.45)^2} = 1.19$, is also inaccurate. For example, Fig. 6 clearly shows that the downburst diminished in the presence of ABL winds and the reduction of wind speed at $z/D = 0.0313$ is 1.32 times (i.e., $\overline{U}_{\mathrm{ABL\ off}}/\overline{U}_{\mathrm{ABL\ on}} = 1.32$). The results in Fig. 6 also demonstrate that the future measurements should include additional measuring heights below $z/D = 0.0125$, as well as more probes in the region of high velocity gradients. In addition, the forthcoming study will also explore this relationship for other radial distances and velocity components.

A generalization of the above results for other azimuth angles is shown in Fig. 7. Notice that the direction of ABL flow is along $\theta = 0°$. The largest discrepancies between two cases of downburst outflows are observed for $\theta = 0°$ at almost all heights.

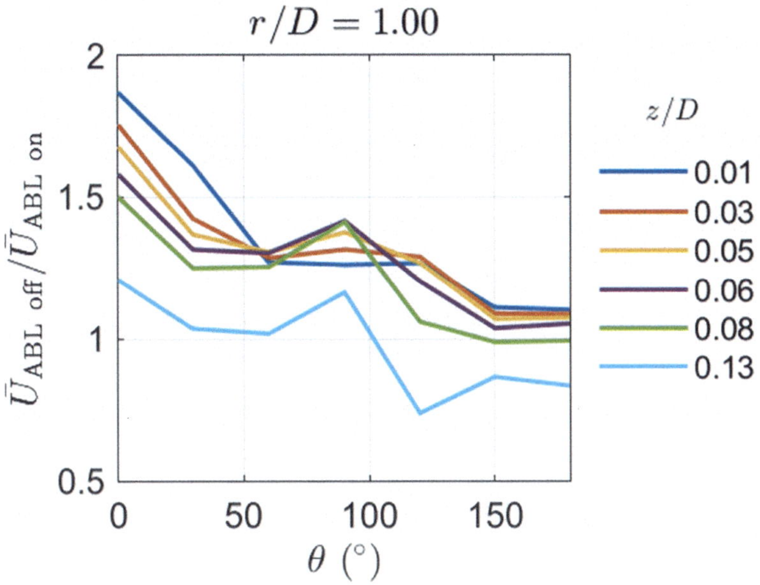

Fig. 8. The ratio of mean downburst wind speeds as function of height (z/D) and azimuth angle (θ) with (blue) and without (red) ABL winds.

The only exception is at $z/D = 0.13$, in which case the largest difference between two downbursts is at $\theta = 120°$. Notice that at $z/D = 0.06$ and 0.08, the downbursts with and without ABL are the same on the far end of ABL winds (i.e., $\theta = 150°$ and $180°$). At the first five heights, the discrepancies between the pure downburst and downburst in ABL diminish with increasing θ.

The relationship between downburst outflows without and with ABL winds is also portrayed in Fig. 8. These results could be used in the future studies to analytically formulate a relationship between the strength of a downburst with ABL winds and downburst without ABL winds. Figure 8 shows that the strength of ABL winds rather linearly diminishes with increasing θ. It is important to recall that this graph is applicable only for a single r/D positions ($r/D = 1$). It is interesting to note that the values of $\overline{U}_{\mathrm{ABL\,off}}/\overline{U}_{\mathrm{ABL\,on}} < 1$ are only observed at the highest investigated level ($z/D = 0.13$) and for $\theta > 100°$. In principle, for $\theta > 90°$ downburst outflow and ABL winds are propagating into the same quadrant, but Figs. 7 and 8 demonstrate that the augmentation of the combined flows is observed only at $z/D = 0.13$. For other investigated heights, the combined flow is attenuated in comparison to the pure downburst. This result is probably due to the combining effects of the log-like profile of ABL winds (i.e., increasing wind speed with height throughout the boundary layer) and the nose shape velocity profile in downburst outflows. At the higher elevations and for $\theta > 100°$, the flow in the downburst leading vortex without ABL winds is reversed in comparison to the radially advancing outflow close to the surface. Such schematic representations of downburst outflows as a rolling vortex are provided in, for example, Fujita (1985) and Hjelmfelt (1988), and more recently in Lompar et al. (2018).

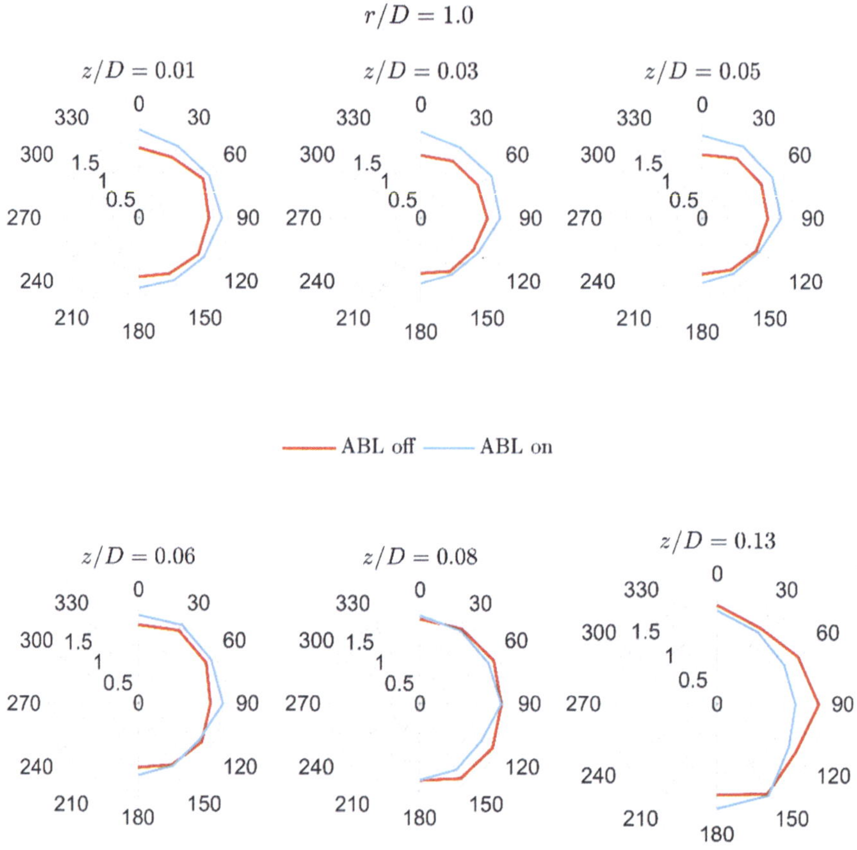

Fig. 9. Same as Fig. 7, but for standard deviation of velocities. Standard deviation is not normalized in this figure.

Here, we hypothesize that the reversed flow in the upper regions of the vortex in the presence of ABL winds is disturbed (i.e., overpowered) by the oppositely directed ABL winds for $\theta > 100°$, hence, resulting in $\overline{U}_{\mathrm{ABL\ off}}/\overline{U}_{\mathrm{ABL\ on}} < 1$. However, this hypothesis needs to be rigorously tested in the future studies. That being said, the future work should extend the results of this article to different r/D values, as well as describe the complete three-dimensional structure and dynamics of the interaction between downburst outflows and ABL winds.

Lastly, Fig. 9 shows the azimuthal profiles of standard deviation of downburst velocities with and without ABL winds. With the exception of the highest two measuring heights, the fluctuations in the combined flows are higher than in the pure downburst outflow. At the top two heights ($z/D = 0.08$ and 0.13), however, the standard deviation (i.e., fluctuations) in the pure downburst is more pronounced than in the case of the combined flows. The same result is also observed for $\theta = 120°$ and $z/D = 0.06$. Also, with the exclusion of the top height, there is an overall (weak) positive trend of standard deviation with increasing θ. The high value of standard deviation at $z/D = 0.13$ and $\theta = 180°$ support, at least in principle and qualitatively, the proposed hypothesis of the dominance of ABL winds in the reversed flow region of the downburst outflow vortex. However, further research on this topic is much needed.

5 Conclusions and Future Work

This article presents the unique set of experiments performed in the WindEEE Dome, at Western University in Canada, with the goal to investigate the interaction between atmospheric boundary layer (ABL) and downburst winds. Most numerical and all experimental simulations of downbursts neglect the existence of background ABL winds. In other cases, the ABL winds are linearly superimposed to downburst outflows through a simple vector or sometimes algebraic summations. The results from this paper shed some light on this subject.

Statistical properties of time series of ABL winds with and without downburst flow differ significantly, in particular when it comes to higher order statistics. On the other hand, downburst without ABL wind is stronger than in the presence of ABL winds. This result is observed at almost all heights and in particular for the mean wind speeds. However, there are exceptions to this finding; such as for the azimuth angles $\theta > 120°$ at the highest measuring level. The instantaneous peaks are sometimes higher in the cases when two flows are combined. As one of the conclusions from these results, we demonstrate that the simple summation (vector or algebraic) of ABL winds and downburst outflows, which has sometimes been proposed in literature, cannot be accurate.

The future work on this subject will focus on extending the results from this paper to other radial positions (this study investigated only one radial position) as well as different feature of the flow. Moreover, the dependency between the flows will rigorously be formulated as a function of height, radial distance, and azimuth positions. The experimental results will be compared and combined with the current analytical models of downburst winds. In addition, the downburst records in this study are statistically steady time series after the initial ramp-up, while the real downbursts are highly transient. Thus, the unsteady and transient portions of the experimental time series that resembles the full scale downburst events will also be investigated in the future work.

Acknowledgements. The authors thank Julien LoTufo for his technical help in conducting the experiments in the WindEEE Dome. Constructive comments from two anonymous reviewers are highly appreciated.

References

Abd-Elaal E-S, Mills JE, Ma X (2014) Empirical models for predicting unsteady-state downburst wind speeds. J Wind Eng Ind Aerodyn 129:49–63

Burlando M, Romanic D, Solari G, Hangan H, Zhang S (2017) Field data analysis and weather scenario of a downburst event in Livorno, Italy, on 1 October 2012. Mon Weather Rev 145:3507–3527

Elawady A, Aboshosha H, El Damatty A, Bitsuamlak G, Hangan H, Elatar A (2017) Aero-elastic testing of multi-spanned transmission line subjected to downbursts. J Wind Eng Ind Aerodyn 169:194–216

Chay MT, Albermani F, Wilson R (2006) Numerical and analytical simulation of downburst wind loads. Eng Struct 28:240–254

Fujita TT (1981) Tornadoes and downbursts in the context of generalized planetary scales. J Atmos Sci 38:1511–1534

Fujita TT (1985) The downburst: microburst and macroburst. Satellite and Mesometeorology Research Project, Department of the Geophysical Sciences, University of Chicago, Chicago, Illinois, United States

Gutmark E, Wolfshtein M, Wygnanski I (1978) The plane turbulent impinging jet. J Fluid Mech 88:737–758

Hall JW, Ewing D (2005) The development of the large-scale structures in round impinging jets exiting long pipes at two Reynolds numbers. Exp Fluids 38:50–58

Hangan H (2010) Current and future directions for wind research at western: a new quantum leap in wind research through the wind engineering, energy and environment (WindEEE) Dome. Wind Eng JAWE 35:277–281

Hangan H, Refan M, Jubayer CM, Romanic D, Parvu D, LoTufo J, Costache A (2017) Novel techniques in wind engineering. J Wind Eng Ind Aerodyn 171:13–33

Hjelmfelt MR (1988) Structure and life cycle of microburst outflows observed in Colorado. J Appl Meteorol 27:900–927

Ho C-M, Nosseir NS (1981) Dynamics of an impinging jet. Part 1. The feedback phenomenon. J Fluid Mech 105:119–142

Holmes JD, Oliver SE (2000) An empirical model of a downburst. Eng Struct 22:1167–1172

Holmes JD, Hangan HM, Schroeder JL, Letchford CW, Orwig KD (2008) A forensic study of the Lubbock-Reese downdraft of 2002. Wind Struct 11:137–152

Jubayer CM, Hangan H (2018) A hybrid approach for evaluating wind flow over a complex terrain. J Wind Eng Ind Aerodyn 175:65–76

Jubayer CM, Romanić D, Hangan H (2017) Effect of a large scale impinging jet on a standard tall building. 7th European and African conference on wind engineering (EACWE 2017), Liège, Belgium, 3–6 July 2017

Jubayer CM, Elatar A, Hangan H (2016) Pressure distributions on a low-rise building in a laboratory simulated downburst. In: Bluff body aerodynamics and applications, Boston, Massachusetts, United States, 7–11 June 2016

Kim J, Hangan H (2007) Numerical simulations of impinging jets with application to downbursts. J Wind Eng Ind Aerodyn 95:279–298

Lompar M, Ćurić M, Romanic D (2018) Implementation of a gust front head collapse scheme in the WRF numerical model. Atmos Res 203:231–245

Refan M, Hangan H (2018) Near surface experimental exploration of tornado vortices. J Wind Eng Ind Aerodyn 175:120–135

Romanic D, Parvu D, Hangan H (2017) Influence of background winds and storm motion on downburst outflow. In: 7th European and African Conference on Wind Engineering (EACWE 2017), Liège, Belgium, 3–6 July 2017

Solari G, Repetto MP, Burlando M, De Gaetano P, Pizzo M, Tizzi M, Parodi M (2012) The wind forecast for safety management of port areas. J Wind Eng Ind Aerodyn 104–106:266–277

Solari G, De Gaetano P, Repetto MP (2015) Thunderstorm response spectrum: fundamentals and case study. J Wind Eng Ind Aerodyn 143:62–77

Field Measurement of Gas Dispersion in the Atsugi Area

R. Sasaki[1(✉)], K. Miyashita[1], and R. Yoshie[2]

[1] Wind Engineering Institute Co., Ltd.,
3-29, Kanda-Jimbo-cho, Chiyoda-ku, Tokyo 101-0051, Japan
{sasaki,miya}@wei.co.jp
[2] Tokyo Polytechnic University,
1583 Iiyama, Atsugi, Kanagawa 243-0297, Japan
yoshie@arch.t-kougei.ac.jp

Abstract. The ability to predict the dispersion of gases is highly important for the environmental assessment of cities and industrial areas, where gaseous emissions have become a matter of global concern. Our study aims to understand the extent to which theoretical models of gaseous dispersion make accurate predictions when compared to actual measurements from real-world geographical areas and to investigate its limitations. With this in mind, we conducted a field study on gas dispersion in the urban area of Atsugi city in Japan. The release point for the gas was set on the rooftop of a building. The sampling points were located on the ground. We compared the distribution of dispersed gases in terms of concentration obtained from the field measurements with our calculations based on the plume model. Both sets of results were in good agreement with each other, except when there was atmospheric instability in an area where dispersion was measured. Meteorological parameters of wind direction fluctuation and vertical wind profile were studied to further understand the areas of atmospheric instability.

Keywords: Gas dispersion · Field measurement · Plume model · Atmosphere stability

1 Introduction

Prediction of gas dispersion patterns is important not only for environmental assessment of gas emissions but also to confront the threat of toxic substances released into the atmosphere. Field measurements of gas dispersion have been carried out in some areas including St. Louis (McEloy and Pooler 1968), Salt Lake City (Allwine et al. 2002), Oklahoma City (Allwine et al. 2004), London (Arnold et al. 2004), and New York City (Hanna et al. 2006). Nevertheless, the prediction method involving wind tunnel tests and computational fluid dynamics (CFD) has not yet been validated because of the lack of adequate measured data regarding wind speed and gas concentrations in urban areas.

For the purpose of obtaining such validation data, our team based at Tokyo Polytechnic University conducted a preliminary field measurement of gas dispersion

F. Ricciardelli and A. M. Avossa (Eds.): INVENTO 2018, LNCE 27, pp. 665–678, 2019.
https://doi.org/10.1007/978-3-030-12815-9_50

over a 300-m wide area in 2014. We found that wind speed and gas concentration data agreed well with the values calculated using wind tunnel tests and CFD (Miyashita et al. 2015). As the next step of this study, we have carried out a similar field measurement with an aim to investigate dispersion characteristics over a wider area. A vertical profile of wind speed was attempted to measure using a Doppler lidar in order to apply an inflow condition to CFD. We then compared gas concentrations obtained from the field measurements with theoretical model predictions. Furthermore, we investigated the relationship between gas dispersion and atmospheric stability from wind speed and solar radiation. To incorporate more detail into our investigation of the atmospheric condition, fluctuation of wind direction and the vertical profile of wind speed and direction were studied in cases of atmospheric instability, where measurements were not completely in agreement with theoretical predictions.

2 Field Measurements and Experimental Conditions

2.1 Field Site

Field measurement was conducted in the Atsugi area, which is approximately 5 km wide, as shown in Fig. 1. Atsugi is located approximately 40 km to the southwest of the Tokyo metropolitan area, and the population is approximately 200000. The Sagami river flows in a north-south direction and the topography is mostly flat. The surface roughness around the area is category II or III based on the AIJ Recommendations for Loads on Buildings (2015). The measurement was conducted from August 5th to August 8th, 2014.

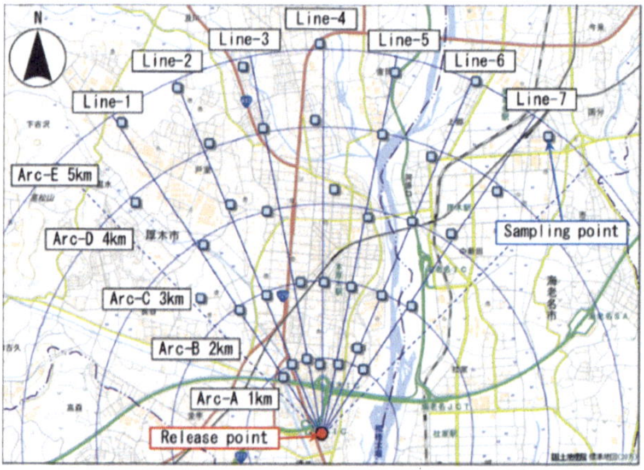

(Geospatial Information Authority of Japan)

Fig. 1. Location of the release and sampling points.

2.2 Release and Sampling

The release point was located to the south of downtown Atsugi. It was set on the rooftop of a five-storied building, which was approximately 20 m high as shown in Fig. 2(a) and (b). It was assumed that the wind at the release point was not influenced by other buildings, because there were few surrounding buildings that rose higher than the release point. In our study, PMCH (perfluoro-methyl cyclohexane, C_7F_{14}, MW = 350) was used as a tracer gas. The release rate was approximately 150 g/h. The measurement schedule is indicated in Table 1. In this study, one Run was defined as a 30-min period of measurement. We performed a total of 32 Run, undertaking 8 Run per day. The gas release was initiated 30 min before the measurement, and ended when the measurement was complete.

The five sampling arcs were approximately 1, 2, 3, 4, and 5 km to the north of the release point. Seven sampling lines were set, with the central line heading north from the release point, three lines to its east, and three to its west at longitudinal intervals of 11.25°. A total of 37 samplers were placed close to the points of intersection between the arcs and lines, as illustrated in Fig. 1. The height at which all samples were taken was 1.5 m, as shown in Fig. 2(c). The collecting tubes in the sampler adsorbed the gas. Afterwards, the samples were analyzed using a gas chromatography apparatus.

(a) Overview of the building (b) Release point (c) Sampling point

Fig. 2. Release and sampling points for collection of gas dispersion data.

Table 1. Measurement schedule with Run numbers specified.

		10		11		12		13		14		15		16	hour
August 5th, 2014					○	1	2	3	4	5	6	7	8		
August 6th, 2014					○	9	10	11	12	13	14	15	16		
August 7th, 2014			○	17	18	19	20	21	22	23	24				
August 8th, 2014			○	25	26	27	28	29	30	31	32				

○ : starting time for gas release

2.3 Meteorological Measurement

In order to record the meteorological conditions during measurement, a 2D ultrasonic anemometer (Vaisala Corp., WXT520), a pyranometer (EKO Instruments Co., Ltd.,

MS-402F), and a Doppler lidar (Mitsubishi Electric Corp., LR-09FLIII) were set up near the release point as shown in Fig. 3. The anemometer was placed on the pole of the penthouse, at an approximate height of 25.5 m. The other instruments were set up on the roof, at an approximate height of 20 m. To investigate the horizontal wind flow characteristics near the ground surface, five 2D ultrasonic anemometers (Vaisala Corp., WXT520) were set up, as illustrated in Fig. 2(c), with the measurement being taken at a height of 2 m. They were placed at Arc-D from August 5th to August 7th, and almost on the central line on August 8th.

In this study, each meteorological measurement took an average time of 10 min. Wind speed and direction data were sampled using the ultrasonic anemometer at the rate of 1 Hz, and a moving average of 3 s was applied to the data. With respect to lidars, the sampling rate was 1 Hz, and several observations were made from between 114.35 m and 826.85 m above ground, at 37.5-m intervals.

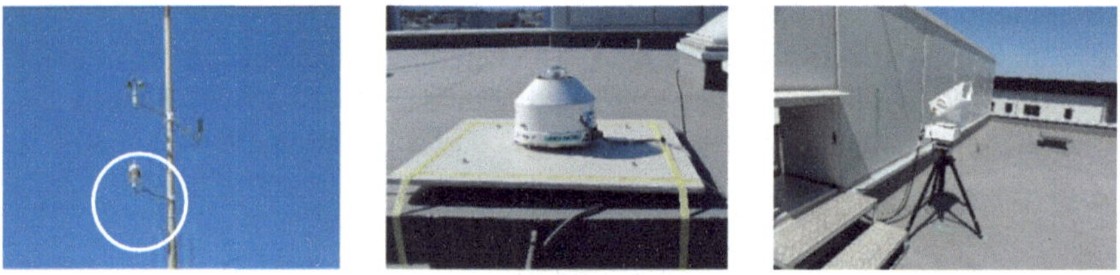

(a) Ultrasonic anemometer (b) Pyranometer (c) Doppler lidar

Fig. 3. Measurement devices placed near the release point.

2.4 Meteorological Conditions and Atmospheric Stability

Figure 4 shows meteorological measurements at the release point for four days. Left and right figures represent temperature and relative humidity, and wind speed and wind direction, respectively, as values averaged over 10 min. During the measurement period, the weather was influenced by the Pacific anticyclone and was almost fine between August 5th and August 7th. Relatively strong winds of approximately 8 m/s, were blowing from the southwest on August 5th and 6th. On the 7th, wind speed decreased slightly, and wind direction changed to south. On August 8th, wind speed became low (less than 4 m/s), and wind direction fluctuated from east-northeast to south-southwest. Furthermore, the temperature was lower and relative humidity was higher than on the other days.

In the previous studies, some parameter, e.g. the Obukhov length, was often used for estimating the atmospheric stability based on multiplicate instruments in vertical line. However, the atmospheric stability was calculated using wind speed and solar radiation in this study, based on the classification shown in Table 2. This is because there was only 2D ultrasonic anemometer in vertical line. Figure 5 shows the atmospheric stability calculated using this method. The wind speed at a height of 10 m was obtained from the value measured at observation height using the equation

$$U_{10} = U_{obs}(10/25.5)^{1/3} \quad (1)$$

where U_{obs} represents wind speed at observation height (25.5 m) and U_{10} denotes wind speed at a height of 10 m.

On the 5th and 6th of August, atmospheric stability was classified as C, C-D, and D, meaning it was almost neutral. It began to destabilize gradually starting August 7th, so that class B was occasionally observed. On August 8th, it became strongly unstable, with class A-B appearing more often than class C.

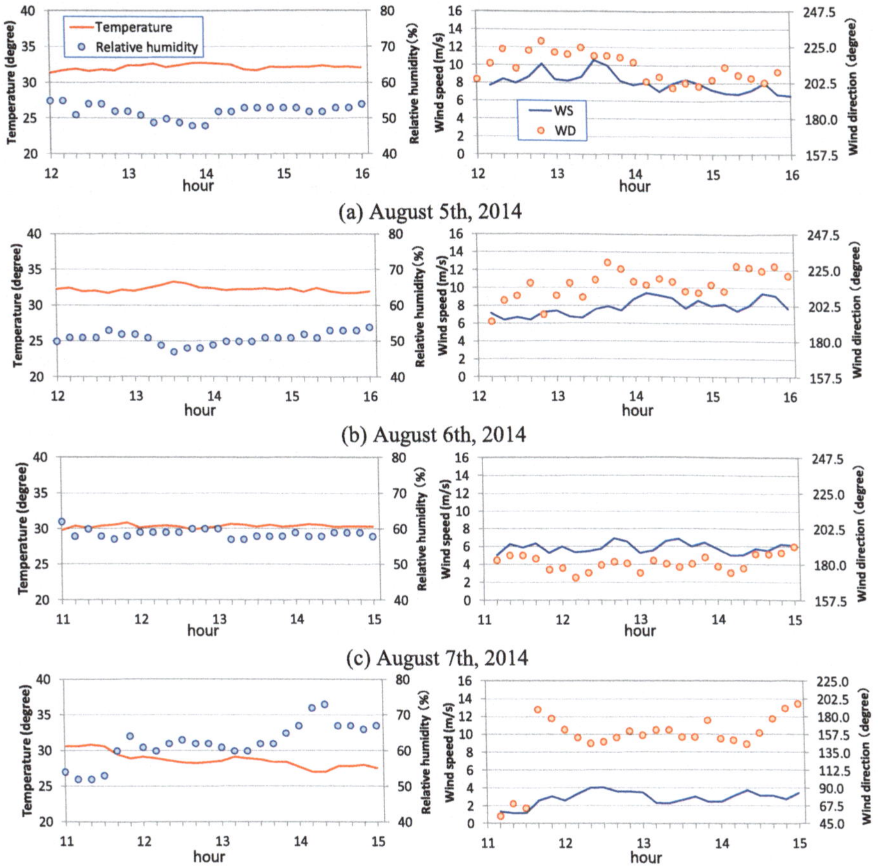

(a) August 5th, 2014

(b) August 6th, 2014

(c) August 7th, 2014

(d) August 8th, 2014

Fig. 4. Meteorological measurements on each day.

Table 2. Classification of atmospheric stability.

Wind speed U_{10} (m/s)	Amount of solar radiation (T) kW/m^2			
	$T \geq 0.60$	$0.60 > T \geq 0.30$	$0.30 > T \geq 0.15$	$0.15 > T$
$U_{10} < 2$	A	A-B	B	D
$2 \leq U_{10} < 3$	A-B	B	C	D
$3 \leq U_{10} < 4$	B	B-C	C	D
$4 \leq U_{10} < 6$	C	C-D	D	D
$6 \leq U_{10}$	C	D	D	D

A: Extremely unstable; B: Moderately unstable; C: Slightly unstable; D: Neutral

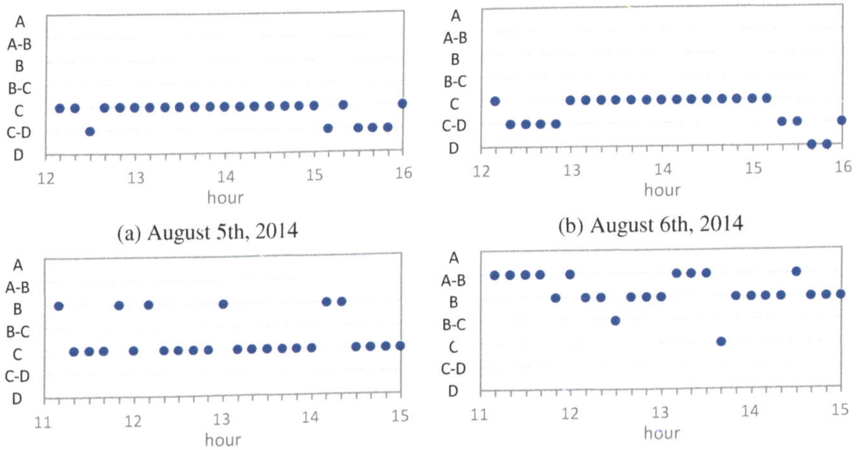

(a) August 5th, 2014 (b) August 6th, 2014

(c) August 7th, 2014 (d) August 8th, 2014

Fig. 5. Atmospheric stability during measurement periods.

3 Results

3.1 Distribution of Gas Concentrations

Figure 6 illustrates the distribution of gases in terms of their concentration over the area covered by all Run. The concentration values are expressed in a logarithmic scale. The arrows denote the wind direction at the release point. It can be seen that the high-concentration area extends in the direction of the wind, as at Run-05 and Run-08, and the concentration decreases with increasing distance from the release point. It is assumed that there is no influence of topographical features such as the Sagami River on the distribution, because the area is not extended along the river.

Figure 7 shows the 30-min averaged wind speed and wind direction at the release point and at five sampling points at ground level. The length and direction indicate wind speed and direction, respectively. It is shown that the wind direction at the ground corresponds well to the direction recorded at one of the release points. The wind speed

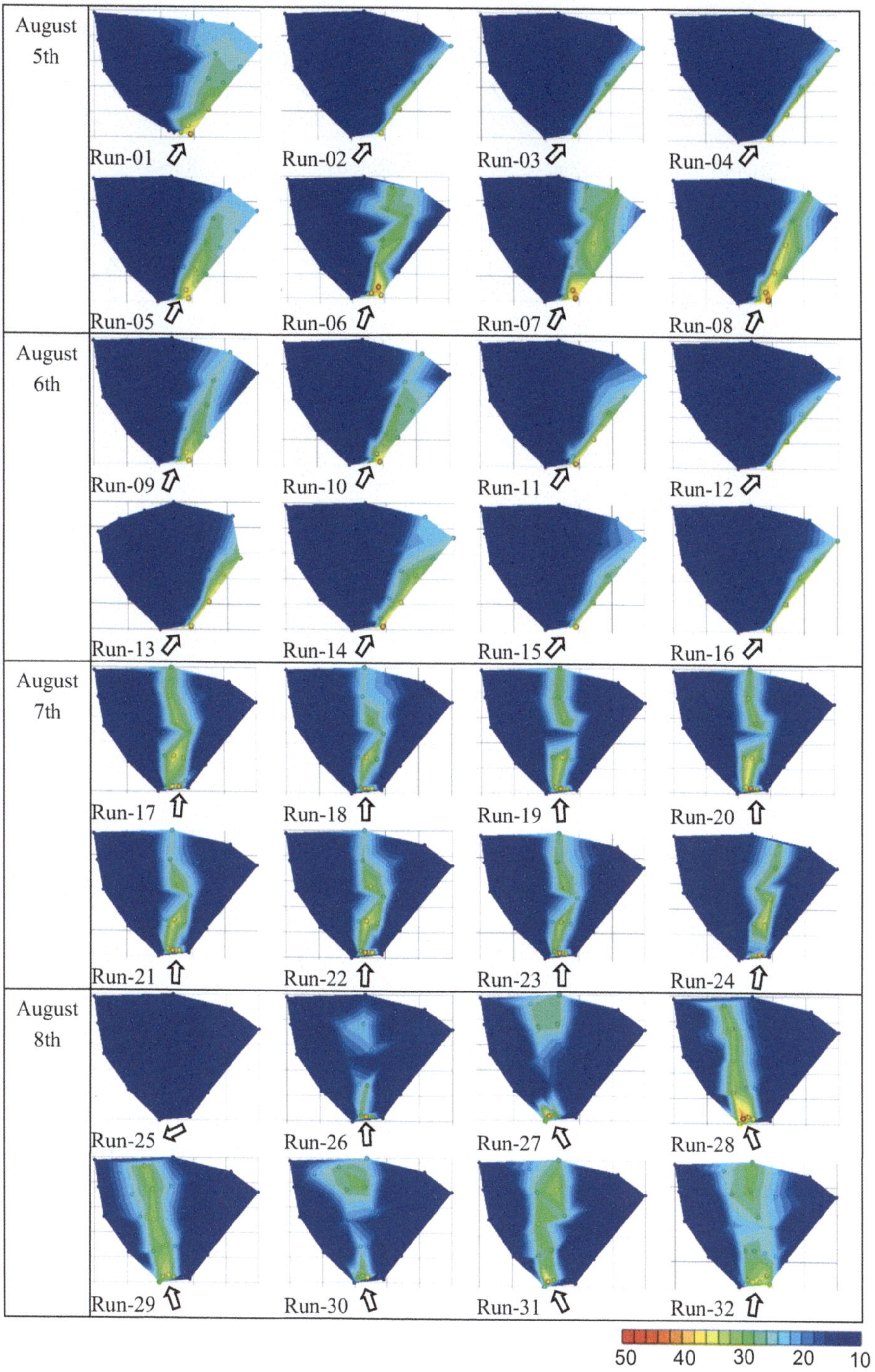

Fig. 6. Distribution of gas in terms of concentration (ppq [Log-scale])

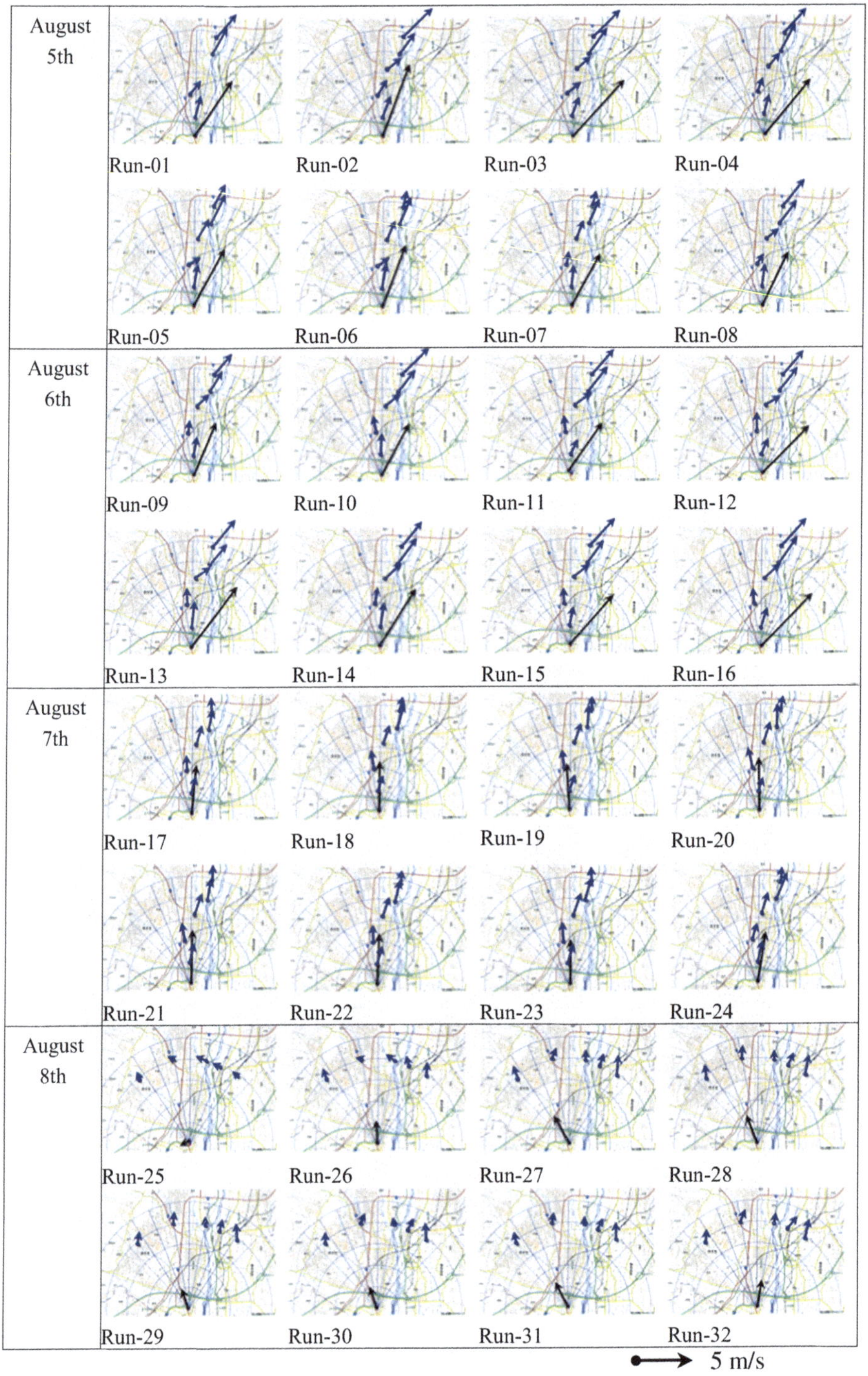

Fig. 7. Wind speed and direction at the ground level.

at the Arc-B sampling point tends lower because this point was in the downtown area and therefore surrounded by several buildings. On August 8th, wind speed at the ground was very low, similar to that at the release point.

Figure 6 shows that there is a gap indicating a low-concentration pocket within the high-concentration area. The gap near Arc-B is probably the result of scattering between sampling points in the case of Run-19 to Run-22. In addition to this, we suspect that the wind speed near the gap is low for reasons relating to the dense infrastructure mentioned above, constraining the gas from properly spreading over the area. By contrast, in the case of Run-27 and Run-30, the gap may be related to atmospheric stability or fluctuation in wind direction.

From the next section onwards, the following runs are selected as representative cases, in order to investigate the characteristics of gas dispersion in more detail: Run-05 to Run-08 on August 5th, when atmospheric stability is relatively high. Run-27 to Run-30 on August 8th, when atmosphere stability is strongly unstable. Run-19 to Run-22 on August 7th when atmospheric stability is between the former two conditions.

3.2 Comparison Between Measured Data and Predicted Values from Theoretical Model

The plume model described below in Eq. (2) was used as the theoretical model in this study (Environmental Research and Control Center, 2000). Equation (3) was used to correct the dispersion width for the horizontal direction σ_y corresponding to the evaluation time in this study. Figure 8 represents a comparative analysis of the gas concentration results obtained from the measured data and the values predicted by the theoretical model. In this figure, measurement results are given as five data points along the line that most closely approximates wind direction at the release point. In case of Run-27, Run-28 and Run-30, the concentration was approximately zero at some measurement points; therefore, five data points were not available and are not shown. The atmospheric stabilities for every 10-min period are shown as well. When the atmospheric stability falls into two classes in one Run, as in Run-07, two lines can be seen.

$$\mathrm{C}(x, y, 0) = \frac{Q}{\pi \sigma_y \sigma_z U} exp\left(-\frac{y^2}{2\sigma_y^2}\right) exp\left(\frac{-H^2}{2\sigma_z^2}\right) \tag{2}$$

Where x is the arc distance from the release point (m), y represents the distance of the across-wind line from the along-wind line (m), C $(x, y, 0)$ denotes gas concentration at the ground at position (x, y), Q is the amount of gas released per second (m^3/s), U is the wind speed averaged over 10-min periods at the release point, and H is the height of the release point (m).

σ_y, σ_z are given by the following table:

Atmospheric stability	$\sigma_y(x) = \gamma_y \cdot x^{\alpha_y}$			$\sigma_z(x) = \gamma_z \cdot x^{\alpha_z}$		
	x(m)	α_y	γ_y	x(m)	α_z	γ_z
A	0 ~ 1,000 1,000 ~	0.901 0.851	0.426 0.602	0 ~ 300 300 ~ 500 500 ~	1.122 1.514 2.109	0.0800 0.00855 0.000212
B	0 ~ 1,000 1,000 ~	0.914 0.865	0.282 0.396	0 ~ 500 500 ~	0.964 1.094	0.1272 0.0570
C	0 ~ 1,000 1,000 ~	0.924 0.885	0.1772 0.232	0 ~	0.918	0.1068
D	0 ~ 1,000 1,000 ~	0.929 0.889	0.1107 0.1467	0 ~ 1,000 1,000 ~ 10,000 10,000 ~	0.826 0.632 0.555	0.1046 0.400 0.811

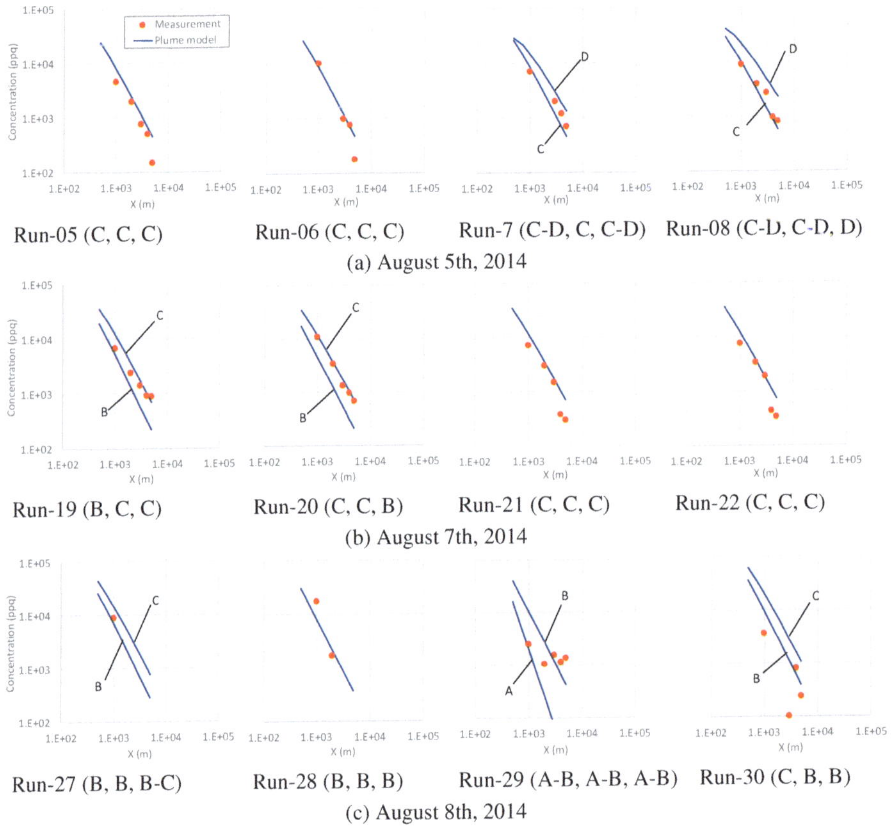

Run-05 (C, C, C) Run-06 (C, C, C) Run-7 (C-D, C, C-D) Run-08 (C-D, C-D, D)

(a) August 5th, 2014

Run-19 (B, C, C) Run-20 (C, C, B) Run-21 (C, C, C) Run-22 (C, C, C)

(b) August 7th, 2014

Run-27 (B, B, B-C) Run-28 (B, B, B) Run-29 (A-B, A-B, A-B) Run-30 (C, B, B)

(c) August 8th, 2014

Fig. 8. Comparison of concentrations obtained from field measurement and from the theoretical model (atmospheric stability class for each 10-min period in brackets).

$$\sigma_{y'} = \sigma_y (t/t_0)^r \tag{3}$$

where $\sigma_{y'}$ is the corrected σ_y, t is the time of evaluation (30 min) t_0 if the time of evaluation based on the Pasquill-Gifford scheme (3 min), and r is a constant (0.2).

Figure 8 indicates that the results from measurements and those from the model used in this study are in agreement, except in the case of Run-29 and Run-30. The strong alignment of results for the most part appears to be the result of the field site selection, specifically that there were no high-rise buildings around the release point. This is, apparently, the ideal condition in which to apply the theoretical model. Furthermore, we found that the wind direction at the release point is important in predicting gas dispersion, because the distance of the across-wind line from the along-wind line has an effect on the accuracy of the model. The theoretical model is least likely to be in agreement with the measured results when the atmospheric stability falls into class A-B and class B that is, when the atmosphere is extremely or moderately unstable.

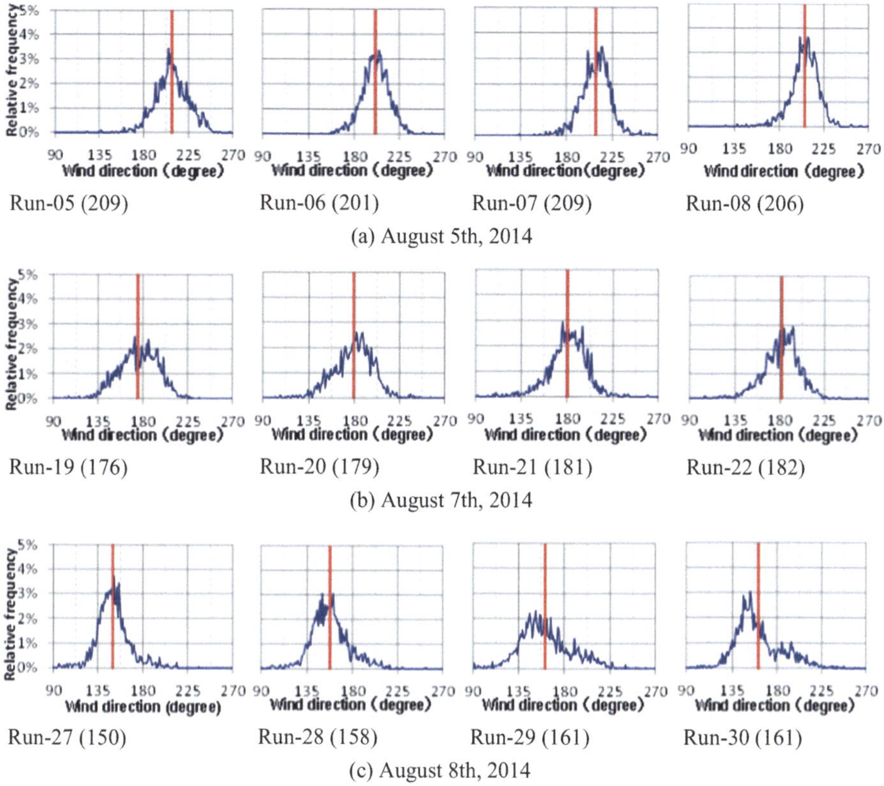

Fig. 9. Relative frequency of wind directions (red line and number in bracket indicate average wind direction).

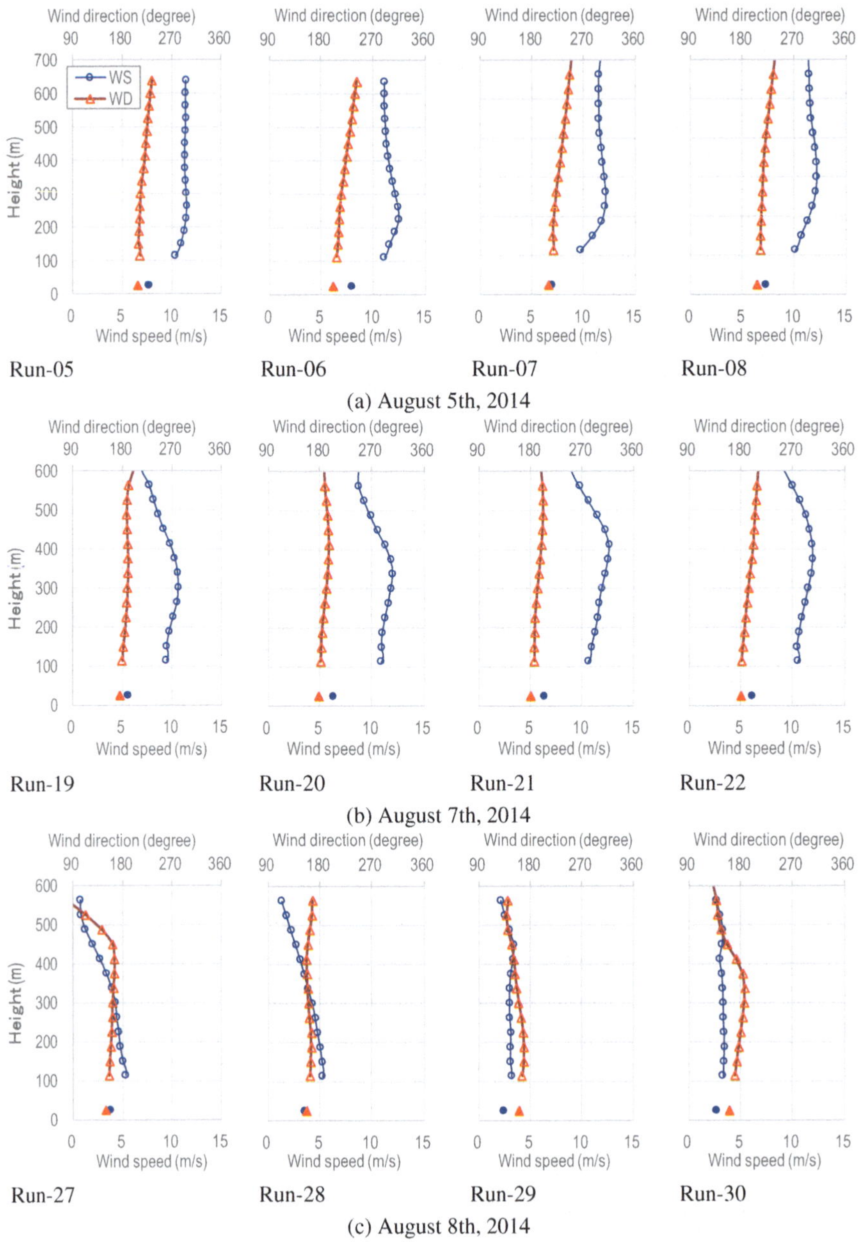

Fig. 10. Vertical profile of wind speed and wind direction.

3.3 Characteristics of Wind Turbulence and Vertical Profile

We found that model predictions do not align well with measurement results when the atmosphere is strongly unstable. To understand the weather conditions at these Run measurements in detail, we investigated the fluctuation of wind direction and the vertical profile of wind speed and direction for each of them.

Figure 9 shows the relative frequency of wind direction changes for each Run. The distribution of wind direction is relatively narrow on August 5th. However, we found that the values are distributed over a wider range on August 7th and 8th, especially at Run-29 and Run-30. Clearly, wind direction tended to fluctuate on these two days.

Figure 10 illustrates the vertical profile of wind speed and direction based on lidar and ultrasonic anemometer measurements at the release point. In this figure, data from the anemometer are represented as filled marks. Figure 10(a) shows that wind speed generally increases with an increase in height from the ground to an approximate height of 200 m and then reaches an almost constant value. On August 7th, wind speed rises with increasing height from the ground. However, it decreases with increasing height past approximately 400 m above the ground. Furthermore, on August 8th, wind speed is seen to have generally decreased with an increase in height at Run-27 and Run-28, and had an almost constant value at Run-29 and Run-30. As for wind direction, it shifts slightly westward with increasing height on August 5th and 7th. By contrast, the profile of wind direction on August 8th is quite different from that seen on the other two days.

Our investigation into wind turbulence and vertical profile indicates that the atmospheric condition on August 8th was strongly unstable. We suggest that this instability resulted in the gap in the high-concentration area seen in Fig. 6.

4 Concluding Remarks

In this study, we conducted a field measurement of gas dispersion in the Atsugi area, approximately 5 km wide. We found that an area of high concentration of the dispersed gas tends to extend along the direction of the wind. Furthermore, the alignment between the measured data and predicted values from the theoretical model of gas concentration was examined. We found that the results agreed well with each other when the atmospheric stability was high. However, some measurements showed a clear gap of low concentration within the high-concentration area. We investigated the meteorological conditions for these measurement points using fluctuation of wind direction and vertical profile of wind speed and direction as parameters. Our results suggest that strongly unstable atmospheric conditions cause the gap.

In future, wind tunnel tests and CFD will be applied to our data to investigate the results in more detail. Especially, the vertical profile based on the measurements of Doppler lidar will be attempt to use as inflow condition of CFD. Furthermore, we plan to conduct another field measurement in the future to improve the credibility of our findings. Measurements in locations with a variety of wind speeds and wind directions at ground level are necessary components of a thorough investigation of the characteristics of gas dispersion.

Acknowledgements. The authors would like to thank Professor Akihito Yoshida of Tokyo Polytechnic University, and the foundation of Sun Mall, Atsugi for permission to use their building rooftop as our release point. We also thank Toshiba Infrastructure Systems & Solutions Corporation and Japan Weather Association for support in performing the measurements, and Mr. Naoki Matayoshi of Japan Aerospace Exploration Agency for his support of measuring using Doppler lidar.

References

Allwine KJ et al (2002) Overview of urban 2000: a multiscale field study of dispersion through an urban environment. Bull Amer Meteor Soc 83:521–536

Allwine KJ et al (2004) Overview of urban 2003: an atmospheric dispersion study in Oklahoma City. In: Eighth symposium on integrated observing and assimilation systems in the atmosphere, oceans, and land surface and symposium on planning, nowcasting, and forecasting in the urban zone, Seattle, WA, Amer. Meteor. Soc., J7.1

Architectural Institute of Japan (2015) Recommendations for Loads on Buildings (in Japanese)

Arnold SJ et al (2004) Introduction to the DAPPLE air pollution project. Sci Total Environ 332 (1–3):139–153

Environmental Research and Control Center (2000) Chisso sankabutsu soryo kisei manual: Nitrogen oxide regulation of discharge of harmful industrial waste manual (new publication) (in Japanese)

Hanna S et al (2006) Analysis of Joint Urban 2003 (JU2003) and Madison Square Garden 2005 (MSG05) meteorological and tracer data. In: Sixth symposium on the urban environment, Atlanta, GA, Amer. Meteor. Soc., J7.1

McEloy JL, Pooler F (1968) St. Louis Dispersion Study. Report AP-53. US Public Health Service, National Air Pollution Control Administration

Miyashita et al (2015) Full scale experiment of gas dispersion in an urban area and its prediction by wind tunnel experiment and large eddy simulation, Part 1-4. Wind Eng JAWE 40(2):87–94 (in Japanese)

Definition of Wind Loads on the Riyadh Western Metro Station

P. Schito(✉), L. Rosa, and A. Zasso

Department of Mechanical Engineering, Politecnico di Milano,
Via La Masa 1, 20156 Milan, Italy
{paolo.schito,lorenzo.rosa,alberto.zasso}@polimi.it

Abstract. Design of large modern structures can be strongly affected by wind actions, in terms of dynamic loading as well as peak façade cladding load. The wind forcing can be described by means of CFD analysis as well by wind tunnel tests using a scaled model instrumented with pressure taps. The density and the distribution of pressure taps on the building surface is critical, as well as the averaging of the surface pressure time history for the definition of the wind actions. Dynamic actions are calculated using a modal approach, while pressure excitation is computed according to time equivalent area averaging. The definition of the time equivalent area averaging is carefully analysed, in order to provide pressure values consistent with the forcing area.

Keywords: Wind loads · Wind tunnel tests · Dynamic response · TMD

1 Introduction

The design of modern large structures is often driven by the choice of architectural solutions that create emotion and wonder on the sightseers. The development of new construction technology by using more advanced materials, allows to designers to present structures with increasing complexity and innovative contents. Novel designs show complex shapes that sometimes cannot be reconducted to the building typologies reported in standard codes (i.e. Eurocode [1], ASCE [2]).

The design of current buildings uses several different materials for the different components of the building, up to the external surface, where the wind loads are transferred to the main structure. The size of the structure and its flexibility can be such that the wind pressure loads may induce deformations of the structure, leading to inertial accelerations that need to be considered in the structural design. Typically, in case of structures with very large dimensions, wind tunnel tests are required in order to consider the actual wind actions on the structure [3].

Wind tunnel tests on buildings are generally performed on rigid models, since, typically, the wind induced loads do not change significantly the shape of the building itself and its aerodynamic performances. Wind loads can be assessed by using an integral measurement (typically the forces on the building foundation) or by assessing the wind pressure acting on the building envelope. If both measurements are available, a verification of the reliability of the methodology can be conducted. Wind pressure is measured by means of pressure taps located on the building surface: the distribution of

F. Ricciardelli and A. M. Avossa (Eds.): INVENTO 2018, LNCE 27, pp. 679–688, 2019.
https://doi.org/10.1007/978-3-030-12815-9_51

the pressure taps must be suitable for the description of the wind action on the building. The calculation of the wind induced loads is generally performed taking into consideration the transient characteristics of the turbulent incoming wind with the dynamics of the structure.

Many strategies can be adopted for the calculation of wind loads on structures. CFD calculation can be used for the definition of the aerodynamic actions on the structure and the Standard recommended approach can be adopted to compute the wind loads, using the numerical model to extract the pressure coefficient distribution on the building surface. In this paper wind loads are extracted through wind tunnel tests using the high frequency pressure integration technique (HFPI), [4]. The practical basis of this approach is to develop an analytical modal approach considering as input the pressure data acquired simultaneously over the surface of the structure and the modal information (modal masses, mode shapes and frequencies) obtained from a finite element model of the structure. The calculation of the wind induced loads is generally performed taking into consideration the transient interaction of the turbulent incoming wind with the dynamics of the structure. The transient pressure distribution on the building surface is measured according to the incoming turbulent wind characteristics, that depend on the terrain characteristics and on the existing surrounding structures. The terrain features are estimated according to the area where the building will be erected (e.g. flat terrain – built environment) and indicates the main characteristics of the incoming wind in terms of vertical wind profile, vertical turbulence intensity and vertical turbulence length scale.

Fig. 1. Rendering of Riyadh Western Metro Station.

The structure that has been considered is the Riyadh Western Metro Station (see Fig. 1): it is a very light structure, mainly in steel, with length larger than 300 m and a height up to 25 m. The peculiar shape of the building and in particular of the smaller dune that acts as a cantilever produces very particular dynamic response of the structure.

2 Experimental Model

The wind tunnel model of Riyadh Western Metro Station has been realized with a geometrical scale of 1:100. Measurements were carried out for 36 wind directions spaced 10° apart, simulating the natural wind profile expected at the construction stage. Atmospheric boundary layer wind tunnel tests consider the characteristics of the natural wind, correctly scaled. In particular, the wind reproduces the variation of the mean velocity with the height from the ground and its turbulence characteristics. The simulation of the natural wind in the test section is achieved using passive turbulence generators placed at the beginning of the test section and roughness elements on the wind tunnel floor upstream the turntable (see Fig. 2). The model is designed to measure the pressures distribution by means of about 400 pressure taps simultaneously acquired. Pressure measurements are performed using the high-speed scanning pressure equipment Initium and the miniature pressure scanners ESP (*500* Hz sampling frequency). Pressure taps have been distributed on all surfaces of the dunes, including roof and façades (see Fig. 3). The model of the dunes was CNC made using resin. The external surface was left intentionally rough in order to simulate a higher Reynolds number and consequently a flow more similar to the full scale. The experimental setup considers not only the buildings, but also the most important surrounding buildings and the orography of the site.

Fig. 2. Wind tunnel model of Riyadh Western Metro Station.

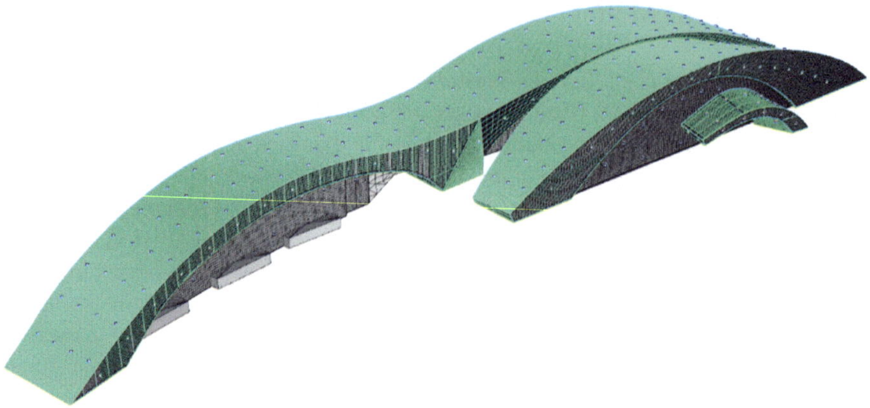

Fig. 3. Indication of pressure taps locations for the wind tunnel model.

3 Numerical Model

The computation of the wind loads is performed using the high frequency pressure integration technique (HFPI): the pressure data acquired simultaneously over the surface of the structure are coupled with the modal information (modal masses, mode shapes and frequencies) from a finite element model of the structure, to develop an analytical modal model of the building.

According to the modal approach, considering n modes, the vector of the nodal displacements $\boldsymbol{X}$ is expressed as:

$$\boldsymbol{X}(t) = \boldsymbol{\Phi}\underline{q}(t) \tag{1}$$

Where $\boldsymbol{\Phi}$ is the matrix of the eigenvectors and $\underline{q}(t)$ is the vector of generalized displacements.

The equation of motion can be written as:

$$\boldsymbol{M}\ddot{\boldsymbol{X}}(t) + \boldsymbol{C}\dot{\boldsymbol{X}}(t) + \boldsymbol{K}\boldsymbol{X}(t) = \boldsymbol{F}(t) \tag{2}$$

$$\boldsymbol{\Phi}^T\boldsymbol{M}\boldsymbol{\Phi}\ddot{\underline{q}}(t) + \boldsymbol{\Phi}^T\boldsymbol{C}\boldsymbol{\Phi}\dot{\underline{q}}(t) + \boldsymbol{\Phi}^T\boldsymbol{K}\boldsymbol{\Phi}\underline{q}(t) = \boldsymbol{\Phi}^T\boldsymbol{F}(t) \tag{3}$$

$$\boldsymbol{M}_q\ddot{\underline{q}}(t) + \boldsymbol{C}_q\dot{\underline{q}}(t) + \boldsymbol{K}_q\underline{q}(t) = \boldsymbol{Q}(t) \tag{4}$$

Where the forcing term $\boldsymbol{F}(t)$ is represented by the pressures acting on the cladding multiplied by its tributary area.

Displacements and accelerations are calculated by integrating, step by step, the equations of motion of the structure (Eq. 4).

Dynamic loads $\boldsymbol{F}_{IN}$ on each mass are calculated as:

$$\boldsymbol{F}_{IN}(t) = -\boldsymbol{m}\ddot{\boldsymbol{X}}(t) \quad (5)$$

The output of this analysis is the time history of the total loads (pressure loads on the cladding and inertial loads on the masses) on the entire structure. For the determination of the most critical situation, some strain indicators have been developed (e.g. maximum foundation moment).

The definition of the pressure loads is computed according to the instantaneous pressure measured on the experimental pressure taps, filtered according to the pressure tap reference area.

4 Spatial Averaging and the Aerodynamic Admittance Function

During wind tunnel tests, pressures are measured at specific points along the model. These punctual measurements are then used to compute the wind loads on finite areas. This passage, from pressure acting on a point to pressure acting on an area, needs a processing of the point pressure to account for the “spatial averaging” effect.

Indeed, fluctuating pressures acting on finite areas have smaller peak values than the pressure of a single point in that area. This occurs because, for a given area, the high frequency components of each punctual pressure are not fully correlated, and when these pressures are averaged over the finite area, the averaged fluctuations are smaller.

The “spatial averaging” effect is simulated by a “time filtering”, using an appropriate filter function or “aerodynamic admittance” function, [5].

For standard structures, a moving average filter is generally used, with a time window equal to $\tau = KL/V$ being V the mean wind speed, L a characteristic length

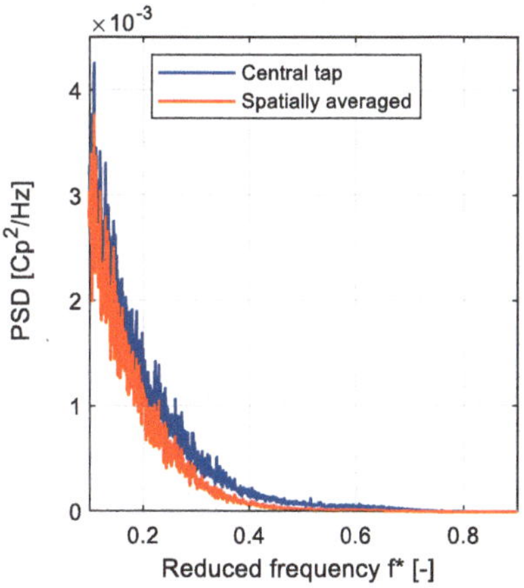

Fig. 4. PSD comparison between the central pressure tap in the tile and the spatially averaged signal, $\alpha = -135°$.

(e.g. diagonal of the tributary area), and K a number that depends on the structure (in general between 1–4, as described by Holmes [5]).

For complex structures like these dunes, specific tests should be performed to assess which is the effect of the "spatial averaging" and what is the appropriate processing of the time histories to simulate it. According to this method, the local pressure spectrum is multiplied by an "aerodynamic admittance" function that was experimentally evaluate. To evaluate the "aerodynamic admittance" the pressure was measured on a spatially dense array of pressure taps (hereafter called tile) in 3 significant finite areas of the surface (*20* taps for a *6* m^2 tile full scale), averaged, and finally compared with the single point measurement.

Signals acquired by the pressure taps in the tile were averaged at each time instant in order to obtain a new equivalent spatial averaged signal for each incoming wind direction. The new spatially averaged signal is assumed to be representative of the tile area.

As an example Fig. 4 shows the spectra of the time history of pressure acquired by the tap placed at the center of the tile and the spatially averaged signal for the same exposure angle. As one can see the spatially averaged signal has a lower energy content, especially for high reduced frequencies f^* ($f^* = f \cdot L/V$ where L is the diagonal of the tile and V is the design wind speed). It can be noted that for reduced frequencies higher than 0.5 the signals have no more energy content.

4.1 Experimental Admittance

Assuming that the spatial averaged signal is representative for the tile area, the transfer function (TF) between the signal of the tap placed at the center of the tile and the equivalent spatial averaged signal is computed for each exposure angle. This TF is called admittance function. Focusing on the most significant wind exposure angles in terms of dynamic contents, the transfer functions are plotted in Fig. 5, where the experimental TF (continuous line) is compared with the fitted curve (dashed line).

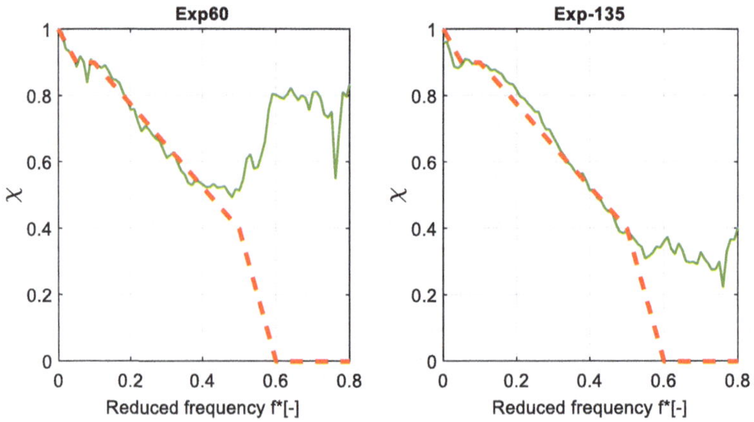

Fig. 5. Transfer Function module for wind exposure angles 60° and −135°.

TFs have been expressed as function of the reduced frequency f^*. It should be noted that for reduced frequencies higher than 0.5, the fitted curve approaches zero values, since we do not have energy content anymore (see Fig. 4) and thus it is no more significant since the signal contain only noise. The fitted aerodynamic admittance curve considers all the tiles and the wind directions, taking the most conservative case. The filter obtained with the previous procedure has been expressed as function of the reduced frequency. This makes possible to extend it to a generic surface of size L, with the hypothesis that the coherence between two points in an area it is not affected by the extent of the area, which is current practice.

5 Dune 0 Response to the Wind

The dynamic response of Dune 0 has been evaluated from the wind tunnel measured wind excitation. This is the most challenging structure since it is a standalone cantilevered structure almost *50* m long, see Fig. 6. Figure 7 shows the first and second vibrations mode of Dune 0, whereas in Fig. 8 the end section of Dune 0 is reported with its reference system.

The analysis considers a structural damping ratio of 0.7% and a mean incoming wind with return period of 100 years equal to V = 44 m/s. The numerical simulations were done for each exposure angles tested in the wind tunnel and the exposure angle $\alpha = 0°$ showed the worst dynamic response.

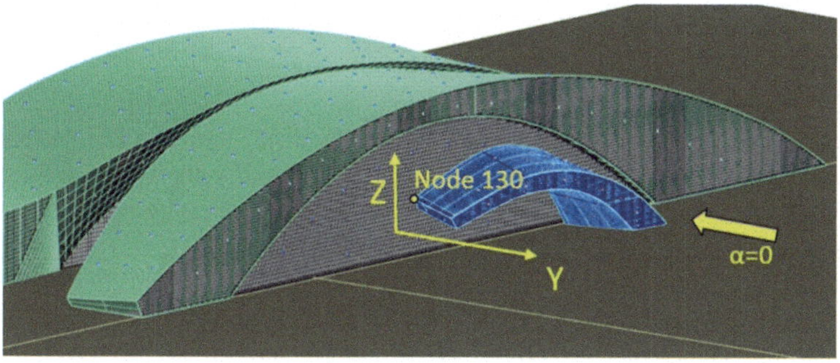

Fig. 6. Schematic model of Dune 0 and reference system.

The displacement of the dune 0 free end (monitoring the structural FEM node 130) shows large displacements and high accelerations along y-direction and z-direction, mainly imputable to the first mode shape of the structure. The displacement of node 130 showed values larger than ± 200 mm in the horizontal and vertical directions, while acceleration showed values that exceed *5* m/s^2. Figures 9 and 10 show the time histories of displacement and acceleration along y-direction (blue line), whereas the trend along z-direction (not shown here) was very similar. Figures 11 and 12 show the spectra of the previous time histories along y and z direction respectively (blue line): as one can see the contribution of the first vibration mode is prevalent and the intensity of the vibration along y and z directions is very similar.

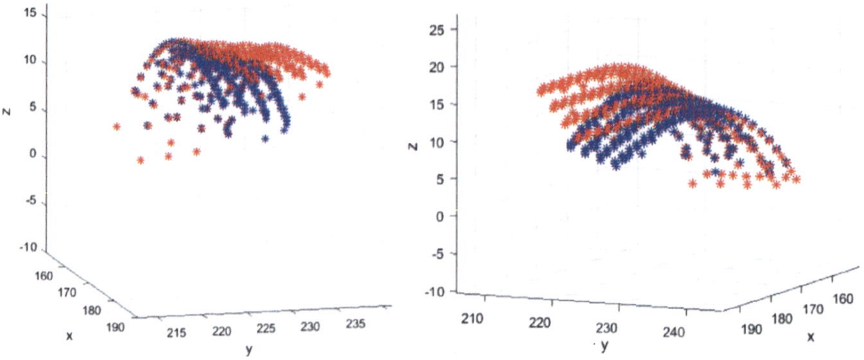

Fig. 7. First (sx) and second (dx) vibration mode of Dune 0. Blue points: not deformed geometry. Red points: mode shape.

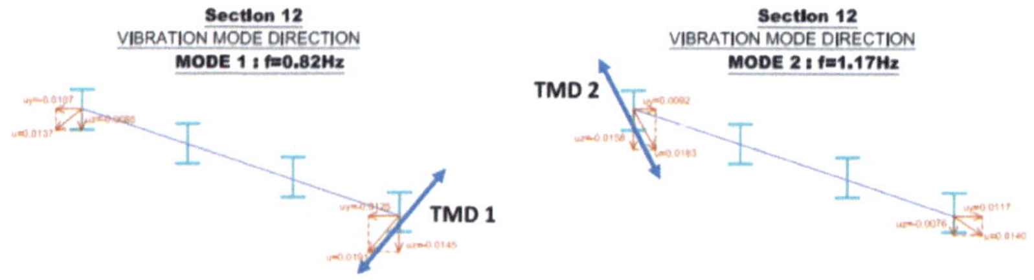

Fig. 8. Schematic model of the end section of Dune 0 and locations and direction of action for the installation of TMD.

The analysis of the data shows that both the displacements and accelerations evaluated were too high. Since it was not possible to modify the structure in order to increase its stiffness, it was necessary to increase the total damping of the structure by means of a tuned mass damper.

A preliminary study, by applying a damping ratio of *4%* on the first two modes, was conducted and the results are reported in Figs. 9 and 10 (red line). The displacements show values that exceed ± 75 mm, and accelerations slightly over *1.5* m/s^2. The signal spectrum (Figs. 11 and 12, red line) is largely reduced for the modes taken into account for the damping ratios of *4%*.

The reduction in oscillations and in accelerations seem satisfying with such damping ratio, therefore it was suggested to install two TMDs (each installed on the tip of the dune, and synchronized on its mode frequency). Each TMD should be working following the direction of the mode direction (following the maximum displacement value, that is located on the dune tip).

The modal mass of mode 1 was estimated in *30* tons, therefore a TMD with a weight of *0.6* tons is sufficient to guarantee the *4%* damping ratio. The modal mass of mode 2 is estimated in *50* tons, therefore a TMD with a weight of *1* ton is sufficient to guarantee the *4%* damping ratio. The location and direction of installation of TMDs is reported in Fig. 8.

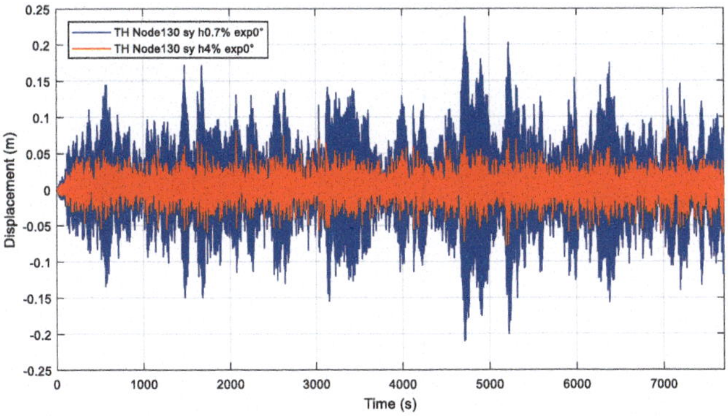

Fig. 9. Time history displacement FE node 130, y-direction, h = 0.7% and h = 4%.

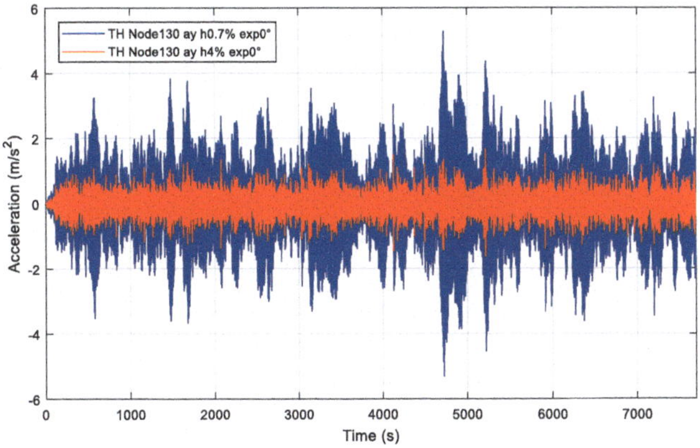

Fig. 10. Time history acceleration FE node 130, y-direction, h = 0.7% and h = 4%.

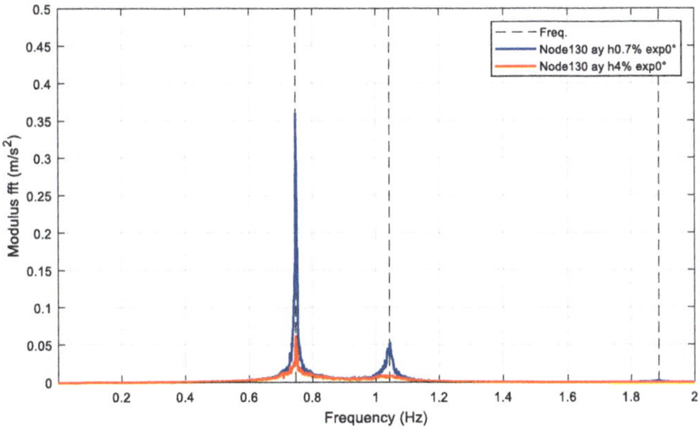

Fig. 11. Spectra acceleration FE node 130, y-direction, h = 0.7% and h = 4%.

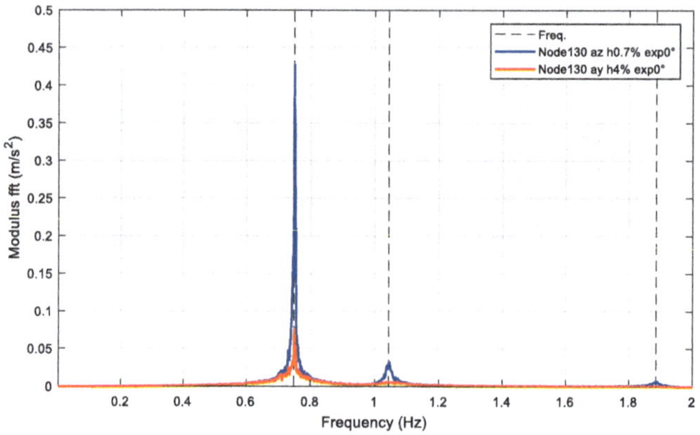

Fig. 12. Spectra acceleration FE node 130, z-direction, h = 0.7% and h = 4%.

As an alternative option, it would be possible to have each damper vertically guided, by increasing, roughly proportionally, the overall damping masses.

6 Conclusions

The method described in this article shows how it is possible to combine a modal approach analysis, with experimental wind tunnel test in order to simulate the dynamic of a structure. Results achieved with a rigid model are used to load a numerical model of the structure in order to evaluate the dynamic response of the structure. The wind excitation is extrapolated from measured wind tunnel data, by deriving the loading on the cladding surface of the structure, by taking into consideration the extension of the area and the equivalent loading, through an admittance function. The response of the building is evaluated and, by applying an equivalent damping to the structure, the displacements and the accelerations of the structures are limited. The required damping system, using a TMD technology is evaluated, and an estimation of the requirements of the system are presented.

References

1. Eurocode1 (2005) UNI EN 1991-1-4. Actions on structures - Part 1-4: General actions - Wind actions, April 2005
2. ASCE 7-05. Minimum Design Loads for Buildings and Other Structures
3. Irwin PA (2009) Wind engineering challenges of the new generation of super-tall buildings. J Wind Eng Ind Aerodyn 97(7–8):328–334
4. Rosa L, Tomasini G et al (2012) Wind-induced dynamics and loads in a prismatic slender building: a modal approach based on unsteady pressure measurements. J Wind Eng Ind Aerodyn 107–108:118–130
5. Holmes JD (1997) Equivalent time averaging in wind engineering. J Wind Eng Ind Aerodyn 72:411–419

Characterization of Mean Wind Profiles and Surface Roughness Assessment from Wind LIDAR Measurements

V. Sepe[1], F. Rizzo[1], F. Ricciardelli[2], and A. M. Avossa[2(✉)]

[1] Department of Engineering and Geology, Università "G. D'Annunzio" di Chieti-Pescara, Viale Pindaro 42, 65127 Pescara, Italy
{v.sepe, fabio.rizzo}@unich.it

[2] Department of Engineering, Università della Campania "Luigi Vanvitelli", Via Roma 9, 81031 Aversa, Italy
{friccia, albertomaria.avossa}@unicampania.it

Abstract. To evaluate wind loading on medium to high-rise structures, an accurate representation of the mean wind profile and of the structure of turbulence is required. Wind models are currently incorporated in Codes of Practice, which have been theoretically derived and experimentally calibrated. Nevertheless, the limited amount of data makes such calibration questionable at times. Wind LIDARS have recently appeared and are becoming more and more common for the assessment of wind power and for the monitoring of wind characteristics. They provide a direct measurement of the wind velocity at the site and at the height of interest and offer the possibility of analysing the spatial distribution of mean and fluctuating wind speed. In this paper, wind LIDAR measurements are used to characterize the vertical wind profile, to calibrate surface roughness to be used in wind models and to discuss their accuracy.

Keywords: Wind LIDAR · Mean wind profile · Surface roughness · Wind models · Neutral atmosphere

1 Introduction

To evaluate wind loading on medium- to high-rise structures, an accurate representation of the characteristics of strong winds in the Atmospheric Boundary Layer (ABL) is needed. On the other hand, lower winds contribute to fatigue damage accumulation, and are therefore also of interest for Structural Engineers. Several theoretical and empirical models have been proposed to describe the vertical distribution of mean wind speeds, the calibration of which is based on the knowledge of the underlying surface roughness. Although the topic is well established, there is still no common consensus on what is the most appropriate wind model, especially for Code of Practice implementation, and on how to calibrate surface roughness. Recently, observations of wind speed profiles through Doppler Lidar anemometry are becoming available (e.g. Drew *et al.* 2013; Lane *et al.* 2013; Wood *et al.* 2013; Lim *et al.* 2016; Demartino *et al.* 2017; Kent *et al.* 2018). These provide a direct measurement of the wind velocity at the

F. Ricciardelli and A. M. Avossa (Eds.): INVENTO 2018, LNCE 27, pp. 689–702, 2019.
https://doi.org/10.1007/978-3-030-12815-9_52

site and height of interest, in a range of several hundred meters above the ground, and with a sampling rate up to 1 Hz or more. In addition, wind LIDARS, and in particular wind scanners, also offer the possibility of analysing the spatial distribution of mean and fluctuating wind speed. In spite of the inaccuracies associated with space averaging and with non-synchronous measurement of continuous beam LIDARS, wind LIDAR measurements are becoming a useful tool in Wind Engineering and Wind Energy.

In this paper, wind LIDAR measurements over a period of about nine months are used to assess the accuracy of the logarithmic profile (Lumley and Panofsky 1964; Tennekes 1973; Simiu and Scanlan 1986) and of the linearized Deaves and Harris wind model (Deaves and Harris 1978; Cook 1985). The former is used by several Codes of Practice, among which is Eurocode 1 (CEN 2005), and is known to lose accuracy at around 200 m of elevation; the latter, obtained by adding to the logarithmic law only the linear term of the polynomial correction of the Deaves and Harris model, has the advantage of keeping a rather good accuracy up to 300 m of elevation.

2 LIDAR Measurements

For the measurement of the atmospheric wind field, a WindCube® V2 wind LIDAR manufactured by Leosphere was used. This device, which operates as a pulsed Doppler anemometer, is a ground-based, small and compact, and the measurements of the three wind speed components are derived by a laser beam of 1.54 μm wavelength with a fixed focus. In particular, to obtain the three-dimensional wind vector, the beam is inclined by 30° from vertical direction and measurements are taken under four different azimuth angles, each separated by 90°, rotating on a cone. When the prism holds still the LIDAR sends a stream of pulses (5,000 to 10,000) in a given direction, and the measurement is taken simultaneously at different heights, by considering the time elapsed between sending and receiving the beam. Having sent the required number of pulses, the prism rotates to the next azimuth angle to be scanned. A full rotation takes about 6 s. The length of the laser pulses of about 26 m defines the vertical extension of the measurement (up to 260 m). The three-dimensional wind velocity vector is then derived from four successive measurements.

At the beginning of October 2015 the wind LIDAR was installed at 10 m of elevation at the Aversa site; the device is located on the roof of a 2-story building belonging to the School of Science and Technology of the University of Campania "Luigi Vanvitelli" (Fig. 1). The Aversa site (40°58′00″N, 14°12′00″E) is located 15 km North of Napoli, at an elevation of 50 m and at a minimum distance from sea of 15 km (Fig. 2). The surroundings are characterised by a flat orography and densely built areas with low- to medium-rise constructions. Measurements started being taken on October 9, 2015 and were stopped at the end of 2017. In particular, the measurements taken until July 27, 2016 have been used for the analyses. The device is able to measure simultaneously the volume-averaged instantaneous three components of the wind speed at 12 elevations from 40 m to 300 m of vertical distance from the instrument, with a sampling rate of 1 Hz. For this particular experiment elevations of 50, 60, 70, 80, 90, 100, 110, 120, 150, 200, 220 and 250 m were selected. Ten-minute averages of the horizontal speed were calculated from the measurements; these are 144 for each day of measurements, with a total of about 39300 for the 273 days considered.

Fig. 1. WindCube® V2 installation

Fig. 2. Aversa site (40°58′00″N, 14°12′00″E), (©2018 Google-Imagery ©2018 DigitalGlobe).

3 Vertical Profiles of Mean Wind Velocity

For each ten-minute time interval the average wind speed $V(z_i)$ and the average wind angle $\alpha(z_i)$ have been evaluated at the twelve selected elevations z_i, so obtaining the mean wind speed and wind angle profiles.

Each one of the 39333 profiles is represented by the pair (V_{50}, α_{50}), where $V_{50} = V(z_1)$ and $\alpha_{50} = \alpha(z_1)$ denote the mean wind speed and wind angle at the lowest elevation $z_1 = 50$ m.

As expected, for each profile the average wind angle $\alpha(z_i)$ changes with the elevation z_i and such a variation decreases for an increasing wind speed V_{50} of the profile; this result is shown in Fig. 3, where Standard Deviation σ_α of the angles $\alpha(z_i)$ of each profile as a function of the profile wind speed V_{50} is given. It is observed that for wind speeds $V_{50} < 4$ m/s there is a strong variability of the angle of attach with the elevation, indicating that thermal effects are dominant; for 4 m/s $< V_{50} <$ 10 m/s there is still a pronounced variability of the wind direction, larger than what it could be ascribed to a Ekman spiral effect; for $V_{50} >$ 10 m/s the wind direction variability becomes small, and the profiles are those corresponding to a neutral atmospheric stratification.

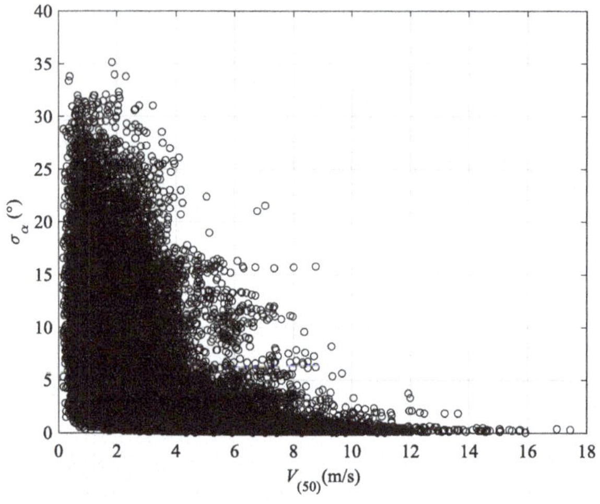

Fig. 3. Standard Deviation σ_α of the angles $\alpha(z_i)$ of each profile as a function of the profile wind speed V_{50} (39333 data points)

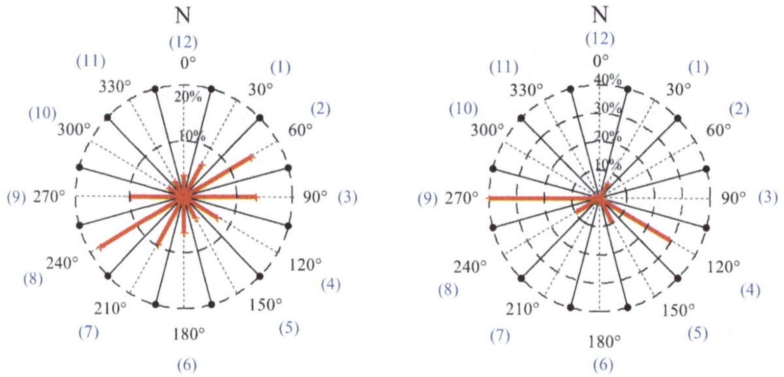

Fig. 4. Frequency roses for the entire dataset of 10-min profiles (39333 profiles, left), and for $V_{50} \geq$ 12 m/s (65 profiles, right).

Frequency roses in Fig. 4 show the occurrence of the wind direction α_{50} for the 12 sectors of 30° degrees each, for the entire dataset of ten-minute profiles (39333 records, left) and for profiles with $V_{50} \geq$ 12 m/s (65 records, right).

It is noticed that the prevailing sectors do not coincide with those of stronger winds, which can be identified as α_{50} = 120° ± 15° (sector 4) and α_{50} = 270° ± 15° (sector 9).

Figure 5 reports the measured wind speed profiles (19 and 25 profiles, respectively) for sectors 4 and 9.

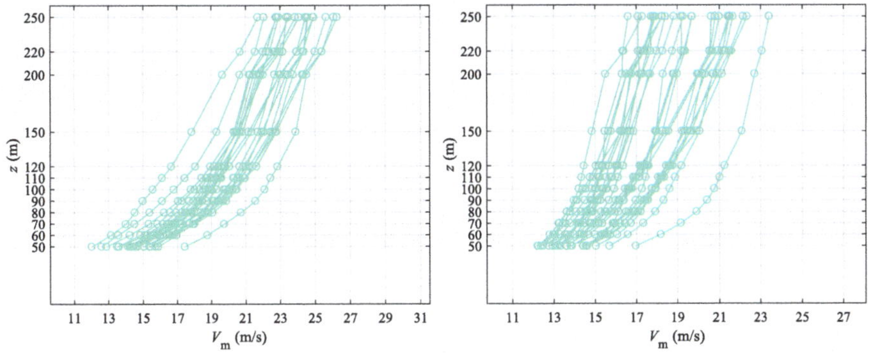

Fig. 5. Measured profiles with $V_{50} > 12$ m/s; left: sector 4 (120° ± 15°), right: sector 9 (270° ± 15°)

The measured profiles of Fig. 5 have been fitted with logarithmic profile (CEN 2005):

$$V_m(z) = v_o k_r \ln\left(\frac{z}{z_0}\right);\ k_r = 0.19\left(\frac{z_0}{z_{0,\mathrm{II}}}\right)^{0.07} \tag{1}$$

and with the linearized Deaves and Harris profile:

$$V_m(z) = v_o c_x \frac{ln\left(\frac{z}{z_0}\right) + 5.75\frac{z}{z_g}}{ln\left(\frac{10}{z_0}\right)};\ z_g = \frac{k_r v_o}{15f};f = 1.454 \cdot 10^{-4} \sin\varphi \tag{2}$$

where v_o is the reference wind velocity, i.e. the 10-min. average velocity at 10 m of elevation in flat country of exposure category II, z_0 is the roughness length, k_r is the terrain factor, $z_{0,\mathrm{II}}$ = 0.05 m is the roughness length for category II exposure, c_x is the exposure factor, z_g is the gradient height, f is the Coriolis parameter and φ is the latitude of the site.

In the curve fitting of measured data the coefficient c_x has been expressed as a function of z_0 and z_g (see Fig. 6), through the equation (Cook 1985):

$$c_x = \frac{\left[ln\left(\frac{z_g}{z_{0,\mathrm{II}}}\right) + 2.79\right] \cdot ln\left(\frac{10}{z_0}\right)}{\left[ln\left(\frac{z_g}{z_0}\right) + 2.79\right] \cdot ln\left(\frac{10}{z_{0,\mathrm{II}}}\right)} \tag{3}$$

In Cook (1985), Eq. (3) is used with z_g = 2550 m and $z_{0,\text{II}}$ = 0.03 m, values which are found to well apply to the UK territory. In that case Eq. (3) is found to depend only on z_0. On the other hand, when Eq. (2) is used to evaluate z_g, then Eq. (3) depends also on latitude and v_o. However, it is found that the variability of Eq. (3) with v_o is very negligible, whereas some little dependency with the value chosen for $z_{0,\text{II}}$ does exist. This is shown in Fig. 6, where Eq. (3) is represented as given in Cook (1985), and for a latitude of 40°58′00″N (where the LIDAR is installed) and for v_o = 10, 20 and 30 m/s, both for $z_{0,\text{II}}$ = 0.03 m and $z_{0,\text{II}}$ = 0.05 m.

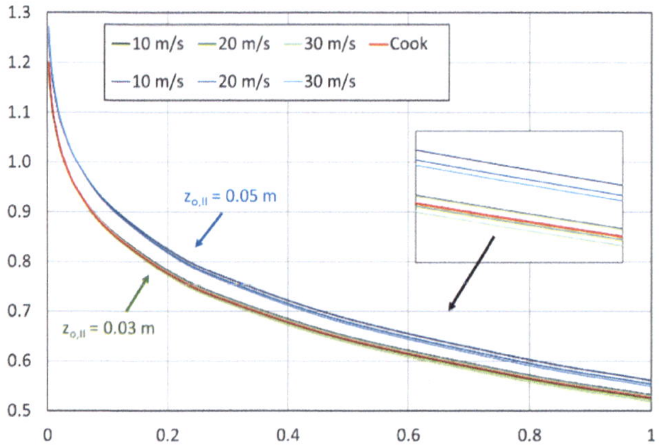

Fig. 6. $c_x(z_0)$ as given in Cook (1985), and for a latitude of 40°58′00″N and for v_o = 10, 20 and 30 m/s, both for $z_{0,\text{II}}$ = 0.03 m and $z_{0,\text{II}}$ = 0.05 m

Figure 7 shows a sample of the curve fitting of profiles in the sectors 4 and 9, respectively, with both the logarithmic and linearized Deaves and Harris profiles. In both cases the optimal curve fitting has been obtained by searching for the minimum RMS error between the measured data and the analytical wind speed at the 12 available elevations, with v_o and z_0 varying between 0 and 30 m/s (step 0.05 m/s) and between 0 and 8 m (step 0.005 m), respectively, using Eqs. (1), (2) and (3). The approximation of the experimental points (Table 1) is rather good with both approximations. The results obtained for the two selected profiles are representative of the entire dataset, as qualitatively the same observations can be drawn from analysis of the other profiles.

Table 1. Parameters of the interpolated profiles of Fig. 7.

	Sector 4		Sector 9	
	Log	D&H	Log	D&H
v_o (m/s)	24.1	19.6	18.2	14.4
z_0 (m)	3.84	2.32	1.84	0.69
k_r	0.257	–	0.245	–
c_x	–	0.380	–	0.626
z_g (m)	–	3407		2299
RMS error (m/s)	0.254	0.352	0.120	0.207

In general it is observed that the approximation of the logarithmic profile is better than that of the Deaves and Harris profile (lower RMS error) and that the RMS error is very low compared with v_o.

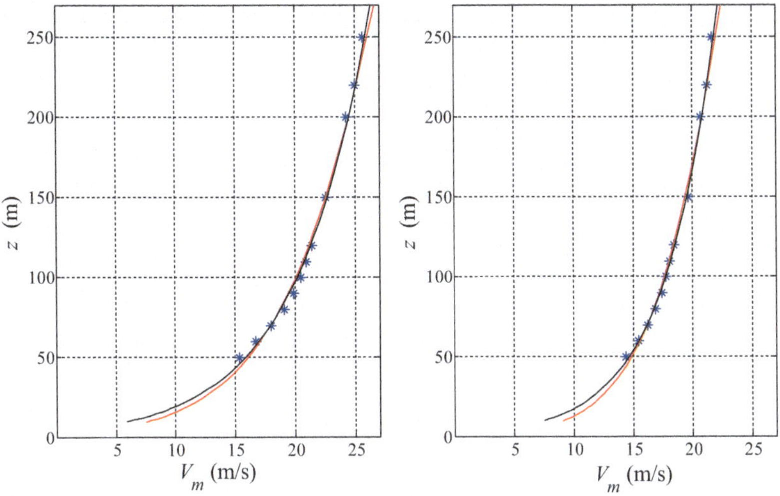

Fig. 7. Sample curve fitting of measured profiles in sector 4 (left) and sector 9 (right): black line is the logarithmic approximation and red line is the Deaves and Harris approximation

Given the satisfactory result in approximating the measured data through the logarithmic and Deaves and Harris profiles for the entire dataset, in Fig. 8 the RMS error contour plots in the $z_0 - v_o$ plane are shown for the same records of Fig. 7. Contour lines are shown corresponding to 1.1, 1.2 and 1.5 times the minimum RMS error and to both the logarithmic and Deaves and Harris models.

Although the sensitivity of RMSE to the parameters z_0 and v_o is different depending on model (logarithmic or Deaves and Harris) and on wind direction (sectors 4 or 9), for both sample profiles of Fig. 7, as well as for most of the available profiles for $V_{50} > 12$ m/s shown in Fig. 5, the contour plots suggest a satisfactory identification of the optimal pair $z_0 - v_o$.

For sectors 4 and 9, Table 2 reports the statistics of the relevant quantities for the logarithmic and Deaves and Harris models as obtained from the processing of the entire dataset, which confirm the observations already made with reference to the profiles shown in Fig. 7. Comparison of the mean RMS error associated with the logarithmic and with the Deaves and Harris approximations (line 1) suggests that the former better interpolates the measured profiles; this is also confirmed by the standard deviation, minimum and maximum values of the same quantities (lines 2, 3 and 4), which are also always lower for the logarithmic profile than for the Deaves and Harris profile. On the other hand, comparison of the mean RMS error (line 1) with the mean reference wind velocity among different profiles (line 9) suggests that the approximation is anyways good, being the RMS error always lower than 2.5% of the reference mean wind velocity. It also appears that the identification of the roughness length of the two profiles is rather scattered; in fact comparison between the standard deviation and mean values of the identified roughness length (lines 6 and 5) reveal that the Coefficient of Variation of such quantity is 0.33 and 0.42 for sector 4 for the logarithmic and Deaves and Harris profiles, respectively, and of 0.70 and 1.15 for the same profiles in sector 9; similar conclusion is found comparing the minimum and maximum values of the same quantities (lines 7 and 8).

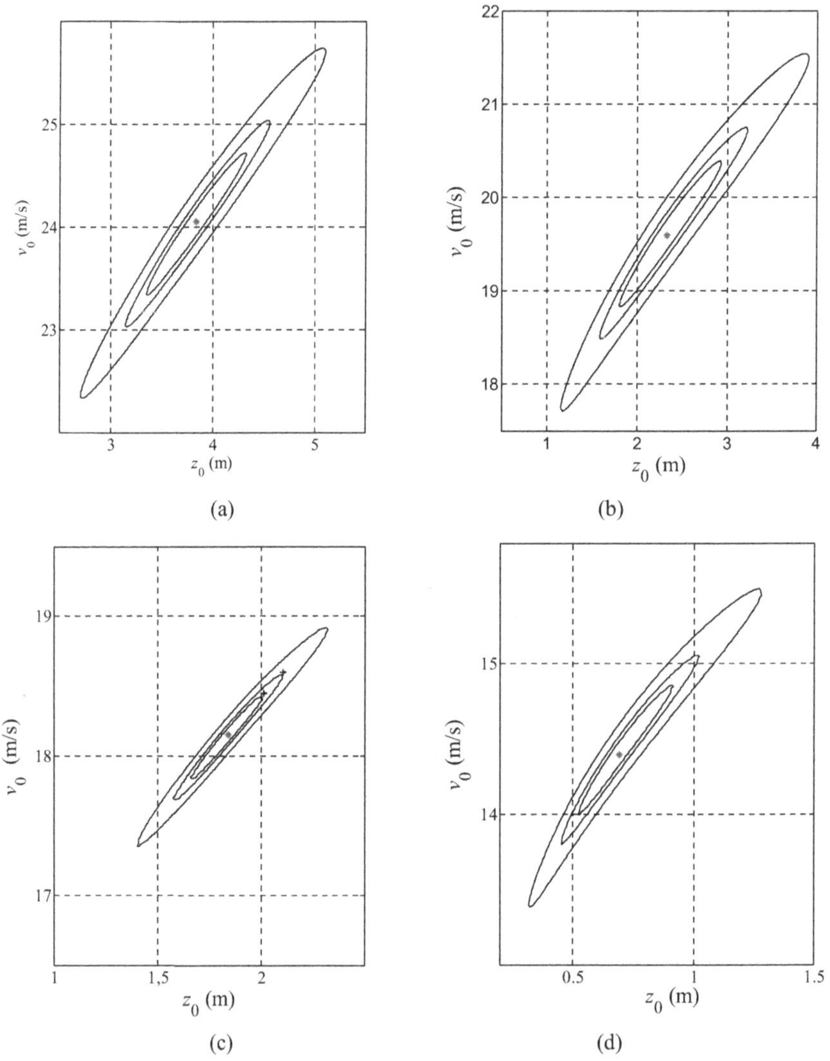

Fig. 8. Contour plots of the RMS error of the approximated profiles in the $z_0 - v_o$ plane. Sector 4: logarithmic profile (a) and Deaves and Harris profile (b); Sector 9: logarithmic profile (c) and Deaves and Harris profile (d). The three contours correspond to 1.1, 1.2 and 1.5 times the minimum RMS error.

Given the large scatter, also in this case it appears that the logarithmic approximation is more accurate than the Deaves and Harris approximation, as it gives rise to more stable values of the identified roughness length; finally, it also appears that identification of the roughness length is more stable when the basic wind velocity is higher (sector 4) than when it is lower (sector 9).

Table 2. Statistics of v_o and z_0 and of the RMS error for the logarithmic and Deaves and Harris models for sectors 4 and 9, profiles with $V_{50} > 12$ m/s.

		Sector 4		Sector 9	
		Log	D&H	Log	D&H
RMS error (m)	mean	0.29	0.36	0.21	0.27
	σ	0.11	0.14	0.08	0.09
	min	0.12	0.17	0.08	0.13
	max	0.50	0.61	0.37	0.47
z_0 (m)	mean	3.90	2.31	1.10	0.33
	σ	1.29	0.98	0.77	0.38
	min	1.42	0.60	0.08	0.005
	max	6.55	4.30	2.98	1.40
v_o (m/s)	mean	22.4	18.0	15.1	11.9

Another observation that can be drawn from Table 2 is that both the values of the roughness length z_0 and of the reference wind velocity v_o identified through the Deaves and Harris model are smaller than those identified through the logarithmic model; this was already observed by Deaves (1981).

As a consequence of what previously observed, if the same values of z_0 and v_o are taken for the logarithmic and for the Deaves and Harris modes, then the latter would overestimate the measured wind speed.

Finally, a by-eye calibration of the roughness length to be used for design is attempted. In Fig. 9, the c_r coefficient obtained using the logarithmic and the Deaves and Harris models, with the roughness length values given in Eurocode 1(CEN, 2015) are shown. In particular, Fig. 9a and b show the experimental profiles for sector 4 compared with the theoretical curves deriving from application of the logarithmic (Fig. 9a) and Deaves and Harris (Fig. 9b) models, with values of the roughness length $z_0 = 0.003, 0.01, 0.05, 0.3$ and 1.0 m (categories from 0 to IV). Figure 9c and d are the equivalent for sector 9.

The figures confirm that application of the logarithmic and Deaves and Harris wind models to the data measured in sector 4 would require a value of the roughness larger than that applying to exposure category IV, and that such value is larger when used with the logarithmic model that with the Deaves and Harris model. As to the data coming from sector 9, it is observed that a value of the roughness length in the order of 1.0 m would be reasonable for the logarithmic model, whereas the Deaves and Harris model shows very scattered values of z_0. All this is consistent with the values presented in Table 2.

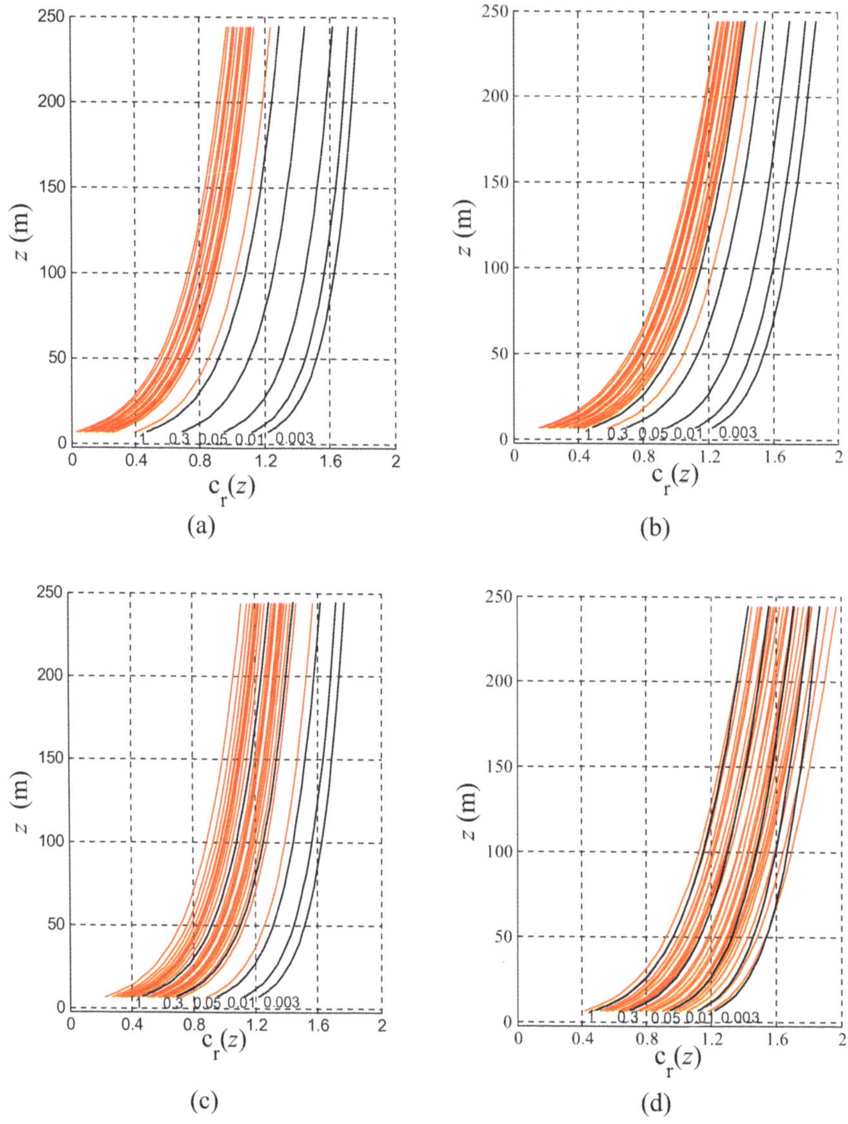

Fig. 9. Experimental c_r profiles compared with the theoretical curves deriving from application of the logarithmic and Deaves and Harris models: sector 4 logarithmic (a), sector 4 Deaves and Harris (b), sector 9 logarithmic (c), sector 9 Deaves and Harris (d); the experimental data (red curves) are for $V_{50} > 12$ m/s

It is noticed that the values of the roughness length provided by the CORINE Land Cover project (European Environment Agency 1985) for urban areas lies between 1.0 and 4.0 m; this turns out to be consistent with the results produced for sector 4 using both the logarithmic and the Deaves and Harris approximations; it is still consistent with what found for sector 9 using the logarithmic approximation, whereas it tends to overestimate roughness when the Deaves and Harris model is used.

4 Turbulence Models

Besides the variation of the mean wind velocity with height, wind models contain a description of the variation of the fluctuating longitudinal wind component. In particular, together with the logarithmic approximation of the mean wind profile, the assumption is usually made (CEN 2005) that the standard deviation of the longitudinal component of turbulence is constant with height; this brings the turbulence intensity profile to be the inverse of the mean wind velocity profile, and in flat orography this becomes:

$$I_v(z) = \frac{1}{\ln\left(\frac{z}{z_0}\right)} \tag{4}$$

On the other hand, a different vertical profile of the intensity of turbulence is associated with the Deaves and Harris wind model, given as (Cook 1985):

$$I_v(z) = \frac{3\left(1 - \frac{z}{z_g}\right)\left(0.538 + 0.09 ln \frac{z}{z_o}\right)^{\left(1-\frac{z}{z_g}\right)^{16}}}{ln \frac{z}{z_o}\left(1 + 0.156 ln \frac{6z_g}{z_o}\right)} \tag{5}$$

depending on geostrophic height and not giving rise to a constant vertical distribution of the longitudinal fluctuating wind component.

Indeed, experimental evidence indicates that the assumption of a constant longitudinal fluctuating wind component is not exact; as an example in Fig. 10 two samples of eight records each, of the vertical profiles of the RMS longitudinal wind component are shown, for sectors 4 and 9, respectively, showing a decreasing fluctuating wind component with increasing height between 50 and 250 m.

Finally, in Fig. 11 the vertical profiles of the longitudinal intensity of turbulence are shown, for sectors 4 and 9. The thin blue lines represent the experimental data, the solid red line is their mean and the dashed red lines are the mean values plus and minus one standard deviation. The green lines represent Eq. (4), with $z_0 = 3.90$ m and $z_0 = 1.10$ m for the two sectors, respectively; these are the optimum values obtained fitting the mean wind profiles with the logarithmic model. Finally, black lines represent Eq. (5), with $z_0 = 2.31$ m and $z_0 = 0.33$ m for the two sectors, respectively, which are the optimum values obtained fitting the mean wind profiles with the Deaves and Harris model. It is observed that both models tend to overestimate the turbulence intensity, and that such overestimate depends on sector, therefore on roughness length. This difference can be partly ascribed to the space averaging and limited sampling frequency of Lidar measurements.

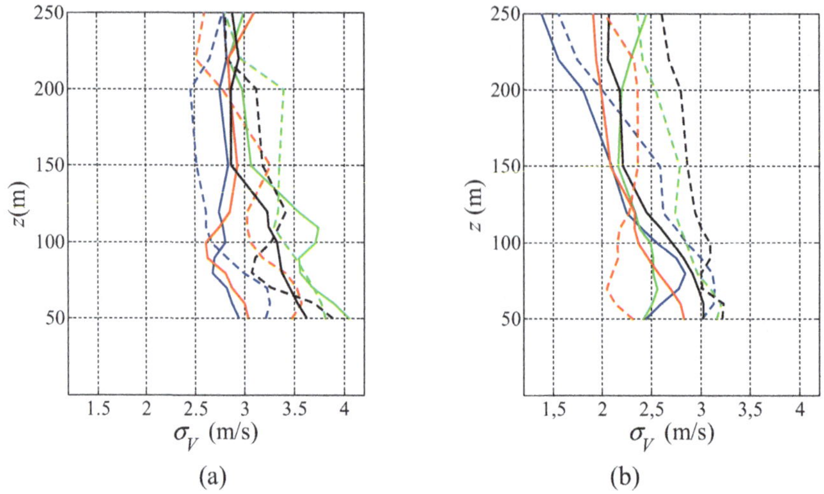

Fig. 10. Vertical profiles of the longitudinal fluctuating wind component: 8 samples for sector 4 (a) and 8 samples for sector 9

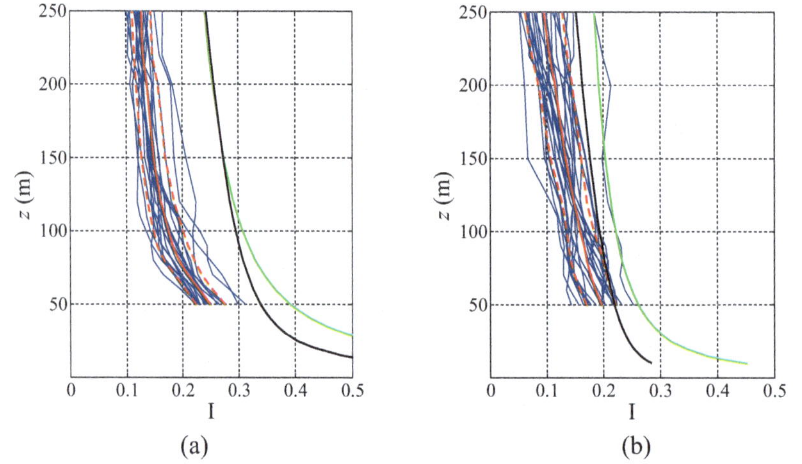

Fig. 11. Vertical profiles of the longitudinal intensity of turbulence for sectors 4 (a) and 9 (b). Blue lines are the experimental data, solid red line is the mean and dashed red lines are the mean values plus and minus one standard deviation; green lines are Eq. (4) with z_0 = 3.90 m and z_0 = 1.10 m; black lines are Eq. (5) with z_0 = 2.31 m and z_0 = 0.33 m

5 Conclusions

In this paper some results of wind LIDAR measurements over a flat orography have been presented and discussed. The two dominating directions of 120° and 270° have been observed, featuring the strongest winds. For these two directions the 10-min records showing a mean wind velocity at 50 m larger than 12 m/s have been selected for discussion. These records prove to correspond to neutral atmospheric stratification, as they feature a minor variation of the direction along the height (whereas a very large variation is observed in records with lower mean values) and a mean profile well approximated by a logarithmic law. These experimental profiles have been fitted with the logarithmic law and with the Deaves and Harris wind model, by finding the optimum model parameters, i.e. the parameters that best fit the experimental data. It is found that both model fit very well the experimental data, with a slightly better performance of the logarithmic model, and that the logarithmic model require larger values of the reference wind velocity and of the roughness length. However, it is also observed a rather large scatter of the identified values of the roughness length, in the sense that different profiles within the same sector correspond to rather different values of roughness length. These values are also larger than those given in EN1991-1-4 (CEN 2005). Finally it is observed that the measured vertical profiles of the longitudinal fluctuating wind component are not constant, as assumed in the wind model of EN1991-1-4; when the experimental vertical profiles of the longitudinal intensity of turbulence are compared with those deriving from application of the EN1991-1-4 and of the Deaves and Harris models, it is found that both models tend to overestimate turbulence, and therefore the fluctuating component of the wind action.

Acknowledgements. The results presented in this paper have been produced in the framework of the research project GELMINACAL, founded by the Italian Ministry for University (MIUR). The research has also received financial support from the University "G. D'Annunzio" of Chieti-Pescara.

References

Comité Européen de Normalisation, CEN (2005) Eurocode 1: Actions on structures - Part 1–4: General actions - Wind actions, EN1991-1-4

Cook NJ (1985) The designer's guide to wind loading of building structures. Part 1 background, damage survey, wind data and structural classification. Building Research Establishment, Butterworths, London

Demartino C, Avossa AM, Ricciardelli F, Calidonna CR (2017) Wind profiles identification using wind LIDARS: an application to the area of Lametia Terme. In: Proceedings of 7th european and african conference on wind engineering, EACWE, 4–7 July 2017, Liège, Belgium

Deaves DM (1981) Terrain-dependence of longitudinal R.M.S. velocities in the neutral atmosphere. J Wind Eng Ind Aerodyn 8:259–274

Deaves DM, Harris RL (1978) A mathematical model of the structure of strong winds. Construction Industry Research and Information Association, Report 76

Drew DR, Barlow JF, Lane SE (2013) Observations of wind speed profiles over Greater London, UK, using a Doppler lidar. J Wind Eng Ind Aerodyn 121:98–105

European Environment Agency (1985). https://www.eea.europa.eu/publications/COR0-landcover

Kent CW, Grimmond CSB, Gatey D, Burlow JF (2018) Assessing methods to extrapolate the vertical wind-speed profile from surface observations in a city centre during strong winds. J Wind Eng Ind Aerodyn 173:100–111

Lane SE, Barlow JF, Wood CR (2013) An assessment of a three-beam doppler lidar wind profiling method for use in urban areas. J Wind Eng Ind Aerodyn 119:53–59

Lim KEW, Watkins S, Clothier R, Ladani R, Mohamed A, Palmer JL (2016) Full-scale flow measurement on a tall building with a continuous-wave Doppler Lidar anemometer. J Wind Eng Ind Aerodyn 154:69–75

Lumley JL, Panofsky HA (1964) The structure of atmospheric turbulence. Wiley/Interscience, New York

Simiu E, Scanlan RH (1986) Wind effects on structures, 2nd edn. Wiley, Hoboken

Tennekes H (1973) The logarithmic wind profile. J Atmos Sci 30:234–238

Wood CR, Pauscher L, Ward H, Kotthaus S, Barlow J, Gouvea M, Lane S, Grimmond CSB (2013) Wind observations above an urban river using a new lidar technique, scintillometry and anemometry. Sci Total Environ 442:527–533

Design and Experimental Optimisation of a cm-Scale Wind Turbine for Energy Harvesting

G. Tomasini[1(✉)], D. Tarsitano[1], S. Giappino[2], and L. Bernini[1]

[1] Department of Mechanical Engineering, Politecnico di Milano, via La Masa, 1, 20156 Milan, Italy
{gisella.tomasini,davide.tarsitano, luca.bernini}@polimi.it

[2] GVPM, Politecnico di Milano, via La Masa 34, 20156 Milan, Italy
stefano.giappino@polimi.it

Abstract. Wind turbine is a good choice for energy harvesting on trains. The air flow is generated by the train movement and it is quite simple to find places with high flow speed near the wireless sensor nodes location which are typically mounted on bogies and wheelsets. The requested amount of energy is relatively small, in the order of 0.01–0.05 W, allowing for wind turbines with very small diameter, between 2 to 6 cm. In the present work a newly designed turbine is described. The turbine is placed in a duct with protection purposes and it has a symmetrical design that results in power production independently from the train movement direction.

Keywords: Duct wind turbines · Energy harvesting · Sensor nodes · Wind tunnel test · BEM approach

1 Introduction

This work deals with the design and the optimisation by means of wind tunnel tests of a centimetre scale wind turbine to be used as energy harvester in autonomous sensor nodes for diagnostic/monitoring of freight trains.

The sensor node has to be installed in correspondence of the axle-box of the railway vehicle: it is composed by one or more sensors (digital accelerometers, temperature sensors, etc.), a micro-controller, a wireless transmission system (MiWi) and a rechargeable battery (Milani et al. 2014). The consumption of such sensorial node depends on the type of monitoring and, mainly, on the ratio between the sleep mode time (a very slow consumption mode, when the node is not working) and the full power mode time, when the microcontroller runs at its full capability, acquiring data from the sensors, performing data analysis and sending the results. For the considered application, that consists in monitoring very slow phenomena as crack propagation or wear progression, the node is set for measuring the acceleration (post-processing and sending the data) for a 1-s time-window (full power mode time) every 10–30 min (sleep mode time): with these parameters, the mean consumption is about 4–8 mW (Tomasini et al. 2016).

F. Ricciardelli and A. M. Avossa (Eds.): INVENTO 2018, LNCE 27, pp. 703–715, 2019.
https://doi.org/10.1007/978-3-030-12815-9_53

Different energy sources may be adopted for the development of energy harvesting systems: solar energy, mechanical energy, wind energy (associated to the relative motion between train and air), etc.

Within the research presented in this paper, a cm-scale wind turbine has been chosen for powering the rechargeable battery of a sensor node. The objective of the paper is to design, realize and test, by wind tunnel tests, a cm-scale wind turbine in order to define the best configuration.

The state of the art in the field of small scale wind turbines for energy harvesting is focused on the performance analysis of off-the-shelf cm-scale turbines using commercially available components small scale rotors (Xu et al. 2010, 2014). Few recent studies (Howey et al. 2011) analyse the aerodynamics of the blades, using the Blade Element Momentum based code to design and build a customized cm-scale turbine to optimize its efficiency for a low wind speed air-ventilation application.

With respect to the cited works, in this research the aerodynamic design of a centimetre scale blade is carried out using an inhouse BEM code and the layout is specifically studied to be adopted in railways applications (wide range of wind speed, bidirectional operational conditions, as required by the trains operational conditions).

Different prototypes of rotors are designed considering the specific requirements of the final application. Three rotor prototypes connected to a generator (26 mm diameter motor) are built and tested in wind tunnel by means of active control tests, where the rotational velocity of the motor/generator is imposed, and passive tests, where the generator is connected to a passive circuit with variable electrical loads. Open field tests are performed to verify the performances of the prototypes with respect to the numerical results found with the BEM method. Finally, tests with ducted turbine are carried out with various layouts of ducts to evaluate the effects of different parameters as the presence of a contraction cone/diffuser and the wind direction, upwind or downwind the rotor.

2 Problem Definition and Initial Design

The target of the paper is to design a wind turbine to be installed on freight trains, in correspondence of the axle-box: thus, for this application, the wind velocity is represented by the train speed if running in still air or, more in general, by the relative speed between the train and the ambient air. The range of velocity in which the turbine should extract energy is ranged between 10 m/s and 30 m/s that are, respectively, the minimum velocity below that it is acceptable that the turbine cannot recover energy and the maximum velocity of a freight train. In particular, the turbine should be optimized at 10 m/s but it should be able to harvest energy up to 30 m/s.

To be installed in correspondence of the axle-box, the turbine should have a small size. The reduction in rotor diameter leads to an increase in rotational speed (on equal tip speed) but the angular rotor velocity has to be limited for selecting a commercial generator. By setting the Tip Speed Ratio (TSR) equal to 1 (that is considered the lower limit to obtain optimum power coefficient and, as reported in the literature, Rancourt et al. (2007), it represents a feasible value for centimeter wind turbines) and considering generators with a maximum rotational speed ranged between 5000 and 15000 RPM, the two selected diameters are D = 30 mm and D = 40 mm. Choosing a maximum chord of about 15 mm, compatible with standard fabrication processes, the Reynolds number is about Re = 2 * 10^4 and the minimum number of blades is four.

Defined these parameters in agreement with the specific application, the final rotors are designed by means of the BEM code. This code allows to calculate the steady loads and, thus, also the thrust and the power for different settings of the wind speed, rotor angular speed and pitch angle. Providing as input the number of blades, the rotor diameter, the TSR and the maximum angular speed, it is possible to use the algorithm to identify the optimal angles of attack and, consequently, the optimal twist and chord distribution along the radius, which maximise the extracted power.

Two turbines, respectively 30 mm and 40 mm of rotor diameter, employing a low Reynolds number airfoil, SD7032 (Lyon et al. 1997), were optimized with the BEM code: the SD7032 airfoil has a maximum thickness equal to 9.6% and a maximum camber of 3.4% of the chord length. The main parameters of the designed rotors are resumed in Table 1.

In addition to the two optimised rotors, having only one leading edge, one 40 mm symmetrical rotor with constant blade pitch fixed at 45° and flat plate (rectangular) blade section was designed. This symmetrical rotor was not optimized using the BEM code for lack of knowledge about the flat plate response at the considered conditions. The symmetrical design has been investigated for its ability to work in both the train running directions.

Table 1. Main characteristics of the designed turbines.

Turbine	Airfoil	Diameter [mm]	Maximum chord [mm]	Twist angle [deg]
D30	SD7032	30	13	55–20
D40	SD7032	40	17	55–20
D40-sym	Flat plate	40	17	45

All the rotors prototypes have been produced in bronze with rapid prototyping technique with high precision and good mechanical strength characteristics. Figure 1 shows the designed rotor D30 (a) and D40-sym (b) and the corresponding real prototype D30 (c).

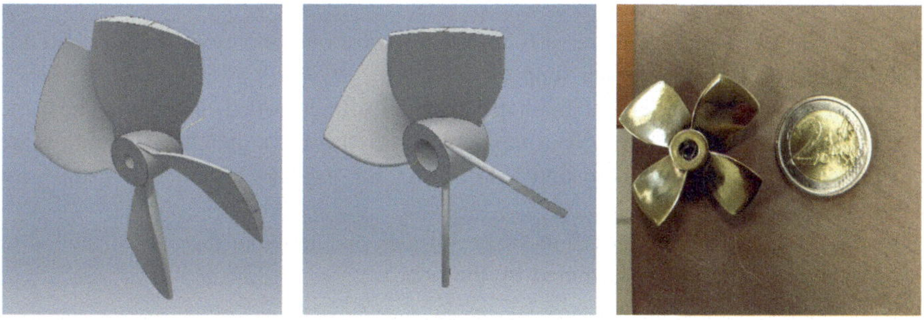

(a) (b) (c)

Fig. 1. D30 BEM designed rotor (a), D40-sym designed rotor (b) and D30 real prototype (c).

The performances of the two optimised turbines, according to the BEM numerical code, are presented in Table 2.

Table 2. Maximum power and power coefficient for different wind speed (BEM).

Wind speed	D40		D30	
[m/s]	P [W]	Cp	P [W]	Cp
10	0.29	0.378	0.16	0.380
20	2.36	0.383	1.32	0.380
30	8.11	0.390	4.50	0.385

2.1 Ducted Turbines

Since the turbines must be mounted on freight trains, they should be able to correctly work in hard environment conditions as storm, snow, dust, etc. For this reason, and in order to increase the turbine performance, considering that the flow direction in this application is always perpendicular to the rotor plane, a specific duct, where the turbine is inserted, was designed and tested. In order to maximize the power production of the ducted turbine, a diffuser was studied, starting from the findings presented in Jafari and Kosasih (2014) (Fig. 2a). The principle behind the study is related to the geometry of the diffuser, which forces the flow downstream to expand through the diffuser. The expansion reduces the pressure at the outlet of the diffuser and causes an increase in the mass flow at the inlet, due to the higher pressure drop between inlet and outlet. An increase of a mass flow through the rotor leads to an increase of power production. This effect is called "back pressure effect". In Jafari and Kosasih (2014), the identified optimum ranges for the two main ratios between length, height and diameter (Fig. 2a) are L/D = 0.03–0.04 and H/D = 0.1–0.15.

The final design of the sensorial node shall include the generator, the electronic board and the rotor. By positioning the board under the turbine, the total length of the duct is conditioned by the length of the board and it is fixed. On the other hand, in order to evaluate the effects, in terms of performances, of different ratios L/D and H/D and of different positions of the rotor within the duct, the prototype of the ducted turbine was designed with a modular approach. The complete duct is realized by combining three different types of modules as represented in Fig. 2b:

- module B: cylindrical in shape, it is designed to host only the motor support and it is connected to the wind tunnel support;
- modules D and E: divergent in shape, they are dimensioned in order to host the rotor and they have a length comparable with the rotor height. The ratios are H/D = 0.1 for both of them and L/D = 0.03 for module D and L/D = 0.04 for module E;
- modules A and C: cylindrical in shape, of different lengths, they are designed to reach the fixed length determined by the board.

The generator selected for its short dimensions and performance in terms of maximum rotational velocity is the 16ECP36 – 8B Ultra ECTM Portescape three phase brushless motor.

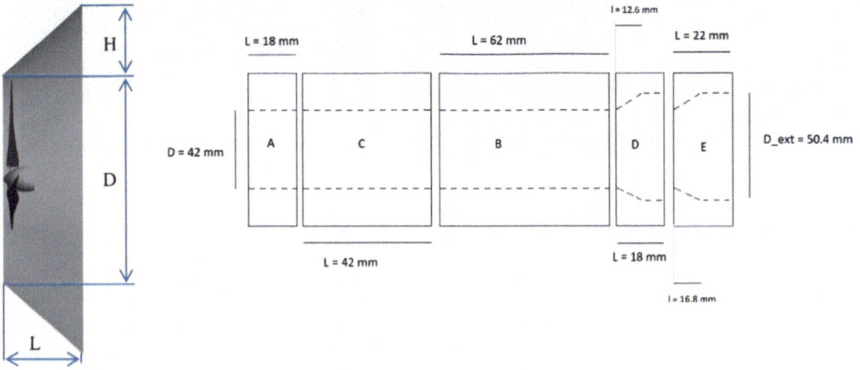

(a) (b)

Fig. 2. Parameters of the diffuser (a), geometrical characteristics of the five modules that can be combined to compose the full duct.

3 Wind Tunnel Tests

Wind tunnel tests have been performed in the 'Sergio De Ponte' POLIMI wind tunnel, section of 1 × 1.5 m, with a maximum wind speed equal to 55 m/s.

3.1 Measurement Setup

The experimental tests are carried out in two configurations (Fig. 3):

- open field, rotor set in the wind tunnel without duct (Fig. 3a);
- confined flow, rotor set inside a duct (Fig. 3b).

In the open field tests, two operational conditions are investigated:

- active control test (ACT);
- passive test (PT).

Tests are performed in open field on D30 and D40 prototypes to estimate the performances and to compare them with those numerically evaluated by the BEM code. Furthermore, open field tests are carried out also on the D40-sym to compare it with the other two turbines.

In the confined flow tests, the symmetrical turbine is set within different duct layouts (with/without diffuser, upwind and downwind, …) to identify the best configuration for the final application.

In the active control tests the motor is closed-loop controlled and it rotates at a fixed angular speed. The power generated or absorbed by the motor corresponds to the power introduced by the turbine, except for the bearing dissipations. A block diagram of the

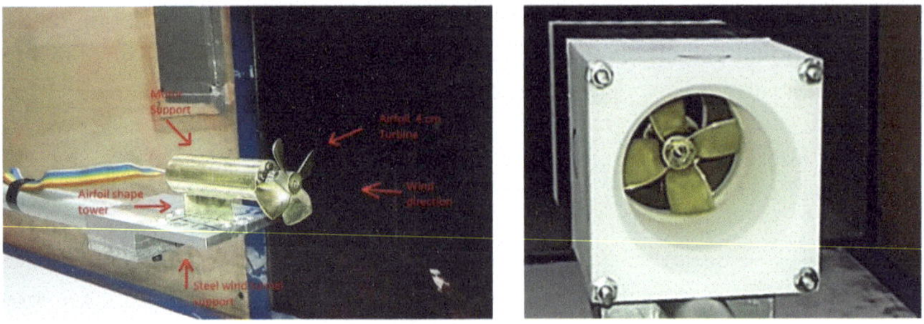

(a) (b)

Fig. 3. Picture of the wind tunnel test setup for open field tests (a) and confined flow tests (b).

experimental setup in the ACT is sketched in Fig. 4a. The parameter Ω_{ref} represents the angular velocity of the motor given in input to the system by the user, Ω is the angular velocity of the motor shaft, I represents the motor current feedback.

The following quantities are measured:

- the current, I, feedback by the motor, proportional to the torque;
- the angular velocity of the motor shaft, Ω (that corresponds to the angular speed of the rotor), by means of an encoder;
- the wind speed, measured by a pitot tube.

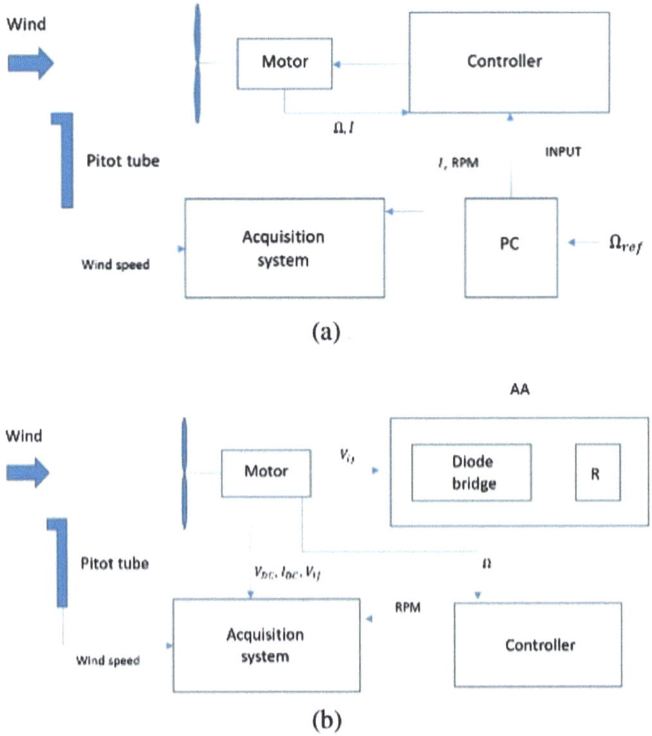

Fig. 4. Block diagram of the experimental setup for ACT (a) and PT (b) configurations.

Passive tests are performed in order to evaluate the actual power available at the circuit accounting for all the mechanical and electrical efficiencies. In Fig. 4b is presented the block diagram of the experimental setup adopted for the PT configuration.

The principal block is the "AA", composed by a three phase diode bridge rectifier and resistive loads. These resistive loads are modifiable in order to define different steady-speed operational conditions.

In the PT configuration, the following quantities are measured:

- the voltage V_{DC};
- the angular velocity of the motor shaft, Ω;
- the wind speed, measured by a pitot tube.

3.2 Open Field Tests

3.2.1 Active Controlled Tests (ACT)

The goal of the active tests is to verify the BEM numerical model by means of a comparison in terms of power produced by the turbine and the corresponding power coefficient. In these tests, the torque applied (if positive) or absorbed (if negative) by the motor to maintain the imposed speed Ω_{ref} corresponds, except for the dissipations in the generator (bearings, etc.), to the aerodynamic moment generated by the wind speed acting on the rotor.

As first step, some preliminary tests have been performed to evaluate the bearing losses: the motor rotates in active control configuration, at different rotational velocities, without any wind flow. The current, necessary to maintain the motor at constant rotational speed (positive), represents the current needed to overcome the bearing losses, I_{diss}.

In open field, with active controlled test configuration, the D30 and D40 turbines are tested. Figure 5 shows the numerical-experimental comparison in non-dimensional terms (C_p coefficient vs TSR, Fig. 5a and b) and dimensional terms (power output vs rotational speed, Fig. 5c and d) for both the turbines D30 and D40.

In terms of comparison with the numerical model results, from the analysis of the C_p coefficient, it is possible to observe that the experimental power coefficients of both the turbines have lower peaks and shifted on the left: the numerical results have an optimum point at about 1.5 of TSR while the experimental ones at about 1. Similar results were found in Howey et al. (2011), where the agreement with the BEM results was discrete in the increasing part of the curve but the maximum C_P value found experimentally was lower and anticipated (at lower Tip Speed Ratio) with respect to the numerical prevision. The main reason of this behaviour is associated to a lack of accuracy in the airfoil aerodynamic coefficients at these very low Reynolds number ($Re = 1\ 10^4$).

Moreover, from Fig. 5a and b it is possible to observe that, in some cases, the power coefficients trend varies with the wind speed, while the three numerically evaluated curves are perfectly superposed (as they have to be according to the theory). Considering that the main discrepancies arise at 10 m/s for the smaller turbine (D30), it is the authors' opinion that this behaviour is not due to a Re number dependency but to a low resolution of the measurement system: at this very low speed the energy introduced by the rotor is of the same order of magnitude of the energy associated to the bearing losses.

Regarding the power, Fig. 5c and d show that, due to the anticipation of the pick value, the power recovered by both the turbines according to the experimental tests is higher than that estimated numerically for rotational speed lower than 10000 RPM and reaches a maximum of 2.4 mW with D30 and nearly 5 mW with D40 at the wind speed 30 m/s. In terms of comparison between the two turbines, D30 and D40, the experimental results confirm the numerical ones: power harvested by the smaller rotor is about 50% of that recovered by the greatest one, at all tested wind speeds.

In conclusion, the agreement with the numerical results, at these low Re numbers, is not excellent, especially for the position of the maximum value, but the BEM approach may be used, for a first approach to the problem, to optimize the blade parameters because it allows to evaluate, with an acceptable approximation, the expected performances.

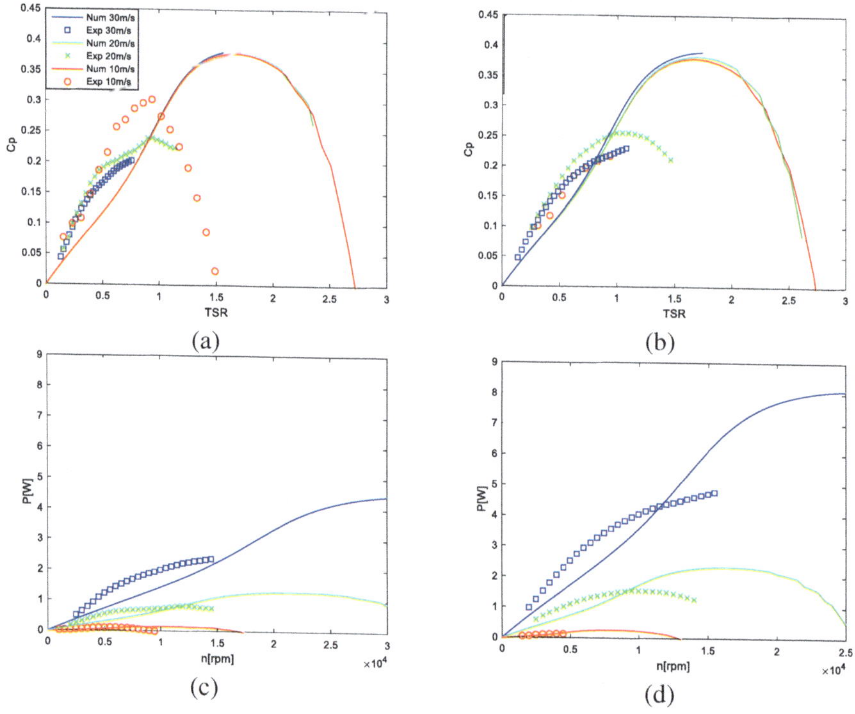

Fig. 5. Open field, active controlled tests, numerical-experimental comparison: power output as a function of the rotational speed with D30 (a) and D40 (b) and C_p and a function of TSR with D30 (c) and D40 (d).

3.2.2 Passive Tests (PT)

The passive tests are carried out because they represent the actual operational condition of the turbine in its final application. In this configuration, the motor has only the function of generator: the turbine recovers wind energy to produce electrical energy and it occurs only when the wind speed is high enough to overcome the internal motor/generator friction resistances. It is important to underline that, in the passive tests, the effective measured power is that available after the rectifier stage, lower than that introduced by the turbine because it is reduced by the mechanical and electrical losses.

Once the turbine rotates, the generator, without any resistive load, reaches its maximum rotational velocity for the specific wind speed. To reach other steady operating velocities, different resistive loads are applied to the electrical circuit. The acquisition is done only when the system reaches the steady-state conditions. In this way it is possible to obtain the performance curves, by applying increasing electrical loads. Knowing the resistance load R_{eq} and the voltage V_{DC}, the electric power P, extracted after the conditioning electrical circuit (accounting for the electrical losses), is calculated as:

$$P = \frac{V_{DC}^2}{R_{eq}} \tag{1}$$

The passive tests have been performed with the three prototypes, with different lengths of the shaft (between the rotor and the generator), in upwind and downwind conditions, in order to quantify the effects of all these parameters.

The main findings of these tests are presented in Fig. 6. A comparison between D40 and D40-sym turbine is performed in upwind conditions, without extension shaft (rotor directly connected to the generator, see Fig. 3a), in order to understand how much energy is wasted adopting a non-optimized blade. First of all, it is possible to observe that at 10 m/s of wind speed the D40-sym turbine is not yet able to generate energy, while the airfoil blade turbine produces a maximum power of 50 mW at about 2500 RPM.

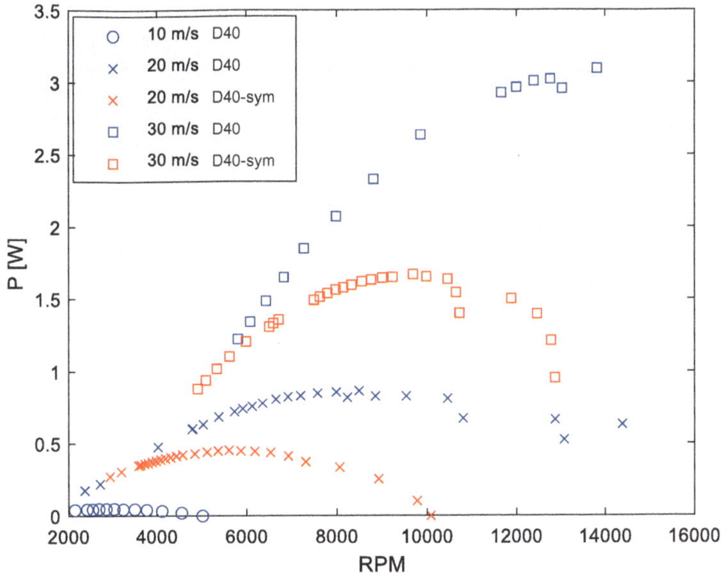

Fig. 6. Open field, passive tests, comparison between D40 and D40-sym turbine: power output as a function of the rotational speed.

The starting wind speed for the non-optimized turbine is about 12.5 m/s. At 20 m/s and 30 m/s, the D40-sym turbine reaches its maximum values at lower rotation velocities and the maximum recovered energy is about 50% lower than that obtained with the D40. Anyway, the harvested energy at these wind speeds is significantly higher than the minimum power required for this application (minimum threshold equal to about 4–8 mW) and, consequently, this turbine was finally selected for the tests with ducted configuration.

3.3 Ducted Symmetric Turbine

The goal of the tests with ducted configuration is the determination of the best duct layout in terms of power generation.

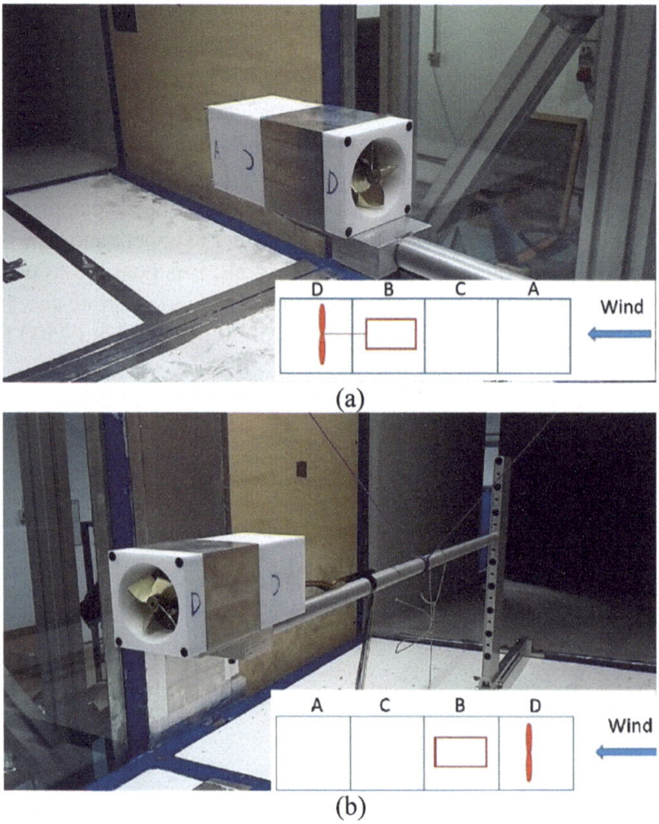

Fig. 7. Open field, passive tests, comparison between D40 and D40-sym turbine: power output as a function of the rotational speed.

Different configurations have been tested to evaluate the effects of:

- the presence of a diffuser (module D or E, see Fig. 2) downstream, Fig. 7a;
- the presence of a contraction cone (module D or E, see Fig. 2) upstream, Fig. 7b;
- the upwind (Fig. 7b) and downwind (Fig. 7a) configuration.

In order to evaluate the effect of the presence of a divergent modulus, Fig. 8c compares the power generated by two identical duct layouts, except for the end modulus which is divergent (E) for the first one (blue points) and cylindrical (A) for the second (red points). For both the configurations the turbine is positioned upwind. The presence of the divergent increases the power at all wind speeds: at 20 m/s and 30 m/s the improvement is about of 30%. On the contrary, no significant effects have been observed by changing the type of diffuser (modulus E or D) or by adding or removing a contraction cone.

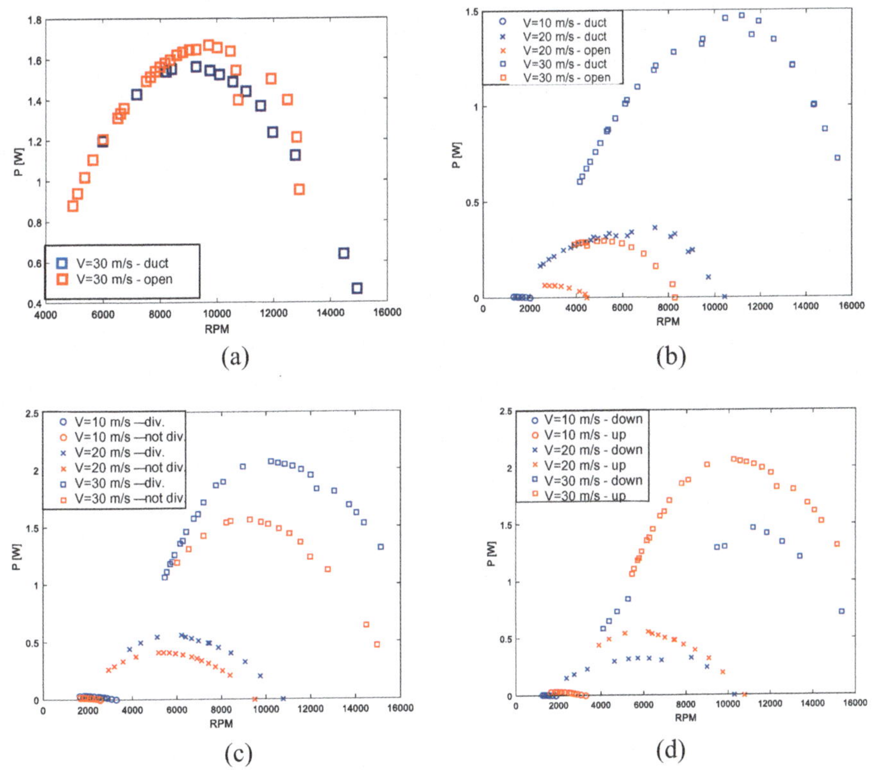

Fig. 8. D40-sym turbine, passive tests, power output as a function of the rotational speed: open-field vs ducted configuration, upwind (a); open-field vs ducted configuration, with divergent, downwind (b); with divergent vs without divergent, ducted configuration, upwind, (c); upwind vs downwind, ducted configuration, with diffuser and contraction cone (d).

Finally, the comparison between the same layout (with contraction cone and diffuser) in upwind and downwind conditions show that the reduction in power measured when the turbine is downstream is acceptable because, at 20 m/s, it results however greater than 250 mW, which is significantly up the limit threshold (8 mW).

4 Conclusions

Different cm-scale turbines, to be used as energy harvesters in sensor nodes, were designed and their performances have been evaluated by wind tunnel tests. Two 4 blade airfoil turbines, with a diameter of 30 mm and 40 mm were designed using the numerical BEM approach; a third 40 mm symmetrical rotor with constant blade pitch and flat plate blade section was built to be able to work with wind coming from two opposite directions, as required by the final application on trains.

Active control tests in open field allowed to verify the goodness of the design methodology applied: the agreement between the numerical and experimental results is discrete and the BEM approach can be used in a first step of design phase to optimize the blade parameters and to evaluate the expected performances also with these very low dimensions and high rotational velocities.

With the passive tests, the real performances of the turbines were studied, accounting on the losses in the conditioning circuit. The comparison between the airfoil and the symmetric turbine in open field showed that the second one leads to a reduction in the harvested power of about 50%; on the other hand, the harvested energy at 20 m/s of wind speeds is significantly higher than the minimum power required for this application (minimum threshold equal to 4–8 mW).

The final tests were performed with the ducted symmetric turbine. The ducted configuration was selected considering that, when the sensor node is mounted on a train, the relative wind speed has fixed direction and it has to correctly work in hard environmental conditions. The different ducted configurations tested have allowed to conclude that the presence of the duct improves the performances with respect to the open field conditions, especially when the turbine is downstream; moreover, the presence of a divergent increases the harvested power of about 30% while the contraction cone has not significant effects.

References

Howey DA, Bansal A, Holmes AS (2011) Design and performance of a centimetre-scale shrouded wind turbine for energy harvesting. Smart Mater Struct 20(8). Article no. 085021

Jafari SAH, Kosasih B (2014) Flow analysis of shrouded small wind turbine with a simple frustum diffuser with computational fluid dynamics simulations. J Wind Eng Ind Aerodyn 125:102–110

Lyon CA, Broeren AP, Giguere P, Gopalarathnam A, Selig MS (1997) Summary of low-speed airfoil data, vol 3

Milani D, Bassetti M, Braghin F, Tomasini G (2014) Design of a wireless sensor powered by a piezoelectric energy harvester. In: ASME 2014 12th biennial conference on engineering systems design and analysis, ESDA 2014

Rancourt D, Tabesh A, Fréchette L (2007) Evaluation of centimeter-scale micro wind mills: aerodynamics and electromagnetic power generation. In: Proceedings of the 7th international workshop on micro and nanotechnology for power generation & energy conversion app's (PowerMEMS 2007), Freiburg, Germany, 27–29 November 2007

Tomasini G, Giappino S, Costa A (2016) Galloping-based piezo-aeroelastic energy harvester for wireless sensors to be installed on freight trains. In: Proceedings of SPIE - the international society for optical engineering, vol 9799. Article no. 97990L

Xu FJ, Yuan FG, Hu JZ, Qiu YP (2014) Miniature horizontal axis wind turbine system for multipurpose application. Energy 75:216–224

Xu FJ, Yuan FG, Hu JZ, Qiu YP (2010) Design of a miniature wind turbine for powering wireless sensors. In: Proceedings of SPIE - the international society for optical engineering, vol 7647. Article no. 764741

Experimental Study of Wind Loads on Domed Free Roofs

Y. Uematsu(✉) and R. Yamamura

Department of Architecture and Building Science,
Tohoku University, Aoba-ku, Sendai 980-8579, Japan
yasushi.uematsu.d8@tohoku.ac.jp

Abstract. Design wind loads on domed free roofs have been investigated in a wind tunnel. Wind pressure distributions both on the top and bottom surfaces of the roof models were measured. The rise-to-span ratio, *f*/*D*, was varied from 0.1 to 0.4. First, the effect of the Reynolds number on the pressure distribution is examined. Then, the effects of *f*/*D* on the wind pressures and forces are made clear. The wind force coefficients for designing the main wind force resisting systems are proposed, in which focus is on the axial forces induced in the columns supporting the roof as the most important load effect, assuming that the roof is rigid and supported by four columns. Finally, the peak wind force coefficients for the design of cladding/components are proposed based on the maximum and minimum peak wind force coefficients irrespective of wind direction.

Keywords: Domed free roof · Design wind load · Wind tunnel experiment · Codification

1 Introduction

Free roofs are widely used for structures providing shade and weather protection in public spaces. Membrane is often used for the cladding. Since free roofs are usually supported by columns and no walls, wind action is directly exerted both on the top and bottom surfaces. Consequently, the flow around the roof is rather complicated. Being light and flexible, they are vulnerable to dynamic wind actions. Therefore, wind resistance is one of the most important technological problems, when designing these roofs. Whereas a large database of knowledge exists for wind pressure distributions on enclosed buildings of various shapes (e.g. Holmes 2001), only a few studies have been made on the wind loading on free roofs, which may be due to difficulties in model making and wind pressure measurements. Many pressure taps are required to be installed both on the top and bottom surfaces of the roof model in order to measure the wind force distributions in detail. In practice, however, the number of pressure taps is limited, because the roof thickness should be as small as possible.

Regarding the planar free roofs, such as gable and mono-sloped roofs, extensive researches have been made by several researchers, e.g. Gumley (1984), Letchford and Ginger (1992), Ginger and Letchford (1994), Natalini et al. (2002) and Uematsu et al. (2007, 2013). Indeed, the wind force coefficients for designing such free roofs are

F. Ricciardelli and A. M. Avossa (Eds.): INVENTO 2018, LNCE 27, pp. 716–729, 2019.
https://doi.org/10.1007/978-3-030-12815-9_54

specified in codes and standards (e.g. AIJ Recommendations for Loads on Buildings, 2015). By comparison, only a few studies have been made of the wind loading on curved free roofs. The Australian/New Zealand Standard specifies the wind force coefficients for H.P.-shaped free roofs. However, the range of rise-to-span (or sag-to-span) ratio is limited to small values. Wind loads on H.P.-shaped free roofs with larger rise-to-span ratios were studied by Uematsu et al. (2014a, b, 2015) experimentally and by Takeda et al. (2014) numerically. Uematsu et al. (2014a, b) proposed wind force coefficients for the design of main wind force resisting systems of such roofs as a function of the rise-to-span ratio, based on a combination of the lift and aerodynamic moment coefficients. Natalini et al. (2013) investigated the wind loads on cylindrical free roofs in a wind tunnel. They measured only the mean wind pressures acting on the roof. Dynamic load effects of turbulent winds are not discussed.

Wind loads on spherical domes of enclosed buildings have been studied by many researchers, e.g. Hongo et al. (1995). Wind pressure coefficients for designing such roofs are specified in the AIJ Recommendations for Loads on Buildings (2015). By comparison, to the authors' best knowledge, no systematic study has been made on the wind loads on domed free roofs, although many structures have been constructed in practice.

The present paper experimentally investigates the design wind loads on domed free roofs. The wind tunnel model is made by a 3D printer, which makes the roof thickness as thin as 2 mm. Note that Natalini et al. (2002) indicated that the thickness significantly affected the pressure distributions on free roofs. In the present models, the pressure taps are drilled both on the top and bottom surfaces of the model, providing the net wind forces acting on the roof. The tubing is installed in the model. Wind pressure distributions both on the top and bottom surfaces are measured in a turbulent boundary layer simultaneously. Based on the results, we propose the design wind force coefficients both for the main wind force resisting systems and cladding/components.

2 Wind Tunnel Experiment

Figure 1a shows a wind tunnel model. The length scale λ_L of the models is assumed 1/100. The rise-to-span ratio, f/D, ranges from 0.1 to 0.4 with the diameter D being fixed to 150 mm (15 m in full scale). The roof of 2 mm thickness is supported by four columns of 6.5 mm external diameter. The mean roof height H (= $h - f/2$) is 80 mm (8 m in full scale) for all models, which means that the eaves height depends on f/D. Our preliminary experiment (Wen et al. 2018) indicated that the mean roof height did not affect the wind forces on the roof significantly, provided that the eaves height was higher than 30 mm (3 m in full scale). Considering the practical situation where the eaves height is generally higher than approximately 3 m, we can understand the general features of wind forces on domed free roofs based on the present experiment.

The coordinate system and notation are shown in Fig. 1b. Nine pressure taps are installed both on the top and bottom surfaces along a centerline (Fig. 1c) to obtain the net wind forces acting on the roof model. A part of tubing system is installed in the roof and columns. The wind direction θ is changed from 0° to 90° at an increment of 5°, providing the net wind force distribution over the whole area.

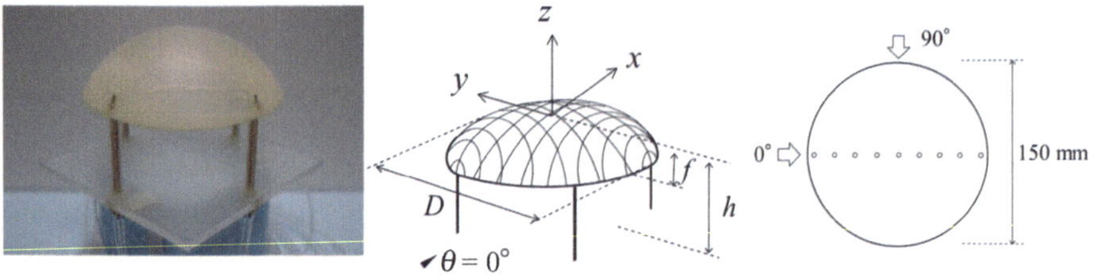

(a) Picture ($f/D = 0.3$) (b) Coordinate system and notation (c) Pressure taps and wind direction

Fig. 1. Experimental model

The wind tunnel flow is a turbulent boundary layer. The power law exponent α of the mean wind speed profile is proximately 0.22. The wind tunnel speed U_H at the mean roof height H is 9 m/s for $f/D = 0.1$ and 0.2 and 10 m/s for $f/D = 0.3$ and 0.4. The velocity scale λ_V is assumed 1/3.0 for $f/D = 0.1$ and 0.2 and 1/2.8 for $f/D = 0.3$ and 0.4. The turbulence intensity I_{uH} at the mean roof height H is approximately 0.16. Wind pressures at all pressure taps are measured simultaneously at a sampling rate of 500 Hz after passing through low-pass filters of 300 Hz. The tubing effect is compensated in the frequency domain by using the frequency response function of the measuring system. Each measurement is made for a time duration of 10 min in full scale. The measurements are repeated 10 times under the same condition. The statistical values of wind forces etc. are evaluated by applying ensemble average to the results of these consecutive 10 runs.

The wind pressures, p_t and p_b, acting on the top and bottom surfaces are normalized by the mean velocity pressure q_H ($=1/2\rho U_H^2$, with ρ being the air density) and expressed as wind pressure coefficients, C_{pt} and C_{pb}, respectively. The net wind force per unit area is provided by the difference between p_t and p_b. Therefore, the wind force coefficient C_f is defined by $C_f = C_{pt} - C_{pb}$; positive downward.

3 Experimental Results and Discussion

3.1 Effect of the Reynolds Number on the Pressure Distribution

It is well accepted that the wind pressures on curved structures are affected by the Reynolds number, *Re*. This is because the location of flow separation from the structure's surface depends on *Re*. In this section, the effect of *Re* on the wind pressure distribution is investigated by changing the wind speed U_H. Here, the Reynolds number is defined as $Re = U_H(2R)/\nu$, where ν is the coefficient of kinematic viscosity of air, and R represents the radius of curvature of the dome. Note that '2*R*' is used for the reference length in order to associate the flow around the domed roof with that around a two dimensional circular cylinder, which has been studied extensively. In those studies the diameter (=two times the radius) is used for defining *Re*.

In the experiment the wind speed U_H was varied from 3 to 11 m/s; the corresponding Reynolds number ranged from 0.33×10^5 to 3.09×10^5. The distribution of the mean wind force coefficients $C_{f,\mathrm{mean}}$ along the centerline when $\theta = 0°$ were measured at various wind speeds. Figures 2a and b show the results for $f/D = 0.2$ and 0.3, respectively. In the figure 's' represents the coordinate along the centerline, normalized by the maximum value s_{max}. When $f/D = 0.1$, not shown here to save space, the distribution did not change with Re. This feature indicates that the wind flows along the roof's top surface without separation. On the other hand, when $f/D \geq 0.2$, the distribution changes with Re, which implies that the flow along the roof's top surface separates at some point and the separation point shifts with Re. The flow separates downward at the windward edge of the roof. However, the wind pressures on the bottom surface are not affected by Re. Therefore, the change in the wind force coefficient distribution is mainly attributed to the wind pressure coefficient distribution on the top surface.

Figure 2 indicates that the distribution is minutely affected by Re when $Re > 1.0 \times 10^5$. This feature is consistent with a finding by Macdonald et al. (1988) for cylindrical structures. They investigated the Reynolds number effects on the flow around circular cylinders by comparing the wind tunnel results obtained at various Reynolds numbers with the results of full-scale measurements on actual silos. According to their investigation, the wind-tunnel results obtained at $Re > 1.0 \times 10^5$ in a turbulent boundary layer simulating natural winds may represent the practical situation in full-scale. Therefore, the measurements are made at $Re > 1.0 \times 10^5$ in the following sections.

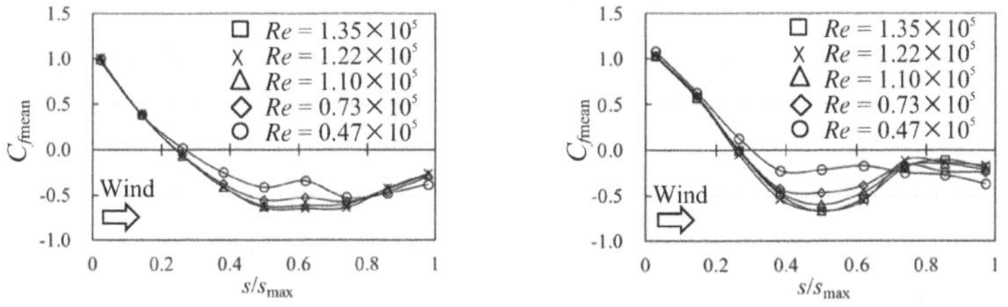

(a) $f/D = 0.2$ (b) $f/D = 0.3$

Fig. 2. Variation of the distribution of mean wind force coefficients $C_{f,mean}$ with Re

3.2 Distributions of Wind Pressure and Force Coefficients

Figure 3 shows the distributions of mean wind pressure coefficients, C_{ptmean}, on the top surface along the centerline for $\theta = 0°$. When $f/D = 0.1$, the value changes smoothly along the line. On the other hand, the curve has an inflection point in the leeward area when $f/D \geq 0.2$. The inflection point shifts windward as the f/D ratio increases. It is located near the leeward edge when $f/D = 0.2$ and at $s/s_{max} \approx 0.7$ when $f/D = 0.4$. This inflection point roughly corresponds to the flow separation point.

The distributions of mean wind pressure coefficients, C_{ptmean} and C_{pbmean}, on the top and bottom surfaces are shown in Figs. 4 and 5, respectively. Although the roof has curvatures both in the along-wind and across-wind directions, the contour lines on the top surface are almost perpendicular to the centerline. This feature indicates that the flow separation line is roughly perpendicular to the centerline. The flow separates

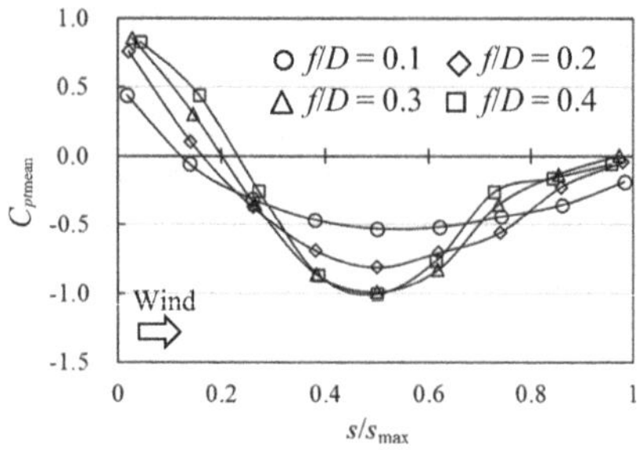

Fig. 3. Distributions of mean wind pressure coefficients along the centerline on the top surface

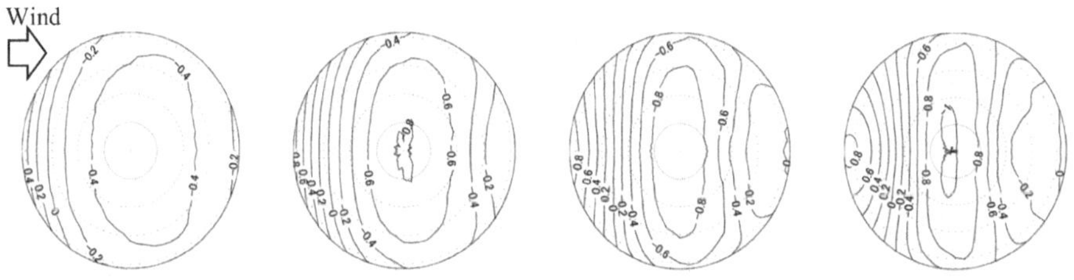

(a) $f/D = 0.1$ (b) $f/D = 0.2$ (c) $f/D = 0.3$ (d) $f/D = 0.4$

Fig. 4. Distribution of mean wind pressure coefficients $C_{pt\mathrm{mean}}$ on the top surface

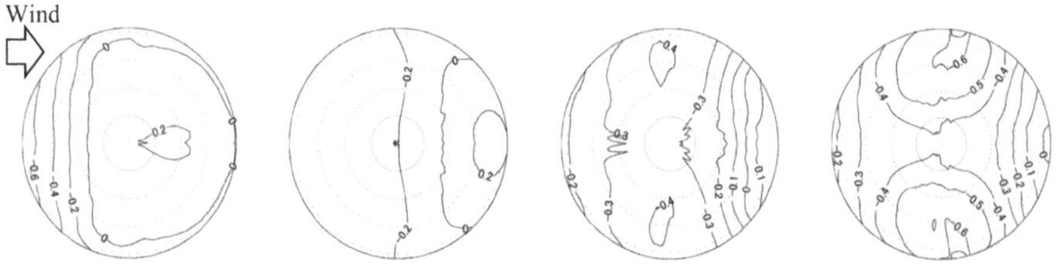

(a) $f/D = 0.1$ (b) $f/D = 0.2$ (c) $f/D = 0.3$ (d) $f/D = 0.4$

Fig. 5. Distribution of mean wind pressure coefficients $C_{pb\mathrm{mean}}$ on the bottom surface

downward at the windward curved edge and reattaches on the bottom surface. The reattachment line is also perpendicular to the centerline. The separation bubble zone expands leeward as the f/D ratio increases. When $f/D = 0.4$, the bottom surface is totally immersed in the separation bubble. The spatial variation of $C_{pb\mathrm{mean}}$ is generally smaller than that of $C_{pt\mathrm{mean}}$. Therefore, the distributions of mean wind force coefficients $C_{f\mathrm{maen}}$, shown in Fig. 6, look similar to those of $C_{pt\mathrm{mean}}$ (Fig. 4). In other words, the wind force distribution is dominated by the wind pressure distribution on the top surface. The roof is subjected to positive wind forces (downward) in the windward area and negative wind forces (upward) in the other area. The negative wind force becomes the largest in magnitude near the roof top.

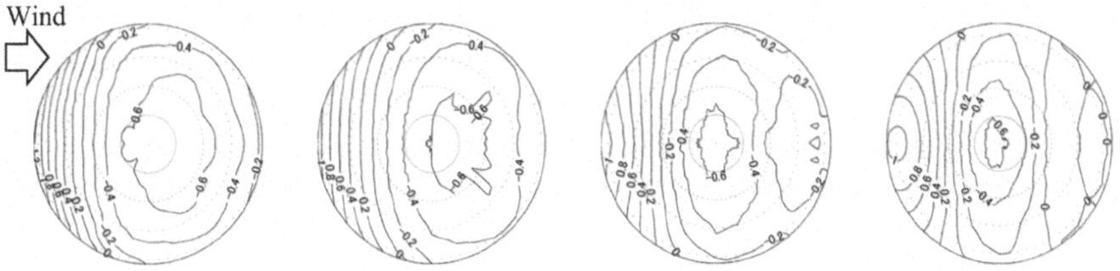

(a) $f/D = 0.1$ (b) $f/D = 0.2$ (c) $f/D = 0.3$ (d) $f/D = 0.4$

Fig. 6. Distribution of mean wind force coefficients $C_{f\text{mean}}$

The distributions of $C_{pt\text{mean}}$ along the centerline for $f/D = 0.1$ and 0.4 are compared with those on enclosed structures with $f/D = 0.1$ and 0.5 obtained by Hongo et al. (1995) in Fig. 7. Note that the conditions of wind tunnel flows employed in the experiments are somewhat different from each other. Nevertheless, the distributions look similar to each other when the f/D ratio is relatively large, such as $f/D = 0.4$, for example, which implies that the flow over the roof is not affected by the wall significantly in such a case. By comparison, the distributions are quite different from each other when the f/D ratio is small, such as $f/D = 0.1$, for example. In the case of enclosed structure, the flow separates upward at the windward edge and then reattaches on the top surface, resulting in large suctions on the top surface in the separation bubble zone. In the free roof case, on the other hand, the flow does not separate upward at the windward edge. Therefore, the windward area is subjected to positive wind forces.

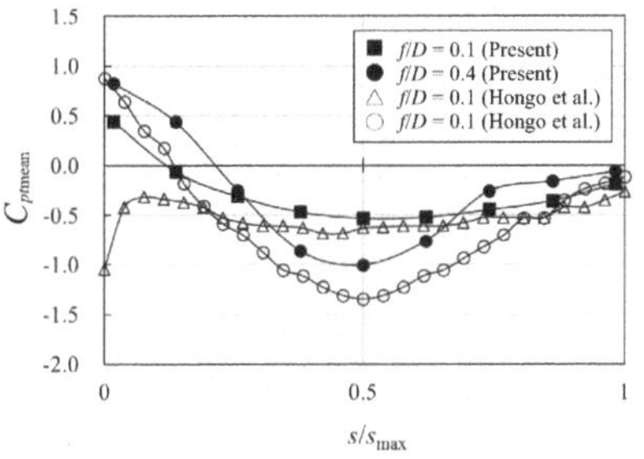

Fig. 7. Comparison between free roof and enclosed structure for the mean wind pressure coefficients along the centerline on the top surface

4 Wind Force Coefficients for the Main Wind Force Resisting Systems

4.1 Basic Concepts for Proposing Wind Force Coefficients

The following assumptions are made for discussing the wind force coefficients for designing the main wind force resisting systems:

(1) The structure is rigid enough to ignore the resonance effects on the wind loads. Therefore, the design wind loads are represented by equivalent static loads based on a quasi-static assumption. In this case, the wind loads are given by the product of velocity pressure q_H, wind force coefficients C_f and a gust effect factor G_f. This assumption is thought to be reasonable for such small structures as considered in the present study.
(2) The roof is supported by four columns (Fig. 1b). Focus is on the axial forces induced in the columns as the most important load effect when discussing the design wind force coefficients. Two sets of wind force coefficients providing the maximum tension and compression (called Loads cases A and B, respectively) are proposed. The same assumption was made in our previous studies for the other roof shapes (Uematsu et al. 2007, 2014a, b; Yamamura et al. 2018).
(3) The dynamic load effect of turbulence is considered in the gust effect factor G_f, which is evaluated based on the load effect. The value of G_f can be approximated by the square of gust factor G_v of approach flow with an appropriate value of the peak factor g_v.
(4) Considering the characteristics of wind force coefficients mentioned above, the roof is divided into three zones, R_1 to R_3, as shown in Fig. 8 and constant wind force coefficients, C_{NW}, C_{NC} and C_{NL}, providing the maximum load effects are specified. Wind forces at any points in each zone are assumed to be perfectly correlated with each other. Therefore, values of C_{NW}, C_{NC} and C_{NL} are determined from the wind force coefficient distribution along the centerline parallel to the wind direction. In practice, the correlation of wind force coefficients at different points in each zone is not perfect, as shown in Fig. 9, which shows the correlation coefficient R between fluctuating wind forces at Point 5 (reference point) and the other point along the centerline perpendicular to the wind direction. It is found that the correlation of wind force fluctuations with respect to that at Point 5 is rather low for points far from the centerline, such as Points 1 and 2, for example. By comparison, the correlation is relatively high for the points located in the range of $y/D = -0.25$ to 0.25, e.g. Points 3–7. The area providing such higher correlation is relatively wide because of the roof shape. Strictly speaking, the above assumption is not correct. However, it may be acceptable from the practical viewpoint, because it overestimates the design wind loads. At present the effect of this assumption on the estimated wind loads is not clear. It is the subject of future study.

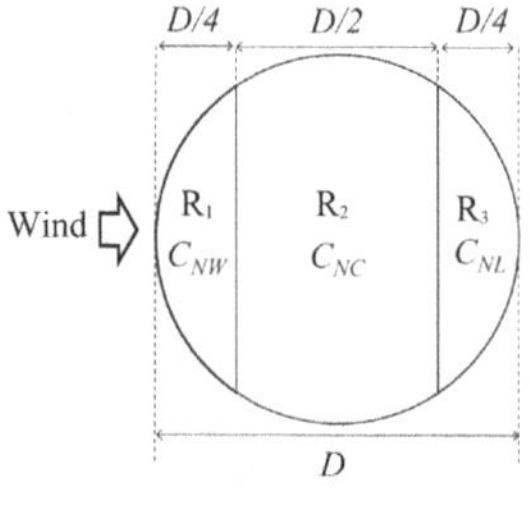

Fig. 8. Roof division for the specification of wind force coefficients

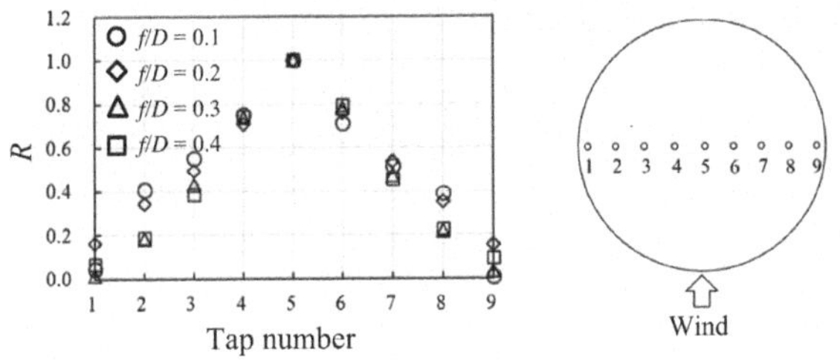

Fig. 9. Correlation coefficient between fluctuating wind forces at Point 5 and the other point

4.2 Wind Force Coefficients Providing the Maximum Load Effects

The axial forces induced in the columns are calculated by using the time history of wind force coefficients C_f at nine points along the centerline when $\theta = 0°$. Applying a conditional sampling technique to the time history of wind force coefficients, we detected the distribution of C_f at an instant when the maximum tension (Load case A) or the maximum compression (Load case B) occurred in the columns. The results for $f/D = 0.2$ and 0.4 are shown in Figs. 10 and 11, respectively. In the figure the results of 10 runs are plotted. It is interesting to note that the scatter of the results is relatively small. This feature implies that the flow field generating the maximum load effect is almost the same for a given f/D ratio. Therefore, ensemble average is first applied to the C_f distributions obtained from the 10 runs, and then the distribution is spatially averaged over each zone. The wind force coefficients, C_{NW}, C_{NC} and C_{NL}, obtained in this way, can be used for the design purpose.

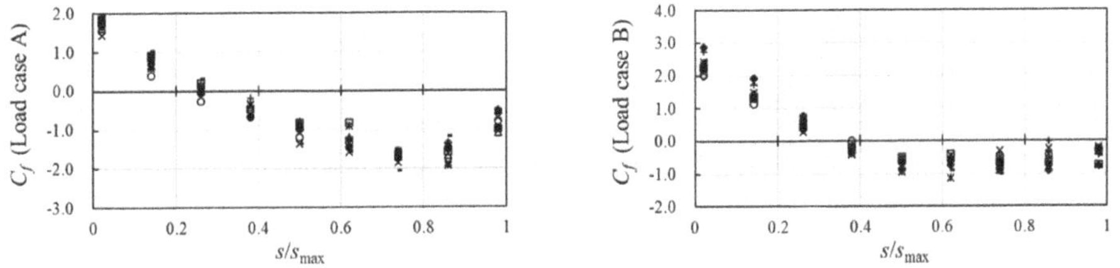

(a) Maximum tension (Load case A)

(b) Maximum compression (Load case B)

Fig. 10. Distributions of C_f along the centerline providing the maximum load effects for $f/D = 0.2$

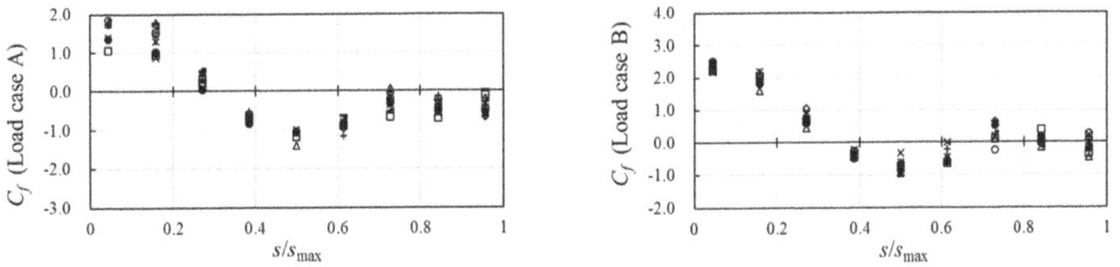

(a) Maximum tension (Load case A)

(b) Maximum compression (Load case B)

Fig. 11. Distributions of C_f along the centerline providing the maximum load effects for $f/D = 0.4$

4.3 Gust Effect Factor Based on the Load Effects

The values of C_{NW}, C_{NC} and C_{NL} obtained in the previous section involve the dynamic load effects of approaching turbulence, which can be evaluated by the gust effect factor G_f. The gust effect factor is defined by the ratio of the maximum or the minimum axial force to the mean value induced in the columns. Figure 12 shows the ratio of the maximum peak value $N_{\rm peak}$ to the mean value $N_{\rm mean}$ of column axial force, plotted against the non-dimensional mean axial force defined as follows:

$$N^*_{mean} = \frac{16N_{mean}}{q_H \cdot (\pi D^2)}. \tag{1}$$

In the figure, the results for cylindrical roofs (Yamamura et al. 2018) are also plotted. Because their wind tunnel experiment was carried out with models of the same size in the same turbulent boundary layer, the dynamic load effect of approaching turbulence is expected to be the same as that for the domed models used in the present study. When the values of $|N^*_{\rm mean}|$ are small, the $N_{\rm peak}/N_{\rm mean}$ ratio (G_f) exhibits a large value. However, as the value of $|N^*_{\rm mean}|$ increases, the $N_{\rm peak}/N_{\rm mean}$ ratios collapse into a narrow range around 1.8. Therefore, the value of $G_f = 1.8$ is used for evaluating the wind force coefficients hereafter.

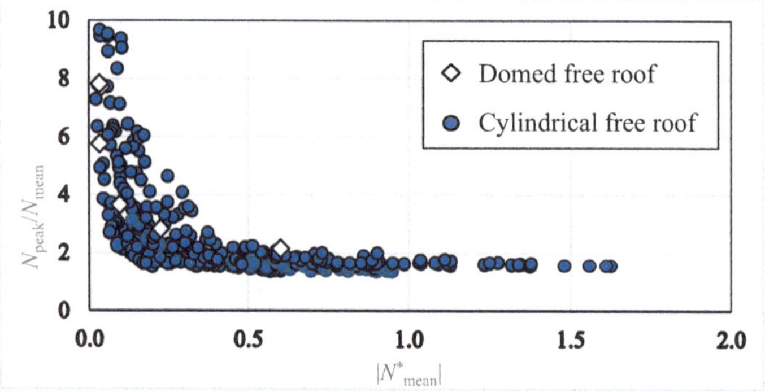

Fig. 12. Ratio of the peak to the mean value of the column axial force

4.4 Design Wind Force Coefficients

The wind force coefficients, C^*_{NW}, C^*_{NC} and C^*_{NL}, for designing the main wind force resisting systems are provided by the area-averaged wind force coefficients, C_{NW}, C_{NC} and C_{NL}, obtained in Sect. 4.2, divided by $G_f = 1.8$. The results are summarized in Table 1. In practical design, the wind loads on domed free roofs can be provided by using these coefficients and a gust effect factor G_f appropriately evaluated with consideration of the terrain roughness around the construction site.

Table 1. Wind force coefficients for designing the main wind force resisting systems

f/D	Load case A			Load case B		
	C^*_{NW}	C^*_{NC}	C^*_{NL}	C^*_{NW}	C^*_{NC}	C^*_{NL}
0.1	0.6	–0.6	–0.6	1.1	–0.3	–0.4
0.2	0.5	–0.5	–0.8	0.8	–0.2	–0.3
0.3	0.5	–0.5	–0.6	0.9	–0.3	–0.1
0.4	0.7	–0.4	–0.2	0.9	–0.2	0

5 Peak Wind Force Coefficients for Cladding/Components

The tributary area of cladding and its immediately supporting structures is assumed 1 m^2 based on the AIJ Recommendations for Loads on Buildings (2015). Regarding the evaluation of peak pressure coefficients, the tributary area A (m^2) of spatial average is related to the averaging period T_a (s) of moving average applied to the time history of pressure fluctuations at a representative point in the area by the following equation (see Uematsu and Isyumov 1998):

$$A = \left(\frac{T_a \cdot U_H}{k_p}\right)^2. \tag{2}$$

where k_p represents a decay constant when the root coherence of pressure fluctuations at two different points in the area is approximated by an exponential function of the non-dimensional frequency, $f\Delta/U_H$, with Δ being the distance between the two points. Assuming that $k_p \approx 5$ as a typical value, we obtain $T_a \approx 0.18$ s. Hence, a moving average with an averaging period of approximately 0.2 s in full scale was applied to the time history of wind force coefficients $C_f(t)$. In practice, consecutive 3 data were used for computing the moving average.

Figures 13 and 14 show the distributions of the maximum and minimum peak values of $C_f(t)$, $C_{f\max}$ and $C_{f\min}$, during a time duration of 10 min in full scale. Large positive values of $C_{f\max}$ are induced near the leading edge. This feature is related to larger $|C_{pb}|$ values due to the downward flow separation at the leading edge. On the other hand, large $|C_{f\min}|$ values are induced in an elliptical area near the roof top. This feature is related to large $|C_{pt}|$ values induced by the accelerated flow near the roof top.

Considering the distributions of $C_{f\max}$ and $C_{f\min}$ (Figs. 13 and 14) together with a fact that the domed free roof is an axisymmetric structure, the roof is divided into three zones, R_a – R_c, as shown in Fig. 15. The definition of zones is the same as that used in the AIJ Recommendations for Loads on Buildings (2015). The positive and negative peak wind force coefficients for designing cladding/components are specified based on the most critical maximum and minimum wind force coefficients irrespective of wind direction in each zone. The specified values are summarized in Table 2. The positive peak wind force coefficient significantly changes from zone to zone; it is the largest in Zone R_a and the smallest in Zone R_c. When f/D = 0.1, the value is generally large, particularly in Zone R_a. As the f/D ratio increases, the difference in $C_{f\max}$ between zones becomes smaller. On the other hand, the negative peak wind force coefficient does not change so much from zone to zone; the magnitude is the smallest in Zone R_a and the largest in Zone R_c. It is interesting to note that the value is the largest in magnitude when $f/D = 0.2$. The values of $C_{f\min}$ in the three zones are almost the same when $f/D = 0.2$ and 0.3.

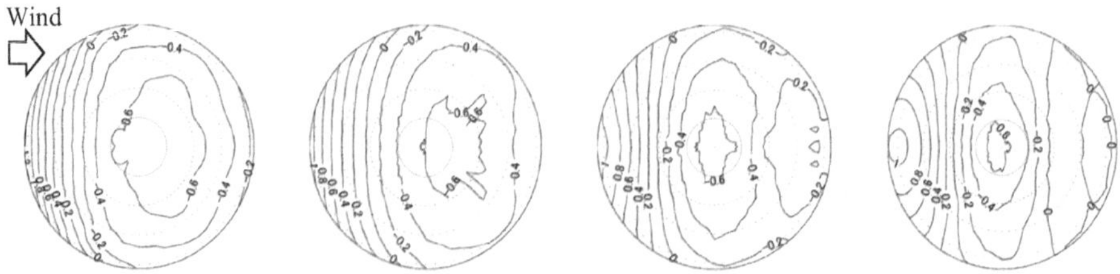

(a) $f/D = 0.1$ (b) $f/D = 0.2$ (c) $f/D = 0.3$ (d) $f/D = 0.4$

Fig. 13. Distribution of the maximum peak wind force coefficients, $C_{f\max}$

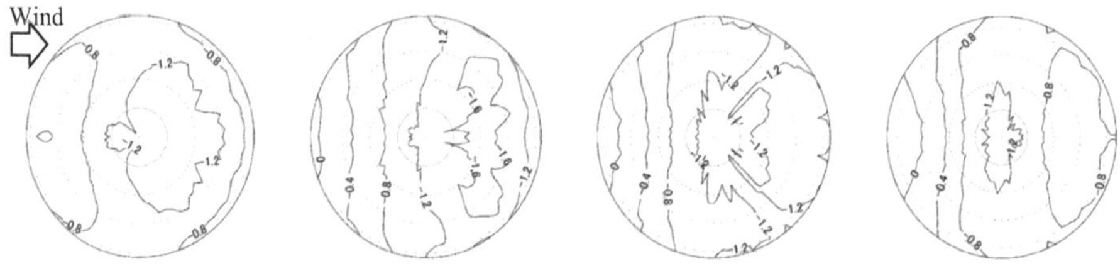

(a) $f/D = 0.1$ (b) $f/D = 0.2$ (c) $f/D = 0.3$ (d) $f/D = 0.4$

Fig. 14. Distribution of the minimum peak wind force coefficients, $C_{f\min}$

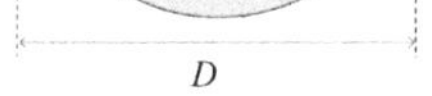

Fig. 15. Roof division for the specification of peak wind force coefficients for cladding/components

Table 2. Specification of the peak wind force coefficients for designing cladding/components

(a) Positive peak wind force coefficients

f/D	Zone R_a	Zone R_b	Zone R_c
0.1	3.2	2.2	1.3
0.2	2.5	1.7	0.9
0.3	2.4	2.0	1.1
0.4	2.4	2.1	1.2

(b) Negative peak wind force coefficients

f/D	Zone R_a	Zone R_b	Zone R_c
0.1	−1.0	−1.2	−1.4
0.2	−1.6	−1.7	−1.7
0.3	−1.2	−1.2	−1.2
0.4	−0.9	−1.1	−1.3

6 Concluding Remarks

The present paper has investigated the wind force coefficients for designing domed free roofs based on wind tunnel experiments. The wind tunnel model of 2 mm thickness was made by a 3D printer and had nine pressure taps on each of the top and bottom surfaces along a centerline. The net wind force per unit area acting on the roof is provided by the difference between the wind pressures on the top and bottom surfaces. The distributions of wind pressure coefficients, C_{pt} and C_{pb}, on the top and bottom surfaces and the wind force coefficients, C_f (=$C_{pt} - C_{pb}$), were obtained by rotating the wind tunnel model.

First, the effect of the Reynolds number on the pressure distribution was examined. The results indicate that the wind-tunnel results obtained at $Re > 1.0 \times 10^5$ in a turbulent boundary layer simulating natural winds represent the practical situation in full-scale.

Second, the characteristics of wind pressure and force coefficients were investigated. It is found that the flow around and the resultant wind pressures on the roof are affected by the rise-to-span ratio, f/D, significantly. The contour lines of wind pressure coefficients were found to be almost perpendicular to the wind direction.

Third, a discussion was made of the wind force coefficients for designing the main wind force resisting systems, in which the roof was assumed rigid and supported by four columns. Focus was on the axial forces induced in the columns by wind loading as the most important load effect and two load cases providing the maximum tension and compression were considered. The roof is divided into three zones, R_1 to R_3, by two lines perpendicular to the wind direction, and two sets of wind force coefficients for these three zones are provided as a function of f/D. A conditional sampling technique was employed to obtain the wind force coefficient distributions producing the maximum tension and compression in the columns. The proposed wind force coefficients are the spatially averaged values of the wind force coefficients over the roof zones.

Finally, the positive and negative peak wind force coefficients for designing cladding/components were provided based on the distributions of the most critical maximum and minimum wind force coefficients irrespective wind direction. Considering that the domed roof is an axisymmetric structure, the roof was divided into three zones, R_a to R_c, by two circles. The positive and negative values of peak wind force coefficients were provided as a function of *f*/*D*.

Acknowledgements. The present study was financially supported by Membrane Structures Association of Japan.

References

Architectural Institute of Japan (2015) Recommendations for Loads on Buildings. (in Japanese)
Ginger JD, Letchford CW (1994) Wind loads on planar canopy roofs, Part 2 fluctuating pressure distributions and correlations. J Wind Eng Ind Aerodyn 51:353–370
Gumley SJ (1984) A parametric study of extreme pressures for the static design of canopy structures. J Wind Eng Ind Aerodyn 16:43–56
Holmes JD (2001) Wind loading of structures. Span Press, London
Hongo T, Suzuki Y, Tsuchiya M (1995) Experimental study of wind forces on spherical roofs – Part 1. Effects of wind profile and turbulence intensity on mean pressure distribution. J Wind Eng (JAWE) 62:23–33. (in Japanese)
Letchford CW, Ginger JD (1992) Wind loads on planar canopy roofs, Part 1: mean pressure distributions. J Wind Eng Ind Aerodyn 45:25–45
Macdonald PA, Kwok KCS, Holmes JD (1988) Wind loads on circular storage bins, silos, and tanks: I point pressure measurements on isolated structures. J Wind Eng Ind Aerodyn 31:165–188
Natalini B, Marighetti JO, Natalini MB (2002) Wind tunnel modeling of mean pressures on planar canopy roof. J Wind Eng Ind Aerodyn 90:427–439
Natalini MB, Morel C, Natalini B (2013) Mean loads on vaulted canopy roofs. J Wind Eng Ind Aerodyn 119:102–113
Takeda F, Yoshino T, Uematsu Y (2014) Design wind force coefficients for hyperbolic paraboloid free roofs. J Phys Sci Appl 4(1):1–19
Uematsu Y, Isyumov N (1998) Peak gust pressures on the roof and wall edges of a low-rise building. J Wind Eng Ind Aerodyn 77 & 78:217–231
Uematsu Y, Iizumi E, Stathopoulos T (2007) Wind force coefficients for designing free-standing canopy roofs. J Wind Eng Ind Aerodyn 95:1486–1510
Uematsu Y, Miyamoto Y, Gavanski E (2014a) Wind loading on a hyperbolic paraboloid free roof. J Civ Eng Archit 8:1233–1242
Uematsu Y, Sakurai H, Miyamoto Y, Gavanski E (2013) Wind force coefficients for designing porous canopy roofs. J Civ Eng Archit 7(9):1047–1055
Uematsu Y, Miyamoto Y, Gavanski E (2014b) Wind loading on a hyperbolic paraboloid free roof. J Civ Eng Archit 8(10):1–19
Uematsu Y, Miyamoto Y, Gavanski E (2015) Effects of porosity on the Wind loads on a hyperbolic paraboloid canopy roof. J Civ Eng Archit 9(6):715–726

Wen L, Yamamura R, Uematsu Y (2018) Characteristics of wind forces on domed free roofs, Summaries of technical papers of annual meeting, Structure I, Architectural Institute of Japan, pp 129–130. (in Japanese)

Yamamura R, Uematsu Y, Takadate Y (2018) Wind force coefficients for designing curved free roofs Part 1: cylindrical roofs, Research paper on membrane structures 2017, Membrane Structures Association of Japan, 31, 11–24. (in Japanese)

Wind Pressure Distribution on a Porous Double Skin Façade System

A. Zasso[1(✉)], F. Perotti[2], L. Rosa[1], P. Schito[1], G. Pomaranzi[1], and N. Daniotti[3]

[1] Department of Mechanical Engineering, Politecnico di Milano, Via G. La Masa 1, 20156 Milan, Italy
{alberto.zasso,paolo.schito}@polimi.it, {lorenzo.rosa,giulia.pomaranzi}@polimit.it

[2] Department of Civil and Environmental Engineering, Politecnico di Milano, Piazza Leonardo da Vinci 32, 20133 Milan, Italy
federico.perotti@polimi.it

[3] Department of Mechanical and Structural Engineering and Material Science, University of Stavanger, 4036 Stavanger, Norway
nicolo.daniotti@uis.no

Abstract. Permeable double skin facades (permeable DSF) are part of modern architectural concepts. An accurate assessment of the wind loads and the porosity effects on this kind of facades is crucial for a correct design and performance evaluation. Measuring and predicting DSF airflow is not a straightforward task due to the interaction between the wind turbulence and the outer porous skin. Moreover, Eurocode and many other National Codes do not supply any prescription about such issue. In this paper the comprehensive experimental study of wind loads acting on the porous double skin facade of the *New Bocconi Campus*, currently under construction in Milan, is reported. Cladding loads assessment has been carried out performing wind tunnel tests on properly scaled rigid models able to reproduce the aerodynamic behaviour of the porous medium. Peak pressures for the inner facade have been estimated and compared to the ones expected forcing a standard glazed facade.

Keywords: Wind tunnel tests · Pressure measurements · Double skin facade · Porous double skin facade · Porous facade

1 Introduction

Nowadays the environmental issue and the energy efficiency optimisation are becoming more and more important: part of the design of civil structure is devoted to the energy performance of the system too. At this purpose, double skin facades (now on DSF) are often adopted, since they may have an impact in the energy savings (Xiaotong and Chen 2010); they are a multilayer structure characterized by two facades (or skins) between the interior and the exterior environment. Particularly, this study deals with Porous Double Skin Facade, where the outer skin is a metallic mesh and the inner one is a standard glazed facade. This cladding system is firstly adopted to control the amount of sun that penetrates the building but the presence of an outer skin may

F. Ricciardelli and A. M. Avossa (Eds.): INVENTO 2018, LNCE 27, pp. 730–741, 2019.
https://doi.org/10.1007/978-3-030-12815-9_55

induce modifications to the wind pressure distribution on the inner facade; since the cladding system can achieve up to 25% of the total building cost (Overend and Zammit 2006), it is important to assess if this kind of facade system may induce a reduced wind pressure distribution with respect to the standard single layer facade. Additionally, Eurocode and many other National Code do not seem to propose specific guidelines dedicated to the assessment of the wind load in this cladding system, especially if the outer skin porosity is higher than 1%.

1.1 Literature Review

The DSF systems have been studied and frequently used as solution to improve the thermal performances of buildings. On the contrary, their beneficial effect on the inner face's pressure distribution has been very little investigated, as pointed out by da Silva and Gomes (2008). Looking at the existing literature, there is a dearth of knowledge about the wind DSF interaction in terms of pressure distribution, especially when outer skin is characterized by high value of porosity, as for the object of this study.

Wind permeable facades have been studied for what concerns the rain screens and porous roofs only. Although they are characterized by lower values of porosity with respect to the ones which this project is interested in, they may give a first idea of what happens behind a wind permeable facade: Gerhardt and Janser (1994) have shown the advantages we may have installing such screens as outer skin in addition to the standard glazed facade, in terms of pressure reduction. Belloli et al. (2014) have evaluated the effects of the porosity of a perforated steel plates as cladding of a tall slender tower.

Results confirm the pressure reduction due to the permeable elements.

Nowadays porous structural elements find several applications in civil engineering also for their aerodynamic behavior: perforated shroudings have been deeply investigated as they may represent a suitable device able to control vortex shedding induced vibration (Belloli et al. 2012).

1.2 Case Study

The main objective of this research is the assessment of the cladding loads for the *New Bocconi Campus*, a complex of buildings under construction in the Southern part of Milan; it consists in six different buildings (Fig. 1(a)), interesting a total area of 36000 m^2. The cladding system for each structure is made by a porous DSF and so the glazed facade is covered by the outer metallic mesh: in this way a gap flow between the two layers can develop. In order to estimate the wind load acting on such peculiar facade, ad-hoc wind tunnel tests have been commissioned in order to study the aerodynamic effect induced by porosity of the external skin on the pressure distribution acting on the inner glazed facade. The main aim of this project is to assess the external peak pressure distribution on the inner glazed facade, when an outer porous mesh is installed. This output is then compared to the peak pressure distribution measured on the standard facade model in order to estimate the expected percentage reduction.

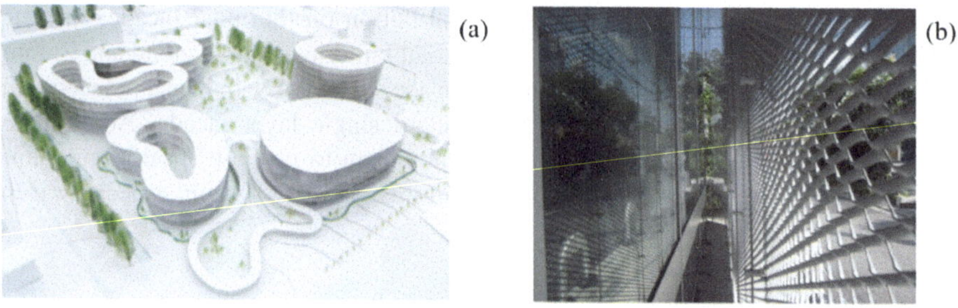

Fig. 1. The *New Bocconi Campus* render (a) and DSF porous system example (b).

2 Wind Tunnel Test Set-up

Wind tunnel tests have been performed in the Boundary Layer Wind Tunnel at Politecnico di Milano. The test section dimensions (4 m high, 14 m wide and 36 m long) allow to choose large geometric scale $\lambda_L (\lambda = \text{model/real})$, attaining Reynolds Number that would give reliable informations on full-scale conditions, while keeping low blockage effects. In order to deeply understand the aerodynamic behavior of the porous medium, this research project is divided in two phases. In the first one, the six buildings are tested as they would have a standard facade system to assess the wind loads acting on the glazed facade without the porous mesh. Once these pressure distributions are known, the attention may be focused on that building characterized by the worst conditions in terms of wind load. Only the aforesaid structure is instrumented and tested during the second stage, when the outer porous skin is properly modeled and reproduced in the model scale. By comparing the two tests outcomes, the beneficial effect of the porous medium in terms of pressure reduction can be highlighted.

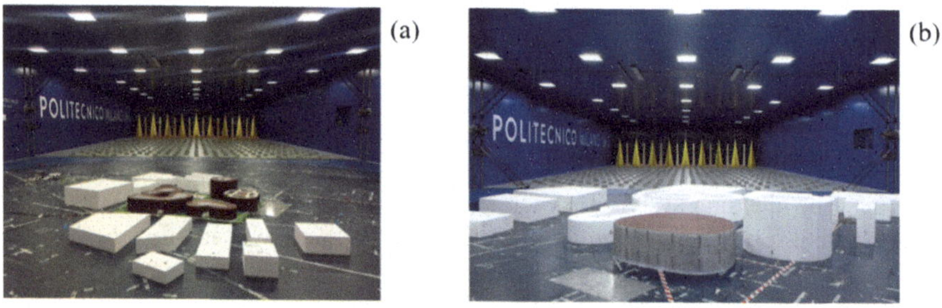

Fig. 2. NoDSF setup (a) and DSF setup (b).

2.1 First Stage: No DSF Model

The first stage of the research considers all the *New Bocconi Campus* buildings: rigid models, able to reproduced the full-scale geometry, are instrumented with pressure taps. Given the curved shapes of the structures, in order to improve local flow similarities, sand papers are applied to the model surface. In this way, the surface roughness is increased in order to avoid Reynolds dependency and to simulate higher Reynolds number flow condition (Fig. 2(a)). The geometric scale is chosen equal to 1:100 and the scale velocity is 1:2 instead. Being the height of the building not relevant (lower than 35 m at full scale), pressure taps are positioned in one single level, running along the perimeter of each building. The pressure measurement system consists of 8 PSI ESP-32 HD high speed pressure scanners, connected to a data acquisition system with 500 Hz sampling frequency. The outcome of each test is a 120 s time history for each pressure tap. A Pitot tube placed 7 m upwind the model measures the reference velocity at 0.35 m height (12 m/s). 36 different inflow directions (spaced 10° apart) are tested, according to the convention depicted in Fig. 3(a) (Table 1).

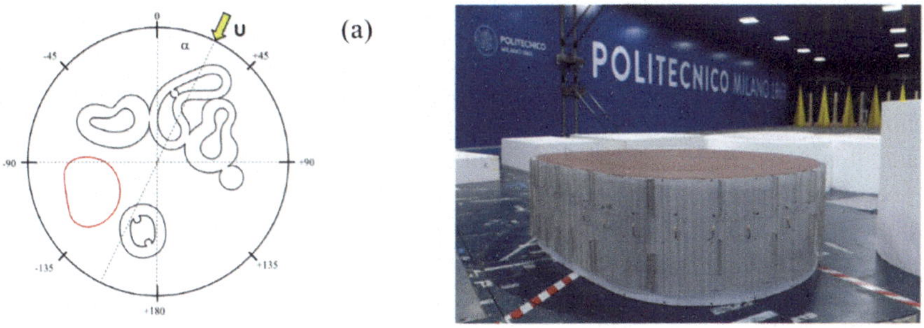

Fig. 3. Wind exposure angles and reference system (a) and DSF building during the second stage of the tests (b).

Table 1. Scale factors for the No DSF stage

	λ
Length scale	1/100
Velocity scale	1/2.53
Time scale	1/39.53

2.2 Scaling Procedure for the Metallic Mesh

In order to test the Double Skin facade buildings, the outer porous mesh has to be reproduced. Experimental results of wind tunnel tests strongly depend on the flow conditions, that should be the same for full and model scale. At this aim, the porous outer skin has to be properly scaled such that the kinematic similitude is fulfilled and this requires a dedicated study. When air flows through a porous panel, a pressure drop in the fluid occurs due to the friction stresses that develop at the contact regions. As a consequence, the usual geometric scale applied to the global structure dimensions cannot be applied. As shown by Belloli et al. (2014) and Letchford et al. (2000), the pressure loss coefficient must be maintained passing from real scale to model scale. It is defined as:

$$k = \frac{\Delta p}{\frac{1}{2}\rho U^2} \tag{1}$$

where Δp is the pressure difference measured before and after the porous screen, ρ is the air density and U is the mean wind speed. This coefficient accounts for the resistance to flow through the porous panel, including the effects of the porosity and the holes shape. The k coefficient can be obtained through experimental tests only: an open circuit conduct properly designed to characterize the porous media is used. It has a circular section (diameter 400 mm), fully covered by the mesh being tested. The parameter has been computed for the real scale mesh first and then compared to the model scale mesh one: as shown in Fig. 4, the two loss coefficients are very close in the range of velocity of interest.

In the real structure, the outer skin is made by a diamond mesh expanded metal, with shorter diagonal and longer diagonal equal to 35 mm and 85 mm respectively; the open area $\boldsymbol{\beta}$ is 33%. Moreover, the outer skin does not simply follow the inner glazed facade because it is bent in the transversal plane such that waves with small curvature radii are reproduced. The model scale mesh is instead characterized by the same mesh shape with shorter diagonal and longer diagonal equal to 3 mm and 6 mm respectively; the small waves are recreated in the wind tunnel model, in order to obtain the reproduction of the real scale outer skin.

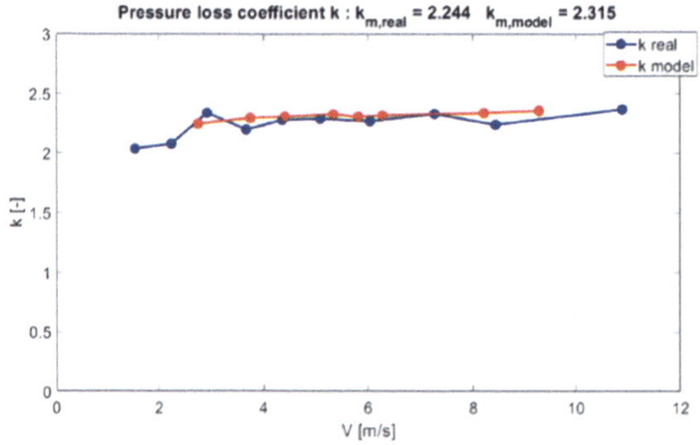

Fig. 4. Experimental pressure loss coefficient with varying velocity: real mesh and model mesh.

2.3 Second Stage: DSF Model

The second stage of the tests interests the most critical building in terms of wind pressure distribution only, while the remaining buildings of the *New Bocconi Campus* are reproduced by styrofoam blocks. The outer mesh found by the previously described test is installed in the model (Fig. 2(b)). The geometrical scale is 1:50 in this phase, such that a gap flow can develop. The velocity scale is instead the same as the first test. In order to obtain a full characterization of the wind pressure distribution on the DSF system, pressure taps are placed both on the inner facade and on the porous skin: the mesh presents pressure taps pointing the gap between the two facades and the external environment in such a way that it is possible to compute also the net pressure on the mesh (Fig. 3(b)). The same pressure measurement system as in the first stage is used. The outcome is a 120 s time series for each pressure tap, for incoming wind direction (Table 2).

Table 2. Scale factors for the DSF stage

	λ
Length scale	1/50
Velocity scale	1/2.13
Time scale	1/23.47

2.4 Flow Configuration

Wind tunnel tests have been performed in a correctly scaled atmospheric boundary layer. In order to reproduce it during both the wind tunnel test stages, a group of nine spires and a set of bricks have been used: the former has the function to recreate wind structures as big as possible and to fasten the creation of the boundary layer while the ground roughness slows down the wind near the floor in order to reproduce the effect due to the presence of the trees and buildings at the full scale. Figures 5 and 6 show the experimental wind tunnel mean velocity, turbulence intensity profiles and the power spectra compared to the ones prescribed by the Eurocode.

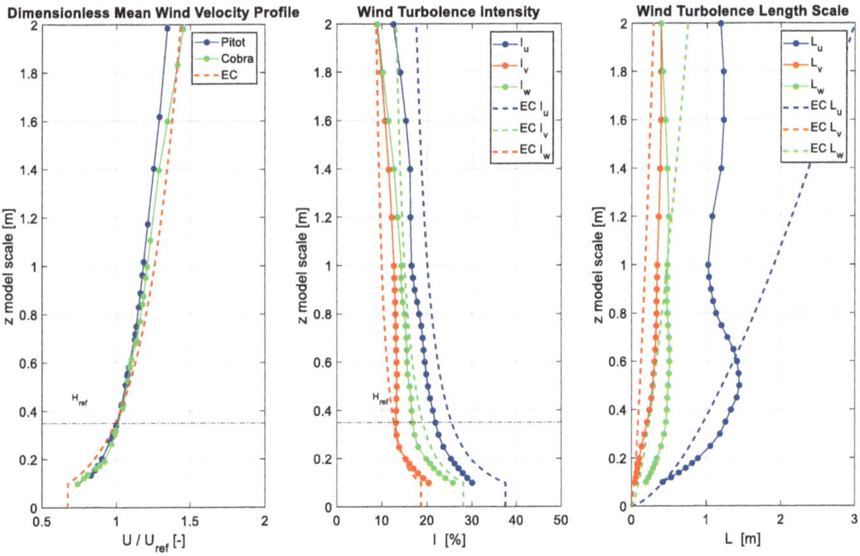

Fig. 5. Mean velocity profile, turbulence intensity profile and integral length scales.

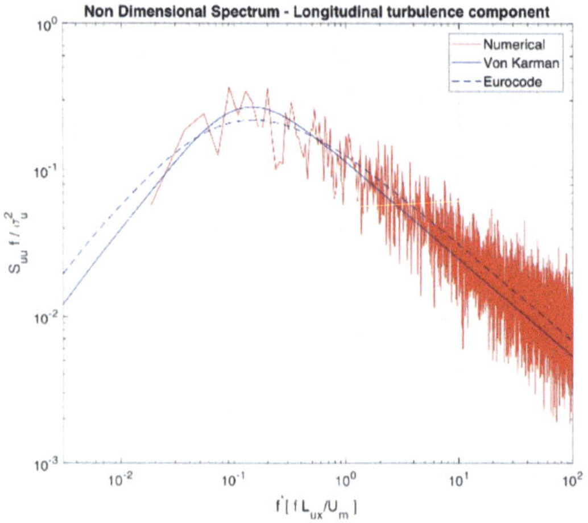

Fig. 6. Dimensionless power spectra: longitudinal fluctuating component u.

3 Experimental Results

The following sections show the experimental results in terms of pressure distributions: the first test (where no DSF structures have been tested) outcomes is compared to the DSF test results focusing on just one building, characterized by the worst condition in terms of wind load.

3.1 Coefficients Definition

The measured pressures are shown as non-dimensional coefficient C_p, Eq. 2, where $p^k(t)$ is the local pressure at the kth pressure tap, p_s is the ambient (static) pressure in the test section and $q_H = \frac{1}{2}\rho U^2$ is the wind reference dynamic pressure.

$$C_p^k(t) = \frac{p^k(t) - p_s}{q_H} \tag{2}$$

From the definition itself, positive pressure loads represent thrusts while negative pressure loads represent suctions (inward and outward arrows respectively in the pressure distributions figures). When referring to the outer porous skin, the net pressure acting on it is computed as the difference in the time domain between the external (ext) and the internal (int) pressure coefficients acquired at the kth tap:

$$C_{p,net}^k(t) = C_{p,ext}^k(t) - C_{p,int}^k(t) \tag{3}$$

Acquired pressure signals have been statistically analyzed in order to estimate the extreme pressure coefficient values (both positive and negative) associated to a given exceedence probability. Peak values (design values) have been statistically estimated according to the so-called Cook and Mayne Method, fully exploited in the reference Cook (1986). As current practice in wind engineering design facade generally recommends, before proceeding with the extreme value analysis, a low pass moving average filter with an averaging time equal to 3 s real scale (the so-called 3 s gust) is applied to the signals in order to account for decorrelation of gust over large areas in such a way that the effect of eddies smaller than the loaded area cancels out.

3.2 No DSF Model and DSF Model Comparison

A direct comparison of the negative peak pressure distributions from the first test and the inner façade of the DSF system is shown in Fig. 7(b), when the incoming wind direction is −140° (Fig. 3(a)). The strongest suctions exhibited in the standard facade system do not occur anymore in the glazed facade of the DSF. It seems that the presence of the metallic mesh is able to strongly reduce the peak pressures. Focusing on the DSF system, Fig. 7(a) shows negative peak pressure distribution on the glazed facade (inner ring) and on the metallic mesh (outer ring), where the arrows indicate the net pressure on the porous panels. It emerges that the net pressure, even if it is

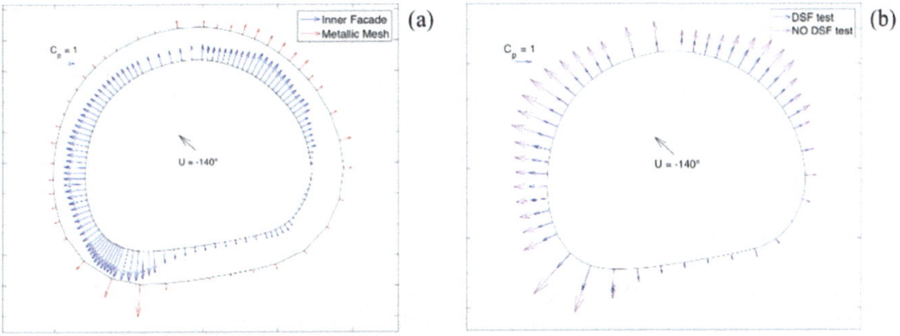

Fig. 7. C_p Positive (a) and negative (b) peak values distributions for the glazed facade with wind exposure angle −140°: noDSF test and DSF test comparison.

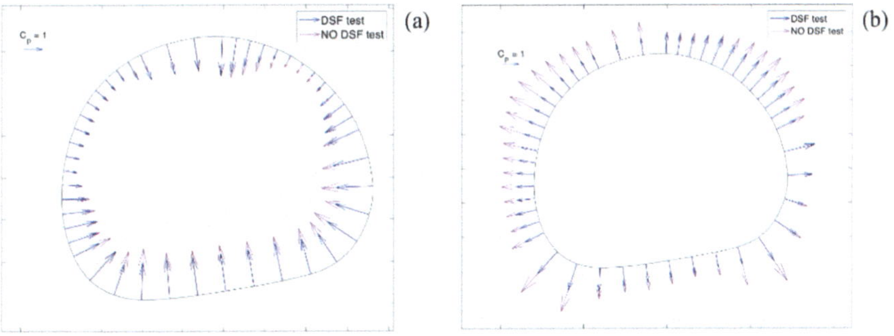

Fig. 8. C_p Positive (a) and negative (b) envelope distributions for the glazed facade: no DSF test and DSF test comparison.

characterized by smaller values, is related to the one acting on the inner facade. As far as the cladding design is concerned, the most interesting diagrams are the envelope ones: Fig. 8 shows positive and negative envelope distributions from the two tests. Especially for the latter, the presence of an outer skin results in a beneficial effect in terms of wind loads, which are actually lower in modulus.

3.3 Filtering Effect of the Mesh

This research does not focus on the peak value estimation only, but further investigations have been conducted both in the time and frequency domain in order to deeper understand the role of a porous medium in front of a glazed facade.

Let us analyze two signals recorded by pressure taps placed in the same position during the two stages, as shown in Fig. 9(a), where they are plotted together. It appears that the time histories recorded in the DSF model is shifted towards lower (in absolute terms) mean value and it exhibits lower standard deviation; this implies a lower dynamic content than the blue signal (first stage), characterized by sharp and concentrated peaks instead. Such different nature is due to the mesh, which is able to cut off high frequency content of pressure fluctuations.

This kind of behavior is even clearer if we switch to the frequency domain by analyzing the PSD of the aforementioned signals, aiming to underline their different energy content. The power spectral densities are plotted according to the logarithmic full scale frequency in such a way that the frequency contents coming from different scales can be directly compared. As expected, the energy content of the no DSF

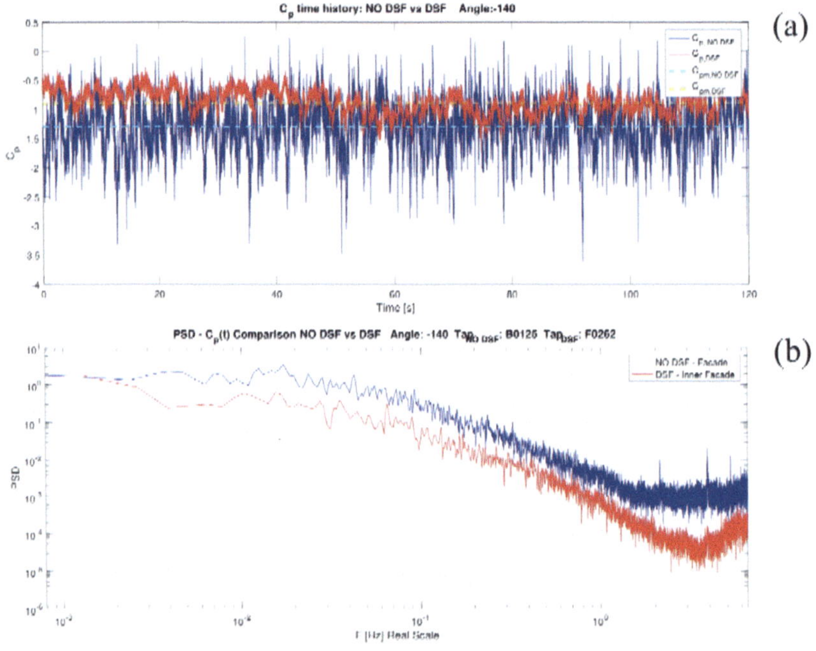

Fig. 9. C_p time histories (a) and PSD of pressure fluctuations (b).

configuration is higher than the one recorded in the corresponding point on the inner facade of the DSF case within the entire frequency axes. It is worth pointing out how at about 3 Hz full-scale an increase of the energy content occurs. Such phenomenon is generally not present in the PSD of pressure fluctuations acting on the standard facade, where the typical trend is monotonically decreasing with increasing frequency up to a value beyond which it remains constant. Basically there is a lack of energy of the signal within such frequency band which might be connected to the presence of the mesh, which alters somehow the incoming wind turbulence resulting into a pressure signal which is of course different from the one which might be recorded on a plain surface.

4 Reduction of the Peak Values

Once the peak pressure distributions from both tests are known, starting from the peak pressures obtained from the first test, a reduction factor can be computed by considering the corresponding pressure taps in the inner facade of the second test model (we remind that pressure taps in the second stage have been placed such that we have a correspondence between the 2 models, despite the different length scale λ_L).

Making reference to both positive ($\hat{C}_p$) and negative ($\check{C}_p$) peak pressures, the reduction factors are respectively defined as follows:

$$\hat{\eta} = \frac{\hat{C}_{pNoDSF} - \hat{C}_{pDSF}}{\hat{C}_{pNoDSF}} [\%] \tag{4a}$$

$$\check{\eta} = \frac{\check{C}_{pNoDSF} - \check{C}_{pDSF}}{\check{C}_{pNoDSF}} [\%] \tag{4b}$$

The reduction factors want to give an idea of the peak pressure reduction recorded between the two wind tunnel measurements. In order to obtain significant values, the η factor is computed only if the starting pressure (i.e. the one from the no DSF test) is in modulus higher than 1.5 C_p. Looking at the reduction factor of the envelope diagrams (Fig. 10), it appears that the porous double skin system may play a very important role in the wind load decrease, since highest η values are exhibited where the no DSF envelopes have the strongest peak pressures leading to high percentage value of $\hat{\eta}$ and $\check{\eta}_p$. The meaningful pressures exhibit reduction greater than 40% and in particular the strongest (dimensioning) suction from the no DSF test is equal to 63%, since Cp increases from −3.21 to −1.18 (Table 3).

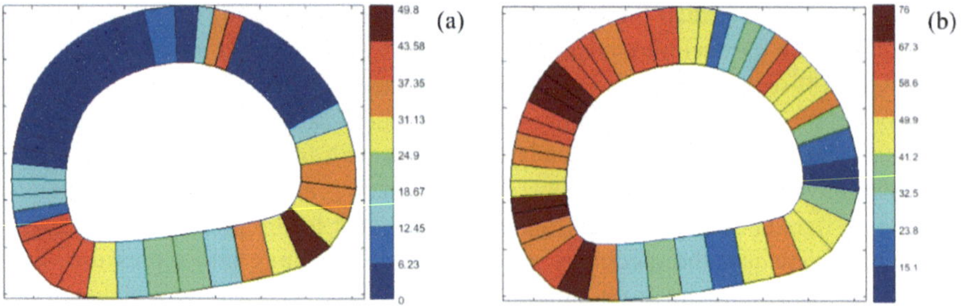

Fig. 10. Reduction factors for the positive (a) and negative (b) pressure envelopes for the tested building.

Table 3. Extreme values from envelopes of No DSF test and corresponding reduction factors.

	NoDSF	DSF	η [%]
$\check{C}_p$	−3.214	−1.189	63
$\hat{C}_p$	2.509	1.581	37

5 Conclusions

The main purpose of this research project has been the assessment of the wind pressure distribution on the porous double skin facade of the *New Bocconi Campus* buildings. Due to the dearth of reference case studies in literature and specific provisions in the Codes, wind tunnel tests have been commissioned in order to compute cladding loads relying on the estimated peak pressures from experimental measurements. The tests have been performed in two different stages: the first one, which has involved all the buildings, has aimed to the estimation of the expected wind pressures as in the case of a standard glazed facade; after that, a metallic mesh able to reproduce the real scale outer skin effect has been selected and installed in the model during the second stage.

By comparing the outcomes from the two tests, a pressure reduction has been effectively observed switching from the standard facade system to the inner skin of the DSF. As far as peak distribution on the inner facade is concerned, the aerodynamic effect induced by the porosity of the outer skin has resulted in strong reduction with respect to the most severe peak values (which happened to be suctions) recorded on the no DSF model. Moreover, it has been highlighted that the presence of an outer porous skin is able to cut off high frequency content of pressure fluctuations.

References

Belloli M, Rocchi D, Rosa L, Zasso A (2012) Wind tunnel studies on the effects of porous elements on the aerodynamic behavior of civil structures. In: The Seventh International Colloquium on Bluff Body Aerodynamics and Applications (BBAA7), pp 1132–1141

Belloli M, Rosa L, Zasso A (2014) Wind loads and vortex shedding analysis on the effects of the porosity on a high slender tower. J Wind Eng Ind Aerodyn 126:75–86

Cook NJ (1986) Designers guide to wind loading of building structures. Part 1
da Silva FM, Gomes MG (2008) Gap inner pressures in multi-storey double skin facades. Energy Build 40(8):1553–1559
Gerhardt H, Janser F (1994) Wind loads on wind permeable facades. J Wind Eng Ind Aerodyn 1–2:37–48
Letchford C, Row A, Vitale A, Wolbers J (2000) Mean wind loads on porous canopy roofs. J Wind Eng Ind Aerodyn 84(2):197–213
Overend M, Zammit K (2006) Wind loading on cladding and glazed façades. In: Proceedings
Xiaotong P, Chen L (2010) Cfd modeling for thermal behavior of open-type double-skin facades. In: 2010 International Conference on Intelligent Computation Technology and Automation (ICICTA), vol 1. IEEE, pp 918–921

Correction to: Cyclist Aerodynamics: A Comparison Between Wind Tunnel Tests and CFD Simulations for Helmet Design

S. Giappino, S. Omarini, P. Schito, C. Somaschini,
M. Belloli, and M. Tenni

Correction to:
Chapter "Cyclist Aerodynamics: A Comparison Between Wind Tunnel Tests and CFD Simulations for Helmet Design" in: F. Ricciardelli and A. M. Avossa (Eds.):
***Proceedings of the XV Conference of the Italian Association for Wind Engineering*, LNCE 27,**
https://doi.org/10.1007/978-3-030-12815-9_27

In the original version of this chapter, the following belated correction has been incorporated: The author's name was corrected from "S. Somaschini" to "C. Somaschini". The correction chapter has been updated with the change.

The updated version of this chapter can be found at
https://doi.org/10.1007/978-3-030-12815-9_27

F. Ricciardelli and A. M. Avossa (Eds.): INVENTO 2018, LNCE 27, p. C1, 2019.
https://doi.org/10.1007/978-3-030-12815-9_56

Author Index

F. Ricciardelli and A. M. Avossa (Eds.): INVENTO 2018, LNCE 27, pp. 743–745, 2019.
https://doi.org/10.1007/978-3-030-12815-9